METZGEN

10×10
or
20×20

437 26
93% 26

P9-DWN-533

Materials and Processes in Manufacturing

E. PAUL DeGARMO

Registered Professional Engineer
Professor of Industrial Engineering and Mechanical Engineering Emeritus
University of California, Berkeley

J TEMPLE BLACK

Registered Professional Engineer
Professor/Chairman of Industrial and Systems Engineering
University of Alabama, Huntsville

RONALD A. KOHSER

Associate Professor of Metallurgical Engineering
University of Missouri-Rolla

MATERIALS AND PROCESSES IN MANUFACTURING

Sixth Edition

Macmillan Publishing Company

New York

Collier Macmillan Publishers

London

Copyright © 1984, Macmillan Publishing Company, a
division of Macmillan, Inc.

Printed in the United States of America

All rights reserved. No part of this book may be
reproduced or transmitted in any form or by any means,
electronic or mechanical, including photocopying,
recording, or any information storage and retrieval
system, without permission in writing from the Publisher.

Earlier edition © 1957 and 1962 by Macmillan
Publishing Company, copyright © 1969 by E. Paul
DeGarmo, and copyright © 1974 and 1979 by Darvic
Associates, Inc.

Macmillan Publishing Company
866 Third Avenue, New York, New York 10022

Collier Macmillan Canada, Inc.

Library of Congress Cataloging in Publication Data

DeGarmo, E. Paul, Date
  Materials and processes in manufacturing.

  Bibliography: p.
  Includes index.
  1. Manufacturing processes.   2. Materials.
I. Black, J Temple.   II. Kohser, Ronald A.   III. Title.
TS183.D4   1984   671   83-5386
ISBN 0-02-328620-2 (Hardcover Edition)
ISBN 0-02-946130-8 (International Edition)

Printing: 4 5 6 7 8   Year: 5 6 7 8 9 0 1

ISBN   0-02-328620-2

PREFACE

This sixth edition of *Materials and Processes in Manufacturing* is the latest revision of the well-received text first introduced by E. Paul DeGarmo in 1957 and subsequently revised in 1962, 1969, 1974, and 1979. Here again, Professor DeGarmo has maintained overall direction of the edition, but has delegated much of the writing to two new co-authors: Dr. Ronald A. Kohser and Dr. J T. Black. Dr. Kohser assumes responsibility for the sections on materials, casting and forming processes, and joining processes. Dr. Black prepared sections relating to machining processes, and processes and techniques relating to manufacturing.

Looking back, we find that the trends noted in the fifth edition—namely increased automation and computer control plus emphasis on productivity with quality—have manifested themselves to the point where the United States now finds itself locked in a world-wide productivity battle, the likes of which have not been seen since World War II. Specifically, countries like Japan have made significant inroads into many of the commercial and industrial market areas once considered the sole domain of the United States. They have developed modern manufacturing systems significantly different in philosophy, methodology, and complexity. Strategies such as "just-in-time" production and "total quality control" have revolutionized the field. Set-up times are reduced to the point where it becomes economical to produce small batches. Worker motivation and productivity improve and everyone in the company assumes responsibility for quality improvement. The demonstrated success of these methods and philosophies are forcing their implementation world-wide.

Computers and microprocessors are literally invading the world of manufacturing and we expect to see continued development and application. Software control systems will operate all manner and type of machines. Computer-aided design and computer-aided manufacturing will become closely coupled with computer-aided testing and inspection to place the entire production system under computer control. Industrial robots will perform operations with extreme precision and reliability.

This edition reflects these changes in the hope that it will contribute to the needed progress through its use in engineering schools and colleges as well as industry in both the United States and numerous foreign countries. Considerable space has been devoted to robotics, computer-aided manufacturing, and new manufacturing management philosophies (such as Group technology, just-in-time manufacturing, and total quality control) as the authors recognize these techniques and tools as the methods and machines of the factory of the future.

Engineers are involved in the total manufacturing process in a variety of ways—the development of materials, the development and design of processing equipment, and the processing of materials. The vast majority of engineers, however, are concerned with materials and processes as the means whereby their designs are brought into reality. At the design stage many decisions are made that determine which materials are to be used and what processes must be employed to process them. At some stage in the design-material selection-processing sequence someone must make decisions regarding the material to be used and the processing that will be employed. These decisions always affect the cost of the product, and they may vitally affect its functioning. Therefore, it is most desirable that the designer either make these decisions or, at least, be in on the making of them. If not, either costs or functionality, or both, may suffer. The design, material, and processing constitute a system that should be considered as an entity.

New, and usually more specialized, materials continue to be developed to meet special requirements. These often require special, more precise processing in order for their properties to be effectively utilized. Traditional materials are now recognized to exist in limited quantities. Therefore, the economics of scarcity, the reduction of waste, and the necessity of recycling will assume increasingly more important roles during material selection. Similarly, sophisticated and more versatile machines have become commonplace. In most cases, however, their full potential can be utilized only if the designer has an understanding of their capabilities and limitations. Therefore, although the basic objective of this edition remains the same as for prior editions, even more attention is given to the design-material selection-processing relationship. Chapter 10, "Material Selection," has been rewritten to place additional emphasis on this relationship. The basic processes continue to be stressed, but with greater emphasis on the manner in which they integrate into modern multifunctional manufacturing and production systems.

The use of digital readout controls and numerical-, tape-, and computer-control systems is emphasized. Special attention is given to the processes that permit components to be formed into final, or semifinal, shape with little or no waste of materials.

Although considerable effort has been made to include all the significant and promising new developments in both materials and processes, the major emphasis remains on fundamentals, which provide an enduring basis for understanding both existing phenomena and those not yet in use. Thus the chapters on materials are designed to emphasize why they are suitable for certain applications, why they react as they do when subjected to certain processing, and why they must be treated in a specific manner to obtain desired results. How they are processed is an important, but secondary, objective. Similarly, with respect to machine tools, the primary emphasis is on what they will do, how they do it, their accuracy, and their relative advantages and limitations—particularly economic. Although some attention must be given to details regarding their construction and operation, this is solely for the purpose of providing a better understanding of the relationship of the tools to the foregoing objectives.

In the fifth edition, case studies were introduced at the ends of various chapters. This edition contains thirty-nine case studies, taken, for the most part, from actual industrial situations experienced by one of the authors. It is emphasized that they do not necessarily involve only the subject matter dealt with in the chapter that they follow. Rather, all the subject matter required for their solution is contained in chapters up to that point. (In some cases, the student will have to consult standard data sources for needed information, as would be necessary in actual practice.) Such case examples are extremely useful in making students aware of the great importance of properly coordinating design, material selection, and manufacturing in order to achieve a satisfactory and failure-free product.

As in previous editions, great care has also been given to the illustrations. The photographs have been selected to instruct, not to advertise a particular product. Many were made especially for this text, and numerous companies have been most cooperative in this regard. However, it should be understood that in many instances safety guards have been removed from equipment so as to show important details, and the personnel shown are not wearing certain items of safety apparel that would be worn in normal operation.

The book continues to be organized so that it can be used either in courses that cover both materials and processes, or in courses covering only manufacturing processes. For the combined type of course, the use of all chapters will give comprehensive coverage of both materials and processes. For use in courses covering only manufacturing processes, Chapters 2 through 10 can be omitted. However, they are available as a ready reference for explanation of why materials behave as they do when processed.

The authors wish to acknowledge the assistance and cooperation of wives and families during the preparation of this sixth edition. Discussions with students and colleagues have been quite helpful and were influential in revising several key chapters. The encouragement and constructive criticism of many are truly appreciated.

E. Paul DeGarmo
J Temple Black
Ronald A. Kohser

CONTENTS

I MATERIALS

II CASTING AND FORMING PROCESSES

III MACHINING PROCESSES

IV JOINING PROCESSES

V PROCESSES AND TECHNIQUES RELATED TO MANUFACTURING

CASE STUDIES

PART I

MATERIALS

CHAPTER 1

4 resources
MAN POWER
MATERIALS
MACHINERY
MONEY

JOB SHOP

MACHINERY GROUPED BY FUNCTION
CAN PRODUCE A WIDE VARIETY
CANNOT " " LARGE AMOUNT OF
Q
LABOR IS HIGH COST   6 - 2000 PES
MACH IS LOW COST   flexibility

Process plant
usually liquids

MASS PRODUCTION SHOP   specialized machinery
low skills large Q high efficient low flexibility

Introduction   AUTOMOTIVE INDUSTRY

Materials, manufacturing, and the standard of living. The standard of living in any civilization is determined, primarily, by the goods and services that are available to its people. In most cases, materials are utilized in the form of manufactured goods. These goods are typically divided into two classes: consumer goods and producer goods. Producer goods are those manufactured for other companies to use to manufacture either producer or consumer goods. Consumer goods are those purchased directly by the consumer or the general public. For example, someone has to build the rolling mill used to roll the sheets of steel which are then formed and become the fenders of your car. Similarly, many service industries depend heavily on the use of manufactured products, just as the agricultural industry is heavily dependent on the use of large farming machines for efficient production.

The more efficiently we can produce and convert materials into usable manufactured products while avoiding waste and achieving the desired function with the prescribed quality, the greater will be our productivity and the better will be our standard of living.

The history of man has been linked with his ability to work with materials, beginning with the stone age and ranging through the copper and bronze eras, the iron age, and recently the age of steel, with our sophisticated ferrous and nonferrous materials. We are now entering the age of tailor-made materials such as composites, as indicated in Figure 1-1, which details the alloys that have

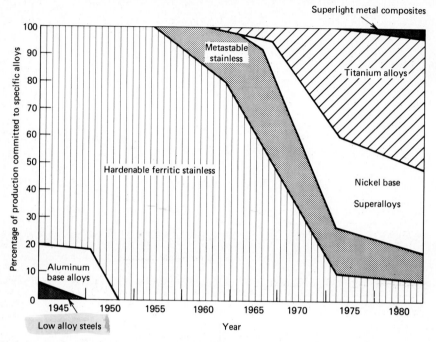

FIGURE 1-1 Changing trends in the mix of alloys used for compressor airfoils in jet engines. (*From E. E. Weismantel, AEG-265-8/68(700), Aircraft Engine Group, General Electric Company.*)

been used in the manufacture of compressor airfoils since 1945. As the materials became more sophisticated, having greater strength and lighter weight, they also became more difficult to manufacture with existing production methods. The tools wear out so rapidly or are so excessively costly that the material and process combination ceases to be economical. Quite often the most hostile environment the material ever encounters in its lifetime is the processing environment.

Although we are no longer dependent on using materials only in their natural state or in modified forms, there is obviously an absolute limit to the amounts of many materials available here on earth. So while the variety of materials continues to increase, we must use these resources efficiently and recycle as much as we can of those materials which are fast depleting. Figure 1-2 shows the effectiveness of recycling. Of course, recycling only postpones the exhaustion date. A low annual increase rate of usage is more significant. Tin (Sb), for example, has an annual growth rate of 2% and a current recycling rate of 20%.

Like materials, processes have also proliferated greatly in the last 30 years with new processes being developed to handle the new materials more efficiently and with less waste. Advances in manufacturing technology probably account for 40% of our improvements in productivity.

Materials, men, and equipment are interrelated factors in manufacturing that must be combined properly in order to achieve economical production. This important concept is indicated in Figure 1-3. What may be a proper combination

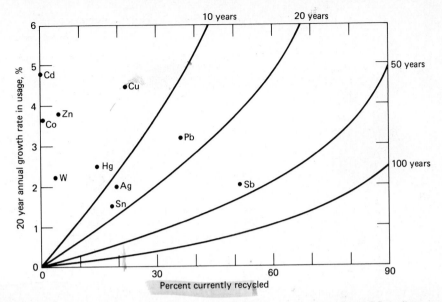

FIGURE 1-2 Effectiveness of recycling. Numbers 10, 20, 50, and 100 indicate years metal exhaustion is postponed by recycling. (*From A. Hurlich,* Metal Progress, *Oct. 1977.*)

for one product may not be optimal for another. The optimal combination for producing a small quantity of a given product may be very inefficient for a larger quantity of the same product. The proper combination for one product may be entirely wrong for a different product. Consequently, a systems approach, taking into account all the factors, must be used. This requires a sound and broad understanding of materials and processes and equipment on the part of those who make all the decisions involved.

The roles of engineers in manufacturing. Many engineers have as their function the designing of products that are to be brought into reality through the processing or fabrication of materials. In this capacity they are a key factor in the material selection–manufacturing procedure. A design engineer, better than any other person, should know what he or she wants a design to accomplish. He knows what assumptions he has made about service loads and requirements, what service environment the product must withstand, and what appearance he wants the final product to have. In order to meet these requirements he must select and specify the material(s) to be used. In most cases, in order to utilize the material and to enable the product to have the desired form, he knows that certain manufacturing processes will have to be employed. In many instances, the selection of a specific material may dictate what processing must be used. At the same time, when certain processes are to be used, the design may have to be modified in order for the process to be utilized effectively and economically. Certain dimensional tolerances can dictate the processing, and some processes require certain tolerances. In any case, in the sequence of converting the design into reality, such decisions must be made by someone. In most

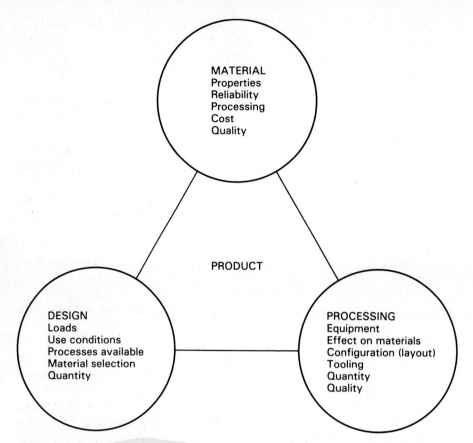

FIGURE 1-3 Interrelationships among materials, design, and processing in manufacturing a product.

instances they can be made most effectively at the design stage, by the designer if he has a reasonably adequate knowledge concerning materials and manufacturing processes. Otherwise, decisions may be made that will detract from the effectiveness of the product, or the product may be needlessly costly. It is thus apparent that design engineers are a vital factor in the manufacturing process, and it is indeed a blessing to the company if they can design for producibility—that is, for efficient production.

Manufacturing engineers select and coordinate specific processes and equipment to be used, or supervise and manage their use. Some design special tooling that is used so that standard machines can be utilized in producing specific products. These engineers must have a broad knowledge of machine and process capabilities and of materials, so that desired operations can be done effectively and efficiently without overloading or damaging machines and without adversely affecting the materials being processed. These tool or manufacturing engineers also play an important role in manufacturing.

A relatively small group of engineers design the machines and equipment used in manufacturing. They obviously are design engineers and, relative to their products, they have the same concerns of the interrelationship of design, materials, and manufacturing processes. However, they have an even greater concern regarding the properties of the materials that their machines are going to process and the interreaction of the materials and the machines.

Still another group of engineers—the materials engineers—devote their major efforts toward developing new and better materials. They, too, must be concerned with how these materials can be processed and with the effects the processing will have on the properties of the materials.

Although their roles may be quite different, it is apparent that a large proportion of engineers must concern themselves with the interrelationship between materials and manufacturing processes.

As an example of the close interrelationship of design, material selection, and the selection and use of manufacturing processes, consider the electrical appliance plug shown in Figure 1-4. This plug was purchased at a retail store for $1.40, and the manufacturer probably received about 85 cents for it. As shown in Figure 1-4, it consists of 10 parts. Thus the manufacturer had to produce, assemble, and sell the 10 parts for less than 85 cents—an average of 8½ cents per part—in order to make a profit. Only by giving a great deal of attention to design, selection of materials, selection of processes, selection of the equipment used for manufacturing (tooling), and utilization of personnel could such a result be achieved.

The appliance plug is a relatively simple product, yet the problems involved in its manufacture are typical of those with which manufacturing industries must deal. The elements of design, materials, and processes mentioned are all closely related. Each has its effect on the others. For example, if the two plastic shell components were to be fastened together by screws and nuts, instead of by the two U-shaped clips, entirely different machines, processes, and assembly procedures would be required. Similarly, the success of the plug is dependent on the proper material being selected for the clips. The material had to be sufficiently ductile to permit it to be bent without breaking, yet it had to be sufficiently strong and stiff to act as a spring and cause the clips to hold the shells together firmly. It is apparent that both the material and the processing had to be considered when the plug and the clips were designed in order to assure a satisfactory product that could be manufactured economically.

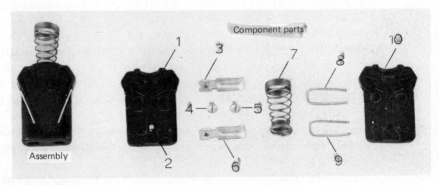

FIGURE 1-4 Appliance plug, assembled and disassembled.

A list of all the problems that had to be solved to enable the appliance plug to be produced for 85 cents would be quite long. Imagine the magnitude of a similar list for an automobile or a space rocket. The fact that a modern automobile can be purchased for under $6000 is one proof that industry has learned to deal effectively with the multitude of problems that attend the design and production of complex, modern products. Solutions to these problems require engineers who have a fundamental and comprehensive knowledge concerning materials and manufacturing processes and their interrelationship, and the application of this knowledge at all stages, from the conceptual design through the supervision of the production equipment and facilities.

Manufacturing and production systems. The systems used to manufacture goods are typically laid out in one of three methods: the job shop, the flow shop, and the project shop. A fourth type, the process or continuous process layout, is common in the chemical industry and will not be further elaborated on here because it primarily deals with liquids (such as an oil refinery) rather than solids.

The most common of these layouts is the *job shop*, characterized by large varieties of components, general-purpose machines, and a functional layout (see Figure 38-2). This means that machines are collected by function (all lathes together, all milling machines together, etc.) and the parts are routed around the shop in small lots to the various machines.

Flow shops are characterized by larger lots, special-purpose machines, less variety, and more mechanization. Flow shop layouts are typically either continuous or interrupted. If *continuous*, they basically run one large-volume complex item in great quantity and nothing else. The appliance plug was made this way. A transfer line producing an engine block is another typical example. If *interrupted*, the line works on large lots but is periodically changed over to run a similar but different component.

The *project shop* is characterized by the immobility of the item being manufactured. In the construction industry, bridges and roads are good examples; in the manufacturing of goods, large airplanes or locomotives are manufactured this way. It is necessary that the men, machines, and materials come to the site. The number of end items is not large, so the lots sizes of the components going into the end item are not large. Thus the job shop and the project shop are often linked together, with the job shop making components for the project shop in small lots.

Naturally, there are many hybrid forms of these manufacturing systems, but the job shop is the most common system and may continue to be because of

1. The proliferation in numbers and varieties of products, which means that lot sizes will decrease as variety increases.
2. Continued increase in the variety of materials with widely diverse properties.

Because of its design, the job shop has been shown to be the least cost efficient of all the systems. The hub of the problem is that component parts in a typical job shop spend only 5% of their time in machines and the rest of the time waiting or being moved from one functional area to the next. Once the part is on the machine, it is actually being processed (i.e., having value added by changing its shape) only about 30% of the time. The rest of the time it is being loaded, unloaded, inspected, and so on. The advent of programmable machines serves to improve the percentage of time the machine is making chips as tool movements are programmed and the machines can automatically change tools or load and unload parts. However, there are a number of trends which are forcing manufacturing management to consider means by which the job shop system itself can be redesigned to improve its overall efficiency. These trends have forced manufacturing companies to explore the benefits of concepts such as *group technology* (GT). The application of GT allows one to reorganize the job shop into totally new types of manufacturing systems, called flexible manufacturing systems and cellular manufacturing systems. In these systems, machines are grouped (laid out) so that a family of component parts (parts that display similar processing needs) can be handled. See Chapter 38 for a detailed discussion.

Many innovative advances in industrial manufacturing and management have been successfully implemented by the Japanese. They have a *Just-In-Time* (JIT) production objective, in which they try to operate their production in very small batches. They try to eliminate setup time so that it becomes economical to have small lot sizes, which in turn pays great benefits in quality, worker motivation, productivity, and inventory costs and control. The JIT system prevents large lots of defective items from ever being produced and has been able to implement a *total quality control* (TQC) system wherein the worker is primarily responsible for the quality. It appears that most of the management and manufacturing concepts embodied in JIT and TQC are readily transferable to American industry. These production systems will be discussed in Chapter 39. The JIT system lends itself to much greater job enlargement and job involvement for the worker. It is a natural for the efficient implementation of *computer-aided design and computer-aided manufacturing* (CAD/CAM) techniques.

Flow shops have been characterized as having specialized equipment designed to produce large quantities of products. This tends to be the accepted view of mass-production manufacturing: using workers that are "less skilled" in a limited range of operations. Actually, mass production involves the manufacture of large quantities of standardized products produced through the use of division of labor or labor specialization.

The project shop is characterized by the movement of the necessary men, machines, equipment, and materials to the item being manufactured, that item typically being too large to move easily, such as a plane or a locomotive, or actually stationary, such as a bridge or building. Building a house can be considered a project. When built in the context of a development, where many houses are being constructed in one location, crews move from house to house, taking it through various stages of construction. Generally speaking, lot sizes are small and the end items expensive in the project shop. Of course, even houses can be built on factory assembly lines and trucked to the site with significant reductions in cost. Characteristics of the basic manufacturing systems are summarized in Chapter 38.

Configuration analysis. In the manufacturing of *hardware*, the primary objective is to produce a component having a desired configuration, size, and finish. Every component has a shape that is bounded by various types of surfaces of certain sizes that are spaced and arranged relative to each other. Consequently, a component is manufactured by producing the surfaces that bound the shape. Surfaces may be

1. Plane or flat.
2. Cylindrical: external or internal.
3. Conical: external or internal.
4. Irregular: curved or warped.

Figure 1-5 illustrates how a shape can be analyzed and broken up into these basic, bounding surfaces. Parts are manufactured by using processes that will either (1) remove portions of a rough block of material so as to produce and leave the desired bounding surfaces, or (2) cause material to form into a stable configuration that has the required bounding surfaces. Consequently, in designing an object, one delineates and specifies the shape, size, and arrangement of the bounding surfaces. Next, the designed configuration must be analyzed to determine what materials will provide the desired properties and what processes can best be employed to obtain the end product at the most reasonable cost. This is called *designing for producibility.*

Production terms. Production terms that are in common usage today have a rank or order that is important to understand. The rank order shown in Table 1-1 is not absolute but relative, with overlap because of inconsistent popular usage. In this text a *production system* will refer to the total company and will include within it the *manufacturing systems.*

An obvious problem exists here in the terminology of manufacturing and production. The same term can refer to many different things. For example, "drill" can refer to the machine tool that does these kind of operations; the operation itself, which can be done on many different kinds of machines; or the cutting tool, which exists in many different forms. It is therefore important to use modifiers whenever possible: "Use the *radial* drill *press* to drill a hole with a 1-inch-diameter spade drill." The emphasis of this book will be directed toward the understanding of the processes, machines, and tools required for manufacturing and how they interact with the materials being processed.

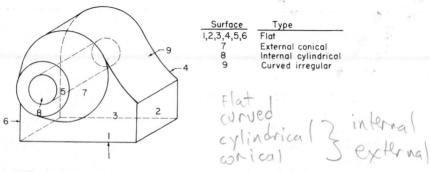

Surface	Type
1,2,3,4,5,6	Flat
7	External conical
8	Internal cylindrical
9	Curved irregular

FIGURE 1-5 Object composed of seven geometric surfaces. Dashed lines are hidden surfaces.

TABLE 1-1 Production Terms

Term	Meaning	Examples
System (production)	All aspects of men, machines, materials, and information considered collectively needed to manufacture parts or products; integration of all units of system critical (see Figure 38-4)	Company that makes engines, assembly plant, glassmaking factory, foundry
Process or sequence of operation	A series of manufacturing operations resulting in specific end products; *manufacturing system* is an arrangement or layout of many processes, as in the job shop or flow shop	Injection molding, rolling steel plates, spot-welding auto bodies, series of connected operations
Machine or machine tool	A specific piece of equipment designed to accomplish specific processes; often called a *machine tool;* machine tools link together to make a manufacturing process; term is often interchanged with *process*	Spot welder, milling machine, lathe, drill press, forge, drop hammer, die caster
Job	A collection of operations done on machines or a collection of tasks performed by one man at one station (location) on an assembly line	Operate lathe, inspector, final assembly, forklift driver
Operation (sometimes called a process)	A specific action or treatment, the collection of which make up the job of a worker	Drill, ream, bend, solder, turn, face, mill, extrude, heat-treat
Tools or tooling	Refers to the implements used to hold, cut, shape, or deform the work materials; called *cutting tools* if referring to machining; can refer to *jigs and fixtures* in workholding and *punches and dies* in metal forming	Grinding wheel, drill bit, tap, end milling cutter, die, mold, clamp, three-jaw vise

The basic manufacturing processes.   Manufacturing processes can be classified as

1. Casting or molding processes.
2. Forming or metalworking processes.
3. Machining (material removal) processes.
4. Joining/assembly.
5. Finishing.
6. Heat treating.
7. Other.

These classifications are not mutually exclusive. For example, some finishing processes involve a small amount of metal removal or metal forming. A laser can be used for either joining or metal removal or heat treating. Occasionally, we have a process such as shearing, which is really metal cutting but is viewed as a (sheet) metal-forming process. So our categories of process types are far from perfect.

Casting and *molding* involve introducing liquid, granular, or powdered material into a previously prepared mold cavity. Liquid material (usually molten metal) takes the shape of the cavity and solidifies; it retains the desired shape of the mold cavity after it is removed, either by the mold being opened or broken away. Where granular or powdered material is involved, the application of considerable pressure is required in order to cause it to conform to the shape of the mold cavity and to acquire the desired density. Heat often is applied in addition to pressure. When the material has permanently attained the desired shape and density, the mold is opened and the part is removed.

An important advantage of casting and molding is that, in a single step, materials can be converted from a crude form into a desired shape. In most cases, a secondary advantage is that excess, or scrap material can easily be recycled. Figure 1-6 illustrates schematically the basic concepts of these processes.

Casting processes commonly are classified into two types, based on whether the mold is permanent and can be used repeatedly or nonpermanent so that a new mold must be prepared for each casting made. These processes will be discussed in detail in Chapter 11. Molding processes usually are classified according to the material being molded, commonly metals, plastics, or glass. The powder metallurgy process is discussed in Chapter 12 and the molding of plastics in Chapter 9.

Forming and *shearing* operations are extensive in number and utilize material (metal or plastics) that previously has been cast or molded. In many cases the materials pass through a series of forming or shearing operations, so the form of the material for a specific operation may be the result of all the prior operations. The basic purpose of forming and shearing is to modify the shape and size and/or physical properties of the material. Often the concurrent modifications of the properties that may occur are not desired since, as will be discussed in a later chapter, they may limit what can be accomplished in a process or may make additional processing necessary.

The basic forming and shearing processes are listed in Table 1-2. Some of the forming and shearing processes are shown in Figure 1-7.

Metalforming and shearing operations are done both "hot" and "cold," a reference to the temperature of the material at the time it is being processed with respect to the temperature at which this material can recrystallize (i.e., grow new grain structure). Above this temperature, we do hot working; below it, cold. Most metalworking processes are done either way, with some exceptions such as coining, always done cold. These processes generally require large dies or containers to hold or shape the metal, operate under high-pressure conditions, and typically involve friction, which wastes a lot of energy. Therefore, heavy and costly equipment is required.

Machining or *metal removal processes* refer to the removal of certain selected areas from a part, to obtain a desired shape or finish. We classically think of

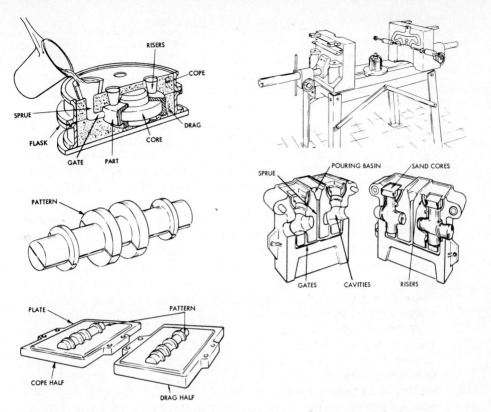

FIGURE 1-6 Casting processes. Sand casting. Patterns are used to produce the cavity in sand. (*Right*) Molds are permanent metal type and can be reused. (*From Manufacturing Producibility Handbook; courtesy General Electric Company.*)

these processes as "chipmaking" processes, but in recent years many chipless machining processes have been developed and we can include "cutting with heat" in this family as well.

Chips are formed by interaction of a cutting tool with the material being machined. The chip formation process is basically a shear process, resulting in a chip being separated from the workpiece and rubbing processes between the tool and work and the tool and the chip. Figure 1-8 shows a chip being formed by a single-point cutting tool.

Cutting tools are used to perform the basic and related machining processes which are shown schematically in Figure 1-9. The cutting tools are mounted in machine tools, which provide the required movements of the tool with respect to the work (or vise versa) to accomplish the process desired.

To accomplish the seven basic machining processes—shaping, drilling, turning, milling, sawing, broaching, and abrasive machining—eight basic types of machine tools have been developed. These are: shapers (and planers), drill presses, lathes, boring machines, milling machines, saws, broaches, and grinders. Most of these machine tools are capable of performing more than one of the basic machining processes. This obvious advantage has led to the develop-

TABLE 1-2 Basic Metal Forming and Shearing Processes

Process Term	Meaning	Examples
Rolling	Reducing thickness or imparting specific cross section to sheets or bars, by passing them between pairs of rollers	Rolling shapes round, hex, square; rolling rails; roll-forming chair legs, window channels, automobile frame members, radiator sections
Extruding	Forcing hot or cold metals or plastics through dies to develop long materials of desired cross-sectional form	Aluminum window channels, shaving cream tubes, seamless tubing, plastic garden hose, synthetic fibers, moldings, and trims
Drawing	Pulling ribbons of metal (skelp) through dies to form tubes; pulling rods or wires through dies for sizing and for reduction of diameter	Tubing, pipe, wire, through dies; metal strip through rolls
Stretching	Elongating materials, to remove kinks, harden, reduce thickness, or to stretch form over dies	Stretching rods, wire, tubes, strip, for straightening; tension winding of wire and strip on-and-off of reels and coils; tensioning rods and wires for prestress of concrete
Pressing or deep drawing	Applying large forces with or without impact, for forming a workpiece, or to force one piece into another; or to hold workpieces under compression while undergoing heating, cementing, or laminating	Pots and pans, lamp reflectors, automotive fenders, typewriter covers, refrigerator or furnace panels; pressing electric motor shaft into laminations, handles into hammer heads
Coining	Embossing or imprinting a surface pattern on cold metal or forcing a part to size by applying dies at high pressures	Medallions, nameplates, scale markings, knobs, buttons, trim, electrical contacts, welding projections
Hammering	Repeated blows, for heavy vibration, for rough forging, or for crushing	Forging, nailing, shaking out parts or cores, swaging (forming end on workpiece), upsetting (shortening and thickening workpiece), heading nail and bolt heads
Forging	Stamping, pressing, or hammering metal, usually while hot, between dies, to form rough workpieces for finishing	Crankshafts, connecting rods, hammers, socket wrenches, axles, rough balls (for ball bearings), gears, axes, knives, forks, pry-bars
Bending	Deflecting materials beyond the elastic limit for the purpose of causing a permanent change of shape or form	Pipe, steel plates, rails, ornamental iron, angles, reinforcing iron, brackets, structural members
Braking	Bending sheet metal over straight dies for sharp folds	Cabinet panels, furnace covers, fiber and metal boxes, heat ducts

TABLE 1-2 **(continued)**

Process Term	Meaning	Examples
Spinning	Developing a "form of revolution" from a sheet metal disk by stretching and bending it while rotating; force is applied at progressively increasing radii, causing disk to conform to backing form	Access covers, hoppers, mixing tubs, body panels, aircraft sections, food machine pans, processing vessels
Seaming–flanging	Joining or attaching by pressing together folded edges of sheet metal	Drums, boxes, cans, heating ducts, flexible pipe
Roll forming	Rolling sheets into cylinders; or making longitudinal folds, bends, and seams, in long strip	Drums, boilers, tanks; decking, concrete forms; corrugated roofing, gutter pipe; steel furniture members
Winding	Coiling or wrapping wire or strip to form the desired configuration	Coils, springs, metal cylinders, shafts, plywoods, laminations
Shearing	Parting sheet metal or rods or bar stock by placing material in the acute angle formed between two dies sliding across each other	Sheet metal shears, rod and bolt cutters, wire nippers, bar cropping, plate notching, angle shearing, cutting sheet to size, trimming off surplus, paper cutting
Stamping or piercing and blanking	Applying large forces to the workpiece, by impact; for forming or punching sheet metal; causing die to pass into or onto another die, to form the metal or to cause a hole to be formed in the intervening sheet metal material	Washers, motor laminations, slots in laminations, container covers, switch plates, thumb tacks; electrical hardware connectors, lugs, terminals; razor blades, speed nuts; bolt holes in channels, beams, and angles; socket holes for electronic chassis, business machine cards, perforating sheets for ornamentation or expansion, holes in leather, paper, fiber
Nibbling	Cutting sheet metal by making a series of overlapping small punchings at high speed	Trimming sheets to size or line, cutting circles, cutting sheet metal to contours, notching

ment of *machining centers* specifically designed to permit several of the basic processes, plus some of the related processes, to be done on a single machine tool with a single workpiece setup.

Cutting tools specifically designed to perform these processes are held and driven by tool-holding devices in the machine tools. The cutting tools are often segregated according to whether they are of single-point geometry, have multiple cutting edges or teeth, or use abrasive grits, as in grinding. Referring again to Figure 1-9, it is observed that some of the basic processes are always performed

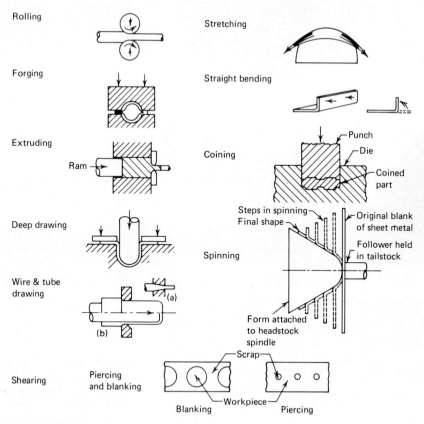

FIGURE 1-7 Common forming and shearing process.

by single-point tools (turning, shaping), while others are always performed by multiple-tooth cutters (milling, drilling). Some typical cutting tools are shown in Figure 1-10. This is just a small sample of the great variety of cutting tools commercially available. While they may seem quite different to the novice, they all essentially remove metal by the same basic chip formation mechanism.

It is necessary for the engineer who determines how the product is to be made to select the tooling needed to machine the part and to specify the machining speed and feed rate as these directly control the production rates in these machines. Even in automated equipment, this will still be necessary. Complicating this decision is the great variety of cutting tool materials now available for both single- and multiple-point tools. These materials range from sophisticated tool steels through many grades of carbides, ceramics, and more recently, cubic boron nitride and synthetic diamonds. Each tool material will display a different tool life when cutting different work material.

Within the collection of processes known as *machining processes,* we can include those wherein metal is removed by chemical, electrical, electrochemical, or highly concentrated heat sources. Generally speaking, these nontraditional or

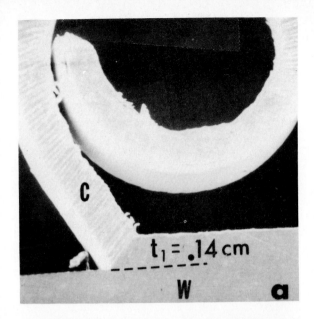

FIGURE 1-8 Series of electron micrographs of chip formation. (a) Low magnification chip formation process as seen inside a scanning electron microscope. The workpiece is a rectangular plate of high-purity gold, polished on the sides so that the plastic deformation of the shear process can be readily observed. The specimen was prepared by a "quick stop mechanism." (b) Higher magnification view of the shear area close to the tool tip region. The shear fronts have progressed from the tool tip toward the free surface. The shear angle is φ.

"chipless" processes have evolved to fill a specific need when conventional processes were too expensive or too slow when machining very hard materials. There are usually no high reaction forces with these processes to distort the workpiece, so precise, accurate cuts in delicate parts are possible. However, most of these processes have very low metal removal rates compared to con-

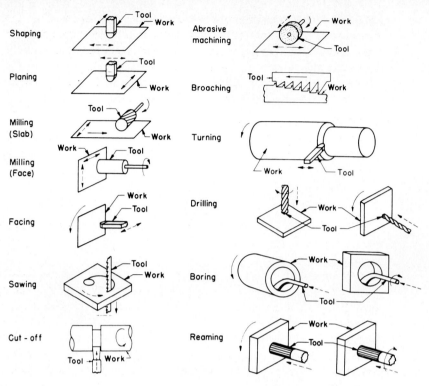

FIGURE 1-9 Schematic representation of the basic machining processes.

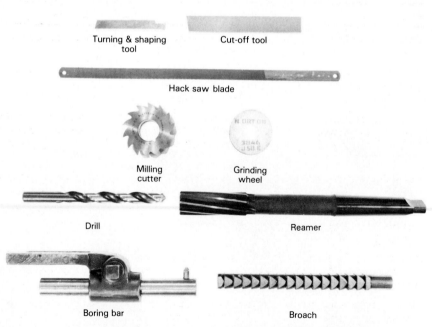

Turning & shaping tool

Cut-off tool

Hack saw blade

Milling cutter

Grinding wheel

Drill

Reamer

Boring bar

Broach

FIGURE 1-10 Typical cutting tools used in common machining operations.

MATERIALS

ventional processes and are thus best employed in special
first uses of a laser was to machine holes in ultra-high-s
being used today to drill tiny holes in turbine blades for je
shows the more common chipless machining processes.

Perhaps the largest collection of processes, both in term
quantity, are the *joining processes*, which include the follov

1. Mechanical fastening.
2. Soldering and brazing.
3. Welding.
4. Press fitting.
5. Shrink fitting.
6. Adhesive bonding.

Mechanical fastening includes semipermanent methods, such as bolting, and
permanent methods, such as riveting and staking. A wide variety of special
devices has been developed to meet specific needs.

Soldering and *brazing* are used to make semipermanent connections between
metal parts by means of solders or braze metals that have a lower melting point
than the metals being joined. When sufficient heat is applied to the base metal
to melt the solder or braze metal, they form an alloy with the surface of the
parent metal and, upon solidification, form a joint. Such a joint is commonly
thought to be permanent, but it must be remembered that the parts can be
unfastened, purposely or accidentally, by reheating; the parent parts are not
destroyed.

TABLE 1-3 Nontraditional Chipless Machining Processes

Process	Metal-Removal Mechanism	Examples
Chemical machining, milling or blanking	Chemical etching	Photoengraving
Electrochemical machining (ECM) or drilling or grinding	High-intensity "reverse electroplating" using high current densities	Machine cavities in dies
Ultrasonic machining (drilling or welding)	Abrasive slurry with grits vibrated into work by ultrasonic means; actually forms chips	Tool and die work (in nonconductors)
Electrodischarge machining (EDM)	Spark erosion of metals by local heating and melting	Drill holes in very hard tool and die materials
Laser beam machining (LBM) or drilling or heat treating	High-energy laser melts and vaporizes metal	Drill holes in turbine blades
Electron beam machining (EBM) or welding or cutting	High-energy electron beam melts and vaporizes metal	Microhole drilling in integrated-circuit boards
Plasma jet machining or cutting or welding	Ionic plasma, very high temperature jets to melt materials	Rapid cutting of plates

lding, of both metals and plastics, involves either melting the materials being joined at the interface or combining temperature and pressure so as to cause localized coalescence. Consequently, in most instances higher temperatures are involved than for brazing and soldering and the union is permanent The heating is provided by gas–oxygen flames, electric arcs, or by the electrical resistance of the interface, as in resistance welding. Other sources of heat for welding include electron beams, laser beams, combustion of fine aluminum powder and even friction heating between rotating parts. However, if the surfaces are clean enough and placed in very intimate contact (close atomic proximity), they can weld with just pressure. In this vein, *explosive welding* and *ultrasonic welding* have been developed.

In *press fitting,* mating parts, on which the dimension of the interior member is the same as or slightly greater than the interior dimension of the exterior member, are forced together. Obviously, there is a slight displacement of metal on the two parts. Such a joint normally will be permanent unless a suitable amount of force is used to press the parts apart. In *shrink fitting,* on the other hand, there is a substantial amount of interference between the interior and exterior parts, and they can be joined only by expanding the exterior part by heating or by contracting the interior part by cooling. Once the joint has been made, it ordinarily can be disassembled only by corresponding heating or cooling of one part.

Adhesive bonding produces joints by means of various adhesive agents. Most such joints are permanent unless loads too great are applied so as to produce failure. The use of adhesive bonding has increased greatly in recent years.

Finishing processes are yet another class of processes typically employed for cleaning, removal of burrs left by machining or providing a protective and/or decorative surface on workpieces. These processes include

1. Cleaning.
   a. Chemical.
   b. Mechanical
2. Deburring.
3. Painting.
4. Plating.
5. Buffing.
6. Galvanizing.
7. Anodyzing.

Cleaning removes such foreign substances as dirt, grease, and scale, which result from various manufacturing operations or handling. Frequently, it must be done in preparation for subsequent finishing or manufacturing operations. Machining, casting, and shearing operations often leave sharp, and possibly dangerous, edges that are removed by *deburring.*

Buffing, sometimes called *polishing,* reduces the roughness of a surface by actual smearing of the material, reducing microscopic protrusions and filling in small hollows. Cleaning, deburring, and buffing primarily improve the appearance of workpieces. However, although customers usually prefer such improved appearance, they seldom will consciously pay an extra price for it, and these operations must be done at minimum cost. As pointed out in Chapter 37, the designer can do much to eliminate the necessity for such operations or to permit them to be done economically.

Galvanizing and *anodyzing* are done almost exclusively to provide corrosion resistance, although anodyzing, in some cases, is used to provide an improved surface for later painting. In galvanizing, a coating of zinc is built up on the surface of steel, either by dipping in a molten zinc bath or by electrolytic deposition. In anodyzing, on the other hand, the surface of the metal is converted, extending into the metal for a few thousandths of an inch. Consequently, it causes no appreciable change in dimensions. It most commonly is used on aluminum alloys, providing a surface that is corrosion-resistant in numerous media.

Painting and *plating* are, of course, the adding of protective and decorative materials to the surface of a workpiece. Although paints and lacquers can be applied by brush, in manufacturing they almost always are applied by dipping, spraying, or by an electrolytic process. Metals are plated on to surfaces by the electroplating process. However, some plating is done by a process wherein metal is melted in an arc or an oxyacetylene flame and caused to impinge on the surface of the workpiece.

A modification of the electroplating process can be used to produce certain desired shapes by plating thick deposits, up to ¾ inch, on a mandrel that has an external shape coinciding with the desired inside shape of the workpiece. After the desired thickness of the plated material has been obtained, the mandrel is removed. This process is called *electroforming*.

Heat treatment is the heating and cooling of a metal for the specific purpose of altering its metallurgical and mechanical properties. Because the changing and controlling of these properties is so important in the processing and performance of metals, heat treatment is a very important manufacturing process. Each type of metal reacts differently to heat treatment. Consequently, a designer should know not only how a selected metal can be altered by heat treatment but, equally important, how a selected metal will react, favorably or unfavorably, to any heating or cooling that may be incidental to a manufacturing process. Through proper knowledge and use of heat treatment, less expensive metals can often be substituted in place of more costly materials, adverse effects from processing can be avoided, or less costly processing can be employed.

There are some other fundamental manufacturing operations other than the processes which we must consider. The first of these is *inspection*, which helps us to determine whether we have achieved the desired objectives stated by the designer in the specifications. This activity generally falls into the quality control department and provides feedback to the design and manufacturing engineers with regard to the process behavior. Essential to this inspection function are measurement activities. This material is covered in Chapters 16 and 17.

Another fundamental area of manufacturing is *materials handling* or *conveyance* of the product from one process to another. This is a very critical operation in that it always adds cost to the production process but not value. The means of conveyance are tied directly to design or layout of the processes, so that different manufacturing systems require very different materials handling systems. Programmable material handling systems, such as robots, are radically changing this function on the factory floor.

Finally, the manufacturing engineer may be involved in *packaging* of the finished products or final testing of the finished products to determine their reliability. All of these operations are important, but space does not permit us to deal with them in this book.

Historical development of machine tools. In 1962, Amber & Amber[1] presented their *Yardstick for Automation*. The chart that they developed has been updated and is included here as Table 1-4 in a somewhat abbreviated form. The key to the chart is that each level of automation is keyed to the human attribute which is being replaced (mechanized or automated) by the machine. Therefore, the A(0) level of automation, in which no human attribute was mechanized, covers the stone age through the iron age. Two of the earliest machine tools were the crude lathes for making Etruscan wooden bowls in 700 B.C. and the windlass-powered broaching of rifling grooves into rifle barrels for guns used over 300 years ago.

The first industrial revolution can be tied to the development of powered

[1] Amber & Amber, Anatomy of Automation, Prentice-Hall, Inc. Englewood Cliffs, N.J., 1962.

TABLE 1-4   Yardstick of Automation

Orders of Automation	Human Attribute Replaced	Examples
A(0)	*None:* lever, screw, pulley, wedge	Hand tools, manual machine
A(1)	*Energy:* muscles replaced	Powered machines and tools, Whitney's milling machine
A(2)	*Dexterity:* self-feeding	Single-cycle automatics
A(3)	*Diligence:* no feedback	Repeats cycle; open-loop numerical control or automatic screw machine; transfer lines
A(4)	*Judgment:* positional feedback	Closed loop; numerical control; self measuring and adjusting
A(5)	*Evaluation:* adaptive control; deductive analysis; feedback from the process	Computer control; model of process required for analysis and optimization
A(6)	*Learning:* by experience	Limited self-programming; some artificial intelligence (AI)
A(7)	*Reasoning:* exhibits intuition; relates causes from effects	Inductive reasoning; advanced AI
A(8)	*Creativeness:* performs design unaided	Originality
A(9)	*Dominance:* supermachine, commands others	Machine is master (Hal from ''2001, A Space Odyssey'')

Source: Amber & Amber, *Anatomy of Automation,* Prentice-Hall, Inc., Englewood Cliffs, N.J., 1962. Used by permission of Amber & Amber.

Outboard bearing

FIGURE 1-11 Model of Wilkinson's horizontal boring machine. (*British Crown Copyright, Science Museum, London.*)

machine tools, dating from 1775, when John Wilkinson in England, constructed a horizontal boring machine for machined internal cylindrical surfaces, such as in piston-type pumps. In Wilkinson's machine, a model of which is shown in Figure 1-11, the boring bar extended through the casting to be machined and was supported at its outer end by a bearing. Modern boring machines still employ this basic design. Wilkinson reported that his machine could bore a 57-inch-diameter cylinder to such accuracy that nothing greater than an English shilling (about 1/16 inch or 1.59 mm) could be inserted between the piston and the cylinder.

The next A(1) machine tool came along in 1794, when Henry Maudsley developed an engine lathe with a practical slide tool rest. This machine tool, shown in Figure 1-12, was the forerunner of the modern engine lathe. The lead

FIGURE 1-12 Maudsley's screw-cutting lathe. (*British Crown Copyright, Science Museum, London.*)

screw and change gear, which enabled threads to be cut, were added about 1800. The first planer was developed in 1817 by Richard Roberts in Manchester, England, and the first horizontal milling machine is credited to Eli Whitney in 1818 in New Haven, Connecticut (see Figure 1-13). The development of machines that could not only make specific products but could also produce other machines to make other products was fundamental to the first industrial revolution.

→ interchangeable parts

While the early work in machine tools and precision measurement was done in England, the earliest attempts at interchangeable manufacturing apparently occurred almost simultaneously in Europe and the United States. These, basically, involved the use of filing jigs, with which duplicate parts could be hand-filed to substantially identical dimensions. In 1798, Eli Whitney, using this technique, was able to obtain and eventually fulfill a contract from the U.S. government to produce 10,000 army muskets, the parts of each being interchangeable. However, this truly remarkable achievement was accomplished primarily by painstaking handwork and not by specialized machines.

precision measurement

Joseph Whitworth, starting about 1830, accelerated the use of Wilkinson's and Maudsley's machine tools by developing precision measuring methods. Later he developed a measuring machine using a large micrometer screw. Still later he worked toward establishing thread standards and made plug and ring gauges. His work was valuable because precise methods of measurement were a prerequisite for developing interchangeability, a requirement for later mass production.

The next significant machine tool was the drill press with automatic feed developed by John Nasmyth in 1840 in Manchester, England. Surface grinding machines came along about 1880 and the era was completed with the invention of the bandsaw by Leighton Wilkie in 1933. In total there were eight basic machine tools in the first industrial revolution for machining: lathe, milling machine, drill press, broach, boring mill, planer (shaper), grinder, and saw.

The A(2) level of automation was clearly delineated when machine tools became single-cycle, self-feeding machines, displaying dexterity. Many examples of this level of machine are given in Chapters 19 through 26, as they still exist in great numbers in many factories. The A(3) level requires the machine to be *diligent* or repeat cycle automatically. These machines are open loop, meaning that they do not have *feedback,* and are controlled by either an internal fixed program, such as a cam, or externally programmed with a tape or more recently a computer. A(3), A(4), and A(5) levels are basically superimposed on A(2)-level machines, which must be A(1) by definition. The A(3) level includes robots and numerical control (NC) machines which have no feedback, and many special-purpose machine tools.

The A(4) level of automation required that *judgment* in the human being be replaced by a capability in the machine to measure and compare results with desired position or size and make adjustments to minimize errors. This is feedback or closed-loop control. The first numerical control machine was developed in the early 1950s at MIT. It had positional feedback control and is generally recognized as the first A(4) machine tool. By 1958, the first NC machining center was being marketed by Kearney and Trecker. This machining center was a compilation of many machine tools capable of performing many processes; in

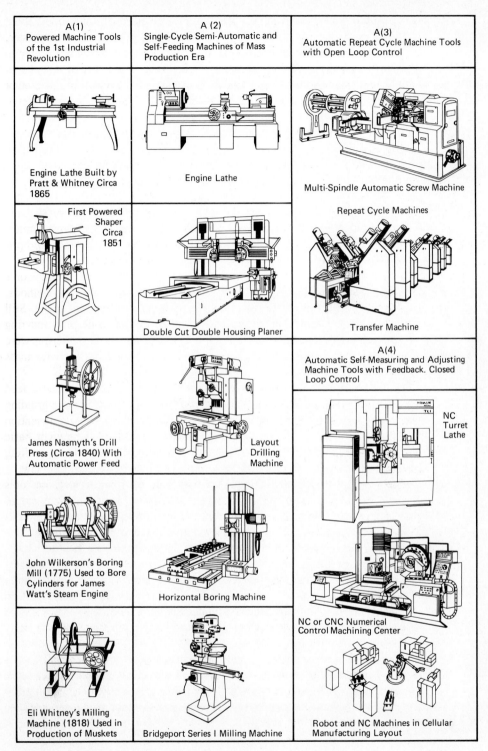

A(1) Powered Machine Tools of the 1st Industrial Revolution	A (2) Single-Cycle Semi-Automatic and Self-Feeding Machines of Mass Production Era	A(3) Automatic Repeat Cycle Machine Tools with Open Loop Control
Engine Lathe Built by Pratt & Whitney Circa 1865	Engine Lathe	Multi-Spindle Automatic Screw Machine
First Powered Shaper Circa 1851	Double Cut Double Housing Planer	Repeat Cycle Machines Transfer Machine
James Nasmyth's Drill Press (Circa 1840) With Automatic Power Feed	Layout Drilling Machine	A(4) Automatic Self-Measuring and Adjusting Machine Tools with Feedback. Closed Loop Control NC Turret Lathe
John Wilkerson's Boring Mill (1775) Used to Bore Cylinders for James Watt's Steam Engine	Horizontal Boring Machine	NC or CNC Numerical Control Machining Center
Eli Whitney's Milling Machine (1818) Used in Production of Muskets	Bridgeport Series I Milling Machine	Robot and NC Machines in Cellular Manufacturing Layout

FIGURE 1-13 Machine tools of the first industrial revolution (A1), the mass production era (A2), and examples of A(3) and A(4) levels of automation.

this case, milling, drilling (tapping), and boring (see Figure 1-13). It could also automatically change tools to give it greater flexibility. Within 10 years, the NC machine tools became computer numerical control (CNC) machine tools. Thus these machines had their own microprocessor and could be programmed directly or could be driven with a tape prepared externally to the machine. In either event, CNC machines are still A(4) machines.

With the advent of the NC type of machine and more recently the programmable robot, two types of automation were defined. *Hard* or *fixed automation* is exemplified by transfer machines or automatic screw machines, and *flexible* or *programmable automation* is typified by NC machines or robots we can teach or program externally via computers.

The A(5) level requires that machines perform *evaluations* of the process itself. Thus the machine must be cognizant of the multiple factors on which the machining processes performance is predicated, evaluate the current setting of the input parameters versus the outputs from the process, and then determine how to alter the inputs to optimize the process. This is called *adaptive control* (AC).

There are very few examples of A(5) machines on the factory floor and virtually none at the A(6) level wherein the machine control has artificial intelligence capability. Levels A(3) through A(6) will be discussed in greater detail in Chapter 38. Levels A(7) through A(9) will be left to the whims of the science fiction writer.

Automation involves machines, or integrated groups of machines, which automatically perform required machining, forming, assembly, handling, and inspection operations, and through sensing and feedback devices, automatically make necessary corrective adjustments. There are relatively few completely automated production units, but there are numerous examples of highly automated individual machines. While the potential advantage of a completely automated plant are tremendous, in practice, step-by-step automation of individual operations is required. Thus it is important to have a piecewise plan to convert from the classical job shops to the automated integrated factory of the future. This will be discussed in more detail in Chapter 38. However, the most serious limitation in automation will be available capital, as the initial investments for automation equipment and installations will be large. Proper engineering economics analysis must be employed to evaluate these investments, so students who anticipate getting into manufacturing should consider a course in this area a firm requirement.

The problem of continually reducing production costs does, and always will, remain. New materials are constantly being sought. We are truly entering the age of composites. New processes are needed to deal with new materials which do not pollute the environment. Despite the great advances of recent years, even greater progress may be expected in the future. More attention will be given to eliminating the wastage of materials and improvement in quality. The consumer wants products to have high reliability. Thus the pressures are many and careful planning of production is required.

Planning for production.   Low-cost manufacture does not just happen. There is a close and interdependent relationship between the design of a product,

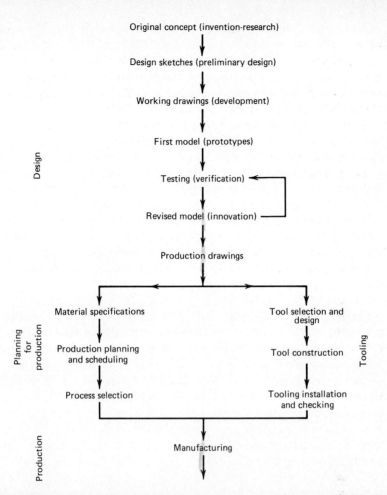

FIGURE 1-14 Traditional steps required to convert an idea into a finished product.

selection of materials, selection of processes and equipment, layout of processes, and tooling selection and design. Each of these steps must be carefully considered, planned, and coordinated before manufacturing starts. This lead time, particularly for complicated products, may take months, even years, and the expenditure of large amounts of money may be involved. Typically, the lead time for a completely new model of an automobile is about 5 years; for a modern aircraft it may be 4 years.

Figure 1-14 shows the steps involved in getting one product from the original idea stage to the point where it was coming off the assembly line. Note that most of the steps are closely related to the others. For example, the design of the tooling was conditioned by the design of the parts to be produced. It is often possible to simplify the tooling if certain changes are made in the design of the parts or the design of the manufacturing systems. Similarly, the selection of the material to be used will affect the design of the tooling or the processes selected. On the one hand, it frequently is desirable to change the design of a part so as to enable it to be produced with tooling already on hand and thus avoid the purchase of new equipment. Close coordination of all the various phases of manufacture is essential if economy is to result. All mistakes and "bugs" should be eliminated during the preliminary phases, because changes become more and more costly as work progresses.

With the advent of computers and machines that can be controlled by either tapes made by computers or by the computers themselves, we are entering a new era of production planning. The integration of the design function and the manufacturing function through the computer is called CAD/CAM (computer-aided design/computer-aided manufacturing). The design is used to determine the manufacturing process planning and the programming information for the manufacturing processes themselves. Detailed drawings can also be made from the central data base used for the design and manufacture, and programs can be generated to make the dies as needed. In addition, extensive computer-aided testing and inspection (CATI) of the manufactured parts is taking place. There is no doubt that this trend will continue at ever-accelerating rates as computers become cheaper and smarter.

Organization for production. The most important factor in economical, and successful, manufacturing is the manner in which the resources—men, materials, and capital—are organized and managed so as to provide effective coordination, responsibility, and control. The success of the JIT production system could be attributed to a different management approach. This approach was characterized by a wholistic approach to people and includes

1. Consensus decision making by management.
2. Vertical integration of company.
3. Mutual trust, integrity, and loyalty.
4. Working in teams or groups or clusters.
5. Incentive pay in the form of bonuses for company performance.
6. Stable (even lifetime) employment for 35 to 50% of the work force.
7. Large pool of part-time temporary workers.

There are many companies in the United States that employ some or all of these elements and, obviously, there are many different ways a company can be organized and managed. For the sake of argument, a simple organization chart is given in Figure 1-15, indicating the relationship between the various departments and the people involved. In this simple U.S. company, coordination at the design state is provided by a committee composed of all the vice presidents, the manager of planning and scheduling, the production engineer, and the chief industrial engineer. Each can provide vital input as to whether a new product should be made or an existing one altered or discontinued. The vice president of manufacturing must know that the product can be manufactured economically and what equipment will be required. The manager of planning and scheduling must know what material will be required and at what time. The production engineer will know what special tooling and equipment will be required, if any, and can assure its availability when needed, and often he can suggest design modifications to reduce and simplify tooling requirements. The chief industrial engineer must be able to predict the labor costs, determine the

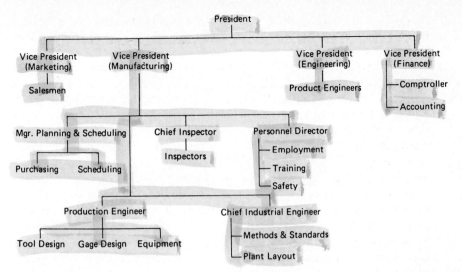

FIGURE 1-15 Organization chart for manufacturing portion of a typical modern business.

methods to be used, and plan the layout of the equipment. With such a systems approach, costly errors can be avoided and changes made on paper, rather than in the factory, and a better and less costly product usually results.

Each person on such a committee obviously will not know all about the regular work of the other members. But it is important that he understand how their functions are interrelated and how decisions within their own department will affect the operations of the others. For example, in designing the special tooling, the production engineer must keep in mind that, usually, the tools will be operated by people, and ease and speed of operation should be given just as much consideration as functional performance.

The type of cooperation and coordination that has just been described calls for engineers who are more than specialists in a given field. They also must possess a broad fundamental knowledge of design, metallurgy, processing, economics, accounting, and human relations. In the manufacturing game, low-cost mass production is the result of a team that cooperates to operate a plant as a coordinated unit. This is the key to producing more goods of better quality at less cost.

REVIEW QUESTIONS

1. What role does manufacturing play relative to the standard of living of a country?
2. How does recycling aide in today's world of manufacturing?
3. Explain the differences between job shop, flow shop, project shop, and cellular shop manufacturing systems.
4. How does a "system" differ from a process? From a machine? From a job? From an operation?
5. Is a cutting tool the same thing as a machine tool?
6. What are the basic manufacturing processes?

7. Why would it be advantageous if casting could be used to produce a complex-shaped part to be made from a hard-to-machine metal?
8. What are the basic factors that distinguish mass production?
9. Why may large-scale production not be the same as mass production?
10. List the forming processes used to make a "tin" can.
11. What deformation process is involved in machining?
12. It is acknowledged that chip-type machining, basically, is an inefficient process, yet it probably is used more than any other to produce desired shapes. Why?
13. What are the basic chip-type processes?
14. List three purposes of finishing operations.
15. What are the basic chipless machining processes?
16. List five assembly processes.
17. How is electrochemical machining related to electroplating?
18. How does brazing differ basically from welding?
19. What is a basic difference between mechanization and automation?
20. Give an example of a machine found in the home for each level of automation (Table 1-4).
21. What difficulties might result if the step "Production Planning and Scheduling" were omitted from the procedure shown in Figure 1-14?
22. A company is considering making automobile bumpers from aluminum instead of from steel. List some of the factors it would have to consider in arriving at its decision.
23. Discuss briefly the relationship of design to production.
24. It has been said that low-cost products are more likely to be more carefully designed than high-priced items. Do you think this is true, and why?

CASE STUDY 1. Economics of Mass Production

Obtain from a store some mass-produced product selling for around $1. It should be possible to disassemble the item. Then list the parts and make a sketch of each part. Note any design features which make the part easy to manufacture or assemble. Try to determine the material(s) used in each part. See if you can determine any way in which any of the parts could be altered to further simplify the manufacture or assembly of the parts without detracting from its functional worth or quality.

CHAPTER 2

Properties of Materials

In selecting a material, the primary concern of engineers is to match the material properties to the service requirements of the component. Knowing the conditions of load and environment under which a component must operate, engineers must select an appropriate material, using tabulated test data as the primary guide. They must know what properties they want to consider, how these are determined, and what restrictions or limitations should be placed on their application. Only by having a familiarity with test procedures, capabilities, and limitations can engineers determine whether the listed values of specific properties are, or are not, directly applicable to the problem at hand, and then use them intelligently to select a material.

Metallic and nonmetallic materials.   Perhaps the most common classification that is encountered in engineering materials is whether the material is *metallic* or *nonmetallic*. The common metallic materials are such metals as iron, copper, aluminum, magnesium, nickel, titanium, lead, tin, and zinc and the alloys of these metals, such as steel, brass, and bronze. They possess the *metallic properties* of luster, high thermal conductivity, and high electrical conductivity; are relatively ductile; and some have good magnetic properties. Some common nonmetals are wood, brick, concrete, glass, rubber, and plastics. Their properties vary widely, but they generally tend to be less ductile, weaker, and less dense than the metals, and they have poor electrical and thermal conductivities.

Although it is likely that metals always will be the more important of the two groups, the relative importance of the nonmetallic group is increasing rapidly, and since new nonmetals are being created almost continuously, this trend is certain to continue. In many cases the selection between a metal and nonmetal is determined by a consideration of required properties. Where the required properties are available in both, total cost becomes the determining factor.

Physical and mechanical properties.   One material can often be distinguished from another by means of *physical properties,* such as color, density, specific heat, coefficient of thermal expansion, thermal and electrical conductivity, magnetic properties, and melting point. Some of these, for example thermal conductivity, electrical conductivity, and density, may be of prime importance in selecting material for certain uses. Those properties that describe how a material reacts to various applied loads, however, are often more important to the engineer responsible for selecting materials.

The *mechanical properties* of materials are determined by subjecting them to standard laboratory tests, so that their reaction to changes in the controlled, influencing conditions can be determined. In using the results of such tests, however, the engineer must remember that they apply *only* to the specific test conditions. Caution should be exercised in the application of results, for the actual service conditions rarely duplicate the conditions of testing.

Stress and strain.   When a load is applied to a mechanism or structure, the material is deformed (*strained*) and internal reactive forces (*stresses*) are produced to resist the applied force. For example, if a weight, W, is suspended from a bar of uniform cross section, as in Figure 2-1, the bar will elongate slightly by an amount ΔL. For a given weight, W, the magnitude of the *elongation,* ΔL, will depend upon the original length of the bar. The amount of deformation of each unit length of the bar, expressed as $e = \Delta L/L$, is called the *unit strain*. Although it is a ratio of a length to another length and is a dimensionless number, it is usually expressed in terms of millimeter per meter, inch per inch, or as a percentage.

Application of the load W also produces reactive stresses within the bar, through which the load is transmitted to the supports. *Stress* is defined as the force or load being transmitted divided by the cross-sectional area transmitting the load. Thus, in Figure 2-1, the stress is $S = W/A$, where A is the cross-sectional area of the supporting bar. It ordinarily is expressed in terms of megapascals (in SI units) or pounds per square inch (in the English system).

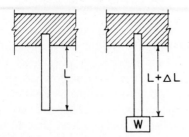

FIGURE 2-1 Tension loading and resulting elongation.

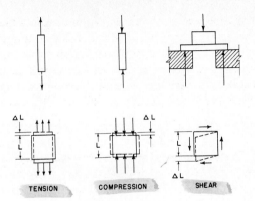

FIGURE 2-2 Examples of tension, compression, and shear loading, and their strain response.

In Figure 2-1, the weight tends to stretch or lengthen the bar, so the strain is known as a *tensile strain* and the stress as a *tensile stress*. Other types of loadings produce other types of stress and strain, as illustrated in Figure 2-2. *Compressive forces* tend to shorten the material and produce *compressive stresses and strains*. *Shearing stresses and strains* result from two forces acting on an element of material offset with respect to each other.

Static properties. When the loads applied to a material are constant and stationary, or nearly so, they are said to be *static*. In many uses the load conditions are essentially static, and it becomes important to characterize the behavior of materials under such conditions. Consequently, a number of standardized tests have been developed as a means to determine and report the *static properties* of materials. The documented test results can be used to select materials, provided that the service conditions are sufficiently similar to those of testing. Even when the service conditions differ from those of testing, the results can be used to qualitatively rate and compare various materials.

The tensile test. Considerable information about the properties of a material can be obtained from a uniaxial *tensile test*. A standard specimen is loaded in tension in a testing machine such as the one shown in Figure 2-3. Standard test conditions assure meaningful and reproducible test results. Standard specimens, the two most common of which are shown in Figure 2-4, are designed to produce uniform uniaxial tension in the central test portion and assure reduced stresses in the sections that are gripped.

A load, W, is applied and measured by the testing machine, while the elongation (ΔL) or strain over the specified gage length is determined by an external measuring device attached to the specimen. The result is a plot of coordinated load–elongation points, producing a curve of the form in Figure 2-5. Since characteristic loads will differ with different-size specimens and elongations will vary with different gage lengths, it becomes desirable to remove the size effects and establish a plot that is characteristic of the material's response to the test conditions. If the load is divided by the *original* cross-sectional area and the elongation is divided by the *original* gage length, the size effects are eliminated

FIGURE 2-3 Hydraulic-type tension and compression testing machine. (*Courtesy Tinius Olsen Testing Machine Co., Inc., Willow Grove, Pa.*)

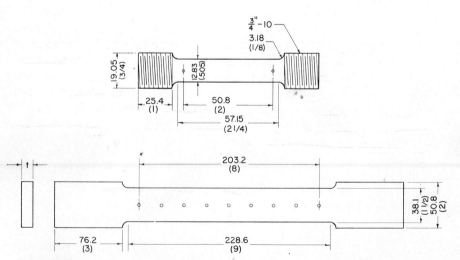

FIGURE 2-4 Two common types of standard tensile test specimens. (*Upper*) round; (*Lower*) flat. Dimensions are in millimeters with inches in parentheses.

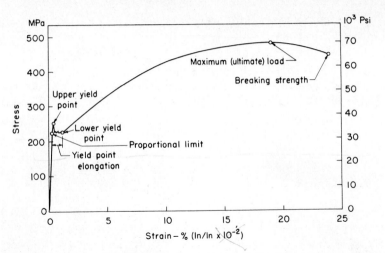

FIGURE 2-5 Stress–strain diagram for a low-carbon steel.

and the plot becomes known as an *engineering stress–strain curve,* as shown in Figure 2-5. This curve is simply a load–elongation curve with the scales of both axes modified to remove size effects.

In Figure 2-5 it will be noted that, up to a certain stress, the strain is directly proportional to the stress. The stress at which this proportionality ceases to exist is known as the *proportional limit.* Up to the proportional limit, the material obeys *Hooke's law,* which states that stress is directly proportional to strain. The proportionality constant, or ratio of stress to strain in this region, is known as *Young's modulus* or the *modulus of elasticity.* It is an inherent and constant property of a given material and is of considerable importance. As a measure of stiffness, it indicates the ability of a given material, for a given cross section, to resist deflection when loaded. It is commonly designated by the symbol *E*.

Up to a certain stress, if the applied load is removed, the specimen will return to its original length. Thus, from zero stress up to this point, the behavior is elastic and this region of the curve is known as the *elastic region.* The maximum stress for which truly elastic behavior exists is called the *elastic limit.* For some materials, the elastic limit and proportional limit are almost identical. In most cases, however, the elastic limit is slightly higher. Neither quantity, however, should be assigned great engineering significance, for the values are quite dependent upon the sensitivity of the test equipment.

The amount of energy that a unit volume of material can absorb within the elastic range is called the *resilience* or, in quantitative terms, the *modulus of resilience.* Because energy is the product of force times distance, the area under the load—elongation curve up to the elastic limit is equal to the energy absorbed by the specimen. In dividing load by original area to produce engineering stress, and elongation by gage length to produce engineering strain, the area under the stress–strain curve becomes the energy per unit volume or the modulus of resilience. This energy is potential energy and is therefore released whenever a member is unloaded.

Elongation beyond the elastic limit becomes unrecoverable and is known as *plastic deformation.* Upon removal of all loads, the specimen will retain a permanent change in shape. The engineer is usually interested in either the elastic or plastic response, but rarely both. For most components, plastic flow, except for a slight amount to permit the redistribution of stresses, represents failure of the component through loss of dimensional and tolerance control. In manufacturing where plastic deformation is used to shape a product, design stresses must be sufficient to put the workpiece into the plastic region. Thus some means is desired to determine the transition from elastic behavior to plastic flow.

Beyond the elastic limit, increases in strain do not require proportionate increases in stress. In some materials, a point may be reached where additional strain occurs without any increase in stress, this point being known as the *yield point* or *yield-point stress.* For low-carbon steels, as in Figure 2-5, two distinct points are significant: the highest stress preceding extensive strain, known as the *upper yield point,* and the lower, relatively constant, "run-out" value, known as the *lower yield point.* The lower value is the one that would appear in tabulated data.

Most materials, however, do not have a well-defined yield point, but have a stress–strain curve of the form shown in Figure 2-6. For such materials, the elastic-to-plastic transition is *defined* by the *offset yield strength.* This is the value of stress that will produce a given and tolerable amount of permanent strain. Deformations used are usually 0.2 or 0.1%, although 0.02% may be used when minute amounts of plastic deformation may lead to component failure. Offset yield strength is then determined by drawing a line parallel to the elastic line, displaced by the offset strain, and reporting the point where it intersects the stress–strain curve, as illustrated in Figure 2-6. The value is reproducible and is independent of equipment sensitivity, but it is meaningless unless reported with the amount of offset used.

As the straining of the material continues into the plastic range, the material gains increasing load-bearing ability. Since load-bearing ability is equal to strength times cross-sectional area, and the cross-sectional area is decreasing with tensile

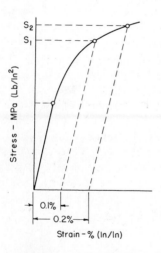

FIGURE 2-6 Stress–strain diagram for a material not having a well-defined yield point, showing the offset method for determining yield strength.

FIGURE 2-7 Standard 0.505-inch-diameter tensile specimen showing a necked portion developed prior to failure.

stretching of the specimen, the material must be increasing in strength. When the mechanism for this strengthening is discussed, in Chapter 3, it will be seen that the strength will always continue to increase with deformation. In the tensile test, however, a point is reached where the drop in area with increased strain dominates the increase in strength and the overall load-bearing capacity peaks and begins to diminish, as in Figure 2-5. The value of this point on the stress–strain curve is known as the *ultimate strength* or *tensile strength* of the material. The weakest point of the tensile bar at that time continues to be the weakest point by virtue of the decrease in area, and deformation becomes localized. This localized reduction of the cross-sectional area, known as *necking,* is shown in Figure 2-7. It is accompanied by a reduction in the amount of load required to produce additional straining, and the stress–strain curve drops.

If straining is continued far enough, the tensile specimen will ultimately fracture. The stress at which this occurs is called the *breaking strength* or *fracture strength.* For relatively ductile materials, the breaking strength is less than the ultimate tensile strength and necking precedes fracture. For a brittle material, fracture usually terminates the stress–strain curve before necking and possibly before the onset of plastic flow.

Ductility and brittleness. The extent to which a material exhibits *plasticity* is significant in evaluating its suitability to certain manufacturing processes. Metal deformation processes, for example, require plasticity; the more plastic a material is, the more it can be deformed without rupture. This ability of a material to be deformed plastically without fracture is known as *ductility.*

One of the major ways of evaluating ductility is to consider the *percent elongation* of a tensile-test specimen. As shown in Figure 2-8, however, materials do not elongate uniformly along their entire length when loaded beyond necking. Thus it has become common practice to report ductility in terms of percent elongation of a specified gage length. The actual gage length used in the evaluation is of great significance. For the entire 8-inch gage length of Figure 2-8, the elongation becomes 31%. If the center 2-inch segment is considered, elongation becomes 60%. Thus quantitative comparison of material ductility through elongation requires testing of specimens with identical gage lengths. A more meaningful measure of ductility may be the *uniform elongation* or *percent elongation prior to necking,* but elongation at fracture is the commonly reported value.

Another indication of ductility is the *percent reduction in area* that occurs in the necked region of the specimen. This is computed as

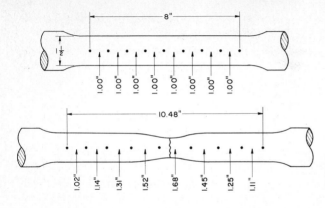

FIGURE 2-8 Elongation in various sections of a tensile test specimen.

$$\frac{A_o - A_f}{A_o} \times 100\%$$

where A_o is the original cross-sectional area and A_f is the smallest area in the necked region, independent of gage length.

Other terms that are often related to the ductility of materials include *malleability, workability,* and *formability.* These terms relate to the ability of a metal to undergo mechanical working processes without rupture. Although plasticity is the controlling property, the terms above relate to the material response to specific processes and as such do not describe a material property.

If the material fails with little or no ductility, it is said to be *brittle.* Thus brittleness can be viewed as the opposite of ductility. Brittleness should not be considered as the lack of strength, however, but simply the lack of significant plasticity.

Toughness. *Toughness* is defined as the work per unit volume required to fracture a material and is commonly expressed as a *modulus of toughness.* One means of measuring toughness is through the tensile test, for the total area under the stress–strain curve represents the energy required to produce a fracture in a unit volume of material.

Caution should be exercised in the use of toughness data, however, for the values can vary markedly with different conditions of testing. As will be seen later, variation in temperature and load-application rate can change the nature of a material's stress–strain curve and, hence, the toughness. Toughness is commonly associated with impact of shock loadings. The values obtained from impact tests, however, often fail to correlate with those from static-type tests.

True stress/true strain curves. The stress–strain curve of Figure 2-5 is a plot of *engineering stress, S,* versus *engineering strain, e,* where S is computed as load (W) divided by the *original* cross-sectional area (A_o) and e is the elongation (ΔL) divided by the *original* gage length (L_o). As noted previously and illustrated in Figures 2-7 and 2-8, the cross section of the test bar changes as the test proceeds, first uniformly and then nonuniformly after necking begins. The actual stress within the specimen, therefore, should be based on the instantaneous cross-sectional area and not the original and will be greater than the engineering stress shown in Figure 2-5. True stress, σ, can be computed by taking simultaneous readings of load and minimum specimen diameter. The true area can be computed and true stress determined as

$$\sigma = \frac{W}{A} \quad \left(\text{as opposed to the engineering stress, } S = \frac{W}{A_o} \right)$$

The determination of the true strain is somewhat more complex. What is desired is the strain at a point, since we have already seen strain variation throughout the specimen. One way is to utilize the engineering strain expression with an infinitesimal gage length, l_o, such that

$$e' = \frac{l - l_o}{l_o} = \frac{l}{l_o} - 1$$

This strain, e', is often called the *mean strain*, or, more correctly, the *zero gage length strain*. Although the infinitesimal gage length, l_o, cannot be measured directly, volume constancy enables determination through measurements of diameter and is valid up to necking.

$$\frac{l}{l_o} = \frac{A_o}{A} = \frac{D_o{}^2}{D^2}$$

A more precise and useful measure for true strain is the *natural strain* or *logarithmic strain*, which is the integral summation of the incremental elements, expressed as

$$\epsilon = \int_{l_o}^{l} \frac{dL}{L} = \ln \frac{l}{l_o} = 2 \ln \frac{D_o}{D}$$

Before the onset of necking, where strain is uniform, the natural strain can be related to the engineering strain by

$$\epsilon = \ln (e + 1)$$

Figure 2-9 shows the type of curve that results when the uniaxial tensile test data are plotted in the form of true stress vs. true strain. It should be noted that the true stress of the material, a measure of the material strength at that point,

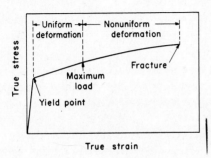

FIGURE 2-9 True stress–true strain curve.

FIGURE 2-10 Section of a tensile test specimen stopped just prior to failure, showing a crack already started in the necked region. (*Courtesy E. R. Parker.*)

continues to rise throughout the test, even after necking. Data beyond the point of necking should be used with extreme caution, however, for once the geometry of the neck begins to form, the stress state in that region becomes a triaxial tension instead of the uniaxial tension assumed for the test. Voids or cracks, such as in Figure 2-10, tend to open in the necked region as a preface to failure. Diameter measurements no longer reflect the true load-bearing area and the data are further distorted.

Repeated static loads. In many applications, static loads are applied more than once. In order to understand how a ductile metal, such as steel, behaves when subjected to repeated slow loading and unloading, refer to the stress-strain diagram of Figure 2-11. Unloading and reloading within the elastic range results in simply cycling up and down the linear portion of the diagram between O and A. However, if unloading takes place from point B in the plastic region, the unloading curve follows the path BeC, which is approximately parallel to OA. The permanent set at this point would be OC. Reloading from C would tend to follow the curve CfD, a slightly different path from that of unloading. The area within the loop formed by the two paths is called a *hysteresis loop*. The energy

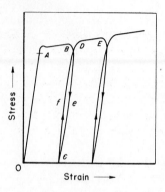

FIGURE 2-11 Stress–strain diagram obtained by unloading and reloading a specimen.

represented by this loop corresponds to the energy per unit volume that was transformed to heat within the material during the unloading and reloading cycle.

When most materials are plastically deformed, they *work-harden;* that is, they become harder and the yield-point stress is raised. This is a progressive phenomenon, with the result that, as the applied load is increased to produce plastic deformation, a greater load will be required to produce further deformation. In our example above, the work-hardening properties of the material have raised the yield point to *D* and elastic behavior is experienced up to this point upon reloading. Beyond the new yield point, *D,* additional plastic deformation takes place. If unloading then took place at *E,* with subsequent reloading as indicated, another hysteresis loop would be formed, and further raising of the yield point would be observed. From this example it may be seen that beyond the elastic region the true stress/true strain curve actually represents the locus of the yield stress for various amounts of strain.

Damping capacity.   The hysteresis loop, discussed in the previous section, was caused when some of the mechanical energy that was put into the material during the loading and unloading cycle was converted into heat energy. This process produces *mechanical damping,* and materials that possess this property to a high degree are able to absorb mechanical vibrations or damp them out rapidly. This is an important property of materials for certain uses, such as crankshafts and engine bases. Gray cast iron is used in many applications because of its high damping capacity. Materials with lower damping capacities, such as brass and steel, will continue to ring when struck by a blow.

Hardness.   *Hardness* is a very important, yet difficult to define, property of materials. Numerous tests have been developed around the definition of resistance to permanent indentation under static or dynamic loading. Other tests evaluate resistance to scratching, energy absorption under impact loading, wear resistance, or even resistance to cutting or drilling. Clearly, these phenomena are not the same. Thus, although hardness can be measured by a variety of well-standardized methods, there may be no correlation between the results obtained from the various tests. Caution should be exercised so that the test selected clearly evaluates the phenomenon of interest.

Brinell hardness test.   One of the earliest standardized methods of measuring hardness is the *Brinell test.* A hardened steel ball 1 centimeter in diameter is pressed into a smooth surface of material by a standard load of 500, 1500, or 3000 kilograms. The load and ball are removed and the diameter of the spherical indentation is measured, usually by means of a special grid or traveling microscope. The Brinell hardness number is equal to the load divided by the spherical surface area of the indentation expressed in kilograms per square millimeter:

$$\text{Brinell hardness number (BHN)} = \frac{\text{load}}{\text{surface area of indentation}}$$

In actual practice, the Brinell hardness number is determined from tables that tabulate the number versus the diameter of the indentation.

The Brinell test is subject to several limitations, however:

1. It cannot be used on very hard or very soft materials.
2. The test may not be valid for thin specimens. Preferably, the thickness of the material should be at least 10 times the depth of the indentation. Standards specify minimum hardnesses for which tests on thin specimens are valid.
3. The test is not valid for case-hardened surfaces.
4. The test should be conducted on a location far enough removed from the edge of the material so that no edge bulging results.
5. The noticeable indentation may be objectionable on finished parts.
6. The edge of the indentation may not always be clearly defined or may be rather difficult to see on material of certain colors.

Nevertheless, the test is not difficult to conduct and has the advantage that it measures the hardness over a relatively large area and, therefore, does not reflect small-scale variations. It is used to a large extent on iron and steel castings. Figure 2-12 shows a standard Brinell tester. Relatively small, portable testers are also available.

Rockwell hardness test. The widely used *Rockwell hardness test* is similar to the Brinell test in that the hardness value is a function of the indentation of a test piece by an indentor under static load. The nature of the test can be explained in connection with Figure 2-13. A small indentor, either a ¹⁄₁₆-inch ball or a diamond cone called *a brale*, is first seated firmly in the material by

FIGURE 2-12 Brinell hardness tester. (*Courtesy Tinius Olsen Testing Machine Co., Inc.*)

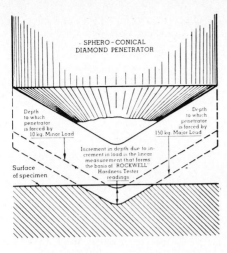

FIGURE 2-13 Operating principle of the Rockwell hardness tester. (*Courtesy Wilson Instrument Division, ACCO, Inc.*)

the application of a ''minor'' load of 10 kilograms, causing a very small indentation. The indicator on the dial of the tester, shown in Figure 2-14, is set at zero and a ''major'' load is then applied to the indentor to produce a deeper indentation. After the indicating pointer has come to rest, the major load is removed. With the minor load still applied, the pointer now indicates the Rockwell hardness number on the appropriate scale of the dial. The number is a measure of the *plastic* or *permanent penetration* produced by the major load.

FIGURE 2-14 Rockwell hardness tester having digital readout. (*Courtesy Wilson Instrument Division, ACCO, Inc.*)

TABLE 2-1 Loads and Indentors for Rockwell Hardness Tests

Test	Load (kg)	Indentor
A	60	Brale
B	100	$\frac{1}{16}$-in. ball
C	150	Brale
D	100	Brale
F	60	$\frac{1}{16}$-in. ball
G	150	$\frac{1}{16}$-in. ball

Different combinations of major loads and indentors are available and are used for materials of various degrees of hardness. Table 2-1 provides a partial listing of available Rockwell hardness scales. Since multiple scales exist, Rockwell hardness numbers must always be accompanied by a letter indicating the particular combination of load and indentor used in the test. The notation R_C60 (Rockwell C 60) indicates that the brale indentor was used with a major load of 150 kilograms and a reading of 60 was obtained. The B and C scales are used more extensively than the others.

The standard Rockwell tester should not be used on materials less than $\frac{1}{16}$ inch thick, on rough surfaces, or on materials that are not homogeneous, such as gray cast iron. As a result of the small size of the indentation, localized variations of roughness, composition, or structure can greatly influence the results. For thinner materials or purposes where a very shallow indentation is desired, the *Rockwell superficial-hardness test* is used. Operating on the same Rockwell principle, the test employs smaller major and minor loads and uses a more sensitive depth-measuring device. The test was designed primarily for determining the hardness of thin sheet metal and the surface hardness of materials that have received such surface treatments as nitriding or carburizing.

In comparison with the Brinell test, the Rockwell test offers the benefit of direct readings in a single step. Since it can be conducted rapidly, it is suitable for routine tests of hardness in mass production. Furthermore, it has the additional advantage that the smaller indentation is often not objectionable to the appearance of the component or is more easily removed in a later operation.

Vickers hardness test. The *Vickers hardness test* is similar to the Brinell test, with a square-based diamond pyramid being used as the indentor. As in the Brinell test, the Vickers hardness number is the ratio of the load to the surface area of the indentation in kilograms per square millimeter. An advantage of the Vickers machine is the increased accuracy in determining the diagonal of a square as opposed to the diameter of a circle and the assurance that even light loads will produce plastic deformation.

Numerous advantages make the indentation hardness methods quite popular: (1) simple to conduct, (2) little time involved, (3) little surface preparation required, (4) can be done on location, (5) is relatively inexpensive, and (6) often

FIGURE 2-15 Microhardness tester. (*Courtesy LECO Corporation.*)

provides results that can be correlated to quality control, material strength, and so on.

Microhardness tests. Various microhardness tests are available for use when it is necessary to determine hardness over a very small area of a material. The *Tukon tester,* shown in Figure 2-15, was developed for this purpose. The position for the test is selected under high magnification. A small diamond penetrator is then loaded with a predetermined load of from 25 to 3600 grams. The hardness number, known as a *Knoop hardness number,* is then obtained by dividing the load (in kilograms) by the projected area of the now diamond-shaped indentation (in square millimeters). The length of the indentation is determined using a microscope, because the mark is very small.

Durometer hardness test. For testing very soft, elastic materials, such as rubbers and nonrigid plastics, a *Durometer* is often used. This instrument, shown in Figure 2-16, measures the resistance of the material to elastic penetration by

FIGURE 2-16 Durometer hardness tester. (*Courtesy Shore Instrument & Mfg. Company, Inc.*)

FIGURE 2-17 Scleroscope hardness tester. (*Courtesy Shore Instrument & Mfg. Company, Inc.*)

a spring-loaded conical steel indentor. No permanent deformation occurs. A similar test is used to evaluate molding sands in the casting industry.

Scleroscope hardness test—a dynamic test. In the *Scleroscope test,* hardness is measured by the rebound of a small, diamond-tipped "hammer" that is dropped from a fixed height onto the surface of a material being tested, using a test instrument such as in Figure 2-17. Obviously, this test measures the resilience of the material, and the surface on which the test is made must have a fairly high polish to obtain good results. Scleroscope hardness numbers are comparable only among similar materials. A comparison between steel and rubber, therefore, would not be valid.

Scratch hardness tests. As mentioned previously, hardness can also be defined as the ability of a material to resist being scratched. At least two types of tests have been developed to determine hardness by this method.

The *Mohs hardness scale* arranges 10 minerals in order of ascending hardness as follows:

1. Talc.
2. Gypsum.
3. Calcite.
4. Fluorite.
5. Apatite.
6. Feldspar.
7. Quartz.
8. Topaz.
9. Sapphire or corundum.
10. Diamond.

According to this scale, a given material should be able to scratch any material having a lower Mohs number. In this manner, any substance can be assigned an approximate number on the Mohs scale. Ordinary glass, for example, would be about 5.5; hardened steel, about 6.5. This test is sufficiently crude that it is not suitable for manufacturing purposes, but it is quite useful in mineral identification.

Another crude, but often useful, hardness test is the *file test,* wherein one determines whether the material can be cut with a file. This test can be either a pass–fail test using a single file, or a semiquantitative evaluation using a series of files pretreated to various levels of known hardness.

Relationships among the various hardness tests.   Since the various tests tend to evaluate somewhat different material phenomena, there is no simple relationship between the several types of hardness numbers that can be determined. Approximate relationships have been developed, however, by testing the same material on various devices. Table 2-2 gives a comparison of hardness values for plain carbon and low-alloy steels. It may be noted that for Rockwell C numbers above 20, the Brinell numbers are approximately 10 times the Rockwell numbers. Also, for hardnesses below 320 Brinell, the Vickers and Brinell numbers agree closely. Since the relationships will vary with material, mechanical treatment, and heat treatment, tables such as Table 2-2 should be used with caution.

Relationship of hardness to tensile strength.   Table 2-2 also shows a comparison of hardness and tensile strength for steel. For plain carbon and very low alloy steels, the tensile strength (in psi) can be determined fairly well by multiplying the Brinell hardness number by 500. This provides a simple, and very useful, method of determining the approximate tensile strength of a steel by means of a hardness test. For other materials, the relationship may be too variable to be dependable. For example, for duraluminum the ratio is about 600, whereas for soft brass it is about 800.

Compression tests.   When a material is subjected to compressive forces, the relationships between stress and strain are similar to those for a tension test. Up to a certain value of stress, the material behaves elastically; beyond it plastic flow occurs. In general, however, the compression test is more difficult and more complex than the standard tensile test. Test specimens must have larger cross-sectional areas to resist bending or buckling. As deformation proceeds, the cross section of the specimen tends to increase producing a substantial increase in required load (true stress versus true strain is substantially the same for both cases). Frictional effects between the testing machine surfaces and the end surfaces of the specimen will tend to alter the results if not properly considered. The selection of the tension or compression mode of testing, however, is largely determined by the type of service to which the material is to be subjected.

Failure under compressive loading is generally by buckling or by shear along a plane at 45° to the axis of loading. Figure 2-18 shows a compression failure of a wood specimen.

TABLE 2-2 Hardness Conversion Table FOR STEEL

Brinell Number	Vickers Number	Rockwell Number C	Rockwell Number B	Scleroscope Number	Tensile Strength MPa	Tensile Strength 1000 psi
	940	68		97	2537	368
757[a]	860	66		92	2427	352
722[a]	800	64		88	2324	337
686[a]	745	62		84	2234	324
660[a]	700	60		81	2144	311
615[a]	655	58		78	2055	298
559[a]	595	55		73	1903	276
500	545	52		69	1765	256
475	510	50		67	1703	247
452	485	48		65	1641	238
431	459	46		62	1462	212
410	435	44		58	1407	204
390	412	42		56	1351	196
370	392	40		53	1303	189
350	370	38	110	51	1213	176
341	350	36	109	48	1138	165
321	327	34	108	45	1069	155
302	305	32	107	43	1007	146
285	287	30	105	40	951	138
277	279	28	104	39	924	134
262	263	26	103	37	883	128
248	248	24	102	36	841	122
228	240	20	98	34	800	116
210	222	17	96	32	738	107
202	213	14	94	30	683	99
192	202	12	92	29	655	95
183	192	9	90	28	627	91
174	182	7	88	26	600	87
166	175	4	86	25	572	83
159	167	2	84	24	552	80
153	162		82	23	524	76
148	156		80	22	510	74
140	148		78	22	490	71

TABLE 2-2   Hardness Conversion Table (*Continued*)

Brinell Number	Vickers Number	Rockwell Number		Scleroscope Number	Tensile Strength	
		C	B		MPa	1000 psi
135	142		76	21	469	68
131	137		74	20	455	66
126	132		72	20	441	64
121	121		70		427	62
112	114		66			58

ªTungsten carbide ball; others standard ball.

Dynamic properties.   In many engineering applications, materials are subjected to dynamic loadings. Such cases may involve components that (1) experience sudden loads or loads which rapidly vary in magnitude, (2) are loaded and unloaded repeatedly, or (3) undergo frequent changes in loading modes, such as from tension to compression. For such service conditions, the engineer is concerned with properties other than those determined by the static tests.

Unfortunately, the dynamic tests are not as well standardized, or controlled as easily, as the static tests. In addition, many of the dynamic tests do not give results that can be used directly in design. In such cases, the tests merely classify materials relative to each other regarding their behavior when subjected to certain loading conditions. Nevertheless, they can serve a very useful purpose, provided that one remembers their limitations.

The impact test.   Several tests have been developed to evaluate the fracture resistance of a material under rapidly applied dynamic loads, or *impacts*. Of those tests that have become common, two basic types have emerged: (1) bending impact tests, which include the standard Charpy and Izod tests, and (2) tension impact tests.

The bending impact tests utilize specimens that are supported as beams. In the *Charpy test*, the specimen contains either a V, keyhole, or U notch—the keyhole and V being most common. As show in Figure 2-19, the Charpy test specimen is supported as a simple beam, and the impact is applied to the center, behind the notch, to complete a three-point bending. The *Izod test* specimen is

$$\frac{12}{95} = \frac{14}{99}$$

FIGURE 2-18   Failure of wood under compressive loading.

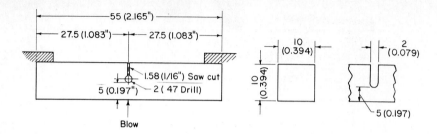

FIGURE 2-19 Standard Charpy impact specimens and mode of loading. Illustrated are keyhole and U notches. Standard V-notch specimen has notch geometry shown in Figure 2-20.

supported as a cantilever beam and is impacted on the end as in Figure 2-20. Standard testing machines, such as the one shown in Figure 2-21, apply a predetermined impact energy by means of a swinging pendulum. After breaking or deforming the specimen, the pendulum continues to swing with an energy equal to its original minus that absorbed by the broken specimen. This loss is measured by the angle attained by the pendulum on its upward swing.

Test specimens for bending impacts must be prepared with careful precision to assure consistent and reproducible results. Notch profile, particularly the radius at the root of V-notch specimens, is extremely critical, for the test measures the energy required to both initiate and propagate a fracture. The effect of notch profile is shown dramatically in Figure 2-22. Here, two specimens have been made from the same piece of steel with the same reduced cross-sectional area. The one with the keyhole notch fractures and absorbs only 58 joules (43 ft-lb) of energy, while the other specimen resists fracture and absorbs 88 joules (65 ft-lb) during impact.

Additional cautions should be placed on the use of impact test data for design purposes. These tests only indicate the impact resistance of materials that contain a standardized notch. Changes in the form of the notch or minor variations from standard geometry can produce significant changes in the results. The test also evaluates a standard specimen under only one condition of impact rate. Under modified test conditions using faster rates of loading or wide specimens with a higher degree of constraint, many materials will behave in a more brittle fashion. Bending impact tests are valuable in ranking materials as to their sensitivity to notches and the multiaxial stresses that exist around a notch. In addition, testing

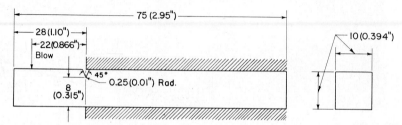

FIGURE 2-20 Izod impact specimen and mode of loading.

FIGURE 2-21 Impact testing machine. (*Courtesy Tinius Olsen Testing Machine Co., Inc.*)

temperature can be varied to enable evaluation of the fracture resistance of a material as a function of temperature. Such information can provide invaluable information to the engineer involved in material selection.

The *tensile impact test,* illustrated schematically in Figure 2-23, avoids many of the objections inherent in the Charpy and Izod tests, but is more difficult to perform. The behavior of ductile materials under uniaxial impact loading can be studied without the complications introduced by the use of a notched speci-

FIGURE 2-22 Notched and unnotched impact specimens before and after testing. Both specimens had the same cross-sectional area.

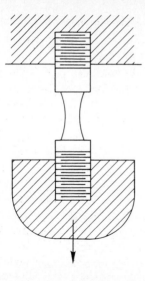

FIGURE 2-23 Tensile impact test schematic.

men. Various methods, such as drop-weight, modified pendulum, and variable-speed flywheel, have been used to supply the impact.

Metal fatigue and endurance limit. Metals may also fracture when subjected to repeated applications of stress, even though all stresses lie below the ultimate tensile strength and usually below the yield strength determined by a tensile specimen. This phenomenon, known as *metal fatigue,* may result from the repetition of a particular loading cycle or from an entirely random variation of stress. Since such failures probably account for more than 90% of all mechanical fractures, it is important for the engineer to know how materials will react to fatigue conditions.

Although there are an infinite number of possible repeated loadings, the periodic, sinusoidal mode is most suitable for experimental reproduction and subsequent analysis. Restricting conditions even further by considering only equal-magnitude tension-compression reversals, curves such as that of Figure 2-24 can be developed. If this material were subjected to a normal static tensile test, it would break at about 480 MPa (70,000 psi). However, if it were repeatedly subjected to a reversing stress of ±380 MPa (55,000 psi)—considerably below

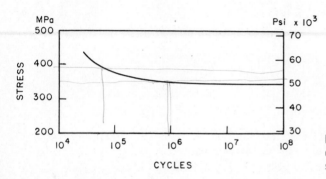

FIGURE 2-24 Typical *S–N* or endurance limit curve for steel.

its breaking stress—it would fail when the load was repeated about 100,000 times. Similarly, if a stress of ±350 MPa (51,000 psi) were applied, 1,000,000 cycles could be sustained prior to failure. By reducing the applied stress to below ±340 MPa (49,000 psi), this steel will not fail regardless of the number of stress applications. Such a curve is known as a stress versus number of cycles or *S–N curve*. Any point on the curve is the *fatigue strength* corresponding to the given number of cycles of loading. The limiting stress level, below which the material will not fail regardless of the number of cycles of loading, is known as the *endurance limit* or *endurance strength* and is an important criterion in many design applications.

A different number of cycles of loading is required to reveal the endurance limit of different materials. For steels, about 10 million (10^7) cycles are usually sufficient. Several of the nonferrous metals require 500 million (5×10^8) cycles. Some aluminum alloys require an even greater number, such that no endurance limit is apparent under typical test conditions.

The apparent fatigue strength of materials may be affected by several factors. One of the most important of these is the presence of stress raisers such as small surface cracks, machining marks, and so on. Data for *S–N* curves are obtained from polished specimens, and the observed lifetime is the cumulative number of cycles required to initiate a fatigue crack and propagate it to failure. If a part contains a surface crack or flaw, the cycles required for crack initiation can be markedly reduced. In addition, stresses concentrate at the tip of the crack producing an accelerated rate of crack growth. Great care should be taken to eliminate stress raisers and surface flaws on parts subjected to cyclic loadings. Proper design and manufacturing practices are often more critical than material selection and heat treatment for fatigue applications (see Chapter 39).

Another factor worthy of consideration is the temperature of testing. Figure 2-25 shows the shifts in the *S–N* curve for Inconel alloy 625 (Ni–Cr–Fe alloy) as temperature is varied. Since most test data are generated at room temperature, caution should be exercised when the application involves elevated service temperatures.

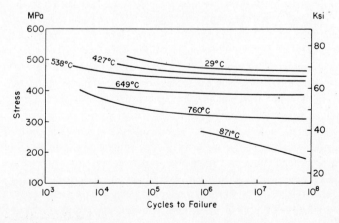

FIGURE 2-25 Fatigue strength of Inconel alloy 625 at various temperatures. (*Courtesy Huntington Alloy Products Division, The International Nickel Co., Inc.*)

TABLE 2-3   Ratio of Endurance Limit to Tensile Strength for Various Materials

Material	Ratio
Steel, AISI 1035	0.46
Steel, screw stock	0.44
Steel, AISI 4140 normalized	0.54
Wrought iron	0.63
Copper, hard	0.33
Beryllium copper (heat-treated)	0.29
Aluminum	0.38
Magnesium	0.38

Fatigue lifetime can also be altered by changes in the environment. When metals are subjected to corrosion during the repeated loadings, a situation known as *corrosion fatigue,* specimen lifetime and the endurance limit are significantly reduced. Special corrosion-resistant coatings, such as zinc or cadmium, may be required for these applications. More recently, tests conducted in a vacuum have produced significantly different results from those conducted in air and variations have been observed with different humidity levels. Direct application of test data, therefore, should be done with caution. If the magnitude of the applied stress is varied during service, a condition common to many components, the fatigue response of the metal becomes quite complex. While low-stress cycles are less damaging to the material, a few high-stress cycles may substantially reduce the expected lifetime. Such variations and their response are of significant importance to design engineers.

Table 2-3 shows the approximate ratio of the endurance limit to ultimate tensile strength for several metals.

Fatigue failures.   Metal components that fail as the result of repeated applications of load and the fatigue phenomenon are commonly called *fatigue failures.* These fractures form a major part of a larger classification known as *progressive fractures.* If the fracture surface of Figure 2-26 is examined closely, two points of fracture initiation can be located. These points usually correspond to discontinuities, in the form of a fine surface crack, a sharp corner, machining marks, or even "metallurgical notches" such as an abrupt change in metal structure. Once started, the crack propagates through the metal upon repeated application of load, crack growth being due to the stress at the tip of the crack exceeding the strength of the material. Crack propagation continues until the remaining section of metal no longer has sufficient area to sustain the applied load, at which time complete failure of the remaining section occurs. The section of metal involved in this final failure will have a relatively coarse, granular appearance; whereas the fatigued section between the origin of the crack and the coarse-appearing area will be relatively smooth.

FIGURE 2-26 Progressive fracture of an axle within a ball-bearing ring, starting from two points (arrows).

The smooth areas of the fracture may often contain a series of crescent shaped ridges radiating outward from the origin of the crack. These markings, however, may not be observable under ordinary visual examination. They may be very fine, they may have been obliterated by a rubbing action during repeated cyclic loading, or there may only be a few such marks if failure occurred after only a few cycles of loading ("low-cycle fatigue"). Electron microscope studies of the fracture surface can often reveal these small parallel ridges, or *striations,* which are characteristic of progressive failure. Figure 2-27 presents a typical fatigue fracture at high magnification.

Because the final area of fracture has a crystalline appearance, it has often been said that such failures are due to the metal having "crystallized." Since solid metals are always crystalline, such a conclusion is obviously erroneous and the term should not be applied.

Another common misnomer is to apply the term "fatigue failure" to all fractures having the characteristic progressive failure appearance. The general appearance of fractures where fatigue is a major factor is often the same as for other fractures where fatigue may be only a minor contributor. Also, the same

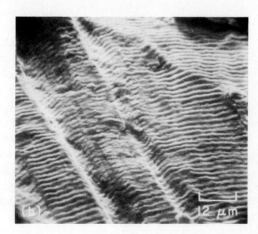

FIGURE 2-27 Fatigue fracture of AISI type 304 stainless steel viewed in a scanning electron microscope at 810×. Well-defined striations are visible. (*From "Interpretation of Scanning-Electron-Microscope Fractographs,"* Metals Handbook, *Vol. 9, 8th ed., American Society for Metals, Metals Park, Ohio, 1970, p. 70.*)

fracture phenomena may lead to different general appearances depending on the specific conditions of load magnitude, load type (torsional, bending, tension), temperature, and so on. Correct failure analysis requires far more information than can be provided by the examination of a fracture surface.

A final fact regarding failure by fatigue relates to the misconception that failure is time dependent. The failure of materials under repeated loads below their static strengths is not a function of time, but is dependent on the history of loading. High cyclic frequencies can produce failure in relatively short time intervals.

Temperature effects.   It cannot be overemphasized that test data used in design and engineering decisions should be obtained under conditions that best simulate the conditions of service. Engineers are frequently being confronted with the design of structures, such as aircraft, space vehicles, gas turbines, and nuclear power plants, that require operation under temperatures as low as $-130°C$ ($-200°F$) or as high as $1250°C$ ($2300°F$). Consequently, it is imperative for the designer to know both the short-range and long-range effects of temperature on the mechanical and physical properties of a material being considered for such applications. From a manufacturing viewpoint, the effects of temperature variations are equally important. Since numerous manufacturing processes involve the use of heat, the processing may tend to alter the properties in a favorable or unfavorable manner. Often a material can be processed successfully, or economically, only because its properties can be changed by heating or cooling.

To a manufacturing engineer, the most important effects of temperature on materials are those relating to the tensile and hardness properties. Figure 2-28 illustrates changes in key data for the case of a medium-carbon steel. Similar effects are shown for magnesium in Figure 2-29. In general terms, an increase in temperature tends to promote a drop in strength and hardness properties and an increase in elongation. For forming operations, these trends are of consid-

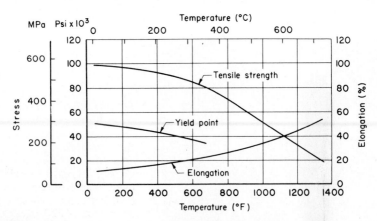

FIGURE 2-28 Some effects of temperature on the tensile properties of a medium-carbon steel.

MATERIALS

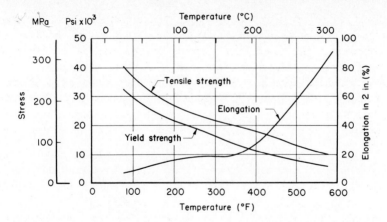

FIGURE 2-29 Effects of temperature on the tensile properties of magnesium.

erable importance because they permit forming to be done more readily at elevated temperatures where the material is weaker and more ductile.

Figure 2-30 adds another dimension by showing the effects of both temperature and strain rate on the ultimate tensile stress. From this graph it can be clearly seen that the rate of deformation can strongly influence mechanical properties. Room-temperature, standard-rate tensile test data will be of little use to the engineer concerned with the behavior of a material being hot rolled at speeds of 1200 meters per minute (4600 feet per minute). The effects of strain rate on the more important yield strength value are more difficult to evaluate, but follow the same trends as tensile strength.

The effect of temperature on impact properties became the subject of intense study when ships, structures, and components fractured unexpectedly in cold environments. Figure 2-31 shows the effect of decreasing temperature on the impact properties of two low-carbon steels. Although of similar compositions, the steels show distinctly different response. The steel indicated by the solid

FIGURE 2-30 Effects of temperature and strain rate on the tensile strength of copper. (*From A. Nadai and M. J. Manjoine, J. Appl. Mech., Vol. 8, 1941, p. A82, courtesy ASME.*)

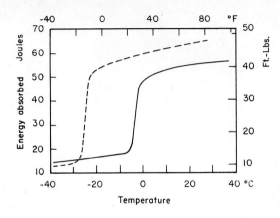

FIGURE 2-31 Effect of temperature on the impact properties of two low-carbon steels.

line becomes brittle at temperatures below −4°C (25°F), while the other steel retains its fracture resistance down to −26°C (−15°F). The temperature at which the response goes from high to low energy absorption is known as the *transition temperature* and is useful in evaluating the suitability of materials for certain applications. All steels tend to exhibit the rapid transition in impact strength when temperature is decreased, but the temperature at which it occurs varies with the material. Special steels with high nickel contents and several other alloys have been developed for cryogenic applications requiring retention of impact resistance to −195°C (−320°F).

Creep. The long-term effect of temperature is manifest in a phenomenon known as *creep*. If a tensile-type specimen is subjected to a fixed load at an elevated temperature, it will elongate continuously until rupture occurs, even though the applied stress is below the yield strength of the material at the temperature of testing. Although the rate of elongation is small, it is sufficient to be of great importance in the design of equipment such as steam or gas turbines, power plants, and high-temperature pressure vessels that operate at high temperatures for long periods of time.

If a single specimen is tested under fixed load and fixed temperature, a curve such as that of Figure 2-32 is generated. The curve contains three distinct stages:

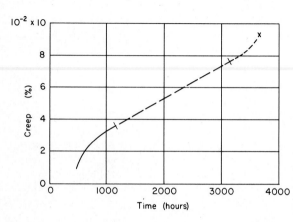

FIGURE 2-32 Creep curve for a single specimen at a fixed temperature, showing three stages of creep.

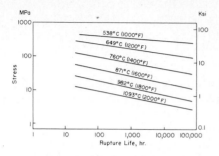

FIGURE 2-33 Stress-rupture diagram of a solution-annealed Incoloy alloy 800 (Fe-Ni-Cr alloy). (*Courtesy Huntington Alloy Products Division, The International Nickel Company, Inc.*)

a short-lived initial stage, a rather long second stage of rather linear elongation rate, and a short-lived third stage leading to fracture. From each test, two significant pieces of engineering data are obtained, the rate of elongation in the second stage, or *creep rate,* and the total elapsed time to rupture. Tests conducted at higher temperatures or with higher applied loads would produce higher creep rates and shorter rupture times.

A very useful engineering tool where creep is a significant factor is a *stress–rupture diagram,* such as the one in Figure 2-33. Rupture-time data from a number of tests at various temperatures and stresses are plotted on a single diagram. Creep-rate data can also be plotted to show the effects of temperature and stress as in Figure 2-34.

In general, the alloying elements of nickel, manganese, molybdenum, tungsten, vanadium, and chromium are helpful in lowering the creep rate of steel. At high temperatures, coarse-grained steels seem to be more creep resistant than fine-grained steels. Below the lowest recrystallization temperature (a property discussed in Chapter 3), however, the reverse is true, with a fine-grained structure being preferred. Killed steels show superior creep-resisting properties when compared to rimmed steels.

Machinability. Material properties relating to the response of a material to various processing techniques are some of the most difficult properties to define. The terms "malleability," "workability," and "formability" have already been introduced as measures of a material's suitability to plastic deformation processes. *Machinability* is another such term, important from the production viewpoint, but hard to define in that it depends not only on the material involved, but also on the specific process and aspect of interest. In some cases one is concerned with how easily a material cuts, with little regard for surface finish. At other

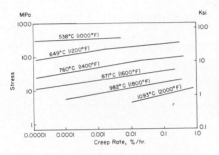

FIGURE 2-34 Creep-rate properties of solution-annealed Incoloy alloy 800. (*Courtesy Huntington Alloy Products Division, The International Nickel Company, Inc.*)

times surface finish may be of prime importance. The formation of fine chips, so as to facilitate chip removal, may be another desirable feature. Also, tool life may be a consideration. Thus the general term "machinability" may involve several properties of a material, each of varying importance.

Perhaps the best approach to machinability is to consider that one is usually interested in removing the greatest amount of material in the shortest time, without requiring tool redressing or replacement, producing a satisfactory surface finish, and maintaining a low overall cost. Thus, good machinability is associated with the removal of material with moderate forces, formation of rather small chips, minimization of tool abrasion, and production of good surface finish. Chapter 17 will deal more fully with the relationships between machinability and the properties of materials.

The fracture mechanics approach.   This chapter would not be complete without mention of the many tests and design concepts based on the fracture mechanics approach. Using the premise that all materials contain flaws, material characterization tends to focus on three quantities: (1) the size of the largest or most critical flaw, (2) the applied stress, and (3) the *fracture toughness*—a property that describes the conditions necessary for flaw growth or propagation leading to fracture. If nondestructive testing or quality control checks are applied, the size of the largest flaw that might go undetected can be determined. Assuming this flaw to be in the most critical location and knowing the stress in that location, the designer can specify a material with sufficient fracture toughness that the flaw would not propagate to a failure in service. Conversely, if the material and stress conditions were defined, the size of the maximum permissible flaw that would not propagate could be determined. The approach has proved invaluable in many areas where fracture could be catastrophic and has shown great refinement and increased acceptance in recent years.

REVIEW QUESTIONS

1. In general, how do the properties of nonmetallic materials differ from those of the metals?
2. What are some of the common physical properties of materials?
3. In general, how are the mechanical properties of materials determined?
4. Why should an engineer or designer use caution when applying test results to actual products?
5. What is engineering stress? Engineering strain?
6. Of what significance to the designer is the modulus of elasticity?
7. What is the major significance of the yield point or yield strength?
8. What is the significance of the modulus of resilence?
9. How is the elastic-to-plastic transition generally reported for materials that do not have a well-defined yield point?
10. Why is a manufacturing engineer generally more concerned with the plastic response of a metal than with its elastic properties?
11. Why is ultimate tensile strength not a good material property to use for design purposes?
12. Why might an engineer be concerned about the ductility of a given material?
13. How is ductility generally measured?
14. Why might an engineer be concerned about the toughness of an engineering material?
15. What is the difference between true stress and engineering stress?
16. What is meant by the term "work hardening" when applied to metals?

17. Why might an engineer be interested in the damping capacity of an engineering material?
18. What are the two major types of hardness tests applied to metals?
19. Why must Rockwell hardness test results always be reported with a letter such as A, B, or C?
20. How might the hardness of a very thin specimen be determined?
21. What are some of the advantages of the indentation hardness methods?
22. Explain the lack of correlation between the various types of hardness tests?
23. How can hardness be an indication of material strength?
24. What are the two major types of impact tests?
25. Why should a designer use extreme caution when applying impact test data for design purposes?
26. What is metal fatigue?
27. What is an endurance limit or endurance strength?
28. Why are stress raisers or surface flaws so critical in a part that will experience fatigue-type loading?
29. What are the two major regions observable on the fracture surfaces of most fatigue fractures?
30. How might the major engineering properties of a metal change when it is heated to elevated temperature?
31. Why is there a concern when a steel product is expected to receive impacts at low temperature?
32. Under what conditions might the creep properties of a material be of concern to a designer?
33. Why might it be difficult to develop a universally acceptable test for machinability?
34. What is the basic premise of the fracture mechanics approach to testing and design?
35. What is fracture toughness?

CASE STUDY 2.   The Mixed-Up Steel

You are an engineer employed by the ALO Company. Because of a shipping accident, 100 bars each of AISI 1020 and AISI 1040 hot-rolled steel have become mixed during shipment to the company's warehouse in Alaska. It is essential that the bars be correctly identified. You are being sent to the Alaska warehouse to identify each bar. The only equipment available at the warehouse that can be used in the identification is a 227-kN (60,000-lb) tensile testing machine, a Brinell hardness tester, and the equipment in a small machine shop.

Determine the best procedure to use and justify your decision, making use of standard data-source references, such as Volume 1 of the ASM Handbook.

CHAPTER 3

The Nature of Metals and Alloys

The structure–property relationship.   The fundamental engineering *properties* of materials presented in Chapter 2 are the direct result of the *structure* of the particular material. Moreover, such properties as strength and ductility are often sensitive to minute variations of structure, some of which are macroscopic, others microscopic, and still others are on the atomic scale. For the engineer to control the properties of materials and intelligently use them to their optimum, he first must have a working knowledge of material structure.

The basic structure of materials.   Since all materials are composed of the same basic components—*protons, neutrons,* and *electrons*—it is amazing that such a variety of materials exists with such widely varying properties. Variation is explained when one considers the many possible combinations of these units in a macroscopic assembly. The subatomic components, listed above, combine in different arrangements to form the various elemental *atoms,* each having a *nucleus* of protons and neutrons surrounded by the proper number of electrons to maintain charge neutrality. Atoms then combine in distinctive arrangements to form *molecules* or *crystals.* These units can then be assembled in differing amounts and configurations to form a microscopic scale structure, or *microstructure,* and ultimately an engineering component. The engineer, therefore, has at his disposal a wide variety of metals and nonmetals that possess an almost unlimited range of properties. Because the properties of materials depend upon

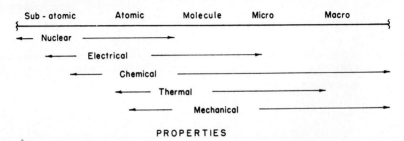

FIGURE 3-1 General relationship of structural level to engineering properties.

all levels of structure, as shown schematically in Figure 3-1, it is important for the engineer to understand the entire structure spectrum, from atomic to macroscopic.

Atomic structure. Experiments have revealed that atoms consist of a relatively dense nucleus composed of positively charge protons and neutral particles of nearly identical mass, known as neutrons. Surrounding the nucleus are the negatively charged electrons, which have only $\frac{1}{1839}$ the mass of a neutron and appear in numbers equal to the protons to maintain a net charge balance. Distinct particle groupings produce the known elements, ranging from the relatively simple hydrogen atom to unstable transuranium atoms over 250 times as heavy. Except for density and specific heat, however, the weight of atoms has relatively little influence on engineering properties.

The light electrons that surround the nucleus play a far more significant role in determining properties. Again, experiments reveal that the electrons are arranged in a characteristic structure consisting of shells and subshells, each possessing a distinctive energy. Upon absorbing a small amount of energy, an electron can jump from a low-energy shell near the nucleus to a higher-energy shell farther out. The reverse jump can occur with the release of a distinct amount or *quantum* of energy.

Each of the various shells and subshells contains only a limited number of electrons. The first shell, nearest the nucleus, can contain only two. The second shell can contain eight; the third, 32. Each shell and subshell is most stable when it is completely filled. For atoms containing electrons in the third shell and beyond, however, relative stability is associated with eight electrons in the outermost layer.

If, in its outer shell, a normal atom has slightly less than the number of electrons required for stability (for example, seven in its third shell), it will readily accept an electron from another source. It will then have one electron more than the number of protons and becomes a negatively charged atom or *negative ion*. Extra electrons may cause the formation of ions having negative charges of 1, 2, 3, and so on. If an atom has a slight excess of electrons beyond the number required for stability (such as sodium, with one electron in the third shell), it will readily give up the excess electron and become a *positive ion*. The remaining electrons become more strongly bound, making the removal of electrons progressively more difficult.

The number of electrons surrounding the nucleus of a neutral atom is called the *atomic number*. More important, however, are those electrons in the outermost shell (or subshell) known as *valence electrons*. These are influential in determining chemical properties, electrical conductivity, some mechanical properties, the nature of interatomic bonding, atom size, and optical characteristics. Elements with similar electron configurations in their outer shells will tend to have similar properties.

Atomic bonds. Atoms are rarely found as free and independent units, but usually are linked or bonded to other atoms in some manner as a result of interatomic forces. The electronic structure of the atoms influences the nature of the bond, which may be classed as *primary* (strong) or *secondary* (weak).

· The simplest type of primary bond is the *ionic bond*. Electrons break free of atoms with excesses in their valence shell, producing positive ions, and unite with atoms having an incomplete outer shell to form negative ions. The positive and negative ions have a natural attraction for each other, producing a strong bonding force. Figure 3-2 illustrates the process for a bond between sodium and chlorine. In the ionic type of bonding, however, the atoms do not unite in simple pairs. All positively charged atoms attract all negatively charged atoms. Thus, for example, sodium ions surround themselves with negative chlorine ions, and chlorine ions surround themselves with positive sodium ions. The attraction is equal in all directions and results in a three-dimensional structure, such as in Figure 3-3, rather than the simple link of a single bond. For stability in the structure, total charge neutrality must be maintained, thereby requiring equal numbers of positive and negative charges. General characteristics of materials joined by ionic bonds include moderate to high strength, high hardness, brittleness, high melting point, and electrical insulating properties (all charge transport must be through ion movement).

A second type of primary bond is the *covalent* type. Here the atoms being linked find it impossible to produce completed shells by electron transfer, but achieve the same goal by electron sharing. Adjacent atoms share outer-shell electrons so that each achieves a stable electron structure. The shared negative electrons locate between the positive nuclei to form the bonding link. Figure 3-4 illustrates this type of bond for chlorine, where two atoms, each containing seven valence electrons, share a pair to form a stable *molecule*. Stable molecules can also form from the sharing of more than one electron from each atom, as

FIGURE 3-2 Mechanism of ionization of sodium and chlorine, producing stable outer shells by electron transfer.

FIGURE 3-3 Three-dimensional structure
of the sodium chloride molecule. Note how
ions are surrounded by ions of opposite
charge.

in the case of nitrogen in Figure 3-5a. Atoms need not be identical (as in HF in Figure 3-5b), the sharing need not be equal, and an atom may share with more than one other atom. For elements such as carbon (four valence electrons), one atom may share valence electrons with each of four neighboring carbon atoms. The resulting structure becomes a network of bonded atoms, Figure 3-5c, instead of a finite, well-defined molecule. Like the ionic bond, the covalent bond tends to produce materials with high strength and high melting points. Atom movement within the material (deformation) requires the breaking of distinct bonds, thereby making the material characteristically brittle. Electrical conductivity depends upon bond strength, ranging from conductive tin (weak covalent bond) through semiconductive silicon and germanium to insulating diamond. Engineering materials possessing ionic and covalent bonds tend to be ceramic (refractories or abrasives) or polymeric in nature.

A third type of primary bond can result when a complete outer shell cannot be formed by either electron transfer or electron sharing, and is known as the *metallic bond*. If there are only a few valence electrons (1, 2, or 3) in each of a grouping of atoms, these electrons can be removed relatively easily while the remainder are held firmly to the nucleus. The result is a structure of positive ions (nucleus and nonvalence electrons) surrounded by a wandering assortment of universally shared valence electrons (electron cloud or gas), as in Figure 3-6. These highly mobile ''free'' electrons account for the observed high electrical and thermal conductivity as well as the opaque optical properties (free electrons can absorb light radiation energies). Moreover, they provide the ''cement'' necessary to produce the positive–negative–positive attractions necessary for bonding. Bond strength, and therefore material strength, varies over a wide range. More significantly, however, is the observation that the positive ions can move within the structure without the breaking of distinct bonds. Materials

FIGURE 3-4 Formation of a chlorine
molecule through a covalent bond.

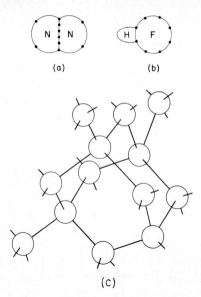

(a)   (b)

(c)

FIGURE 3-5 Examples of covalent bonding in N_2, HF, and diamond.

bonded by metallic bonds therefore can be deformed by atom movement mechanisms and produce a deformed material every bit as strong as the original. This is the basis of metal plasticity, ductility, and many of the shaping processes used in metal fabrication.

Secondary bonds. Weak or secondary bonds, known as *van der Waals forces*, can link molecules that possess a nonsymmetric distribution of charge. Some molecules, such as hydrogen fluoride and water, can be viewed as electric dipoles, in that certain portions of the molecule tend to be more positive or negative than others (an effect referred to as *polarization*). The negative part of one molecule tends to attract the positive part of another to form a weak bond.

Another weak bond can result from momentary polarization caused by random movements of the electrons and the resulting momentary electrical unbalance. This random and momentary polarization leading to attractive forces is called the *dispersion effect.*

A third type of weak bond is the *hydrogen bridge,* where a small hydrogen nucleus is simultaneously attracted to the negative electrons of two different atoms, thereby forming a three-atom link. Such bonds play a significant role in

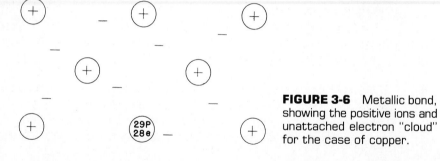

FIGURE 3-6 Metallic bond, showing the positive ions and unattached electron "cloud" for the case of copper.

biological systems but, as with all secondary bonds, are rarely of engineering significance.

Interatomic distances. Since the space occupied by the electron shells is very large compared to the size of the actual electrons, most of the total volume of an atom is vacant space, and the "size of an atom" becomes somewhat undefined. The interatomic or bonding forces tend to pull the atoms together; at the same time, there are repelling forces between the positive nuclei. There exists some equilibrium distance where the forces of attraction and repulsion are equal. The atoms will assume this separation, and to increase or decrease this distance will require energy, such as thermal energy or electrical or mechanical forces.

Thus atoms can be assigned a distinct size, the equilibrium distance between the centers of two neighboring atoms being considered to be the sum of the atomic radii. The atomic radius is not a constant, however. Temperature can change its value, producing an observable thermal expansion. Removal of electrons from the outer shell will decrease the radius and the addition of electrons will increase it. Consequently, a negative ion is larger than its base atom, and a positive ion is smaller. Atomic radius also changes with the number of adjacent or nearest-neighbor atoms. With more neighbors, there is less attraction to any single neighbor atom, and the interatomic distance is increased. Thus iron has a slightly different atomic size in its two different crystal forms, as will be discussed in a later chapter.

Atom arrangements in materials. Moving to the next level of material structure, we find that the arrangement of atoms in a material has a significant effect on its properties. Depending on the manner of atomic grouping, materials are classified as having *molecular structures, crystal structures,* or *amorphous structures.*

Molecular structures have a distinct number of atoms that are held together by primary bonds, but they have only relatively weak bonds with other similar groups of atoms. Typical examples of molecules include O_2, H_2O, and C_2H_4 (ethylene). Each molecule is free to act more or less independently, giving these materials relatively low melting and boiling points. Molecular materials tend to be weak, since the molecules can easily move past one another. Upon changes of state from solid to liquid or gas, the molecules remain intact as distinct entities.

Crystal structures are assumed by solid metals and most minerals. Here atoms are arranged in a regular geometric array known as a *space lattice.* These lattices are describable through a unit building block which is essentially repeated throughout space in a periodic manner. Such blocks are known as *unit cells.*

In amorphous structures, such as glass, the atoms have a certain degree of local order but, when viewed as an aggregate, have a more disorganized atom arrangement than the crystalline solids.

Crystal structure of metals. From a manufacturing viewpoint, metals are the most important class of materials. Most often, they are the materials being processed and are used in the machines performing the processing. Consequently, in order to perform manufacturing operations intelligently, it is essential to have a basic knowledge of the fundamental nature of metals and their behavior when subjected to mechanical or thermal treatment.

More than 50 of the known chemical elements are classed as metals, and about 40 have commercial importance. These materials are characterized by a metallic bond and possess certain distinguishing characteristics: strength, good electrical and thermal conductivity, luster, the ability to be deformed permanently to a fair degree without fracturing, and a relatively high specific gravity, as compared with nonmetals. The fact that some metals possess properties different from the general characteristics simply expands their engineering utility.

When metals solidify by cooling, they assume a crystalline structure; that is, the atoms arrange themselves in a space lattice. Most metals exist in only one lattice form. A few, however, can exist in the solid state in two or more lattice forms, the particular form depending on the conditions of temperature and pressure. These metals are said to be *allotropic,* and the change from one lattice form to another is called an *allotropic change.* The most notable example of such a metal is iron, where the property makes possible the use of heat-treating procedures to produce a wide range of characteristics. It is largely due to its allotropy that iron is the base of our most important alloys.

Metals are known to solidify into 14 different crystal structures. However, nearly all of the important commercial metals solidify into one of three types of lattices, these being body-centered cubic, face-centered cubic, and hexagonal close-packed. Table 3-1 shows the lattice structure of a number of common metals at room temperature. Figure 3-7, compares these structures to each other and to the easily visualized, but rarely observed, simple cubic structure.

The simple cubic structure of Figure 3-7a can be constructed by placing single atoms at all corners of a cube and subsequently linking identical cubes together. Assuming that the atoms are spheres with atomic radii touching each other, computation reveals that only 52% of available space is occupied. Each atom has only six nearest neighbors. Both of these observations are unfavorable to the metallic bond, where atoms desire the greatest number of nearest neighbors and high-efficiency packing.

TABLE 3-1 Types of Lattices of Common Metals at Room Temperature

Metal	Lattice Type
Aluminum	Face-centered cubic
Copper	Face-centered cubic
Gold	Face-centered cubic
Iron	Body-centered cubic
Lead	Face-centered cubic
Magnesium	Hexagonal
Silver	Face-centered cubic
Tin	Body-centered tetragonal
Titanium	Hexagonal

	Lattice Structure	Unit Cell Schematic	Ping-Pong Ball Model	Number of Nearest Neighbors	Packing Efficiency	Typical Metals
a	Simple cubic	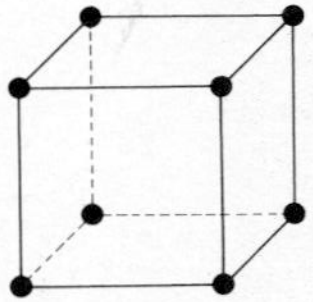		6	52%	None
b	Body-centered cubic	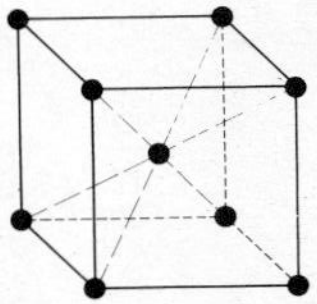		8	68%	Fe, Cr, Mn, Cb, W, Ta, Ti, V, Na, K
c	Face-centered cubic	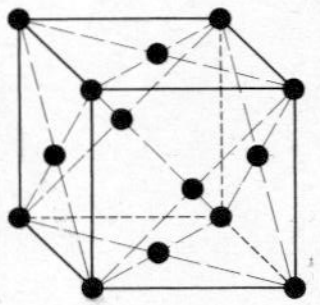		12	74%	Fe, Al, Cu, Ni, Ca, Au, Ag, Pb, Pt
d	Hexagonal close-packed	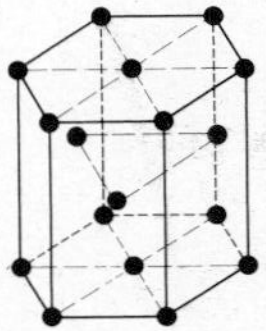		12	74%	Be, Cd, Mg, Zn, Zr

FIGURE 3-7 Comparison of crystal structures: simple cubic, body-centered cubic, face-centered cubic, and hexagonal close-packed.

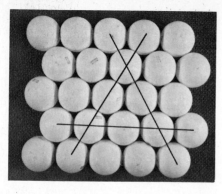

FIGURE 3-8 Close-packed atomic plane showing three directions of closest packing.

If the cube is expanded somewhat to allow the insertion of an additional atom in the center, the *body-centered-cubic* (bcc) structure results, as in Figure 3-7b. Each atom now has eight nearest neighbors and 68% of the space is occupied. Such a structure is more favorable to metals and is observed in the elements Fe, Cr, Mn, and so on, as listed in Figure 3-7b.

If Ping-Pong balls, used to simulate atoms, were placed in a box and agitated until a stable arrangement was produced, we would find the structure to consist of layered *close-packed planes*, where each plane looks like Figure 3-8. Two different structures can result, depending upon the sequence in which the various planes are stacked, but both are identical in nearest neighbors (12) and efficiency of occupying space (74%).

One of these sequences produces a structure that can be viewed as an expanded cube with an atom inserted in the center of each face, the *face-centered-cubic* (fcc) structure of Figure 3-7c. Such a structure occurs in many of the most important engineering metals and tends to produce high formability (ability to be permanently deformed without fracture).

The second of the structures is known as *hexagonal-close-packed* (hcp), wherein the close-packed planes can be clearly identified (see Figure 3-7d). Metals having this strucutre tend to have poor formability and often require special processing procedures.

Development of metallic grains. As metals solidify, a small particle of solid forms in the liquid, having the lattice structure characteristic of the given material. This particle then acts as a *seed* or *nucleus* onto which other atoms in the vicinity tend to attach themselves. The resulting arrangement is a crystal composed of repetitions of the same basic pattern throughout space, as illustrated in Figure 3-9.

In actual solidification it is expected that many seed or nuclei particles would form independently at various locations in the liquid mass and have random orientations. Each then grows until it begins to interfere with its neighbors, as illustrated in two dimensions in Figure 3-10. Since adjacent lattice structures have different alignments, growth cannot produce a single continuous structure. The small continuous segments of solid are known as *crystals* or *grains*, and the surfaces that divide them (i.e., the surfaces of crystalline discontinuity) are known as *grain boundaries*. The process through which the grain structure is produced is one of *nucleation and growth*.

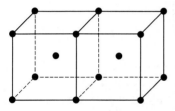 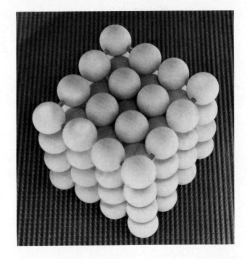

FIGURE 3-9 Growth of crystals to produce an extended lattice. (a) Line schematic; (b) ping-pong ball model.

Grains are the smallest structural units of metal that are observable with ordinary light microscopy. If a piece of metal is polished to a mirror finish with a series of abrasives and then exposed to an attacking chemical for a short time (etched), the grain structure can be seen. The atoms on the grain boundaries are more loosely bonded and tend to react with the chemical more readily than those that are part of the grain interior. When subsequently viewed under reflected light, the attacked boundaries appear dark compared to the relatively unaffected (still flat) grains, as in Figure 3-11. Occasionally, grains are large enough to be seen by the naked eye, as on some galvanized steels, but usually magnification is required.

The number and size of the grains in a metal are a function of two factors: the rate of nucleation and the rate of growth. The greater the rate of nucleation, the smaller the resulting grains. Similarly, the greater the rate of growth, the larger the grain size. Because the overall *grain structure* will influence certain mechanical and physical properties, it is an important property for the engineer

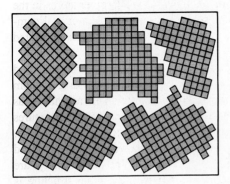 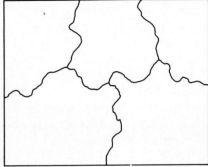

FIGURE 3-10 Schematic representation of the growth of crystals.

FIGURE 3-11 Photomicrograph of alpha ferrite (essentially pure iron) showing grain boundaries; 1000×. (*Courtesy United States Steel Corporation.*)

to be able to both control and specify. One such specification scheme is the ASTM *grain-size number,* defined by

$$N = 2^{n-1}$$

where N = number of grains per square inch visible in a prepared specimen at $100\times$

n = ASTM grain-size number

Higher numbers correspond to smaller grain sizes. Materials with ASTM grain size 7–9 are often desired when good formability is required.

Elastic deformation of single crystals. To a great extent, the mechanical behavior of materials is dependent upon their crystal structure. Therefore, to understand mechanical behavior, it is important for the engineer to have some understanding of the way crystals react when subjected to mechanical loading. Much of what is known about this subject has been obtained from the study of carefully prepared single crystals that may be several inches long. In general, observation reveals that the behavior of metal crystals depends upon (1) the lattice type, (2) the interatomic forces, (3) the spacing between planes of atoms, and (4) the density of atoms on various planes.

If the applied loads are relatively low, the crystal responds by simply stretching or compressing the distance between atoms, as in Figure 3-12. The lattice unit does not change, and the atoms retain their basic positions. The applied

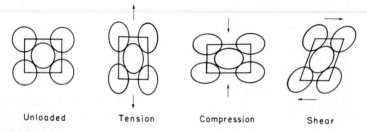

| Unloaded | Tension | Compression | Shear |

FIGURE 3-12 Distortion of the crystal lattice in response to elastic loadings.

load serves only to disrupt the force balance of the atomic bonds in a manner so as to transmit the applied load through the body. When the load is removed, balance is restored and the lattice resumes its original size and shape. The response to such loads is *elastic* in nature. The amount of stretch or compression (strain) is proportional to the applied load or stress.

Elongation or compression in one direction in response to an applied force also produces an opposite change in dimensions at right angles to that force. The ratio of lateral contraction to axial tensile strain under uniaxial tensile loading is known as *Poisson's ratio*. This ratio is always less than 0.5 and usually is about 0.3

Plastic deformation in a single crystal. As the magnitude of applied load is increased, the distortion increases to a point where the atoms must either (1) break bonds to produce a fracture, or (2) slide over one another to produce a permanent shift of atom positions. For metallic materials, the second phenomenon generally requires lower loads and thus occurs preferentially in nature. The result is a plastic deformation, wherein a permanent change in shape occurs without a concurrent deterioration in properties.

Investigation reveals that the mechanism of plastic deformation is the shearing of atomic planes over one another to produce a net displacement. Conceptually, this is similar to the distortion of a deck of playing cards when one card slides over another. As we shall see, the actual mechanism, however, is a progressive one rather than all atoms in a plane shifting simultaneously.

Recalling that a crystal structure is a regular and periodic arrangement of atoms in space, it becomes possible to link atoms into flat planes in a nearly infinite number of ways. Planes having different orientations with respect to the basic lattice will have different atomic densities and different spacing between adjacent parallel planes, as illustrated in Figure 3-13. Given the choice of all possibilities, plastic deformation tends to take place along planes having the highest atomic density and greatest parallel separation. The reason for this may be seen in the simplified Figure 3-14. Planes A and A' have higher density and greater separation than planes B and B'. In visualizing relative motion, the atoms of B and B' would interfere significantly with one another, whereas planes A and A' do not experience this difficulty.

Within the preferred planes are also preferred directions. If sliding occurs in a direction corresponding to close packing of atoms in the plane (as in Figure 3-8), atoms can simply follow one another rather than each having to negotiate its own path. Thus, plastic deformation occurs by the preferential sliding of

FIGURE 3-13 Schematic diagram showing crystalline planes with different atomic densities and interplanar spacings.

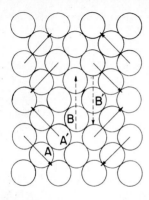

FIGURE 3-14 Planar schematic representing the greater deformation resistance of planes of lower atomic density and closer interplanar spacing.

maximum-density planes (close-packed planes if present) in directions of closest packing.

The ease with which a metal may be deformed depends on the ease of shearing one atomic plane over an adjacent one and the favorability with which the plane is oriented with respect to the load. For example, a deck of playing cards will not "deform" when laid flat on a table and pressed from the top, or stacked on edge and pressed uniformly. Only if the deck is skewed with respect to the applied load will sliding be produced.

With this understanding, let us now consider the properties of the various crystal structures:

Body-centered-cubic: In the body-centered-cubic structure, there are no close-packed planes. Slip therefore occurs on planes with large interplanar spacings (six of which are illustrated in Figure 3-15) in directions of closest packing (that is, the cube diagonals). If each combination of plane and direction is considered as a slip system, we find that 48 such systems exist. The probability that one of these systems will be oriented for easy shear is great, but the force necessary to effect deformation is rather large. Materials with this structure generally possess high strength with only moderate deformation capabilities. (See the typical metals in Figure 3-7).

Face-centered-cubic: In the face-centered-cubic structure, each unit cube possesses four close-packed planes, as illustrated in Figure 3-15. Each plane contains three close-packed directions (the face diagonals), giving 12 possible slip

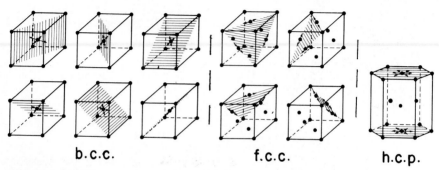

b.c.c.　　　　f.c.c.　　　　h.c.p.

FIGURE 3-15 Slip planes in the various lattice types.

systems. Again the probability of one system being favorably oriented for shear is great, and the required force is relatively low. Face-centered metals possess excellent ductilities, as an inspection of Figure 3-7 will reveal.

Hexagonal-close-packed: The hexagonal lattice also contains close-packed planes, but only one such plane exists for the lattice. Although this plane contains three close-packed directions and the force required to produce deformation is rather low, the probability of favorable orientation to the applied load is rather small. Metals with the hcp structure tend to have low ductilities and often appear to be brittle.

Dislocation theory of slippage. A theoretical calculation of the strength of metals (resistance to plastic deformation) based on the sliding of atomic planes over one another predicts yield strengths on the order of 20,000 MPa (3 million psi). Observed strengths are usually 100 to 150 times less than this value, indicating a discrepancy between theory and reality.

Explanation is provided by the fact that plastic deformation does not occur by all the atoms in one plane slipping simultaneously over all atoms in an adjacent plane. Instead, the motion takes place by the progressive slippage of a localized disruption, the disruption being known as a *dislocation.* Consider an analogy. An individual wants to move a carpet a short distance in a given direction. One approach would be to pull on one end and try to "shear the carpet over the floor." This would require a large force acting over a small distance. An alternative approach would be to form and work a wrinkle across the floor to produce a net shift of the whole carpet—a low-force-over-large-distance approach to the same task. In the region of the wrinkle, there was excess carpet with respect to the floor below, and motion of this excess was relatively easy.

It has been shown that metal crystals do not have all their atoms in perfect arrangement, but contain various localized imperfections. Two such imperfections are the *edge dislocation* and *screw dislocation,* as illustrated in Figure 3-16. Edge dislocations are the ends of extra half-planes of atoms. Screw dislocations correspond to a partial tearing of crystal planes. In each case, the dislocation is a disruption to the regular, symmetrical arrangement of atoms which can be moved about with rather low applied forces. It is the motion of these

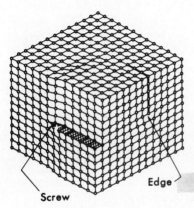

Edge

Screw

FIGURE 3-16 Schematic representation of screw and edge dislocations.

microscopic dislocations under applied loads that produces the observed macroscopic plastic deformation.

All engineering metals contain dislocations, usually in abundant quantities. The ease of deformation, therefore, depends on the ease of producing dislocation movement. Barriers to dislocation motion tend to increase the overall strength of the metal. These barriers take the form of crystal imperfections and may be of the point type (missing atoms or *vacancies,* extra atoms or *interstitials,* or substituting atoms of a different variety as may occur in an alloy), line type (another dislocation), or surface type (crystal grain boundary).

Strain hardening or work hardening.   Many metals possess a unique property, in that after undergoing some deformation, the metal possesses greater resistance to further plastic flow. In essence, metals become stronger when plastically deformed, a phenomenon known as *strain hardening* or *work hardening.*

Understanding of this phenomenon can come from further consideration of the carpet analogy. Suppose this time that the goal is to move the carpet diagonally. The best way would be to move a wrinkle in one direction and then a second one 90° to the first. But suppose that both wrinkles were started simultaneously. We would find that wrinkle 1 would impede the motion of wrinkle 2, and vice versa. In essence, the device that made deformation easy can also serve to impede the motion of other, similar devices. Returning to metals, we find that plastic deformation is accomplished by the motion of dislocations. As dislocations move, they are more likely to encounter and interact with other such dislocations, thereby producing resistance to further motion. Moreover, mechanisms exist to markedly increase the number of dislocations in a metal undergoing deformation, the effect being an increased probability of interaction.

This phenomenon becomes significant when one considers mechanical working processes operating in the cold-working range. Strength properties of metals can be increased markedly by deformation, thereby enabling a deformed inexpensive metal to often substitute for a stronger, undeformed, but more costly one.

Experimental evidence strongly confirms the current theory. When a load is applied to a metal crystal, deformation will commence on the slip system most favorably oriented. The net result is often an observable slip and rotation in a macroscopic (but still single crystal) specimen, as illustrated in Figures 3-17 and 3-18. Dislocation motion on one slip system may become blocked as strain hardening produces increased resistance. Slip stops on the first system and, as load increases, further deformation will take place through alternative systems offering less resistance. This phenomenon, known as *cross slip,* has also been observed.

Plastic deformation in polycrystalline metals.   Thus far, only the deformation of single crystals has been considered. Metals, as normally encountered, are polycrystalline. Within each crystal of a polycrystalline metal, deformation proceeds in the manner just described. However, since adjacent grains do not have their lattice structures aligned in the same orientation, an applied loading will cause different deformations within the various grains. This type of response

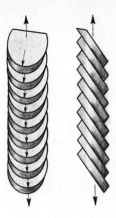

FIGURE 3-17 Schematic representation of slip and rotation resulting from deformation.

is shown in Figure 3-19, where the slip lines in the various grains can be seen. It may be noted that the slip lines do not cross over from one grain to another. The grain boundaries act as barriers to dislocation motion. Finer grain structure, that is, more grains per unit area, generally tends to produce greater strength and hardness coupled with increased impact resistance. This "universal improvement of properties" is strong motivation for controlling grain size during processing.

Grain deformation and fiber structure. When a metal is deformed a considerable degree, the grains become elongated in the direction of metal flow, as can be seen in Figure 3-20. At moderate magnification, a cross section of such a deformed metal may appear fibrous. Concurrent with nonuniformity of structure is a nonuniformity of properties. Because of strain hardening and the fact that the intergranular boundaries are no longer randomly oriented, the strength and other mechanical properties will not be the same in all directions. Electrical and magnetic properties may also show directional variation.

The possibility of employing this increase in strength in certain directions is important to both the designer and manufacturing engineer. Certain processes,

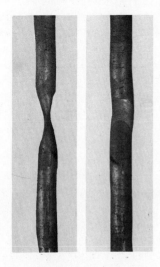

FIGURE 3-18 Front and side views of a single zinc crystal that has been elongated in uniaxial tension. (*Courtesy E. R. Parker.*)

FIGURE 3-19 Slip lines in a polycrystalline material. (*From Richard Hertzberg*, Deformation and Fracture Mechanics of Engineering Materials; *courtesy John Wiley & Sons, Inc.*)

such as forging, can be designed to utilize directional properties. Caution should be used, however, for improvement of properties in some directions usually is accompanied by a decline in properties in other directions. Moreover, directional structures may impose serious difficulties in some operations, such as sheet metal drawing.

Fracture of metals. Under certain conditions of load, temperature, impact, and so on, metals may respond by fracture. *Brittle fracture* is the most cata-

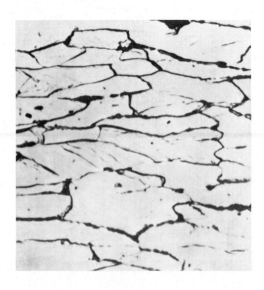

FIGURE 3-20 Deformed grains in cold-worked 1008 steel after 50% reduction by rolling; 1000×. (*From* Metals Handbook, *8th ed. 1972*, American Society for Metals, Metals Park, Ohio.)

strophic, for it occurs without the prior warning of plastic deformation and propagates rapidly through the metal. Such fractures usually are associated with metals having the bcc or hcp crystal structure. *Ductile fracture* generally occurs when plastic deformation is extended too far.

The actual mechanism and type of fracture will vary depending on the material, temperature, state of stress, and rate of loading. Shear or slip fractures are due to extensive shear on slip planes. Intergranular fracture occurs by separation along the grain boundaries. Cleavage fracture results from the pulling apart of metal grains along distinct "cleavage planes" due to tensile loads within the crystal.

Recrystallization. Plastic deformation increases the energy of a metal by means of creating many additional dislocations and increasing the surface area of the grain boundaries through distortion. If a polycrystalline metal is heated to a high enough temperature after being plastically deformed, new, equiaxed, unstrained crystals will form from the original distorted structure as in Figure 3-21. This process of reducing the internal energy is known as *recrystallization*. The temperature at which recrystallization takes place is different for each metal and varies with the amount of prior deformation. In general, the greater the amount of deformation, the lower is the recrystallization temperature. However, there is a practical lower limit below which recrystallization will not take place in a reasonable length of time. Table 3-2 gives the short-time recrystallization temperatures of several metals.

Noting that metals may fracture if deformed too much, we find it common practice to recrystallize material after certain initial amounts of cold work. Ductility is restored and the material is ready for further deformation. This process, known as *recrystallization annealing*, enables deformation to be carried out to great lengths without danger of fracture and is important to many manufacturing processes. If metals are deformed above the recrystallization temperature, working and recrystallization take place simultaneously and large deformations are made possible.

Having already noted the desirability of fine grain size in improving properties, we find recrystallization to be a means of grain-size control. In metals that do not undergo allotropic changes, a coarse grain structure can be converted to a fine grain structure through recrystallization. The material must first be plastically deformed to provide the driving force for recrystallization. Control of the recrystallization process then establishes the final grain size.

Grain growth. The recrystallization process tends to produce uniform grains of comparatively small size. If the metal is held at or above the recrystallization temperature for any appreciable time, however, the new grains will start to "grow." In effect, some grains become larger at the expense of their neighbors. Since properties tend to diminish with increased grain size, control is of prime importance here.

Hot and cold working. When metals are deformed plastically below the recrystallization temperature, the process is called *cold working*. The metal strain hardens and the structure consists of distorted grains. When deformation takes

FIGURE 3-21 Recrystallization of 70–30 brass. (a) Cold worked 33%; (b) heated at 580°C (1075°F) for 3 seconds, (c) 4 seconds, and (d) 8 seconds; 45×. (*Courtesy J. E. Burke, General Electric Company.*)

place above the recrystallization temperature, the process is called *hot working*. A recrystallized structure continually forms, and no strain hardening is apparent. The temperature above which hot forming can be performed depends on the material being worked, as shown in Table 3-2.

Alloys. Up to this point, the discussion in this chapter has been confined to the nature and behavior of pure metals. For most manufacturing applications, however, metals are not used in their pure form, but in the form of alloys. An *alloy* can be defined as a material composed of two or more elements, at least

TABLE 3-2 Lowest Recrystallization Temperature of Common Metals

Metal	Temperature [°C (°F)]
Aluminum	150 (300)
Copper	200 (390)
Gold	200 (390)
Iron	450 (840)
Lead	Below room temperature
Magnesium	150 (300)
Nickel	590 (1100)
Silver	200 (390)
Tin	Below room temperature
Zinc	Room temperature

one of which is a metal, which possesses metallic properties. The addition of a second element to a first to form an alloy usually results in a change of properties. Knowledge of alloys is important to the intelligent selection of material for given applications.

Alloy types. The response of a metal to an alloy addition may follow any of three mechanisms. The first, and probably the simplest response occurs when *the two components are insoluble in each other in the solid state.* In this case, the base metal and the alloying element each maintain their individual identities, structures, and properties. The alloy, in effect, assumes a composite structure consisting of two types of building blocks in an intimate mechanical mixture.

The second possibility occurs when *the two elements are soluble in each other in the solid state.* They thus form *a solid solution,* with the alloying element being dissolved in the base metal. These solid solutions may be of two types: (1) *substitutional* and (2) *interstitial.* In the substitutional type, some atoms of the alloy element occupy sites normally occupied by atoms of the host or base metal. Replacement is random in nature, the alloy atom being free to occupy any atom site in the base lattice. In the interstitial type, the alloy element atoms squeeze into the "unoccupied" spaces within the base metal lattice.

The third possibility occurs where *the elements combine to form intermetallic compounds.* In this case, atoms of the alloying element combine with atoms of the base metal *in definite proportions and in definite geometric relationships.* Bonding is primarily of the nonmetallic variety (i.e., ionic or covalent), and the lattice structures are more complex than for metallic materials. Such compounds tend to be hard, brittle, high-strength materials.

Even though alloys are composed of more than one type of atom, their structure is one of lattices and grains just as in pure metals. Their behavior when subjected to loading, therefore, should be similar to that of pure metals, with due provision for the structural modifications. Plastic flow along atomic planes may be impeded by the presence of unlike atoms. Grains in composite-type mixtures may show different responses to the same loading, reflecting the different properties of the component units.

Atomic structure and electrical properties. As with mechanical properties, the structure of materials strongly influences their electrical properties. Electrical conductivity involves the movement of valence electrons through the crystalline lattice to produce a net transport of charge. The more perfect the atomic arrangement is in a metal, the higher the conductivity. Conversely, the more lattice imperfections or irregularities, the higher the resistance to electrical conduction.

Electrical resistance depends largely on two factors: (1) lattice imperfections, and (2) temperature. Vacant atomic sites, interstitial atoms, substitutional atoms, dislocations, and grain boundaries all act as disruptions to the regularity of a crystalline lattice. Temperature becomes important when one considers the associated atomic vibrations. We have already seen that mechanical energy can displace atoms from their equilibrium positions and stretch bonds. Thermal energy causes atoms to vibrate about their equilibrium positions. This vibration interferes with electron transport, reducing conductivity at higher temperatures. At low temperatures, resistivity becomes primarily a function of crystal imperfections. Thus the best conductors are pure (defect-free) crystalline solids at low temperatures.

The conductivity of metals is primarily a result of the "free" electrons in the metallic bond. Materials with covalent bonds require bonds to be broken to provide electrons available for conduction. Thus the electrical properties of these materials depend on bond strength. Diamond, for instance, is a strong insulator. Silicon and germanium, however, have weak bonds that can easily be broken by thermal energy. These materials, when pure, are known as intrinsic *semiconductors*, since moderate amounts of applied energy can enable the materials to conduct small amounts of electricity.

The conductivity of the nonmetallic semiconductors can be substantially improved by a process known as *doping*. Both silicon and germanium have four valence electrons and four covalent bonds. If one of these elements is replaced with an atom containing five valence electrons, such as phosphorus, the four bonds form, leaving one excess electron, as in Figure 3-22. The extra electron

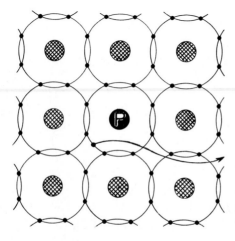

FIGURE 3-22 Schematic representation of an *n*-type semiconductor, with the excess electron of the phosphorus atom being free for conduction.

is free to move about and conduct electricity. Such materials are known as *n*-type extrinsic semiconductors.

A similar effect can be created by substituting an atom with three valence electrons, such as aluminum. An electron is missing from a bond, creating an electron "hole." As an electron jumps into this hole, it creates a hole in the spot it vacated. Movement of electron holes is equivalent to a countermovement of electrons, and thereby produces improved conductivity. These materials are known as *p*-type semiconductors. The control of conductivity through semiconductor devices underlies much of the current advance in solid-state devices and circuitry.

REVIEW QUESTIONS

1. Why might an engineer concerned with the macroscopic mechanical properties of a material be interested in its microscopic structure?
2. What are the two major alternatives for atoms to combine into a larger structural unit?
3. Why might atoms prefer to exist as ions?
4. What are the three types of primary bonds?
5. What are some general characteristics of ionically bonded material?
6. What types of atoms tend to form covalent bonds?
7. Why are ionic and covalently bonded materials characteristically brittle?
8. Why are metals opaque?
9. Why are metals generally good electrical conductors?
10. Why might we expect metallically bonded materials to exhibit plasticity?
11. What are some general characteristics of molecular-bonded material?
12. What is the difference between molecular, crystal, and amorphous structures?
13. What are some of the general characteristic of metallic materials?
14. What is an allotropic material?
15. What are the three most common crystal lattices found in metals?
16. Why is the simple cubic structure not favorable to metals?
17. What is the efficiency of filling space with spheres in the body-centered-cubic arrangement? Face-centered-cubic? Hexagonal-close-packed?
18. What is the difference between fcc and hcp when viewed in terms of stacked close-packed planes?
19. Why are most engineering metals polycrystalline aggregates?
20. What is a grain boundary?
21. What is the basic mechanism for elastic deformation in a metal?
22. Why do bcc metals tend to have high strengths?
23. Why are fcc metals generally weak and ductile?
24. Why do metals not exhibit their theoretical strength of approximately 3 million pounds per square inch?
25. How might one increase the strength of a material that contains abundant quantities of dislocations?
26. Why might a metal with small grain size be stronger than the same metal with larger grains?
27. Why does a deformed metal want to recrystallize?
28. Why might an engineer want to induce recrystallization in a metal?
29. Why is grain growth generally undesirable?
30. What is an alloy?
31. What might one do to improve the electrical conductivity of a metal?
32. What structural feature is responsible for the semiconductor characteristics of silicon and germanium?

CASE STUDY 3. The Misused Test Data

Supreme Sheet Metal Company is a supplier of formed sheet metal panels for various applications. One of its customers has approached the company with a request to convert one such panel from 1008 steel sheet (a low carbon/high ductility steel) to a thinner-gage, high-strength material as a means of reducing the weight of its product. It was also its desire to retain the same design and, thereby, the same forming dies, if at all possible.

John Doakes, a young engineer with Supreme Sheet Metal, was assigned the task of determining the feasibility of the substitution. He first pulled a uniaxial test specimen from the new material and observed a 6% uniform elongation. Next he placed a grid on one of the 1008 steel sheets, deformed it into the desired shape and observed a maximum tensile strain of only 4% during forming. Concluding that the new material had adequate formability for the particular component, he reported favorably on the substitution.

When production began, numerous rupture-type failures occurred in the region of maximum strain, indicating insufficient ductility. What had John overlooked?

CHAPTER 4

Production and Properties of Common Engineering Metals

Engineering metals possess a wide range of usable properties. Some of these are inherent to the particular metal, but many can be varied by controlling the manner of production and processing. Metals, to a large extent, are history-dependent materials, with the final properties being affected by the specific details of the processing history. Thus it is helpful for the engineer to have a working knowledge of how metals are produced and processed and the effects of production and processing on final properties and engineering utility.

IRON

Iron is the fourth most plentiful element in the earth's crust and, for centuries, has been the most important of the basic engineering metals. It is rarely found in the metallic state, but occurs in mineral compounds known as ores. The metallic iron or steel metal produced from these ores has played a central role in the development of civilization, and it appears that it will continue to do so in the foreseeable future. New advances in technology of iron and iron alloys continue to expand their utility in engineering application.

Pig iron. *Pig iron* is the first product in the process of converting iron ore into useful metal and is produced in a blast furnace. These furnaces, illustrated

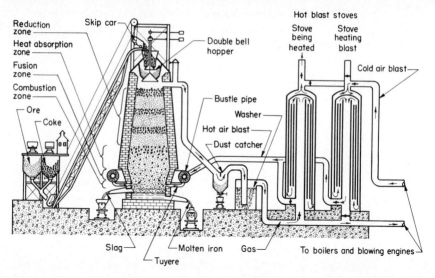

Reduction zone
Heat absorption zone
Fusion zone
Combustion zone
Ore
Coke
Skip car
Double bell hopper
Bustle pipe
Washer
Hot air blast
Dust catcher
Slag
Tuyere
Molten iron
Gas
Hot blast stoves
Stove being heated
Stove heating blast
Cold air blast
To boilers and blowing engines

FIGURE 4-1 Schematic diagram of a blast furnace and its associated equipment.

schematically in Figure 4-1, are large, round, costly steel structures from 30 to 40 meters (98 to 131 feet) tall, lined with refractory firebrick. The diameter varies with position on the furnace, but at the largest point, called the *bosh,* they usually are about 10 meters (33 feet) in diameter.

Iron ore, which basically is an oxide of iron with companion impurities, is processed in a manner so as to break the iron-oxygen bonds and produce metallic iron. Approximately two parts of iron ore, one part of coke, and one-half part of limestone are put into the blast furnace through the double bell hopper at the top. The process is then controlled so as to reduce the iron oxide to iron and simultaneously regulate the reactions involving the *gangue,* the earthy ore contaminant which may contain silica, alumina, calcium oxide, magnesium oxide, water, phosphorus, and sulfur. Control of these components is important because they may substantially affect the product material. The overall process is known as *chemical reduction.*

Four raw materials are involved in the reactions. Ore is the source of the metallic iron and may be of various forms and compositions, among them being (1) *hematite,* a mixture of Fe_2O_3 and gangue, and (2) *magnetite,* the Fe_3O_4 form of iron oxide.

Often the ore is processed into spherical *pellets* or irregular-shaped agglomerates known as *sinter* prior to addition to the furnace. These processes facilitate handling, enable fine material to be used, and increase the efficiency of the furnace. Limestone is added for several reasons, but primarily as a fluxing agent to enable the gangue material and coke ash to enter into a fusible liquid. A second purpose is to provide a material with which undesirable elements or compounds will combine in preference to the metallic iron being produced. The coke serves as a source of the heat necessary to cause the reducing processes to occur and produce molten iron. Air is the fourth raw material. It is preheated to 600 to 700°C (1100 to 1300°F) and is blown into the furnace through nozzles, called *tuyeres,* near the base of the furnace. As the air passes up through the incandescent coke, large volumes of carbon monoxide gas are formed. This gas, together with the carbon in the coke, enables the reduction of the iron oxide through several possible reactions:

$$Fe_2O_3 + 3CO \rightleftharpoons 2Fe + 3CO_2$$

$$Fe_2O_3 + 3C \rightleftharpoons 2Fe + 3CO$$

The actual reduction process does not take place in as simple a manner as indicated by these equations, but involves several sequential steps with the same net result. Although the reactions above are reversible, they are forced to go in the desired direction by regulating the charge, temperature, and amount of air.

As the input material settles near the bosh and the temperature increases, the iron is reduced to a spongy incandescent mass. In this state the iron absorbs more carbon, which lowers its melting point until it finally melts and drips down over the unburned incandescent coke into the base or hearth of the furnace.

Meanwhile, the limestone, which aids in reducing the iron oxides, combines with the calcium and magnesium oxides, alumina, and silica to form a fusible *slag*. Being lighter than the molten iron, the slag floats on top of it and is drained off (tapped) periodically through a hole (slag notch) located at a level above the iron tap hole. Although once discarded, now this material is processed in a variety of manners, among them being (1) solidification and crushing for road-bed aggregate, and (2) water spray atomizing to form a sandlike material for construction purposes.

Blast furnace opeation is a continuous process with ore, limestone, and coke being fed in steadily and the molten pig iron being tapped off every 4 to 6 hours. Although most iron is now sent directly to steelmaking facilities in the form of molten metal, earlier processing involved the solidification into small molds known as pigs, from whence the name "pig iron." From 1000 to 4000 tons can be produced in a 24-hour period, depending on the furnace. Furnace operation is continuous, being terminated only during production lags and for relining or repair. Modern furnaces can usually be operated from 4 to 6 years without being shut down for repair.

For each ton of pig iron tapped from a blast furnace, about 3 tons of hot or combustible gases are obtained. These are used to preheat the air blast by means of stoves, part of the auxiliary equipment shown in Figure 4-1. Ordinarily, four stoves are associated with each blast furnace. These are rotated in operation so that three receive heat from the gases while the fourth is giving up its heat to the air blast. By-product gases may also be used to generate power for operation of additional equipment, such as compressors and blowers.

Recent advances in blast furnace technology tend to center on increased productivity, energy conservation, energy-form substitutions, and the use of lower-grade materials. *Taconite*, a low-grade ore that contains from 20 to 27% recoverable iron, is now being used extensively. Beneficiation treatments are applied at or near the mine site to produce pellets that contain about 63% iron and are suitable for use in the blast furnace.

Composition of pig iron.   Along with the iron ore, other oxides in the input feed may be reduced in the blast furnace. All the phosphorus and most of the manganese enter into the pig iron. Oxides of silicon and sulfur are only partially reduced. Calcium, magnesium, and aluminum oxides are fluxed by the calcium carbonate in the limestone, go into the fusible slag, and are removed. Consequently, the resulting pig iron contains from 3 to 4.5% carbon, all the phosphorus that was present, and most of the manganese. The silicon and sulfur content is determined to some extent by the raw materials, and also by controlling the chemistry of the slag and the temperature of the furnace. All of the reduced elements enter the melt, while those elements that are oxidized go into the slag. Pig iron therefore has roughly the following analysis:

Carbon	3.0–4.5%
Manganese	0.15–2.5%
Silicon	1.0–3.0%
Sulfur	0.05–0.1%
Phosphorus	0.1–2.0%

Types and uses of pig iron. Pig iron is important primarily as the raw material for other processes. A small portion of the total output is cast into final shapes as it comes from the blast furnace, but most iron is transfered in the molten state and fed into various types of furnaces to be made into steel. Several types of pig iron are made, each varying somewhat in composition and properties.

Direct reduction of iron ore. As a result of the desire to employ lower-grade ores and readily available fuels that are currently unsuitable for blast furnace operations, many attempts have been made to develop a process that would at least partially replace the blast furnace as a source of metallic iron. Such processes are referred to as *direct reduction processes* and have as their goal either (1) production of steel directly from iron ore, (2) manufacture of a product equivalent to blast-furnace pig iron for use in current steelmaking processes, or (3) the production of a low-carbon iron to be used in a similar manner to scrap in the current steelmaking processes. Almost every type of apparatus suitable for the purpose has been adapted, including pot furnaces, reverberatory furnaces, regenerative furnaces, shaft furnaces, rotary and stationary kilns, retort furnaces, electric furnaces, and fluidized-bed reactors. Reducing agents include coke, coal, graphite, char, distillation residues, fuel oil, tar, combustible hydrocarbon gases, and hydrogen. Many processes have passed through the pilot-plant stage and several have become economically favorable as a result of local resource conditions. To date, however, no such process has shown sufficient promise that it would challenge the blast furnace as the chief source of iron for the steelmaking industry.

Interest in direct reduction processes is high because they can be used for low-grade ores and with fuels other than coking-grade coal. Moreover, there is an economic advantage in that a direct reduction plant is probably less expensive than the traditional blast furnaces. Plants currently in successful operation, however, are competing largely due to favorable supplies of natural resources in their immediate vicinity. Many processes require extensive amounts of natural gas, a rapidly diminishing fuel material.

Cast iron.   *Cast iron* is essentially the same as pig iron, but is the term applied to the metal when cast in product form. Ordinarily, pig iron and scrap are melted and then cast into prepared molds. The metallurgical properties of cast iron are discussed in Chapter 5 and its melting and utilization in the casting process in Chapter 11.

Wrought iron.   Although no longer produced in any substantial quantity, *wrought iron* was the most important structural metal prior to 1855. Pig iron was further refined to a very pure iron and was then combined with siliceous slag and processed to form an iron matrix with fibrous particles of slag. The product was strong, ductile, and corrosion resistant, but has been almost entirely replaced by steel and other materials.

STEEL

The manufacture of *steel* is essentially an oxidation process that decreases the amount of carbon, silicon, manganese, phosphorus, and sulfur in a mixture of molten pig iron and steel scrap. In 1856, the *Kelly–Bessemer process* essentially opened up an industry by enabling the manufacture of commercial quantities of steel. The *open-hearth process* surpassed the Bessemer process in tonnage produced in 1908 and was producing over 90% of all steel in 1960. *Oxygen furnaces* of a variety of types and *electric furnaces* now produce most of our commercial steels.

The Bessemer process.   The Kelly–Bessemer process, or simply the *Bessemer process,* made use of the fact that air passed through molten pig iron enables exothermic reactions to occur that refine the metal into steel. Carbon oxidizes to produce gaseous CO or CO_2. Action of the air also forms large quantities of iron oxide, which further reacts to form oxides of silicon and manganese, taking these elements into the slag. In this process, however, the reactions do not affect the phosphorus and sulfur contents of the metal and leave these somewhat undesirable elements in the steel.

About 25 tons could be produced in a 15-minute blow, but chemical control of the product was difficult to achieve. High nitrogen contents were also introduced into the metal as the air passed through the melt.

The open-hearth process.   The *open-hearth furnace,* developed commercially around 1870, was basically a shallow refractory-lined "dish" up to 5 by 11 meters (16 by 36 feet) in surface and about ⅔ meter (2 feet) deep. Approximately 150 to 250 tons of selected material was heated to a molten state, raised to elevated temperatures, and processed to obtain a product of desired chemical composition. Heated air and fuel gas entered one end of the hearth, passed over the charge of pig iron, scrap steel, and limestone, and finally heated a heat reservoir (checker) on the other side of the furnace with the waste gases. Upon reversal of flow, the checkers yield their stored heat to the incoming air to continue the removal of manganese, phosphorus, silicon, and sulfur from the melt.

Approximately 7 to 10 hours were required for a melting and refining cycle, enabling periodic samplings and additions to the melt. The process was easy to control, produced uniform-quality material with low phosphorus and sulfur content, had high flexibility, and could use up to 80% scrap in its charge. While more recent advances have speeded up the process either by blowing oxygen over the surface of the molten metal or by introducing it directly into the molten metal through a water-cooled pipe, a process known as *bottom lancing*, most open-hearth opeations have been phased-out in lieu of basic oxygen furnaces.

Basic-oxygen process. In 1952, the *basic-oxygen process* was developed, taking advantage of the availability of large quantities of pure oxygen. Approximately 200 to 400 tons of steel scrap and molten pig iron are charged into a large, cylindrical, open-mouth vessel that has a basic refractory lining. The furnace is returned to its upright position and burned lime and flux are poured in through a chute. A water-cooled oxygen lance, roughly 18 meters (65 feet) long and 25 cm (10 inches) in diameter, is then lowered to within 2 meters of the bath and high-purity oxygen is blown onto the surface of the metal. The oxygen, which is ejected under considerable pressure at rates in excess of 1 ton per minute, blows the slag aside and reacts violently with the exposed iron to form iron oxide. Part of the oxide reacts with the flux to form a basic slag, while the remainder is mixed with the bath, through turbulence, and oxidizes the impurities. The oxygen treatment requires only 15 to 20 minutes of the total 40- to 45-minute cycle. Samples are taken and analyzed and adjustments are made to produce desired compositions prior to tapping. Figure 4-2 shows the process in schematic.

Production capacity for a single furnace offers a tenfold increase over the open-hearth furnace. Capital investment and operating expenses are lower for the oxygen furnace provided that a sufficient supply of hot metal is available. Improved ingot quality, lower energy requirements, and easier environmental control all favor the oxygen process over the open-hearth process. The major disadvantage is the limited amount of scrap input (30 to 35%) and large amount of hot metal required. Also, with the reliance on only a few furnaces, prolonged downtime can seriously affect an entire plant.

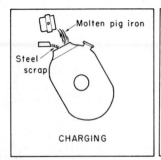 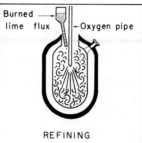 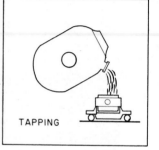

FIGURE 4-2 Schematic diagram of the basic-oxygen steelmaking process.

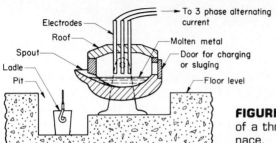

FIGURE 4-3 Schematic diagram of a three-phase electric arc furnace.

Current development efforts relate primarily to increasing the amount of scrap in the charge and improving the energy efficiency of the process. The energy required to produce a ton of steel from scrap is only 26% of the energy required to produce the same steel from iron ore. In addition, air pollution effluents, water pollution, and mining wastes are all reduced by more than 75%. Bottom lancing modifications have also been recently introduced.

Electric furnace processes.   A substantial tonnage of steel is also produced in electric furnaces. Although once associated with small quantities of specialty materials, such as tool steel, stainless steel, die steels, and aircraft-quality metal, these furnaces are now widely used when smaller quantities and high alloy contents are desired or where their ability to utilize up to 100% scrap provides a distinct advantage.

In one type of furnace, an *electric arc* is maintained between electrodes and the metal being melted. Three electrodes are normally used in conjunction with a three-phase power source. High currents flowing through the conductive metal provide the heat necessary for the process. Various slag baths are used to control the chemistry of the product. Commercial electric furnaces of the types illustrated in Figures 4-3 and 4-4 range in capacity from 5 tons to more than 150 tons.

The *induction-type furnace,* illustrated in Figure 4-5, usually has a capacity of less than 5 tons. They are primarily used for making high-quality or high-alloy steels or for remelting metals such as cast iron in an iron foundry.

Major disadvantages of the electric furnace process relate to the inefficiency of the energy system once meltdown has been completed, and the pickup of undesired gases by the melt during processing.

The production of ingots.   Regardless of the method by which steel is made, it must undergo a change of state from liquid to solid before it becomes a usable product. The liquid can be converted directly into a steel casting or into a form suitable for further processing. In most cases, the latter option is exercised through either *continuous casting* or the forming of *ingots* to produce the raw material for subsequent forging and rolling operations.

In casting ingots, the desire is to obtain metal as free of flaws as possible. The molten steel is tapped from the furnace into pouring ladles, nearly all of which are of the bottom-pouring type illustrated in Figure 4-6. By extracting the metal from the bottom of the ladle, slag and floating matter are not transferred

FIGURE 4-4 Electric arc furnace, tilted for pouring. (*Courtesy Pittsburgh Lectro-melt Furnace Corporation.*)

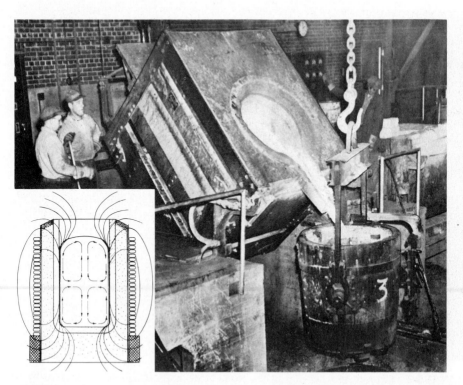

FIGURE 4-5 Pouring molten metal from an electric induction furnace. Inset shows the principle of this type of furnace. (*Courtesy Ajax Magnethermic Corporation.*)

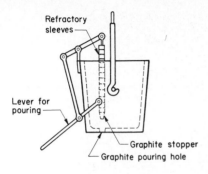

FIGURE 4-6 Schematic diagram of a bottom-pouring ladle.

Labels on figure: Refractory sleeves; Lever for pouring; Graphite stopper; Graphite pouring hole

to the ingots. When poured into the ingot molds, the metal may enter from either the top or the bottom. Although the majority of ingots are poured from the top, there are several disadvantages to the procedure. Hot metal may be splashed onto the side walls of the mold, solidify, and later become part of the ingot when the pouring level passes that point. This produces a disruption to the continuity of ingot's internal structure. Slag entrapment is also more likely in top poured ingots. To avoid these difficulties, ingots are sometimes poured from the bottom. In this procedure, illustrated in Figure 4-7, the bottom of several outer ingots is connected to the bottom of a central pouring ingot by

FIGURE 4-7 Pouring of steel ingots by the bottom-pouring process. The bottom of the center mold being poured is connected to the bottom of the remaining molds in the cluster by means of tile channels. (*Courtesy United States Steel Corporation.*)

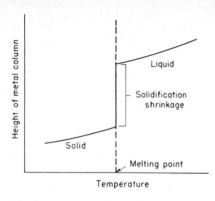

FIGURE 4-8 Height of a column of metal as a function of temperature, showing solidification shrinkage.

ceramic tile tunnels to form a spiderlike arrangement. Hot metal poured into the center is conveyed through the tunnels to fill the outer molds from the bottom. Bottom pouring is more costly than top pouring so it is usually used only for high-quality steels.

Figure 4-8 schematically shows the contraction of a metal undergoing cooling, the jump at the melting point being known as the *solidification shrinkage*. When metals solidify, it is expected that a shrinkage will be observed in the region of the last material to be liquid. In ingots, solidification proceeds inward from the mold walls and upward from the bottom. Shrinkage takes the form of a *pipe* coming in from the top as illustrated in Figure 4-9. Since the pipe surface has been exposed to the atmosphere at elevated temperature, oxides and surface contaminants form which prevent the metal from welding back together during subsequent processing. That portion of the ingot containing the pipe must be recycled as scrap and may be a substantial fraction of the ingot.

FIGURE 4-9 Section of an ingot, showing a "pipe" (top) and "segregation" (dark areas). (*Courtesy Bethlehem Steel Corporation.*)

Although the amount of shrinkage cannot be changed, the shape and location can be greatly controlled. One procedure to reduce the amount of pipe is to use a ceramic *hot top* on the top of the ingot mold. By retaining heat at the top, the liquid reservoir at the end of solidification is more of a uniform layer on top of the ingot, thereby minimizing the depth of shrinkage. Variation of the shape of the mold controls not only the solidification shape, but also the structure of the ingot structure. Tapered ingots having the big end up are commonly used for greatest soundness.

Several new processes have been developed to overcome some ingot-related difficulties, such as piping, entrapped slag, and structure variation, as well as to reduce or replace the associated transport, mold stripping, and reheating operations required for further processing. A variety of *continuous-casting tech-niques* have been used extensively. Figure 4-10 illustrates the most common procedure, wherein liquid metal flows from a ladle, through a tundish, into a bottomless, water-cooled mold, usually made of copper. Cooling is controlled so that the outside has solidified before the metal exits the mold. The material is further cooled by direct water sprays to assure complete solidification. The cast solid is then either bent and fed horizontally through a reheat furnace for further processing or cut to desired lengths. Mold shape, and thus the shape of the cast product, may vary such that products may be cast with cross section close to the desired final shape.

Continuous casting virtually eliminates the problems of piping and mold spat-

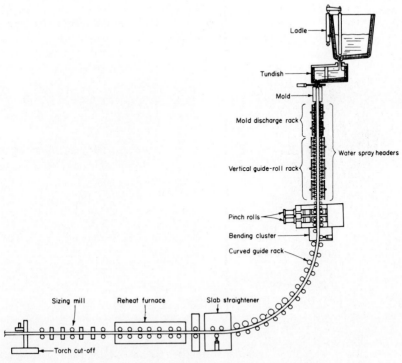

FIGURE 4-10 Schematic representation of the continuous casting process for producing billets, slabs, and bars. (*Courtesy* Materials Engineering.)

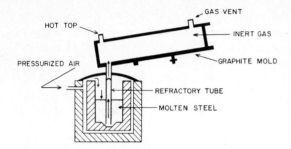

FIGURE 4-11 Schematic representation of the pressure-pouring process for casting ingots and slabs.

ter, as well as the cost of ingot molds, handling, and stripping. Some small surface defects, such as small surface cracks, may occur and have to be removed where use involves critical applications. The steel, copper, and aluminum industries are all producing material by continuous casting.

Molten steel may also be cast into slabs suitable for further processing by means of the pressure-pouring process depicted in Figure 4-11. Molten metal is forced up into a graphite mold by air pressure and solidifies. The mold is hinged for easy slab removal and can be used repeatedly. Because the metal is taken from the bottom of the ladle and is introduced through the bottom of the slab mold, excellent slab quality is obtained.

Degassification of ingots.   During the oxidation that takes place in the making of steel, considerable amounts of oxygen can dissolve in the molten metal. When this molten metal is then cooled to produce solidification, oxygen and other gases are rejected from the solid as the saturation levels decrease (see the schematic of Figure 4-12). The rejected oxygen links with atomic carbon to produce a CO gas evolution. A porous structure results, where the gas bubble-induced *porosity* has the form of either small, dispersed pores or large blowholes. Pores that are totally internal can be welded shut during subsequent hot forming, but if they are exposed to the air at elevated temperatures, the pore surfaces oxidize and will not weld. Cracks and internal defects may then appear in the finished product.

In many cases, it is desirable to avoid the porosity difficulties by removing the oxygen or rendering it nongaseous prior to solidification. Where high-quality steel is desired or where subsequent deformation may be inadequate to produce

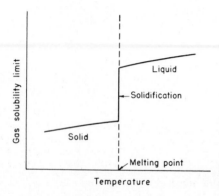

FIGURE 4-12   Gas solubility in a metal as a function of temperature.

welding of the pores, the metal is usually fully *deoxidized* (or *killed*), to produce a killed steel. Aluminum, ferromanganese, or ferrosilicon is added to the molten steel while it is in the ladle to provide material with a higher affinity for oxygen than the carbon. This process is particularly useful with higher-carbon steels. Rejected oxygen simply reacts to produce solid metallic oxides dispersed throughout the structure. Full shrinkage is observed as this material solidifies and a large portion of scrap may be generated when the exposed pipe region is cut off the solidified ingot. This may be somewhat offset by the use of the heated refractory "hot top" described earlier which slows the rate of solidification at the top, giving a more uniform shrink instead of a pipe.

For steels with lower carbon contents, a partial deoxidation may be employed to produce *a semikilled steel*. Enough deoxidant is added to partially suppress bubble evolution, but not enough to completely eliminate the effect of oxygen. Some pores still form in the center of the ingot, their volume serving to cancel some of the solidification shrinkage, thereby reducing the extent of piping and scrap generation.

For steels of sufficiently low carbon (usually less than 0.2% carbon), a process, known as *rimming*, may be employed. The steel is only partially deoxidized prior to solidification. When the material is poured into the ingot mold, the first metal to solidify is almost pure iron, being very low in carbon, oxygen, sulfur, and phosphorus. These elements are rejected from the solid into the liquid ahead of the solidification front. When the concentration of dissolved gases exceeds the saturation point, a layer of CO bubbles evolves and the effervescence tends to drive entrapped particles toward the center of the ingot. Further solidification produces additional porosity in the inner portions of the ingot and, if properly controlled, the rimming action can produce a porosity level that will approximately compensate for solidification shrinkage.

The outside of the ingot is clean, defect-free metal which provides an excellent blemish-free surface when the ingot is rolled into flat strips or similar product. The holes on the inside of the rimmed ingots have bright, clean surfaces, because they have not been exposed to the air. When hot working is performed, the surfaces weld together to produce a sound product. Figure 4-13 presents a schematic comparison of killed, semikilled, and rimmed ingot structures.

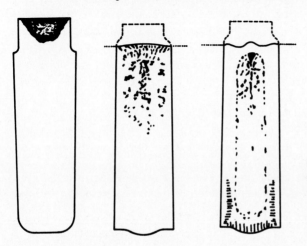

FIGURE 4-13 Schematic comparison of killed, semikilled, and rimmed ingot structures. (*Courtesy American Iron and Steel Institute, Washington, D.C.*)

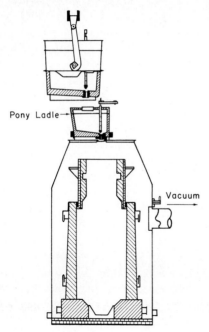

Pony Ladle→

Vacuum →

FIGURE 4-14 Method of vacuum degassing steel while pouring ingots.

In addition to oxygen, small amounts of other dissolved gases, particularly hydrogen and nitrogen, have deleterious effects on the performance of steels. This is particularly important in the case of alloy steels because several of the major alloying elements, such as vanadium, columbium, and chromium, tend to increase the solubility of these gases. Consequently, several methods have been devised for *degassing* steels, and considerable quantities of degassed steel are now used for critical applications, such as turbine rotor shafts. Figure 4-14 illustrates a method that is widely used to produce degassed ingots, known as *vacuum degassing*. The ingot mold is enclosed in an evacuated chamber and the metal stream passes through the vacuum during pouring, the vacuum serving to remove the dissolved gases. An alternate procedure is to perform the melting in an induction furnace and enclose both the furnace and the ingot mold in a vacuum chamber.

When exceptional purity is required, a *consumable-electrode remelting* process may be employed, as illustrated in Figure 4-15. A solidified metal electrode is remelted by an electric arc in an evacuated chamber and resolidifies into a new form. This process, known as *vacuum arc remelting* (VAR) [or *vacuum induction melting* (VIM) if induction heating replaces the electric arc], is highly effective in removing all dissolved gases, but does not remove nonmetallic impurities from the original metal.

If extremely gas-free and clean metal is required, it may be obtained by using the *electroslag remelting process* (ESR), illustrated in Figure 4-16. In this process, an electrode is melted and recast by means of an arc with the surface of the melted metal now covered by a thick blanket of molten flux. Nonmetallic impurities are collected in the flux blanket and there is no danger of arc spatter collecting on the mold walls and becoming part of the ingot.

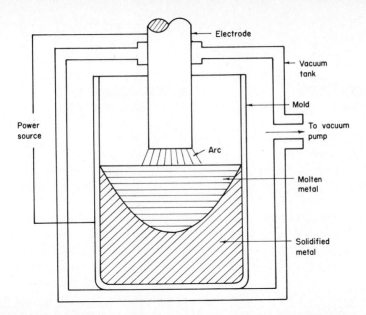

FIGURE 4-15 Method of producing degassed ingots by consumable-electrode vacuum remelting.

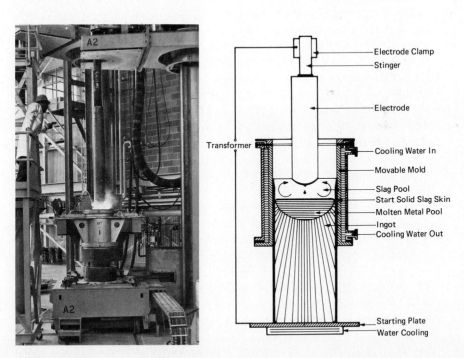

FIGURE 4-16 (*Left*) Production of an ingot by the electroslag remelting process. (*Right*) Schematic representation of this process. (*Courtesy Carpenter Technology Corporation.*)

COPPER

Production. *Copper* is an important engineering metal that has been in use for over 6000 years. Used in its pure state, copper is the backbone of the electrical industry. It is also the major metal in several highly important engineering alloys—brass and bronze.

Most copper ore is in the form of sulfides or oxides, with major U.S. deposits being found in Arizona, Utah, New Mexico, and Montana. With the depletion of the best domestic ore, deposits in South America and Africa have recently assumed great importance.

Even the best ore contains a rather low percentage of copper, so the first step in production is a concentration process. Sulfide ores are crushed, ground, and concentrated by flotation to yield a product that is about 50% copper. Oxide ores cannot be concentrated by flotation, so an acid leach is used followed by either precipitation on scrap iron or electrowinning. The concentrate may then be roasted to reduce the sulfur and arsenic. It is then melted with suitable fluxes in reverberatory or electric furnaces, a process known as *smelting*. Lighter impurities combine and float to the top as a slag while copper, iron, sulfur, and any precious metals form a product known as *matte* in the lower part of the furnace.

The molten matte (containing 40 to 80% copper) is transferred to a converter that is similar to those used in making Bessemer steel. Air blown through the matte oxidizes and blows out the sulfur and oxidizes the iron for removal into the slag. The product is known as *blister copper* and is approximately 99% pure.

Blister copper may be upgraded to refined copper by melting in a furnace and removing the principal impurity, oxygen, in a reducing atmosphere formed by the use of green logs thrust into the melt (poling) or by injection of a reducing gas such as methane. Most blister copper, however, undergoes only a partial furnace refining and is then cast into copper anodes for electrolytic refining. These anodes and thin copper starting sheets, or cathodes, are suspended in tanks containing a solution of copper sulfate and sulfuric acid. An electric current passed through the solution dissolves the copper anodes and refined copper is deposited on the cathode. Gold, silver, and other valuables are recovered from the sludge on the bottom of the tanks.

The resulting refined copper contains between 0.02 and 0.05% oxygen and is called *electrolytic tough-pitch* (ETP) *copper*. If the intended use of the copper is as the base for an alloy, refinement to low oxygen content may not be necessary and the poling process may be eliminated.

Copper with lower oxygen content may be obtained by two other methods. The first uses an inert-gas atmosphere in the reverberatory furnace and produces an *oxygen-free high-conductivity* (OFHC) *copper*. The second procedure is to use deoxidation with a strong reducing agent such as phosphorous or silicon. This approach has the disadvantage that the electrical conductivity is reduced by 10 to 20%. The use of calcium, lithium, or boron as the reducing agent causes less reduction in the conductivity.

The most pressing problem currently facing the copper industry is environmental. Smelting produces large quantities of sulfur compounds, trace elements, and particulate emissions that must be contained before release to the atmosphere. Existing practices must be modified or new techniques developed as a solution to the problem. Similar problems also confront the lead industry, where the principal ore is also a sulfide. Processing techniques are in a state of dynamic change. Further modifications relate to the use of leaching techniques to recover copper from low-grade ore or previous waste material, economics now favoring a recovery process.

Properties. The wide use of copper is based, primarily, on three important properties: its high electrical *conductivity*, high *ductility*, and *corrosion resistance*. Obviously, its excellent conductivity accounts for its importance to the electrical industry. The better grades of conductor copper have a conductivity rating of about 102%, reflecting the better oxidation procedures now available as opposed to 1913, when the standard was established.

While copper in the soft pure state has a tensile strength of only about 200 MPa (30,000 psi), its elongation in 2 inches is about 60%. By cold working, it can be hardened and the tensile strength increased to above 450 MPa (65,000 psi) with a decrease in elongation to about 5%. Its relatively low strength and high ductility make it a very desirable material where forming operations are necessary. Furthermore, the hardening effects of cold working may easily be removed because the recrystallization temperature is less than 260°C (500°F).

Copper, as a pure metal, is not used extensively in manufactured products except in electrical equipment. More often, it is the base material for some alloy to which it imparts its good ductility and corrosion resistance. In nonelectrical uses, it is usually one or both of these properties that accounts for its selection.

If copper is stressed at high temperatures over long periods of time, it is subject to intercrystalline failure at about half its normal room-temperature strength. Material containing more than 0.3% oxygen is also subject to hydrogen embrittlement when exposed to reducing gases above 400°C (750°F).

ALUMINUM

Production. Although *aluminum* has been a commercial metal for less than 100 years, it now ranks second to steel in both worldwide quantity and expenditure. It is important in virtually all segments of the world economy, with principal uses in transportation, construction, electrical applications, containers, consumer durables, and mechanical equipment.

Aluminum is the most abundant metallic element in the earth's crust. Supply, therefore, is limited only by the economics of mining, extracting, and processing. Under current conditions, bauxite is the major source, bauxite being $Al_2O_3 \cdot nH_2O$ with impurity oxides of iron, silicon, and titanium. Most ore contains about 50% aluminum oxide (alumina) and is sufficiently inexpensive that the major portion of the ore cost lies in transportation. Kaolin-type clays become an alternative source of aluminum if bauxite is limited.

The first stage of the processing is the separation of the alumina from the impurity oxides in the ore. If not performed, the reduction process that produces aluminum from its oxide would also reduce the other metallic oxides. Most industries use modifications of the Bayer process (developed in 1888) wherein the bauxite is digested in a caustic soda leach at elevated temperature and pressure. The alumina dissolves out as a solution of sodium aluminate, is separated, selectively precipitated as hydrated aluminum oxide, and finally converted to pure Al_2O_3 by calcination.

Further processing involves the reduction of the oxide to molten metal. Because alumina has a very high melting temperature (2045°C or 3720°F), it cannot be reduced by the usual furnace techniques used for iron. However, in 1886, Charles M. Hall and Paul Heroult discovered that if Al_2O_3 were dissolved in molten cryolite, the aluminum could be deposited at the cathode of an electrolytic cell.

Industry now carries out the electrolysis in cells made of steel shells lined with carbon, which acts as the cathode. The cells are filled with molten cryolite into which about 16% of alumina is dissolved. Carbon anodes dip into the electrolyte and introduce the current. The separated aluminum is deposited on the bottom of the cell, being heavier than molten cryolite, and is drawn off periodically as it collects. Powdered alumina is added to the bath to replace the aluminum drawn off. Electrolytic refinement requires 15 to 18 kilowatt-hours of electricity per kilogram of aluminum. A new process to produce aluminum by converting alumina to aluminum chloride and then reducing the chloride electrolytically has the potential of significantly reducing the energy requirement. In addition, the significant energy saved by remelting scrap is a substantial motivation for recycling aluminum.

Properties and uses.   The properties of aluminum that make it of engineering significance are its *workability, light weight, corrosion resistance,* and good electrical and thermal *conductivity*. Aluminum has a specific gravity of 2.7 as compared to 7.85 for steel. In addition, when exposed to air, aluminum forms an adherent surface oxide that possesses good corrosion resistance. Therefore, aluminum is an excellent material for those applications where relatively good corrosion resistance must be combined with light weight, or in areas which utilize the high electrical and thermal conductivities.

Probably the most serious weakness of aluminum from an engineering viewpoint is its relatively *low modulus of elasticity,* roughly one-third that of steel. (Under identical loading, aluminum will deflect three times as much as steel.). This factor, coupled with the higher cost of aluminum, makes it necessary to use sections that place the metal properly so as to obtain adequate stiffness while using minimal material. Fortunately, this often can be done with relative ease because of aluminum's good workability.

In its pure state aluminum is soft, ductile, and not very strong. As a result, it is not often used in this condition. Its electrical conductivity is about 60% that of copper. For this reason it often is used for electrical transmission lines, but it is usually reinforced by a steel core so as to obtain adequate strength. In most other applications it is alloyed with copper, manganese, magnesium, or silicon to produce increased strength and hardness. Thus, aluminum serves as the base metal of a series of extremely useful alloys that are constantly increasing in importance. These alloys will be discussed in detail in Chapter 8.

TITANIUM

Titanium is a strong, lightweight, corrosion-resistant metal that has been of commercial importance only since 1948. Because its properties are between those of steel and aluminum, its importance is increasing rapidly. Yield strength is about 415 MPa (60,000 psi) and can be raised to 1300 MPa (190,000 psi) by alloying—a strength comparable to many alloy steels. Density, on the other hand, is only 56% that of steel, and the modulus of elasticity is about one-half that of steel. Mechanical properties are retained well up to temperatures of 480°C (900°F). From the negative side, titanium and its alloys suffer from high cost, fabrication difficulties, and a high reactivity at elevated temperatures (above 480°C).

Production.   Titanium is difficult to produce. One method involves the reduction of titanium tetrachloride with magnesium in an inert atmosphere. The resulting magnesium chloride and free magnesium are then leached out with hydrochloric acid, leaving sponge or powder titanium. The sponge or powder is then melted in a vacuum unit and chilled to form solid metal. Another method of production involves the electrolysis of titanium tetrachloride in a fused salt bath. Processing problems are numerous, but titanium is now being produced routinely and is readily available on the metals market.

Properties and uses.   Commercially pure titanium has typical properties as follows:

	Annealed	Hard-Rolled
Yield strength	275 MPa	655 MPa
Ultimate strength	415 MPa	760 MPa
Elongation in 2 inches	25%	10%

Numerous alloys have been developed to improve these properties and are grouped into three classes on the basis of stable structure at room temperature. Some of these alloys have yield strengths of 1380 MPa (200,000 psi) and ultimate strengths of 1520 MPa (220,000 psi) at room temperature in the heat-treated condition.

Uses of titanium relate primarily to the *high strength-to-weight ratio* and *corrosion resistance* as well as *retention of these properties at elevated temperatures*. Applications are primarily in the area of aerospace but also include chemical and electrochemical processing equipment, marine implements, and ordnance equipment. The metal can be cast, forged, rolled, or extruded to desired shape with special process modifications or control. Some applications relating to bonding utilize the unique property that titanium wets glass and some ceramics. Titanium carbide is used where a very hard material is required, such as in cutting-tool tips.

MAGNESIUM

Production. *Magnesium* is the lightest of the commercially important metals, having a specific gravity of about 1.75. Two major processes are used to produce magnesium: electrolysis of molten anhydrous magnesium chloride and the thermic reduction of dolomite with ferrosilicon. The magnesium chloride for the first process may be obtained by processing various mineral deposits or by processing seawater, which contains about 1.07 kilograms of magnesium per cubic meter (1 pound per 15 cubic feet).

Properties and uses. Like aluminum, magnesium is relatively weak in the pure-state and for engineering purposes is almost always used as an alloy. Its modulus of elasticity is even less than that of aluminum, being only about one-fifth that of steel. Therefore, it is usually necessary to require considerable thickness or deep sections to obtain adequate stiffness. In various types of castings, this restriction is easily accommodated. Sand, permanent mold and die casting are all well developed for magnesium. While forming behavior is poor at room temperature, most conventional processes can be performed when the material is heated to temperatures between 230 and 370°C (450 to 700°F).

Wear, creep, and fatigue properties are rather poor. Corrosion resistance is such that paint or some other type of surface protection is often required. Rapid oxidation or burning of the metal can also occur during elevated-temperature processing or operations that may generate high temperature, such as machining. A protective, oxygen-free, atmosphere may be required during these operations.

These restrictions, coupled with high cost, limit magnesium to applications where light weight is very important. Aside from the possibility of burning chips, the machining characteristics are sufficiently good that in many applications the savings in machining costs more than compensate for the increased material expense.

ZINC

Production. *Zinc* ore is most commonly a sulfide (zincblende), with zinc contents ranging from 2 to 15%. For further processing, the ore must first be concentrated by a process of crushing and grinding followed by either "gravity" separation or flotation. The zinc sulfide concentrate, containing from 48 to 60% zinc, is then put through a roasting process in which the sulfur is burned to SO_2 and the zinc converted to ZnO. The SO_2 is generally processed into sulfuric acid.

Metallic zinc is then produced from the oxide by either a carbon reduction process using a furnace or by an electrolytic process. In the furnace process, a mixture of zinc oxide and coal is briquetted or sintered and fired at a temperature in excess of 1000°C. Zinc is reduced and vaporized (since the temperature is in excess of the boiling point of zinc) and is then liquefied in a condenser. All grades of zinc, except special high purity (99.99% Zn) can be produced by the furnace process. If further purity is desired, the impure zinc can be subjected to fractional distillation to produce 99.99 + % zinc.

In the electrolytic process, the roasted ore is dissolved in sulfuric acid and the zinc-bearing solution is filtered and purified. Electrolysis plates the zinc onto prepared aluminum cathodes, from which it is periodically stripped, melted, and cast into slabs. Purity exceeds the 99.99% level.

Properties and uses.   As a pure metal, zinc has only one important use: *galvanizing* iron and steel. This process, in which steel is acid-cleaned and then coated with a layer of zinc either by dipping in a bath of molten metal or by electrolytic plating, provides excellent corrosion resistance even when the coating is scratched or marred. Galvanizing accounts for about 35% of all zinc used. *Sherardizing* is a similar process in which zinc is applied to the surface of steel components in the form of a diffusion coating. The goal again is corrosion resistance, but the form of the coating and process is different.

The second major use of zinc is in alloys used in the *die-casting* process. About 40% of the annual consumption of zinc is for this process, which requires a fluid metal and low melting point. In its pure state, zinc softens at relatively low temperatures but has rather low strength and is brittle. The die-casting metals are alloys designed to retain the low melting point but have considerably improved strength properties. They will be discussed at greater length in Chapter 8. Most zinc die castings are subsequently plated or painted.

The third major use for zinc is as the principal alloying element in *brass* (copper–zinc alloys). Brass making utilizes about 15% of the total zinc produced.

CHROMIUM, MOLYBDENUM, NICKEL, COBALT, TIN, AND LEAD

Chromium, molybdenum, nickel, cobalt, and tin are of considerable importance as engineering metals but seldom are used in their pure forms. *Chromium, molybdenum,* and *nickel* have been used for years as alloying elements in steels. Nickel is used as the base metal of a number of corrosion-resistant nonferrous alloys. Nickel or *cobalt* form the basis for the *superalloys*, which can maintain useful strengths and properties at temperatures slightly in excess of 1100°C (2000°F). Cobalt is also used as a binder metal in various powder-based components and sintered carbides.

Tin is used primarily as a coating on steel to provide corrosion resistance, in combination with copper to produce bronze, or in certain alloys that are used as bearing materials.

Lead is often used in the pure state as a corrosion-resisting material. In most uses, however, it is used as an additive to improve corrosion resistance or the machining characteristics of alloy metals. Principal uses, such as storage batteries, paint pigment, and cable covering, consume about 60% of the annual output.

"UNUSUAL METALS"

Several uncommon metals have achieved importance in modern technology as a result of their somewhat unique properties. *Hafnium, thorium,* and *beryllium* are used in nuclear reactors because of their low neutron-absorption characteristics. High-temperature applications often require metals with high melting points, such as *niobium* (2470°C), *molybdenum* (2610°C), *tantalum* (3000°C) *rhenium* (3170°C), and *tungsten* (3410°C). Depleted *uranium,* because of its very high density (19.1 grams per cubic centimeter), is useful in special applications where maximum weight must be put into a limited space, as in counterweights. *Zirconium* is used for its outstanding corrosion resistance to most acids, chlorides, and organic acids. When alloyed with a small percentage of hafnium, it has a yield strength of about 579 MPa (84,000 psi) and a tensile strength of about 620 MPa (90,000 psi). With extensive interest in space, nuclear, high-temperature, and electronic applications, a considerable amount of research and development work is being done on these uncommon metals.

GRAPHITE

Properties and uses.   While technically not a metal, *graphite* is a material that, in improved forms, has considerable importance as an engineering material. It possesses the unique property of having increased strength at higher temperatures. Recrystallized, polycrystalline graphites have mechanical strengths, as measured by modulus of rupture, up to 70 MPa (10,000 psi) at room temperature, which double at 2500°C (4500°F).

Large quantities of graphite are used as electrodes in arc furnaces, but other uses are developing rapidly. The addition of small amounts of borides, carbides, nitrides, and silicides greatly lowers the oxidation rate of graphite at high temperatures and also improves its mechanical strength. This makes it highly suitable for use in rocket-nozzle inserts and permanent molds for casting various products. It can be machined quite readily to excellent surface finishes.

The use of graphite fibers in composite materials will be discussed in Chapter 9.

REVIEW QUESTIONS

1. Why is it helpful for an engineer to have a knowledge of how metals are produced and processed?
2. What are the four raw materials involved in the extraction of iron from ore?
3. What is the role of limestone in a blast furnace?
4. In addition to iron, what are the major chemical elements in pig iron?
5. What are the possible benefits of direct reduction of iron ore?
6. What is the goal of steelmaking processes with regard to the amounts of carbon, silicon, manganese, phosphorus, and sulfur?
7. What was the major disadvantage of the open-hearth steelmaking process?
8. What are the major disadvantages of the basic oxygen process of steelmaking?

9. Why might electric furnaces be desirable when large quantities of steel scrap are available?
10. What is the advantage of extracting molten steel from the bottom of a ladle?
11. What is solidification shrinkage?
12. Why might an engineer want to degas molten steel prior to solidification?
13. What is a "killed" steel?
14. What is a "rimmed" steel?
15. Why do the gas holes in a rimmed material not appear in a hot-worked product?
16. Why might a metal be vacuum degassed?
17. When might an engineer be able to justify the additional cost of vacuum arc remelting or electroslag remelting?
18. What is the principal impurity in blister copper?
19. What are the three properties of copper that make it attractive as an engineering metal?
20. What is the major engineering motivation for the recycling of aluminum?
21. What properties of aluminum make it attractive as an engineering metal?
22. What may be the most serious weakness of aluminum from an engineering viewpoint?
23. What are the major attributes of titanium? Its major disadvantages?
24. Compare the specific gravities of steel, aluminum, and magnesium.
25. Why might magnesium products have to be painted?
26. What are the three major uses of zinc?
27. Why are zinc alloys attractive materials for die casting?
28. Which metals form the basis for the superalloys?
29. What is the unique feature of the superalloy materials?
30. What metals might be classified as high-temperature or "refractory" metals?
31. What is one unique property of graphite?

CASE STUDY 4.   The Broken Wire Cable

Figure CS-4 shows one end of a wire cable which broke, permitting the large boom of a crane to fall. Enlarged views of two typical individual wire ends are also shown. The company that manufactured the cable said the cable was completely normal and had been overloaded, whereas a man who was injured by the falling boom insisted that no overload had been applied and that the cable was defective.

Who was right? Why is such a wire cable not likely to be defective when new?

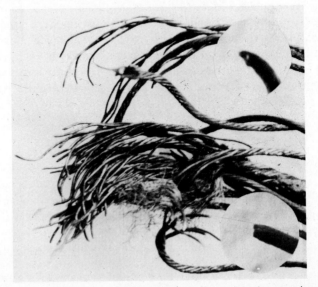

FIGURE CS-4   Wires at one end of a broken wire cable. Insets show enlarged views of two individual wire ends, typical of all wires in the cable.

CHAPTER 5

Equilibrium Diagrams

As our study of engineering materials becomes more focused on specific metals and alloys, it is increasingly important that the natural characteristics and properties of the material under various environments be known. In what condition is the material? Is the composition uniform throughout? If not, how much of each component is present? Is something present that may give undesired properties? What will happen if temperature is increased or decreased; pressure is changed; or chemistry is varied? The answers to these and other important questions can be obtained through use of equilibrium phase diagrams.

Phases.   Before moving to a discussion of these diagrams, it is imperative that a working definition of the term *phase* be developed. As a first-order definition, a phase is simply a form of material possessing a single characteristic structure and associated characteristic properties. Uniformity of chemistry, structure, and properties is assumed throughout the phase. More rigorously, a phase is *any physically distinct, chemically homogeneous, and mechanically separable portion of a substance*. In layman's terms, this requires a unique structure, uniform composition, and well-defined boundaries or interfaces.

A phase can be continuous, like the air in a room; or discontinuous, like grains of salt in a shaker. A phase can be solid, liquid, or gas. In addition, a phase can be a pure substance or a solution, provided that the structure and composition are uniform throughout. Alcohol and water mix in all proportions and will therefore form a single phase when combined. Oil and water tend to form isolated regions with distinct boundaries and must be considered as two distinct phases.

Equilibrium phase diagram. The *equilibrium phase diagram* is a *graphical mapping* of the natural tendencies of a material system, assuming infinite time at the conditions specified. Areas of the diagram are assigned to the various phases, with the boundaries indicating the equilibrium conditions of transition.

With the background just developed, let us now consider the types of phase mappings that may be useful. Three *primary variables* are at our disposal: *temperature, pressure, and composition*. The simplest type of diagram is a pressure–temperature (P-T) diagram for a fixed composition material.

For simplicity, consider the P-T diagram for water presented in Figure 5-1. With composition fixed, the diagram enables the determination of the stable form of water at any condition of temperature and pressure. Holding pressure constant and varying temperature, the transition boundaries locate the melting and boiling points. Still other uses can be presented. Locate a temperature where the stable phase at atmospheric pressure is the solid (ice). Now maintain that temperature and begin to decrease the pressure. A transition is encountered wherein solid goes directly to gas with no liquid intermediate. The process is that of freeze drying and is employed in the manufacture of numerous dehydrated products.

Temperature–composition diagrams. The water diagram serves as an example of the use of phase diagrams. In engineering applications, however, the P-T phase diagram is rarely used. Most processes are conducted at atmospheric pressure, with variations coming primarily in temperature and composition. The most useful mapping, therefore, would be a *temperature–composition phase diagram* at atmospheric pressure. For the remainder of the chapter, it is this second form of diagram that will be considered.

For mapping purposes, temperature is placed on the vertical axis and composition on the horizontal. Figure 5-2 shows the form of such a mapping for

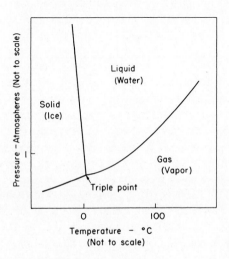

FIGURE 5-1 Schematic pressure–temperature diagram for water.

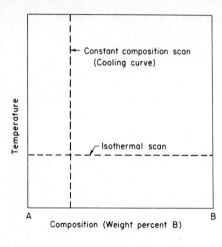

FIGURE 5-2 Temperature–composition equilibrium phase diagram mapping.

the A–B system where the left-hand vertical corresponds to pure material A and the percentage of B increases as we move toward pure B at the right of the diagram. Experimental investigations to fill in the details of the diagram basically take the form of vertical or horizontal scans designed to locate transition points.

Cooling curves. Considerable information can be obtained from vertical scans through the diagram in which a fixed composition material is heated and subsequently slow-cooled by removing heat at a uniformly slow rate. Transitions in structure appear as characteristic points in a temperature versus time plot of the cooling cycle, known as a *cooling curve*.

For the system composed of sodium chloride (common table salt) and water, five different cooling curves are presented in Figure 5-3. Curve (a) is for pure

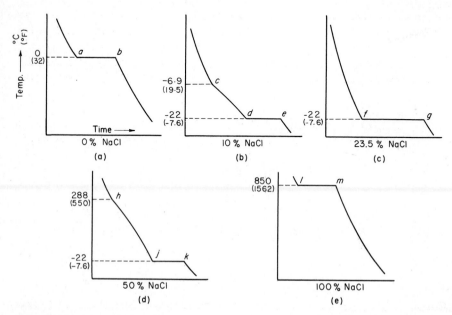

FIGURE 5-3 Cooling curves for various compositions of NaCl-H₂O solutions.

water being cooled from the liquid state. A smooth continuous line is observed for the liquid, the extraction of heat producing a concurrent drop in temperature. When the freezing point is reached (point *a*), the material changes state and releases heat energy equivalent to the liquid-to-solid transition. Heat is still being extracted from the system, but its source is the change in state and not a decrease in temperature. Thus an isothermal hold (*a-b*) is observed until solidification is completed. From this point, the newly formed solid experiences a smooth drop in temperature as heat extraction continues. Such a curve is characteristic of pure metals or substances with a distinct melting point.

Curve (b) of Figure 5-3 is the cooling curve for a 10% solution of salt in water. The liquid region undergoes a continuous cooling down to point *c*, where the slope abruptly decreases. At this temperature small particles of ice begin to form and the slope change is due to the energy released in this transition. The formation of these ice particles leaves the remaining solution richer in salt and imparts a lower freezing temperature to it. Further cooling must take place for additional solid to form. The formation of more solid leaves the remaining liquid richer in salt and the freezing point is progressively lowered. Instead of possessing a distinct melting point or freezing point, the material is said to have a *freezing range*. When the temperature of point *d* is reached, the remaining liquid solidifies into an intimate mixture of solid salt and solid water (to be discussed later) and an isothermal hold is observed. Further heat extraction from the solid material produces a continuous drop in temperature.

For a solution of 23.5% salt in water, a distinct freezing point is observed, as shown by curve (c) of Figure 5-3. Compositions with richer salt concentrations show phenomena similar to those presented earlier but with solid salt being the first substance to form from the liquid melt.

The key transition points can be transferred to a temperature–composition diagram with the designating letters corresponding to those on the cooling curves. Figure 5-4 presents such a map with several key lines being drawn. Line *a-f-l* denotes the lowest temperature at which the material is totally liquid and is known as the *liquidus line*. Line *d-f-j* is the highest temperature at which a material is completely solid and is called the *solidus line*. Between the liquidus and solidus two phases coexist, one being liquid and the other being solid.

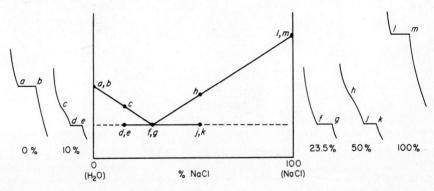

FIGURE 5-4 Partial equilibrium diagram for NaCl and H_2O derived from cooling curve information.

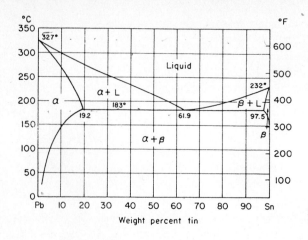

FIGURE 5-5 Lead–tin equilibrium diagram.

Cooling-curve studies have thus enabled the determination of various bits of key information regarding the system being studied. An equilibrium phase diagram, therefore, can be viewed as a collection of cooling curve data taken over an entire range of alloy compositions.

Solubility studies. The observant reader will note that the ends of the diagram still remain undetermined. Both pure materials have only one transition point below which they appear as a single-phase solid. Can ice retain some salt in solid solution? If so, how much? Can solid salt hold water and still remain a single phase? Completion of the diagram, therefore, requires several horizontal scans to determine any *solubility limits,* the conditions of saturation at various temperatures.

These isothermal scans with variable composition will usually require analysis of the specimens by X-ray technique, microscopy, or other investigative approaches to determine the composition at which transitions occur. As the scan moves away from the pure metal, the first line encountered (provided that the temperature is in the all-solid region) denotes the solubility limit and is known as the *solvus line.* Figure 5-5 presents the equilbrium phase diagram for the lead–tin system, using the conventional notation wherein Greek letters are used to label the various single-phase solids. The upper portion closely resembles the salt–water diagram, but solubility of one metal in the other can be seen at both ends of the diagram.

Having now been exposed to the concepts of equilibrium diagrams, let us move on to consider several specific examples. Presentation will move from the simple to the more complex.

Complete solubility in both liquid and solid states. If two metals are each completely soluble in the other in both the liquid and solid states, a rather simple diagram results, as illustrated in Figure 5-6 for the copper–nickel system. At temperatures above the liquidus line, the two materials are in liquid solution no matter what the composition. Similarly, below the solidus, the materials from a solid solution at all compositions. Between the liquidus and solidus is a two-phase region where liquid and solid solutions coexist.

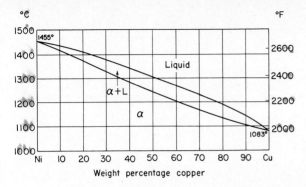

FIGURE 5-6 Copper–nickel equilibrium diagram, showing complete solubility.

Partial solid solubility. As might be expected, many materials exhibit neither complete solubility nor complete insolubility in the solid state. Each is soluble in the other up to a certain limit or saturation point, the value of this limit being a function of temperature. Such a diagram has already been observed in the lead–tin system of Figure 5-5.

Consideration of Figure 5-5 shows that the maximum solubility of tin in lead in the solid state is 19.2% by weight. Similarly, tin will dissolve up to 2.5wt % lead in solid solution. If the temperature is decreased from this point of maximum solubility, the amount of substance capable of being held in solution generally decreases. Thus, if a saturated solution of tin in lead is cooled from 183°C, the material moves from a single-phase region into a two-phase region. Some tin-rich second phase must precipitate from solution. This fact is used to control the properties of many engineering alloys.

Insolubility. If one or both of the components are insoluble in the other, the diagrams also reflect this phenomenon. Figure 5-7 illustrates the case where component A is completely insoluble in component B. Figure 5-8 presents the extreme case where the materials are completely insoluble in each other in both the liquid and solid states.

Utilization of diagrams. Before considering some of the more complex components of phase diagrams, let us return to a complete solubility diagram (Figure

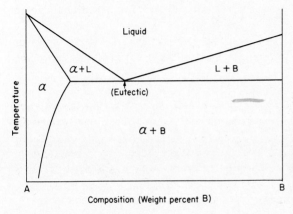

FIGURE 5-7 Equilibrium diagram of two materials, one of which is partially soluble in the other in the solid state.

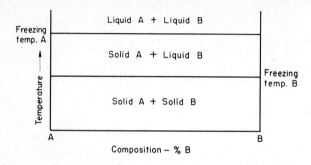

FIGURE 5-8 Equilibrium diagram of two materials that are completely insoluble in each other in both the liquid and solid states.

5-9) and pursue the utilization of these diagrams to obtain useful information. Three basic pieces of information can be obtained at each point on the diagram.

1. *The phases present.* By simply locating the point of consideration on the temperature–composition mapping and identifying the region of the diagram in which it appears, the stable phases can be determined.
2. *The composition of each phase.* If the point lies in a single-phase region, the composition of the phase is, by necessity, the overall composition of the material being considered. If the point lies in a two-phase region, a *tie-line* is drawn. A tie-line is simply an isothermal line drawn through the point of consideration, terminating at the boundaries of the single-phase regions on either side. The composition at which the tie-line intersects the neighboring single-phase regions determines the compositions of these phases in the two-phase mixture. For example, consider point a in Figure 5-9. The tie-line for this point runs from S_2 to L_2. The point S_2 is the intersection of the tie-line with the solid phase, and thus the solid in the two-phase mixture at a will have the composition of point S_2. Similarly, the liquid will have the composition of point L_2.
3. *The amount of each phase present.* If the point lies in a single-phase region, the amount must be 100% of that phase. If the point lies in a two-phase region, the relative amounts can be determined by a *lever-law* calculation using the previously drawn tie-line.

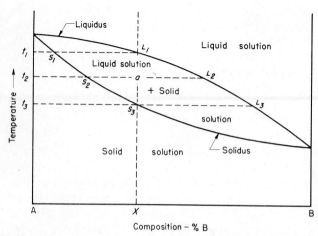

FIGURE 5-9 Equilibrium diagram of two materials that are completely soluble in each other in both the liquid and solid states.

Consider the cooling of alloy X in Figure 5-9 in a manner sufficiently slow so as to approximate equilibrium. At temperatures above t_1, the material is in a single-phase liquid state. Now go to temperature t_1 and draw a tie-line from S_1 to L_1. Any solid forming will be of composition S_1, but common sense tells us that almost all of the material will still be liquid of composition L_1. The entire tie line lies to the left of composition X. Now drop to temperature t_3, where the material is entirely solid of composition S_3. The tie-line lies to the right of composition X. Extrapolating these observations to intermediate temperatures, such as t_2, we predict that the amount of liquid will be proportional to the relative length of the tie-line to the left of point a. Namely, the amount of liquid at t_2 is

$$\frac{a - S_2}{L_2 - S_2} \times 100\%$$

Similarly, the amount of solid corresponds to the proportion of the tie-line to the right of point a. Since the calculations consider the tie-line as a lever with the phases at each end and a fulcrum at the composition line, they are called lever-law calculations.

Other applications of phase diagrams relate to an overall view of the system or the location of the transition points for a specific alloy. For instance, the temperature necessary to put an alloy into a given phase field can easily be determined. Changes that may occur upon the slow heating or slow cooling of a given material can be predicted. In fact, most of the questions posed at the beginning of this chapter can now be answered.

Solidification of Alloy X. Using the tools discussed above, it now becomes relatively simple to follow the solidification of alloy X in Figure 5-9. At temperature t_1, the first minute amount of solid forms with the chemistry of point S_1. As temperature drops, more solid forms, but the chemistry of both solid and liquid phases is shifting in accordance with the tie-line end points. Finally, at t_3, solidification is complete, and the composition of the single-phase solid is that of alloy X, as required.

The composition of the final solid is different from that of the first solid that was formed. If cooling is sufficiently slow (equilibrium conditions are approached) the composition of the entire mass of solid tends to become uniform at the value predicted by the tie-line. Compositional differences are removed by the phenomenon of diffusion, wherein atoms migrate from point to point in the crystal lattice under the energy impetus of elevated temperature. If the cooling rate is rapid, a nonuniform material may result, with the initial solid that formed retaining a composition different from the latter portions of the solid. This structure is referred to as being *cored*.

Three-phase reactions. Several of the diagrams previously presented contain a distinct feature in which phase regions are separated by a horizontal line. These lines are further characterized by either a **V** intersecting from above or an inverted **V** intersecting from below, denoting the location of a *three-phase equilibrium reaction*.

One common type of three-phase reaction, known as a *eutectic*, has already been observed in Figures 5-4, 5-5, and 5-7. Understanding of such reactions is possible by the use of the tie-line and lever-law concepts and will be presented using the lead–tin diagram of Figure 5-5. Consider any alloy containing between 19.2 and 97.5 wt % tin at a temperature just above the 183°C horizontal line. Tie-line and lever-law computations show that the material contains either a lead-rich or tin-rich solid and remaining liquid, the liquid having a composition of 61.9 wt % tin. [Note that any liquid will always have a composition of 61.9 wt % at 183°C (361°F), regardless of the overall composition of the alloy.] If we now focus on the liquid and allow it to cool to just below 183°C (361°F), a transition occurs wherein liquid of composition 61.9% tin goes to a mixture of lead-rich solid with 19.2% tin and a tin-rich solid containing 97.5% tin. The relative amounts of the two components maintain the overall chemistry. The form for this eutectic transition is similar to a chemical reaction:

$$\text{liquid} \rightarrow \text{solid}_1 + \text{solid}_2$$

Since the two solids have chemistries on either side of the intermediate liquid, a separation must have occurred within the system. Such a separation in a solidifying melt results from two metals that are soluble in the liquid state but only partially soluble in the solid state. Separation requires atom movement, but the distances between the two solids cannot be great. The resulting eutectic structure is an intimate mixture of two single-phase solids and assumes a characteristic set of physical and mechanical properties. Alloys of the eutectic composition have the lowest melting points of all alloys in a given system and are often used as casting alloys or as filler metal in soldering or brazing applications.

Figure 5-10 summarizes some other types of three-phase reactions that may occur in engineering systems. These include the *peritectic, monotectic,* and *syntetic,* the suffix —*ic* denoting that at least one of the three phases in the reaction is a liquid. If the suffix —*oid* is used, it simply denotes that all phases involved are solids; the form of the reaction remains the same. Two all-solid reactions can occur: the *eutectoid* and *peritectoid.* These reactions tend to be a bit more sluggish since all changes must occur in the solid state.

Intermetallic compounds. A final feature occurs in systems wherein the bonding attractions of the component materials is sufficiently strong that compounds tend to form. These compounds are single-phase solids and tend to break the diagram into recognizable subareas. If components A and B form a compound A_xB_y and the compound cannot tolerate any deviation from that fixed ratio, the intermetallic is known as a *stoichiometric intermetallic* and appears as a single vertical line in the diagram. If some degree of deviation is tolerable, the vertical line expands into a region and the compound is a *nonstoichiometric intermetallic.* Figure 5-10 shows schematic representations for both types of intermetallic compounds.

In general, intermetallics tend to be hard, brittle materials, these properties being related to their ionic or covalent bonding. If they are present in large quantities or lie along grain boundaries, the overall alloy can be extremely brittle. If the same intermetallic can be uniformly distributed throughout the structure in small particles, the effect can be considerable strengthening of the alloy.

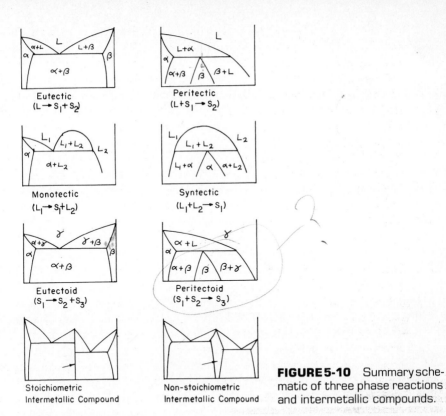

Eutectic
$(L \to S_1 + S_2)$

Peritectic
$(L + S_1 \to S_2)$

Monotectic
$(L_1 \to S_1 + L_2)$

Syntectic
$(L_1 + L_2 \to S_1)$

Eutectoid
$(S_1 \to S_2 + S_3)$

Peritectoid
$(S_1 + S_2 \to S_3)$

Stoichiometric
Intermetallic Compound

Non-stoichiometric
Intermetallic Compound

FIGURE 5-10 Summary schematic of three phase reactions and intermetallic compounds.

Complex diagrams. Most equilibrium diagrams of actual alloy systems will be one of the basic types just discussed or combinations thereof. Often, these diagrams appear to be quite complex and formidable to the manufacturing engineer. By focusing on the particular alloy in question and analyzing specific points using the tie-line and lever-law concept, even the most complex diagram can be simply dissected. Knowledge of the properties of the various components then enables predictions about the overall product.

The iron–carbon equilibrium diagram. Because steel, composed essentially of iron and carbon, is such an important engineering material, the iron–carbon equilibrium diagram of Figure 5-11 is by far the most important of those with which the average engineer must deal. Actually, the diagram most frequently used is an iron–iron carbide diagram, for a stoichiometric intermetallic carbide of the form Fe_3C can be used to terminate the useful range of the diagram at 6.67 wt % carbon. Names and notations have evolved historically and will be used in their generally accepted form.

Three distinct three-phase reactions can be identified in the diagram. At 1495°C (2723°F), a *peritectic* occurs for alloys with low weight percent carbon. Because of its high temperature and the extensive single-phase gamma region immediately below it, the peritectic reaction rarely assumes any engineering significance. A *eutectic* is observed at 1148°C (2098°F), with the eutectic point at 4.3% carbon. Alloys containing greater than 2.11% carbon will experience this

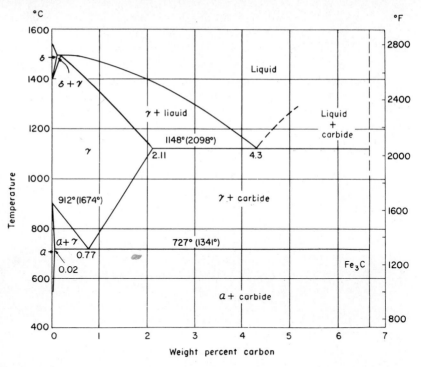

FIGURE 5-11 Iron—carbon equilibrium diagram. α, ferrite; γ, austenite; δ, δ-ferrite; Fe₃C, cementite.

eutectic and are classified by the general term *cast irons*. The final three-phase reaction is a *eutectoid* at 727°C (1341°F), with the eutectoid point at 0.77 wt % carbon. Alloys with less than 2.11% carbon can undergo a transition from a single-phase solid solution (γ) through the eutectoid to a two-phase mixture and are known as *steels*. Thus the point of maximum carbon solubility in iron, 2.11 wt %, forms an arbitrary division between the steels and cast irons.

To further our understanding of the diagram, let us now consider the four single-phase components, three of which occur in pure iron, the fourth being the carbide at 6.67% carbon. Upon solidification, pure iron forms a body-centered-cubic solid that is stable down to 1394°C (2541°F). Known as *delta ferrite*, this phase is only stable at extremely elevated temperatures and has no significant engineering importance. From 1394°C (2541°F) to 912°C (1674°F), pure iron assumes a face-centered-cubic structure known as austenite (γ) (in honor of the famed metallurgist Sir Robert Austen of England). Key features of austenite are the high formability characteristic of the fcc structure and its high solubility of carbon. Hot forming of steel benefits from these features of formability and compositional uniformity. Moreover, most heat treatment of steel begins with the single-phase austenite structure. *Alpha ferrite*, or more commonly just *ferrite*, is the stable form of iron at temperatures below 912°C (1674°F). This body-centered-cubic structure can hold only 0.02 wt % carbon in solution and forces a two-phase mixture in most steels. The only other change upon the further cooling of iron is the nonmagnetic-to-magnetic transition at the Curie

118

point (770°C) (1418°F). Since this is not associated with any change in phase, it does not appear on the equilibrium phase diagram.

The fourth single-phase region is the brittle intermetallic, Fe_3C, which also goes by the name *cementite* or iron carbide. As with most intermetallics, it is quite hard and brittle and care should be exercised in controlling the structures in which it occurs. Alloys with excessive amounts of cementite or cementite in undesirable form tend to have brittle characteristics. Since cementite dissociates prior to melting, its exact melting point is unknown and the liquidus line remains undetermined in the high-carbon region of the diagram.

A simplified iron carbon diagram.   If we focus only on the materials normally known as steels, a simplified diagram is often used. Those portions of the iron–carbon diagram near the delta region and those above 2% carbon content are of little importance to the engineer and are deleted. A simplified diagram, such as the one in Figure 5-12, focuses on the eutectoid region and is quite useful in understanding the properties and processing of steel.

The key transition described in this diagram is the decomposition of single-phase austenite (γ) to the two-phase ferrite plus carbide structure as temperature drops. Control of this reaction, which arises due to the drastically different carbon solubilities of austenite and ferrite, enables a wide range of properties to be achieved through heat treatment.

To begin to understand these processes, consider a steel of the eutectoid composition, 0.77% carbon, being slow cooled along line x-x' in Figure 5-12. At the upper temperatures, only austenite is present, the 0.77% carbon being dissolved in solid solution with the iron. When the steel cools to 727°C (1341°F), several changes occur simultaneously. The iron wants to change from the fcc

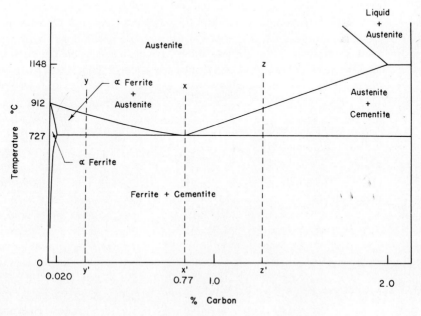

FIGURE 5-12 Simplified iron–carbon diagram.

austenite structure to the bcc ferrite structure, but the ferrite can only contain 0.02% carbon in solid solution. The rejected carbon forms the carbon-rich cementite intermetallic with composition Fe_3C. In essence, the net reaction at the eutectoid is

$$\text{austenite} \rightarrow \text{ferrite} + \text{cementite}$$
$$0.77\% \ C \qquad 0.02\% \ C \qquad 6.67\% \ C$$

Since this chemical separation of the carbon component occurs entirely in the solid state, the resulting structure is a fine mechanical mixture of ferrite and cementite. Specimens prepared by polishing and etching in a weak solution of nitric acid and alcohol reveal the lamellar structure of alternating plates that forms on slow cooling. This structure, seen in Figure 5-13, is composed of two distinct phases, but has its own set of characteristic properties and goes by the name *pearlite,* because of its resemblance to mother-of-pearl at low magnification.

Steels having less than the eutectoid amount of carbon (less than 0.77%) are known as *hypoeutectoid steels.* Consider now the transformation of such a material, represented by cooling along line y-y' in Figure 5-12. At high temperatures, the material is entirely austenite, but upon cooling enters a region where the stable phases are ferrite and austenite. Tie-line and lever-law calculations show that low-carbon ferrite nucleates and grows, leaving the remaining austenite richer in carbon. At 727°C (1341°F), the austenite is of eutectoid composition (0.77% carbon) and further cooling transforms the remaining austenite to pearlite. The resulting structure is a mixture of primary or proeutectoid ferrite (ferrite that formed above the eutectoid reaction) and regions of pearlite. An example of this structure is shown in Figure 5-14.

Hypereutectoid steels are steels that contain greater than the eutectoid amount of carbon. When such a steel cools, as in z-z' of Figure 5-12, the process is similar to the hypoeutectoid case, except that the primary or proeutectoid phase is now cementite instead of ferrite. As the carbon-rich phase forms, the remaining austenite decreases in carbon content, reaching the eutectoid composition at 727°C (1341°F). As before, any remaining austenite transforms to

FIGURE 5-13 Pearlite; 1000×. (*Courtesy United States Steel Corporation.*)

FIGURE 5-14 Photomicrograph of a hypoeutectoid steel showing ferrite (white) and pearlite; 500×. (*Courtesy United States Steel Corporation.*)

pearlite upon slow cooling through this temperature. Figure 5-15 is a photomicrograph of the resulting structure.

It should be remembered that the transitions that have been described by the phase diagrams are for equilibrium conditions, which can be approximated by slow cooling. With slow heating, these transitions occur in the reverse manner. However, when alloys are cooled rapidly, entirely different results may be obtained, because sufficient time is not provided for the normal phase reactions to occur. In such cases, the phase diagram is no longer a useful tool for engineering analysis. Since such processes are often important to the heat treatment of steels and other metals, their characteristics will be discussed in Chapter 6 and new tools will be developed to aid our understanding.

Cast irons. Those alloys of iron and carbon having greater than 2.11% carbon are called *cast irons*. More specifically, these are alloys with sufficient carbon that they will experience the eutectic transformation during cooling. Being relatively inexpensive, with good fluidity and rather low liquidus temperatures, they are readily cast and occupy an important place in engineering applications.

Most commercial cast irons also contain a significant amount of silicon, the general composition being 2.0 to 4.0% carbon, 0.5 to 3.0% silicon, less than 1.0% manganese, and less than 0.2% sulfur. Silicon produces two major effects.

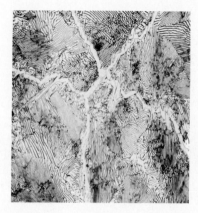

FIGURE 5-15 Photomicrograph of a hypereutectoid steel showing primary cementite along grain boundaries; 500×. (*Courtesy United States Steel Corporation.*)

First, it partially substitutes for carbon, such that use of the phase diagram requires replacing the weight percent carbon scale with a carbon equivalent. Several formulations of carbon equivalent exist, the simplest being percent carbon plus one-third of the percent silicon. As a second effect, silicon tends to promote the formation of graphite as the carbon-rich single phase instead of the Fe_3C intermetallic. Thus the eutectic reaction now has two distinct possibilities as seen in the modified phase diagram of Figure 5-16:

$$liquid \rightarrow austenite + Fe_3C$$

or

$$liquid \rightarrow austenite + graphite$$

The final microstructure, therefore, has two possible extremes: (1) all of the carbon-rich phase in the form of Fe_3C, and (2) all of the carbon-rich phase in the form of graphite. In practice, these extremes can be approached in the various types of cast irons by control of the process variables. Graphite formation is promoted by slow cooling, high carbon and silicon contents, heavy section sizes, inoculation practices, and alloy additions of Ni and Cu. Cementite (Fe_3C) is favored by fast cooling, low carbon and silicon levels, thin sections, and alloy additions of Mn, Cr, and Mo.

Four basic types of cast irons are produced, the most common being *gray cast iron*. In this type, most of the carbon is in the form of graphite flakes formed during the eutectic reaction (some carbide may form at the lower eutectoid reaction). When fractured, the freshly exposed surface has a gray appearance (see Figure 5-17) and a graphite smudge can usually be obtained if one rubs a finger across a freshly fractured or machined surface.

Gray cast iron is the least expensive of the four types and is characterized by those features which promote the formation of graphite. Typical compositions range from 2.5 to 4.0% carbon, 1.0 to 3.0% silicon, and 0.4 to 1.0% manganese.

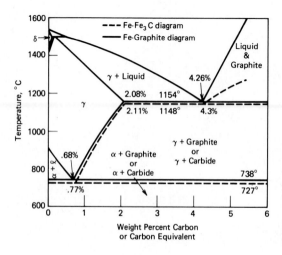

FIGURE 5-16 Iron—carbon diagram showing two possible eutectic reactions. Graphite is solid line; cementite is dashed.

FIGURE 5-17 (*Left to right*) Fractures of gray, white, and malleable iron. (*Courtesy Iron Castings Society*, Rocky River, Ohio.)

The microstructure consists of three-dimensional graphite flakes dispersed in a matrix of ferrite, pearlite, or other iron-based structure (such as those to be encountered in Chapter 6). Figure 5-18 shows a typical cross section. Because the graphite flakes have no appreciable strength, they act essentially as voids in the structure. Moreover, the pointed edges of the flakes act as preexisting notches or crack initiation sites, giving the material a characteristic brittle nature. Size and shape of the graphite flakes have considerable effect on the overall properties of gray cast iron. When maximum strength is desired, small, uniformly distributed flakes are desired with a minimum amount of mutual intersection.

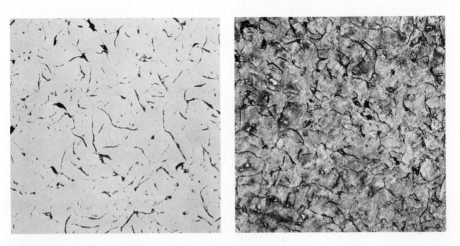

FIGURE 5-18 Photomicrographs of typical gray cast iron; 1000×. (*Left*) unetched; (*right*) etched. (*Courtesy Bethlehem Steel Corporation.*)

A more effective means of controlling strength is through control of the matrix structure. Several distinct classes of gray iron are identified on the basis of tensile strength, the class number corresponding to the minimum tensile strength in thousands of pounds per square inch. Class 20 iron (minimum tensile strength of 20,000 psi) consists of high-carbon equivalent metal with a ferrite matrix. Higher strengths, up to class 40, are attainable with lower-carbon equivalents and a pearlite matrix. To go above class 40, alloying is required to provide solid solution strengthening and heat treatment practices are often employed. Gray irons can be obtained up through class 80, but in all cases, the presence of graphite flakes results in near-zero ductility.

Gray cast irons possess excellent compressive strengths (compressive forces do not promote crack propagation), excellent machinability (graphite acts to break up the chips and lubricate contact surfaces), good wear resistance (graphite flakes self-lubricate), and outstanding vibration damping characteristics (graphite flakes absorb vibration energy). High silicon contents promote good corrosion resistance and the fluidity desired for casting applications. For these reasons, together with its low cost, it is an excellent material for large machinery parts that are subjected to high compressive loads and vibration.

White cast iron has essentially all of the carbon in the form of iron carbide and receives its name from the white surface that appears when the material is fractured (see Figure 5-17). Features promoting its formation are those which favor cementite over graphite: a low-carbon equivalent (1.8 to 3.6% carbon, 0.5 to 1.9% silicon, and 0.25 to 0.8% manganese) and rapid cooling.

By virtue of the large amounts of iron carbide, white iron is very hard and brittle and finds application where extreme wear resistance is required. For these uses, it is common to alloy the material so as to produce the hard, wear-resistant *martensite* structure as the iron-rich phase either upon solidification or by subsequent heat treatment. (This structure will be discussed in Chapter 6.) In this manner, both phases contribute to the wear-resistant characteristics.

Other applications of white iron involve a surface layer with an underlying substrate of other material. Mill rolls that require extreme wear resistance may have a white cast iron surface and steel interior. Variable cooling rates produced by tapered sections or metal chill bars placed in the molding sand can be used to produce a surface of white cast iron on a gray iron casting. An example of this can be found in inexpensive scissors, where white iron forms along the thin cutting edge and gray iron forms in the heavier backup section. A fracture surface will show white iron, a transition region of mixed white and gray iron known as the *mottled zone,* and then gray iron.

Other white cast iron may be put through a controlled heat treatment cycle where the cementite dissociates and some or all of the carbon is converted to irregular graphite spheroids. This product, known as *malleable cast iron,* has appreciably better ductility than that exhibited by gray cast iron because the more favorable graphite shape removes the internal notches. The rapid cooling required to produce the starting white iron structure, however, restricts the size and thickness of malleable iron products, most weighing less than 4.5 kg (10 pounds).

Two types of malleable iron can be produced, depending on the nature of the thermal cycle. If white iron is heated and held for a prolonged time just below

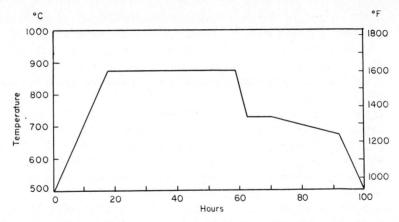

FIGURE 5-19 Typical thermal cycle for manufacture of ferritic malleable cast iron. (*Courtesy Iron Castings Society.*)

the melting point, the carbon in cementite reverts to graphite. Subsequent slow cooling through the eutectoid reaction causes the austenite to go to ferrite and more graphite. The resulting product is known as *ferritic malleable iron* and has properties consistent with its structure of irregular graphite spheroids in a ferrite matrix: 10% elongation, 240-MPa (35-ksi) yield strength, 345-MPa (50-ksi) tensile strength, and excellent impact strength, corrosion resistance, and machinability. Figure 5-19 shows a typical thermal cycle and Figure 5-20 shows the final structure.

If the casting is rapidly cooled through the eutectoid transformation after the first thermal hold, the carbon in the austenite does not form additional graphite but is retained in a pearlite or martensite matrix. These structures are stronger than ferrite and the resulting product, known as *pearlitic malleable iron,* is characterized by higher strength and lower ductility than its ferritic counterpart. Typical properties range from 1 to 4% elongation, 310 to 590 MPa (45 to 85 ksi) yield strength, and 450 to 725 MPa (65 to 105 ksi) tensile strength, with reduced machinability.

Most malleable iron (ferritic-type) has poorer wear resistance than gray cast

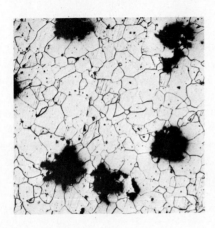

FIGURE 5-20 Photomicrograph of malleable iron. (*Courtesy Iron Castings Society.*)

iron but is as good or better in machinability. Because of its superior ductility, it is widely used for automotive parts, such as axle housings and brackets, and for pipe fittings and other applications where a uniform material of moderate ductility must be machined at rather low cost.

The graphite structure of malleable iron provides quite an improvement in properties, but it would be even better if it could be obtained during solidification rather than by prolonged heat treatment at highly elevated temperatures. Certain materials have been found that promote graphite formation and change the morphology of the graphite product. If sufficient magnesium (added in the form of MgFeSi or MgNi alloy) or cerium is added to the liquid iron just prior to solidification, graphite tends to form as regular spheroids during solidification. The added material is known as a nodulizing agent and the product as *ductile or nodular cast iron*. Subsequent control of the thermal history can produce a wide range of matrix structures, with ferrite or pearlite being the most common (see Figure 5-21). Properties span a wide range from 2 to 18% elongation, 275 to 620 MPa (40 to 90 ksi) yield strength, and 415 to 825 MPa (60 to 120 ksi) tensile strength.

The combination of good ductility, high strength, and castability makes nodular iron a rather desirable engineering material. Unfortunately, the cost of the nodulizer, higher-grade melting stock, better furnaces, and improved process control may make ductile iron almost as expensive as malleable iron. Nevertheless, it is replacing gray iron, malleable iron, and steel castings in numerous applications.

Although existing for over 30 years, a material known as *compacted graphite iron* (or vermicular graphite iron) has begun to attract considerable attention because it has now become far more reproducible. Produced by a method similar to ductile iron, compacted graphite iron is characterized by a graphite structure intermediate to the flake graphite of gray iron and the nodular graphite of ductile iron and tends to possess the desirable properties and characteristics of each. Depending upon composition and section size, tensile strengths of 310 to 520 MPa (45 to 75ksi) and yield strengths of 240 to 415 MPa (35 to 60 ksi) have been reported with elongation values averaging 4%. Machinability, castability,

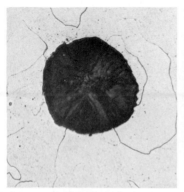 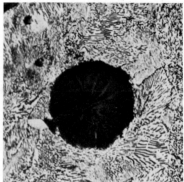

FIGURE 5-21 (*Left*) Ductile iron with ferrite matrix. (*Right*) Ductile iron with pearlite matrix; 500×. Note spheroidal graphite nodule.

damping capacity, and thermal conductivity all approach those of gray iron. Impact and fatigue properties approach those of ductile iron. Areas of potential application tend to be those requiring high strength with good machinability and thermal conductivity, such as engine blocks and brake drums.

REVIEW QUESTIONS

1. What characteristics must be possessed for a structure to be a phase?
2. What is an equilibrium phase diagram?
3. What are the three primary variables generally considered in equilibrium phase diagrams?
4. What type of phase diagram mapping would be most useful for engineering applications?
5. How can cooling curves be used to locate changes in a material's structure?
6. Why does a cooling curve for a pure metal show an isothermal hold at the freezing point?
7. What is a liquidus line? A solidus line?
8. What are solubility limits? What is the name of the phase diagram line that shows their values?
9. How does the solubility of one metal in another generally change with a decrease in temperature?
10. What are the three basic pieces of information that can be obtained for each point on an equilibrium phase diagram?
11. How can the composition of phases in a two-phase mixture be determined from a tie-line?
12. Why might the chemistry of a solid not be uniform if the material was quickly frozen?
13. What is a eutectic reaction?
14. What features in an equilibrium phase diagram indicate the presence of a three-phase reaction?
15. Describe a eutectic structure.
16. What feature of eutectic alloys makes them attractive for casting applications?
17. What are the typical mechanical property characteristics of intermetallic compounds?
18. Why might the presence of an intermetallic compound not be undesirable?
19. What are the three three-phase reactions that appear in the iron–carbon phase diagram?
20. By definition, what distinguishes cast irons from steels?
21. What are some of the key characteristics of austenite?
22. What are the stable phases in steel at room temperature?
23. What feature in the iron–carbon phase diagram is most responsible for the wide variety of properties that can be obtained by steel heat treatment?
24. What is pearlite? Describe its structure.
25. What is a hypoeutectoid steel? A hypereutectoid steel?
26. What are the attractive engineering features of cast irons?
27. What are the two major effects of silicon in cast iron?
28. What new phase is common in cast irons but is not prevalent in steels?
29. What features promote graphite formation?
30. Describe the microstructure of gray cast iron.
31. What structural unit is generally altered to increase the strength of gray irons?
32. What are some of the attractive engineering properties of gray cast irons?
33. What types of applications would be appropriate for white cast iron?
34. What is malleable iron and how is it made?
35. What structural feature is responsible for the high ductility of ductile or nodular cast iron?
36. What is compacted graphite iron? What are its attractive features?

CASE STUDY 5. Improper Utilization of Phase Diagrams

Harry Simon, a production engineer with Missouri Machine Co., needed to reduce the thickness of a standard strip of aluminum–3% copper alloy for use in a shimming application. Noting that the available rolling equipment had limited capacity, he decided to hot-roll the material in an effort to reduce the required forces. In selecting the heating temperature, he consulted the phase diagram in Figure CS-5 and selected 575°C (single-phase α region, 25°C below where liquid would form). Upon rolling at this temperature, however, the strip fragmented, breaking into pieces, rather than deforming uniformly. What had Harry overlooked?

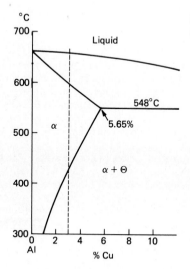

FIGURE CS-5 Aluminum-rich section of the aluminum—copper equilibrium phase diagram.

CHAPTER 6

Heat Treatment

THEORY AND PROCESSES OF HEAT TREATMENT

Heat treatment, by definition, is the *controlled heating and cooling of metals for the purpose of altering their properties*. Because the mechanical and physical properties can be altered so much by heat treatment, it is one of the most important and widely used manufacturing processes, providing a usually simple and low-cost means of obtaining desired properties. However, if performed improperly, more harm than good can result. Thus heat treatment must be understood and correlated with the other manufacturing processes in order to obtain effective results.

Although actual heat treatment applies only to processes where the heating and cooling are done for the specific purpose of altering properties, heating and cooling often occur as incidental phases of other manufacturing operations. (Obvious examples are hot-forming operations and welding.) The effect on the properties of the metal—beneficial or harmful—will be the same, even though no change in properties may be desired. Thus the designer who selects the metal and the engineer who determines its processing must be aware of possible changes and take them into account.

Because proper application of heat treatment requires a thorough understanding of the material's response to various processes, both the theory of heat treatment and the various processes will be considered in this chapter.

PROCESSING HEAT TREATMENTS

While heat treatment often is associated only with those thermal processes designed to increase strength, the definition permits inclusion of numerous processes which, for lack of a better term, we will call *processing heat treatments*. These are performed with a major goal of preparing the material for fabrication, including improving machining characteristics, reducing forming forces and energy consumption, and restoring ductility for further deformation. Thus heat treatment has tremendous capabilities, permitting the same metal to be softened for ease of fabrication, and then by another process, be given a totally different set of properties for service.

Equilibrium diagrams as aids to heat treatment.   Most processing heat treatments involve rather slow cooling or extended times at elevated temperatures, thus tending to approximate equilibrium conditions. The resulting structures, therefore, can reasonably be predicted by the use of the equilibrium phase diagrams. The diagram indicates the temperatures that must be attained to achieve a desired product and the change that will occur upon subsequent cooling. It should be remembered, however, that the diagram is for truly equilibrium conditions, and departures from equilibrium may lead to substantially different results.

Processing heat treatments for steel.   Because most processing heat treatments are applied to plain carbon and low-alloy steels, they will be presented here with the simplified iron–carbon equilibrium diagram of Figure 5-12 serving as a reference guide. Figure 6-1 shows this diagram with the key transition lines being labeled by standard notation. The eutectoid line is designated by the symbol A_1, and A_3 designates the boundary between austenite and ferrite +

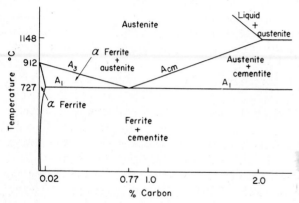

FIGURE 6-1 Simplified iron–carbon diagram for steels with transition lines labeled in standard notation.

austenite. The transition from austenite to austenite + cementite is designated as the A_{cm} line.

A number of process heat-treating operations are classified under the general term *annealing*. These may be employed to reduce hardness, remove residual stresses, improve toughness, restore ductility, refine grain size, reduce segregation, or alter the mechanical, electrical, or magnetic properties of the material. By producing a certain desired structure, characteristics can be imparted that are favorable to the subsequent operations or application. The temperature, cooling rate, and specific details of the process are determined by the material being treated and the objectives of the treatment.

In *full annealing*, hypoeutectoid steels are heated to 30 to 60°C (50 to 100°F) above the A_3 temperature to convert the structure to homogeneous single-phase austenite of uniform composition and temperature, held at this temperature for a period of time, and then slowly cooled at a controlled rate to below the A_1 temperature. A general rule is to provide 1 hour at temperature per inch of thickness of the largest section, but energy savings have motivated a reduction in this time. Cooling from the soaking temperature usually is done in the furnace, decreasing the temperature at a rate of 10 to 30°C (20 to 50°F) per hour to at least 30°C (50°F) below the A_1, followed by air cooling to room temperature. The resulting structure is coarse pearlite (widely spaced lamellae) with excess ferrite in amounts predicted by the phase diagram. The material is quite soft and ductile.

The procedure for hypereutectoid alloys is basically the same, except the original heating is only into the austenite plus cementite region (30 to 60°C above the A_1). If the material is slow-cooled from the pure austenite region, a continuous network of cementite may form on the grain boundaries and make the material brittle. When properly annealed, the structure of a hypereutectoid steel will be coarse pearlite plus excess cementite in dispersed spheroidal form.

Full anneals are time consuming and require considerable energy to maintain the elevated temperatures. When extreme softness is not required and cost savings are desired, *normalizing* may be employed. Here the metal is heated to 60°C (100°F) above the A_3 (hypoeutectoid) or A_{cm} (hypereutectoid), soaked to obtain uniform austenite, then removed from the furnace and allowed to cool in still air. Resultant structures and properties depend upon the subsequent cooling rate. Although wide variations are possible depending on the size and geometry of the metal, fine pearlite with excess ferrite or cementite generally is produced.

Where cold working has severely strain-hardened a metal, it is often desirable to restore the ductility, either for service or to permit further processing without danger of fracture. A *process anneal* is often used for this purpose. The metal is heated to a temperature slightly below the A_1, held long enough to achieve softening, and then cooled at any desired rate (usually in air). Since austenite is not formed, the existing phases simply change their morphology. Process anneals are often used to recrystallize low-carbon steel sheets. Because the material is not heated to as high a temperature as in other processes, a process anneal is somewhat cheaper, more rapid, and tends to produce less scaling.

A *stress-relief anneal* often is employed to remove residual stresses in large steel castings and welded structures. Parts are heated to temperatures below the

A_1 (550 to 650°C; 1000 to 1200°F), held for a period of time and then slow-cooled. Times and temperatures vary with the conditions of the component.

When high-carbon steels must be prepared for machining or forming, a process known as *spheroidization* is employed. The goal is to produce a structure wherein all cementite is in the form of small, well-dispersed spheroids or globules in a ferrite matrix. This can be accomplished by a variety of techniques, including (1) prolonged heating at a temperature just below the A_1 followed by relatively slow cooling, (2) prolonged cycling between temperatures slightly above and slightly below the A_1, or (3) in the case of tool steel or high-alloy steel, heating to 750 to 800°C (1400 to 1500°F) or higher and holding at this temperature for several hours, followed by slow cooling.

Although the selection of a processing heat treatment often depends on the desired objectives, steel composition strongly influences the choice. Low-carbon steels (less than 0.3% carbon) are most often normalized or process annealed. Steels of the medium (0.4 to 0.6%) carbon range are usually full annealed. Above 0.6% carbon a spheroidization treatment generally is required. Figure 6-2 provides a graphical summary of the process heat treatments.

Heat treatments for nonferrous metals. Most nonferrous metals do not have the significant phase transitions observed in the iron–carbon system and, for them, process heat treatments do not play such a significant role. Aside from precipitation hardening, which will be discussed later, nonferrous metals are ordinarily heat-treated for three purposes: (1) to obtain a uniform structure (such as to eliminate coring), (2) to provide stress relief, or (3) to bring about recrystallization. Coring, which may be present in castings that have cooled too rapidly, can be removed by heating to moderate temperatures and holding for a

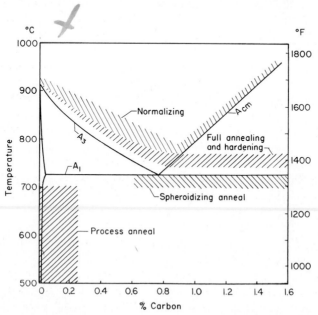

FIGURE 6-2 Graphical summary of process heat treatments for steel on an equilibrium diagram.

sufficient period to allow thorough diffusion to take place. Similarly, stresses that result from forming, welding, or brazing can be removed by heating for several hours at relatively low temperatures. Recrystallization, as discussed in Chapter 3, is a function of the particular metal, the degree of prior straining, and the time provided for completion. In general, the more a metal has been strained, the lower is the recrystallization temperature. If a nonferrous metal has been strained, it is relatively easy to bring about recrystallization, producing new, equiaxed, and stress-free grains. Without straining, no recrystallization will occur and only grain growth will result. It is important to remember that only through recrystallization can new equiaxed grains (of the same or finer grain size as the original) be obtained in nonferrous metals in the solid state.

HEAT TREATMENTS TO INCREASE STRENGTH

Six major mechanisms are available to increase the strength of metals: (1) solid solution hardening, (2) strain hardening, (3) grain size refinement, (4) precipitation hardening, (5) dispersion hardening, and (6) phase transformations. All can be induced or altered by heat treatment but not all can be applied to any given metal.

In *solid solution hardening,* a base metal dissolves other atoms in solid solution, either as *substitutional solutions,* where the new atoms occupy sites on the regular crystal lattice, or as *interstitial solutions,* where the new atoms squeeze into "holes" in the base lattice. The amount of strengthening depends on the amount of dissolved solute and the size difference of the atoms involved. Distortion of the host structure makes dislocation motion more difficult.

Strain hardening, as discusssed in Chapter 3, produces increased strength by plastic deformation under cold-working conditions.

Because grain boundaries act as barriers to dislocation motion, a metal with small grains will tend to be stronger than the same metal with larger grains. Thus *grain-size refinement* can be used to increase strength, except at elevated temperatures where failure is by a grain-boundary diffusion-controlled creep mechanism. Grain-size refinement is one of the few processes capable of improving both strength and ductility.

Precipitation hardening or *age hardening* is a method whereby strength is obtained from a nonequilibrium structure produced by a three step (solution treat–quench–and age) heat treatment.

Strength obtained from distinct second-phase particles in a base matrix is called *dispersion hardening.* To be effective, these second phases should be stronger than the matrix, adding strength through both their reinforcing action and by the additional barriers presented to dislocation motion.

Phase transformation strengthening involves alloys which can be heated to form a single high-temperature phase and subsequently transformed to one or more low-temperature phases upon cooling. Where phase transformation is used to increase strength, the cooling is usually rapid and the phases produced are nonequilibrium in nature.

STRENGTHENING HEAT TREATMENTS FOR NONFERROUS METALS

All of the mechanisms just described can be used to increase the strength of nonferrous metals. Solid solution strengthens single-phase metals. Strain hardening is applicable if sufficient ductility is present. Eutectic-forming alloys possess considerable dispersion hardening. The most effective strengthening mechanism for nonferrous metals, however, is precipitation hardening.

Precipitation or age hardening.   Some alloy systems, mostly nonferrous, possess a sloping solvus line such that certain alloys can be heated into a single-phase solid solution, and, owing to decreasing solubility, will form two distinct phases at lower temperatures. If the heated single phase is rapidly cooled, however, a supersaturated solid solution can be formed wherein the material required to form the second phase is trapped in the base lattice. A subsequent *aging* process permits the excess solute atoms to precipitate out of the supersaturated matrix and produce a controllable nonequilibrium structure.

As an example, consider the silver-rich portion of the silver–copper system shown in Figures 6-3 and 6-4. If an alloy composed of 92.5% silver and 7.5% copper (sterling silver) is slowly cooled from 775°C (1430°F), upon crossing the solvus line *A-B* at 760°C (1400°F) β phase is precipitated out of the α solid solution because the solubility of copper in silver decreases from 8.4% at 780°C (1435°F) to less than 1% at room temperature. If this same alloy were cooled rapidly from 775°C (1430°F), there would not be sufficient time for the normal structure to form and the α phase would be retained in a highly supersaturated state. Because this supersaturated condition is not normal, the excess copper would like to precipitate out of solution and coalesce into β-phase particles. Atom movement or diffusion is required, and therefore time at elevated tem-

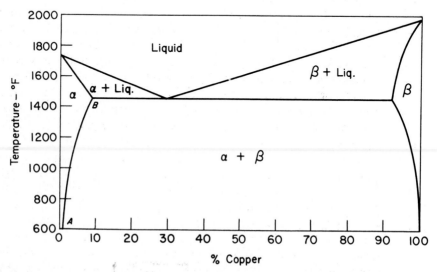

FIGURE 6-3 Silver–copper equilibrium diagram.

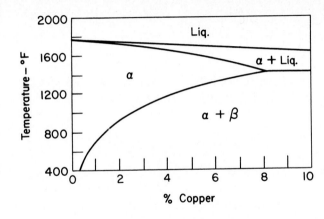

FIGURE 6-4 Enlargement of the copper-rich section of the silver—copper equilibrium diagram (Figure 6-3).

perature may have to be provided. If, in the case of sterling silver, the temperature is raised to around 260°C (500°F), precipitation occurs such that after 30 minutes the hardness will have increased from about 55 Brinell to about 110 Brinell.

Precipitation hardening is a three-step controlled heat treatment. The first step is a heating (*solution treatment*) to put the material into a single-phase solid solution. This heating must not exceed the eutectic temperature so as to avoid melting in a cored structure. After soaking to achieve a uniform single phase, the material is *quenched* (rapidly cooled), usually in water, to prevent diffusion and to produce the supersaturated solid solution. In this state the material is rather soft, possibly softer than in the annealed condition. It can be straightened, formed, or machined while in this soft condition.

At this point, precipitation-hardening materials can be classified into two types: (1) *naturally aging,* in which the required diffusion from the unstable supersaturated to the stable two-phase structure can occur at room temperature, and (2) *artificially aging,* which require elevated temperatures. With the first type the after-quench softness can be retained by refrigeration. (Aluminum alloy rivets are examples. Removed from the refrigeration and headed, they attain full strength in several days.) The properties of the second type can readily be controlled by means of temperature and aging time. Aging is a continuous process which begins by the clustering of solute atoms on distinct planes of the parent lattice. Various transitions may then occur, leading to the ultimate formation of a distinct second phase with its own characteristic crystal structure.

A key concept in this sequence is that of *coherency.* If the crystallographic planes are continuous in all directions, the solute aggregates tend to distort (strain) the adjacent lattice for a sizable distance away and a small aggregate looks like a much larger barrier to dislocation motion. When the aggregates reach a certain size, they tend to break free of the parent structure, forming distinct second-phase particles with interphase boundaries. Coherency is lost and the mechanism of strengthening is *dispersion hardening,* wherein the particles present only their physical dimensions as dislocation blocks. The material is said to be *overaged.* A benefit of artificial aging, therefore, is the ability to stop the process at any stage by simple quenching. The structure and properties of that stage are retained, provided that the material does not subsequently experience elevated temperature conditions. Artificially aging alloys are quite popular because of the ability to tailor properties to needs.

Precipitation hardening is responsible for the engineering strengths of many aluminum, copper, and magnesium alloys. In many cases, this is more than double the strength observed upon conventional slow cooling. By special processing, some age-hardenable ferrous alloys have also been produced.

STRENGTHENING HEAT TREATMENTS FOR STEEL

Iron-base metals have been heat-treated for centuries, and today over 90% of all heat treating is performed on steel. The striking changes that resulted from plunging red-hot steel into cold water or some other quenching medium were awe-inspiring to the ancients. Those who did such heat treatment in making of swords or armor were looked upon as possessing unusual powers, and much superstition arose regarding the process. Because quality was directly related to the act of quenching, great importance was placed on the quenching medium that was used. For example, urine was thought to be a very superior quenching medium, and that from a red-haired boy was deemed particularly effective.

The isothermal transformation diagram.   It has been only within the last hundred years that the art of heat treating has begun to turn into a science. One of the major barriers to understanding was the fact that the strengthening treatments were nonequilibrium in nature. Variations in cooling produced variations in structure and properties.

One of the aids to understanding the nonequilibrium processes was the *isothermal transformation* (IT) or *time–temperature–transformation* (T-T-T) diagram, obtained by heating thin specimens of a given metal to form uniform single-phase austenite, "instantaneously" quenching to a temperature where austenite was not the stable phase, holding for variable periods of time, and observing the resultant product via photomicrographs.

For simplicity, consider a carbon steel of eutectoid composition and the resulting T-T-T diagram (Figure 6-5). Above 727°C (1341°F), austenite is the stable phase. Below this temperature, the face-centered austenite would like to transform to body-centered ferrite and carbon-rich cementite. Two factors control the rate of transition: (1) the motivation or driving force for the change, and (2) the ability to form the desired products (i.e., the ability to move atoms through diffusion). Figure 6-5 can be interpreted as follows. For any given temperature below 727°C (1341°F), time zero corresponds to a sample quenched "instantaneously." The structure usually is unstable austenite. As time passes (moving horizontally across the diagram), a line is encountered representing the start of transformation and a second line indicating completion of the phase change. At elevated temperatures (just below 727°C), diffusion is rapid, but the rather sluggish driving force dominates the kinetics. At lower temperature, the driving force is high but diffusion is quite limited. Kinetics are more rapid at a compromise intermediate temperature than at either extreme, resulting in the often referred to *C-curve* terminology. That portion of the C which extends farthest to the left is known as the *nose* of the T-T-T diagram.

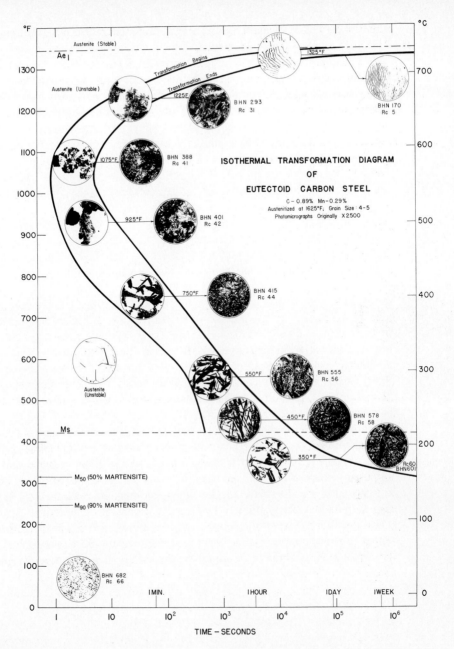

FIGURE 6-5 Isothermal transformation diagram (T-T-T diagram) for eutectoid composition steel. Structures resulting from transformation at various temperatures are shown as insets. (*Courtesy United States Steel Corporation.*)

Now consider the products of the various transformations. If the transformation occurs between the nose of the curve and the A_1 temperature, the departure from equilibrium is not great. Austenite forms ferrite and cementite in the combined structure known as pearlite. Because of diffusion capabilities at higher temperatures, the lamellae spacing will be greater for the higher-temperature product, coarse pearlite, than for that formed nearer the nose, fine pearlite.

If quenched to a temperature between the nose and the temperature designated as M_s, another type of product must form. The process now is a significant departure from equilibrium, and the diffusion required to form lamellar pearlite is no longer available. The metal still has the same goal however—to change its structure from austenite to a mixture of ferrite and cementite. The resulting structure is not that of alternating plates, but rather a dispersion of discrete cementite particles in either a lathlike or needlelike matrix of ferrite. Electron microscopy may be required to resolve the carbides in this structure, known as *bainite*. Because of the fine dispersion of carbide, its strength exceeds that of fine pearlite, and ductility is retained because soft ferrite is the continuous phase.

If the metal is quenched to below the M_s temperature, a different type of transformation occurs. The metal still desires to go from the face-centered to body-centered structure, but it cannot expel the required amount of carbon to form ferrite. As a response to the nonequilibrium conditions, the material undergoes an instantaneous change in crystal structure with no diffusion. The trapped carbon distorts the structure such that a body-centered tetragonal lattice results (a distorted body-centered cubic); the degree of distortion is a function of the amount of trapped carbon. The new structure, shown in Figure 6-6, is known as *martensite* and, with sufficient carbon, is exceptionally strong, hard, and brittle. Dislocation motion necessary for metal flow is effectively blocked by the highly distorted lattice.

As shown in Figure 6-7, the hardness and strength of steel in the martensitic form is a strong function of the carbon content. Below 0.10% carbon, martensite is not very strong, although it is tough. Hardness is typically 30 to $35R_c$ at 0.1% carbon and drops rapidly with decreasing carbon. Since no diffusion occurs in the transformation, higher-carbon-content material forms higher-carbon-containing martensite, with concurrent increase in strength and hardness and decrease in toughness and ductility. Thus from 0.3 to 0.7% carbon, the maximum hardness increases rapidly; above 0.7% carbon, the maximum hardness rises only slightly with increased carbon, a feature related to retained austenite.

The amount of martensite formed upon cooling is a function of the lowest temperature encountered and not the time at that temperature, as shown in Figure 6-8. Returning to Figure 6-5, we see below the M_s, a temperature designated

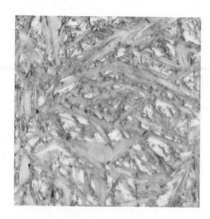

FIGURE 6-6 Photomicrograph of martensite; 1000×. (*Courtesy United States Steel Corporation.*)

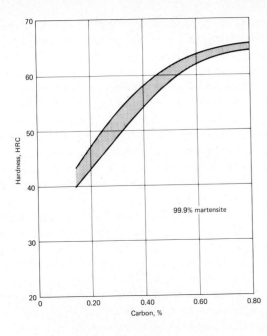

FIGURE 6-7 Effect of carbon on the hardness of martensite.

as M_{50} at which the structure will be 50% martensite and 50% untransformed austenite. At the lower M_{90} temperature, the structure is 90% martensite. If no further cooling is undertaken, the untransformed austenite can remain within the structure indefinitely. This *retained austenite* can cause loss of strength or hardness, dimensional instability or cracking, or brittleness. Since most quenches are to room temperature, retained austenite problems become significant when the martensite finish, or 100% martensite, temperature lies below ambient. Higher carbon contents and alloy additions both decrease all martensite-related temperatures, and these materials may require refrigeration or a quench in liquid nitrogen to obtain full hardness.

Note that all the transformations that occur below the A_1 are one-way transitions, austenite-to-something. These are the only reactions possible, and it is impossible to convert one product to another without first reheating to above A_1 to again form some stable austenite.

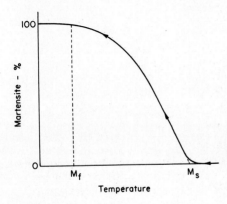

FIGURE 6-8 Schematic representation depicting the amount of martensite formed upon quenching to various temperatures from M_s through M_f.

The T-T-T diagram is quite useful in determining the kinetics of transformation and the nature of the product. The left-hand curve shows the elapsed time at constant temperature before transformation begins and the right-hand curve, the time required for complete transformation at that temperature. If a hypo- or hypereutectoid steel were considered, additional regions would be added to correspond to the primary equilibrium phases. These regions would not extend below the nose, however, since the nonequilibrium bainite and martensite phases do not have to maintain fixed compositions as do the phases in the equilibrium diagram. Figure 6-9 is a T-T-T cruve for a hypoeutectoid steel.

Tempering of martensite. Because it lacks good toughness and ductility, medium- or high-carbon martensite is not a useful engineering microstructure, despite its great strength. A subsequent heating, known as *tempering,* usually is required to restore some desired degree of toughness at the expense of a decrease in strength and hardness.

Martensite, in essence, is a supersaturated solid solution of carbon in alpha ferrite and therefore is metastable. By reheating in the range of 100 to 700°C (200 to 1300°F), carbon atoms will be rejected from solution, and the structure will move to a mixture of the stable ferrite and cementite phases. This decomposition of martensite to ferrite and cementite is a time - and temperature-controlled diffusion phenomenon with a spectrum of intermediate and transitory conditions. The initial stage, which occurs at 100 to 200°C (200 to 400°F), is the precipitation of an intermediate carbide with the composition of $Fe_{2.4}C$, known as epsilon (ϵ) carbide. This allows the matrix to revert to the body-centered-cubic configuration. From 200 to 400°C (400 to 750°F), the structure becomes one of ferrite and cementite. Little is observable in the microscope, however, for the cementite particles are submicroscopic and the original martensite boundaries are retained. Figure 6-10 shows such a structure, which appears as a rather mottled mass with little well-defined structure. Electron microscope studies reveal the fine carbide structure responsible for the softening and improved ductility.

If tempering progresses into the 400 to 550°C (750 to 1000°F) range, the

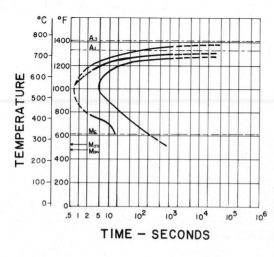

FIGURE 6-9 Isothermal transformation diagram for a hypoeutectoid steel (1050) showing additional region for primary ferrite.

FIGURE 6-10 Eutectoid steel, hardened and tempered at 315°C (600°F); 1000×. (*Courtesy United States Steel Corporation.*)

martensite boundaries disappear and a ferrite structure nucleates and grows. As the precipitated carbides grow in size, the properties move farther in the direction of a weaker but more ductile material. Figure 6-11 shows some of the newly formed ferrite (white) in the tempered material.

Above 550°C (1000°F), the new ferrite grains totally consume the original structure, and the cementite particles become larger and more spheroidal. Figure 6-12 shows such a structure having the highest toughness and ductility and the lowest strength of the tempered-martensite structures. Heating above the A_1 temperature will cause the structure to revert to stable austenite. Thus an infinite range of structures, and corresponding range of properties, can be obtained by quenching steel to obtain 100% martensite and then tempering it to the desired state. This is commonly known as the *quench-and-temper* process and the product is *tempered martensite*.

Continuous-cooling transformations. While the T-T-T diagrams have provided significant information about the structures obtained through nonequilibrium thermal processing, they usually are not applicable to direct engineering utilization because the assumptions of instantaneous cooling followed by constant temperature transformation fail to match reality. A diagram showing the

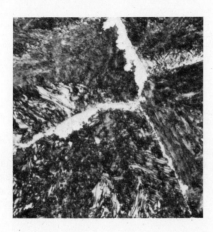

FIGURE 6-11 Eutectoid steel, hardened and tempered at 540°C (1000°F); 1000×. (*Courtesy United States Steel Corporation.*)

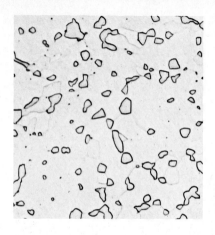

FIGURE 6-12 Photomicrograph of structure obtained by prolonged tempering above 540°C (1000°F), sometimes called spheroidite; 1000×. (*Courtesy United States Steel Corporation.*)

results of continuous cooling at various rates of temperature would be more useful. What will be the result if the temperature is dropped 300°C per second, 30°C per second, or 3°C per second?

A *continuous cooling transformation* (C-C-T) *diagram,* such as is shown schematically in Figure 6-13, can provide answers to these questions and several

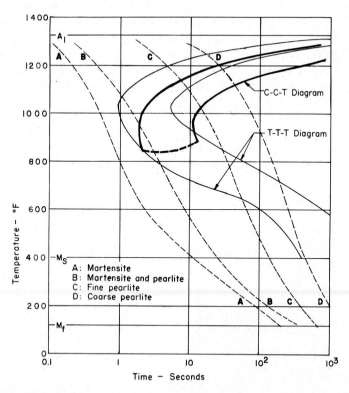

FIGURE 6-13 Schematic C-C-T diagram for a eutectoid composition steel, showing several superimposed cooling curves and the resultant structures. (*Courtesy United States Steel Corporation.*)

others. Critical cooling rates required to obtain products can easily be determined. If cooled fast enough, the structure will be all martensite. A slow cool may produce coarse pearlite and some primary phase. Intermediate rates usually produce mixed structures, the time at any one temperature usually being insufficient for complete transformation. If each structure is considered as providing a companion set of properties, the wide range of possibilities obtainable through controlled heating and cooling of steel becomes evident.

The Jominy test for hardenability. A tool that is commonly used to assist understanding of nonequilibrium heat treatment is the *Jominy end-quench hardenability test* and associated diagrams. In this test, depicted in Figure 6-14, an effort is made to reproduce the entire spectrum of cooling rates on a single specimen by quenching a heated bar from one end. The quench is standardized by specifying specimen geometry, the water temperature (24°C or 75°F), internal nozzle diameter, water pressure (63.5 mm or $2\frac{1}{2}$-inch vertical travel unimpeded), and gap between the nozzle and specimen.

After the specimen has been cooled, a flat region is ground along one side and R_C hardness readings are taken every 1.6 mm ($\frac{1}{16}$ inch) along the bar and plotted as shown in Figure 6-15. The hardness values then can be related to the cooling-rate data so that the cooling rate required to produce a given set of properties in the steel being considered can be determined rather precisely.

Applications of the test assume that identical results will be obtained if the same material undergoes identical cooling histories. If the cooling rate is known for a given location within a part (from experiment or theory), the properties at that location can be predicted as those at the equivalent cooling-rate location in a Jominy test bar. If specific properties are required, the necessary cooling rate for a given material can be determined. If the cooling rate is restricted, an acceptable material can be selected. Figure 6-16 shows Jominy curves of several common engineering steels.

Hardenability. When attempting to understand the heat treatment of steel, several key effects must be understood: the effect of carbon content, the effect

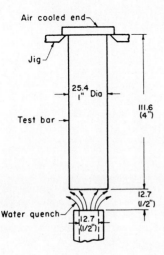

FIGURE 6-14 Schematic diagram of the Jominy hardenability test.

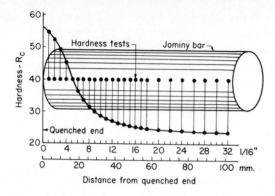

FIGURE 6-15 Typical hardness distribution in Jominy bars.

of alloy additions, and the effect of various quenching conditions. The first two relate to the material and the third to the process.

Hardness is a mechanical property related to strength and is a strong function of the carbon content of a metal. *Hardenability,* on the other hand, is a measure of the depth to which full hardness can be attained under a normal hardening cycle and is related primarily to amounts and types of alloying elements. In Figure 6-16 all the steels have the same carbon content but differ in type and amounts of alloy elements. Maximum hardness is the same in all cases, but the depth of hardening varies considerably. Figure 6-17 shows the results for steels containing the same alloying elements but variable amounts of carbon. Note the change in peak hardness.

The primary reason for adding alloy elements to commercial steels is to increase the hardenability, not to improve the strength properties. Steels with greater hardenability need not be cooled as rapidly to achieve a desired level of strength or hardness, and they can be completely hardened in thicker sections.

Materials selection for steels requires accurate determination of need. Strength and structure tend to be determined by carbon content, the general rule being to stay as low as possible and still meet specifications. Because heat can only be extracted from the surface of a metal, the depth of required hardening sets the conditions for hardenability and quench. For a given quenching condition,

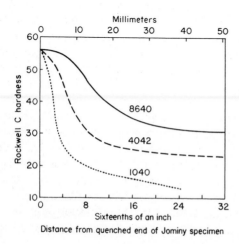

FIGURE 6-16 Jominy hardness curves for engineering steels with the same carbon content and varying types and amounts of alloy elements.

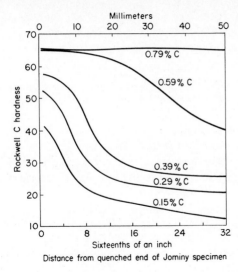

FIGURE 6-17 Jominy hardness curves for engineering steels with identical alloy conditions but variable carbon content.

different alloys will produce different results. Because alloy elements increase the cost of a material, a general rule is to select only what is required to assure compliance with specifications. Money is often wasted by specifying an alloy steel for an application where a plain carbon steel, or steel with lower alloy content (and thus less costly), would be satisfactory. Another alternative when greater depth of hardness is required is to modify the quench conditions such that a faster cooling rate is achieved. Quench changes may be limited, however, by cracking or warping problems, depending on the shape, size, complexity, and precision of the part being treated.

Quench media. Quench media vary in their effectiveness, the variation being best understood by considering the three stages of quenching. When a piece of hot metal is first inserted into a tank of liquid quenchant, that adjacent to the metal vaporizes and forms a gaseous layer separating the metal and liquid. Cooling is slow in this *vapor-jacket stage (first stage)* since all heat transport now must be through a gas. This stage occurs when the metal is above the boiling point of the quenchant. Soon bubbles nucleate and remove the gas, liquid contacts the metal, vaporizes (removing its heat of vaporization from the metal), forms a bubble, and the process continues. This *second stage of quenching* provides rapid cooling as a result of the large quantities of heat removed by the mechanism. When the metal cools below the boiling point of the quenchant, all heat transfer occurs through conduction across the solid-liquid interface, aided by convection or stirring within the quenchant; this is the *third stage*.

Water is a fairly good quenching medium because of its high heat of vaporization and the second stage of quenching extending down to 100°C (212°F), usually well into the martensite transition range or below. Water also is cheap, but the clinging tendency of the bubbles may cause soft spots in the metal. Agitation is recommended when using a water quench.

Brine is a more severe quenching medium than water because the salt nucleates bubbles, forcing a more rapid transition through the vapor-jacket stage.

Unfortunately, brine tends to accentuate corrosion problems unless completely removed. Sodium or potassium hydroxide sometimes is used when very severe quenching is desired and one wishes to obtain good hardness in low-carbon steels. Various degrees of agitation or spraying of the quench can be used to increase the effectiveness of a given medium.

When a slower cooling rate is desired, oil quenches are often used. Various oils are available that have high flash points and different degrees of quenching effectiveness. Since the boiling points are often quite high, the transition to stage three cooling generally precedes the martensite start. The slower cooling through the M_s to M_f temperature range leads to a milder temperature gradient and less tendency to cracking. Molten salt baths are used as quench mediums and possess the property of going directly to the third stage.

Still slower cooling can be obtained by cooling in still air, packing the metal in sand, and a variety of other methods.

The role of design in the heat treatment of steel. Design details and material selection play important roles in the satisfactory and economical heat treatment of parts, and proper consideration of these factors usually does not impose serious limitations. In fact, proper consideration of them usually leads to more simple, more economical, and more reliable products. Failure to relate design and materials to heat-treatment procedure usually produces disappointing or variable results and often service failure.

Undesirable design features are (1) nonuniform sections or thicknesses, (2) sharp interior corners, and (3) sharp exterior corners. Because these may easily find their way into the design of parts, the designer should be aware of their effect in heat-treating operations. Undesirable results may include nonuniform structure and properties, undesirable residual stresses, cracking, warping, and dimensional changes.

Heat can only be extracted from a piece of metal through its exposed surface. Thus, if a piece to be hardened has a nonuniform cross section, the thin portion will cool rapidly and fully harden while the thick region may not harden, except on the surface. This surface may even be tempered somewhat by the heat retained in the center of the heavy section. The shape that might be closest to the ideal from the viewpoint of quenching would be a doughnut, having a uniform cross section with maximum exposed surface.

Residual stresses are the often-complex results of the various dimensional changes that occur during heat treatment. Thermal contraction during cooling is a well-understood phenomenon. In addition, the various phases and structures that may form are often characterized by different densities and, therefore, a volume expansion or contraction accompanies the phase transformations. When austenite transforms to martensite, there is a volume expansion of up to 4%. Austenite transforming to pearlite also experiences a volume expansion, but it is of smaller magnitude.

If all temperature changes occur uniformly throughout the part, all changes in dimension would occur simultaneously and the result would be a component free of residual stresses. However, most parts being heat-treated experience nonuniform temperatures during the cooling or quenching process. Cross sections should be designed so that temperature differences are as low as possible

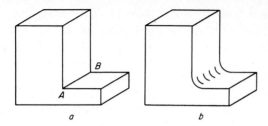

FIGURE 6-18 (*Left*) Shape containing nonuniform sections joined by a sharp interior corner that may crack in quenching. (*Right*) Improved design using a large radius to join sections and to avoid cracking in heat treatment.

and are not concentrated. If this is not possible, slower cooling and a material that will harden in an oil or air quench may be required. Because materials having greater hardenability invariably are more expensive, the design alternative clearly has advantages.

When temperature differences and resultant residual stresses become severe or localized, additional problems, such as cracking or distortion, can result. Figure 6-18a shows an example where a sharp interior corner was placed at a change of cross section. Upon quenching, stresses concentrate along line *A-B* and a crack is almost certain to result. When changes in cross section or other transition must be made, they should be gradual, as in the redesigned Figure 6-18b. Generous fillets at interior corners, radiused external corners, and smooth transitions all reduce problems. Use of a more hardenable material or less severe quenching also will aid in preventing unnecessary problems.

Figure 6-19a shows the cross section of a die that consistently cracked during hardening. Eliminating the sharp corners and adding holes to provide a more uniform cross section during quenching, as in Figure 6-19b, eliminated the difficulty.

One of the ominous features about improperly designed heat-treated parts is the fact that the residual stresses may not produce immediate failure, but may contribute to a failure at a later time. Applied stresses well within the ''safe'' designed limit may couple with residual stresses to produce loads sufficient to cause failure. (See chapter 39 for more details.) Dimensions may change or warping result during subsequent machining or grinding operations. Corrosion reactions may be significantly accelerated in the presence of residual stresses. Time and money are lost that could be saved if proper design practices are employed. Heat treatment is an important manufacturing process that enables better results to be obtained with less costly materials, *if used properly*. The designer, however, must take into account all the facts and conditions related to it when he designs a part, selects a material, and specifies, directly or indirectly, the heat treatment for it.

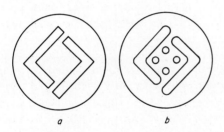

FIGURE 6-19 Improved design techniques to provide more uniform sections.

Techniques to reduce cracking. Two variations of rapid quenching have been designed to minimize the high temperature gradients which often result in cracking. Rapid quenching still is required to prevent transformation to the weaker pearlite structure, but instead of quenching through the martensite transformation, the component is rapidly quenched to a temperature several degrees above the M_s usually in a bath of molten salt. Holding for a brief time at this temperature enables the piece to come to a nearly uniform temperature. If the piece is held at this temperature long enough, the austenite will transform to bainite, a process known as *austempering*. If the piece is stablilized and then slow cooled through the martensite transformation, the process is known as *martempering* or *marquenching*. Here the product is martensite, which must be tempered the same as martensite formed by rapid quenching. Figure 6-20 shows these schematically on a T-T-T diagram (a misuse of the diagram but good for visualization).

Pieces with complicated shapes, undesirable design features, or high precision can all benefit from these modified hardening techniques. Straightening can also be performed after stabilization in the quench bath, before final hardening occurs.

Ausforming. A process often confused with austempering is that of *ausforming*. Certain alloys tend to retard the pearlite transformation far more than the bainite reaction and produce a T-T-T curve such as that shown in Figure 6-21 for 4340 steel. If a metal is heated to form austenite and then quenched to the temperature of the "bay" between the pearlite and bainite regions, it can retain its austenitic structure for a useful period of time. Deformation can be performed here on an austenitic material, a structure that, technically is not stable at the temperatures involved. Benefits include the increased ductility of the face-centered crystal structure, the finer grain size characteristic of recrystallization at a lower temperature, and some degree of possible strain hardening. Following deformation, the material can either be slowly cooled to produce a bainitic

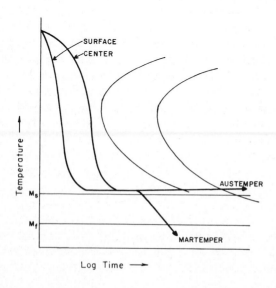

FIGURE 6-20 Schematic representation of the austempering and martempering processes.

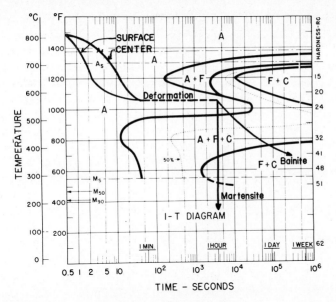

FIGURE 6-21 T-T-T diagram for 4340 steel, showing the "bay" and the ausforming process in schematic. (*Courtesy United States Steel Corporation.*)

structure or quenched to form martensite, which then is tempered. The material that results shows extremely high strength and ductility, toughness, creep resistance, and fatigue life, far greater than if the processes of deformation and transformation were conducted in their normal sequence. Ausforming is an example of combined heat treatment and deformation techniques known as *thermomechanical processing*.

SURFACE HARDENING OF STEEL

It is often desirable for products to have a hard, wear-resistant surface coupled with a tough, impact-resistant core. The methods developed to produce such properties generally fall into three types: selective heating of the surface, altered surface chemistry, and deposition of an additional surface layer.

Selective heating techniques. If a steel has sufficient carbon content to attain the desired surface hardness, the different properties can be obtained by simply varying the thermal histories of the various regions. *Selective heating* generally requires steels of at least 0.3% carbon to be operable. Maximum hardness depends upon the carbon content, while the depth of hardness depends on the material's hardenability.

Flame hardening employs a high-intensity oxyacetylene flame to raise the surface temperature high enough to reform austenite, which is then quenched and tempered to the desired surface properties. Heat input is quite rapid and is concentrated on the surface. Slow heat transfer and short times leave the interior at low temperature and therefore free from any significant changes.

FIGURE 6-22 Surface hardening by the flame-hardening process. (*Courtesy Linde Division, Union Carbide Corporation.*)

Considerable flexibility is provided since the rate and depth of heating can be varied. Depth of hardening can range from thin skins through 0.6 cm ($\frac{1}{4}$ inch). Flame hardening is often employed on large objects where size or shape prohibits the use of alternative techniques. Figure 6-22 shows its application to hardening the teeth of large gears. Equipment varies from crude, hand-held torches to fully automated and computerized units.

Induction heating is particularly well suited to surface hardening, the rate and depth of heating being controlled by the amperage and frequency of the generator. The steel part is placed inside a wound coil which is then subjected to alternating current. The changing magnetic field induces a current in the steel which flows through the surface layers and heats by electrical resistance. Heating rates are extremely rapid and efficiency is high. The process can be adapted to special shapes and offers the benefits of excellent reproducibility, good quality control, and the possibility of automation. Figure 6-23 shows an induction hardened gear where hardening has been applied to those areas subject to wear. Distortion during hardening is negligible since rigidity is provided throughout the process by the cool interior of the product.

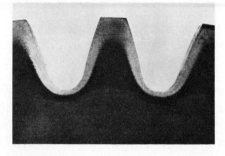

FIGURE 6-23 Section of gear teeth, showing localized surface hardening. (*Courtesy TOCCO Division, Park-Ohio Industries, Inc.*)

Other selective heating techniques employ immersion in a *lead pot* or *salt bath* to heat the surface. *Laser beam hardening* and *electron beam hardening* have been used to produce hardened surfaces on uneven geometries with high speed and little distortion. Beam size, intensity, direction, and speed of travel can all be controlled and programmed for complete automation. Production equipment can perform multiple operations with efficiencies often in excess of 90%.

Techniques using altered surface chemistry. When steels contain insufficient carbon to attain the desired surface properties upon selective heating, an alternative approach is to alter the surface chemistry. The most common technique in this category is *carburizing*. In *pack carburizing*, the parts are packed in a high-carbon solid medium, enclosed in a gastight box, and heated in a furnace for 6 to 72 hours at roughly 900°C (1650°F). The hot carburizing compound produces CO gas which reacts with the metal, releasing carbon which is readily absorbed into the austenite. The boxes are then taken from the furnace and the parts removed and thermally processed. Direct quenching from the carburizing operation then produces the different structures due to the different carbon contents in the metal. A slow cool, reheat, and quench or duplex core-case heat treatment are alternative processing methods. The carbon content of the surface varies from 0.7 to 1.2% depending on the details of the process. Case depth may range from a few hundreths of a millimeter to in excess of 10 millimeters, but cases over 1.5 mm (0.060 inch) are not often employed.

Several problems are encountered in pack carburizing. Heating is inefficient; temperature uniformity is questionable; handling is often difficult; and the process is not readily adaptable to continuous operation. *Gas carburizing* overcomes these difficulties by replacing the solid carburizing compound with a carbon-providing gas, usually containing an excess of CO. While the mechanisms and processing are the same, the operation is faster and more easily controlled. Accuracy and uniformity are increased, and continuous operation is possible. Special types of furnaces are required, however.

In *liquid carburizing* or *cyaniding*, the carbon is supplied by immersing the part in molten cyanide salt. Both carbon and nitrogen are added to the steel surface in this process, used primarily to put thin cases on small parts. Safety concerns associated with the toxic cyanide fumes nearly eliminated this technique; however, some nontoxic liquid carburizing media have recently been introduced.

Nitriding produces surface hardening by the formation of alloy nitrides in the surface layers of special steels which contain nitride forming elements such as aluminum, chromium, molybdenum, or vanadium. The parts are heat-treated and tempered at 525 to 675°C (1000 to 1250°F) prior to nitriding. After cleaning and removal of any decarburized surface material, they are heated in a dissociated ammonia atmosphere (containing nitrogen and hydrogen) for 10 to 40 hours at 500 to 625°C (950 to 1150°F). Nitrogen diffusing into the steel will then form alloy nitrides, hardening the metal to a depth of about 0.65 mm (0.025 inch). Very hard cases are formed and distortion is low. No subsequent thermal processing is required. Although the surface hardness is higher than for other hardening methods, the long times at elevated temperatures, coupled with the thin case and associated dimensional precision, restrict the application of nitriding mainly to parts requiring high quality cases.

Plasma processes, such as *ionitriding,* have recently emerged as attractive alternatives to the conventional methods. Parts to be treated are placed in an evacuated "furnace" and a direct current potential of 500 to 1000 volts is applied between the parts and the furnace walls. Low-pressure nitrogen gas is introduced into the chamber and becomes ionized. The ions are accelerated toward the product surface, become embedded, and generate sufficient heat to promote inward diffusion. This is the only heat associated with the process; the furnace acts only as a vacuum container and electrode. Advantages include shorter cycle times, reduced consumption of gases, significantly reduced energy costs, reduced space requirements and the possibility of total automation. Product quality is improved over that of conventional nitriding and the process is applicable to a wider range of materials.

HEAT TREATMENT EQUIPMENT

Furnaces. To facilitate production heat treatment, many types of heating equipment have been developed in a wide range of sizes. Furnaces can be either *batch-type* or *continuous-type.* Batch furnaces are those in which the workpiece remains stationary throughout its time in the furnace and may be of either horizontal or vertical design. Continuous furnaces move the components through the heat treatment operation at rates selected to be compatible with other manufacturing operations.

Horizontal batch-type furnaces are often called *box furnaces,* because they resemble a rectangular box. As shown in Figure 6-24, a door is provided on one end to permit work to be inserted and removed. Gas or electricity provides the source of heat. For large or long workpieces, *car-bottom furnaces* are used, such as shown in Figure 6-25. The work is loaded onto a refractory-topped flatcar which is moved on rails into the furnace.

Horizontal furnaces are relatively easy to construct in any size, are easily insulated, and are thermally efficient. However, it is difficult to heat-treat long, slender work in them because of the sagging or warping that is likely to occur. Vertical *pit furnaces,* as shown in Figure 6-26, have been designed for such work. These are cylindrical chambers sunk into the floor with a door on top that can be swung aside to permit the work to be lowered into the furnace. Suspended in this manner, long workpieces are less likely to warp. This type of furnace is also used to heat batches of small parts which can be loaded into baskets and lowered into the furnace.

Another furnace type is the *bell furnace,* shown in Figure 6-27. The heating elements are contained within a bottomless bell that is lowered over the work. An airtight inner shell is often employed to contain a protective atmosphere during the heating and cooling cycles and thereby reduce tarnish or oxidation. After the work is heated, the furnace unit can be lifted off and transferred to another batch, the inner shell retaining the controlled atmosphere during cooling.

FIGURE 6-24 Box-type electric heat-treating furnace. (*Courtesy Lindberg, A Unit of General Signal.*)

FIGURE 6-25 Car-bottom box-type furnace. (*Courtesy Hevi Duty Electric Company.*)

FIGURE 6-26 Vertical pit furnace heat treating a 6.7-meter (22-feet)-long mill roll. (*Courtesy Lindberg, A Unit of General Signal.*)

Sometimes an insulated cover may be placed over the heated work if slower cooling is desired, as when annealing large batches of steel or other metal.

The *elevator-type furnace* is a modification of the bell furnace, in which the bell is stationary and the work is raised up into it by means of a movable platform that forms the bottom of the furnace. An interesting variation of this furnace is one for which there are three vertical positions. In the middle position the work is loaded onto the platform elevator. In the upper position the work is in the bell furnace, and in the lower position it is in a quench tank. Such furnaces are used where work must be quenched as soon as possible after being removed from the furnace.

Continuous furnaces are used for large production runs of the same or similar parts, where a steady flow of workpieces is moved through the furnace by some type of conveyor or push-mechanism, the conveyor often ending such that the workpiece falls into a quench tank to complete the treatment. These furnaces can be designed to conduct complex heating, holding, and quenching or cooling cycles in an exact and repeatable manner with very low labor cost. Circular

FIGURE 6-27 Bell-type heat-treating furnace. (*Courtesy Surface Combustion Corporation.*)

continuous furnaces, where the workpieces move on a rotating hearth, are convenient when a single workstation is used to load and unload the furnace.

Nearly all the above furnaces, if properly designed, can employ artificial gas atmospheres which can serve to prevent scaling or tarnishing, prevent decarburization, or supply carbon or nitrogen for surface modification. Most furnace atmospheres are generated from natural gas. Motivated by fuel and energy shortages, however, many users have found nitrogen-based atmospheres to offer cost savings, energy savings, increased safety, and environmental attractiveness.

When a liquid heating medium is preferred over gas, *salt bath furnaces* may be employed. Electrically conductive salt can be heated by passing a current between two electrodes suspended in the bath. The electrical currents also cause the bath to circulate and thus maintain uniform temperature. Nonconductive salt baths are heated by some form of immersion heater. In addition to being a source of uniform heat, the salt can also be selected to provide prevention of scaling or decarburization. A similar furnace is the *lead bath,* where molten lead replaces salt as the heating medium.

The heating rates of gas atmosphere furnaces can be made comparable to liquid baths by incorporation of the *fluidized bed* concept. These furnaces utilize a container of mobile inert particles, such as aluminum oxide, which are heated and fluidized (suspended) in a flowing stream of gas. Items to be heated are immersed directly into the bed. Any fluidizing gas can be used and high heat transfer rates, high thermal efficiency, and low fuel consumption have been observed.

The use of *electrical induction heating* has simplified many heat-treating operations. Small parts can be through-heated and hardened as in other styles of furnaces. Local or selective hardening is possible at a very rapid production rate. Furthermore, a standard induction unit can be adapted to a wide variety of products by simply changing the induction coil and adjusting equipment settings.

Furnace controls. All heat-treating operations should be carried out under rigid controls if uniformity is desired. For this reason, most heat-treating furnaces are equipped with indicating and controlling pyrometers. Some furnaces are equipped with controllers that also regulate the rate of heating or cooling. It should be remembered that it is the temperature of the workpiece and not that of the furnace that controls the result. Since the temperature of the work may be many degrees different from the furnace, ample time must be allowed to bring the work to uniform temperature. While heat treatment consumes large amounts of energy, its use may actually be an energy conservation measure, because it enables the manufacture of a higher-quality, more durable product. Higher strengths may permit the use of less material for a comparable part.

REVIEW QUESTIONS

1. What is heat treatment?
2. How might heating and cooling associated with other manufacturing operations affect material properties?
3. What are processing heat treatments?
4. How might equilibrium phase diagrams be used to aid processing heat treatment?
5. What are some of the goals of annealing operations?
6. What structure results from full annealing a hypoeutectoid steel?
7. What is the attractive feature of normalizing compared to full annealing?
8. What types of materials may be spheroidized? Why?
9. What are the six major mechanisms available to increase the strength of metals?
10. What is solid solution hardening?
11. What are the stages of a precipitation hardening treatment?
12. What is the difference between a coherent precipitate and the second-phase particles in dispersion hardening?
13. What is the most effective strengthening mechanism for nonferrous metals?
14. What are the two types of precipitation hardening materials? How do they differ?
15. What was one of the major barriers to the understanding of the strengthening heat treatments?
16. Describe the transition that occurs when austenite goes to martensite.
17. What is retained austenite? Why is it undesirable?
18. Why is martensite tempered before put in use?
19. What is the benefit of a C-C-T diagram over a T-T-T diagram?
20. How can "equivalent-cooling rates" be used to predict the properties of a heat-treated product?
21. How does hardenability differ from hardness?
22. What is the primary reason for adding alloy elements to commercial steels?
23. What are the three stages of quenching?
24. Why is brine or salt water a better quench than water when rapid heat extraction is desired?

25. What concerns exist when heat-treating a part with nonuniform sections?
26. Why might residual stresses be undesirable?
27. How do austempering and martempering act to reduce the cracking tendencies that occur when quenching a metal?
28. What are some of the selective heating techniques of surface hardening?
29. What are several techniques for surface hardening of steel that involve altered surface chemistry?
30. Why are protective atmospheres often required during heat treatment?
31. Why might the use of heat treatment actually be an energy conservation measure?

CASE STUDY 6.   A Flying Chip from a Sledgehammer

Industrial sledgehammers are used throughout JCL Industries, most having a 15-pound head of AISI 1060 steel. To reduce tool replacement costs, the company machine shop periodically gathers hammers with heavily deformed heads (mushroomed) and grinds off the deformed segment.

A reground hammer was placed back in use. Upon striking a metal plate, a chip flew off a corner of the hammer head and lodged in the eye of a worker. A lawsuit resulted.

Investigation showed that the head of a new hammer should have a bulk hardness between R_C 44 and 55. The chip, however, had a hardness of R_C 65 on the fracture surface. Inspection of other hammers showed numerous chipped regions, all on redressed hammers.

What do you suspect to be the problem? How would you alter the procedures or policies of JCL Industries to eliminate a possible recurrence, yet minimize expense?

CHAPTER 7

Alloy Irons and Steels

Without alloy irons and steels, the state of technology would be set back considerably. Many varieties of alloys have been developed to meet the specific needs of an advancing civilization. However, the availability of many varieties has often resulted in poor selection or excess cost for an unnecessary and expensive alloy material. It is the responsibility of the design and manufacturing engineer to be knowledgeable in this area and to make the best selection from the available alternatives.

Plain-carbon steel. Steel theoretically is an alloy of iron and carbon. When produced commercially, however, certain other elements—notably manganese, phosphorus, sulfur, and silicon—are present in small quantities. When these four foreign elements are present in their normal percentages, the product is referred to as *plain-carbon steel*. Its strength is primarily a function of its carbon content. Unfortunately, the ductility of plain-carbon steel decreases as the carbon content is increased, and its hardenability is quite low. In addition, the properties of ordinary carbon steels are impaired by both high and low temperatures, and they are subject to corrosion in most environments.

Plain-carbon steels are generally classed into three subgroups, based on carbon content. *Low-carbon steels* have less than 0.30% carbon, possess good formability and weldability, but not enough hardenability to be hardened to any significant depth. Their structures usually are ferrite and pearlite, and the material generally is used as it comes from the hot-forming or cold-forming process. *Medium-carbon steels* have between 0.30 and 0.80% carbon, and they can be quenched to form martensite or bainite if the section size is small and a severe water or brine quench is used. The best balance of properties is attained at these carbon levels, the high fatigue and toughness of the low-carbon material being in good compromise with the strength and hardness that comes with higher carbon content. These steels are extremely popular and find numerous applications. *High-carbon steels* have more than 0.80% carbon. Toughness and formability are quite low, but hardness and wear resistance are high. Severe quenches can form martensite, but hardenability is still poor. Quench cracking is often a problem when the material is pushed to its limit.

Plain carbon steels are the lowest-cost steel material and should be considered for many applications. Often, however, their limitations become restrictive. When improved material is required, steels can be upgraded by the addition of one or more alloying elements.

Alloy steels. The differentiation between "plain carbon" and "alloy" steel is often somewhat arbitrary. Both contain carbon, manganese, and usually silicon. Copper and boron also are possible additions to both classes. Steels containing more than 1.65% manganese, 0.60% silicon, or 0.60% copper are designated as alloy steels. Also, a steel is considered to be an alloy steel if a definite amount or minimum of other alloying element is specified or required. The most common alloy elements are chromium, nickel, molybdenum, vanadium, tungsten, cobalt, boron, and copper, as well as manganese, silicon, phosphorus, and sulfur in amounts greater than are normally present.

Effects of the alloying elements. In general, alloying elements are added to steel in small percentages—usually less than 5%—to improve strength or hardenability, or in much larger amounts, often up to 20%, to produce special properties such as corrosion resistance or stability at high or low temperatures. Certain additions may be made during the steelmaking process to remove dissolved oxygen from the melt. Manganese, silicon, and aluminum are frequently used for this deoxidation. Aluminum and, to a lesser extent, vanadium, columbium, and titanium are used to control austenitic grain size. Machinability can be enhanced by additions of sulfur, lead, selenium, and tellurium. Other elements may be added to improve the strength or toughness properties of the product metal. Manganese, silicon, nickel, and copper add strength by forming solid solutions in ferrite. Chromium, vanadium, molybdenum, tungsten, and other elements increase strength by forming dispersed second-phase carbides. Columbium, vanadium, and zirconium can be used for ferrite grain-size control. Nickel and copper are added to low-alloy steels to provide improved corrosion resistance.

For constructional alloy steels, the principal effect of alloying elements is to increase hardenability. The commonly used elements, in order of decreasing effectiveness, are manganese, molybdenum, chromium, silicon, and nickel. Small quantities of vanadium are quite effective, but the response drops off as quantity is increased. Boron is also extremely significant in steels with less than 0.65% carbon.

Still other characteristics of the individual elements are worthy of note.

Manganese is present in most plain-carbon steels to combine with sulfur and form soft manganese sulfides. This prevents formation of iron sulfide along grain boundaries, which would lead to brittleness of the metal. In alloy steels, it increases hardenability, slightly strengthens ferrite, and lowers the martensite transformation temperatures. It often is added in amounts greater than 1%. Manganese steel having about 1 to 1.5% manganese and 0.9 to 1.0% carbon has desirable nondistorting qualities when being hardened and is often used in die work. Sometimes 0.5% chromium and vanadium are also added. When manganese is used in large percentages (11 to 14%), an austenitic alloy known as *Hadfield steel* is produced. Its high hardness with good ductility, high strain hardening capability, and excellent wear resistance make it ideal for mining tools and similar applications.

As mentioned previously, *sulfur* usually is not desired in steel because of the embrittling effect of iron sulfide. In the form of manganese sulfide, however, sulfur is not harmful provided that the sulfides are not in large quantities and are well dispersed. If manganese sulfide is present in large quantities and in proper form, it can impart desirable machinability properties. Therefore, some *free-machining steels,* which are to be machined automatically and are to be used for parts that will not be subjected to much impact, have 0.08 to 0.15% sulfur added. The manganese content is usually increased to make certain that no iron sulfide is formed.

Nickel is added primarily for its increase in toughness and impact resistance, particularly at low temperature. It also lessens distortion in quenching, improves corrosion resistance, lowers the critical temperatures, and widens the temperature range for successful heat treatment. It is used in amounts of 2 to 5%, often combined with other alloying elements to improve toughness. When 12 to 20% nickel is used in steel with low carbon content, good corrosion resistance is provided. A steel with 36% nickel has a thermal expansion coefficient of almost zero. Commonly known as *Invar,* this metal is used for measuring devices. Because of its high cost, nickel should only be used where it is uniquely effective, as in providing low-temperature impact resistance for cryogenic steels.

Although large percentages of *chromium* can impart corrosion resistance and heat resistance, in the amounts used in low-alloy steels, these effects are minor. For these materials, chromium serves to increase hardenability and increase strength. Less than 2% chromium is generally employed, and often chromium and nickel are used together in a ratio of about 1 part chromium to 2 parts nickel. Chromium carbides often are desired for their superior wear resistance.

Molybdenum, as used in ordinary alloy steels, improves hardenability and increases strength properties, particularly under dynamic and high-temperature conditions. It tends to form stable carbides that persist at elevated temperatures, thereby retaining fine grain size and unusual toughness. Molybdenum steels are somewhat resistant to tempering and therefore maintain their strength and creep-resistant properties at elevated temperatures. Resistance to temper embrittlement is also noted.

Molybdenum is often used in conjunction with chromium, its amount seldom being in excess of 0.3%. Molybdenum is used in larger amounts in tool steels because of its effect, similar to tungsten, of imparting hardness that persists at red heat. Within the temperature ranges where it is effective, it is about twice as potent as tungsten and much cheaper. It is commonly used in forging dies that must resist impact and abrasion at elevated temperature.

Vanadium is another alloying element that forms strong carbides that persist at elevated temperature. The carbides do not readily go into solution when the metal is heated prior to quenching, and therefore inhibit grain growth. The resulting effect of 0.03 to 0.25% vanadium is an increase in strength properties, particularly the elastic limit, yield point, and impact strength, with almost no loss in ductility.

Tungsten is more effective than molybdenum in producing hardness at very high temperatures, again attributable to stable carbides. It thus is a primary alloying element in tool steels that must maintain their hardness at high operating temperatures. It also serves as a principal alloying element in some air-hardening steels.

Although the resistance of *copper* to atmospheric corrosion has been known for centuries, only recently has it been used as an addition to steel (in amounts from 0.10 to 0.50%) to provide this property. It now is used extensively in low-carbon sheet steel and in structural steels. Surface quality and hot-working behavior deteriorate, however.

Silicon in small percentages has somewhat the same effect on steel as does nickel, increasing the strength properties, especially the elastic limit, with little loss in ductility. It is an important alloying element (usually 0.2 to 0.7%) in certain high-yield-strength structural steels. It also is used in spring steels (which contain about 2% silicon, 0.8% manganese, and 0.6% carbon) and to promote the large grain size desirable for steels used for magnetic applications in electrical equipment.

Boron is a very powerful hardenability agent, being from 250 to 750 times as effective as nickel, 75 to 125 times as effective as molybdenum, and about 100 times as powerful as chromium. Only a few thousandths of a percent are sufficient to produce the desired effect in low-carbon steels, but the results diminish rapidly with increasing carbon content. Since no carbide formation or ferrite strengthening is produced, improved machinability and cold-forming capability often result from the use of boron in place of other hardenability agents.

In addition to its use as a deoxidizer, *aluminum* may be added to steels in amounts of 0.95 to 1.30% to produce a nitriding steel, as discussed in Chapter 6. *Titanium* and *niobium* (columbium) are other carbide formers. Steels with 0.15 to 0.35% *lead* show substantially improved machinability. *Zirconium, cerium* and *calcium* control the shape of inclusions and thereby promote toughness.

Table 7-1 shows the basic effects of the common alloying elements. A working knowledge of the information contained in this table is useful to the design engineer in selecting an alloy steel for a given requirement. Of course, alloying elements are often used in combination, resulting in a large variety of alloy steels being available. To simplify matters, a classification system has been developed and is in general use in a variety of industries.

The AISI-SAE classification system. Undoubtedly the most important group of alloy steels, from the manufacturing viewpoint, is that designated by the AISI identification system. This system, which classifies alloys by chemistry, was started by the Society of Automotive Engineers (SAE) to provide some standard-

TABLE 7-1 Principle Effects of Major Alloying Elements in Steel

Element	Percentage	Primary Function
Manganese	0.25–0.40 > 1	Combine with sulfur to prevent brittleness Increase hardenability, by lowering transformation points and causing transformations to be sluggish
Sulfur	0.08–0.15	Free-machining properties
Nickel	2–5 12–20	Toughener Corrosion resistance
Chromium	0.5–2 4–18	Increase hardenability Corrosion resistance
Molybdenum	0.2–5	Stable carbides; inhibits grain growth
Vanadium	0.15	Stable carbides; increases strength while retaining ductility; promotes fine grain structure
Boron	0.001–0.003	Powerful hardenability agent
Tungsten		Hardness at high temperatures
Silicon	0.2–0.7 2 Higher percentages	Increases strength Spring steels Improve magnetic properties
Copper	0.1–0.4	Corrosion resistance
Aluminum	0.95–1.30	Alloying element in nitriding steels
Titanium	—	Fixes carbon in inert particles Reduces martensitic hardness in chromium steels
Lead	—	Improves machinability

ization of steels used in the automotive industry. It was later adopted and expanded by the American Iron and Steel Institute (AISI) and has become the most universal system in the United States. Both plain-carbon and low-alloy steels are identified by a four-digit number, the first number indicating the major alloying elements and the second number indicating a subgrouping of the major alloy system. Groupings by the first two digits is according to an arbitrary table. The last two digits indicate the approximate carbon content of the metal in "points" of carbon, where one point is equivalent to 0.01% carbon. Table 7-2 presents the basic composition classification. As examples, a 1080 steel would be a plain-carbon steel with 0.80% carbon. Similarly, a 4340 steel would be a Mo–Cr–Ni alloy with 0.40% carbon.

A letter prefix may be used to indicate the process employed to produce the steel, such as basic open-hearth (C) or electric furnace (E). An X prefix is used to indicate permissible variations in the range of manganese, sulfur, or chromium. The letter B between the second and third digits indicates that the base metal has been supplemented by addition of boron. Similarly, a letter L in this position indicates a lead addition for enhanced machinability.

The *H-grade AISI steels* are designated by the letter H as a suffix to the standard designation. These steels are for use where hardenability is a major requirement, with slightly broader variations in steel chemistry being permitted. The steel is supplied to meet hardenability standards as specified by the customer in terms of hardness values at specific locations from the quenched end of a Jominy hardenability specimen.

Other systems of designation, such as the American Society for Testing and Materials (ASTM) and U.S. government (MIL and federal) specification systems focus more on specific applications. Acceptance for a given grade may be based more on physical or mechanical properties than on chemistry of the metal. Many low-carbon structural and alloy steels are referred to by their ASTM designation.

Balanced alloy steels. From the previous discussion, it is apparent that two or more alloying elements may produce similar effects. Thus, it is possible to obtain steels having almost identical properties although their chemical compositions are substantially different. This fact is strikingly demonstrated in Figures 7-1 and 7-2. Here the test data show that several steels of quite different compositions have almost identical property ratios *when heat-treated properly*. This fact should be kept in mind by all who select and specify the use of alloy steels. It is particularly important when one realizes that some alloying elements are much more costly than others, and that some may be in short supply in certain countries (the United States, for one) in times of emergency or as a result of political constraints. Overspecification is often employed to guarantee success in spite of sloppy manufacturing or heat-treatment practices. In most cases, the correct steel to use is the least expensive one that can be heat-treated to satisfactorily achieve the desired properties. This usually means taking advantage of the effects provided by *all* the elements in a steel through ''balanced'' compositions that avoid needlessly large amounts of expensive elements.

An excellent example of what can be achieved is seen in the series of *EX steels* originated in 1963 by the Society of Automotive Engineers (SAE). The system designates new grades of wrought alloy steels on a temporary basis. Grades are removed from the list when they are either promoted to full SAE standard steel status or dropped for lack of interest. More than 50 of these EX steel compositions have been approved since 1963. These have been designed to reduce the need for expensive alloying elements or those which are in short supply, notably nickel and chromium, or to improve a particular attribute of a standard grade of alloy steel. Table 7-3 lists the compositions and equivalent standard grades (on the basis of hardenability) for several EX steels. By comparing the compositions of the EX steel and the equivalent composition from Table 7-2, the alloy savings become readily apparent.

In selecting alloy steels, it is important to keep usage in mind. For example, for one use it might be permissible to increase the carbon content in order to obtain greater strength. For a different usage, for example in welding, it would be better to keep the carbon content low and use a balanced amount of alloying elements to obtain the required strength without the cracking or fracture problems associated with possible higher-carbon martensitic structures. Steel selection often involves the defining of required properties, the selection of the best microstructure to provide those properties, and finally the selection of the steel with the proper carbon content and hardenability characteristics to achieve

TABLE 7-2 Some AISI–SAE Standard Steel Designations

AISI Number	Type	Alloying Elements (%)					
		Mn	Ni	Cr	V	Mo	Other
1xxx	Carbon steels						
10xx	Plain carbon						
11xx	Free cutting (S)						
12xx	Free cutting (S) and (P)						
15xx	High manganese						
13xx	High manganese	1.60–1.90					
2xxx	Nickel steels		3.5–5.0				
3xxx	Nickel–chromium		1.0–3.5	0.5–1.75			
4xxx	Molybdenum						
40xx	Mo					0.15–0.30	
41xx	Mo, Cr			0.40–1.10		0.08–0.35	
43xx	Mo, Cr, Ni		1.65–2.00	0.40–0.90		0.20–0.30	
44xx	Mo					0.35–0.60	

Type		Ni	Cr	Mo	V	Si
46xx	Mo, Ni (low)	0.70–2.00		0.15–0.30		
47xx	Mo, Cr, Ni	0.90–1.20	0.35–0.55	0.15–0.40		
48xx	Mo, Ni (high)	3.25–3.75		0.20–0.30		
5xxx	Chromium					
50xx			0.20–0.60			
51xx			0.70–1.15			
6xxx	Chromium–vanadium					
61xx			0.50–1.10		0.10–0.15	
8xxx	Ni, Cr, Mo					
81xx		0.20–0.40	0.30–0.55	0.08–0.15		
86xx		0.40–0.70	0.40–0.60	0.15–0.25		
87xx		0.40–0.70	0.40–0.60	0.20–0.30		
88xx		0.40–0.70	0.40–0.60	0.30–0.40		
9xxx	Other					
92xx	High silicon					1.20–2.20 Si
93xx	Ni, Cr, Mo	3.00–3.50	1.00–1.40	0.08–0.15		
94xx	Ni, Cr, Mo	0.30–0.60	0.30–0.50	0.08–0.15		

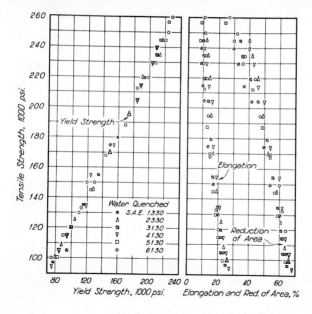

FIGURE 7-1 Straight-line relationship of mechanical properties of properly heat-treated SAE alloy steels. (*Courtesy American Society for Metals, Metals Park, Ohio.*)

that goal. By not specifying exact composition, but by purchasing on the basis of properties, savings can often be made. The producer is free to supply any material that will meet the desired properties and can take advantage of the residual elements in increased amounts of recycled scrap. Reduced material costs result in lower steel prices.

High strength/low alloy (HSLA) structural steels. There are two general categories of alloy steels: the high strength/low alloy types, which rely largely

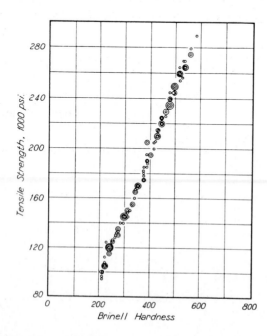

FIGURE 7-2 Relationship of hardness and tensile strength for a group of standard alloy steels. (*Courtesy American Society for Metals, Metals Park, Ohio.*)

TABLE 7-3 Compositions and Equivalents of Several EX Steels

EX Number	C	Mn	Cr	Mo	Other	Equivalent AISI Grade
15	0.18–0.23	0.90–1.20	0.40–0.60	0.13–0.20	—	8620
24	0.18–0.23	0.75–1.00	0.45–0.65	0.20–0.30	—	8620
31	0.15–0.20	0.70–0.90	0.45–0.65	0.45–0.60	0.70-1.00Ni	4817

on chemical composition to develop the desired mechanical properties in the as-rolled or normalized condition, and the constructional alloy steels, in which desired properties are developed by thermal treatment. Many manufactured products require steels with good hardenability, ductility, and fatigue strength—the constructional alloy steels. For structural applications, however, high yield strength, good weldability, and corrosion resistance are most desired, with only limited ductility and virtually no hardenability. Development of steels possessing these properties in the as-rolled condition have made possible substantial cost and weight savings in automobiles, trains, bridges, and buildings.

The low-alloy structural steels have about twice the yield strength of the plain-carbon structural steels. This increase in strength, coupled with resistance to martensite formation in a weld zone, is obtained by adding low percentages of several elements, notably manganese, silicon, niobium (columbium), and vanadium, as well as several others. About 0.2% copper is usually added to improve corrosion resistance. Although numerous types of these steels have been obtained by the addition of alloying elements in various combinations and quantities, four of the more common ones are listed in Table 7-4.

These steels represent a significant contribution to the field of structural materials. Through their use, weight savings of 20 to 30% have been achieved

TABLE 7-4 Typical Compositions and Strength Properties of Several Groups of Low-Alloy Structural Steels

Group	Chemical Composition[a](%)					Strength Properties				Elongation in 2 inches (%)
						Yield		Tensile		
	C	Mn	Si	Cb	V	MPa	ksi	MPa	ksi	
Columbium or vanadium	0.20	1.25	0.30	0.01	0.01	379	55	483	70	20
Low manganese–vanadium	0.10	0.50	0.10		0.02	276	40	414	60	35
Manganese–copper	0.25	1.20	0.30			345	50	517	75	20
Manganese–vanadium–copper	0.22	1.25	0.30		0.02	345	50	483	70	22

[a]All have 0.04% P, 0.05% S, and 0.20% Cu.

without any sacrifice in strength or safety. They currently are produced in sufficient tonnages that their cost is little more than that of ordinary grades of structural steel.

Quenched-and-tempered structural steel.   The need for even stronger structural steels that can be welded, notably for use in submarines and pressure vessels, has led to the development of several alloys that are always used in the quenched-and-tempered condition. These steels have yield strengths in the range 550 to 1050 MPa (80 to 150 ksi); tensile strengths of 650 to 1400 MPa (95 to 200 ksi), and elongations in 2 inches of 13 to 20%. The chemical compositions of most of these steels fall into two groups, one using nickel and low manganese and the other high manganese and silicon and some zirconium and boron but less chromium and no nickel. Typical compositions are:

Element	Group A	Group B
Carbon	0.18–0.20%	0.15–0.21%
Manganese	0.10–0.40%	0.80–1.10%
Phosphorus	0.025%	0.035%
Sulfur	0.025%	0.04%
Silicon	0.15–0.35%	0.40–0.90%
Nickel	2.00–3.50%	
Chromium	1.00–1.80%	0.50–0.90%
Molybdenum	0.06%	0.28%
Zirconium		0.05%
Boron		0.0025%

Such steels are typically water-quenched from about 900°C (1650°F) and tempered at 625 to 650°C (1150 to 1200°F) to produce a tempered martensite structure. When welded they are still tough, even though not tempered after welding, because the resulting martensite has a very low carbon content. They have excellent impact resistance at low temperatures and good atmospheric corrosion resistance. Because of their superior strength properties, they usually permit considerable weight saving, which can offset the added material cost.

Free-machining steels.   The increased use of high-speed machining, particularly on automatic machine tools, has spurred the development of several varieties of *free-machining steels*. These steels machine readily and form small chips so as to reduce the rubbing against the cutting tool and associated friction and heat. Formation of small chips also reduces the likelihood of chip entanglement in the machine and makes chip removal much easier.

These steels are basically carbon steels that have been modified in one of two ways. The first method is to increase the sulfur to a value of 0.08 to 0.33% and the manganese to 0.7 to 1.6%. The second modification utilizes 0.25 to 0.35% sulfur and 0.15 to 0.35% lead, with small amounts of tellurium, selenium, or bismuth. The additives cause the formation of inclusions, such as MnS, which act as discontinuities in the structure to form broken chips. The inclusions may also provide a built-in lubricant that prevents the formation of a built-up edge on the tool which alters the cutting geometry (see Chapter 18). It should be remembered, however, that these steels have somewhat reduced ductility and impact properties.

Alloy steels for electrical applications.   There are two groups of alloy steels that are widely used in the electrical industry. The addition of small amounts of silicon, 0.5 to 5%, in the *silicon steels* results in increased resistivity and permeability. Increased resistivity decreases eddy-current losses, while improved permeability decreases hysteresis losses. When such steel is used in electrical motors, generators, and transformers, the power losses and associated heat problems are reduced. For this reason, silicon steel frequently is used for the magnetic circuits of electrical equipment. For armature cores the silicon content is about 0.5%, whereas in transformers about 5% silicon is used. Silicon causes the steel to be brittle, so the amount used should be kept low where the steel is subjected to high rotative speeds.

Cobalt increases the magnetic saturation point in steel when it is used in amounts up to about 36%. For this reason *cobalt alloy steels* are used in electrical equipment where high magnetic densities must exist. For example, it is often used for the pole pieces of electromagnets. High-cobalt alloys are also used for most permanent magnets.

Corrosion-resistant or stainless steels.   *Corrosion-resistant or stainless steels* contain sufficient amounts of chromium that they can no longer be considered low-alloy steels. The corrosion resistance is imparted by the formation of a strongly adherent chromium oxide on the surface. Good resistance to many corrosive media encountered in the chemical industry may be obtained by the addition of 4 to 6% chromium to low-carbon steel. Usually, 0.4 to 0.8% silicon and 0.5% molybdenum are also added.

Where improved corrosion resistance and outstanding appearance are required, materials are designed to utilize a superior chromium oxide that forms when the amount of atomic chromium in solution (excluding chromium carbides, etc.) exceeds 12%. In this category, a variety of stainless steels exist. One major classification method is on the basis of microstructural characteristics. The AISI designation scheme for these metals is a three-digit code that identifies family and particular alloy within the family, as outlined in Table 7-5.

Chromium is a ferrite stabilizer, the addition of chromium tending to decrease the temperature range over which austenite is stable. If sufficient chromium is added to iron, a material can be obtained that is ferritic at all temperatures below solidification. *Ferritic stainless steels* are therefore low carbon/high chromium alloys. They possess rather poor ductility or formability because of the bcc crystal structure, but they are readily weldable. No martensite can form in these materials, because there is no possibility of austenite that can transform.

If the austenite region is not completely eliminated, stainless metals can be produced that are austenite at high temperature and ferrite at low. Such a metal can be quenched to form martensite and is known as a *martensitic stainless*

TABLE 7-5 AISI Designation Scheme for Stainless Steels

Series	Alloys	Structure
200	Chromium, nickel, manganese, or nitrogen	Austenitic
300	Chromium and nickel	Austenitic
400	Chromium only	Ferritic or martensitic
500	Low chromium (< 12%)	Martensitic

steel. Carbon content can vary to produce the desired strength level, but chromium must be adjusted to assure more than 12% in solution. In addition to the higher base cost, approximately 1½ times that of the ferritic material, martensitic stainlesses require more costly processing. They are usually annealed for fabrication and hardened by a full austenitize-quench-stress relieve-temper cycle.

Nickel is an austenite stabilizer, and with sufficient amounts of both chromium and nickel, it is possible to produce a stainless steel in which austenite is the stable phase at room temperature, an *austenitic stainless steel*. From 3.5 to 22% nickel is used and in some cases small amounts of molybdenum or titanium. The cost of this material may be more than double that of the cheaper ferritic stainless. Manganese and nitrogen are also austenite stabilizers and may be substituted for some of the nickel to produce a lower cost, somewhat lower-quality austenitic stainless steel (the 200 series alloys).

Austenitic stainless steels are nonmagnetic and are highly corrosion resistant in almost all media except hydrochloric acid and other halide acids and salts. In addition, they may be polished to a mirror finish and thus combine attractive appearance with good corrosion resistance. Formability is outstanding (characteristic of the fcc crystal structure), and they respond quite well to strengthening by cold work. The response of the popular 18–8 (18% chromium and 8% nickel) grade to a small amount of cold work is as follows:

	Water Quench	Cold-Rolled 15%
Yield strength	260 MPa (38 ksi)	805 MPa (117 ksi)
Tensile strength	620 MPa (90 ksi)	965 MPa (140 ksi)
Elongation in 2 inches	68%	11%

One should note that these materials are often water-quenched to retain the alloys in solid solution—no transformation occurs since the stable phase is austenite for all temperatures involved.

Austenitic stainless steels are expensive metals and, although produced in large tonnages, they should not be specified where the less expensive ferritic or martensitic alloys would be adequate or where a true stainless steel is not re-

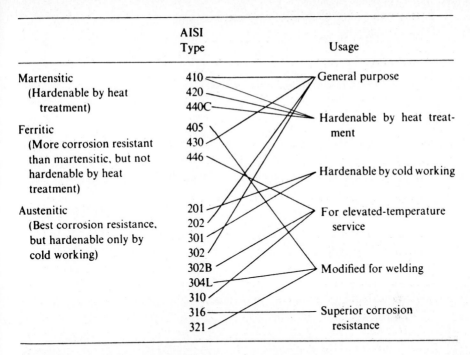

FIGURE 7-3 Classification and uses of stainless steels.

quired. Figure 7-3 lists several grades from each of the three major classifications and notes some key properties. Table 7-6 presents the typical compositions of the basic stainless classes.

Problems with the basic stainless grades generally relate to loss of corrosion resistance (sensitization) when the level of chromium in solution drops below 12%. Since chromium depletion is usually caused by the formation of chromium carbides along grain boundaries, especially at elevated temperatures, one method of prevention is to keep the carbon content as low as possible, usually less than

TABLE 7-6 Typical Compositions of the Ferritic, Martensitic, and Austenitic Stainless Steels

Element	Ferritic	Martensitic	Austenitic[a]
Carbon	0.08–0.20%	0.15–1.2%	0.03–0.25%
Manganese	1–1.5%	1%	2%(5.5–10%)
Silicon	1%	1%	1–2% (0%)
Chromium	11–27%	11.5–18%	16–26%
Nickel			3.5–22%
Phosphorus and sulfur			Normal (0%)
Molybdenum			Some cases
Titanium			Some cases

[a]Values in parentheses are for one type.

0.10%. Another approach is to tie up the carbon with small amounts of "stabilizing" elements such as titanium or columbium, alloys that have a stronger affinity to form carbides. Rapidly cooling these metals through the range 480 to 820°C (900 to 1500°F) also retards carbide formation.

Other stainless steels have been developed for special uses. Ordinary stainless steels are difficult to machine because of their work-hardening properties and the fact that they "seize" and thus prevent clean cutting. *Free-machining varieties*, produced by the addition of sulfur or selenium, are available that machine nearly as well as a medium-carbon steel.

Another class of stainless material is that of the *precipitation-hardening stainless steels*. These alloys are basically martensitic or austenitic types with low carbon, modified by the addition of alloying elements that permit age hardening at relatively low temperatures. Properties such as a 1790-MPa (260-ksi) yield strength and a 1825-MPa (265-ksi) tensile strength with 2% elongation can be attained with none of the distortion associated with quenching. These grades, however, are among the most expensive stainlesses and should be used only when required.

The maraging steels. When 18 to 25% nickel and significant amounts of cobalt, molybdenum, and titanium are added to very low-carbon steel, a material results that air-quenches from about 815°C (1500°F) to form martensite, which, in turn, will age-harden at 480°C (900°F) to produce yield strengths in excess of 1725 MPa (250 ksi) with elongations in excess of 11%. This material is very useful in such applications as rocket-motor cases, where high strength and good toughness are important.

A typical maraging steel in the 1725-MPa (250-ksi) strength class has a composition of:

0.03% C	0.10% Al
18.5% Ni	0.003% B
7.5% Co	0.10% Si maximum
4.8% Mo	0.10% Mn maximum
0.40% Ti	0.01% S maximum
0.01% Zr	0.01% P maximum

This steel can be hot-worked from 760 to 1260°C (1400 to 2300°F). Air cooled from 815°C (1500°F), it has a hardness of about $30R_c$ and a structure of soft, tough martensite. It is easily machined and, because of its low work-hardening rate, can be cold-worked to a high degree. Aging at 480°C (900°F) for 3 to 6 hours, followed by air cooling, raises the hardness to about $52R_c$ and produces full strength. Maraging steels can be welded, if welding is followed by the full solution and aging treatment. As expected, the cost of this material is quite high.

Steels for high-temperature service. Continued developments in rocketry, missiles, jet aircraft, and nuclear power have increased the need for metals that have good strength characteristics, corrosion resistance, and, particularly, creep resistance at high temperatures. Much work has been done to produce both ferrous and nonferrous alloys that have the desired properties at temperatures from 550 to 950°C (1000 to 1750°F). The nonferrous alloys will be discussed in a later chapter.

The ferrous alloys, which ordinarily are used below 760°C (1400°F), are low-carbon materials with less than 0.1% carbon. One alloy is a modified 18–8 stainless steel stabilized with either columbium or titanium. A 1000-hour rupture stress of 40 to 50 MPa (6000 to 7000 psi) is observed at 760°C (1400°F), with considerably higher strengths at lower temperatures. Iron also is a major component of other high-temperature alloys, but when amounts become less than 50%, the metal can hardly be classified as ferrous in nature. High strength at high temperature usually requires the more expensive nonferrous materials.

Tool steels. *Tool steels* are metals designed to provide wear resistance and toughness combined with high strength. They are basically high-carbon alloys, where the chemistry provides the balance of toughness and wear desired.

Several classification or breakdown systems have been applied to tool steels, some using chemistry as a basis and others employing hardening method and major mechanical property desired. The standard AISI–SAE designation system identifies letter grades by basic principles such as quenching method, primary application, special characteristic, or specific industry involved. Table 7-7 lists the seven basic families of tool steel and corresponding AISI–SAE grades. Individual alloys are then listed numerically within the grade to produce a letter–number identification system.

Water-hardening carbon tool steels (W grade) account for a large percentage of all tool steels used. They are the least expensive and are used for a wide variety of parts that are usually quite small and not subjected to severe usage or elevated temperatures. Because strength and hardness are functions of the carbon content, a wide range of these properties can be obtained through com-

TABLE 7-7 Basic Types of Tool Steel and Corresponding AISI–SAE Grades

Type	AISI–SAE Grade	
1. Water hardening	W	
2. Cold work	O	Oil-hardening
	A	Air-hardening medium-alloy
	D	High carbon/high chromium
3. Shock resisting	S	
4. High speed	T	Tungsten base
	M	Molybdenum base
5. Hot work	H	H1–H19: chromium base
		H20–H39: tungsten base
		H40–H59: molybdenum base
6. Plastic mold	P	
7. Special purpose	L	Low alloy
	F	Carbon–tungsten

position variation. These steels must be quenched in water to obtain high hardness and, because their hardenability is low, they can be used only for relatively light sections if full depth of hardness must be obtained. They also are rather brittle, particularly at higher hardness.

The uses of plain carbon steels, according to carbon content, are somewhat as follows:

0.60–0.75% carbon: machine parts, chisels, set screws, and the like, where medium hardness with considerable toughness and shock resistance is required.

0.75–0.90% carbon: forging dies, hammers, and sledges.

0.90–1.10% carbon: general-purpose tooling requirements, which require good balance of wear resistance and toughness, such as drills, cutters, shear blades, and other heavy-duty cutting edges.

1.10–1.30% carbon: small drills, lathe tools, razors, and similar light-duty applications, where extreme hardness is necessary without great toughness.

In applications where improved toughness is desired in conjunction with high strength and hardness, small amounts of manganese, silicon, and molybdenum are often added. Vanadium additions of about 0.20% are often used to form strong stable carbides that retain fine grain size during heat treating. One of the main weaknesses of carbon tool steels is the fact that they do not hold their hardness at elevated temperatures. Prolonged exposure to temperatures over 150°C (300°F) results in undesired softening. This places a considerable limitation on their use.

When larger parts must be hardened or distortion must be minimized, *oil- or air-hardening grades* (O and A designations, respectively) are often employed. Because of the higher hardenability, they can be hardened by less severe quenches, maintaining close dimensional tolerances and minimizing the tendency to cracking. Manganese tool and die steels form one segment of this group. These contain 0.75 to 1.0% carbon and 1.0 to 2.0% manganese and are moderate in cost. In some cases, manganese is reduced and chromium, silicon, and nickel are added to give greater toughness. These steels are not as hard as the plain manganese types, but have less tendency to crack because of the greater hardenability and less severe quench requirement.

Chromium tool and die steels are also in this class. The low-chrome tool steels are much the same as plain carbon tool steels, with chromium added to produce the desired hardenability and toughness. Chromium levels are usually between 0.5 and 5.0%. High-chromium tool steels are designated by a separate letter (D) and contain between 10 and 18% chromium. These steels ordinarily must be annealed before they can be machined. After machining, they are hardened and tempered and will usually retain the final hardness at temperatures up to 425°C (800°F). They are used for tools and dies that must withstand hard usage over long periods, often at elevated temperatures. Forging dies, die-casting die blocks, and drawing dies are often made from such steel.

Shock-resisting tool steels (S designation) have been developed for both hot and cold impact applications. Low carbon contents (approximately 0.5%) are used to provide toughness, with carbide-forming alloying elements being added to supply the desired abrasion resistance, hardenability, and hot-work characteristics.

High-speed tool steels are used primarily for cutting tools that machine metal and other special applications where retention of hardness at red heat is required. The most common type of high-speed steel is the tungsten-based T1 alloy, also known as 18-4-1 because the analysis contains 0.7% carbon, 18% tungsten, 4% chromium, and 1% vanadium. This alloy has a balanced combination of shock resistance and abrasion resistance and is used for a wide variety of cutting applications. Other high-speed tool steels have cobalt added to improve hardness at elevated temperatures. Toughness diminishes and forming problems increase.

The *molybdenum high-speed steels* (M designation) were developed to reduce the amount of tungsten and chromium required to produce the desired properties. The M2 variety is now the most widely used high-speed steel, its higher carbon content and balanced analysis producing properties applicable to all general-purpose high-speed uses.

Hot-work tool steels (H designation) are designed to perform adequately under environments of prolonged high temperature. All employ additions of carbide-forming alloying elements: H1 to H19 are chromium-base types with about 5.0% chromium; H20 to H39 are tungsten-base types with 9 to 18% tungsten coupled with 3 to 4% chromium; and H40 to H59 are molybdenum-based types.

Other types of tool steels include (1) the *plastic mold steels* (P designation) made specifically for the requirements of zinc die casting and plastic molding dies, (2) the *low-alloy special-purpose tool steels* (L designation), such as the L6 extreme toughness variety; and (3) the *carbon–tungsten type* of special-purpose tool steels (F designation), which are water hardening but substantially more wear-resistant than the plain-carbon tool steels.

Alloy cast steels and irons. The effects of alloying elements in steel are the same regardless of the method used to produce the final shape. To couple the attributes of the casting process and the benefits of alloy material, many alloy cast steels have been produced. In the case of cast iron, however, alloys often perform functions in addition to those previously cited.

Whereas alloy steels are usually heat-treated, many alloy cast irons are not, except for stress relieving or annealing. If a cast iron is to be hardened, chromium, molybdenum, and nickel are frequently added to improve hardenability. In addition, chromium tends to offset the undesirable quenching effects of high carbon, and molybdenum and nickel offset the high silicon.

If alloy cast irons are not to be heat-treated, the alloy elements are often selected to alter properties through affecting the formation of graphite or cementite, modifying the morphology of the carbon-rich phase, or simply strengthening the matrix material. Alloys are often added in small amounts to improve strength properties or wear resistance. High-alloy cast irons are often designed to provide corrosion resistance, particularly at high temperatures such as are encountered in the chemical industry.

Nickel promotes graphite formation and tends to produce a finer graphite structure. Chromium, on the other hand, retards graphitization and stabilizes cementite. They are frequently used together in the ratio of 2 to 3 parts of nickel to 1 part of chromium.

If the silicon content of gray cast iron is lowered, the strength is increased considerably. However, if this is carried very far, hard, white cast iron results. The addition of about 2% nickel will minimize the formation of white cast iron. By adjusting the silicon content and adding a small amount of nickel, a cast iron of good strength is obtained without sacrificing machinability.

Molybdenum strengthens gray cast iron to some extent and often forms carbides. In addition, it is used to control the size of the graphite flakes. From 0.5 to 1.0% is ordinarily added.

Among the high-alloy cast irons, *austenitic gray cast iron* is quite common. This contains about 14% nickel, 5% copper, and 2.5% chromium. It possesses good corrosion resistance to many acids and alkalis at temperatures up to about 800°C (1500°F).

Alloy cast irons and steels are generally identified by their ASTM specification number.

REVIEW QUESTIONS

1. What is plain-carbon steel?
2. What are some of the limitations of plain-carbon steels?
3. What are the three classes of plain-carbon steels and the characteristics of each?
4. What constitutes an alloy steel?
5. What are the most common alloy elements added to steel?
6. What are some of the benefits obtained from adding alloy elements to steel?
7. How might the machinability of steels be improved?
8. Why might a high-nickel steel be attractive for low-temperature applications?
9. For what reason is copper often added to low-carbon sheet and structural steels?
10. What is the major reason for a boron addition?
11. What is the significance of the last two digits in the AISI–SAE designation system?
12. What are H-grade AISI steels?
13. What are EX steels?
14. What is the goal of the HSLA steels?
15. What are free-machining steels?
16. Why are silicon steels useful for electrical applications?
17. How do stainless steels get their corrosion resistance?
18. What are the three families of stainless steels?
19. What are some of the outstanding characteristics of austenitic stainless steels?
20. What causes sensitization of a stainless steel?
21. What are two of the special types of stainless steels?
22. What properties are desired in tool steels?
23. What is the major limitation of the W-grade tool steels?
24. What special characteristics are desired in the high-speed tool steels?
25. What are some of the reasons alloy elements may be added to cast irons that are not to be heat-treated?
26. How are alloy cast irons and cast steels generally identified?

CASE STUDY 7. Heat-Treated Axle Shafts

The high strength and fatigue requirements of automobile axle shafts generally require a heat-treated steel with a tempered martensite structure and a surface hardness of approximately R_C 50. The shafts are approximately 35 mm (1⅜ inches) in diameter and 1.06 meters (3½ feet) in length, and distortion or warpage must be minimized. In the 1960s, these requirements were met by oil-quenching bars of medium-carbon alloy steels, such as 4140, 8640, and 5140. Now they are frequently manufactured from the less-expensive plain-carbon steels, such as 1038 and 1040. How would you propose to heat-treat these lower hardenability metals to the desired hardness and structure and still limit distortion?

A further cost saving can be obtained if a small segment of the axle shaft surface can serve as the inner race of a bearing. This, however, requires a surface hardness of R_C 60 in this region. How would this additional requirement modify your material selection? Assuming that plain-carbon steel is still used, how might this region of high surface hardness be obtained?

CHAPTER 8

Nonferrous Alloys

Nonferrous metals and alloys are playing increasingly important roles in modern technology. Because of their number and the fact that the properties of the individual metals vary widely, both in their relatively pure form and as base metals for alloys, they provide an almost limitless range of properties for the design engineer. Even though they are not produced in as great tonnages and are more costly than iron and steel, they make available certain important properties or combinations of properties that cannot be obtained in steels, notably:

1. Resistance to corrosion.
2. Ease of fabrication.
3. High electrical and thermal conductivity.
4. Light weight.
5. Color.

While it is true that corrosion resistance can be obtained in certain ferrous alloys, several of the nonferrous alloys possess this property without requiring special and expensive alloying elements. Nearly all the nonferrous alloys possess at least two of the qualities listed above, and some possess all five. For many applications, certain combinations of these properties are highly desirable, and the availability of materials that provide them directly is a strong motivation for the use of nonferrous alloys.

In most cases, the nonferrous alloys are inferior to steel in respect to strength.

Also, the modulus of elasticity may be considerably lower, a fact that places them at a disadvantage where stiffness is a necessary property. Fabrication, however, is usually easier than for steel. Those alloys with low melting points are often easy to cast, either in sand molds, permanent molds, or dies. Many alloys have high ductility coupled with low yield points, the ideal conditions for easy cold work and high formability. High machinability is characteristic of several nonferrous alloys. Fabrication savings can often overcome the higher cost of the nonferrous material and favor its use in place of steel. The one fabrication area in which the nonferrous alloys are somewhat inferior to steel is weldability. Due to recent developments, however, it is often possible to produce satisfactory weldments from the viewpoint of both quality and economy.

COPPER-BASE ALLOYS

As was discussed in Chapter 4, copper is seldom used in its pure state except in the electrical industry. For other engineering applications, it is almost always used in the form of an alloy with elements such as zinc, tin, nickel, aluminum, silicon, or beryllium. These alloys are commonly identified through a system of copper alloy numbers developed by the Copper Development Association (CDA). Table 8-1 presents a breakdown of this system, which has been adopted by the ASTM, SAE, and the U.S. government. Alloys numbered from 100 to 190 are mostly copper with less than 2 percent alloy. Numbers 200 to 799 are other wrought alloys. The 800 and 900 series are all casting alloys.

TABLE 8-1 Standard Designations for Copper and Copper Alloys (CDA System)

	Wrought Alloys		Cast Alloys
100–155	Commercial coppers	833–838	Red brasses and leaded red brasses
162–199	High-copper alloys		
200–299	Copper–zinc alloys (brasses)	842–848	Semi-red brasses and leaded semi-red brasses
300–399	Copper–zinc–lead alloys (leaded brasses)	852–858	Yellow brasses and leaded yellow brasses
400–499	Copper–zinc–tin alloys (tin brasses)	861–868	Manganese and leaded manganese bronzes
500–529	Copper–tin alloys (phosphor bronzes)	872–879	Silicon bronzes and silicon brasses
532–548	Copper–tin–lead alloys (leaded phosphor bronzes)	902–917	Tin bronzes
600–642	Copper–aluminum alloys (aluminum bronzes)	922–929	Leaded tin bronzes
		932–945	High-leaded tin bronzes
647–661	Copper–silicon alloys (silicon bronzes)	947–949	Nickel–tin bronzes
		952–958	Aluminum bronzes
667–699	Miscellaneous copper–zinc alloys	962–966	Copper–nickels
		973–978	Leaded nickel bronzes
700–725	Copper–nickel alloys		
732–799	Copper–nickel–zinc alloys (nickel silvers)		

Copper–zinc alloys. Zinc is by far the most popular alloying addition to copper, the alloys of copper and zinc being commonly known as *brasses*. If the zinc content is not over 36%, brass is a single-phase solid solution of zinc in copper. Since this is identified as the alpha phase, these alloys are often called *alpha brasses*. They are quite ductile and formable, these characteristics increasing with the zinc content up to 36%. Above 36% zinc the alloys enter a two-phase region involving a brittle beta phase. Cold-working properties are rather poor for the high-zinc brasses, but deformation is easy when performed hot. Like copper, brass is hardenable by cold working and, in commercial grades, is available in various degrees of hardness.

Table 8-2 lists some of the most common copper–zinc alloys, their compositions, properties, and typical uses. Brasses range in color from copper to nearly white, the lower-zinc brasses being more coppery than those with more zinc. The addition of a third element, however, can change color considerably.

Most brasses have good corrosion resistance. In the 0 to 40% zinc region, the addition of a small amount of tin produces excellent resistance to seawater corrosion. Cartridge brass with tin becomes admiralty brass; muntz metal with tin is known as naval brass. Brasses with 20 to 36% zinc are subject to a selective corrosion, known as *dezincification,* when in acid or salt solutions. Another corrosion problem in brasses with more than 15% zinc is season cracking or stress-corrosion cracking. Both stress and exposure to corrosive media are required for this failure to occur. Thus brasses must often undergo a stress relief to remove the residual stresses induced by cold working prior to being put into service.

Where high machinability is needed, as in automatic screw-machine stock, 2 to 3% lead is added to brass to assure production of free-breaking chips.

Many uses of brass relate to the high electrical and thermal conductivities coupled with adequate strength. Plating characteristics are outstanding and make it an excellent base for decorative chrome or similar coatings. A unique property of alpha brass is its ability to have rubber vulcanized to it without any special treatment except thorough cleaning. It is widely used in mechanical rubber goods because of this property.

An alloy containing from 50 to 55% copper and the remainder zinc is often used for filler metal in brazing. It is an effective agent for joining steel, cast iron, brasses, and copper, producing joints that are nearly as strong as those obtained by welding.

Copper–tin alloys. Alloys of copper and tin, commonly called tin *bronzes,* are considerably more expensive than the brasses because of the high price of tin. Consequently, they have now been replaced to a considerable degree by less expensive nonferrous alloys, but are used in certain applications because of special properties.

The term "bronze" is somewhat confusing, since some alloys that contain no tin are called bronzes because of their color. The true bronzes usually contain less than 12% tin. Strength increases with tin content up to about 20%, beyond which the alloys become brittle. Copper–tin alloys are characterized by good strength, toughness, wear resistance, and corrosion resistance. They are often

TABLE 8-2 Composition, Properties, and Uses of Some Common Copper–Zinc Alloys

CDA Number	Common Name	Composition (%)					Condition	Tensile Strength		Elongation in 2 inches (%)	Typical Uses
		Cu	Zn	Sn	Pb	Mn		MPa	ksi		
220	Commercial bronze	90	10				Soft sheet Hard sheet Spring	262 441 503	38 64 73	45 4 3	Screen wire, hardware, screws, jewelry
240	Low brass	80	20				Annealed sheet Hard Spring	324 517 627	47 75 91	47 7 3	Drawing, architectural work, ornamental
260	Cartridge brass	70	30				Annealed sheet Hard Spring	365 524 634	53 76 92	54 7 3	Munitions, hardware, musical instruments, tubing
270	Yellow brass	65	35				Annealed sheet Hard	317 524	46 76	64 7	Cold forming, radiator cores, springs, screws
280	Muntz metal	60	40				Hot rolled Cold rolled	372 551	54 80	45 5	Architectural work, condenser tube
443–445	Admiralty metal	71	28	1			Soft Hard	310 655	45 95	60 5	Condenser tube (salt water), heat exchangers
360	Free-cutting brass	61.5	35.5		3		Soft Hard	324 427	47 62	60 20	Screw-machine parts
675	Manganese bronze	58.5	39	1		0.1	Soft Bars, half hard	448 579	65 84	33 19	Clutch disks, pump rods, valve stems, high-strength propellers

used for bearings, gears, and fittings that are subjected to heavy compressive loads. When used as bearing material, up to 10% lead is often added.

The most popular wrought bronze is phosphor bronze, which usually contains from 1 to 11% tin. Alloy 521 (CDA) is typical of this class and contains 92% copper, 8% tin, and 0.15% phosphorus. Hard sheet of this material has a tensile strength of 758 MPa (110 ksi) and an elongation in 2 inches of 3%. Soft sheet has a 379-MPa (55-ksi) tensile strength and 65% elongation. It is used for pump parts, gears, springs, and bearings.

Alloy 905 is a commonly used cast bronze containing 88% copper, 10% tin, and 2% zinc. In the cast condition, the tensile strength is about 310 MPa (45 ksi), with an elongation of 25% in 2 inches. It has very good resistance to seawater corrosion and is used on ships for pipe fittings, gears, pump parts, bushings, and bearings.

Copper–nickel alloys. Copper and nickel have complete solid solubility as seen in Figure 5-6, and a wide range of alloys have been developed. High thermal conductivity coupled with corrosion resistance make these materials a good choice for heat exchangers, cookware, and other heat-transfer applications. *Cupro-nickels* contain 2 to 30% nickel. *Nickel silvers* have 10 to 30% nickel and at least 5% zinc. An alloy with 45% nickel is known as *constantan,* and the 67% nickel alloy is called *Monel.*

Other copper-base alloys. The alloys previously discussed acquire their strength primarily through solid-solution strengthening and cold work. In the copper alloy family, three alloying elements produce materials that are precipitation hardenable: aluminum, silicon, and beryllium.

Aluminum bronze alloys usually contain between 6 and 12% aluminum and often 2 to 5% iron. With aluminum contents below 8%, the alloys are very ductile. With more than 9%, hardness approaches that of steel, and higher aluminum contents result in brittle, but very wear resistant, materials. A cast aluminum bronze having 86.2% copper, 10.2% aluminum, and 3.3% iron has a tensile strength of 480 to 550 MPa (70 to 80 ksi) and 18 to 22% elongation in 2 inches. By varying the aluminum content and the heat treatment, the tensile strength may be varied from about 415 to 860 MPa (60 to 125 ksi). Large amounts of shrinkage occur in parts cast in aluminum bronze. Castings of this material should be designed with this in mind.

Silicon bronzes contain up to 4% silicon and 1.5% zinc (higher zinc contents are used when the material is cast). Strength, formability, machinability, and corrosion resistance are quite good. Tensile strengths can approach 900 MPa (130 ksi) with cold work, whereas the soft material has a tensile strength of about 380 MPa (55 ksi) and 65% elongation. Uses include boiler, stove, and tank applications requiring high strength with corrosion resistance.

Copper–beryllium alloys can be age-hardened to produce the highest strengths of the copper-based metals. They ordinarily contain less than 2% beryllium, but are quite expensive. When annealed the material has a yield strength of 172 MPa (25 ksi), tensile strength of 482 MPa (70 ksi), and elongation of 50%. After heat treatment, these properties can rise to 1100 MPa (160 ksi), 1245 MPa

(180 ksi), and 5%, respectively. The modulus of elasticity is about 125,000 MPa (18,000 ksi) and the endurance limit is around 275 MPa (40 ksi). These properties make the material excellent for springs, but cost limits application to small components requiring long life and high reliability. Other applications relate to the unique properties: strength of steel, nonsparking, nonmagnetic, and conductive.

Looking to the future, we see a continuing demand from the electronics and telecommunications industry for new alloys combining ductility, conductivity, and strength. In the high-strength area, the demand will be for non-beryllium-containing alloys. Yield strengths of 700 to 1000 MPa (100 to 150 ksi) with high conductivities may be possible using combined techniques such as fiber reinforcement, splat cooling, dispersion strengthening, and precipitation hardening. Tensile strengths of 2050 MPa (300 ksi) and fatigue limits of 550 MPa (80 ksi) may be attainable with copper-base alloys.

ALUMINUM ALLOYS

Aluminum has become the most important of the nonferrous metals. Pure aluminum is outstanding for its light weight, high thermal and electrical conductivity, and corrosion resistance. In the annealed condition, however, pure aluminum has only about one-fifth the strength of ordinary structural steel, and its modulus of elasticity—a property that can be modified only slightly by alloying and not at all by heat treatment—is only one-third that of steel. Although it costs more per pound than steel, it is only a little over one-third as heavy, thereby making it cheaper per unit volume. Corrosion resistance is far superior to that of ordinary steel.

Electrical-conductor grade aluminum is used in large quantities and has replaced copper in many electrical transmission lines and for bus bars. This grade, commonly designated by the letters EC, contains a minimum of 99.45% aluminum and has an electrical conductivity 62% that of copper for the same size wire and 200% that of copper on an equal-weight basis.

Aside from its electrical uses, most aluminum is used in the form of alloys. These have much greater strength than pure aluminum, yet retain the advantages of light weight, good conductivity, and corrosion resistance. Some alloys are available that have tensile properties, except for ductility, superior to those of low alloy/high yield strength structural steel. On a strength-to-weight basis, most of the aluminum alloys are superior to steel, but wear, creep, and fatigue properties are usually somewhat poorer. The selection between steel and aluminum for any given application is largely a matter of cost, although in many cases the advantages of reduced weight or corrosion resistance may justify additional expense. In most cases, aluminum replaces steel or cast iron, where the need for lightness, corrosion resistance, low maintenance expense, or high thermal or electrical conductivity offsets the added cost. Aluminum is playing a major role in downsizing and weight reduction of motor vehicles. While bar stock and sheet metal are only marginally competitive, extensive use of aluminum castings is expected in areas such as manifolds, blocks, and transmission cases.

Corrosion resistance of aluminum and its alloys. Pure aluminum is very reactive and forms a tight, adherent oxide coating on the surface as soon as it is exposed to air. This oxide is resistant to many corrosive media and serves as a corrosion-resistant barrier to protect the underlying metal. When alloying elements are added to aluminum, oxide formation is somewhat retarded, so the alloys, in general, do not have quite the superior corrosion resistance of pure aluminum.

The oxide coating on aluminum alloys causes some difficulty in relation to its weldability. In resistance welding, it usually is necessary to remove the oxide immediately before welding in order to obtain consistent results. In fusion welding processes, the aluminum oxidizes so readily that it is necessary to use special fluxes or protective inert-gas atmospheres. Suitable welding techniques have been developed, however, to enable aluminum to be welded with complete success, from both quality and cost viewpoints.

Wrought aluminum alloys. *Wrought aluminum-base alloys* can be divided into two basic types: those that achieve strength by *solid solution alloying and work hardening* and those that can be *precipitation-hardened*. Table 8-3 lists some common wrought aluminum alloys, using the standard four-digit designation system for aluminums. The first digit indicates the major group as follows:

Major Alloying Element	
Aluminum, 99.00% and greater	1xxx
Copper	2xxx
Manganese	3xxx
Silicon	4xxx
Magnesium	5xxx
Magnesium and silicon	6xxx
Zinc	7xxx
Other element	8xxx

The second digit indicates modifications of the original alloy or impurity limits, and the last two digits identify the particular alloy or indicate the aluminum purity.

Having specified alloy chemistry, clarification of alloy condition is done through the *temper* designation, a letter or letter–number suffix. Symbols and their meanings are as follows:

-F: as fabricated, as in casting.
-H: strain-hardened.
   -H1: strain-hardened by working to desired dimensions; a second digit, 1 through 9, indicates the degree of hardness, 8 being commercially full-hard and 9 extra-hard.

-H2: strain-hardened by cold working, following by partial annealing, second-digit numbers 2 through 8, as above.
-H3: strain-hardened and stabilized.
-O: annealed.
-T: thermally treated (heat treated).
-T1: cooled from hot working and naturally aged.
-T2: cooled from hot working, cold-worked, and naturally aged.
-T3: solution-heat-treated, cold-worked, and naturally aged.
-T4: solution-heat-treated and naturally aged.
-T5: cooled from hot working and artificially aged.
-T6: solution-heat-treated and artificially aged.
-T7: solution-heat-treated and stabilized.
-T8: solution-heat-treated, cold-worked, and artificially aged.
-T9: solution-heat-treated, artificially aged, and cold-worked.
-T10: cooled from hot working, cold-worked, and artificially aged.
-W: solution-heat-treated only.

Additional digits beyond those listed above indicate variations of the basic temper.

It can be noted from Table 8-3 that the work-hardenable alloys are primarily those in the 1000 (pure aluminum), 3000 (aluminum-manganese), and 5000 (aluminum-magnesium) series. Within these series the 1100, 3003, and 5052 alloys tend to be the most popular. Strength tends to increase with increasing alloy number ($1100\rightarrow 3003\rightarrow 5052$), but ductility decreases with increasing strength.

Because of their higher strengths, the precipitation-hardenable alloys are more numerous and are found primarily in the 2000, 6000, and 7000 series. Alloy 2017, the original *duralumin,* is probably the oldest hardenable aluminum. The 2024 aluminum alloy is stronger than 2017 and is widely used in aircraft applications. Also, the ductility of the 2000 series alloys does not decrease significantly with the strength increases produced by heat treatment. The more recently developed precipitation-hardenable alloys are of the 7075 type. In the heat-treated condition, these alloys have yield strengths that approach or exceed those of the high-yield-strength structural steel. Ductility, however, is less than that for steel, and fabrication is more difficult than for the 2024 alloy. Nevertheless, these alloys are widely used in aircraft.

The heat-treatable alloys tend to have poorer corrosion resistance than either pure aluminum or the work-hardenable alloys. Thus, where both high strength and superior corrosion resistance are needed, the wrought aluminum is often produced as *Alclad.* A thin layer of corrosion-resistant aluminum is bonded to one or both surfaces of the high-strength metal during rolling and the material is further processed as a composite. Galvanic protection retards corrosion even when the metal is severely scored.

Because only moderate temperatures are required to lower the strength so that plastic flow occurs readily, aluminum-alloy extrusions and forgings are relatively easy to produce and are manufactured in large quantities. Deep drawing operations can also be carried out easily. In general, the high ductility and low yield strength of the aluminum alloys make them appropriate for almost all forming

TABLE 8-3 Compositions, Typical Properties, and Designations of Some Wrought Aluminum Alloys

| Designation[a] | Composition (%) Aluminum = Balance | | | | | Form Tested | Tensile Strength | | Yield Strength[b] | | Elongation in 2 inches (%) | Brinell Hardness | Uses and Characteristics |
	Cu	Si	Mn	Mg	Others		ksi	MPa	ksi	MPa			
Work-hardening alloys—not heat-treatable													
1100–0	0.12				99 Al	$\frac{1}{16}$-in. sheet	13	90	5	34	35	23	Commercial Al: good forming properties
1100–H14						$\frac{1}{16}$-in. sheet	16	110	14	97	9	32	Good corrosion resistance, low yield strength
1100–H18						$\frac{1}{16}$-in. sheet	24	165	21	145	5	44	Cooking utensils; sheet and tubing
3003–0	0.12		1.2			$\frac{1}{16}$-in. sheet	16	110	6	41	30	28	Similar to 1100
3003–H14						$\frac{1}{16}$-in. sheet	22	152	21	145	8	40	Slightly stronger and less ductile
3003–H18						$\frac{1}{16}$-in. sheet	29	200	27	186	4	55	Cooking utensils; sheetmetal work
5052–0				2.5	0.25 Cr	$\frac{1}{16}$-in. sheet	28	193	13	90	25	45	Strongest work-hardening alloy
5052–H32						$\frac{1}{16}$-in. sheet	33	228	28	193	12	60	High yield strength and fatigue limit
5052–H36						$\frac{1}{16}$-in. sheet	40	276	35	241	8	73	Highly stressed sheetmetal products
Precipitation-hardening alloys—heat-treatable													
2017–0	4.0	0.5	0.7	0.6		$\frac{1}{16}$-in. sheet	26	179	10	69	20	45	Duralumin, original strong alloy
2017–T4						$\frac{1}{16}$-in. sheet	62	428	40	276	20	105	Hardened by quenching and aging
2024–0	4.4		0.6	1.5		$\frac{1}{16}$-in. sheet	27	186	11	76	20	42	Stronger than 2017
2024–T4						$\frac{1}{16}$-in. sheet	64	441	42	290	19	120	Used widely in aircraft construction

Alloy and temper[a]	Composition, %	Form						Remarks
2014–0	4.4 0.8 0.8 0.5	½-in. extruded shapes	27	186	14	97	12	Strong alloy for extruded shapes
2014–T6		Forgings	65	448	55	379	10	Strong forging alloy
2014–T6		1/16-in. sheet	70	483	60	413	8	Higher yield strength than Alclad 2024
Alclad 2014–T6	4.5 1.0 0.8 0.4	1/16-in. sheet	63	434	56	386	7	Clad with heat-treatable alloy[c]
7075–0	1.6 0.2 2.5 { 0.3 Cr, 5.6 Zn }	1/16-in. sheet	33	228	15	103	17	Alloy of highest strength
7075–T6		1/16-in. sheet	76	524	67	462	11	Lower ductility than 2024
Alclad 7075–T6		1/16-in. sheet	76	524	67	462	11	Strongest Alclad product
7075–T6		½-in. extruded shapes	80	552	70	483	6	Strongest alloy for extrusions
6061–T6	0.28 0.6 1.0 0.20 Cr	½-in. extruded shapes	42	290	40	276	12	Strong, corrosion resistant
6063–T6	0.4 0.7	½-in. rod extruded	35	241	31	214	12	Good forming properties and corrosion resistance
6151–T6	0.9 0.6 0.25 Cr	Forgings	48	331	43	297	17	For intricate forgings
2025–T6	4.5 0.8 0.8	Forgings	55	379	30	207	18	Good forgeability, lower cost
2018–T6	4 0.7 2 Ni	Forgings	55	379	40	276	10	Strong at elevated temperatures; forged pistons
4032–T6	0.9 12.2 1.1 0.9 Ni	Forgings	55	379	46	317	9	Forged aircraft pistons
2011–T3	5.5 (0.5 Bi) 0.5 Pb	½-in. rod	55	379	43	297	15	Free cutting, screw-machine products

[a] O, annealed; T, quenched and aged; H, cold rolled to hard temper.
[b] Yield strength taken at 0.2% permanent set.
[c] Cladding alloy: 1.0 Mg, 0.7 Si, 0.5 Mn.

operations. Good dimensional tolerances and fairly intricate shapes can be produced with relative ease.

The machinability of aluminum-base alloys varies greatly. Most cast alloys are machined easily. For the wrought alloys, with the exception of a few special types, special tools and techniques are desirable if large-scale machining is to be done. Free-machining alloys, such as 2011, have been developed for screw-machine work. They can be machined at very high speeds and have replaced brass screw-machine stock in many cases.

Aluminum casting alloys.   Although its low melting temperature tends to make it suitable for casting, pure aluminum is seldom cast. Its high shrinkage and susceptibility to hot cracking cause considerable difficulty and scrap is high. By adding small amounts of alloying elements, however, very suitable casting characteristics are obtained and strength is increased. Large amounts of aluminum alloys are cast, the principal alloying elements being copper, silicon, and zinc. Table 8-4 lists some common commercial aluminum casting alloys and employs the designation system of the Aluminum Association. The first digit indicates the alloy group as follows:

Major Alloying Element	
Aluminum, 99.00% and greater	1xx.x
Copper	2xx.x
Silicon with Cu and/or Mg	3xx.x
Silicon	4xx.x
Magnesium	5xx.x
Zinc	7xx.x
Tin	8xx.x
Other elements	9xx.x

The second and third digits identify the particular alloy or aluminum purity, and the last digit, separated by a decimal point, indicates the product form (that is, casting, ingot, etc.). A modification of the original alloy is indicated by a letter before the numerical designation.

Alloys are designed for both properties and process. Where strength requirements are low, as-cast properties are employed. High-strength castings usually require the use of an alloy that can be subsequently heat-treated. Sand casting has the fewest process restrictions. The aluminum alloys used for permanent-mold castings are designed to have lower coefficients of thermal expansion (or contraction) because the molds offer restraint to the dimensional changes that occur upon cooling. Die-casting alloys require high degrees of fluidity and "castability," because they are often cast into thin sections. Moreover, since die castings ordinarily are not heat-treated, the alloys used are designed to produce rather high "as-cast" strength under rapid cooling conditions. Several of the permanent-mold and die-casting alloys have tensile strengths above 275 MPa (40 ksi).

MAGNESIUM-BASE ALLOYS

The primary reason for the extensive, but specialized, use of *magnesium-base alloys* is their light weight—about 1.74 grams per cubic centimeter, compared with 2.7 for aluminum and 7.8 for iron or steel. Whereas aluminum alloys are most suitable for strength members of mechanically motivated structures, such as airplanes, trains, and trucks, magnesium alloys are best suited for those applications where lightness is the first consideration, and strength is a secondary requirement. Of course, the applications of aluminum and magnesium overlap to some degree.

The designation system for magnesium alloys is not as well standardized as in the case of steels or aluminums, but most producers follow a system using one or two prefix letters, two or three numerals, and a suffix letter. The prefix letters designate the two principal alloying metals according to the following format developed in ASTM specification B275.

A	aluminum	H	thorium	Q	silver
B	bismuth	K	zirconium	R	chromium
C	copper	L	beryllium	S	silicon
D	cadmium	M	manganese	T	tin
E	rare earth	N	nickel	Z	zinc
F	iron	P	lead		

Aluminum, zinc, zirconium, and thorium promote precipitation hardening; manganese improves corrosion resistance; and tin improves castability. Aluminum is the most common alloying element. The numerals correspond to the rounded-off percentages of the two main alloy elements and are arranged in the same order as the letters. The suffix letter distinguishes between different alloys with the same percentage of the principal alloying elements, proceding alphabetically as compositions become standard. Temper designation is much the same as in the case of aluminum, using -F, -O, -H1, -H2, -T4, -T5, and -T6. Some of the more common magnesium alloys are listed in Table 8-5 together with their properties and uses.

Most of the magnesium alloys are rather easily cast and are particularly well suited for die casting. Their machinability is by far the best of any metal. However, it is necessary to keep tools sharp and provide ample space for the chips.

Most alloys can be cold-worked, but deformation is not very good (as one might expect from the hcp crystal structure). If the temperature is raised to the region between 160 and 400°C (325 to 750°F), forming and drawing characteristics improve measureably. Because these temperatures are relatively low and easily attained, many formed and drawn magnesium parts are produced.

Magnesium alloys can be spot-welded nearly as easily as aluminum. Scratch brushing or chemical cleaning is necessary before spot welding. Fusion welding is carried out most easily by using an inert shielding atmosphere of argon or helium gas.

From a use viewpoint, magnesium alloys can be characterized by low wear, creep, fatigue, and corrosion resistance. A low elastic modulus makes it nec-

TABLE 8-4 Composition, Designations, and Properties of Aluminum Casting Alloys

Alloy Desig-nation[a]	Process[b]	Cu	Si	Mg	Zn	Fe	Other	Temper	Tensile Strength MPa	ksi[c]	Elon-gation in 2 inches (%)	Uses, Characteristics, etc.
208	S	4.0	3.0		1.0	1.2		F	131	19	1.5	General-purpose sand castings, can be heat treated
242	S & P	4.0		1.6		1.0	2.0 Ni	T61	276	40	—	Withstands elevated temperatures
295	S	4.5	1.0			1.0		T6	221	32	3.0	Structural castings, heat-treatable
296	P	4.5	2.5			1.2		T6	241	35	2.0	Permanent-mold version of 295
308	P	4.5	5.5		1.0	1.0		F	166	24	—	General-purpose permanent mold
319	S & P	3.5	6.0		1.0	1.0		T6	214	31	1.5	Superior casting characteristics
354	P	1.8	9.0					—	—	—	—	High-strength, aircraft
355	S & P	1.3	5.0					T6	221	32	2.0	High strength and pressure tightness
C355	S & P	1.3	5.0					T61	276	40	3.0	Stronger and more ductile than 355
356	S & P		7.0					T6	207	30	3.0	Excellent castability and impact strength
A356	S & P		7.0					T61	255	37	5.0	Stronger and more ductile than 356
357	S & P		7.0					T6	310	45	3.0	High strength-to-weight castings
359	S & P		9.0					—	—	—	—	High-strength aircraft usage
360	D		9.5			2.0		F	303	44[d]	2.5[d]	Good corrosion resistance and strength

Alloy	Process						Temper				Comments
A360	D		9.5		2.0		F	317	46[d]	3.5[d]	Similar to 360
380	D	3.5	8.5	3.0	2.0		F	317	46[d]	2.5[d]	High strength and hardness
A380	D	3.5	8.5	3.0	1.3		F	324	47[d]	3.5[d]	Similar to 380
383	D	1.5	10.5	3.0	1.3		F	310	45[d]	3.5[d]	High strength and hardness
384	D	3.75	11.3	1.0	1.3		F	331	48[d]	2.5	High strength and hardness
413	D	1.0	12.0		2.0		F	297	43[d]	2.5[d]	General purpose, good castability
A413	D	1.0	12.0		1.3		F	290	42[d]	3.5[d]	Similar to 413
443	D	5.25			2.0		F	228	33[d]	9.0[d]	General purpose, good castability
B443	S & P	5.25			2.0		F	117	17	3.0	General-purpose casting alloy
514	S			4.0			F	152	22	6.0	High corrosion resistance
518	D			8.0	1.8		F	310	45[d]	5.0[d]	Good corrosion resistance, strength, and toughness
520	S			10.0			T4	290	42	12.0	High strength with good ductility
535	S			6.9			F	241	35	9.0	Good corrosion resistance and machinability
712	S		5.8				F	234	34	4.0	Good properties without heat treatment
713	S & P		7.5		1.1		F	221	32	3.0	Similar to 712
771	S		7.0				T6	290	42	5.0	Aircraft and computer components
850	S & P	1.0				6.3 Sn 1.0 Ni	T5	110	16	5.0	Bearing alloy

[a]Aluminum Association.
[b]S, sand cast; P, permanent mold cast; D, die cast.
[c]Minimum figures unless noted.
[d]Typical values.

TABLE 8-5 Composition, Properties, and Uses of Common Magnesium Alloys

Alloy	Temper	Composition (%)							Tensile Strength[a]		Yield Strength[a]		Elongation in 2 inches (%)	Uses and Characteristics
		Al	Rare Earths	Mn	Th	Zn	Zr		MPa	ksi	MPa	ksi		
AM60A	F	6.0		0.13					207	30	117	17	6	Die castings
AM100A	T4	10.0		0.1					234	34	69	10	6	Sand and permanent-mold castings
AZ31B	F	3.0				1.0			221	32	103	15	6	Sheet, plate, extrusions, forgings
AZ61A	F	6.5				1.0			248	36	110	16	7	Sheet, plate, extrusions, forgings
AZ63A	T4	6.0				3.0			234	34	76	11	7	Sand and permanent mold castings
AZ80A	T5	8.5				0.5			234	34	152	22	2	High-strength forgings, extrusions
AZ81A	T4	7.6				0.7			234	34	76	11	7	Sand and permanent-mold casting

Alloy	Temper												Characteristics
AZ91A	F	9.0	0.7					234	34	159	23	3	Die castings
AZ92A	T4	9.0	2.0					234	34	76	11	6	High-strength sand and permanent mold castings
EZ33A	T5		2.6	0.7		3.2		138	20	97	14	2	Sand and permanent mold castings
HK31A	H24			0.7	3.2			228	33	166	24	4	Sheets and plates; castings in T6 temper
HM21A	T5				2.0		0.8	228	33	172	25	3	High-temperature (425°C) sheets, plates, forgings
HZ32A	T5		2.1	0.7	3.2			186	27	90	13	4	Sand and permanent-mold castings
ZH62A	T5		5.7	0.7	1.8			241	35	152	22	5	Sand and permanent-mold castings
ZK51A	T5		4.6	0.7				234	34	138	20	5	Sand and permanent-mold castings
ZK60A	T5		5.5	0.45				262	38	138	20	7	Extrusions, forgings

[a]Properties are minimums for the designated temper.

essary to use thick sections to provide adequate stiffness. Fortunately, the alloys are so light that it often is possible to use the thicker sections required for rigidity and still have a lighter structure than can be obtained with any other metal. Cost per unit volume is low, so the use of thick sections does not push the cost out of line. Moreover, since a large portion of magnesium components are cast, the thick sections actually become a desirable feature. Corrosion resistance is moderate unless exposed to salt water, salt air, or an unfavorable galvanic couple. Adequate corrosion resistance can usually be provided by enamel or lacquer finishes.

Another problem with magnesium alloys is the limited ductility. Here, designers should be aware of the brittle failures possible when components are loaded beyond the assumed conditions. Magnesium automobile wheels are a notable example.

Considerable misinformation has existed regarding the fire hazard that exists in processing magnesium alloys. It is true that magnesium alloys are highly combustible when they are in a finely divided form, such as powder or fine chips, and this hazard should not be ignored. Above 425°C (800°F) a noncombustible atmosphere is required to suppress burning. Castings require additional precautions due to the reactivity of magnesium with sand and water. In sheet, bar, extruded, or cast form, magnesium alloys present no fire hazard.

ZINC-BASE ALLOYS

Zinc-base alloys are of primary importance for their use in die castings. Zinc is low in cost, has a melting point of only 380°C (715°F), does not affect steel dies adversely, and can be made into alloys that have good strength properties and good dimensional stability. Two of the most widely used zinc alloys are characterized in Table 8-6. Alloy AG40A is used widely because of its excellent dimensional stability. Alloy AC41A offers higher strength and better corrosion resistance. A newer alloy, Alloy 7, provides better castability through a lower magnesium content. All alloys have good tensile strengths coupled with exceptional impact resistance. Their development has been responsible for the extensive use of zinc die castings.

Die-cast zinc has a strength greater than all other die-cast metals except the copper alloys. The alloys lend themselves to casting within close dimensional limits, permitting the thinnest sections yet produced, and are machinable with a minimum of cost. Resistance to surface corrosion is adequate for a wide range of applications. Prolonged contact with moisture results in the formation of white corrosion products, but surface treatments can be applied to prevent this corrosion.

The attractiveness of the zinc alloys has recently been enhanced by the development of several zinc–aluminum casting alloys with rather high amounts of aluminum (ZA-8, ZA12, and ZA27). Initially developed for sand, permanent mold, and graphite mold casting, these alloys can also be die-cast to achieve higher performance characteristics than obtained with the conventional alloys. Strength, hardness, and wear resistance are all improved and several of these

TABLE 8-6 Characteristics of Two Zinc Die-Casting Alloys

	ASTM AG40A (SAE 903) (Zamak 3)	ASTM AC41A (SAE 925) (Zamak 5)
Composition (%)		
Copper	0.25	0.75–1.25
Aluminum	3.5–4.3	3.5–4.3
Magnesium	0.02–0.05	0.03–0.08
Iron, maximum	0.1	0.1
Lead, maximum	0.005	0.005
Cadmium, maximum	0.004	0.004
Tin, maximum	0.003	0.003
Zinc	Remainder	Remainder
Properties		
Tensile strength, as cast [MPa (ksi)]	283 (41)	328 (47.6)
Tensile strength, 10 years of aging	241 (35)	271 (39.3)
Elongation in 2 inches, as cast (%)	10	7
Charpy impact strength, as cast (ft-lb)	43	48
Melting point (°F)	717	717

alloys have demonstrated excellent bearing properties. Considering also the lower melting and casting costs, these alloys are attractive alternatives to the conventional aluminum, brass, and bronze casting alloys as well as cast iron.

NICKEL-BASE ALLOYS

Nickel-base alloys are noted for their outstanding strength and corrosion resistance, particularly at high temperatures. *Monel* metal, containing about 67% nickel and 30% copper, has been used for years in the chemical and food-processing industries because of its outstanding corrosion resistance. It probably has better overall corrosion resistance to more media than any other alloy. It is particularly resistant to saltwater corrosion, sulfuric acid, and it even resists high-velocity, high-temperature steam. For the latter reason, Monel has been used for steam turbine blades. It can be polished to have an excellent appearance, similar to stainless steel, and is often used for ornamental trim and household ware. In its common form, Monel has a tensile strength of from 480 to 1170 MPa (70 to 170 ksi), depending on the amount of cold working. The elongation in 2 inches varies from 50 to 2%.

There are three special grades of Monel that contain small amounts of added alloying elements. K Monel contains about 3% aluminum and can be precipitation-hardened to a tensile strength of 1100 to 1240 MPa (160 to 180 ksi). H

Monel has 3% silicon added and S Monel has 4% silicon. They are used for castings and can be precipitation-hardened. To improve upon the machining charactristics of Monel, a special free-machining alloy known as R Monel is produced with about 0.35% sulfur.

Another use for nickel-base alloys is as electrical resistors. Being primarily nickel-chromium alloys, they are often called *Nichromes*. One alloy contains 80% nickel and 20% chromium. Another has 60% nickel, 16% chromium, and 24% iron. They have excellent resistance to oxidation while retaining their strength at red heats.

The nickel-base alloys that have been developed for extreme-high-temperature service will be discussed in the next section.

Most of the nickel alloys are somewhat difficult to cast, but they can be forged and hot-worked. The heating, however, usually must be done in controlled atmospheres to avoid intercrystalline embrittlement. Welding can be performed with little difficulty.

NONFERROUS ALLOYS FOR HIGH-TEMPERATURE SERVICE

As discussed in Chapter 4, titanium and titanium alloys are often used at moderately elevated temperatures. In addition to being strong, lightweight, and corrosion resistant, these alloys are often stronger than steel at temperatures up to 540°C (1000°F). Increased use is currently focused on many aircraft applications.

Rapid developments in the jet engine, gas turbine, rocket, and nuclear fields have stimulated, and have been made possible by, the development of a number of nonferrous alloys that have high strength, creep resistance, and corrosion resistance at temperatures up to and in excess of 1100°C (2000°F). Several of the more common of these *superalloys* are listed in Table 8-7. It will be noted that nickel, iron and nickel, or cobalt form the base metal in these alloys. Most are precipitation-hardenable, and yield strengths above 690 MPa (100 ksi) are readily attained. The nickel-base alloys tend to have higher strengths at room temperature with yield strengths up to 1200 MPa (175 ksi) and ultimate strengths up to 1450 MPa (210 ksi), as compared with 790 MPa (115 ksi) and 1170 MPa (170 ksi), respectively, for the cobalt-base alloys. The 1000-hour rupture strengths of the nickel-base alloys at 815°C (1500°F) are also higher than those of the cobalt-base alloys, up to 450 MPa (65 ksi) versus 228 (33 ksi). Materials such as TD-nickel (a nickel alloy containing 2% dispersed thoria) and columbium give promise of operating at service temperatures above 1100°C (2000°F).

Many of the superalloys are very difficult to machine, so methods such as electrodischarge, electrochemical, or ultrasonic methods are often utilized or they are produced in the form of investment castings. Powder metallurgy techniques are also being used extensively in the manufacture of superalloy components. Because of their ingredients, all these alloys are expensive, and this limits their use to small or critical parts or applications where cost is not the determining factor.

TABLE 8-7 Some Nonferrous Alloys for High-Temperature Service

Alloy	C	Mn	Si	Cr	Ni	Co	Mo	W	Cb	Ti	Al	B	Zr	Fe	Other
							Composition (%)								
Nickel base															
Hastelloy X	0.1	1.0	1.0	21.8	Balance	2.5	9.0	0.6	—	—	—	—	—	18.5	—
IN-100	0.18	—	—	10.0	Balance	15.0	3.0	—	—	4.7	5.5	0.014	0.06	—	1.0 V
Inconel 601	0.05	0.5	0.25	23.0	Balance	—	—	—	—	—	1.4	—	—	14.1	0.2 Cu
Inconel 718	0.04	0.2	0.2	19.0	Balance	—	3.0	—	5.0	0.9	0.5	—	—	18.5	0.2 Cu
M-252	0.15	0.5	0.5	19.0	Balance	10.0	10.0	—	—	2.6	1.0	0.005	—	—	—
Rene 41	0.09	—	—	19.0	Balance	11.0	10.0	—	—	3.1	1.5	0.01	—	—	—
Rene 80	0.17	—	—	14.0	Balance	9.5	4.0	4.0	—	5.0	3.0	0.015	0.03	—	—
Rene 95	0.15	—	—	14.0	Balance	8.0	3.5	3.5	3.5	2.5	3.5	0.01	0.05	—	—
Udimet 500	0.08	—	—	19.0	Balance	18.0	4	—	—	3.0	3.0	0.005	—	0.5	—
Udimet 700	0.07	—	—	15.0	Balance	18.5	5.0	—	—	3.5	4.4	0.025	—	0.5	—
Waspaloy B	0.07	0.75	0.75	19.5	Balance	13.5	4.3	—	—	3.0	1.4	0.006	0.07	2.0	0.1 Cu
Iron–nickel base															
Illium P	0.20	—	—	28.0	8.0	—	2.0	—	—	—	—	—	—	Balance	3.0 Cu
Incoloy 825	0.03	0.5	0.2	21.5	42.0	—	3.0	—	—	0.9	0.1	—	—	30	2.2 Cu
Incoloy 901	0.05	0.4	0.4	13.5	42.7	—	6.2	—	—	2.5	0.2	—	—	34	—
16-25-6	0.08	1.35	0.7	16.0	25.0	—	6.0	—	—	—	—	—	—	Balance	0.15 N
Cobalt base															
Haynes 150	0.08	0.65	0.75	28.0	—	Balance	—	—	—	—	—	—	—	20.0	—
MAR-M322	1.00	0.10	0.1	21.5	—	Balance	—	9.0	—	0.75	—	—	2.25	—	4.5 Ta
S-816	0.38	1.20	0.4	20.0	20.0	Balance	4.0	4.0	4.0	—	—	—	—	4.0	—
WI-52	0.45	0.5	0.5	21.0	1.0	Balance	—	11.0	2.0	—	—	—	—	2.0	—

FIGURE 8-1 Photomicrograph of lead babbitt metal; 75×. Square or triangular white masses are SbSn; small white particles are CuSn.

LEAD–TIN ALLOYS

Lead and tin are nearly always used together in alloys of engineering importance, the two major uses being as bearing materials or as solders. One of the oldest and best bearing metals, composed of about 84% tin, 8% copper, and 8% antimony, is called genuine or tin *babbitt*. Because of the high cost of tin, a more widely used babbitt metal is the lead babbitt composed of 85% lead, 5% tin, 10% antimony, and 0.5% copper. For high speeds and fairly heavy loads, the lead-base babbitts prove unsatisfactory; for slow speeds and moderate loads, they are quite adequate.

Figure 8-1 shows a photomicrograph of a lead babbitt metal. The antimony combines with the tin to form hard particles in the softer lead matrix, a structure typical of many bearing metals. The shaft rides on the harder particles with little friction while the softer matrix acts as a cushion that can distort sufficiently to take care of misalignment and assure a proper fit between the two surfaces.

Soft *solders* are basically lead–tin alloys near the eutectic composition (61.9% tin). A variety of compositions exist, each with a characteristic melting range. Because of the higher price of tin, however, a 50–50 composition or alloys of the lead-rich variety are often used.

REVIEW QUESTIONS

1. What are some of the important properties that may be possessed by nonferrous metals?
2. In what ways are nonferrous alloys often inferior to steel?
3. What are some of the attractive features of nonferrous metals that relate to fabrication ease?
4. What is the most popular alloying addition to copper?
5. What is dezincification? Season cracking?
6. Why would a copper–nickel alloy be a good choice for heat-transfer applications?
7. What are some of the characteristic properties of copper–beryllium alloys?
8. What are some of the attractive engineering properties of aluminum?
9. In what ways might aluminums be mechanically inferior to steel?
10. What is the mechanism by which aluminum alloys possess excellent corrosion resistance?
11. Why might it be difficult to weld aluminum?
12. What are the two basic types of wrought aluminum alloys?
13. What is the temper designation for aluminum alloys? What does it tell us?
14. What is the major asset of precipitation-hardenable aluminum alloys?
15. Why is pure aluminum seldom cast?

16. For what types of applications would the magnesium alloys be attractive?
17. What fabrication processes are particularly compatible with magnesium alloys?
18. What are some of the mechanical limitations of magnesium alloys?
19. What is the principal use of the zinc-base alloys?
20. What are the distinctive engineering characteristics of the nickel-base alloys?
21. What are some of the primary engineering characteristics of titanium alloys?
22. What are the unusual engineering properties of the superalloys?
23. What are the two major uses of lead–tin alloys?

CASE STUDY 8. The Substitute Aluminum Connecting Rods

Winning Racing, Inc., is a manufacturer of high-performance automotive components, specially designed for racing applications. One highly successful product line is a series of specially designed connecting rods, made of forged alloy steel.

Noting the successful use of aluminum alloys in certain racing applications, Team Rabbit has requested Winning Racing to produce a special set of lightweight aluminum connecting rods, using their highly successful existing design. The rods for three engines were made on a special run, using the existing dies, and were put into the engines for testing. During dynamometer testing, however, the engines failed in under 30 minutes, and the failures were attributed to the connecting rods, although none of the rods broke. What had been overlooked that caused the trouble?

CHAPTER 9

Nonmetallic Materials: Plastics, Elastomers, Ceramics, and Composites

A number of nonmetallic materials have substantial importance in manufacturing. Consequently, it is imperative for the design engineer to have an understanding of their natures, properties, advantages, and limitations so he may know when and how they may be used advantageously in his designs. Except in furniture manufacturing, where wood is of prime importance, these materials are plastics, elastomers, ceramics, and composites. Most of these are man-made, permitting a wide range of properties to be obtained, and entirely new materials, and variations of them, are being created almost continuously. As a result, it is difficult for one to keep abreast of all the individual materials that are available at a given time, and no attempt will be made in this chapter to give detailed information about all of them. Instead, the emphasis will be on the basic nature, properties, and processing of these materials, so that the reader may have a good idea as to whether they are potential materials for use in products. For more detailed information about specific materials of these types, texts, handbooks, and compilations that deal exclusively with these materials should be consulted.

PLASTICS

It is difficult to give a precise definition of the term *plastics*. Basically, it covers a group of materials characterized by large molecules that are built up by joining

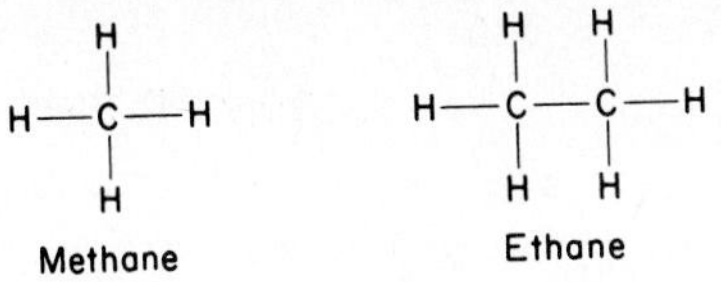

FIGURE 9-1 Linking of hydrogen and carbon in methane and ethane molecules.

small molecules, usually artifically. Practically, it is sufficient to say that they are natural or synthetic resins, or their compounds, that can be molded, extruded, cast, or used as films or coatings. Most of them are organic substances, usually containing hydrogen, oxygen, carbon, and nitrogen.

The molecular structure of plastics. It is helpful to have an understanding of the basic molecular structure of plastics. Most are based on hydrocarbons, in which carbon and hydrogen combine in the relationship C_nH_{2n+2}, known as *paraffins*. Theoretically, these hydrocarbons can be linked together indefinitely to form very large molecules, as illustrated in Figure 9-1. The bonds between the atoms are single pairs of covalent electrons. Because there is no provision for additional atoms to be added to the chain, such molecules are said to be *saturated*. The molecules have strong intramolecular bonds, but the intermolecular attractions are much weaker.

Carbon and hydrogen atoms also can form molecules in which the carbons are held together by double or triple covalent bonds. Ethylene and acetylene are examples, illustrated in Figure 9-2. Because such molecules do not have the maximum possible number of hydrogen atoms, they are said to be *unsaturated* and are important in *polymerization*, where small molecules join into large ones having the same constituents.

In organic compounds, four electron pairs surround each carbon atom, and one electron pair is shared jointly by each hydrogen atom. However, other kinds of atoms can be substituted, such as chlorine or a benzene ring in place of hydrogen and oxygen, sulfur, or nitrogen in place of carbon. Consequently a wide range of organic materials can be created.

Isomers. The same kind and number of atoms can unite in different structural arrangements, thus forming different compounds with completely different properties. Figure 9-3 shows such an example, the compounds being known as *isomers*. These are analogous to allotropism or polymorphism in the case of crystalline materials, such as iron which can exist in both body-centered-cubic and face-centered-cubic structures. A number of plastics are isomers.

Forming molecules by polymerization. Polymerization of large molecules in plastics takes place by either *addition* or *condensation*. Figure 9-4 illustrates polymerization by addition, where a number of basic units (*monomers*) are added

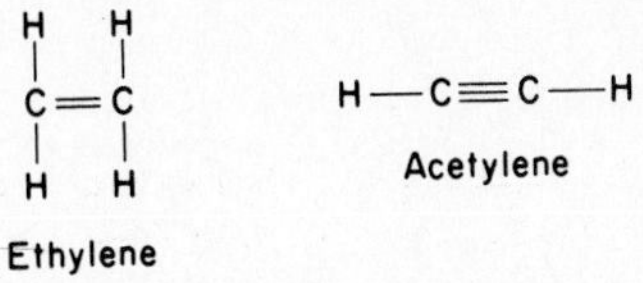

FIGURE 9-2 Covalent bonds in ethylene and acetylene molecules.

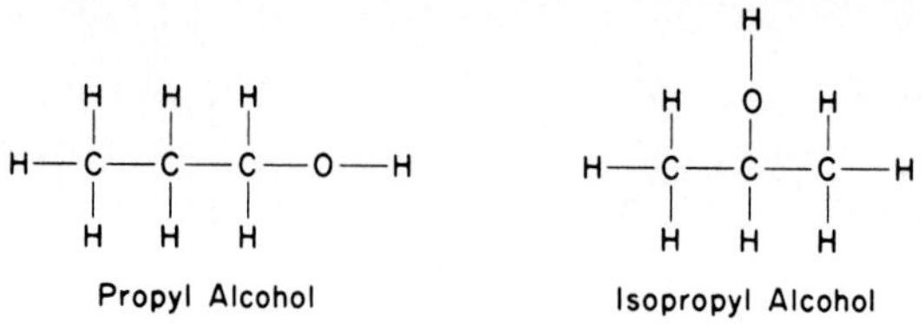

H—C—C—C—O—H (Propyl Alcohol with H atoms)

Propyl Alcohol

H—C—C—C—H (Isopropyl Alcohol)

Isopropyl Alcohol

FIGURE 9-3 Linking of eight hydrogen, three carbon, and one oxygen atoms to form two isomers, propyl and isopropyl alcohol.

together to form a large molecule (*polymer*) in which there is a repeated unit (*mer*). Activators, like $H_2O_2 \rightarrow 2OH$, initiate and terminate the chain. The amount of activator determines the average molecular weight or length of chain. The average number of mers in the polymer material is known as the *degree of polymerization*, and ranges from 75 to 750 for most commercial plastics. *Copolymers* are a special category of polymer where two or more types of mers combine by addition in the chain. This process, illustrated in Figure 9-5, greatly expands the possibilities of creating new types of plastics with improved physical or mechanical properties.

In contrast to polymerization by addition, where all of the component atoms appear in the product, condensation polymerization results in the production of a small by-product molecule, often water, as illustrated in Figure 9-6. The resultant condensation structure is often a three-dimensional framework with all atoms being linked by strong, primary bonds.

Thermosetting and thermoplastic materials. The terms "thermosetting" and "thermoplastic" refer to the material's response to elevated temperature as determined by its material structure. Addition polymers can be viewed as a long chain of tightly bonded carbon atoms with strongly attached pendants such as hydrogen, fluorine, chlorine, or benzene rings. All bonds within the molecules are strong primary bonds. The attraction between neighboring molecules, however, is only by the much weaker van der Waals forces. For these materials, the mechanical and physical properties are largely determined by the intermolecular forces. Because these "secondary bonds" are weakened by elevated temperature, plastics of this type soften with increasing temperature and become harder and stronger when cooled. The softening and hardening of these *thermoplastic* materials can be repeated as often as desired and no chemical change is involved. Because they contain molecules of different sizes, thermoplastic materials do not have a definite melting temperature, but instead, soften over a temperature range.

Since the bonding forces between molecules are much weaker than those within the molecule, deformation occurs by slippage between adjacent molecular chains. Methods to increase the strength of thermoplastics, therefore, focus on

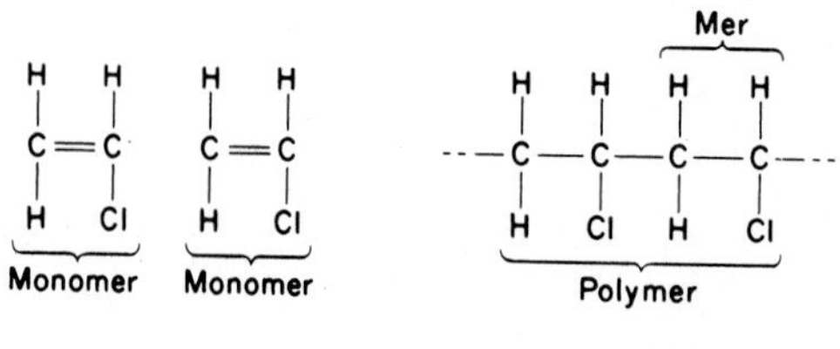

Monomer Monomer   Polymer

FIGURE 9-4 Polymerization by addition: the uniting of monomers.

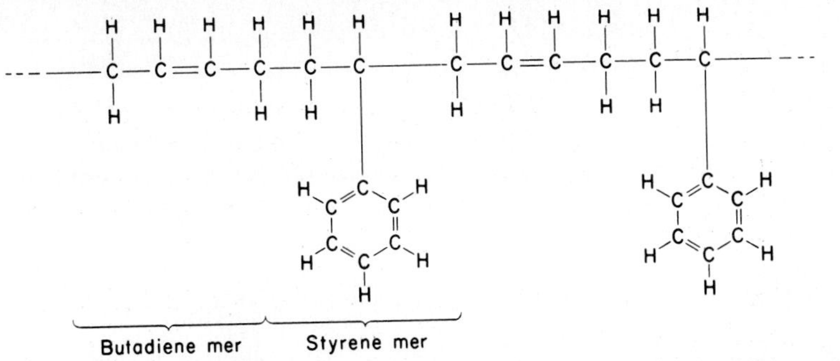

FIGURE 9-5 Polymerization by the addition of two kinds of mers: copolymerization.

restricting intermolecular slippage. Longer chains have less freedom of movement and are therefore stronger. Polymers with large side groupings, such as chlorine or, better yet, benzene groups, instead of hydrogens, will be stronger. Branched polymers, in which the chains divide in a Y with all primary bonds within the chains, offer considerable strength. Connecting adjacent chains with primary bond cross-links, as in the sulfur links when vulcanizing rubber, can significantly impede deformation. Finally, since the secondary bond strength is inversely related to the separation distance of the molecules, methods such as deformation or "crystallization" that produce a parallel alignment of adjacent molecules should increase strength.

Thermosetting plastics, on the other hand, are those with a three-dimensional framework structure in which all atoms are connected by strong, covalent bonds. These materials generally result from condensation polymerization where elevated temperature tends to promote the reaction, hence, the term "thermosetting." Once set, however, additional heatings do not produce softening and they maintain their mechanical properties up to the temperature at which they char or burn. Deformation requires the breaking of primary bonds, so these plastics tend to be strong, but brittle. As a class, thermosetting plastics are significantly stronger than thermoplastics.

Whether a plastic is thermosetting or thermoplastic is of great importance to the person who is selecting it for use, because considerable indication is given not only to its behavior in service but also as to how it must be processed. Thermoplastics are easily molded. However, after the material is formed to shape in a mold at an elevated temperature (and ordinarily under considerable

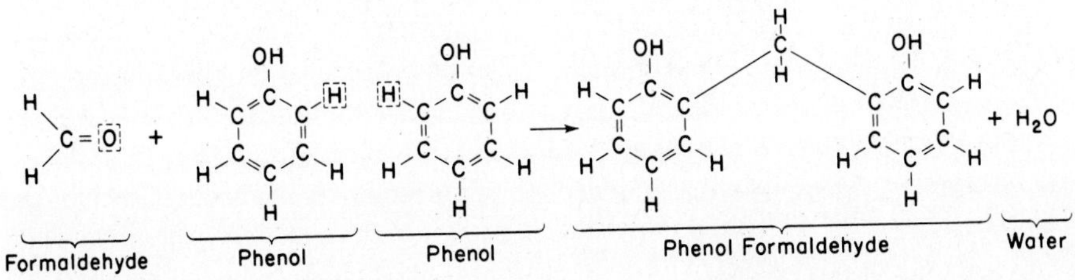

FIGURE 9-6 Formation of phenol–formaldehyde (Bakelite) by condensation polymerization.

pressure), the mold must be cooled to cause the plastic to harden and thus retain its shape when removed. In producing products from thermosetting plastics, the mold remains at an elevated temperature throughout the molding cycle, and the material hardens as a result of the temperature and the pressure. It can then be removed without cooling the mold. The molding procedure is controlled by the type of the plastic.

Types of plastics and their properties. Because there are so many plastics with new ones becoming available almost continuously, it is helpful to have a knowledge of the general properties which they possess and the properties of the several basic types.

1. *Light weight.* Most plastics have specific gravities between 1.1 and 1.6, compared with about 1.75 for magnesium. Thus they are the lightest of the engineering materials.
2. *Corrosion resistance.* Many plastics perform well in hostile, corrosive environments.
3. *Electrical resistance.* They are widely used as insulating materials.
4. *Low thermal conductivity.* They are relatively good heat insulators.
5. *Wide range of colors, transparent, or opaque.* Many plastics have an almost unlimited color range and the color goes throughout, not just on the surface.
6. *Surface finish.* Excellent surface finishes can be obtained by processes used to convert the raw material to the final shape. No added operations are required.
7. *Formability.* Objects can frequently be produced from plastics in only one operation—raw material to final shape—by processes such as casting, extrusion, and molding.
8. *Comparatively low cost.*

The general properties discussed thus far are desirable ones. The inferior properties of plastics have to do with their strength. None of them has strength properties that approach those of the engineering metals. The impact strength of most of them is low, but several—ABS, high-density polyethylene, polycarbonate, and cellulose propionate—have very good impact strengths. Because of their low weight, however, their strength-to-weight ratio is fair. As a class they are not suitable for applications that require high strength unless special strengthening filler materials are added. The dimensional stability of most of them is greatly inferior to metals, and radiation, both ultraviolet and particulate, can markedly alter properties.

Table 9-1 lists properties of a number of common plastics. From the previous discussion of properties and consideration of this table, it is apparent that plastics are best suited for applications that require materials of only low or moderate strength, low electrical and/or thermal conductivity, obtainable in a wide range of colors, and easily transformed from the raw to the finished state. In no other material but plastics can this combination of properties be obtained. Thus a large percentage of all plastics are used as "packaging" or container materials. This classification includes such items as radio cabinets, clock cases, and household appliance housings, which, primarily, serve as containers for the interior mechanisms. Applications, such as insulators in electrical equipment and handles for hot articles, captialize on the low electrical and thermal conductivities. Soft pliable foamed plastics are used extensively as cushioning material. Rigid foams are used inside sheet metal structures, such as airplane and rocket stabilizers, to provide compressive strength. Plastics find application as adhesive or bonding agents when assembling a multitude of objects, as discussed in Chapter 33, and may also serve as inexpensive pressworking tooling where pressures and wear demands are low.

There are many cases where only one or two of the properties of plastics are sufficient to dictate their use and sometimes these properties must be combined with other characteristics not normally found in plastics. Increasingly, fabric- or fiber-reinforced plastics are being used in applications where considerable tensile strength is required, the strength coming primarily from the reinforcement and not the plastic. To a large extent, the use of plastics is due to the fact that they provide a *combination* of several desirable properties not found in other engineering materials. When assessing the suitability of plastics for a particular application and then selecting a specific type, one should seek the best combination of properties.

The following are some comments about several types of plastics listed in Table 9-1 which may be helpful in selecting them for use:

Phenolics: oldest of the plastics, but still widely used; hard, relatively strong, low cost, and easily molded; opaque, but wide color range; wide variety of forms—sheets, rods, tubes, and laminates.

Urea formaldehyde: similar properties to phenolics, but available in lighter colors; useful for containers and housings, but not outdoors; used in lighting fixtures because of translucence in thin sections.

Melamines: excellent resistance to heat, water, and many chemicals; full range of translucent or opaque colors; excellent electrical-arc resistance; tableware, but stained by coffee; used extensively in treating paper and cloth to impart water-repellent properties.

Epoxides: good toughness, elasticity, chemical resistance, and dimensional stability; used as coatings, cements, and ''potting'' materials for electrical components; easily compounded to cure at room temperatures; widely used in tooling applications.

Silicones: semiorganic (spine molecules alternating silicon and oxygen atoms); heat resistant; low moisture absorption; high dielectric properties.

ABS: contain acrylonitrile, butadiene, and styrene; low weight, good strength, and very tough; good under severe service conditions.

Acrylics: highest optical clarity, transmitting over 90% of light; common trade names are *Lucite* and *Plexiglas;* high impact, flexural, tensile, and dielectric strengths; wide range of colors; stretch rather easily.

Cellulose acetate: wide range of colors; good insulating qualities; easily molded; high moisture absorption in most grades and affected by alcohols and alkalies.

Cellulose acetate butyrate: higher impact strength and moisture resistance than cellulose acetate; will withstand rougher usage.

Ethyl cellulose: high electrical resistance and impact strength; retains toughness at low temperatures.

TABLE 9-1 Properties and Major Characteristics of Common Types of Plastics

Material	Specific Gravity	Tensile-Strength (1000 lb/in²)	Impact Strength Izod (ft-lb/in. of notch)	Top Working Temperature [°C (°F)]	Dielectric Strength[b] (volts/mil)	24-hour Water Absorption (%)	Weatherability	Colorability	Optical Clarity	Chemical Resistance	Injection Molding	Extrusions	Formable Sheet	Film	Fiber	Compression or Transfer Moldings	Castings	Reinforced Plastics Moldings	Industrial Thermosetting Laminates	Foam
Thermoplastics																				
ABS material	1.02–1.06	4–8	1.3–10.0		300–400	0.2–0.3	0	×	0	0	✓	✓	✓							
Acetal	1.4	10	1.5	121 (250)	1200	0.22		×		0	✓	✓								
Acrylics	1.12–1.19	5.5–10	0.2–2.3	93 (200)	400–530	0.2–0.4	×		×	0	✓	✓	✓	✓			✓			
Cellulose acetate	1.25–1.50	3–8	0.75–4.0	127 (260)	300–600	2.0–6.0		×	×		✓	✓	✓	✓						
Cellulose acetate butyrate	1.18–1.24	2–6	0.6–3.2	54 (130)	250–350	1.8–2.1	×	×	×		✓	✓	✓	✓						
Cellulose propionate	1.19–1.24	1–5	0.8–9	60 (140)	300	1.8–2.1	×	×	×		✓	✓	✓	✓						
Chlorinated polyether	1.4	6	3.3	149 (300)	400	0.01				×	✓	✓								
Ethyl cellulose	1.16	3–6	1.8–4.0	66 (150)	350	1.6–2.2	×				✓	✓								

TFE-fluorocarbon	2.1–2.3	1.5–3	2.5–4.0	260 (500)	450	0
CFE-fluorocarbon	2.1–2.15	4.5–6	3.5–3.6	199 (390)	550	0
Nylon	1.1–1.2	8–10	2	121 (250)	385–470	0.4–5.5
Polycarbonate	1.2	9.5	14	121 (250)	400	0.15
Polyethylene	0.96	4	10	93 (200)	440	0.003
Polypropylene	0.9–1.27	3.4–5.3	1.02	110 (230)	520–800	0.03
Polystyrene	1.05–1.15	5–9	0.3–0.6	88 (190)	400–600	<0.2
Modified polystyrene	1.0–1.1	2.5–6	0.25–11.0	100 (212)	300–600	0.03–0.2
Vinyl	1.16–1.55	1–5.9	0.25–2.0	104 (220)	25–500	0.2–1
Thermosetting plastics						
Epoxy	1.1–1.7	4–13	0.4–1.5	163 (325)	500	0.1–0.5
Melamine	1.76–1.98	5–8		177 (350)	460	0.1
Phenolic	1.2–1.45	5–9	0.25–5	149 (300)	100–500	0.2–0.6
Polyester (other than molding compounds)	1.06–1.46	4–10	0.18–0.4	149 (300)	340–570	0.5
Polyester (alkyd, DAP)	1.6–1.75	3.2–8	3.6–8			0.16–0.67
Silicone	2.0	3–5	0.2–3.0	288 (550)	250–350	0.4–0.5
Urea	1.41–1.80	4–8.5	0.2–0.5	85 (185)	300–600	1–3

[a] × denotes a principal reason for its use; 0 indicates a secondary reason.
[b] Short-time ASTM Test.

Fluorocarbons: inert to most chemicals; high temperature resistance; very low coefficients of friction (Teflon), used for nonlubricated bearings and nonstick coatings for cooking utensils and electric irons.

Nylon: good abrasion resistance and toughness; excellent dimensional stability; used as bearings with little lubrication; available as monofilaments for textiles, fishing lines, ropes, and so on; expensive (specialized applications only).

Polycarbonates: high strength and outstanding toughness.

Polyethylenes: tough; high electrical resistance; used for bottle caps, unbreakable kitchenware, and electrical wire insulation.

Polystyrenes: high dimensional stability and low water absorption; best all-around dielectric; burns readily and is adversely affected by citrus juices and cleaning fluids.

Vinyls: wide range of types, from thin, rubbery films to rigid forms; tear resistant; good aging properties; good dimensional stability and water resistance in rigid forms; used for floor and wall coverings, upholstery fabrics, and lightweight water hose.

Oriented plastics. Because the intermolecular bonds in thermoplastics are much weaker than the internal atomic bonds, these plastics can be processed to provide high strength in a given direction by aligning the molecules parallel to the applied load. This *orienting* process is accomplished by either stretching or extrusion, as is illustrated in Figure 9-7. The material usually is heated somewhat during the orienting process to aid in overcoming the internal forces and is cooled immediately afterward to "freeze" the molecules in the desired orientation. Uniaxial or biaxial orientations may be imparted.

Orienting may increase the tensile strength by more than 50% but a 25% increase is more typical. The elongation may be increased several hundred percent. One shortcoming is that if oriented plastics are reheated, they tend to return to their original shape, owing to the phenomenon of *elastic memory.*

Additive agents in plastics. For most uses, other materials are added to plastics to (1) improve their properties, (2) reduce the cost, (3) improve their moldability, and/or (4) impart color. Such additive constituents usually are classified as *fillers, plasticizers, lubricants,* or *coloring agents.*

Ordinarily fillers comprise a large percentage of the total volume of a molded plastic product, being added to improve the strength or to decrease cost. To a large degree they determine the general properties of a molded plastic. They also may aid in controlling shrinkage and in improving moldability, although they more commonly reduce the latter property. Whenever possible, fillers are

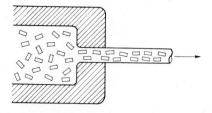

FIGURE 9-7 Schematic representation of the aligning of plastic molecules in the orienting process.

used that are much less expensive than the plastic resin. The most common fillers, and the properties they impart, are:

1. *Wood flour:* the general-purpose filler; low cost with fair strength; good moldability.
2. *Cloth fibers:* improved impact strength; fair moldability.
3. *Macerated cloth:* high impact strength; limited moldability.
4. *Glass fibers:* high strength; dimensional stability; translucence.
5. *Asbestos fiber:* heat resistance; dimensional stability.
6. *Mica:* excellent electrical properties and low moisture absorption.

Other fillers are being used increasingly, particularly for imparting high strength, often at elevated temperatures. "Whiskers" of various metals and nonmetals, such as boron, stainless steel, columbium, tantalum, titanium, zirconium, and silicon carbide, are used. These are from 1 to 5 μm (39 to 197 microinches) in diameter, 30 to 1000 μm (0.0012 to 0.039 inch) in length and have high moduli of elasticity and tensile strengths up to 21,000 MPa (3,000,000 psi). More common is the use of filaments of glass, graphite, or boron, which usually are less than 0.1 mm in diameter but of any desired length. These can provide tensile strengths up to 2450 MPa (350,000 psi) with moduli of elasticity up to 420 000 MPa (60 million psi). Glass-fiber cloth is a very commonly used material (see discussion of composite materials, this chapter, for further details).

When fillers are used, the resin acts as the binding material surrounding the filler particles and holding the mass together. Thus the surface of a molded plastic part is almost pure resin with no filler exposed.

Coloring agents may either be dyes, which actually alter the color of the resin, or colored pigments that through their presence impart a desired color. Most of the fillers do not in themselves produce attractive colors, so a dye is usually needed.

Plasticizers are added in small amounts to increase and control the flow of the plastic during molding. The amount needed for a given resin is governed by the intricacy of the mold. As a rule, the amount of plasticizer is held to a minimum because it is likely to affect the stability of the finished product through gradual loss during aging.

Lubricants are added in small amounts to improve the moldability and to facilitate removal of parts from the molds. Wax, stearates, and occasionally soaps are used for this purpose. They also are held to a minimum because they affect the properties adversely.

Plastics as adhesives. The use of plastics as adhesives is highly developed, and their use for this purpose has expanded tremendously in recent years. This application of plastics is discussed in Chapter 33.

Ablative coatings. Plastics form the base of many ablative coating materials used on rockets and missile motors to provide short-time protection from intense heating conditions, such as the heating experienced by space vehicles in re-entering the earth's atmosphere. Ablation involves the thermal decomposition of high polymers into low-molecular-weight gaseous products and porous carbon char. Some of the heat absorption also may occur by sublimation and chemical decomposition.

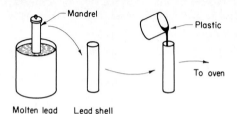

FIGURE 9-8 Steps in the casting of plastic parts.

Molten lead   Lead shell

Production processes for plastic products. Not only does the designer have a large number of plastics available from which he can select, there are also a number of different processes by which a chosen plastic can be converted into a desired product. *Casting, blow molding, hot-compression molding, injection molding, transfer molding, extrusion, laminating,* and *cold molding* are all used extensively. Each has certain advantages and limitations that bear on part design, material selection, and final cost. Not every plastic is suitable for each process. Because it usually is desirable to convert the material into the finished product in a single-process operation, it is important to have an understanding of the various processes so that the material-process selection will be optimal.

Casting is the simplest of the processes because no fillers are used and no pressure is involved. Of course, a mold is required. For certain simple shapes, a model of the product can be made, usually of steel, which is then dipped into molten lead until a thin sheath of lead is formed over the model. The model or mandrel, then is pulled out of the lead sheath, leaving a thin, lead mold into which the liquid plastic is poured. The resin then is cured, usually in an oven at low temperatures (65.6 to 93°C; 150 to 200°F), as indicated in Figure 9-8. After removal from the oven, the lead shell is stripped from the finished product. Some plastics can be cured at room temperatures.

Because cast plastics contain no fillers, they have a distinctive lustrous appearance. The process is inexpensive because no expensive dies or equipment are involved. However, it is limited to small objects of rather simple shape. Small radio cabinets, jewelry, and ornamental objects are commonly made by casting.

Blowmolding is used extensively for making hollow products, such as bottles and other containers. The steps in this process are illustrated in Figure 9-9.

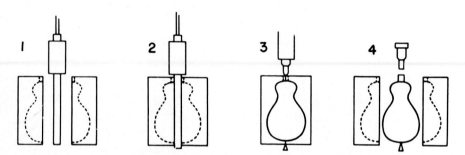

FIGURE 9-9 Steps in "blow molding" plastic parts: (1) tube of heated plastic is placed in open mold, (2) mold closes over extruded tube, (3) air forces tube against mold sides, and (4) mold opens to release product.

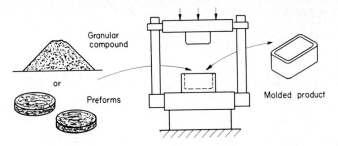

FIGURE 9-10 Schematic representation of the production of plastic parts by the hot-compression molding process.

In *hot-compression molding,* indicated schematically in Figure 9-10, granules or preformed tablets of the raw, mixed plastic material are loaded into the cavity of an open, heated mold. The plunger (male member) of the mold, usually attached to the upper portion of the press, descends, closing the mold and creating sufficient pressure to force the plastic, as it becomes fluid, into all portions of the cavity. After the material has set, or cured, the mold is opened and the part is removed. Usually, a number of cavities are contained within a single mold. The process is simple, but its use is restricted almost exclusively to thermosetting materials. Alternate heating and cooling of the mold is required for thermoplastic materials and this is not economical.

In order to contain the material within the die cavity and enable pressure to be built up, some type of seal is required on hot-compression molding dies. Three types of seals are employed, as illustrated in Figure 9-11. In the *flash type,* excess material is provided which, during the final stage of mold closing, is squeezed out of the cavity. The resulting *flash* must be removed from the finished product, usually requiring an additional operation. This type of mold is relatively inexpensive to make, and it is not necessary to control the amount of raw material closely. However, the dimensions of the product across the mold-opening plane and the density will vary somewhat.

In the plunger or positive type, no material can escape from the mold. Thus to obtain close dimensional control, the raw material must be measured accurately. By controlling the pressure on the plunger, good density control is obtained. The *landed-plunger type* mold is most commonly used, providing good pressure and a definite cutoff to assure accurate dimensions. Figure 9-12 shows a hot compression mold in operation.

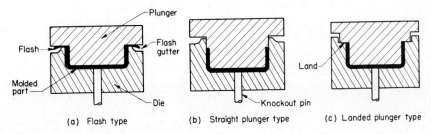

FIGURE 9-11 Three types of molds used in hot-compression molding.

FIGURE 9-12 Typical hot-compression molding press; mold being placed between platens. (*Courtesy Pennwalt Chemicals Corporation.*)

In order to avoid the turbulence and uneven flow that often result from the nonuniform, high pressures in hot-compression molding, *transfer molding* sometimes is used. The raw material is placed in the plunger cavity, where it is heated until it is melted. The plunger then descends, forcing the molten plastic into the die cavities. Because the material enters the cavities as a liquid, there is little pressure until the cavity is completely filled. This makes it easier to obtain thin sections, excellent detail, and good tolerances and finish. The process is particularly useful when fragile inserts are to be incorporated into the product. These are inserted into and maintained in position in the cavity during molding. An undercut on the plunger face causes the sprue to be withdrawn with the plunger when the mold is opened. This procedure is illustrated in Figure 9-13.

Injection molding, illustrated schematically in Figure 9-14, is used to produce more thermoplastic products than any other process. Raw material is fed by gravity from a hopper into a pressure chamber ahead of a plunger. As the plunger advances, the plastic is forced into a heating chamber, where it is preheated. From the preheating chamber it is forced through the *torpedo* section, where it is melted and the flow regulated. It leaves the torpedo through a nozzle that seats against the mold and allows the molten plastic to enter the die cavities through suitable gates and runners. In this process the die remains cool, so the

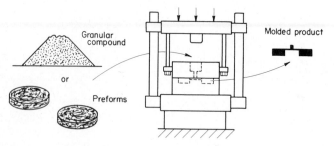

FIGURE 9-13 Schematic diagram of the transfer molding process.

MATERIALS

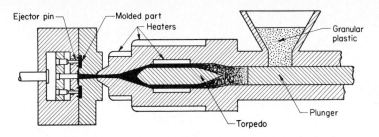

FIGURE 9-14 Schematic diagram of the injection-molding process.

plastic solidifies almost as soon as the mold is filled. To ensure proper filling of the cavity, the material must be forced into the mold rapidly under considerable pressure; premature solidification would cause defective products. While the mold is being opened and closed and the part ejected, the material for the next part is being heated in the torpedo. The complete cycle requires only a few seconds. Figure 9-15 shows a typical injection molding machine.

Because thermosetting plastics must be held in the mold under temperature and pressure for a sufficient time to permit the curing to be completed, a modification of the injection molding process must be used for this type of plastic. In the *jet molding process* shown in Figure 9-16, the plastic is preheated in the feed chamber to about 93°C (200°F) and then further heated to the polymeri-

FIGURE 9-15 Injection-molding machine. Inset shows plastic part being removed from mold. (*Courtesy Pennwalt Chemicals Corporation.*)

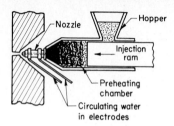

FIGURE 9-16 Jet molding process for injection-molding thermosetting plastics.

zation temperature as it passes through the nozzle. The mold is held at an elevated temperature to complete the curing process. As soon as the charge for one cycle has nearly filled the die cavity, the nozzle is cooled to prevent the plastic in the nozzle from hardening and clogging the machine. Due to economic factors resulting from the relatively long cycle time, little injection molding of thermosetting plastics is done.

Plastic products with long, uniform cross sections are readily produced by *extrusion,* as depicted in Figure 9-17. The plastic material is fed from a hopper into a screw chamber from whence the rotating screw conveys it through a preheating section, where it is compressed, and then forces it through a heated die and onto a conveyor belt. As the plastic passes onto the belt it is cooled by air or water sprays to harden it sufficiently to preserve the shape imparted to it by the die. It continues to cool as it passes along the belt. It is either cut to length, in the case of rigid plastics, or coiled, in the case of flexible plastics. In addition to providing a cheap and rapid method of molding, the extrusion process makes it possible to produce tubing and shapes having reentrant angles.

In the *laminating process,* sheets of paper or cloth made from glass or other types of fibers are impregnated with thermosetting liquid resin and placed together to build up a desired thickness. The resulting "sandwich" is cured, usually under considerable pressure and at an elevated temperature. Such products can be produced to have unusual strength properties, which primarily are the result of the sheet filler that is used. Because the surface is a thin layer of pure resin, laminates usually possess a smooth, attractive appearance. By using transparent resins, the sheet filler can be made visible, so that various decorative effects can be obtained by using cloth, wood, or other suitable materials.

Laminated plastics are produced as sheets, tubes, and rods. Flat sheets are produced as indicated in Figure 9-18. Figure 9-19 illustrates the method that is used to produce laminated tubing. The impregnated sheet stock is wound on a mandrel of the proper diameter. It then is cured in a molding press, after which

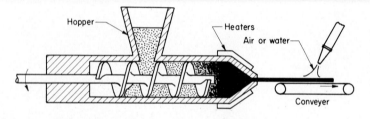

FIGURE 9-17 Extrusion process for producing plastic parts.

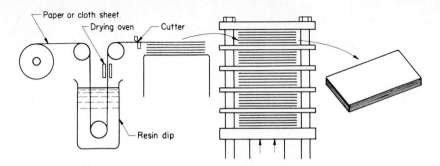

FIGURE 9-18   Method of producing flat laminated plastic sheets.

the mandrel is removed. Rods are produced in a similar manner by using a small mandrel that is removed prior to curing.

Because of their excellent strength qualities, plastic laminates find a wide variety of uses. Some laminated sheets can be blanked (see Chapter 15) and punched readily. Gears frequently are machined from thick laminated sheets; these have unusually quiet operating characteristics when matched with metal gears.

Many laminated-plastic products, which are not flat and contain relatively simple curved shapes, such as boats, automobile bodies, and safety helmets, are made using only moderate or no pressure and low temperatures, often supplied by heat lamps. Usually, a simple female mold is used, made from metal, hard wood, or particle board. The laminating material, often in the form of fabric or glass cloth dipped in the liquid plastic resin, is placed in the mold in layers until the desired thickness is obtained. The mold, containing the laminated material, is then placed in a bag from which the air is evacuated. The external air pressure holds the laminate against the mold during curing in live steam or under heat lamps. The vacuum bag often can be eliminated when plastics are used that cure at room temperature or at temperatures that can readily be obtained from a heat lamp; the pliable resin-dipped material is merely placed in the mold or over a form. Because of the low tooling costs, these processes make it possible to produce economically single items or small numbers of a large part.

Vacuum forming or thermoforming is used extensively to form shapes from thermoplastic materials. As indicated in Figure 9-20, a sheet of plastic is placed

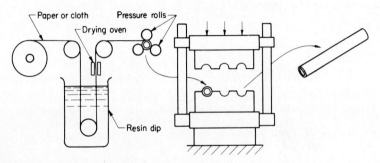

FIGURE 9-19   Method of producing laminated plastic tubing.

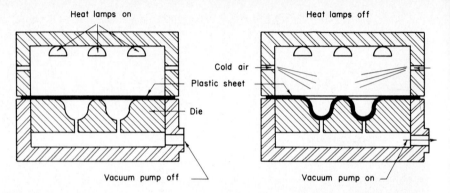

Cold air

Plastic sheet

Die

Vacuum pump off   Vacuum pump on

FIGURE 9-20   Method of molding plastic shapes by heat and vacuum (thermoforming).

over a die or form and heated until it becomes soft. A vacuum then is drawn between the sheet and the form so that the plastic takes the desired shape. It then is cooled, the vacuum dropped, and the part removed from the mold. The entire cycle usually requires only a few minutes. This process is quite economical and is used for producing a wide variety of products, ranging from panels for light fixtures to pages of Braille for the blind.

In the *cold-molding* process, depicted in Figure 9-21, the raw material is pressed to shape while cold, then removed from the mold and cured in an oven. The process is economical, but the resulting products do not have very good surface finish or dimensional tolerances. It is used, primarily, for compounds that are somewhat refractory.

Filament-wound products.   The availability of plastic-coated, high-strength filaments of various materials, such as glass, graphite, and boron, has made it possible to produce containers of various shapes that have exceptional strength-to-weight ratios. The filaments are wound over a form, using longitudinal, circumferential, or helical patterns, or a combination of these, to take advantage of their highly directional strength properties and thus provide directional strength as needed in the product. Figure 9-22 shows a large tank being made by this process. Such tanks can be made in virtually any size, some as large as 4572 mm (15 feet) in diameter and 19,812 mm (65 feet) long, with fair production quantities. The process has been highly mechanized so that uniform quality can be maintained. Because the special tooling for a new size or design is relatively

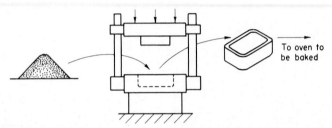

To oven to be baked

FIGURE 9-21   Schematic diagram showing the steps in cold molding.

FIGURE 9-22 Large tank being made by filament winding. (*Courtesy Rohr Corporation.*)

inexpensive, the process is economical and flexible. A variety of plastics is used, with epoxies being very common.

Foamed plastics. *Foamed plastics* have become an important and widely used form. A foaming agent is mixed with the plastic resin and releases gas when the combination is heated during molding. The resulting products have very low densities, ranging from 32 to 641 kg/m^3 (2 to 40 lb/ft^3). Both rigid and flexible foamed plastics can easily be produced. The rigid type is useful for structural applications, packaging and shipping containers, as patterns in the full-mold casting process (see Chapter 11) and for providing rigidity to thin-skinned metal components, such as aircraft fins and stabilizers, by being foamed in place in their interiors. Flexible foams are used primarily for cushioning. Quite a number of the basic plastics, both thermoplastic and thermosetting, have been used in making foams.

It is possible to produce plastic products that have a solid, rigid outer skin and a rigid foam core. Such a product, made in a single molding process using two injection molding machines connected to a single mold, is shown in Figure 9-23.

FIGURE 9-23 Plastic gear having solid outer skin and rigid foam core. (*Courtesy* American Machinist.)

Plastics for tooling. Because of their wide range of properties, their ease of conversion into desired shapes, and their excellent properties when loaded in compression, plastics are widely used in tooling to make jigs, fixtures and forming-die components. Both thermoplastics and thermosetting plastics, particularly cold-setting types, are used for drill and trim jigs, routing and assembly jigs and fixtures, and form blocks for forming. Thermoplastic materials are widely used for punch-press and drop-hammer punches, with the punches frequently formed in the female form block. Thermosetting and cold-setting plastics are used for stretch-press dies and tube-bending mandrels. The use of plastics in these tooling applications usually results in less costly tooling, permitting small quantities to be produced more economically. In addition, the tooling usually can be produced in a much shorter time so that production can get under way at an earlier date.

DESIGN AND SELECTION CONSIDERATIONS FOR PLASTICS

Plastics are not direct substitutes for metals and, as a result, material selection and design principles will often be different. In many cases, plastics are used closer to their design limits and fabrication can often convert raw material to finished products in a single step. The range of properties and processes is so great, however, that if proper selection and design is employed, plastics can be used successfully in many more applications than generally thought.

If one is to successfully design a plastic product, it is essential that four factors be considered: (1) the user requirements, particularly as to temperature, operating environment, and aging; (2) the material, with particular concern as to how well its properties are known; (3) the design; and (4) the production process required. Plastics offer the potential for substituting a single part for several, and various fabrication and fastening techniques usually are available. Obviously, the possibility of integral color, corrosion resistance, light weight, and thermal and/or electrical insulating properties offers unique advantages. But these seldom can be optimized unless one considers all four of the previously mentioned factors simultaneously. The designer must keep abreast of the spectrum of plastics that are available and their properties, while also understanding how design details are related to the processing.

Design factors related to molding. In every casting process in which a fluid or semifluid material is introduced into a mold cavity and permitted to solidify in a desired shape, certain basic problems are encountered. First, the proper amount of material must be introduced and caused to completely fill the mold cavity. Second, any entrapped material within the cavity, usually air, must be removed. Third, any shrinkage of the material that occurs during solidification and/or cooling must be taken into account. Finally, it must be possible to remove the part from the mold after it has solidified; if the mold is to be reused, the mold must be opened. Unless the parts are designed properly, these requirements will not be met. The primary design consideration, of course, must be that the part satisfies its functional requirements. Thus a material must be selected that has the requisite properties in respect to tensile strength, impact strength, dimensional stability, color, and so on. This usually requires close cooperation between the designer and the molder. Often special attention must be given to appearance details, because plastics frequently are used for goods where consumer acceptance is of great importance.

Careful attention must be given to the problem of removing the part from the mold. Because metal molds are rigid, provision must be made so they can be opened. A small amount of unidirectional taper must be provided on each side of the mold parting plane to facilitate withdrawal of the part. Undercuts should be avoided whenever possible; they prevent removal unless special mold sections are provided that move at right angles to the opening motion of the major mold halves. Such mold construction is costly to make and to maintain.

As in all cast products, it is important to provide adequate fillets between adjacent sections to assure smooth flow of the plastic into all sections of the mold and also to eliminate stress concentrations at sharp interior corners. They also make the mold less expensive to produce and lessen the danger of mold breakage where thin, delicate mold sections are encountered. It is even desirable to round exterior edges slightly where permissible. A radius of 0.26 to 0.38 mm (0.010 to 0.015 inch) is scarcely noticeable but will do much to prevent an edge from chipping. Where plastics are used for electrical applications, sharp corners should be avoided because they increase voltage gradients and may lead to failure.

Wall section thickness is very important in plastic products. The curing time is determined by the thickest section. Thus it is desirable to keep sections as nearly uniform in thickness as possible. Primarily, the minimum wall thickness is determined by the size of the part and to some extent by the type of plastic used:

Minimum recommended	0.64 mm (0.025 in.)
Small parts	1.27 mm (0.050 in.)
Average-size parts	2.16 mm (0.085 in.)
Large parts	3.18 mm (0.125 in.)

Thick corners should be avoided because they are likely to lead to gas pockets, undercuring, or cracking. Where extra strength is desired at corners, this can usually best be accomplished by ribbing.

Economical production is greatly facilitated by appropriate dimensional tolerances. A minimum tolerance of ± 0.08 mm (0.003 inch) should be allowed in the direction parallel with the parting line of the mold. In the direction at right angles to the parting line a minimum tolerance of 0.26 mm (0.010 inch) is desirable. In both cases, increasing the tolerance by 50% will reduce manufacturing difficulties and cost appreciably.

Design factors related to finishing. In designing plastic parts, a prime objective should be to eliminate any necessity for machining after molding. This is especially important where machined areas would be exposed; these are poor in appearance, and they also absorb moisture. Parting surfaces of molds are difficult to maintain in perfect condition so that they mate properly. Radii or

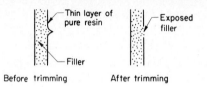

Before trimming   After trimming

FIGURE 9-24 Effect of trimming the flash from a plastic part containing filler.

curved surfaces where parting lines meet make it even more difficult to maintain perfect mating and should be avoided. The result of poor parting-line fit is a small fin or "flash," as illustrated in Figure 9-24. When fillers are used, as they are in most plastics, the exterior surface is a thin film of pure plastic without any filler, providing the smooth high luster that is characteristic of plastic parts. If the flash is trimmed off, a line of exposed filler is produced, which may be objectionable. If parting lines are located at sharp corners, it not only is easier to maintain satisfactory mating of mold sections, but any exposed filler resulting from the removal of a fin will be confined to a corner, where it will be less noticeable.

Because plastics have low moduli of elasticity, large flat areas are not rigid and should be avoided whenever practicable. Ribbing or doming, as illustrated in Figure 9-25, are helpful in providing required stiffness. Flat surfaces also may reveal flow marks from molding and scratches that are bound to result from service. External ribbing can serve the dual function of increasing strength and rigidity and also of preventing scratches from showing. Dimpled, or textured, surfaces often provide a pleasing appearance and do not reveal scratches.

Holes that must be formed by pins in the mold should be given special consideration. In compression-type molding, such pins are subjected to considerable bending influence during mold closing. Where these pins are supported at only one end, their length should not exceed twice the diameter. In transfer-type molds the length can be five times the diameter without maintenance becoming excessive.

Holes that are to be threaded after molding, or are to be used for self-tapping screws, should be countersunk slightly. This facilitates starting the tap or screw and avoids chipping at the outer edge of the hole. If the threaded hole is to be less than 6.35 mm ($\frac{1}{4}$ inch) in diameter, it is best to cut the thread after molding by means of a thread tap. For diameters above 6.35 mm ($\frac{1}{4}$ inch) it usually is better to mold the thread or to use an insert, which will be discussed later. If threads are molded, either a section of the mold must be removable, to permit later unscrewing from the part, or the part must be such that it can be unscrewed from the mold. Both procedures, particularly the latter, are not economical because of the mold delay that results.

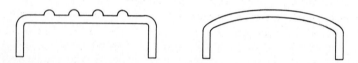

FIGURE 9-25 Method of providing stiffness in large surfaces of plastic parts by the use of ribbing and doming.

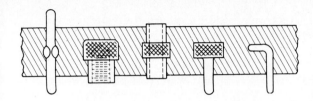

FIGURE 9-26 Typical metal inserts for use in plastic parts.

Inserts. Because of the difficulty of molding threads in plastic parts and the fact that cut threads tend to chip, tapped or threaded inserts generally are used where considerable strength is required or frequent disassembly of the parts may occur. Several types of inserts are shown in Figure 9-26. The use of inserts requires attention to design details to obtain satisfactory results and economy. Inserts usually are made of brass or steel and are held in the plastic only by a mechanical bond. Therefore, it is necessary to provide suitable knurling or grooving so that the insert may be gripped firmly and not become loose in service. A medium or coarse knurl is quite satisfactory to resist torsional loads and moderate axial loads. A groove is excellent for axial loads but offers little resistance to torsional stresses.

If an insert is to act as a boss for mounting or is an electrical terminal, it should protrude slightly above the surface of the plastic in which it is embedded. This permits a firm connection to be made without creating an axial load that would tend to pull the insert from the compound. On the other hand, if it is desired to use the insert to hold two matting parts closely together, the insert should be flush with the surface. In this way the parts can be held together snugly without danger of loosening the insert. Where it is necessary to keep the surface of an insert entirely free from any plastic, a shouldered design is most satisfactory. However, if an insert is used to fasten mating parts that must fit closely, a depression must be made in the mating part to provide clearance for the shoulder. Similarly, a depression has to be provided in the mold. Both operations add to the cost.

Inserts must have adequate support. The wall thickness of the surrounding plastic must be sufficient to support any load that may be transmitted by the insert. For small inserts the wall thickness should be at least half the diameter of the insert. Above 12.7 mm ($\frac{1}{2}$ inch) in diameter, the wall thickness should be at least 6.35 mm ($\frac{1}{4}$ inch).

Machining plastics. Although most plastics are readily machined, their properties vary so greatly that it is impossible to give instructions that are exactly correct for all. It is very important, however, to remember some general characteristics that affect their machinability. First, all plastics are poor heat conductors. Consequently, little of the heat that results from chip formation will be conducted away through the material or be carried away in the chips. As a result, cutting tools run very hot and may fail more rapidly than when cutting metal. Carbide tools frequently are more economical to use than high-speed steel tools if cuts are of moderately long duration or if high-speed cutting is to be done.

Second, because considerable heat and high temperatures do develop at the

FIGURE 9-27   Straight-flute drill (*left*) and "dubbed" drill (*right*) used for drilling plastics.

point of cutting, thermoplastics tend to soften, swell, and bind or clog the cutting tool. Thermosetting plastics give less trouble in this regard.

Third, cutting tools should be kept very sharp at all times. Drilling is best done by means of straight-flute drills or by "dubbing" the cutting edge of a regular twist drill to produce a zero rake angle; these are shown in Figure 9-27. Rotary files and burrs, saws, and milling cutters should be run at high speeds so as to improve cooling, but with the feed carefully adjusted to avoid jamming the gullets. In some cases coolants can be used advantageously if they do not discolor the plastic or cause gumming. Water, soluble oil and water, and weak solutions of sodium silicate in water are used.

Fourth, filled and laminated plastics usually are quite abrasive and may produce fine dust which may be a health hazard.

Finishing plastic parts.   In the majority of cases plastic parts can be designed to require very little finishing or decorative treatment, thus promoting economy. In some cases fins and rough spots can be removed and smoothing and polishing can be done by barrel tumbling with suitable abrasives or polishing agents. Required decoration or lettering can be obtained by etching the mold. These procedures produce letters or designs that protrude from the surface of the plastic only about 0.1 mm (0.004 inch). When higher relief is desired, the mold must be engraved, which adds significantly to mold cost.

Whenever possible, depressed letters should be avoided. Such letters must be raised above the surrounding surface of the mold, requiring the entire remaining mold surface to be cut away, at considerable expense. The cost frequently can be reduced by setting the letters in a small area raised above the main plastic surface. This requires only a small amount of die metal to be undercut.

Many plastic parts now are electroplated, as will be discussed in Chapter 37.

RUBBER AND ARTIFICIAL ELASTOMERS

Elastomers have unique characteristics in that, at room temperature, they can be stretched up to at least twice their original length and upon immediate release of the stress will return quickly to approximately their original dimensions.

Although they are elastic over a wide range, they do not obey Hooke's law. It is their structure, rather than their composition, that produces their elastic properties. They have remarkable capacity for storing energy, and they can be tailored to provide a wide range of stress-strain characteristics.

Natural rubber is the oldest, and still a widely used, elastomer. However, numerous artificial elastomers now are available that have been developed to meet specific needs and, in total, are used more than rubber.

The elastic characteristics of nonelastomers are usually due to the change in distance between adjacent atoms as the result of applied loads. The interatomic forces return the atoms to their normal positions when the load is removed. In elastomers, on the other hand, the elastic properties are due primarily to the fact that in the unstrained condition the basic molecule is in the form of a coil that, like a coil spring, can be stretched. When the load is removed, the stretched coil returns to its normal shape. Thus elastomers exhibit very large degrees of elasticity.

Rubber. Rubber is obtained from the *Hevea brasiliensis* tree, a native of Brazil but grown for commercial purposes primarily in the East Indies and Africa. The trees are tapped to obtain the sap, called *latex,* which consists of about 65% water and 35% rubber. While some latex is shipped in the liquid form for use in certain dipped products, most is coagulated with acetic acid, squeezed to remove the water, and the coagulate milled into sheets and dried. The dried sheet is known as *pale crepe,* or, if smoked after drying, as *smoked sheet.* The sheets are pressed into bales for shipment.

Rubber is used to only a limited extent in the crude form. It is an excellent adhesive and is thus used in many cements made by dissolving the crude rubber in suitable solvents.

The extensive use of rubber as an engineering material dates from 1839, when Charles Goodyear discovered that it could be vulcanized by adding about 30% sulfur and heating it at a suitable temperature. Sulfur causes sufficient cross-linking between the chains of molecules to restrict movement between them, and thus it imparts strength. Subsequently, it was found that the properties of vulcanized rubber could be greatly improved by adding certain pigments, notably carbon black, which would act as stiffeners, tougheners, and antioxidants. Certain *accelerators* have been found that greatly speed the vulcanization process with reduced amounts of sulfur, so most modern rubber compounds contain less than 3% sulfur. As in the compounding of plastics, softeners and fillers are usually added to facilitate processing and to add bulk.

Rubber can be compounded over a wide range, from soft and gummy to very hard, such as ebonite. Where high strength is required, textile cords or fabrics are coated with rubber to withstand the applied loads, the rubber largely serving to insulate the cords from each other and thus prevent chafing and friction. For severe service, steel wires may be coated with rubber and used as the load-carrying medium. This is done in some tires and heavy-duty conveyor belts.

Natural rubber compounds are outstanding for their high flexibility, good electrical insulation, low internal friction, and resistance to most inorganic acids, salts, and alkalies. However, their resistance to petroleum products, such as oil, gasoline, and naptha, is poor. In addition, they lose their strength at elevated temperatures, so it is not advisable to operate them at temperatures above 79 to 82°C (175 to 180°F). They also deteriorate fairly rapidly in direct sunlight unless specially compounded.

Artificial elastomers.   The uncertainty of both the supply and price of natural rubber to the highly industrialized countries in time of war led to the development of a number of artificial elastomers which have had great commercial importance. One, polyisoprene, appears to have the same molecular structure as natural rubber and equal or superior properties. Some of the others are inferior to natural rubber, while others have distinctly different and, frequently, superior properties, thus extending their usefulness for specific applications. Table 9-2 lists the most widely used artificial elastomers, with natural rubber shown for comparison, and gives their typical properties and uses. It must be remembered, however, that the properties can vary considerably, depending on how the materials are compounded and processed.

Processing of rubber and elastomers.   Rubber products are made by several processes. The simplest is where they are formed from a liquid preparation or compound. These commonly are called *latex products.* Dipped products are made by immersing a form repeatedly into the latex compound, causing a certain amount of the liquid to adhere to the surface of the form each time. After each dipping, the film is allowed to dry, usually in air. Dipping is continued until the desired thickness is obtained. After vulcanization, usually in steam, the products are stripped from the forms.

Most latex products now are made by the *anode* process, a process utilizing negative electrical charges on the latex particles. A coagulant is deposited on the form or mold that releases postively charged ions when dipped into the latex. These ions neutralize the charges on the adjacent latex particles and cause them to be deposited on the form. The process goes on continuously, so any desired thickness can be deposited.

When products are to be made from solid elastomers, the first step is the compounding of the elastomers, vulcanizers, fillers, autioxidants, accelerators, and other pigments. This usually is done in a *Banbury mixer,* which breaks down the elastomers and permits mixing in some of the other components to form a homogeneous mass.

Usually, the mix is then put on a *mill,* such as is shown in Figure 9-28, in which chilled iron rolls rotate toward each other at different speeds. They are cooled by the circulation of water through their interiors to remove the heat generated by the milling action, thus preventing the start of vulcanization. The sulfur and accelerators usually are added at this stage.

Rubber compounds and plastics are made into sheet form on *calenders,* such as is shown in Figure 9-29. The sheet coming from a calender is rolled with a fabric liner to prevent the adjacent layers from sticking together.

When cord or square-woven fabric is to be covered with rubber, this also is done on a three- or four-roll calender. On a three-roll calender only one side of the fabric can be coated at each pass. A four-roll calender, such as is indicated schematically in Figure 9-30, makes it possible to coat both sides of the fabric at a single pass.

TABLE 9-2 Properties and Uses of Common Elastomers

Elastomer	Specific Gravity	Durometer Hardness	Tensile Strength (psi)		Elongation (%)		Service Temp. [°C (°F)]		Resistance to:[a]			Typical Applications
			Pure Gum	Black	Pure Gum	Black	Min.	Max.	Oil	Water Swell	Tear	
Natural rubber	0.93	20–100	2500	4000	750	650	−54 (−65)	82 (180)	P	G	G	Tires, gaskets, hose
Polyacrylate	1.10	40–100	350	2500	600	400	−18 (0)	149 (300)	G	P	F	Oil hose, O-rings
EDPM (ethylene propylene)	0.85	30–100	1	3		500	−40 (−40)	149 (300)	P	G	G	Electric insulation, footware, hose. belts
Chlorosulfonated polyethylene	1.10	50–90	4	2		400	−54 (−65)	121 (250)	G	E	G	Tank linings, chemical hose, shoe soles and heels
Polychloroprene (neoprene)	1.23	20–90	3500	4000	800	550	−46 (−50)	107 (225)	G	G	G	Wire insulation, belts, hose, gaskets, seals, linings
Polybutadiene	1.93	30–100	1000	3000	800	550	−62 (−80)	100 (212)	P	P	G	Tires, soles and heels, gaskets, seals
Polyisoprene	0.94	20–100	3000	4000		600	−54 (−65)	82 (180)	P	G	G	Same as natural rubber
Polysulfide	1.34	20–80	350	1000	600	400	−54 (−65)	82 (180)	E	G	G	Seals, gaskets, diaphragms, valve disks
SBR (styrene butadiene)	0.94	40–100	2		1200		−54 (−65)	107 (225)	P	G	G	Molded mechanical goods, disposal pharmaceutical items
Silicone	1.1	25–90		1200		450	−84 (−120)	232 (450)	F	E	P	Electric insulation, seals, gaskets, O-rings
Epichlorohydrin	1.27	40–90		2		325	−46 (−50)	121 (250)	G	G	G	Diaphragms, seals, molded goods, low-temperature parts
Urethane	0.85	62–95	5000		700		−54 (−65)	100 (212)	E	F	E	Caster wheels, heels, foam padding
Fluoroelastomers	1.65	60–90	1	3		400	−40 (−40)	232 (450)	E	E	F	O-rings, seals, gaskets, roll coverings

[a]P, poor: F, fair: G, good: E, excellent.

FIGURE 9-28 Twin rubber mill installation. Rubber is being fed to a tuber from the right-hand mill. (*Courtesy Adamson United Company.*)

Many rubber products, such as inner tubes, garden hose, tubing, and moldings, are produced by extrusion. The compound from a mill is forced through a die by a screw device similar to the plastic extruder of Figure 9-17.

Excellent adhesives have been developed which permit bonding of rubber and artificial elastomers to metal, usually brass or steel. Tanks of all sizes are made by this procedure for transporting and storing a wide variety of corrosive liquids. Usually only moderate pressures and temperatures are required to obtain excellent adhesion.

Elastomers for tooling. When an elastomer is confined, it will act as a fluid, transmitting force quite uniformly in all directions. This phenomenon often makes it possible to substitute an elastomer for one half of a metal die set in connection with metal forming. This procedure also makes it possible to do bulging and forming of reentrant sections, which would be impossible with steel dies except by very costly multipiece dies. Also, because elastomers can be

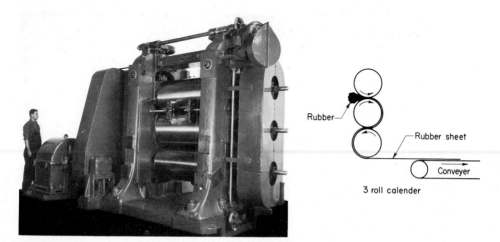

FIGURE 9-29 (*Left*) Three-roll calender used for producing rubber or plastics in sheet form. (*Courtesy Farrel-Birmingham Company, Inc.*) (*Right*) Schematic diagram showing the method of making sheets of rubber in a three-roll calender.

MATERIALS

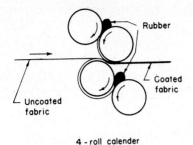

4 - roll calender

FIGURE 9-30 Arrangement of the rolls, fabric, and coating material for coating both sides of fabric in a Z-type four-roll calender.

compounded to range from very soft to very hard, hold up very well when subjected to compression loading, and can quickly and economically be made into a desired shape, they are being used increasingly as tooling materials. Urethanes are currently most popular for these applications.

CERAMICS

Long used in the electrical industry because of their high electrical resistance, in recent years ceramics have assumed considerable engineering importance because of their ability to withstand high temperatures (refractories and refractory coatings), to provide a variety of electrical properties (solid-state electronics), and to resist wear (coated cutting tools). The latter application is one of much current effort with coatings involving titanium carbide, titanium nitride, aluminum oxide, and others (see Chapter 18).

Ceramics are compounds of metallic and nonmetallic elements. Consequently, because there are many possible combinations of metals and nonmetals, there are a multitude of ceramic materials. Also, the same combination of metal and nonmetal may exist in more than one structural arrangement, thereby producing polymorphism. For example, depending on the temperature, silica can exist in three forms—*quartz, tridymite,* and *cristobalite.* The subject of ceramics is too large to be treated in detail here, but some details of their basic nature and properties will be presented so as to indicate their possible uses.

Structure of ceramics. Most ceramics have crystal structures. However, unlike metals, they do not have large numbers of free electrons, the electrons being captive in covalent or ionic bonds. The absence of free electrons makes ceramics poor electrical conductors and results in their being transparent in thin sections. Because ionic bonds tend to produce high stability, ceramics have high melting temperatures.

The crystal structure of ceramics must accommodate atoms of different size. Figure 3-3 illustrates such an arrangement. Several basic crystal structures exist, the cubic and tetrahedral being very common.

In numerous cases ceramic crystals grow into chains, similar to plastic molecules. The chains are now held together by ionic bonds instead of by weak van der Waals forces. However, the bonds between chains are not as strong as those within the chains. Consequently, when forces are applied, cleavage occurs between the chains.

In some cases the atoms form *sheets* and result in *layered* structures; these have relatively weak bonds between the sheets.

Properties of ceramics.   Most *mechanical ceramics* have specific gravities in the range 2.3 to 3.85. By comparing their structure and bonding with that of metals, their behavior under load can be predicted. Metals have considerable ductility because they have lower shear resistance than tensile resistance. Ceramics, on the other hand, have stronger interatomic bonding and higher shear resistance. Thus, they have low ductility with high compressive strength. Theoretically, ceramics could also have high tensile strengths, but ordinarily they do not because small cracks and pores act as stress concentrators and are not reduced through ductility and plastic flow. Failure thus occurs at low average stress values, typical tensile strengths ranging from 21 to 210 MPa (3000 to 30,000 psi). By using special techniques to eliminate the cracks, very high strengths can be obtained; some glass fibers have strengths above 7000 MPa (1,000,000 psi). Some ceramics have melting points above 1649°C (3000°F).

Cermets.   Cermets are combinations of metals and ceramics, bonded together in the same manner in which powder metallurgy parts are produced. They combine some of the high refractory characteristics of ceramics and the toughness and thermal-shock resistance of metals. Oxide cermets usually are chromium–alumina- or chromium–molybdenum–alumina–titania-based. Carbide cermets are based on tungsten, titanium, chromium, or tungsten–titanium carbides.

Parts are produced from cermet materials by pressing the powders in molds at pressures ranging from 70 to 280 MPa (10,000 to 40,000 psi) and then sintering them in controlled-atmosphere furnaces at about 1649°C (3000°F).

Cermets are used principally as crucibles and as nozzles for jet engines or other high-temperature devices.

COMPOSITE MATERIALS

A composite material is a heterogeneous solid consisting of two or more components that are mechanically or metallurgically bonded together. Each of the various components retains its identity in the composite and has its characteristic structure and properties. However, by combining the components, resultant characteristics are imparted that are properties of the composite.

Although there are many types of composites and several methods of classification, one means of division is on the basis of geometry. Three families emerge: *laminar* or layer composites, *particulate* composites, and *fiber-reinforced* composites. Properties of the composite then depend upon (1) the properties of the individual constituents; (2) the relative amounts of the constituents; (3) the size, shape, and distribution of the discontinuous components; (4) the degree of bonding between components; and (5) the orientation of the various components. Considerable freedom exists so that composite materials can often be designed to possess desired sets of properties.

Laminar composites are those having alternating layers of material bonded together in some manner. Plywood is probably the most common engineering material in this category. Here, layers of wood, are bonded with their grain orientations at various angles to one another. Strength and fracture resistance are improved and the product has minimized swelling and shrinkage tendencies. Bimetallic strip consists of two metals with different coefficients of thermal expansion bonded together. Changes in temperature produce flexing or curvature in the product which may be employed in thermostat and other heat-sensing applications. Clad materials are laminar composites. Alclad metal consists of high-strength, heat-treatable aluminum, clad on the exterior with the more corrosion resistant, non-heat-treatable alloys. U.S. coinage material is yet another example of a laminar composite.

Particulate composites consist of discrete particles of one material surrounded by a matrix of another material. Concrete is a classic example, consisting of sand and gravel particles surrounded by cement. Sintered carbide is another example, where tungsten carbide particles are bound in a cobalt metal matrix. Many other powder metallurgy products are particulate composites.

The most popular variety of composite materials is the fiber-reinforced geometry, where thin fibers of one material are embedded in a matrix of another. The matrix supports and transmits loads to the fibers, which, in turn, supply strength to the structure. Wood and bamboo are two natural fiber composites consisting of cellulose fibers in a lignin matrix. Bricks of straw and mud may well have been the first man-made material of this variety. Steel reinforced concrete is actually a particulate composite matrix reinforced with steel "fibers."

The glass-fiber reinforced resins were the first of the "fibrous composites" to be developed, shortly after World War II, in an attempt to produce lightweight materials with high strength and high stiffness. Glass fibers, about 10 μm in diameter, were bonded in a variety of polymers, generally epoxy resins. Limitations generally were related to stiffness since the glass fibers had a modulus of only 70,000 to 90,000 MPa (10 to 13×10^6 psi). Advancements, therefore, focused on either improving the modulus or developing new fibers. Boron–tungsten fibers (boron deposited on a tungsten core) have an elastic modulus of 380,000 MPa (55×10^6 psi) with tensile strengths in excess of 2750 MPa (400,000 psi) and can be cast into a metal matrix if desired. Silicon carbide filaments (SiC on tungsten) have an even higher modulus of elasticity.

Recently, the most popular fibers have been graphite and Kevlar.[1] Graphite can be either the PAN type, produced by the thermal pyrolysis of synthetic organic fibers such as viscose rayon or polyacrylonitrile, or pitch type (from petroleum pitch). They have low density and a range of high tensile strengths and elastic moduli. Kevlar is an organic aramid fiber with 3100 MPa (450 ksi) tensile strength, 131,000 MPa (19×10^6 psi) elastic modulus, and a density approximately one-half that of aluminum. Moreover, it is transparent to radio signals, a characteristic desirable for many aerospace applications.

Fine whiskers of sapphire, silicon carbide, and silicon nitride have also been used to reinforce metallic matrices.

[1]Trade name by DuPont.

TABLE 9-3 Comparison of the Properties of Fiber Composites and Lightweight or Low-Thermal-Expansion Structural Metals

Material	Specific Strength[a] (10^6 in.)	Specific Stiffness[b] (10^6 in.)	Density (lb/in^3)	Thermal Expansion Coefficient [in./(in.-°F)]	Thermal Conductivity [Btu/(hr-ft-°F)]	Raw Material Cost[c] ($/lb)
Graphite–epoxy: high strength (unidirectional)	5.4	400	0.056	−0.3	3	50
Graphite–epoxy: high modulus (unidirectional)	2.1	700	0.063	−0.5	75	300
Glass–epoxy (woven cloth)	0.7	45	0.065	6	0.1	10
Kevlar–epoxy (woven cloth)	1	80	0.05	1	0.5	15
Boron–epoxy	3.3	457	0.07	2.2	1.1	300
Aluminum	0.7	100	0.10	13	100	2
Beryllium	1.1	700	0.07	7.5	120	400
Titanium	0.8	100	0.16	5	4	400
Invar[d]	0.2	70	0.29	1	6	22

[a]Strength divided by density.
[b]Elastic modulus divided by density.
[c]As of 1980.
[d]A low-expansion metal containing 36% Ni and 64% Fe.

Fabrication methods applied to the fibrous composites include simple compression molding, filament winding, pultrusion (bundles of coated fibers are drawn through a heated die), cloth laminations, and autoclave curing (where pressure and elevated temperature are applied simultaneously). Design involves selection of the fiber, matrix, and fabrication method. With the wide variety of each, great flexibility is available and it is possible to tailor a material to a given application. Table 9-3 presents a comparison of various composites and "lightweight" engineering metals.

Limitations of such composites often relate to their relative brittleness and high cost. Graphite fibers, costing $400 to $500 per pound in the late 1960s are expected to be available for about $10 per pound in the near future, but even that cost is considerably higher than many alternative materials. Manufacture tends to be labor intensive and there is a lack of established design guidelines or methods of quality control and inspection.

Nevertheless, these materials have become well established in several major areas. Aerospace applications are extremely common. Sporting equipment, such as golf club shafts, fishing rods, tennis rackets, bicycle frames, and skiis are now available in fibrous composites. Potential automotive applications include drive shafts, springs, and bumpers, if costs reduce to a competitive level. Advancing technology and its concurrent demands have made materials with strengths like steel at one-fifth the weight quite attractive.

Another relatively new area of development is the injection molding of fiber-reinforced plastics. Chopped fibers, up to 6 mm ($\frac{1}{4}$ inch) in length, are mixed with plastic resin (often nylon) and oriented by the flow in the molding process. Both glass and carbon fibers can be used, the latter producing a composite that is stiffer than steel, yet capable of being molded to close tolerances. This process competes with zinc die casting in many applications.

Sheet molding compounds (SMC) are comprised of chopped glass fibers and polyester resin in sheets approximately 2.5 mm (0.1 inch) thick. With strengths of 35 to 70 MPa (5 to 10 ksi) and the ability to be press-formed, they are likely to compete with sheet metal in applications where their light weight and corrosion resistance are attractive.

As one example of the popularity of these materials, manufacturers of trucks in North America used over 34 million kilograms (75 million pounds) of glass-fiber-reinforced plastics in 1982. Applications were in place of steel and aluminum in cab shells and bodies, oil pans, fan shrouds, instrument panels, and engine covers.

REVIEW QUESTIONS

1. What is a plastic?
2. What is a "saturated" molecule?
3. What is an isomer?
4. What are the two distinctly different means of polymerization?
5. What is degree of polymerization and what is its value for most commercial plastics?
6. What causes the difference between thermosetting and thermoplastic materials?
7. How does deformation occur in a thermoplastic material?
8. How might thermoplastic materials be strengthened?

9. Why are thermosetting plastics characteristically brittle?
10. What are some of the general properties of plastics that account for their use as engineering materials?
11. What are some of the limiting properties of plastics?
12. What is the benefit of an oriented plastic?
13. Why are plastic resins seldom used in their pure form?
14. Why should the addition of lubricants in compounding plastics be held to a minimum?
15. What are seven common processes for producing parts from plastics?
16. Why are thermoplastics not as suitable for hot compression as are thermosetting materials?
17. Why is the transfer-molding process used?
18. Why are thermosetting materials more difficult to injection-mold than thermoplastics?
19. What type of starting material would be desirable for a vacuum-forming operation?
20. What are some of the applications for rigid foamed plastics?
21. What are the four major considerations when one is to successfully produce a plastic product?
22. Where is the best location for the mold parting line on a plastic part?
23. Why should the walls in plastic parts be kept as uniform as possible in practice?
24. Explain what results when the flash is trimmed from a plastic part.
25. When should threaded inserts for screws be used in plastic parts?
26. Why should an insert that acts as a mounting boss on a plastic part protrude slightly above the surface in which it is embedded?
27. Why is it desirable for lettering on a plastic part to be raised above the surface?
28. Why do elastomers exhibit elastic properties?
29. What is the purpose of adding an accelerator to an elastomer compound?
30. Which type of artifical rubber has the same molecular structure as natural rubber?
31. Explain how vulcanization increases the strength of elastomers.
32. Under what conditions might elastomers be attractive for tooling applications?
33. How do ceramic crystals differ basically from metal crystals?
34. Explain why there can be so many ceramic materials.
35. Why do most ceramics not have high tensile strengths?
36. What are cermets?
37. What are two common uses for cermets?
38. What is a composite material?
39. What are the three primary geometries of composite materials?
40. What factors influence the properties of composite materials?
41. Give an example of a laminar, particulate, and fibrous composite material.
42. What are some of the modern reinforcing fibers used in fiber-reinforced composites?
43. What are sheet-molding compounds(SMC)?

CASE STUDY 9.   The Repaired Bicycle

The frame of a high-quality, ten-speed bicycle was made of cold-drawn alloy steel tubing, taking advantage of the cold-working characteristics to provide added strength. As a result of excessive abuse, the frame was broken, and a repair was made by conventional arc welding. The repair seemed adequate but, shortly thereafter, the frame again broke, this time adjacent to the repair weld. The break appeared to be ductile in nature, showing evidence of metal flow prior to fracture. What was the probable cause of the second failure?

CHAPTER 10

Material Selection

The objective of any practical work dealing with the manufacture of products is to produce components that will adequately perform their designated tasks. This implies the manufacture of components from *selected engineering materials* with the *required geometrical shape and precision* and *companion material structures* that are optimized for the service environment that the components must withstand. The ideal design is one that will just meet all requirements. Anything better tends to waste money or material. Anything worse, and we have failed to manufacture an adequate product.

During recent years the selection of engineering materials has assumed great importance. Moreover, the process should be one of continual reevaluation. New materials often become available and there may be a decreasing availability of others. Concerns regarding environmental pollution, recycling, and worker health and safety often impose new constraints. The desire for weight reduction or energy savings may dictate the use of different materials. Pressures from domestic and foreign competition, increased serviceability requirements, and customer feedback may all promote materials reevaluation. The extent of product liability actions, often the result of improper material use, has had a marked impact. In addition, the interdependence between materials and their processing has become better recognized. The development of new processes often forces reevaluation of the materials being processed. Therefore, it is imperative that design and manufacturing engineers exercise considerable care in selecting, specifying, and utilizing materials if they are to achieve satisfactory results at reasonable cost and still assure quality.

Most modern products are relatively complex and often utilize a variety of materials. The vacuum cleaner assembly shown in Figure 10-1 is typical with nine different materials being used in the assembly. As shown in Table 10-1, of the 13 components, the materials for 12 have been changed completely from those used originally and that for the thirteenth has been modified. Eleven different reasons were given for the changes. Thus, we see that material selection is both a complex and continuing process.

Recognizing that engineering materials includes virtually all metals and alloys, ceramics, glasses, plastics, elastomers, electrical semiconductors, concrete, composite materials, and many others, it is not surprising that a single individual will have great difficulty in keeping abreast of, and making the necessary decisions concerning, the materials in even a single, fairly simple product. Nevertheless, it is the responsibility of the design engineer, working in conjunction with materials engineers, to select the materials for use in converting designs into reality.

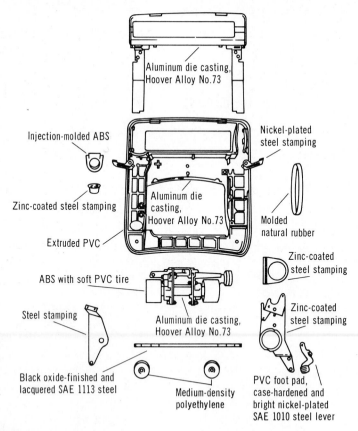

FIGURE 10-1 Materials used in various parts of a vacuum cleaner assembly. (*Courtesy Metal Progress.*)

TABLE 10-1 Examples of Materials Selection in Modern Vacuum Cleaners

Part	Former Material	Present Material	Benefits
Bottom plate	Assembly of steel stampings	One-piece aluminum die casting	More convenient servicing
Wheels (carrier and caster)	Molded phenolic	Molded medium-density polyethylene	Reduced noise
Wheel mounting	Screw-machine parts	Preassembled with a cold-headed steel shaft	Simplified replacement, more economical
Agitator brush	Horsehair bristles in a die-cast zinc or aluminum brush back	Nylon bristles stapled to a polyethylene brush back	Nylon bristles last seven times longer and are now cheaper than horsehair
Switch toggle	Bakelite molding	Molded ABS	Breakage eliminated
Handle tube	AISI 1010 lock seam tubing	Electric seam-welded tubing	Less expensive, better dimensional control
Handle bail	Steel stamping	Die-cast aluminum	Better appearance, allowed lower profile for cleaning under furniture
Motor hood	Molded cellulose acetate (replaced Bakelite)	Molded ABS	Reasonable cost, equal impact strength, much improved heat and moisture resistance; eliminated warpage problems
Extension tube spring latch	Nickel-plated spring steel, extruded PVC cover	Molded acetal resin	More economical
Crevice tool	Wrapped fiber paper	Molded polyethylene	More flexibility
Rug nozzle	Molded ABS	High-impact styrene	Reduced costs
Hose	PVC-coated wire with a single-ply PVC extruded covering	PVC-coated wire with a two-ply PVC extruded covering separated by a nylon reinforcement	More durability, lower cost
Bellows, cleaning tool nozzles, cord insulation, bumper strips	Rubber	PVC	More economical, better aging and color, less marking

Source: Metal Progress, by permission.

The design process. The first step in the manufacture of any product is design, which usually takes place in several distinct stages: (1) conceptual, (2) functional, and (3) production design. During the *conceptual-design* stage, the designer is concerned primarily with the functions the product is to fulfill. Usually, several concepts are visualized and considered, and a decision is made either that the idea is not practical or that the idea is sound and one or more of the conceptual designs should be developed further. Here, the only concern for materials is that materials exist than can provide the desired properties. If no such materials are available, consideration is given as to whether there is a reasonable prospect that new ones could be developed within cost and time limitations.

At the *functional-* or *engineering-design* stage, a practical, workable design is developed. Fairly complete drawings are made, and materials are selected and specified for the various components. Often a prototype or working model is made that can be tested to permit evaluation of the product as to function, reliability, appearance, serviceability, and so on. Although it is expected that such testing might show that some changes may have to be made in materials before the product is advanced to the production-design stage, this should not be taken as an excuse for not doing a thorough job of material selection. Appearance, cost, reliability, and producibility factors should be considered in detail, together with the functional factors. There is much merit to the practice of one very successful company which requires that all prototypes be built with the same materials that will be used in production and, insofar as possible, with the same manufacturing techniques. It is of little value to have a perfectly functioning prototype that cannot be manufactured economically in the expected sales volume, or one that is substantially different from what the production units will be in regard to quality and reliability. Also, it is much better for design engineers to do a complete job of material analysis, selection, and specification at the development stage of design rather than to leave it to the production-design stage, where changes may be made by others, possibly less knowledgeable about all of the functional aspects of the product.

At the *production-design* stage, the primary concern relative to materials should be that they are specified fully, that they are compatible with, and can be processed economically by, existing equipment, and that they are readily available in the needed quantities.

As manufacturing progresses, it is inevitable that situations will arise that may require modifications of the materials being used. Experience may reveal that substitution of cheaper materials can be made. In most cases, however, changes are much more costly to make after manufacturing is in progress than before it starts. Good selection during the production-design phase will eliminate the necessity for most of this type of change. The more common type of change that occurs after manufacturing starts is the result of the availability of new materials. These, of course, present possibilities for cost reduction and improved performance. However, *new materials must be evaluated very carefully to make sure that all their characteristics are well established.* One should always remember that it is indeed rare that as much is known about the properties and reliability of a new material as about those of an existing one. A large proportion of product failure and product-liability cases have resulted from new materials being substituted before their long-term properties were really known.

A procedure for material selection. The selecting of an appropriate material and then converting it into a useful product with desired shape and properties is a complex process. Nearly every engineered item goes through the sequence design → material selection → fabrication → evaluation → redesign. The number of engineering decisions are many. Several methods exist for approaching a design and selection problem.

The *case-history method* assumes that something has worked successfully before and that similar components might be made with the same engineering material and method of manufacture. While the approach is useful, minor variations in service requirements may well require different materials or manufacturing operations. Moreover, this approach precludes the use of new technology, new materials, and other manufacturing advances that have occurred since the formulation of the previous solution. It would be equally unwise, however, to totally ignore the wealth of past experience.

Still other activities involve *improvement of an existing product,* generally seeking to reduce costs or improve quality. Efforts here generally begin with an evaluation of the current product and its present method of manufacture. A frequent pitfall, however, is to lose sight of one of the original design requirements. The forthcoming section on material substitution will present several examples.

The safest and most thorough approach is to consider the task to be the *development of an entirely new product.* Here the full sequence of design, material selection, and development of a manufacturing sequence is to be followed.

The *first step* in any materials selection problem is to *define the needs of the product.* Without any prior biases as to material or method of manufacture, the engineer should form a clear picture of all of the characteristics necessary for this part to adequately perform its intended function. These requirements fall into three major areas: (1) shape or geometry considerations, (2) property requirements, and (3) manufacturing concerns.

The area of shape considerations primarily influences selection of the method of manufacture. While such concerns are somewhat obvious, they may be more complex than one might first imagine. Typical questions include:

1. What is the relative size of the component?
2. How complex is its shape? Are there axes or planes of symmetry? Uniform cross sections? Do you want to consider making it in more than one piece?
3. How many dimensions must be specified?
4. How precise must these dimensions be? (Tolerances) Are all precise? How many are restrictive and which ones?
5. How does the component interact with other components? (Allowances)
6. What surface characteristics are needed? Which surfaces need to be smooth? Hard? Which need to be finished? Which do not?
7. How much can a dimension change by wear or corrosion and the part still perform adequately?

8. Could a minor change in the design shape significantly improve the suitability of the part (increase strength, reliability, fracture resistance, etc.)?

The defining of property requirements is often a far more complex task. Some aspects that should be considered include:

Mechanical properties:

1. What are the needs with regard to static strength?
2. Is the component more likely to fail by deformation or fracture? Do you have a preference?
3. Can you envision impact loadings? If so, of what type and magnitude?
4. Can you envision cyclic loadings? If so, of what type and magnitude?
5. Is wear resistance needed? How much? How deep?
6. Over what temperature range must these properties be present?
7. How much can the material deflect or bend and still function properly?

Physical properties:

1. Are there any electrical property requirements?
2. Are any magnetic properties desired?
3. Are thermal properties significant? Thermal conductivity? Change of dimension with change of temperature?
4. Are there any optical requirements?
5. Is weight a significant factor?
6. What about appearance?

Another important area to evaluate is the service environment of the product throughout its lifetime:

1. What is the lowest, highest, and normal operating temperature for the component?
2. Are all of the desired properties required over this range of temperatures?
3. What is the most severe environment anticipated as far as corrosion or deterioration of material properties?
4. What is the desired service lifetime of the product?
5. What is the anticipated maintenance for this component?
6. What is the potential liability if the product should fail?
7. Should the product be manufactured with recycling in mind

A final area of concern is to determine various factors that would influence method of manufacture.

1. Have standard components and sizes been specified wherever possible?
2. Has the design addressed the requirements which will facilitate ease of manufacture? Machinability? Weldability? Formability? Hardenability? Castability?
3. How many components are to be made? At what rate?
4. What are the maximum and minimum section thicknesses?
5. What is the desired level of quality compared to similar products on the market?
6. What are the anticipated quality control and inspection requirements?

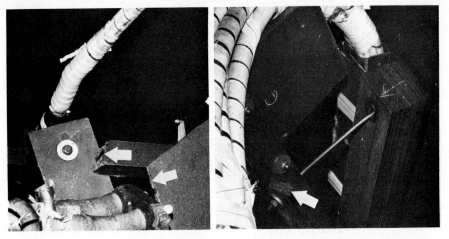

FIGURE 10-2 Failures in the lead-support structure of a large electrical transformer. Material selection failed to consider stresses resulting from shipping.

7. Are there any assembly concerns, relationships to mating parts, and so on, that should be noted?

Although there is a tendency to want to jump to "the answer," the time spent here determining requirements will be well rewarded and it is important that *all* factors be listed and *all* service conditions and uses be considered. Many failures and product liability claims have resulted from simple engineering oversights or the designer not anticipating reasonable use for a product or conditions that the product would have to encounter outside of the narrow, specific function for which he designed it. Figure 10-2 shows such an example. The designer of a large electrical transformer substituted a plastic laminate in place of maple for the members of the structure that supported the heavy, copper leads. In service, certain of these members would have been only in compression, and the laminate was excellent for this type of loading. However, the designer failed to consider that during *shipping,* some of the members might be subjected to bending loads, and the material was weak in this regard. As a consequence, several supporting members broke during shipping, with a domino effect, producing extensive failure and damage.[1] Similarly, the most severe corrosion environment for a military component may well be during storage on the deck of an aircraft carrier (salt spray environment) rather than the environment of actual service.

After the list of required properties is complete, it is often helpful to indicate the relative importance of the various needs. Some requirements may be "absolutes," while others may be relative. Absolutes are those about which there can be no compromise. For example, if ductility is a "must," gray cast iron

---

[1]Another example was the complete failure of a very large steel storage tank during its hydrostatic proof test. The tank would have been operated at a temperature of about 38°C (100°F), but the proof test was to be conducted by filling it with water at ambient temperature. The water and ambient temperatures were about 7°C (45°F), and the tank failed with a brittle fracture, originating in a heat-affected zone adjacent to a weld, with the loss of one life. The designer neglected to consider the brittle fracture properties of the material at the test temperature, and the steel was found to have a ductile-to-brittle transition temperature of 21°C (70°F).

would be ruled out. If it must have electrical conductivity, plastics are out. On the basis of absolutes many materials are quickly eliminated from consideration.

Looking ahead to the actual material selection, it is apparent that, in many cases, no one material will emerge as the obvious choice. Often several materials will, to differing degrees, meet the specific requirements. Compromise and opinion enter into the decision making and it is important that the final choice not overlook a major requirement. The listing and ranking of required properties will go a long way to assure that the person making the selection has considered all of the necessary factors. Should no material meet the requirements and the compromises appear to be too severe, it may be necessary to redesign the product.

Additional factors to consider. When attempting to evaluate candidate materials, the engineer is often forced to utilize "handbook-type data." As noted earlier in the text, it is important to note the specific conditions of the standard tests and question their relevance to the current product. Also, significant data gaps may appear. Here, the best advice would be to consult with the material producers or qualified materials engineers. One should also keep in mind that handbook values represent an average or mean and that actual material properties will vary somewhat.

The introduction of *cost* as a factor is appropriate at this point. Cost is not a service requirement and has not been considered thus far because we have adopted a philosophy that the material must first meet the property requirements to be considered as a candidate. Cost, however, is an important part of the selection process—both material cost and the cost of fabricating the selected material. Often, the final decision involves a compromise between cost, producibility, and service performance or quality. It is desirable for anyone who selects and specifies materials to be familiar with the concepts of value analysis.[2]

Materials availability is another consideration. Often the desired material may not be available in the quantity, size, or shape desired, or may not be available in any form. This may well force the use of a less desirable substitute.

Still other factors should be considered when making material selections:

1. What is the possible misuse of the product by the user? If the product is to be used by the general public, one should anticipate the worst.
2. Have there been prior failures of this product or component or similar ones? Often failure analysis information fails to get back into the hands of designers.
3. Has this material or class of materials established a favorable or unfavorable performance record?
4. Can I take advantage of materials standardization?

Although one should not sacrifice function, reliability, and appearance for standardization, neither should one overlook the possible savings and other benefits that often can be obtained without sacrificing any of the required properties.

[2]See L. D. Miles, *Techniques of Value Analysis and Engineering*, McGraw-Hill Book Company, New York, 1961; or E. P. DeGarmo, J. R. Canada, and W. G. Sullivan, *Engineering Economy*, 6th ed., Macmillian Publishing Co., Inc., New York, 1979, Chapter 4.

Looking ahead to manufacturing considerations. On the basis of the previous considerations, tentative selection of several candidate materials should be made. If one is clearly outstanding, it may be selected at this point; but this is rarely the case.

Further refinement of the selection requires consideration of the possible fabrication processes and the suitability of each "prescreened" material to each process. Familiarity with the manufacturing alternatives is a must along with a knowledge of the associated limitations, economics, product quality, surface finish, precision, and so on.

Selection of an appropriate manufacturing system is yet another engineering judgment and will be treated later. What is significant here, however, is the knowledge that all processes are not compatible with all materials (e.g., steels cannot be economically die cast). Moreover, the designer who implies the nature of the manufacturing process in the specifics of design (i.e., cast versus forge, etc.) may well be eliminating the best particular combination.

Thus the real objective is to arrive at a combination of material and manufacturing process that is the "best" solution for your particular product, that is, a *manufacturing system*. While aspects have been presented in a sequential manner, they are all interrelated. Additional production requirements tend to restrict candidate materials. Specific materials tend to limit fabrication possibilities. The fabrication method influences material properties. Processes designed to alter certain properties may adversely affect other properties. Economics, environment, recycling, serviceability, inspection, energy efficiency, and other concerns influence decisions. The engineer working in this area must be capable of exercising sound judgment. He must understand the product, the materials, the processes, and all of the interrelations. The development of new materials, technological advances in processing methods, increased restrictions in environment and energy, and the demand for products which continue to push the limits of capability further challenge these individuals and assure their value to employers.

Materials substitution. Often in response to advancing technology or market pressures, a new material is substituted without substantially altering the design. While the net result is often improved quality or reduced cost, it is possible to overlook certain features and cause more harm than good.

Consider the drive to produce a lighter-weight, more-fuel-efficient automobile. The development of high-strength low-alloy steel sheets (HSLA) offered the ability to match existing strength levels with thinner sheets and thereby reduce weight. Having overcome some processing problems, the substitution appeared to be a natural one. However, the engineer should be aware of certain compromises. While strength was increased, corrosion resistance and elastic stiffness were essentially unaltered. Thinner sheets would corrode in shorter time and undesirable vibrations would be more of a problem. Design modifications might be necessary.

Aluminum castings have been proposed to replace cast iron transmission housings. Strength and corrosion resistance may be adequate and weight savings are substantial. However, aluminum tends to transmit noise and vibration, whereas cast iron is a damping material. Sound isolation material may have to be added.

While numerous examples may be cited, it should be obvious that the responsible engineer should first reconsider all of the design requirements before authorizing a substitution. Approaching a design or material modification as thoroughly as a new problem may well avoid costly errors.

The effect of product liability on material selection. Product liability actions and court awards, particularly in the United States, have made it imperative that designers and companies employ the very best procedures in selecting materials. While many people would agree that the situation has gotten almost out of control, there unfortunately have been too many instances where sound procedures were not used in selecting materials. No company designer can afford to do so under today's conditions. The five most common faults have been: (1) failure to know and use the latest and best information available about the materials utilized; (2) failure to foresee, and take into account the *reasonable uses* for the product (where possible, the designer is further advised to foresee and account for misuse of the product, as there have been many product liability cases in recent years where the claimant, injured during misuse of the product, has sued the manufacturer and won); (3) the use of materials about which there was insufficient or uncertain data, particularly as to its long-term properties; (4) inadequate, and unverified, quality control procedures; and (5) material selection made by people who are completely unqualified to do so.

An examination of the faults above will lead one to conclude that there is no good reason why they should exist. Consideration of them provides guidance as to how they can be eliminated. While following the very best methods in material selection may not eliminate all product-liability claims, the use of proper procedures by designers and industries can greatly reduce their numbers.

Aids to material selection. From the previous discussion in this chapter, it is apparent that those who select materials should have a broad, basic understanding of the nature and properties of materials and their processing. This is, of course, a primary purpose of this book. However, the number of materials is so great, as is the mass of information that is available and useful in specific situations, that a single book of this type and size cannot be expected to furnish all the information required. Anyone who does much work in materials selection needs to have ready access to other sources of data.

One very useful source is the "Materials Selector" issue of *Materials Engineering*. This is a monthly magazine, and one issue each year, usually in November, contains only tabulated data and advertising about all the common, current engineering materials.

Another "must" is the materials-related volumes of the *Metals Handbook,* published by the American Society for Metals (ASM). This would be Volumes 1 through 3 of the new Ninth Edition (1978–1980). These volumes deal solely with the properties and selection of metals. They are voluminous and detailed— not as easy to use for quick reference as the "Materials Selector," but very complete and authoritative.

The *ASM Metals Reference Book* (1981) provides extensive data about metals and metalworking, all in tabular or graphic form.

The annual "Databook Issue" of *Metals Progress* magazine, published by

ASM, contains a vast amount of useful data regarding materials, process engineering, and fabrication technology.

In addition to the above, persons selecting material should have available several of the handbooks published by the various technical societies and trade associations. They may be material-related (the Aluminum Association's *Aluminum Standards and Data,* the Copper Development Association's *Standards Handbook: Copper, brass, and bronze*), process-related (*Steel Castings Handbook* by the Steel Founders' Society of America, *Heat Treater's Guide* by the American Society for Metals), or profession-related (*SAE Handbook,* Society of Automotive Engineers; *ASME Handbook,* American Society of Mechanical Engineers). These may be supplemented by a variety of supplier-published references. While these are low-cost, excellent references, the user should recognize that they are clearly focused on, and often promote, the product of the supplier.

Another area of necessary information is that of comparative costs of materials. Since these tend to fluctuate, it may be necessary to use a publication such as the weekly *American Metal Market* newspaper. Costs associated with various processing operations are more difficult to obtain and vary greatly from company to company. These costs may be available within a given firm or may have to be estimated. A good source here is the *American Machinist Manufacturing Cost Estimating Guide.*

A comparative rating chart, such as the one shown in Figure 10-3, can be a useful tool when selecting materials. Various desired properties are weighted as to their significance and candidate materials are evaluated on a scale of 1 to 5 relative to each property. Thus, candidate materials are compared by a means that could conceivably be computerized and utilize a large data base of material

RATING CHART FOR SELECTING MATERIALS

Material	Go-No-Go** Screening			Relative Rating Number (†Rating Number x *Weighting Factor)								Material Rating Number
	Corrosion	Weldability	Brazability	Strength (5)*	Toughness (5)	Stiffness (5)	Stability (5)	Fatigue (4)	As-Welded Strength (4)	Thermal Stresses (3)	Cost (1)	$\frac{\Sigma \text{ Rel Rating No.}}{\Sigma \text{ Rating Factors}}$

*Weighting Factor (Range = 1 Lowest to 5 Most Important)
†Range = 1 Poorest to 5 Best
**Code = S = Satisfactory
          U = Unsatisfactory

FIGURE 10-3 Rating chart that may be used for comparing materials.

information. In addition, the designer, in making his chart, places all requirements on a single sheet of paper, thereby assuring that a major need will not be
overlooked.

REVIEW QUESTIONS

1. What are the three usual phases of product design and how does the consideration of materials differ in each?
2. What are some of the pitfalls of the "case-history" approach to materials selection?
3. Why is it important to do a thorough job of defining product needs before progressing on to materials selection?
4. What is the difference between "absolute" and "relative" requirements?
5. Why is it important to consider the type of user when designing and selecting materials for a product?
6. Why must the engineer consider fabrication processes when selecting material for a product?
7. Why should one exercise caution when substituting a new material in an existing design?
8. What are some of the most common faults that have resulted in product liability cases?
9. Why is a thorough and established routine or procedure desirable in selecting materials?
10. Why might a single-page comparative rating chart be a useful tool in materials selection?
11. Why is it important to maintain good records of how material-selection decisions were made?
12. What are several good sources of data on engineering materials.
13. Three materials, X, Y, and Z, are available for a certain usage. Any material selected must have good weldability. Tensile strength, stiffness, stability, and fatigue strength are required, with fatigue strength being considered most important and stiffness the least important of these factors. The three materials are rated as follows in these factors:

	X	Y	Z
Weldability	Excellent	Poor	Good
Tensile strength	Good	Excellent	Fair
Stiffness	Good	Good	Good
Stability	Good	Excellent	Good
Fatigue strength	Fair	Good	Excellent

Using the rating chart shown in Figure 10-3, which material should be selected?

Special Problem   This problem is proposed as an exercise to correlate the various aspects of material selection.

Select a common home, office, or garage item, such as a pair of scissors, appliance part, automobile component, cooking utensil, hand tool, or projector case.

For the selected part, consider and answer the following questions:

a. What is the part supposed to do? What functions must it perform? Under what conditions must it operate?
b. What properties must the material possess in order to perform adequately?

c. What material or materials might you suggest for this part? Why?
d. How would you suggest that this part be fabricated? Why?
e. Would the part require heat treatment? What type?
f. Would the part require surface treatment? Why? What type?

CASE STUDY 10. The Underground Steam Line

An underground steam line for a military base in Alaska, approximately 1.6 km (1 mile) in length, utilized the units depicted in Figure CS-10 each unit being 9.14 meters (30 feet) in length. As shown, the steam line for each unit was enclosed in a cylindrical conduit made of sheet steel, and weighed approximately 356 kg (785 pounds). The conduit was to serve as the return drain line and to provide insulation. Each length of steam line was supported within the conduit by means of three U-shaped legs, made by cold-bending hot-holled steel bar stock. See section A-A for a cross section. These legs were welded to the pipe about 2 meters from each end.

The units were fabricated in California and were transported to Alaska by a sequence of truck, barge, and railroad. Because of an early winter, only one-half of the line was installed the first summer. Before work was resumed the following spring, someone decided to test the line and found that there were numerous holes in the conduit. The holes (slits in casing) were located where the edges of the U-shaped supporting legs contacted the outside conduit (casing) on the tops of the casing. As the result of these holes, the project was delayed and a cost litigation ensued.

1. Why do you think these holes occurred? Was it a design error? A fabrication error? A service error? A service environment error? When did the defects actually develop?
2. What action would you recommend to avoid this problem in the future?
3. What should be done about the part of the line already installed?

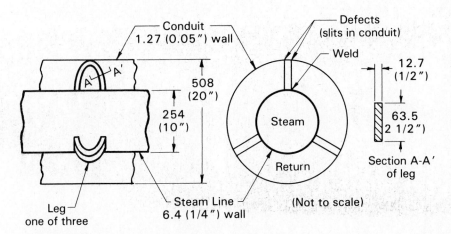

FIGURE CS-10 Schematic diagram showing method of supporting steam line in drain conduit for an underground steam line.

CASTING AND FORMING PROCESSES

CHAPTER 11

Casting Processes

Materials processing is the science and technology by which a material is converted into a final useful shape possessing the necessary structure and properties for its intended use. More loosely, processing is "that which is done to convert stuff into things." Formation of the desired shape is a major portion of processing and casting is a popular means to achieve this goal.

In casting, a solid is melted, heated to a proper temperature, and treated to produce a desired chemical composition. The molten material, generally metal, is then poured into a cavity or mold which contains it in proper shape during solidification. Thus, *in a single step,* simple or complex shapes can be made from any metal that can be melted, with the resulting product having virtually any configuration the designer desires for best resistance to working stresses, minimal directional properties, and usually, a pleasing appearance.

Although some nonmetals are cast, the process is of primary importance in the production of metal products, and only metal casting will be considered in this chapter. The metals most frequently cast are iron, steel, aluminum, brass, bronze, magnesium, and certain zinc alloys. Of these, iron, because of its fluidity, low shrinkage, strength, rigidity, and ease of control, is outstanding for its suitability for casting and is used more than all others.

Cast parts range in size from a few millimeters and weighing a fraction of a gram, such as the individual teeth in a zipper, to 10 or more meters (32.8 feet) and weighing many tons, such as the huge propellors and stern frames of ocean liners. Its use is not restricted to those types of parts, for casting has marked advantages in the production of complex shapes, parts having hollow sections or internal cavities, parts that contain irregular curved surfaces (except those made from thin sheet metal), very large parts, and parts made from metals that are difficult to machine. Because of these obvious advantages, casting is one of the most important of the manufacturing processes. It is currently the sixth largest industry in the United States, involving nearly 4500 companies.

Today it is nearly impossible to design anything that cannot be cast by means of one or more of the available casting processes. However, as with other manufacturing processes, best results and economy can be achieved if the designer understands the various casting processes and adapts his designs so as to use the process most efficiently.

In all casting processes, six basic factors are involved. These are as follows:

1. A mold cavity, having the desired shape and size and with due allowance for shrinkage of the solidifying metal, must be produced. Any complexity of shape desired in the finished casting must exist in the cavity. Consequently, the mold material must be such as to reproduce the desired detail and also have a refractory character so that it will not be significantly affected by the molten metal that it must contain. Either a new mold must be prepared for each casting, or it must be made from a material that can withstand being used for repeated castings, the latter being called *permanent molds*. Inasmuch as permanent molds must be made of metal or graphite and are costly to make, considerable effort has been devoted to methods for economically producing single-usage molds that will enable castings to be made with good accuracy.

2. A suitable means must be available for melting the metal that is to be cast, providing not only adequate temperature but also satisfactory quality and quantity at low cost.

3. The molten metal must be introduced into the mold in such a manner that all air or gases in the mold, prior to pouring or generated by the action of the hot metal upon the mold, will escape, and the mold will be completely filled. A quality casting must be dense and free from defects, such as air holes.

4. Provision must be made so that the mold will not cause too much restraint to the shrinkage that accompanies cooling after the metal has solidified. Otherwise, the casting will crack while its strength is low. In addition, the design of the casting must be such that solidification and solidification shrinkage can occur without producing cracks and internal porosity or voids.

5. It must be possible to remove the casting from the mold. Where the casting is done in molds made from materials such as sand, and the molds are broken up and destroyed after each casting is made, there is no serious difficulty. However, in certain processes where molds of a permanent nature are used, this may be a major problem.

6. After removal from the mold, finishing operations may need to be performed to remove extraneous material that is attached to the casting as the result of the method of introducing the metal into the cavity, or is picked up from the mold through contact with the metal.

CASTING AND FORMING PROCESSES

Much of the development that has taken place in the foundry industry has been directed toward meeting these six objectives with greater economy. Seven major casting processes currently are used. These are:

1. Sand casting.
2. Shell-mold casting.
3. Permanent-mold casting.
4. Die casting.
5. Centrifugal casting.
6. Plaster-mold casting.
7. Investment casting.

Sand casting accounts for, by far, the largest proportion of the total tonnage of castings produced. However, the use of permanent-mold, die, investment, and shell-mold castings has expanded rapidly in recent years.

SAND CASTING

Sand casting uses sand as the mold material. The sand grains, mixed with small amounts of other materials to improve the moldability and cohesive strength, are packed around a pattern that has the shape of the desired casting. Because the grains will pack into thin sections and can be used economically in large quantities, products covering a wide range of sizes and detail can be made by this method. A new mold must be made for each casting, and gravity usually is employed to cause the metal to flow into the mold. Except in the full-mold process, after the sand has been packed firmly around it, the pattern must be removed to leave a cavity of the desired shape. Consequently, the mold must be made in at least two pieces. An opening, called a *sprue hole,* is provided from the top of the mold through the sand and connected to the cavity through a system of channels, called *runners*. The molten metal is poured into the sprue hole and enters the cavity through the runners and an opening called a *gate,* which controls the rate of flow. These essential steps and components are illustrated in Figure 11-1.

Patterns. The first requirement in sand casting is the design and making of a pattern. This is a duplicate of the part that is to be cast but modified in accordance with the basic requirements of the casting process and the particular molding technique that is to be employed. The pattern material is determined primarily by the number of castings to be made. Wood patterns are relatively easy to make and are frequently used when small quantities of castings are required. Wood, however, it is not very dimensionally stable, as it may warp or swell with changes in humidity. Metal patterns are more expensive, but are more dimensionally stable and longer lasting. Hard plastics, such as urethanes, are being used more frequently, particularly with some of the strong, organically bonded sands that tend to stick to other pattern materials. In the full-mold process, expanded polystyrene (EPS) is used, but here each pattern can be used only once.

FIGURE 11-1 Essential steps in sand casting. (a) Bottom (drag) half of pattern in place on mold board between halves of flask ready to receive sand. (b) Drag half of mold completed, ready for turning over. (c) Top (cope) half of pattern and sprue and riser pins in place. (d) Cope half of mold packed with sand.

The modifications that must be incorporated into a pattern are called *allowances*. Probably the most significant of these is the *shrinkage allowance*. Following solidification, the casting continues to contract as it cools, the amount of contraction being as much as 2% or $\frac{1}{4}$ inch per foot. Thus the pattern must be made slightly larger than the desired casting as a means of compensation. The exact allowance is dependent upon the metal that is to be cast. The following allowances are typical.

Cast iron	0.8–1.0% ($\frac{1}{10}$–$\frac{1}{8}$ in./ft)
Steel	1.5–2.0% ($\frac{3}{16}$–$\frac{1}{4}$ in./ft)
Aluminum	1.0–1.3% ($\frac{1}{8}$–$\frac{5}{32}$ in./ft)
Magnesium	1.0–1.3% ($\frac{1}{8}$–$\frac{5}{32}$ in./ft)
Brass	1.5% ($\frac{3}{16}$ in./ft)

The patternmaker often incorporates these allowances into the pattern by using special *shrink rules,* which are longer than a standard rule by the desired shrink allowance. Some caution should be exercised, however, for thermal contraction may not be the only factor affecting dimensions after solidification. Various phase transformations can often bring about significant expansions or contrac-

(e)

(e')

(f)

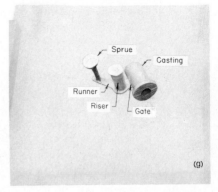
(g)

(e) Mold opened, showing parting surface of drag half, with pattern drawn (removed) and runner and gate cut. (e') Parting surface of cope half of the mold, with pattern and pins removed. (f) Mold closed, ready for pouring metal. (g) Casting removed from mold.

tions. These include eutectoid reactions, martensitic reactions, and graphitization.

In the casting processes where the pattern must be withdrawn from the mold, the mold must be made in two or more sections. This requires that consideration be given to the *parting line* or *surface* where one section fits against the sand of the other section, and to the *draft* or *taper* that must be provided on the pattern to facilitate its withdrawal. These are illustrated in Figure 11-2. If the surfaces of the pattern normal to the parting line were all exactly parallel with the direction in which the pattern had to be moved for withdrawal; the friction between the pattern and the sand, and any movement of the pattern perpendicular to this direction, would tend to cause sand particles to be broken away from the mold. This would be particularly severe at corners between the cavity and the parting surface. By providing a slight taper on all the surfaces parallel with the direction of withdrawal, this difficulty is minimized. Because of this draft, as soon as the pattern is withdrawn a slight amount, it is free from the sand on all surfaces, and it can be withdrawn without damaging the mold.

The amount of draft is determined by the size and shape of the pattern, the depth of the draw, the method used to withdraw the pattern, pattern material,

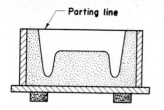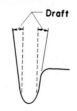

FIGURE 11-2 Relationships of draft to the mold parting line in castings.

mold material, and the molding procedure. Draft seldom is less than 1° or 1:100 ($\frac{1}{8}$ inch per foot),[1] with a minimum of about 1.6 mm ($\frac{1}{16}$ inch) on any surface. On interior surfaces where the opening is small, such as a hole in the center of a hub, the draft should be increased to about 1:24 ($\frac{1}{2}$ inch per foot). Because draft always increases the size of the pattern, and thus the size and weight of the casting, it is always desirable to keep it to the minimum that will permit satisfactory pattern removal. Modern molding procedures, which provide higher strength to the molding sand before the pattern is withdrawn, and the use of molding machines, which substitute mechanical for manual pattern drawing, have permitted substantial reductions in draft allowances. These facilitate the production of light castings with thinner sections, thus saving weight and machining.

When machined surfaces must be provided on castings, it is necessary to provide a *finish allowance*. The amount of the finish allowance depends to a great extent on the casting process and the mold material. Ordinary sand castings have rougher surfaces than shell-mold castings. Die castings are sufficiently smooth that very little or no metal has to be removed, and investment castings frequently do not have to be machined. Consequently, the designer should relate the finish allowance to the casting process and also remember that draft may provide part or all of the extra metal needed for machining.

Some shapes require an allowance for *distortion*. For example, the arms of a U-shaped section may be restrained by the mold, while the base of the U is free to shrink, resulting in the arms sloping outward. Long, horizontal sections tend to sag in the center unless adequate support is provided by suitable ribbing. This type of distortion depends greatly on the particular configuration, and the designer must use experience and judgment in providing any required distortion allowance.

The manner in which the various allowances are included in the pattern for a simple shape is illustrated in Figure 11-3. In general, allowances tend to increase the weight of the casting and the amount of metal that has to be removed by machining. Modern mechanical molding and pattern-drawing equipment, and processes that harden and strengthen the sand before the pattern is withdrawn, have become increasingly popular because of their ability to permit reduced pattern allowances.

Types of patterns. A number of basic types of patterns are in common use, the particular type being determined primarily by the number of duplicate castings required and the complexity of the part.

---

[1] In this instance, and in similar cases that follow throughout the book, where the values are only typical of customary practice and thus are not specified precisely, the SI and English values given are not exact conversions.

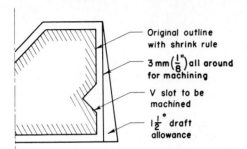

Original outline
with shrink rule

3 mm ($\frac{1}{8}$") all around
for machining

V slot to be
machined

$1\frac{1}{2}°$ draft
allowance

FIGURE 11-3 Various allowances provided on patterns.

One-piece or *solid patterns,* such as are shown in Figure 11-4, are the simplest and cheapest type to make. Essentially, such a pattern is a duplicate of the part to be cast, except for the provision of the various allowances and the possible addition of core prints. (See discussion of core prints in this chapter.) Although this type of pattern usually is inexpensive to construct, in most cases the mold-making process is slow and thus is used only where one or a few duplicate castings are to be made.

Unless a one-piece pattern is quite simple in shape and contains a flat surface that can be placed on the follow board to form a plane parting surface, it may be necessary for the molder to cut out an irregular parting surface by hand. This is time-consuming and costly and requires a skilled workman. This can be avoided by using a special follow board that is dug out so that the one-piece pattern fits down into the follow board up to the depth of the parting-line location. This type of follow-board pattern is illustrated in Figure 11-5. The follow board determines the parting surface, which is usually a plane, but may be curved.

Split patterns are used where moderate quantities of duplicate castings are to be made and also to permit the molding of more complex shapes without resorting to forming the parting plane by hand or using cut-out follow boards. The pattern is split into two sections along a single plane, which will correspond to the parting plane of the mold, as shown in Figure 11-6. One half of the pattern forms the cavity in the lower, or drag, portion of the mold, and the other half serves a similar function in the upper, or cope, section. Tapered pins in the

FIGURE 11-4 Single-piece pattern for a pinion gear.

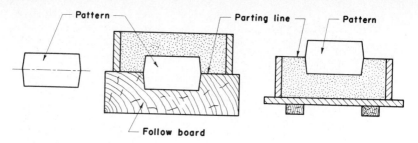

FIGURE 11-5 Method of using a follow-board pattern.

cope half of the pattern fit into corresponding holes in the drag half to hold the halves in proper position while the cope is being filled with sand.

Match-plate patterns, such as are shown in Figure 11-7, are widely used in modern foundry practice because of their suitability for use with modern molding machines to produce large quantities of duplicate castings. In these, the cope and drag sections are fastened to opposite sides of a wood or metal match plate that is equipped with holes or bushings that mate with pins or guides on the halves of the mold flask. The match plate is fitted between the two flask sections, and the sand is packed into the flask to complete the cope and the drag. The entire match-plate pattern is then removed by separating the mold sections. Upon closing the mold, the cavities in the cope and drag will be in proper alignment with respect to each other, because the two halves of the pattern were correctly positioned on the two sides of the match plate, and the guide holes and pins assure identical alignment when the mold is reassembled.

In most cases, the necessary gate and runner system is included on the match plate. This serves the double purpose of eliminating the necessity for the molder to cut the gates and runner by hand and also of assuring that they will be uniform and of the proper size in each mold, thus reducing the likelihood of defects. The gate and runner system can be seen as the dark center section on the match-

FIGURE 11-6 Split pattern, showing the two sections together and separated. Light-colored portions are core prints.

FIGURE 11-7 Match-plate pattern for molding two parts. (*Left*) Cope side; (*right*) drag side.

plate pattern in Figure 11-7, and on the cope section of Figure 11-8. These patterns also illustrate the common practice of having more than one pattern on a match plate. Core prints also are provided when required.

When large quantities are to be produced, or when the casting is quite large, it may be desirable to have the cope and drag halves of split patterns attached to two separate match plates instead of being attached to the opposite sides of a single plate. This permits large molds to be handled more easily, or for two workers, on two machines, to simultaneously produce the two portions of a mold. Such patterns are called cope-and-drag patterns. Figure 11-8 shows a pattern of this type.

FIGURE 11-8 Cope and drag pattern for molding two heavy parts. (*Left*) Cope section; (*right*) drag section.

FIGURE 11-9 Loose-piece pattern for molding a large worm. After sufficient sand is packed around the pattern halves to hold the pieces in position, the wooden pins are withdrawn. The remaining sand is then rammed around the pattern, the mold is opened, and the pattern is removed.

When an object to be cast has protruding sections arranged such that neither a one-piece pattern nor one split along a single parting plane can be removed from the molding sand, *a loose-piece pattern* sometimes can be used. Such a pattern is shown in Figure 11-9. Loose pieces are held to the remainder of the pattern by beveled grooves or by pins. This construction permits all of the pattern except the loose pieces to be withdrawn directly from the sand, after which there is space within the cavity so that the loose pieces may be moved the necessary direction and amount to permit their removal. If the loose pieces cannot be held to the main portion of the pattern by grooves or stationary pins, a long, sliding pin may be used. After the sand is rammed, the pin is withdrawn, thereby freeing the loose pieces from the remainder of the pattern, permitting it to be removed. Obviously, loose-piece patterns are expensive to make, require careful maintenance, slow the molding process, and increase molding costs. They do make possible the casting of complex shapes that otherwise could not be cast, except by the full-mold or investment processes. However, it is desirable to eliminate their necessity by design changes when practicable.

Simple *sweeps* sometimes can be used in place of three-dimensional patterns. Ordinarily, they are used where the shape to be molded can be formed by the rotation of a curved-line element about an axis, as shown in Figure 11-10. The simple sweep eliminates the necessity for making a large and expensive pattern.

It is very important on all castings that intersecting surfaces be joined by a small radius, called a fillet, instead of permitting them to intersect in a line. Fillets avoid shrinkage cracks at such intersections and also eliminate stress concentrations. As a general rule, designers should make fillets generous in size,

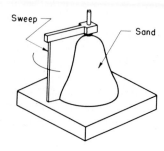

FIGURE 11-10 Method of making a mold using a "sweep."

6.35- and 3.18 mm ($\frac{1}{4}$- and $\frac{1}{8}$-inch) radii commonly being used. Fillets are added to wood patterns by means of wax, leather, or plastic strips of the desired radius, which are glued to the pattern or pressed into place with a heated fillet tool.

Sand conditioning and control. Sand used to make molds must be carefully conditioned and controlled in order to give satisfactory and uniform results. Ordinary silica (SiO_2), zircon, or olivine (forsterite and fayalite) are compounded with additives to meet four requirements:

1. *Refractoriness:* the ability to withstand high temperatures.
2. *Cohesiveness* (referred to as *bond*): the ability to retain a given shape when packed in a mold.
3. *Permeability:* the ability to permit gases to escape through it.
4. *Collapsibility:* the ability to permit the metal to shrink after it solidifies.

Refractoriness is provided by the basic nature of the sand. Cohesiveness, bond, or strength is obtained by coating the sand grains with clays, such as bentonite, kaolite, or illite, that become cohesive when moistened. Collapsibility is obained by adding cereals or other organic materials such as cellulose, that burn out when exposed to the hot metal, thereby reducing the volume of solid bulk and decreasing the strength of the restraining sand. Permeability is primarily a function of the size of the sand particles, the amount and types of clays or other bonding agents, and the moisture content.

Good molding sand always represents a compromise between conflicting factors, the grain size of the sand particles, the amount of bonding agent (such as clay), the moisture content, and the organic matter being selected so as to obtain a satisfactory combination of the four requirements listed previously. Sand composition must be carefully controlled to assure satisfactory and consistent results. A typical green sand mixture consists of 89% silica sand, 8% clay, and 3% water. In addition to the requirements above, the desire to reclaim and reuse the mold material requires control of its temperature during the pour.

Each grain of sand should be coated uniformly with the additive agents. To achieve this, the ingredients are put through a *muller,* which provides the necessary mixing. Figure 11-11 shows a modern, continuous-type muller. The sand is often discharged from the muller through an *aerator,* which fluffs the sand so that it does not tend to pack too hard in handling.

In modern foundry practice many molds and cores are made of sand that is given increased strength through the addition of about 4% silicate of soda or various furfurals or furfural alcohols. Sand mixed with silicate of soda remains soft until exposed to CO_2 gas, after which it hardens in a few seconds. The setting is due to the reaction

$$Na_2SiO_3 + CO_2 \rightarrow Na_2CO_3 + SiO_2 \quad \text{(colloidal)}$$

Sands of this type can be mixed with the silicate of soda in a muller and handled in the normal manner, inasmuch as they are not gassed with CO_2 until after they have been packed around the pattern in the mold flask. Furfurals, with a catalyst, usually will cause hardening within a few minutes. Consequently, they are added and mixed with the sand by special equipment just prior to delivering the sand to the molding station.

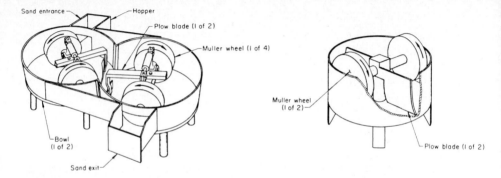

FIGURE 11-11 Schematic diagram of sand mullers. (*Left*) Continuous muller; (*right*) conventional batch muller. Plow blades move the sand and the muller wheels mix the components. (*From ASM Committee on Sand Molding, "Sand Molding," Metals Handbook, Vol. 5, 8th ed., Taylor Lyman, ed., American Society for Metals, Metals Park, Ohio, 1970, p. 163.*)

The major advantages of CO_2–sodium silicate method are that the CO_2 gas is nontoxic and odorless and therefore need not be contained and no heating is required for curing. When hardened, however, these sands have very poor collapsibility, making shakeout and core removal difficult. Unlike other sands, the heating experienced during the pour actually makes the sand stronger (as in firing a ceramic material).

Sand control.   Although sand control is of little concern to the designer of castings, it is a matter of great concern to the foundry worker, who is expected to deliver castings of good and consistent quality. Standard tests and procedures have been developed to maintain consistent sand quality by evaluating *grain size, moisture content, clay content, mold hardness, permeability, and strength.*

Grain size is determined by shaking a known amount of dry silica grains downward through a set of 11 standard sieves having increasing fineness. After shaking for 15 minutes, the amount remaining in each sieve is weighed, and the weights are converted into an AFS[2] number.

Moisture content most commonly is determined by a special device which measures the electrical conductivity of a small sample of sand that is compressed between two prongs. Another device provides a continuous measure of the moisture content, by emission from a radioactive source, as the sand passes along a conveyor belt. A third method is to measure the direct weight loss from a 50-gram sample when it is subjected to a flow of air at about 110°C (230°F) for 3 minutes.

Clay content is determined by washing the clay from a 50-gram sample of molding sand in water that contains sufficient sodium hydroxide to make it alkaline. After several cycles of agitation and washing in such a solution, the clay will have been removed. The remaining sand is dried and then weighed to determine the proportion of the original sample that was clay.

Permeability, and strength tests are conducted on a standard specimen, 2

[2]American Foundrymen's Society.

FIGURE 11-12 Sand rammer for preparing a standard rammed foundry sand specimen. (*Courtesy Harry W. Dietert Company.*)

inches in diameter and 2 inches in height, that is prepared by use of the device shown in Figure 11-12. The sand is compacted inside a steel tube by means of a weight that is dropped three times from a predetermined height.

Permeability is a measure of how easily gases can pass through the narrow voids between the sand grains. Air in the mold before pouring, plus steam produced when the hot metal vaporizes the moisture in the sand, must be allowed to escape, rather than be trapped in the casting as porosity or blow holes. For the permeability test, the sample tube containing the rammed sand specimen is placed on the device shown in Figure 11-13 and subjected to an air pressure of 10 grams per square centimeter. By means of either a flow rate determination or measurement of pressure between the orifice and the sand, an AFS permea-

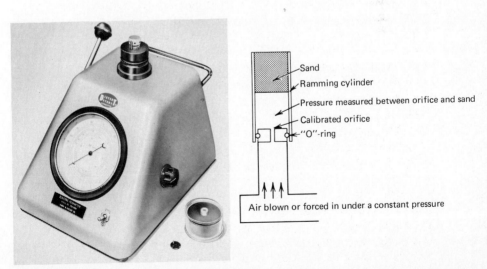

Sand
Ramming cylinder
Pressure measured between orifice and sand
Calibrated orifice
"O"-ring
Air blown or forced in under a constant pressure

FIGURE 11-13 (*Left*) Permeability tester for foundry sand. Standard sample in sleeve is sealed by O-ring on top of unit. (*Courtesy Harry W. Dietert Company.*) (*Right*) Schematic of permeability tester in operation.

FIGURE 11-14 Universal strength-testing machine for foundry sand. (*Courtesy Harry W. Dietert Company.*)

bility number is determined.[3] Most devices provide for direct readout in permeability number.

The compressive strength of the sand is determined by removing the rammed specimen from the tube and placing it in a Universal sand tester such as shown in Figure 11-14. A compressive load is then applied until the specimen breaks, generally in the range 0.07 to 0.2/mPa (10 to 30 psi). The same tester can also be used to measure the transverse bend strength of baked or cured sand specimens.

The hardness to which sand is compacted in a mold is very important because it affects the strength–permeability relationship. It commonly is measured by means of the instrument shown in Figure 11-15, which measures the resistance of the sand to penetration by a 5.08-mm (0.2-inch)-diameter, spring-loaded steel ball.

The making of sand molds. Except in very small foundries or when only a very few castings of a given design are to be made, virtually all sand molds now are made with the use of various types of molding machines. These greatly reduce the labor and skill required and produce castings with better dimensional accuracy and consistency.

Molding machines basically vary in the way the sand is packed within the molding flask, whether mechanical assistance is provided for turning and/or handling the mold, and whether a flask is required. The type of machine shown

[3]An AFS permeability number is used as a measure. This is

$$\frac{V \times H}{P \times A \times T}$$

being the volume of air, in cm³, that will pass through a specimen 1 cm² in area and 1 cm high under a pressure of 1 g/cm², where V is the volume of air (2000 cm³), H the height of the specimen (5.08 cm), P the pressure (10 g/cm²), A the cross section of specimen (20.268cm²), and T the time in seconds required for a flow of 2000 cm³. Because of the constants, the permeability number = 3007.2/T.

CASTING AND FORMING PROCESSES

FIGURE 11-15 Mold hardness tester. (*Courtesy Harry W. Dietert Company.*)

in Figure 11-16 is a common one. With such machines, a match-plate pattern usually is employed, being placed between the two halves of a flask, such as is shown in Figure 11-17. These usually are constructed of aluminum or magnesium and may be either rigid, straight-walled containers with guide pins, or removable jackets. The flask in Figure 11-17 is a snap type that opens slightly to permit it to be slid off the mold after packing is completed.

Packing of the sand is generally done by using one or a combination of several principles. In one, sand is placed on top of the pattern and the pattern, flask and sand are then lifted and dropped several times as shown in Figure 11-18. The kinetic energy of the sand itself produces optimum packing at the pattern. These *jolt-type machines* can be used on the first half of a matchplate pattern or both halves of a cope-and-drag operation.

FIGURE 11-16 Jolt-squeeze, roll-over molding machine. (*Left*) In jolting and squeezing position; (*right*) in roll-over position. (*Courtesy Osborn Manufacturing Company.*)

Squeezing machines pack the sand by the squeezing action of either an air-operated squeezing head; as shown in Figure 11-18, a flexible diaphragm; or small, individually actuated squeeze heads. Squeezing packs the sand firmly near the squeezing head, but the density diminishes as the distance from the head increases. A high-pressure, flexible-diaphragm type of squeezing machine, commonly called a *Taccone machine,* tends to give a more uniform density around all parts of an irregular pattern than can be obtained with a regular flat-plate squeezing machine, as illustrated in Figure 11-19.

Commonly, a combination of jolting and squeezing is used to obtain more uniform density throughout the mold depth. On these the flask, with a match-plate pattern in place, is placed upside down on the table of the machine. After parting dust has been sprinkled on the pattern, the flask is filled with sand. The table and flask then are jolted the desired number of times to pack the sand around the pattern in the lower portion of the flask. The squeeze head then is lowered and pressure is applied to pack the sand in the upper portion of the flask. Both cope and drag may be made on the same machine by rolling the flask over and repeating the operations on the cope half, or the cope and drag portions may be made on separate machines, using cope-and-drag patterns.

Except for very small molds, jolt and/or squeeze machines usually provide mechanical assistance for turning over the heavy mold, as illustrated in Figure 11-16. Some patterns for use on molding machines form the sprue hole, but often it is cut by hand. Often a shallow pouring basin—a depression in the sand at the top of the sprue hole into which the metal is poured and then flows into

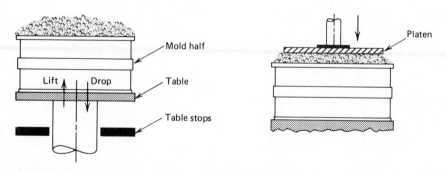

FIGURE 11-18 (*Left*) Jolting a mold half. (*Right*) Squeezing operation.

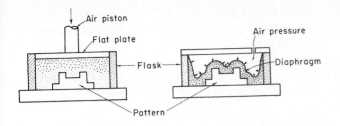

the sprue hole—is formed by a protruding shape on the squeeze board. The pattern will usually include the runner and gate system, but sometimes these are cut by hand. Considerable effort is made in designing and constructing patterns to eliminate the necessity for such handwork.

After a mold is completed, the flask is removed so that it will not be damaged during the pouring of the mold. A *slip jacket*—an inexpensive metal band—may be slid down around the mold to hold the sand in place. Heavy metal weights often are placed on top of the molds to prevent the sections from being separated by the hydrostatic pressure of the molten metal. When used, slip jackets and pouring weights are needed only during pouring and for a few minutes afterward, while solidification occurs. They then can be removed and placed on other molds, so as to keep up with the pouring crew. Thus the amount of expensive equipment is kept to a minimum.

For mass-production molding, a number of machines have been developed that employ one or more of the basic compaction methods, but do this automatically at high compaction pressures. These include automatic matchplate machines and automatic cope and drag machines. Figure 11-20 shows a vertically parted flaskless molding machine. The halves of two separate molds are made at one time, with a complete mold being formed by joining two successive mold blocks. If cores are used, they may be set in place before the halves are closed.

A modification of the CO_2 process is used quite extensively when making certain portions of molds where better accuracy of detail and thinner sections are desired than can be achieved with ordinary molding sand. Sand mixed with silicate of soda is packed around the metal pattern to a depth of about 1 inch,

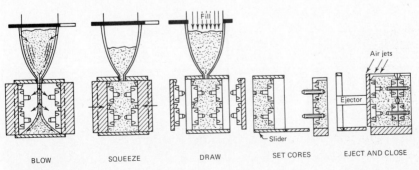

FIGURE 11-20 Method of making molds without the use of a flask. (*Courtesy Belens Corporation.*)

followed by regular molding sand as backing material. After the mold is rammed, CO_2 is introduced through vents in the metal pattern, thereby setting the adjacent sand. The pattern can then be withdrawn with reduced danger of damage to the mold. This procedure is particularly useful where deep draws are necessary. Sand mixed with various "cold-setting" additives, such as furfurals, is used extensively with molding machines of all types to permit thinner sections, deeper draws, and greater accuracy.

For some purposes, molds may be either *skin-dried* or completely dried. This is done to strengthen the mold and also to reduce the amount of gases that will be evolved when hot metal comes in contact with the sand. Molds into which steel is to be cast are nearly always skin-dried just prior to pouring, because much higher temperatures are involved than in the case of cast iron. Such molds may also be given a high-silica wash prior to drying to increase the refractoriness of the surface. Skin drying to the depth of about a half inch is accomplished by means of a gas torch. Frequently, additional binders, such as molasses, linseed oil, or corn flour, must be added to the facing sand when skin drying is to be done.

For castings too large to be made by either bench molding or in the type of molding machines that have been described, large flasks, which rest on the foundry floor, may be employed with various mechanical aids. A *sand slinger*, a mechanized piece of equipment with impeller blades, may be used to fling sand into the flask at high velocity. If skillfully done, uniform compaction of a desired hardness can be achieved. Extra tamping may be done by means of a pneumatic rammer, as illustrated if Figure 11-21.

FIGURE 11-21 Using a pneumatic sand rammer to ram a large floor mold.

FIGURE 11-22 Using dry-sand sections in the construction of a large pit-type mold. (*Courtesy Pennsylvania Glass Sand Corporation.*)

Very large molds usually are made in pits in the floor, as shown in Figure 11-22. Such a mold may be made in essentially the same manner as a floor mold. Large castings usually are complex and portions of the mold cannot be turned over, as can a cope and drag. Moreover the preparation of pit molds is costly and it is desirable to reduce dimensional variations and to provide greater strength and stability to the mold. Thus, molds frequently are built from baked or dried-sand sections. Sections made with the use of furan binders are being used more extensively.

Shell molding. Large numbers of molds now are made by the *shell-molding process*. This process has important advantages in that it provides better surface finish than can be obtained with ordinary sand molding, provides better dimensional accuracy, and less labor is required. In many cases, the process can be completely mechanized. These characteristics make it particularly suitable for mass production. The basic process involves six steps, as illustrated in Figure 11-23.

1. A mixture of sand and a thermosetting plastic binder is dumped onto a metal pattern that is heated to 150 to 232°C (300 to 450°F) and allowed to stand for a few minutes. During this interval the heat from the pattern partially

cures a layer of the sand–plastic mixture, forming a strong, solid-bonded region about 3.2 mm ($\frac{1}{8}$ inch) thick, adjacent to the pattern. The thickness depends upon the pattern temperature and time of contract.

2. The pattern and the sand mixture are inverted to permit all the sand to drop off except the layer of partially cured material adhering to the pattern.
3. The pattern and the partially cured "shell" are then placed in an oven for a few minutes to complete the curing of the shell.
4. The hardened shell is stripped from the pattern.
5. Two shells are glued or clamped together to form a complete mold.
6. Generally, these are placed in a pouring jacket and backed up with shot or sand to provide extra support during the pour.

Because the plastic and sand shell is molded to a metal pattern and is compounded to undergo almost no shrinkage, it has virtually the same dimensional accuracy as the pattern. Tolerances of 0.08 to 0.13 mm (0.003 to 0.005 inch) are readily obtained. The sand used is finer than ordinary foundry sand and, combined with the plastic resin, it results in a very smooth shell and casting

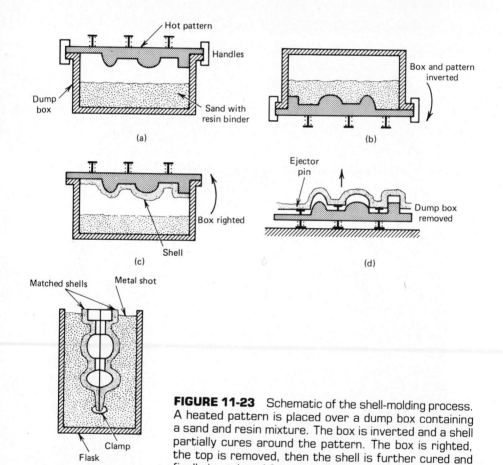

FIGURE 11-23 Schematic of the shell-molding process. A heated pattern is placed over a dump box containing a sand and resin mixture. The box is inverted and a shell partially cures around the pattern. The box is righted, the top is removed, then the shell is further cured and finally is stripped from the pattern. Matched shells are joined and supported in a flask ready for pouring.

surface. Also, the consistency between castings is superior to that obtainable by ordinary sand casting.

Figure 11-24 shows a set of patterns, the two shells before clamping, and the resulting casting. Machines for making shell molds vary from simple types for small operations to large, completely automated types for mass production. The cost of a metal pattern is often rather high and its design must include the gate and runner system. Also, fairly large amounts of expensive binder are required. Nevertheless, the amount of material used to form a thin shell is somewhat small. High productivity, low labor costs, and smooth surfaces and good precision that reduce machining operations all combine to make the process economical for even moderate quantities. In addition, during pouring, the shell becomes very hot, the resin binder burns out, and collapsibility and shakeout characteristics are excellent.

FIGURE 11-24 (*Bottom*) Two halves of a shell-mold pattern. (*Top*) Shell mold and the resulting casting. (*Courtesy Shalco Systems, an Acme-Cleveland Company.*)

The V-process (Vacuum molding). A relatively new sand molding process uses no binder at all, but attains mold strength through use of a specially designed vacuum flask. First, a thin sheet of plastic is draped over a special pattern and is drawn tightly to the pattern surface by a vacuum. Then a vacuum flask is placed over the pattern, the flask is filled with sand, a sprue and pouring cup are formed, and a second sheet of plastic is placed over the sand. A vacuum is then drawn on the flask, making the sand very hard. The pattern vacuum is released and the pattern is then withdrawn. Mold halves are assembled and the mold is poured while mold vacuum is maintained. The sequence is illustrated in Figure 11-25.

The process has numerous advantages. The castings have no moisture-related defects. No binder is used, so binder cost is eliminated and the sand is immediately recyclable. No fumes (binders burning up) are generated during pouring. Any sand or aggregate can be used and shakeout characteristics are exceptional for when the vacuum is released after pouring, the mold virtually collapses. The process is relatively slow, however.

The Eff-set process. In this process, frozen water is the binder. Sand with a small amount of clay and quite a bit of water is packed around a pattern and the pattern is removed. At this point, the mold just has sufficient strength to hold its shape but cannot withstand handling. Liquid nitrogen is then sprayed onto the mold surface and the ice now becomes a firm binder. Molten metal is poured into the mold while it is still frozen. Here again the process benefits from low binder cost and easy shakeout. In addition, the process possesses better than normal fluidity since the vaporizing water tends to provide a "cushion" over which the metal flows.

The full-mold process. The *full-mold process* avoids two bothersome restrictions inherent in the making of ordinary sand molds. These are (1) the cost of

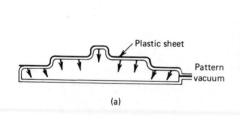

(a)

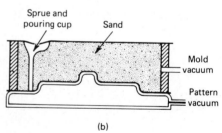

(b)

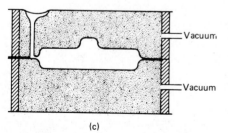

(c)

FIGURE 11-25 Schematic of the V-process or vacuum molding. A vacuum is pulled on a pattern, drawing a plastic sheet tightly against it. A vacuum flask is filled with sand, a second sheet placed on top, and a mold vacuum is pulled. The pattern is withdrawn, mold halves assembled, and the molten metal is poured.

making a relatively expensive wood pattern when only one or a few castings are required, and (2) the necessity for withdrawal of the pattern from the mold, often requiring either some design modification or a complex pattern and/or molding procedure. In the full-mold process, the pattern is made of expanded polystyrene, which remains in the mold *during* the pouring of the metal. When the molten metal is poured, the heat vaporizes the polystyrene pattern almost instantaneously, and the metal fills the space previously occupied by the pattern, as shown in Figure 11-26.

Foamed polystyrene is inexpensive, weighs only about 15.4 kg/m³ (1.2 pounds per cubic foot), is very easily cut, and complex patterns, including the pouring basin, sprue hole, and runner system, can be made by gluing together several simple shapes. Because the pattern is not withdrawn, no draft need be provided. The mold is then completed by packing material around the pattern, taking care not to crush or distort the weak polystyrene. Various sand techniques can be used; vacuum flasks can be employed; or the pattern can be dipped into a refractory wash to provide a solid skin that can be supported by vibrated, loose sand.

The full-mold process can be used for castings of any size. Because of the lessened pattern cost, its application is most economical where only one or a few castings are to be made. It is also useful for complex castings where costly cores or loose-piece patterns might be necessary. Figure 11-27 shows a large styrene pattern and the finished casting ready for machining.

More recently, this has become a large-volume production process. Spheres of polystyrene are blown into heated dies at low pressure where they expand, fill the die, and fuse. The dies can be very complex and large numbers of patterns can be accurately and rapidly produced. Other names for this technique include evaporative pattern casting and the EPS (expanded polystyrene) technique.

Cores and core making. One of the distinct advantages of castings is that hollow or reentrant sections can be included with relative ease. Often such configurations could not be made by any other process. However, to produce such castings it is necessary to use *cores*. Figure 11-28 shows an example of a product that could not be produced by any other process than casting with cores. Of course, cores constitute an added cost, and in many cases the designer can do much to facilitate and simplify their use.

The use of cores can be illustrated by the example of the belt pulley shown in Figure 11-29. The shaft hole could be made in several ways. First, the pulley

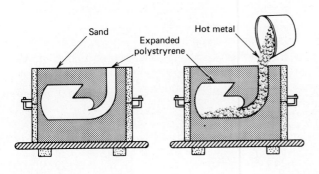

FIGURE 11-26 Schematic of the full-mold process. (*Left*) Expendable pattern in a sand mold. (*Right*) Hot metal vaporizes the pattern and fills the resulting cavity.

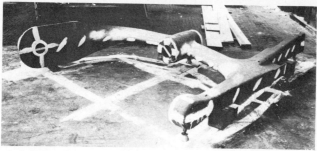

FIGURE 11-27 Polystyrene pattern (*top*) used for making the casting (*bottom*) by the full-mold process. Note that risers, vents, and the sprue have been removed in the lower photo. (*Courtesy Full Mold Process, Inc.*)

could be cast solid and the hole made by machining. However, this procedure would require a substantial amount of costly machining and quite a bit of metal would be wasted. Therefore, it is more economical to cast the pulley with a hole in it of the approximate size desired. One procedure would be to use a split pattern, as shown in Figure 11-29b. Each half of the pattern contains a tapered hole into which green molding sand is packed, just as it is in the remainder of the mold. These sections of sand, which protrude into the hole in the pattern, are called *cores*. In this case, because they are made of green sand, they are called *green-sand cores*. It usually is desirable to make cores of green sand. However, it frequently is impractical to do this because green-sand cores are weak. If they are narrow or long, the pattern cannot be drawn without breaking them, or they will have insufficient strength to support themselves. In addition, the amount of draft that must be provided on the pattern is such that if the core is very long, a considerable amount of machining may have to be done and the advantage of the core is lost. In other cases the shape is such that a green-sand core cannot be used. To overcome these difficulties, it usually is necessary to use *dry-sand cores* as shown in Figure 11-29c and d.

Dry-sand cores can be made in several ways. In each, the sand, mixed with some kind of binding material, is packed into a core box that contains a cavity of the desired shape. The core box, usually made of wood or metal, is analogous

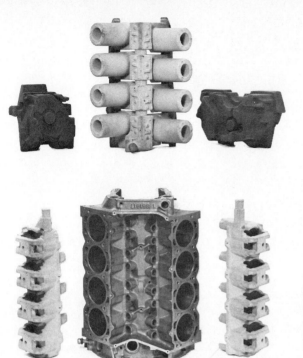

FIGURE 11-28 Dry-sand cores used in the making of a V-8 engine block and the completed casting (*bottom center*). (*Courtesy Central Foundry, Division of General Motors Corporation.*)

to the mold in which a casting is made. The *dump-type core* box, shown in Figure 11-30, is very common. The sand is packed into the box and struck off level with the top surface. Next, a metal plate is placed on top of the box, and the box is turned over and lifted upward, leaving the core resting on the plate.

Some cores can be made in a *split core box,* consisting of two halves that are clamped together with an opening in one or both ends, through which the sand is rammed. The halves of the box are separated to permit removal of the core. Cores that have a uniform cross section throughout their length can be made by a core-extruding machine that is similar to the familiar meat grinder. The cores are cut to length as they come from the machine and placed in core supports for hardening.

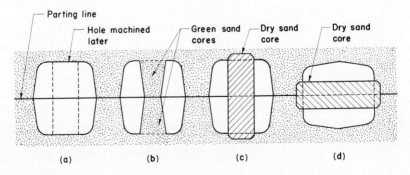

FIGURE 11-29 Four methods of making a hole in a cast pulley.

FIGURE 11-30 (*Clockwise from upper right*) Core box, two core halves ready for baking, and the completed core made by gluing two halves together.

In many cases the sand mix is blown into the core box, using equipment similar to that depicted in Figure 11-31. The oldest and still a common core-making process uses a vegetable or synthetic oil as the binder. Baking is required to polymerize the oil and produce a strong organic bond between the sand grains. In the *hot box* method, the corebox is heated to around 200°C (400°F) so that the surface of the core polymerizes after the sand is blown in. The core can then be removed and be further baked without danger of breakage. In the *cold box* process, the core box is gassed with an organic gas that immediately polymerizes the resin. Here, the gases involved are very toxic, however. Many cores are now made with airset or "no-bake" binders (furan or resins) which harden with little or no heat. The silicate of soda–CO_2 process and the shell-molding method are other alternatives. Shell-molded cores have excellent permeability since they are generally hollow.

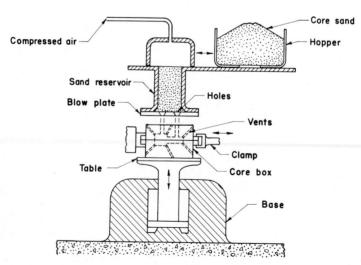

FIGURE 11-31 Method of making cores on a core-blowing machine.

Often cores are molded in halves and assembled after baking or hardening is completed. Rough spots are removed with coarse files or on sanding belts, and the halves are pasted together, often using hot-glue guns. They may then be given a thin coating to produce a smoother surface or greater refractoriness. Graphite, silica, or mica may be used, either sprayed on or brushed on.

To function properly, dry-sand cores must have the following characteristics:

1. Sufficient hardness and strength after baking or hardening to withstand handling and the forces of the molten metal.
2. Sufficient strength before hardening fully to permit any required handling.
3. Adequate permeability to permit the escape of gases.
4. Collapsibility to permit the shrinkage that accompanies cooling of the solidified metal, thereby preventing cracking of the casting and also permitting easy removal from the casting.
5. Adequate refractoriness.
6. A smooth surface.
7. Minimum generation of gases.

Cores occasionally are strengthened by means of wires or rods in order to give sufficient strength for handling and to resist the forces that act against them during pouring. In some cases, particularly in steel casting, where considerable shrinkage is encountered, cores are made hollow at their centers, or straw is put in the center, so they will collapse as the metal cools after solidifying. Failure to provide for free shrinkage may result in cracked castings.

All but very small cores must be vented to enable gases to escape. Vent holes may be produced by pushing small wires into the core or by using waxed strings that burn out and leave vent holes. Coke or cinders sometimes are placed in the center of large cores to provide venting.

When dry-sand cores are used, it usually is necessary to provide recesses in the mold into which the ends of the cores can be placed to provide support and/or to hold them in position. The recesses are known as *core prints*. Figure 11-32 shows molds with core prints containing a vertical and horizontal core respectively.

In some cases the design of a casting does not permit the core to be supported from the sides of the mold. In such instances, the core can be supported, and can be prevented from being moved or floated by the molten metal, by means of small, metal supports called *chaplets*. Figure 11-33 shows some chaplets and

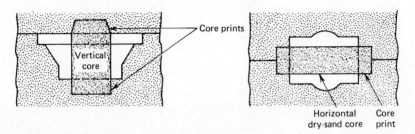

FIGURE 11-32 Mold cavities containing core prints to hold and position dry-sand cores. (*Left*) Vertical core; (*right*) horizontal core.

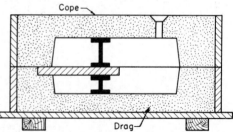

FIGURE 11-33 (*Left*) Typical chaplets. (*Right*) Method of supporting cores by use of chaplets (relative size of chaplets is exaggerated).

illustrates how they are used. The use of chaplets should be minimized because they become a part of the casting and may be a possible source of weakness. When used, chaplets should be of the same, or at least compatible, composition as the casting material. They should be large enough that they do not completely melt and permit the core to float, but small enough that the surface melts and fuses with the metal.

As mentioned previously, dry-sand mold sections are used frequently to facilitate the molding of complex shapes, particularly those containing reentrant sections. Figures 11-34, and 11-35 illustrate how such a procedure may be used advantageously. In Figure 11-34, an intermediate flask section, called a *cheek,* is used to contain a green-sand core. Although considerable skilled handwork is required and the mold must be turned over additional times, such a molding procedure may be advantageous when only a few molds are to be made because it eliminates the necessity of making a special core box. However, if any substantial quantity of the pulley had to be made, the use of a dry-sand core, as illustrated in Figure 11-34, greatly simplifies and speeds the operation. In this case, the pattern is of a different shape, to make a seat for the specially shaped dry-sand core that is set in place to form the reentrant section in the casting. The use of dry-sand cores is especially advantageous where rapid machine molding is used. The procedure, of course, requires less skilled molders. However, the cost of providing a core box and making the cores must be balanced against the saving in labor cost; quantity is a primary factor. If modifications can be

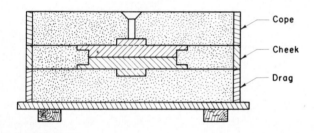

FIGURE 11-34 Method of making a reentrant angle by using a three-piece flask.

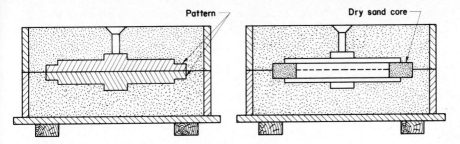

FIGURE 11-35 Molding a reentrant section by using a dry-sand core.

made in the design of a casting that will eliminate the necessity for dry-sand cores or sections and still permit machine molding to be used, this obviously is the best solution.

PERMANENT-MOLD CASTING PROCESSES

Two major disadvantages of the sand-casting processes are the necessity for making a new mold for each casting and the possibility for some dimensional variations. Consequently, much effort has been devoted to developing *permanent-mold casting,* and very substantial success has been achieved. These processes, however, have important limitations which must be considered by designers. The molds are either metal or graphite and, consequently, most permanent-mold castings are restricted to lower melting point nonferrous metals and alloys. However, process modifications exist such that a significant quantity of ferrous castings are now being made by permanent-mold processes.

Nonferrous permanent-mold casting. Of all permanent mold casting, a relatively small amount employs only gravity to introduce the metal. This type often is called *permanent-mold casting,* whereas the other types are given special names. In this process, the molds most commonly are made of fine-grain cast iron or steel, and aluminum-, magnesium-, or copper-base alloys are the metals cast. However, graphite molds are being used increasingly for casting iron and steel. In all cases, the mold halves or sections are hinged so that they can be opened and closed accurately and rapidly by mechanical means.

Permanent molds are heated at the beginning of a run and maintained at a fairly uniform temperature—usually by controlling the casting rate—to avoid too-rapid chilling of the metal. It usually is necessary to coat the cavity surfaces with a thin refractory wash to prevent the casting from sticking to the mold and to prolong the mold life. When used for cast iron, an additional coating of carbon black usually is added, often by means of an acetylene torch. Sand or reusable, retractable metal cores can be used to increase the complexity of the casting.

Numerous advantages can be cited for the process. The mold is reusable. Good surface finish is obtained if the mold is in good condition and dimensional accuracy can usually be within 0.13 to 0.25 mm (0.005 to 0.010 inch). By selectively heating or chilling various portions of the mold, or varying the thickness of the mold wall, the solidification can be controlled to give desired properties. Faster cooling rates generally produce stronger products than result from sand casting.

From the negative side, the process is generally limited to the lower-melting-point alloys. If applied to steels or cast irons, the mold life is extremely short. Even for the low-temperature metals, the mold life is limited due to erosion by the molten metal and thermal fatigue. The actual mold life will vary with the alloy being cast, the mold material, the pouring temperature, mold temperature, and mold configuration. Mold complexity is often restricted because the rigid cavity offers great resistance to shrinkage. Generally, it is good practice to open the mold and remove the casting immediately after solidification, thereby preventing tearing that may occur upon subsequent cooldown. The production cycle must provide sufficient time for full solidification and subsequent cooldown of the mold before another casting is poured. Special provision must be made for venting the molds, because they are not permeable. This usually is accomplished through the slight cracks between the mold halves, or by very small vent holes that will permit the flow of trapped air but not the metal. Mold costs generally require high-volume production to justify the process. Mass production machines are then employed to coat the mold, pour the metal, and remove the casting. Figure 11-36 shows a variety of pistons mass-produced by permanent molding for car and truck engines.

Several variations of permanent-mold casting exist. *Slush casting* is a process wherein the metal is permitted to remain in the mold until a shell, adjacent to the mold cavity, has solidified to the desired thickness. The mold then is inverted, and the remaining molten metal is poured out. The resulting hollow casting has varying wall thickness, but the process is satisfactory for making ornamental objects, such as candlesticks and lamp bases.

Another variation, known as *Corthias casting*, utilizes a plunger that is pushed

FIGURE 11-36 Truck and car pistons, mass-produced by the millions by permanent-mold casting. (*Courtesy Central Foundry, Division of General Motors Corporation.*)

down into the mold cavity, sealing the sprue hole and displacing some of the molten metal into the outer portions of the mold cavity. The positive pressure produces detail and permits thin sections to be cast successfully.

Pressure pouring. The *pressure pouring process,* or low-pressure permanent molding illustrated in Figure 11-37, is used extensively for making railroad-car wheels and steel ingots. A graphite mold is used, and the metal is forced upward into the mold by means of air pressure. When the mold is filled, a plunger is forced downward to seal the sprue hole so that the metal will not drain out of the mold when the pouring pressure is decreased. This requires that the expendable plunger must be at a location where it can be removed from the finished casting as scrap.

Pressure pouring protects the metal entering the mold from the atmosphere, and, as a result, has become quite attractive for aluminum castings. By controlling the air pressure, the mold can be filled from the bottom in a controlled, nonturbulent manner, thereby minimizing gas porosity and dross. Cycle times are generally slow. However, the applied pressure continually feeds metal to compensate for shrinkage and the unused metal in the feed tube simply drops down into the crucible after the pressure is released. Since this metal can be used in subsequent castings, yield may be 90% or more.

Die casting. *Die casting* differs from ordinary permanent-mold casting in that the molten metal is forced into the molds by pressure and held under pressure during solidification. Most die castings are made from nonferrous metals and alloys, but substantial quantities of ferrous die castings now are being produced.

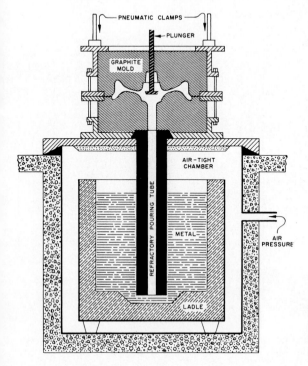

FIGURE 11-37 Schematic diagram of the pressure pouring process. (*Courtesy Amsted Industries.*)

Because of the combination of metal molds or dies, and pressure, fine sections and excellent detail can be achieved, together with long mold life. Special zinc-, copper-, and aluminum base alloys suitable for die casting have been developed which have excellent properties, thereby contributing to the very extensive use of the process.

Because die-casting dies usually are made from hardened tool steel, they are expensive to make. They may be relatively simple, containing only one or two mold cavities, but more often they are complex and may contain eight or more cavities, as shown in Figure 11-38. In order to be opened to remove the castings,

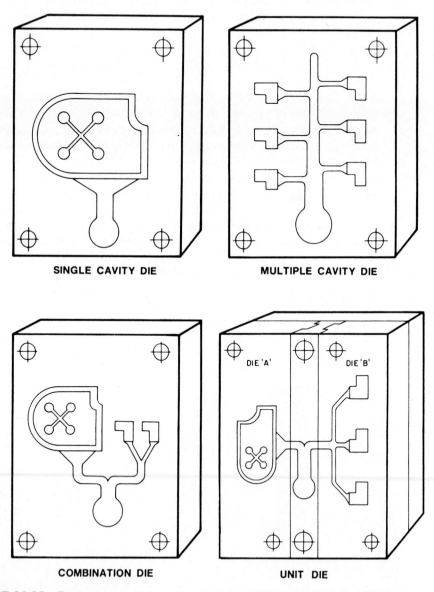

SINGLE CAVITY DIE   MULTIPLE CAVITY DIE

COMBINATION DIE   UNIT DIE

FIGURE 11-38 Some common die-casting die designs. (*Courtesy American Die Casting Institute, Inc., Des Plaines, Ill.*)

the die must be made in at least two pieces; but very often they are much more complicated, containing sections that may move in several directions. In addition, the die sections must contain cooling-water passages and knock-out pins, which eject the casting. Consequently, such dies usually cost in excess of $3000, often over $10,000. It follows, then, that the economical use of die casting is closely related to high production rates, the excellent surface qualities that are obtainable, and the almost complete elimination of machining. The size and weight of die castings are increasing constantly, and parts weighing up to about 9 kg (20 pounds) and measuring up to 610 mm (24 inches) are made routinely.

The die-casting cycle consists of the following steps:

1. Closing and locking the dies.
2. Forcing the metal into the die and maintaining the pressure.
3. Permitting the metal to solidify.
4. Opening the die.
5. Ejecting the casting.

Because the high injection pressures may cause turbulence and air entrapment, the amount and time of applying the pressure varies considerably. A recent trend is toward the use of larger gates and lower injection pressure, followed by much higher pressure after the mold has been completely filled and the metal has started to solidify. This tends to improve the density and reduce porosity. Where the shape of the casting permits, a very high semiforging pressure may be applied when the casting has cooled to forging temperature.

Two basic types of die-casting machines are used. Figure 11-39 illustrates, schematically, the *hot-chamber,* or gooseneck, *machine*. Here, a "gooseneck" is partially submerged in molten metal. Upon each cycle of the machine, a port opens which allows the molten metal to fill the gooseneck. A mechanical plunger or air-injection system then forces the metal out of the gooseneck and into the die, where it rapidly solidifies. This type of machine is fast in operation and has a distinct benefit in that the metal is injected from the same chamber it is melted in. It cannot be used for the higher-melting-point metals, such as brass and bronze, and has a tendency to pick up some iron from the equipment when

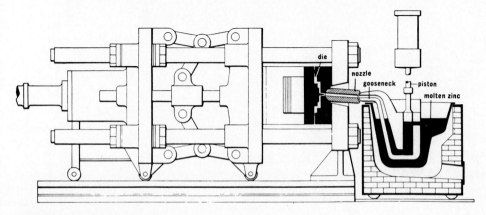

FIGURE 11-39 Schematic of hot-chamber die-casting machine. (*Courtesy Zinc Institute Inc., New York.*)

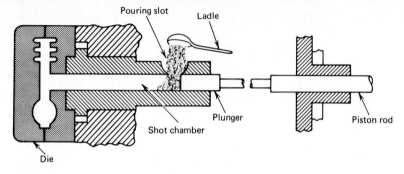

FIGURE 11-40 Schematic diagram of a cold-chamber die-casting machine.

used with aluminum. Thus hot-chamber machines are used primarily with zinc- and tin-base alloys.

Cold-chamber machines, as illustrated in Figure 11-40 and shown in Figure 11-41, are used for most die casting. Here, metal is melted in a separate furnace, then transported to the die casting machine, where a measured quantity is fed into the shot chamber and subsequently driven into the die by a hydraulic plunger. The process can be used for a variety of alloys and is quite popular for aluminum and magnesium. In addition, the pressure cycle can be controlled to produce high-quality castings. Although the cold-chamber machine has a longer operating cycle than do gooseneck machines, because the metal for each cast must be fed individually to the chamber, the productivity still is high, and up to 100 cycles per minute are not uncommon.

FIGURE 11-41 Cold-chamber die-casting machine. Inset shows a close-up of the die and the casting being removed. (*Courtesy Reed-Prentice Corporation.*)

Die-casting dies fill with metal so fast that there is little time for air in the mold cavity to escape and the metal mold has no permeability. Trapped air problems, in the form of blow holes, porosity, or misruns, are quite common. To minimize these, it is crucial that the dies be properly vented, usually by wide, thin (0.13 mm or 0.005 inch) vents at the parting line. Metal filling these escapes must be trimmed off after the casting is removed from the mold. This is usually done with trimming dies that also remove the sprues and runners.

Die castings are generally difficult to make sound due to trapped air and metal turbulence. However, the porosity is generally confined to the center of the casting and the rapidly solidified surface is usually sound and suitable for plating or decorative applications. In the relatively new, *pore-free casting process,* air problems are eliminated by introducing oxygen into the mold before each die-casting shot. When the molten metal is injected, it reacts with the oxygen to form fine, dispersed, oxide particles, virtually eliminating gas porosity. The products can be welded and heat-treated and possess greater strength than do conventional die castings. It has been used with aluminum, zinc, and lead.

Sand cores cannot be used in die casting because the high pressures and flow rates cause the cores to either disintegrate or have excessive metal penetration. As a result, metal cores are used extensively in die castings, and provision must be made for retracting them, usually before the die is opened for removal of the casting. Because a close fit must be maintained between the halves of the die and between any cores and the die sections to prevent the metal from flowing out of the die, both construction and maintenance costs are greatly increased when cores must be used. It is very important that the direction of the core-retracting motions be either a straight line or a circular arc. Otherwise, loose core pieces must be inserted into the die at the beginning of each cycle and then removed from the casting after it is ejected from the die. Such a procedure permits complex shapes to be cast, such as internal or external threads, but only with considerable reduction of the production rate and increased cost. Figure 11-42 is an example of a complex die casting that involves the use of several cores.

Because of the method by which they are produced, die castings tend to have certain distinguishing characteristics. The surfaces tend to be harder than the interior as a result of the chilling action of the metal die on the molten metal. There is some tendency for the inner metal to be porous because of entrapped air. This can be minimized, by proper metal flow, venting, and higher pressures that are timed properly. Good casting design, die design and maintenance, proper procedure, and good equipment make excellent-quality die castings readily obtainable.

It is possible to make special bearing surfaces, or threaded studs or bosses, out of harder metals and cast them into die castings. These must be placed in position within the die before it is closed and the metal injected. Suitable recesses must be provided in the die for holding such parts, and the casting cycle is slowed down.

Excellent dimensional accuracy can be obtained with die casting. For aluminum-, magnesium-, zinc-,. and copper-base alloys, linear tolerances of ± 0.075 mm per 25 mm (± 0.003 inch per inch) of length can be maintained. Minimum section thickness and draft depend on the kind of metal, as follows:

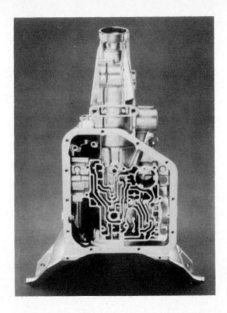

FIGURE 11-42 Die-cast aluminum automatic transmission case, incorporating the bell housing and rear case extension. (*Courtesy Central Foundry, Division of General Motors Corporation.*)

Metal	Minimum Section	Minimum Draft
Aluminum alloys	0.889 mm (0.035 in.)	1:100 (0.010 in./in.)
Brass and bronze	1.27 mm (0.050 in.)	1:80 (0.015 in./in.)
Magnesium alloys	1.27 mm (0.050 in.)	1:100 (0.010 in./in.)
Zinc alloys	0.635 mm (0.025 in.)	1:200 (0.005 in./in.)

As a result of the excellent dimensional accuracy and the smooth surfaces that can be obtained, most die castings require no machining except the removal of a small amount of fin, or flash, around the edge and possibly drilling or tapping of holes. Production rates are high and a set of dies can produce many thousands of castings without significant change in dimensions.

Centrifugal casting. In *centrifugal casting,* the molten metal conforms to the shape of the mold cavity as the result of the centrifugal force that is developed from the mold being rotated about its axial center line, at speeds of from 300 to 3000 rpm, while the molten metal is being introduced. In *true centrifugal casting,* the mold rotates about either a horizontal or vertical axis. Either a dry-sand or metal mold is used, which determines the outer surface of the casting. Although a round shape is most common, hexagonal or other symmetrical but slightly out-of-round shapes can be cast. No mold or core is needed to form the inner surface of the casting. When a horizontal axis is used, as illustrated schematically in Figure 11-43, the inner surface always is cylindrical. If a vertical axis is used, the inner surface is a section of a parabola, as illustrated in Figure 11-44, the exact shape being a function of the speed of rotation.

In true centrifugal casting, the metal is forced against the walls of the mold with considerable force, and it solidifies first at the outer surface. This results

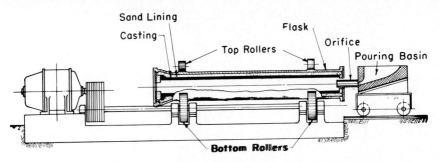

in a dense structure, with all the lighter impurities tending to be at the inner surface, thereby permitting them to be removed readily by a light machining cut, if required.

True centrifugal casting is widely used for the mass production of pipe, pressure vessels, cylinder liners, and brake drums. The equipment required is specialized and is expensive for large castings, but relatively simple and inexpensive equipment is available for small parts. The required permanent molds are relatively costly, but they have quite a long life. When ferrous metals are used, the molds are coated with some type of refractory dust or wash before the metal is introduced, to prolong their life. Yields can be 90% or more since no sprue, gate, or riser is required. Moreover, it is possible to produce composite castings by spinning a second alloy onto the surface of the first material.

In some cases the *semicentrifugal casting* principle, illustrated in Figure 11-45, can be used to advantage. In this case centrifugal force aids the flow of the molten metal from a central reservoir to the extremities of a symmetrical mold. The rotational speeds employed usually are considerably less than the case of true centrifugal casting. Frequently, several identical molds are stacked on top of each other and fed by a single, central reservoir. It is very important that the central feeding reservoir be sufficiently large to assure that the metal in it remains molten until the castings have solidified. The process is best used where the center of the casting is to be removed by machining, since the lighter impurities will concentrate at the center.

Centrifuging, illustrated in Figure 11-46, is another procedure that occasionally is used to provide forced metal flow from a central feeding reservoir into

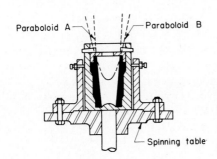

FIGURE 11-44 Vertical centrifugal casting, showing the effect of rotational speed on the shape of the inner surface. Paraboloid A results from fast spinning; paraboloid B from slow.

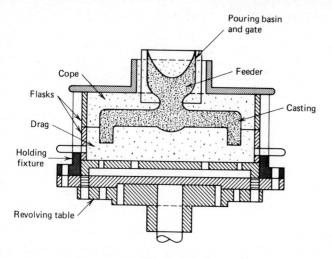

FIGURE **11-45** Semi-centrifugal casting process.

thin, intricate mold cavities. Relatively low rotational speeds are used to produce sound castings with fairly high yields. Investment castings are frequently poured using centrifuging machines.

PLASTER-MOLD CASTING

Two important casting processes utilize plaster molds. These are *investment casting* and the *Shaw process*. In most cases the molds are gypsum plaster with small additions of talc, terra alba, or magnesium oxide, to prevent cracking and to reduce the setting time. Lime or cement may be added to control expansion during baking, and about 25% of fibers can be added to improve the strength. Sand can be added as a filler.

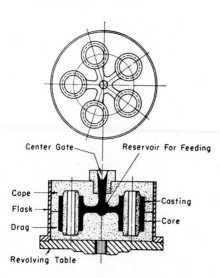

FIGURE **11-46** Method of casting by the centrifuging process. (*Courtesy American Cast Iron Pipe Company.*)

Investment casting. *Investment casting* actually is a very old process. It existed in china for centuries, and Cellini employed a form of it in Italy in the sixteenth century. Dentists have utilized the process since 1897, but it was not until World War II that it attained industrial importance for making jet turbine blades from metals that were not readily machinable. Currently millions of castings are produced by the process each year, its unique characteristics permitting the designer almost unlimited freedom in the complexity and close tolerances he can utilize.

Investment casting involves the following steps:

1. *Produce a master pattern.* The pattern may be made from metal, wood, plastic, or some other easily worked material.
2. *From the master pattern, produce a master die.* This usually is made from low-melting-point metal or from steel, but sometimes is made from wood. When steel is used, the cavity ordinarily is engraved directly in the die. If low-melting-point metals are used, the dies can be cast, using the master pattern. In some cases the dies are machined directly without first making a master pattern, thereby skipping step 1. Rubber molds can also be used.
3. *Produce the wax patterns.* These are made by pouring, or injecting under pressure, molten wax into the master die and allowing it to harden. Plastic and frozen mercury have also been used as pattern material.
4. *Assemble the wax patterns to a common wax sprue.* The individual wax patterns are removed from the master die and several of them are attached to a central sprue and runners by means of heated tools and melted wax. In some cases, several pattern pieces may be united to form a complex, single pattern that, if made in one piece, could not be withdrawn from a master die. The result of this step is a pattern cluster, or *tree*.
5. *Coat the cluster with a thin layer of investment material.* This step usually is accomplished by dipping the cluster into a thin slurry of finely ground refractory. Fine silicaceous material mixed with a special plaster is often used. This step produces a thin but very smooth layer of investment material adjacent to the wax patterns and ensures a smooth surface and good detail in the final product.
6. *Produce the final investment around the coated cluster.* The cluster can be dipped repeatedly in the investment material until the desired thickness is obtained, or it can be placed upside down in a flask and the investment material poured around it.
7. *Vibrate the flask to remove the entrapped air and settle the investment material around the cluster.* This step is necessary only when the investment material is poured around the cluster.
8. *Allow the investment to harden.*
9. *Melt or dissolve the wax pattern to permit it to run out of the mold.* This is generally accomplished by placing the molds upside down in an oven where the wax melts and the residue subsequently vaporizes. The wax is recovered for further use. This step is the most distinctive feature of the process, because it permits a complex pattern to be removed from a mold without the necessity of the mold being made in two or more sections so that it can be opened. Consequently, complicated shapes with reentrant

sections can be cast by this process, and better dimensional accuracy can be achieved. In the early history of the process, when only small parts were cast, the molds were placed in an oven and, as the wax was melted, it was absorbed into the porous investment. Because the wax disappeared from sight, the process was called the *lost-wax* process; this name still is used occasionally.

10. *Preheat the mold preparatory to pouring.* Heating to 538 to 1093°C (1000 to 2000°F) assures that the molten metal will flow more readily to all thin sections. It also gives better dimensional control because the mold and the metal shrink together during cooling.

11. *Pouring the molten metal.* Various methods, beyond simple pouring, are utilized to assure complete filling of the mold, especially where complex, thin sections are involved. Among these are the use of air pressure, evacuation of the air from the mold, and a centrifugal process.

12. *Remove the castings from the mold.* This is accomplished by breaking the mold away from the casting.

Figure 11-47 schematically shows the investment procedure wherein the investment-mold material fills the entire flask. Figure 11-48 shows the shell-mold investment procedure.

Investment casting obviously is a complex process and thus is expensive. Yet its unique advantages make it economically feasible in many cases, particularly since many of the steps can be completely automated. Not only can complex shapes be cast which could not be cast by any other process or be made by machining, but, very thin sections—down to 0.38 mm (0.015 inch)—can be cast. The dimensional tolerances are excellent—0.1 to 1% (0.005 to 0.010 inch per inch) being routinely obtained, combined with the very smooth surfaces. Machining often can be completely eliminated or greatly reduced. Where machining is required, allowances of 0.38 to 1 mm (0.015 to 0.040 inch) usually are ample. These advantages are especially important where difficult-to-machine metals are involved, for the investment casting process can be applied to all metals.

Although most investment castings are less than 76.2 mm (3 inches) in size and weigh less than 0.454 kg (1 pound), there are no specific size limitations, and castings up to 914 mm (36 inches) and weighing 36.3 kg (80 pounds) have been produced. Some typical investment castings are shown in Figure 11-49. It will be noted that complexity of shape is the most common characteristic.

The Shaw process. Obtaining adequate permeability, while retaining the desired smooth mold-cavity surface, is a major problem in plaster-mold casting. The *Shaw process* provides an excellent solution.

In this process, a slurrylike mixture of a refractory aggregate, hydrolyzed ethyl silicate, and a jelling agent are poured over a reusable pattern. This mixture sets to a rubbery jell that allows the pattern to be stripped, but has sufficient strength so that it will return to the exact shape it had while fitted to the pattern. The mold then is ignited to burn off the volatile elements in the mix. Next, it is brought to a red heat in a furnace. This firing makes the mold rigid and hard, but at the same time it causes a network of microscopic cracks to form. This

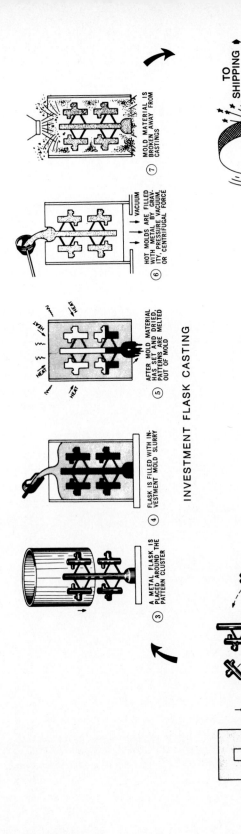

1. WAX OR PLASTIC IS INJECTED INTO DIE TO MAKE A PATTERN
2. PATTERNS ARE GATED TO A CENTRAL SPRUE
3. A METAL FLASK IS PLACED AROUND THE PATTERN CLUSTER
4. FLASK IS FILLED WITH IN-VESTMENT MOLD SLURRY
5. AFTER MOLD MATERIAL HAS SET AND DRIED, PATTERNS ARE MELTED OUT OF MOLD
6. HOT MOLDS ARE FILLED WITH METAL BY GRAV-ITY, PRESSURE VACUUM, OR CENTRIFUGAL FORCE
7. MOLD MATERIAL IS BROKEN AWAY FROM CASTINGS
8. CASTINGS ARE REMOVED FROM SPRUE, AND GATE STUBS GROUND OFF

TO SHIPPING

INVESTMENT FLASK CASTING

FIGURE 11-47 Investment flask-casting procedure (*Courtesy Investment Casting Institute, Dallas, Texas.*)

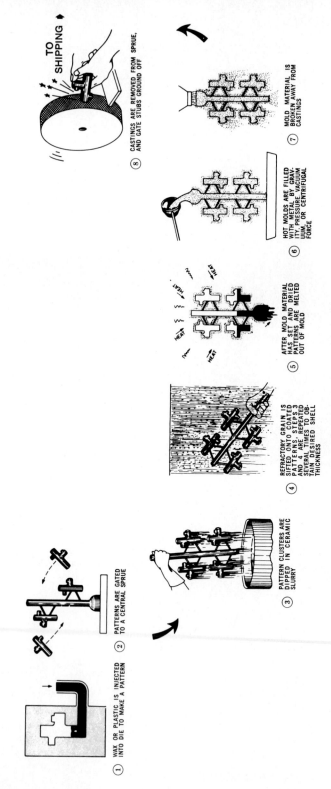

① WAX OR PLASTIC IS INJECTED INTO DIE TO MAKE A PATTERN

② PATTERNS ARE GATED TO A CENTRAL SPRUE

③ PATTERN CLUSTERS ARE DIPPED IN CERAMIC SLURRY

④ REFRACTORY GRAIN IS SIFTED ONTO COATED PATTERNS. STEPS 3 AND 4 ARE REPEATED SEVERAL TIMES TO OBTAIN DESIRED SHELL THICKNESS

⑤ AFTER MOLD MATERIAL HAS SET AND DRIED PATTERNS ARE MELTED OUT OF MOLD

⑥ HOT MOLDS ARE FILLED WITH METAL BY GRAVITY, PRESSURE, VACUUM, OR CENTRIFUGAL FORCE

⑦ MOLD MATERIAL IS BROKEN AWAY FROM CASTINGS

⑧ CASTINGS ARE REMOVED FROM SPRUE, AND GATE STUBS GROUND OFF

TO SHIPPING

INVESTMENT SHELL CASTING

FIGURE 11-48 Investment shell-casting procedure. (*Courtesy Investment Casting Institute, Dallas, Texas.*)

FIGURE 11-49 Group of typical parts produced by investment casting. *(Courtesy Haynes Stellite Company.)*

microcrazing produces fissures that provide excellent permeability and good collapsibility to accommodate the shrinkage of the solidifying metal.

The Shaw process can be used for castings of all sizes and all metals, and it produces excellent surface finish, detail, and dimensional accuracy. Dimensional tolerances between 0.051 mm and 0.25 mm in 25 mm (0.002 to 0.010 inch per inch) are readily obtainable. The molds may be one-piece or multipiece, depending on the type and complexity of the pattern. Figure 11-50 shows some of the wide variety of accurate and complex parts that are cast by this process, and Figure 11-51 shows how the Shaw process occasionally can be combined advantageously with the lost-wax process.

Cores can be used in Shaw-process and other plaster molds much the same as in sand casting. However, as a general rule, thinner cast sections can be obtained in plaster molds because the plaster retards the cooling. This possibility of obtaining thinner sections, better dimensional accuracy, fine details, elimination of machining, and exceptionally fine surface finish with plaster molds must be balanced against their greater cost.

OTHER CASTING PROCESSES

Rubber-mold casting. Several types of artificial rubbers—usually silicone varieties—are available which can be compounded in liquid form and then poured over patterns, forming semirigid molds upon hardening. The molds retain sufficient flexibility to permit them to be stripped from the pattern, thereby permitting quite intricate shapes with reentrant sections to be cast. They are suitable only for quite small castings, such as wax patterns and those of plastics and low-melting-point alloys which melt below about 260°C (500°F).

FIGURE 11-50 (*Left*) Ornamental casting, and (*right*) group of cutters produced by the Shaw process. (*Courtesy Avnet Shaw Division of Avnet, Inc.*)

Expendable graphite mold. For metals, such as titanium, which tend to react with many common mold materials, powdered graphite can be combined with cement, starch, and water and compacted around a pattern. The pattern is removed and the mold is fired at 1000°C (1800°F) to consolidate the graphite into a solid mold. After pouring, the mold is broken to remove the casting.

Continuous casting. Long, continuous shapes, having special cross sections such as shown in Figure 11-52, can be produced readily by the *continuous casting process* discussed in Chapter 4 and illustrated in Figure 4-11. Only a single mold is required, the individual units being cut from a single strand. High-quality metal can be obtained because it is protected from contamination while molten and being poured.

FIGURE 11-51 Method of combining the Shaw process and wax-pattern casting to produce the vanes of an impeller. A wax pattern is added to a metal pattern. After the metal pattern is withdrawn (*center*), the wax pattern is melted, leaving a cavity as shown at the right. (*Courtesy Avenet Shaw Division of Avnet, Inc.*)

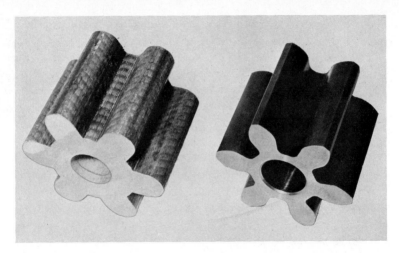

FIGURE 11-52 Gear produced by continuous casting. (*Left*) Piece cut from unfinished casting. (*Right*) After machining. (*Courtesy American Smelting and Refining Company.*)

Electromagnetic casting. This process is a method of casting wherein the metal is contained and solidified in an electromagnetic field rather than in a conventional mold, and it has recently been adapted to continuous casting. There is no sliding contact with the mold wall and the metal can be solidified by direct water impingement. Product surface is smooth and uniform and the process can be directly automated. Several large industries are currently employing it in the continuous casting of aluminum ingot.

MELTING AND POURING

All the casting processes require a furnace for melting the metal. Ideally, such a furnace should (1) provide adequate temperature, (2) minimize contamination, (3) make possible holding the metal at the required temperature without harmful effects so the chemical composition can be altered by alloy additions, (4) be economical, and (5) be capable of control to avoid atmospheric pollution. Except for experimental or very small operations, virtually all foundries use either a cupola, an air furnace, an electric-arc furnace, or an electric induction furnace. Occasionally, in fully integrated steel mills containing a foundry, steel is taken directly from a basic-oxygen furnace and poured into casting molds, but such a practice usually is not done regularly and is reserved for exceptionally large castings. For some experimental work, and in very small foundries, gas-fired crucible furnaces may be used, but these do not have sufficient capacity for most commercial operations.

Selection of the best melting procedure will depend upon such factors as (1) the temperature needed to melt and super-heat the metal, (2) the alloy being melted, (3) the desired melting rate or quantity of metal, (4) the desired quality of the metal, (5) the availability of various fuels, (6) the variety of metals or alloys to be melted, (7) whether melting is to be batch or continuous, (8) the required level of emission control, and (9) the various capital and operating costs.

Cupolas. By far the majority of gray, nodular, and white cast iron is melted in *cupolas,* although the use of electric induction furnaces is steadily increasing. Basically, a cupola is a refractory-lined, vertical steel shell into which alternative charges of coke and iron (either pig iron or scrap) are added in a ratio of 8 to 10 parts of iron to 1 part of coke. Small amounts of limestone or other materials may also be added to act as a flux or to increase the fluidity of the metal, and some alloys may be added to alter the metal composition. When heated under a forced-air draft, the iron charge begins to melt and run down through the coke. Molten metal collects in the lower portion of the cupola and a slag layer floats on top. When a sufficient amount of molten metal has collected, the blast is turned off, the tap hole is opened, and the metal is permitted to flow out into a pouring or holding ladle. The tap hole is then plugged and the blast is again introduced.

Thus, the operation of a cupola is a semicontinuous, batch-type process. As metal is melted and removed, additional charges of coke and iron are added to maintain a uniform charging level.

Cupolas are simple and economical, can be obtained in a wide range of capacities, and can produce excellent-quality cast iron if the proper raw materials and good control are used. Because the hot metal is in intimate contact with the coke, its chemical composition is influenced by what is in the coke. Coke with a too-high sulfur content should be avoided. Some chemistry control problems arise due to the volume of the furnace. By the time the chemistry of the molten metal is determined, a considerable amount of material is working its way through the cupola. Chemical alteration may be performed by use of ladle additions after tapping. Various alloys may be added at this point.

In order to increase the melting rate and give greater economy, a *hot-blast* type of cupola may be employed. Here the stack gases are put through a heat exchanger to preheat the incoming air to temperatures up to 650°C (1200°F). In *water-cooled* cupolas, water is circulated around the shell to provide cooling. Its use may serve to prevent overheating of the lining and prolong its life, or enable elimination of the brick or refractory lining and, thereby, enable a wide variety of slags to be used. *Oxygen-enriched blasts* have been employed to increase temperature and accelerate the rate of melting.

In all modern installations, the large volume of stack gases is passed through dust collectors and various types of pollution-control equipment. Consequently, as indicated in Figure 11-53, a modern cupola installation is quite complex. It should be noted again that the cupola is used exclusively for cast iron.

Indirect fuel-fired furnaces. These include crucibles and pot furnaces where the metal container is heated by an external flame. Use is limited to batch-type melting of nonferrous metals. Stirring action and temperature control are poor but these furnaces offer low capital cost. The crucibles are generally made from clay and graphite, silicon carbide, cast iron, or steel.

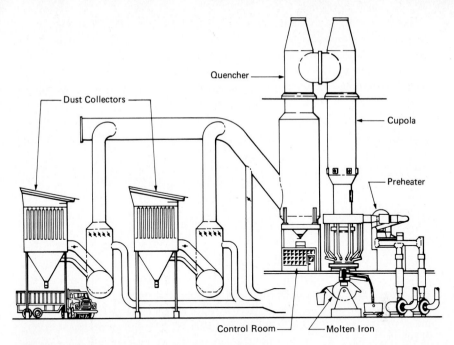

FIGURE 11-53 Modern cupola installation with associated antipollution equipment. (*Courtesy* Foundry Management & Technology.)

Air furnaces or direct fuel-fired furnaces. In these furnaces, illustrated in Figure 11-54, the surface of the metal is heated directly by the burning fuel. They are somewhat similar to small open-hearth furnaces, but have no regenerative equipment and are charged through the removable *bungs*. Capacity is much larger than the crucible furnace, but operation is still primarily limited to batch-type melting of the nonferrous metals (Malleable cast iron is held and adjusted in these furnaces after melting in a cupola.). The rate of heating and melting as well as the temperature and composition of the metal are easily controlled.

Arc furnaces. The use of *arc furnaces* in foundries has increased substantially because of (1) their rapid melting rates, (2) their ability to hold the molten metal for any desired period of time to permit alloying, and (3) the greater ease of providing adequate pollution control.

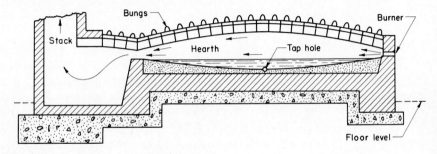

FIGURE 11-54 Section of an air furnace.

The basic features of a *direct-arc* electric furnace are shown in Figure 4-3. In most types the top can be lifted or swung off to permit the charge to be introduced. Heating occurs through lowering the electrodes so that an arc is struck and maintained between the electrodes and the metal charge. The current path is from one electrode across the arc to the metal, through the metal, and back across the arc between the metal and another electrode.

Fluxing materials are added to provide a protective cover over the molten metal. Because the metal is thus protected and the metal maintained at a given temperature as long as desired, high-quality metal of any desired composition can be obtained. Furnaces of this type in capacities up to 200 tons are in use, but those below 25 tons are more common. Up to 50 tons per hour can be conveniently melted in batch-type operations. These furnaces are generally used with ferrous alloys, particularly steel, and provide good mixing and homogeneity to the molten bath. Figure 4-4 shows a typical furnace of this type.

Induction furnaces.   Because of their very rapid melting rates and relative ease of controlling pollution, the use of electric induction furnaces has increased extensively. Two types are used. The *high-frequency type,* or coreless induction furnace, shown in Figure 4-5, consists of a crucible surrounded by a water-cooled coil of copper tubing. A high-frequency electrical current passes through the coil, establishing an alternating magnetic field, which, in turn, induces secondary currents in the metal in the crucible. These secondary currents heat the metal very rapidly.

These furnaces are used for virtually all common alloys, the maximum temperature being limited only by the refractory and efficiency of insulating against heat loss. They provide good control of temperature and composition and are available in capacities through 65 tons. Because there is no contamination from the heat source, they produce very pure metal.

Low-frequency, or *channel-type,* induction furnaces are being used increasingly. As indicated in Figure 11-55, these have an ordinary alternating-current primary coil, but the secondary coil is formed by a loop, or channel, of the molten metal. Because the secondary coil has but one turn, a low voltage/high amperage current is induced in it, which provides the desired heating. Some molten metal, to form the secondary coil, must be used to start such furnaces, but their heating rate is very high, and the temperature is readily controlled. This makes them very useful, and widely used, as holding furnaces, where it is desired to maintain molten metal at a constant temperature for an extended period of time, as in die-casting machines or in mold-pouring systems. Capacities are available up to 250 tons.

Pouring practice.   In order to transfer the metal from the melting furance into the molds, some type of pouring device, or ladle, must be used. Primary considerations are to maintain the metal at the proper temperature for pouring and to assure that only quality metal gets into the molds. The type of ladle used is determined largely by the size and number of castings to be poured. In small foundries, the hand-held, shank-type ladle shown in Figure 11-56 often is used. In large foundries, either bottom-pour or teapot-type ladles, illustrated in Figure 11-57, are used, often in conjunction with a conveyor line on which the molds

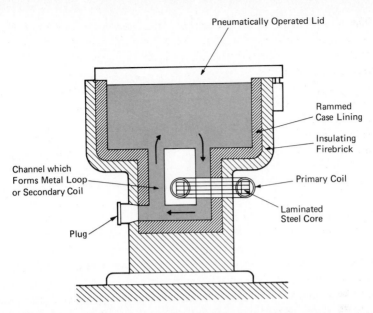

FIGURE 11-55 Principle of the low-frequency induction furnace.

move past the pouring stations. These types of ladles help to assure that slag and oxidized metal do not enter the mold. In modern, mass-production foundries, automatic pouring systems, such as shown in Figure 11-58, are employed. The molten metal is brought from the main melting furnace to a holding furnace by overhead crane. A programmed amount of molten metal is poured into the individual pouring ladles from the holding furnace and, in turn, is poured automatically into the corresponding molds.

Because of its high combustibility at elevated temperatures, special precautions are required in melting and pouring magnesium. During melting, the metal

FIGURE 11-56 Pouring a mold from a shank-type ladle. (*Courtesy Steel Founders' Society of America in Rocky River, Ohio.*)

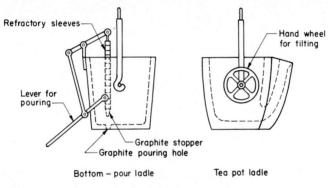

FIGURE 11-57 Types of ladles for use in pouring castings.

must be kept covered by an adequate flux. Prior to pouring, the molten metal should be stirred with a steel rod to free the inpurities, which often are not much lighter than the metal, thereby permitting them to be collected in the flux. The flux then should be skimmed from the metal and a layer of protective material, in the form of powder, immediately put over the surface of the magnesium. Such a protective layer should be maintained at all times, including the time the metal is in the pouring ladle.

Vacuum melting and pouring.   Just as increasing amounts of metals are being melted in a vacuum to remove gases and assure higher purity, increasing num-

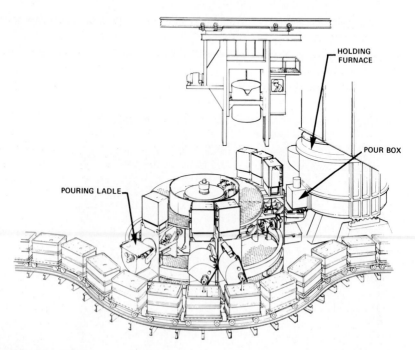

FIGURE 11-58 Machine for automatic pouring of molds on a conveyor line. (*Courtesy Roberts Corporation.*)

bers of castings are being poured in a vacuum to retain their purity in the cast form. This often is done by enclosing both an induction furnace and the mold in a chamber from which the air can be evacuated, and arranging for the metal to be poured directly from the furnace into the mold.

Another method of vacuum melting and pouring, used when extremely high-purity metal is necessary, is carried out by using an arc furnace in which a consumable metal electrode of the metal to be melted is used. This method is used to make titanium ingots. When castings are to be made, the electrode is melted into a crucible and then poured into the mold, all of the equipment being contained in an evacuated chamber. Obviously, these vacuum methods are expensive and should be employed only when the highest quality is essential.

CLEANING, FINISHING, AND HEAT-TREATING CASTINGS

After solidification and removal from the molds, most castings require some cleaning and finishing operations. These may involve all or several of the following steps:

1. Removing cores.
2. Removing gates and risers.
3. Removing fins and rough spots from the surface.
4. Cleaning the surface.
5. Repairing any defects.

The required operations are not always done in the same order, and the particular casting process may eliminate some of them. Because cleaning and finishing operations may involve considerable expense, some consideration should be given to them in designing castings and in selecting the casting method to be used. Often substantial savings can be effected. In recent years much attention has been given to mechanizing these operations.

Sand cores usually can be removed by shaking. Sometimes they must be removed by dissolving the core binder. On small castings, gates and risers can often be knocked off. However, on large castings, and often on small castings, they must be cut off. On nonferrous and cast iron castings this usually is done by means of an abrasive cutoff wheel, power hacksaw or bandsaw. Gates and risers on steel castings, especially large ones, are often removed by an oxyacetylene torch.

After the gates and risers are removed, small castings are often put through tumbling barrels, as shown in Figure 11-59, to remove fins, snags, and sand that adheres to the surface. Tumbling also may be used to remove cores and, in some cases, gates and risers. Frequently, some type of shot or slug material is added to the barrel to aid in the cleaning. Larger castings may be passed through a cleaning chamber on a conveyor, wherein they are subjected to blasts of abrasive or cleaning material, often using the principle illustrated in Figure 15-19. Large castings usually have to be finished manually, using pneumatic chisels, portable grinders, and manually directed blast hoses in separate cleaning rooms.

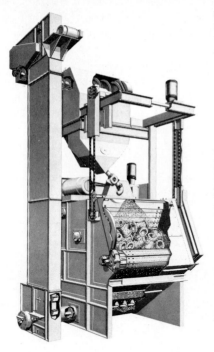

FIGURE 11-59 (*Left*) Tumbling machine for cleaning castings. (*Above*) Uncleaned castings in loading hopper at bottom. Cleaned castings are being discharged at top. (*Courtesy Wheelabrator-Frye Inc.*)

Although it is desireable that castings contain no defects, it is inevitable that some will occur, particularly in large castings where only one or a few of a particular design are made. Some types of defects can be repaired readily and satisfactorily by arc welding. However, it is imperative that the casting be of a material that can be welded satisfactorily, that *all* defective areas be removed down to sound metal by grinding, or chipping, and that a sound repair weld be made.

Heat treatment of castings. The heat treatment of castings has become more common as a means to obtain the full benefit of the alloy additions. Steel castings almost always are given a full anneal to reduce the hardness in rapidly cooled thin sections and to reduce internal stresses that result from uneven cooling. Nonferrous castings are often heat-treated to put them in a softened, stress-relieved condition prior to subsequent machining. For final properties, virtually all of the treatments discussed in Chapter 6 can be applied.

DESIGN CONSIDERATIONS IN CASTINGS

It is very important that the designer of castings give careful attention to several requirements of the process and, if possible, cooperate closely with the foundry if economy and best results are to be obtained. Frequently, minor, and readily permissible, changes in design details will greatly facilitate and simplify the casting of a component and reduce the percentage of defects.

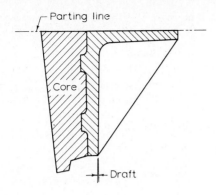

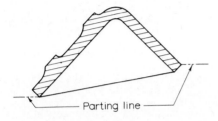

FIGURE 11-60 Elimination of a core by changing the location of the parting line on a casting.

One of the first factors that must be considered by the designer is the location of the parting plane, except in the cases of castings to be produced by the full-mold or investment processes. The location of the parting plane affects the following:

1. The number of cores.
2. Effective and economical gating.
3. Casting weight.
4. Method of supporting cores.
5. Dimensional accuracy.
6. Ease of molding.

In general it is desirable to minimize the use of dry-sand cores. Often a change in the location of the parting plane will assist in this objective, as illustrated in Figure 11-60. Note that the change also reduced the weight of the casting by eliminating the need for draft. Figure 11-61 shows an example of how a simple design change eliminated the need for a dry-sand core. As shown in Figure 11-62, specification of round edges on a part often dictates the location of the parting plane.

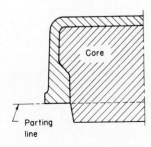

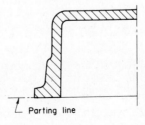

FIGURE 11-61 Elimination of a dry-sand core by change in part design.

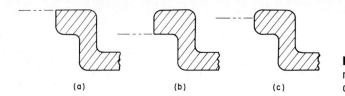

FIGURE 11-62 Effect of rounded edges on the location of the parting line.

(a) (b) (c)

When draft is indicated on a part to be cast, the parting plane may be fixed. More economical molding may be possible if provision for draft is indicated by a note or left to the option of the foundry. Figure 11-63 illustrates some typical options. In addition, the designer should remember that dimensions across a parting plane are subject to more variation than those parallel with the parting plane.

The solidification process is of prime importance in obtaining sound castings, and it is closely related to design. Those portions of a casting that have a high ratio of surface area to volume, and thus are subjected to rapid cooling rates, will tend to be hard or have a hard skin. Heavier sections cool more slowly and, unless special precautions are observed, may contain shrinkage cavities and porosity or have large grain structures.

Ideally, a given casting should have virtually uniform thickness in all sections. In most cases this is not possible. However, when the section thickness must change, such changes should be gradual. Figure 11-64 gives some guides regarding changes in sections.

When sections of castings intersect, two problems arise. The first is the matter of possible stress concentrations. This problem can be minimized by providing generous fillets at all changes of direction to avoid sharp, interior corners. Too large fillets however may cause difficulty because of the second problem—*hot spots*.

When sections of castings intersect, localized thick sections may result, as indicated in Figure 11-65. These thick sections cool more slowly than others

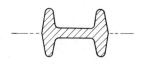

As shown on drawing   As shown on drawing, with draft permitted by note

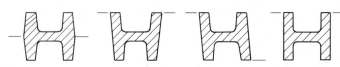

Optional results, with and without draft (exaggerated)

FIGURE 11-63 (*Top left*) Location of parting plane specified by draft. (*Top right*) Part with draft unspecified. (*Bottom*) Various options in producing that part.

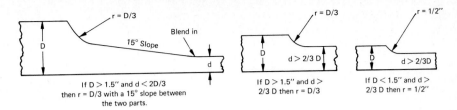

FIGURE 11-64 Guides for section changes in castings.

and result in localized, abnormal shrinkage. When the differences in sections are large, as illustrated in Figure 11-66, the hot-spot areas are likely to result in serious defects in the form of porosity or shrinkage cavities.

Figure 11-67 shows three examples of the incorrect and correct use of radii in reducing hot spots, or shrinkage areas. When a condition exists as shown in Figure 11-67a and b, the metal at the heavy section remains liquid while that in the adjoining legs contracts and exerts tensile strains at the junction. The design in (c) greatly reduces this tendency, but it is not a desirable solution if the fillet is to be stressed in tension, due to the decreased section modulus.

Obviously, voids, porosity, and cracks in castings can result in serious failures. Sometimes cored holes, as illustrated in Figure 11-68, can be used to avoid hot spots. Where heavy sections must exist, an adjacent riser often can be provided to feed the section during shrinkage, as illustrated in Figure 11-69. The riser provides a reservoir of molten metal which, because of its mass, remains molten until after the adjacent portions of the casting have solidified and thus can *feed* the heavy section during shrinkage. A shrinkage cavity occurs only within the riser and is removed when the riser is cut off. Most risers are open to the atmosphere, but *blind risers* (enclosed within the mold) also are used. Both open and blind risers may be seen in Figure 11-27.

Risers add to the cost of castings because extra metal must be melted, and the solidified riser must be removed from the casting. Consequently, it is desirable to eliminate the necessity for them through design changes whenever possible. When their use is required, the design should be optimized by consideration of riser shape, size and location, as well as the nature of the connection between the riser and the casting.

Other means can be used to assist or replace risers in helping to obtain proper solidification. One method is to increase the size of sections that otherwise might cool too quickly, a practice called *padding*. Another procedure is to place pieces of metal, called *chills*, into the molding sand adjacent to heavy sections to

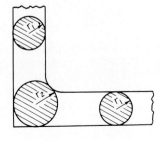

FIGURE 11-65 "Hot spot" at section r_2 caused by intersecting sections.

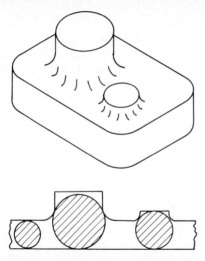

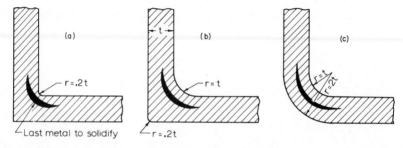

FIGURE 11-66 Hot spot resulting from intersecting sections of various thicknesses.

accelerate the rate of solidification. External chills do not adhere to the casting or may be removed readily. Internal chills are pieces of metal, usually in the form of a coil of wire or a large-headed nail, which are suspended in the mold cavity. They produce a chilling action, but they also become an integral part of the casting. As such they may be the source of internal weakness, and their use is restricted. Still another approach is to retard cooling of the riser by the use of insulated or exothermic toppings and sleeves.

In die casting, the pressure on the metal during solidification tends to overcome the contraction and provide the necessary feeding action. Similarly, in centrifugal and semicentrifugal casting there is a positive pressure to force the metal to flow and thus offset shrinkage.

Proper gating and feeding of a mold can do much to control casting solidification. A satisfactory system will (1) permit the metal to flow rapidly and to fill the mold quickly; (2) allow the metal to flow with minimum turbulence, particularly within the cavity; (3) not cause aspiration of gases; (4) induce solidification at the center with progressive solidification toward the risers; and (5) not cause the surfaces of the cavity or runner system to be eroded by too rapid flow of the metal.

Sprue holes should be adequate in size, decreasing in size toward the bottom

FIGURE 11-67 Last metal to solidify in sections resulting from various radii in the corner.

FIGURE 11-68 Method of eliminating unsound metal at the center of heavy sections in castings by using cored holes.

to prevent aspiration. The *gate,* through which the metal actually flows into the mold cavity, is very important. Frequently, it is located at the parting line of the mold. Ordinarily, more than one gate is used, being connected with the sprue hole by means of a *runner.* In this way the metal can be introduced at several points to aid proper directional solidification. Gates also may be located at the top of a casting, but bottom or side gating usually is preferred because it will cause less turbulent metal flow within the mold cavity. Consequently, if the gate is located at the parting line, the mold cavity must be arranged properly with respect to the parting line to have the metal flow into the cavity in the desired manner.

The wall of the sprue hole is commonly made of exothermic materials, or a stick of exothermic material is added to the sprue after the metal is poured to keep the sprue molten longer and thus aid in feeding the casting.

Intersecting ribs can cause shrinkage problems and should be given special consideration by the designer. Where sections intersect, forming continuous ribs, contraction occurs in opposite directions in each rib. As a consequence, cracking frequently occurs during cooling. By staggering the ribs there is some opportunity for slight distortion to occur so that high stresses are not built up. Figure 11-70 illustrates this situation.

Large unsupported flat areas should be avoided in all types of castings. Such sections tend to warp during cooling. In die castings, good surface appearance of a flat area is difficult to maintain over an extended period of production because the molten metal reacts with the die and any deterioration of the die becomes apparent. Another consideration that is of particular importance in die castings is that the parting line of the die halves be at the corner of the casting whenever possible. Some small amount of fin, or flash, usually will be present at the parting line, particularly as the dies wear. When this flash is removed, or when it is so slight that it does not have to be removed, a small line of demarcation will remain. If this occurs in the middle of a flat surface, it will be visible, whereas at a corner, it will usually go unnoticed.

In designing all types of castings, minimum section thickness must be considered. Exact specifications as to economical and practicable section thicknesses

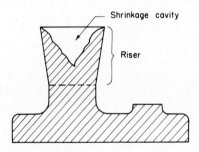

Shrinkage cavity

Riser

FIGURE 11-69 Use of a riser to keep shrinkage cavity out of a casting.

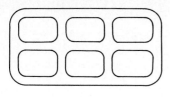

Bad

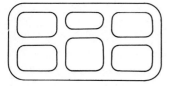

Better

FIGURE 11-70 Method of using staggered ribs to prevent cracking during cooling.

cannot be given because the shape and size of the casting, the kind of metal, the method of casting, and the practice of the individual foundry are all factors that can affect the results obtainable. The following tabulation represents reasonable guidelines:

Material	Minimum		Desirable		Casting Process
	mm	in.	mm	in.	
Steel	4.76	³⁄₁₆	6.35	¼	Sand
Gray iron	3.18	⅛	4.76	³⁄₁₆	Sand
Malleable iron	3.18	⅛	4.76	³⁄₁₆	Sand
Aluminum	3.18	⅛	4.76	³⁄₁₆	Sand
Magnesium	4.76	³⁄₁₆	6.35	¼	Sand
Zinc alloys	0.51	0.020	0.76	0.030	Die
Aluminum alloys	1.27	0.050	1.52	0.060	Die
Magnesium alloys	1.27	0.050	1.52	0.060	Die

In conclusion, one should note that in casting design, probably more than with any other manufacturing process, the designer can gain tremendous benefits by working closely with the supplying foundry.

REVIEW QUESTIONS

1. What is "materials processing"?
2. What can happen to a casting if the mold applies too much constraint during cooling?
3. What are some of the materials used to make casting patterns and their advantages and disadvantages?

4. What types of allowances are incorporated into a pattern?
5. What is a match-plate pattern?
6. What are the four requirements of a molding sand?
7. What is a significant disadvantage of the CO_2–sodium silicate sand?
8. What are some of the properties that are evaluated during mold sand testing?
9. Why is it important that a mold material possess permeability?
10. What are some of the commonly used methods to pack sand within a molding flask?
11. How might the sand adjacent to the pattern be made harder and stronger?
12. What are three major advantages of the shell-molding process?
13. Why must moderate-to-large quantities be required to justify the shell-molding process?
14. What are some of the attractive features of the V-process (vacuum molding)?
15. What is the unusual feature of the Eff-set process?
16. What features make the full-mold process attractive for small quantities of intricate-shaped products?
17. What is the major limitation of green sand cores?
18. What is the attractive feature of air-set or ''no-bake'' core sands?
19. What are some of the characteristics desirable in a dry sand core?
20. What are core prints?
21. What are chaplets? What are some of the restrictions on their use, size, and chemistry?
22. Why are permanent molds heated before a casting is poured?
23. What are some of the major advantages of permanent-mold casting?
24. What are the major restrictions to permanent-mold casting?
25. Why are permanent-mold castings generally removed from the mold immediately after solidification?
26. How does die casting differ from permanent-mold casting?
27. What types of production or product requirements would tend to favor the economical use of die casting?
28. Why has the pressure pouring process become attractive for aluminum castings?
29. What are the two basic types of die-casting machines?
30. For what metals would a hot-chamber die-casting machine be attractive?
31. Why must vents be incorporated into die-casting dies?
32. How does the pore-free casting process eliminate trapped air problems?
33. Why can sand cores not be used in die casting?
34. Why might it be desirable to machine the inner surface of a centrifugal casting?
35. What are common pattern materials for investment casting?
36. Why are investment casting molds preheated prior to pouring?
37. What types of metals can be investment cast?
38. For what materials is the rubber-mold technique appropriate? Expandable graphite molds?
39. What are some of the functions of a casting furnace?
40. What factors might influence the selection of a melting procedure?
41. What is the most popular method for melting cast iron?
42. What features have accounted for the increased popularity of induction furnaces?
43. Why are channel-type induction furnaces useful for holding large amounts of molten metal?
44. What are some of the cleaning and finishing operations commonly performed on castings?
45. Why is the selection of the parting plane a critical step in casting design?
46. What are hot spots?
47. What are risers? Why is it desirable to restrict the size and use of risers?
48. What are some of the design functions of a gating and feeding system?

CASE STUDY 11. The Defective Propellers

The Propco Foundry Co. casts large ship propellers. A new-design propeller, 6.1 meters (20 feet) in diameter and weighing about 32 kN (35 tons), was cast of nickel–aluminum bronze, using the mold–gating–risering arrangement shown schematically in Figure CS-11.[The large end of the propeller hub was approximately 1.52 meters

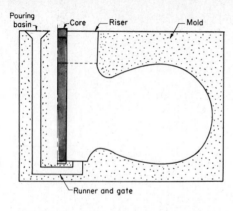

Pouring basin

Core

Riser

Mold

Runner and gate

FIGURE CS-11 Schematic showing gating and risering of mold for a large marine propellor.

(5 feet) in diameter with a cored hole 0.46 meter (1.5 feet) in diameter.] While the ship on which one of the propellers was installed was at sea, an entire blade broke off, resulting in a damage claim of over $250,000. Investigation revealed excessive porosity and shrinkage cavities in the blades of this and two other duplicate propellers. What corrective measures would you recommend to eliminate the difficulty?

CHAPTER 12

Powder Metallurgy

Powder metallurgy is the name given to a process wherein fine metal powders are blended, pressed into a desired shape (compacted), and then heated (sintered) in a controlled atmosphere at a temperature below the melting point of the major constituent for sufficient time to bond the contacting surfaces of the particles and establish desired properties. The process, commonly designated as P/M, readily lends itself to mass production of small, intricate parts of high precision, often without requiring additional machining. There is little material waste; unusual materials or mixtures can be utilized; and controlled degrees of porosity or permeability can be produced. Areas of application tend to be either those for which the process has strong economical advantage or where the product cannot be made by any other process.

A crude form of powder metallurgy appears to have existed in Egypt as early as 3000 B.C., utilizing particles of sponge iron. Around A.D. 1100 the Arabs were making high-quality swords from iron powder. In the nineteenth century, the process was used to produce platinum and tungsten wires. The first significant use in general manufacturing, however, was in the 1920s, when powder metallurgy was used to produce tungsten carbide cutting-tool tips and nonferrous bushings. In recent years, the process has become highly developed and large quantities of P/M products are made annually. From 1960 to 1980, the consumption of iron powder has increased tenfold. Most products are under 50 mm (2 inches) in size, but they have been produced in weights up to 32 kg (70 pounds) with dimensions up to 500 mm (20 inches).

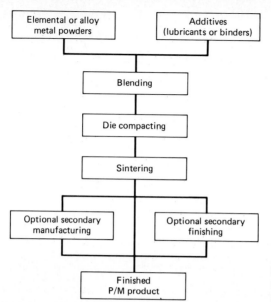

FIGURE 12-1 Schematic of the powder metallurgy process.

The powder metallurgy process normally consists of four basic steps: (1) powder manufacture; (2) blending; (3) compacting; and (4) sintering. Secondary processing may follow to obtain special properties or precision. Figure 12-1 presents a flowchart of the P/M process.

Powder manufacture. The properties of powder metallurgy products are highly dependent on the characteristics of the metal powders that are used. Important properties include: chemistry and purity, particle size, size distribution, particle shape, and the surface texture of the particles. Several processes are used to produce metal powders, the various processes imparting distinct properties and characteristics to the powder and resulting product.

Probably the most common method of making metal powder is by *melt atomization*. In one form, illustrated in Figure 12-2(left), molten metal is atomized by a stream of impinging gas or liquid as it emerges from an orifice. Another method, shown in Figure 12-2(right), utilizes an electric arc impinging on a rapidly rotating electrode within a chamber purged with inert gas. Melt atomization is particularly useful in producing prealloyed powders, since, by using an alloyed melt or electrode, each powder particle is of the alloy composition. Powders of stainless steel, nickel-base alloys such as Monel, titanium alloys, cobalt-base alloys, and some low-alloy steels are available. Other commonly employed methods of powder manufacture include *reduction of oxides or ores* and *electrolytic deposition* from solutions or fused salts. Lesser used processes are: *pulverization or grinding, thermal decomposition of hydrides or carbonyls,* and *condensation of metal vapors*. By one or more of the above methods, almost any metal or alloy can be obtained in powder form. After manufacture, many powders undergo a drying operation and possibly a heat treatment before further processing.

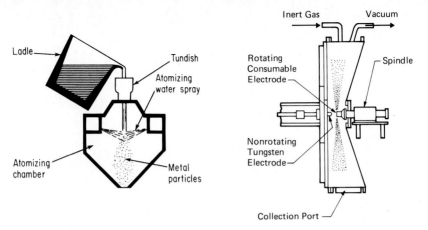

FIGURE 12-2 Two methods for producing metal powders. (*Left*) Melt atomization. (*Right*) Atomization from a rotating consumable electrode.

Powder testing and evaluation.   In addition to evaluation of chemistry, size, shape, and surface texture, metal powders are further evaluated for their adaptability to further processing. *Flow rate,* a measure of the ease by which a powder can be fed and distributed into a die, can significantly influence the cycle time of the pressing operation and, hence, the rate of production. Poor flow characteristics can also result in a pyramiding effect under nozzles and necessitate the use of multiple feed nozzles or a vibrating operation prior to pressing.

Associated with flow characteristics is the *apparent density,* a measure of the powder's ability to fill available space without the application of external pressure. *Compressibility tests* evaluate the effectiveness of applied pressure in raising the density of the powder and *green strength* is used to describe the strength of pressed powders after compacting. Green strength is used to assess the ability to handle products after compacting but before sintering.

Powder blending.   It is rare that a single powder will possess all of the characteristics desired for a given process and product. Most commonly, therefore, the starting material is a mixture of various grades or sizes of powder, or powders of different compositions, with additions of lubricants or binders. It is well known that higher product density results in superior mechanical strength and fracture-resisting properties. While particles of varying sizes should theoretically result in greater density, there is a tendency for the finer sizes to separate and segregate during handling and mixing. Many users, therefore, prefer uniform size powder for their processes.

Product chemistry is often obtained by combining pure metal or nonmetal powders. Unique materials can be produced such as obtaining an intimate distribution of immiscible materials or combining metals and nonmetals into the same products, such as tungsten carbide and cobalt for cutting tool applications. Certain powders can play a dual role, such as graphite which serves as a lubricant during compacting and a source of carbon as it alloys with iron during sintering. Lubricants, such as graphite or stearic acid, improve flow characteristics and compressibility at the expense of reduced green strength. Binders produce the reverse effect.

Blending or mixing opeations can be done either wet or dry, the use of water or solvent often being employed to obtain better mixing, reduce dusting, and lessen explosion hazards. Large lots of powder can be homogenized, often in quantities up to 16 000 kg (35,000 pounds) to assure uniform behavior throughout a production run and a consistent product.

Compacting. One of the most critical and controlling steps in the P/M process is compacting. Here, loose powder is compressed and densified into a shape known as a *green compact*. Product density and the uniformity of that density throughout the compact are important parameters. Generally, uniform high density is desired together with sufficient green strength for in-process handling and transport to the sintering furnace.

Most compacting is done with mechanical presses, although hydraulic and hybrid (combinations of mechanical, hydraulic or pneumatic) presses are often used. The mechanical presses may be of the eccentric or crank type, toggle or knuckle type, or cam type. Figure 12-3 shows a typical P/M press. Pressures of 69 to 690 MPa (5 to 50 tons per square inch) are utilized, 140 to 415 MPa (10 to 30 tons per square inch) being most common. Most presses have capacities of less than 9×10^5 N (100 tons) but an increasing number now have capability

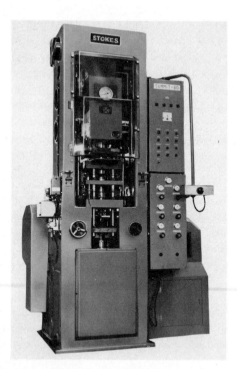

FIGURE 12-3 (*Left*) Typical press for the compacting of metal powders. The removable dieset feature (*right*) allows the machine to be producing parts with one dieset while another is being fitted to produce a second product. (*Courtesy Sharples-Stokes Division, Pennwalt Corporation.*)

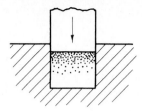

FIGURE 12-4 Compacting with a single plunger, showing resultant nonuniform density.

of 18 to 25 $\times$ 10^5 N (200 to 300 tons) and at least one with a capacity of 27 MN (3000 tons) is in operation. Consequently, most powder metallurgy products have cross sections of less than 2000 mm^2 (3 in.^{2}), but sizes up to 6500 mm^2 (10 in.^{2}) are becoming increasingly common.

In most cases, the prepared powder flows into the die by gravity until there is some excess. The excess is scraped off, providing measurement by volume, and the press closes, compacting the powder. As an alternative, the amount of powder may be controlled by weighing or the desired amount of powder may be preformed into tablets in a tablet-making machine.

During compacting, the powder particles move primarily in the direction of the applied force. The powder does not flow like a liquid, but develops an equal opposing force by means of friction between the particles themselves and the die surfaces. As illustrated in Figure 12-4, when pressure is applied by only one plunger, maximum density occurs below the plunger and decreases as you move down the column. Thus, it is seldom possible to transmit uniform pressures and obtain uniform density throughout the compact. A double-plunger press, as illustrated in Figure 12-5, enables more uniform density to be obtained or thicker products to be compacted. Density variation is a function of both the thickness and width of the part being pressed. Therefore, the ratio t/w should be kept below 2 whenever possible. Ratios above 2 produce considerable variation in density.

Since density is a function of thickness, it is difficult to produce products where more than one thickness is involved. Where nonuniform thickness is desired, more complicated presses or methods are used. Figure 12-6 illustrates the compaction of a dual-thickness part by use of multiple punches in the lower die. When extremely complex shapes are desired, the powder is generally encapsulated in a flexible mold and subjected to uniform high pressure by a liquid or gas medium, a method known as *isostatic compaction*. Other methods to achieve uniform compacting include: the use of a hydraulic press that applies a series of rapid squeezes of controlled, increasing intensity; the technique of

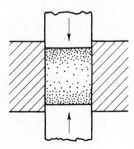

FIGURE 12-5 Density distribution obtained with a double-plunger press.

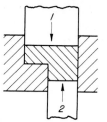
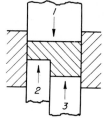

Single lower plunger   Double lower plunger

FIGURE 12-6 Two methods of pressing powders for powder metallurgy parts that have two levels of thickness.

adding powder between applications of pressure to build a uniform compact; or the use of additional lubricant (provided that green strength remains adequate).

Pressing rates vary widely, with about six pieces per minute being the minimum and 100 pieces per minute being quite common. The pressed parts are ejected from the die mechanically. At this stage, the compacts have sufficient strength to withstand a reasonable amount of handling. By means of mechanisms such as bulk movement of powder particles, deformation of particles, and particle fracture or fragmentation, the density of the powder is usually raised to about 80% that of an equivalent melted metal and the nature and uniformity of the remaining porosity has been set.

Dies. Because the powder particles tend to be somewhat abrasive and high pressures are involved, there is considerable wear on the die walls. Consequently, dies are usually made of hardened tool steel. For particularly abrasive powders, or for very high volume work, cemented carbide dies may be employed. Die surfaces should be highly polished and the dies should be heavy enough to withstand the high pressing pressures. Lubricants can be used to reduce die wear. When the lubricant is not incorporated with the powder, it can be sprayed onto the die surfaces prior to filling.

Sintering. In the sintering operation, the compacts are subjected to elevated temperatures in a controlled atmosphere environment. Most metals are sintered at 70 to 80% of their melting point, while certain refractory materials may require temperatures near 90%. When the product is composed of more than one material, the sintering temperature may be above the melting temperature of some of the components. Here, the lower-melting-point materials simply flow into the voids between the higher-melting-point materials.

Most sintering operations involve three stages, and many furnaces employ three distinct zones. The first region, the *burn-off or purge chamber,* is designed to expend air, volatilize and remove lubricants or binders that would interfere with good bonding, and slowly raise the temperature of the compacts in a manner that prevents the buildup of internal pressure from entrapped air and lubricant and the resulting swelling or fracture. If volatile materials are present in appreciable quantities, the final product will tend to be porous and permeable. Certain products take advantage of this possibility. The *high-temperature zone* is the site of actual solid-state diffusion bonding between powder particles. The time must be sufficient to obtain the desired density and final properties, and usually varies from 10 minutes to several hours. Finally, a *cooling zone* is required to lower the temperature and thereby prevent oxidation upon discharge into air and possible thermal shock. Both batch-type and continuous furnaces are used.

All three zones usually operate with a controlled protective atmosphere. This is critical because the fine powder particles have large exposed surface areas and at elevated temperature rapid oxidation can occur and significantly impair the quality of particle bonding. Reducing atmospheres, commonly based on hydrogen, dissociated ammonia, or cracked hydrocarbons, are preferred since they can reduce any oxide already present on the particle surfaces and remove gases liberated during sintering.

During sintering, metallurgical bonds form between powder particles. If different chemistry powders have been blended, alloys may form. In addition, product dimensions will contract, a phenomenon that must be compensated in die design. The product generally contains 10 to 25% porosity.

Spark-discharge sintering is an alternative method of bonding wherein a high energy electrical spark is discharged between the powder particles while they are under pressure in the compacting press. The energy discharge strips the contaminants from the surface of the particles and, because of the intimate contact resulting from the high pressure, bonding results almost instantaneously. This procedure eliminates the separate sintering oven and also provides easier size control by virtually eliminating the shrinkage that occurs during conventional sintering.

Hot isostatic pressing (HIP) combines compaction and sintering in still another manner. Here the powder is hermetically sealed in a flexible, airtight, evacuated can and then subjected to a high-temperature, high-pressure environment, generally around 70 to 100 MPa (10,000 to 15,000 psi) and 1250°C (2300°F). Products emerge 100% dense with uniform, isotropic properties. Near-net shapes are possible, thereby eliminating waste and costly machining operations. A modification of the process produces rods, wire, and small special billets by hot extruding encapsulated powder. Exotic metals such as beryllium, uranium, ziroconium, and high-strength titanium are often processed by HIP techniques. In an area unrelated to P/M, HIP has been employed to heal internal porosity in castings or close and seal internal cracks in products.

Presintering. Powder metallurgy is frequently employed to produce parts from materials that are very difficult to machine. When some machining is desired on such parts, it often can be made easier by employing a *presintering* operation, wherein the compacted parts are heated for a short time at a temperature considerably below the final sintering temperature. This operation imparts sufficient strength to the parts so they can be handled and machined without difficulty. They are then given the final sintering, during which very little dimensional change occurs. Thus, machining after final sintering may be reduced to a minimum or eliminated entirely.

Secondary operations. For many applications, P/M parts are ready for use as they come from the sintering oven. However, many products require one or more secondary operations to provide enhanced precision or special characteristics.

A second pressing operation, known as *repressing, coining, or sizing,* may be used to restore dimensional precision. The part is placed in a die and subjected to pressures equal to or greater than the initial pressing pressure. A small amount of plastic flow takes place, resulting in a very uniform product with respect to size and sharpness of detail. In addition, coining may increase strength by 25 to 50%. Most coining is done cold.

If massive metal deformation occurs in the second pressing, the operation is known as *P/M forging.* Here the powder metallurgy process is used to produce a preform which is one forging operation removed from the finished shape. The normal forging operations of producing a billet or bloom, shearing, reheating, and sequentially deforming it to the desired shape are replaced by the manufacture of a comparatively simple-shape preform by powder metallurgy and a final forging operation to produce a more complex shape, add precision, and provide the benefits of metal flow. The forging operation further densifies the preform, often up to 99% of theoretical density, and significantly improves its properties. While atmospheres or coatings must be used to prevent oxidation during hot forging, the process can significantly reduce scrap or waste and extends the P/M process by improving properties and increasing size and complexity capabilities. The forged products have no segregation, a uniform grain size, and can be made from novel alloys or composites. In addition, subsequent machining operations can be reduced or eliminated. Figure 12-7 shows an example of a part made by conventional forging and then by the P/M forge method. Substantial scrap savings can be obtained as well.

The permeability of P/M products created with controlled porous structures opens up two other possibilities, *impregnation* and *infiltration.* Impregnation refers to the forcing of oil or other liquid into the porous network by either immersing the part in a bath and applying pressure or a combination vacuum-pressure process. The most common application is that of oil-impregnated bearings. Here the bearing itself contains from 10 to 40% oil by volume which will

FIGURE 12-7 Comparison of conventional forging and the use of a powder metallurgy preform to form a gear blank and gear. (*Top*) Sheared stock, rough forging, forged blank, and scrap. (*Bottom*) By using powder-metallurgy preform, finished gear is formed with no scrap. (*Courtesy GKN Forging Limited.*)

be released slowly throughout its lifetime as applied loads and resulting temperatures act upon it.

When the porous structure of a P/M product is not desirable, the part may be subjected to metal infiltration. Here, a molten metal of a lower melting point than the major constituent is forced into the product under pressure or absorbed by capillary action. Engineering properties, such as strength and toughness, are generally comparable to those of solid metal. Infiltration can also be used to seal pores prior to plating, improve machinability, or make components gas- or liquid-tight.

Powder metallurgy products may also be subjected to more conventional finishing operations such as heat treatment, machining, and surface treatment. If the part is of high density or metal impregnated, conventional techniques are employed. Special precautions, however, must be employed when processing low density products. During heat treatment, protective atmospheres must again be employed and certain liquid quenchants should be avoided. When machining, speeds and feeds must be adjusted and care taken to avoid lubricant pickup and associated problems. Nearly all common methods of surface finishing are applicable, again with some modification for porous or low-density parts.

Properties of P/M products. Because the properties of powder metallurgy products depend on so many variables—types and size of powder, amount and type of lubricant, pressing pressure, sintering temperature and time, finishing treatments, and so on—it is difficult to provide generalized information. Properties can range all the way from low-density, high-porous parts with a tensile strength as low as 70 MPa (10,000 psi) to high-density, minimal-porosity pieces with tensile strengths of 1250 MPa (180,000 psi) or more, with strengths of 275 to 350 MPa (40,000 to 50,000 psi) being most common. In general, however, all mechanical properties show a strong dependence on density, with fracture-related properties such as toughness, ductility, and fatigue life showing a stronger dependence than strength and hardness. The strength properties of the weaker metals are often equivalent to the same wrought metal. As alloying elements are added to produce higher strength powder, the resultant properties tend to fall below those of wrought products by varying, but usually substantial amounts. Table 12-1 shows a few powder metallurgy materials and their strength properties, with similar wrought metals shown for comparison. As larger presses and processes such as P/M forging are employed to provide greater density, the strength properties of P/M products will more nearly equal those of wrought materials.

Porosity can also affect physical properties. Corrosion resistance tends to be reduced due to available entrapment pockets and fissures. Electrical, thermal, and magnetic properties all vary with density. On the positive front, the porosity does promote sound and vibration damping and many P/M parts are designed to take advantage of this feature.

Design of powder metallurgy parts. In designing parts that are to be made by powder metallurgy it must be remembered that it is a special manufacturing process and provision should be made for a number of factors that are specific

TABLE 12-1 Comparison of Properties of Powder Metallurgy Materials and Equivalent Wrought Metals

Material	Composition	Condition	Theoretical Density (%)	Tensile Strength MPa	Tensile Strength 10^3 psi	Elongation in 2 inches (%)
Iron [a]		Wrought HR	—	331	48	30
	99% Fe min.	As sintered	89	207	30	9
	99% Fe min.	As sintered	94	276	40	15
Steel[a]	AISI 1025	HR	—	586	85	25
	0.25% C, 99.75% Fe	As sintered	84	234	34	2
Stainless[a] steel	Type 303	Annealed	—	621	90	50
	Type 303	As sintered	82	358	52	2
Aluminum[a]	2014	T6	—	483	70	20
	201AB	T6	94	331	48	2
Aluminum[a]	6061	T6	—	310	45	15
	601AB	T6	94	252	36.5	2
Copper[a]	OFHC & ETP	Annealed	—	234	34	50
		As sintered	89	159	23	8
		Repressed	96	241	35	18
Brass[a]	CDC 260	Annealed	—	303	44	65
	70%Cu–30%Zn	As sintered	89	255	37	26

[a]Equivalent wrought metal shown for comparison. HR = hot rolled; T6 = area hardened.

to this process. The Powder Metallurgy Parts Association recommends six basic rules for designing P/M parts:

1. The shape of the part must permit ejection from the die.
2. The shape of the part should be such that the powder is not required to flow into thin walls, narrow splines, or sharp corners.
3. The shape of the part should permit the construction of strong tooling.
4. The shape of the part should make allowance for the length (thickness) to which thin-walled parts can be compacted.
5. The part should be designed with as few changes in section thickness as possible.
6. Take advantage of the fact that certain forms can be produced by P/M which are impossible, impractical, or uneconomical to obtain by any other method.

Since uniform strength requires uniform density, parts should be designed with uniform cross section and a short thickness compared to cross-sectional area. The ratio of unpressed thickness to final thickness should be kept below 2, if possible. Designs should not contain holes whose axes are perpendicular to the direction of pressing. Multiple-stepped diameters, reentrant holes, grooves, and undercuts should be eliminated. Abrupt changes in section and internal angles without generous fillets should be avoided. Straight serrations can be molded readily, but diamond knurls cannot. The meeting plane between mold

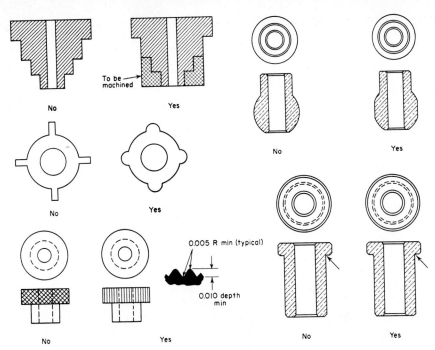

FIGURE 12-8 Some poor and good design details for use in powder metallurgy parts.

punches should be on a cylindrical or flat surface, never on a sphere. Narrow, deep flutes should be avoided. Figure 12-8 illustrates some of these points.

Powder metallurgy products. The products that are commonly produced by powder metallurgy can generally be classified into four groups.

1. *Porous products, such as bearings, filters, and pressure or flow regulators.* Oil-impregnated bearings made from either iron or copper alloys constitute a large volume of P/M products. They are widely used in home appliance and automotive applications since they require no lubrication during their service life. Filters can be made with pores of almost any size, some as small as 0.0025 mm (0.0001 inch).

2. *Products of complex shapes that would require considerable machining when made by other processes.* Large numbers of small gears are made by the powder metallurgy process. Because of the accuracy and fine finish obtainable, many parts require no further processing and others, only a small amount of finish machining. Some gears are oil-impregnated to provide self-lubrication. Other complex shapes, such as pawls, cams, and small activating levers, which ordinarily would involve high machining costs can frequently be made quite economically by powder metallurgy.

3. *Products made from materials that are very difficult to machine.* One of the first modern uses of powder metallurgy was in the production of tungsten carbide cutting tools. This continues to be an important use, with other carbides also being utilized.

4. *Products where the combined properties of two metals, or of metals and*

nonmetals are desired. This unique characteristic of the powder metallurgy process is applied in a number of products. In the electrical industry, copper and graphite frequently are combined in such applications as motor generator brushes, the copper providing high current-carrying capacity with graphite providing lubrication. Similarly, bearings have been made of graphite combined with iron or copper or of mixtures of two metals, such as tin and copper, where a softer metal is placed in a harder metal matrix. Electrical contracts often combine copper or silver with tungsten, nickel, or molybdenum. The copper or silver provides high conductivity while the tungsten, nickel, or molybdenum provides resistance to fusion because of the high arcing temperature of each.

Figure 12-9 shows some typical powder metallurgy products.

Advantages and disadvantages. Like many other manufacturing processes, powder metallurgy has some distinct advantages and disadvantages with which the designer should be familiar in order to use it successfully and economically. Among the important advantages are:

1. *Elimination of machining.* The dimensional accuracy and finish obtainable are such that for many applications all machining can be eliminated. If unusual dimensional accuracy is required, simple coining or sizing operations will give accuracies equal to those obtainable from most production machining.
2. *High production rates.* All the steps of the process are simple and high rates of production can be achieved. Labor requirements are low since the process is often highly automated. Uniformity and reproducibility are among the highest in manufacturing.

FIGURE 12-9 Typical parts produced by the power metallurgy process. (*Courtesy PTX-Pentronix, Inc.*)

3. *Complex shapes can be produced.* Subject to the limitations discussed previously, quite complex shapes can be produced, such as combination gears, cams, and internal keys. Often it is possible to produce parts by powder metallurgy that cannot be machined or cast economically.
4. *Wide variations of compositions can be obtained.* Parts of very high purity can readily be produced, or entirely different materials can be combined. Custom-made compositions are possible.
5. *Wide variations in properties are available.* Density can be varied from low-density porous or permeable products to high-density ones with properties equivalent to wrought counterparts. Damping of noise and vibration can be tailored into a product. Wide variation in strength is also possible.
6. *Scrap is eliminated.* Powder metallurgy is the only common production process wherein no material is wasted, whereas in casting, machining, or press forming the scrap often exceeds 50%. This is important where expensive materials are involved and often makes it possible to use more costly materials without increasing the overall cost of the product.

The major disadvantages of the process are as follows:

1. *Inferior strength properties.* In most cases powder metallurgy parts have mechanical properties less than wrought or cast products of the same material. Their use may be limited when high stresses are involved. However, the required strength can frequently be obtained by using different and usually more costly materials or secondary processing such as P/M forging.
2. *Relatively high die cost.* Because of the high pressures and severe abrasion involved in the process, the dies must be made of expensive materials and be relatively massive. Production volumes of less than 10,000 identical parts are normally not practical.
3. *High material cost.* On a unit weight basis, powdered metals are considerably more expensive than wrought or cast stock. However, the absence of scrap and elimination of machining often offset the higher material cost. Powder metallurgy is usually employed for rather small parts where the material cost is not great.
4. *Design limitations.* The powder metallurgy process simply is not feasible for many shapes. Parts should have essentially uniform sections along the axis of compression. The thickness-to-diameter ratio is limited. Furthermore, the overall size must be within the capacity of available presses.

REVIEW QUESTIONS

1. What are the four basic steps which are usually involved in making products by powder metallurgy?
2. What are five important properties of metal powders that will influence the properties of products made from them?
3. What are several methods of powder manufacture?
4. What is flow rate and why is it important to manufacturing?
5. What is green strength and why is it important?
6. What is the role of a lubricant addition?
7. What is the goal of the compacting operation?
8. How do powder particles flow during compacting?

9. Why should the ratio of thickness to width (t/w) be kept below 2 whenever possible?
10. Why is it desirable to have uniform thickness in a P/M product?
11. What mechanisms may occur during compacting to raise the density to about 80% of an equivalent melted metal?
12. What are the three stages of the sintering process reflected in most furnace designs?
13. Why is a controlled atmosphere necessary during sintering?
14. What changes occur to the compact during sintering?
15. What is HIP?
16. What is the purpose of repressing, coining, or sizing?
17. What benefits can be imparted to a P/M compact by subsequent forging?
18. What is the difference between impregnation and infiltration?
19. Under what conditions would secondary processing operations such as heat treatment, machining, or surface treatment of P/M products be different from conventional?
20. The properties of P/M products are strongly tied to density. Which properties show the strongest dependence?
21. How can P/M parts provide sound and vibration damping characteristics?
22. What are some guidelines for good P/M design?
23. What are the four general categories of powder metallurgy parts?
24. List six advantages of the P/M process.
25. What are the major disadvantages of the P/M process?

CASE STUDY 12. The Short-Lived Gear

A 250-mm (10-inch)-diameter gear was fabricated from AISI 1080 steel. The gear blank was hot-forged, air-cooled, and then full-annealed in preparation for machining. Following finish machining, the gear teeth were surface-hardened by flame hardening and quenching to a hardness of R_C55. After a short period of service, the teeth began to deform in a manner consistent with subsurface flow, and the gear failed to mesh properly. What is a probable cause of the failure?

CHAPTER 13

The Fundamentals of Metal Forming

The ultimate goal of a manufacturing engineer is to produce components of a selected material with a required geometrical shape and a structure optimized for the proposed service environment. Of the above aspects, production of the desired shape is a significant part of the manufacturing process. Four basic alternatives exist: casting, machining, consolidating smaller pieces (welding, powder metallurgy, mechanical fasteners, epoxy, etc.), and deformation processes. Casting processes exploit the fluidity of a liquid as a means of producing the shape. Machining, or more specifically, material removal processes, provide excellent precision and great flexibility, but tend to waste material in the generation of the removed portions. Consolidation processes enable complex shapes to be constructed from simpler components and have a useful domain of application.

Deformation processes, on the other hand, exploit a remarkable property of some materials (usually metals)—their ability to flow plastically in the solid state without deterioration of properties. By simply moving the material to the desired shape, as opposed to removing the unwanted regions, there is little or no waste. The required forces, however, are often high. Machinery and tooling is quite expensive and therefore, large production quantities may be necessary to justify the process.

The overall usefulness of metals in modern society is largely due to the ease with which they may be formed into useful shapes. Nearly all metal products are subjected to metal deformation at some stage of their manufacture. Cast ingots, strands, and slab are reduced in size and converted to basic forms such as sheets, rods, and plates. These forms may then undergo further deformation into wire or the myriad of finished products formed by processes such as forging, extrusion, sheet metal forming, and others. The deformation may be bulk flow in three dimensions, simple shearing, simple or compound bending, or any combination of these and others. The stresses producing these deformations may be tension, compression, shear, or various combinations thereof, illustrated in Table 13-1. Speeds, temperatures, tolerances, surface finishes, and deformation amounts span a wide spectrum. The specific processes are numerous and varied as shown in Table 13-2.

Metal-forming processes: independent variables.   In general, forming processes tend to be complex systems consisting of independent variables, dependent variables and independent–dependent interrelations. *Independent variables* are those aspects of the process over which the engineer has direct control and are generally selected or specified when setting up the process. Consider some of the typical independent process variables:.

1. *Starting material.* The engineer must specify the chemistry and condition of the material to be deformed. In so doing, he is selecting the properties and characteristics of the starting material, which may be chosen for ease in fabrication or may be restricted by the desired final properties of the product.
2. *Starting workpiece geometry.* This may be dictated by previous processing or may be selected by the engineer from a variety of available shapes, often on the basis of economics.
3. *Tool or die geometry.* This is an area of major significance and has many aspects: from the diameter of a rolling mill roll, bend radius in sheet forming, or die angle in drawing or extrusion to the cavity details in forging. Since

TABLE 13-1   Classification of States of Stress

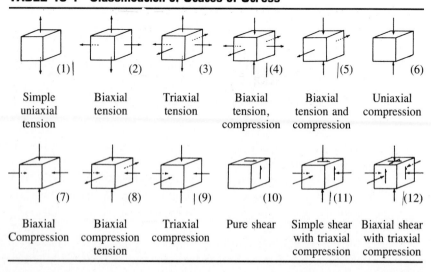

(1)	(2)	(3)	(4)	(5)	(6)
Simple uniaxial tension	Biaxial tension	Triaxial tension	Biaxial tension, compression	Biaxial tension and compression	Uniaxial compression

(7)	(8)	(9)	(10)	(11)	(12)
Biaxial Compression	Biaxial compression tension	Triaxial compression	Pure shear	Simple shear with triaxial compression	Biaxial shear with triaxial compression

TABLE 13-2   Classification of Some Forming Operations

Number	Process	Schematic Diagram	State of Stress in Main Part During Forming[a]
1	Rolling		7
2	Forging		9
3	Extruding		9
4	Shear spinning		12
5	Tube spinning		9
6	Swaging or kneading		7
7	Deep drawing		In flange of blank, 5 In wall of cup, 1
8	Wire and tube drawing		8

[a]See Table 13-1 for key.

TABLE 13-2 Classification of Some Forming Operations (*Continued*)

Number	Process	Schematic Diagram	State of Stress in Main Part During Forming[a]
9	Stretching		2
10	Straight bending		At bend, 2 and 7
11	Contoured flanging	(a) Convex	At outer flange, 6 At bend, 2 and 7
		(b) Concave	At outer flange, 1 At bend, 2 and 7

the tooling will produce and control the metal flow, success or failure of a process often depends upon tool geometry.

4. *Lubrication.* It is not uncommon for friction to account for more than 50% of the power supplied to a deformation process. Since lubricants also act as coolants, thermal barriers, corrosion preventers, and parting compounds, their selection is one of great importance.

5. *Starting temperature.* Temperature is one of the most influential of process variables. Its control may well dictate the success or failure of a process.

6. *Speed of operation.* Most deformation processing equipment has variable speed. Since speed influences lubricant effectiveness, forces required for deformation (see Figure 2-30), and heat transfer, it is obvious that its selection would be significant.

7. *Amount of deformation.* While this may be set by die design, other processes such as rolling permit this to be varied at the discretion of the engineer.

The dependent variables. After the deformation engineer specifies the independent variables, the process then determines the nature and values for a second set of variables, known as *dependent variables*. Examples of these include:

1. *Force or power requirements.* To deform a selected material from a given starting shape to a specified final shape, with a specified lubricant, tooling geometry, speed, and starting temperature, will require a certain amount of force or power. A change in any of the independent variables will bring about a change in the force or power required. The effect, however, is indirect. The engineer cannot directly specify force or power; he can only specify the independent variables and in so doing, indirectly modify the required force.

The ability to predict forces or powers, however, is extremely important, for only by having this knowledge will the engineer be able to specify or select equipment for an operation, select appropriate die materials, compare various die designs or deformation methods, and ultimately optimize processes.

2. *Material properties of the product.* While the engineer can specify the properties of the starting material, the combined deformation and temperature changes imparted by the process will certainly modify them. The customer is not interested in the starting properties, but rather the final properties of the product. Thus, while it is often desirable to select starting properties based on compatibility with the process, it is also necessary to know or be able to predict how the process will alter them.

3. *Exit temperature.* Deformation generates heat. Hot workpieces cool in cold tooling. Properties depend upon both the mechanical and the thermal history of the material. Thus, it is important to know and control the temperature of the material throughout the process.

4. *Surface finish and precision.* Both are product characteristics dependent on the specific details of the process.

5. *The nature of material flow.* Generally deformation processes exert external constraints on the material through control and movement of its surfaces. How it flows or deforms internally depends on the specifics of the process and material. Since properties depend on deformation history, control here is vital. The customer is satisfied only if the desired geometric shape is produced, with the right set of companion properties, and without surface or internal defects.

The metal former's dilemma. As illustrated in Figure 13-1, the problem facing the metal forming engineer becomes quite obvious. On one hand are the independent variables, those aspects of the process over which he has direct and immediate control. On the other are the dependent variables, those aspects over which he must have control but for which his influence is indirect. The dependent variables are determined by the process based on proper selection of the independent variables. If a dependent variable is to be modified, the engineer must determine which independent variable (or variables) is to be changed, in what manner, and by how much. Thus, the engineer must have a knowledge of the independent–dependent variable interrelations.

The link between independent and dependent variables is the most important

Independent variables		Dependent variables
Starting material		Force or power requirements
Starting geometry	–Experience–	
		Product properties
Tool geometry		
	–Experiment–	Exit temperature
Lubrication		
		Surface finish
Starting temperature	–Theory–	
		Dimensional precision
Speed of deformation		Material flow details
Amount of deformation		

FIGURE 13-1 Metal-forming system.

area of knowledge for a metal deformation engineer. Unfortunately, such links are often difficult to obtain. Metal deformation processes are complex systems composed of the material being deformed, the tooling performing the deformation, lubrication at surfaces and interfaces and various other process parameters. The number of different processes and subprocesses is quite large. Various materials often behave differently. Multitudes of different lubricants exist. In fact, some deformation processes may be complex systems of 15 or more interacting independent variables.

The ability to predict and control dependent variables, therefore, comes about in one of three ways:

1. *Experience.* This requires long-time exposure to the process and is often restricted in scope by the realm of past contact.
2. *Experiment.* While possibly the least likely to be in error, direct experiment is both time consuming and costly. Size and speed of deformation are often reduced when conducting laboratory studies. Lubricant performance and heat transfer are generally altered. Thus, while laboratory experiments may be quite valuable, caution should be used when extrapolating their results to different, but similar, conditions.
3. *Theory.* Here one attempts to develop a mathematical model of the process into which he can insert numerical values for the various independent variables and compute a prediction for the dependent variables. Most techniques rely on the applied theory of plasticity with three dimensional stresses. Alternatives vary from crude, first-order approximations, such as slab equilibrium or uniform deformation energy calculations, to sophisticated, computer-based, solutions, such as the finite element or finite difference methods. Solutions may be algebraic equations describing a process and revealing trends or just a numerical answer based on the specific input values.

While the trend is certainly toward mathematical models, both for process design and process computer control, it is important to note that the models can be no better than the accuracy to which the input variables are known. For example, the plasticity behavior (yield strength, ductility, etc.) of the deforming material must be known for the specific conditions of temperature, strain (amount of prior deformation) and strain rate (speed of deformation). Moreover, the same material may behave differently under the same conditions if its microstructure is different (i.e., a 1040 steel annealed to ferrite and pearlite or quenched and tempered to tempered martensite). Microstructure and its effects, however, are difficult to describe in quantitative terms handleable by a computer. The characterization of material behavior under various conditions forms the science of *constitutive relations* and is receiving considerable attention at this time.

Another rather elusive variable is friction. While it is known to be dependent on pressure, area, surface finish, lubricant, speed, and material and often varies for different locations within the same process, most models tend to account for its effect with a single variable of constant magnitude.

While these problems appear to hinder the theoretical approach, it must be noted that the same lack of knowledge hinders the individual trying to document, characterize, and extrapolate experience or experimental results. Theory often reveals process aspects which might otherwise go unnoted, and can be quite useful when extending a process into a previously unknown area or designing a new process.

General parameters. While much metalforming knowledge is specific to a given process, there are certain aspects that are common to all processes and will be treated here.

One important area of knowledge for the metal-forming engineer is *information about the material* being deformed. What is its strength or resistance to deformation at the relevant conditions of temperature, speed of deformation, and amount of prior straining? What are its formability and fracture characteristics? What is the effect of temperature and variations in temperature? To what extent does the material strain harden? What is it recrystallization kinetics? Does it react with various environments or lubricants? These and many other questions must be answered to determine the suitability of a material to a given metal deformation process. Since material properties are quite varied, such details will not be discussed here. The reader is referred to the various chapters on engineering materials and the additional recommended references, such as those cited in Chapter 10.

Several other effects are strongly related to the *speed of deformation*. Material behavior can vary markedly with speed. Some materials may shatter or crack if impact loaded, but will deform plastically under slow-speed loadings. Rate-sensitive materials appear stronger when deformed at faster speed. Thus, more energy is required to produce the same result if we wish to do it faster. Mechanical data obtained from slow-speed tensile tests may not be particularly useful when considering deformation processes at much faster rates of deformation. Moreover, speed sensitivity (the degree to which behavior varies with speed) generally becomes greater when the material is at higher temperature, as is often the case in metal forming.

In addition to mechanical changes related to speed, faster speeds tend to promote improved lubrication efficiency. Also, faster speeds reduce the time for heat transfer and cooling. During hot working, the workpiece stays hotter and tools run cooler.

Other "general variables" include *friction and lubrication* and *temperature*. Both are of sufficient significance that they will be discussed in some detail.

Friction and lubrication under metalworking conditions. An important consideration in metal deformation processes is the friction developed between the workpiece and the forming tool or tools. Sometimes more than 50% of the energy supplied by the equipment is spent in overcoming friction. Product quality aspects, such as surface finish and dimensional precision, are directly related to friction. Further, changes in lubrication can alter the mode of material flow during forming and, in so doing, modify the properties of the final product. Production rates, tool design, tool wear, and process optimization all depend on the ability to determine and control friction. Some processes, such as rolling can only operate with sufficient friction. Friction effects, however, are hard to measure, and since they depend on such variables as contact area, speed, and temperature, they are difficult to scale down for testing or scale up to production conditions.

It should be recognized that friction during metalworking conditions is significantly different from the friction encountered in most mechanical devices. The friction conditions of gears, bearings, journals, and similar components generally involves: two surfaces of similar material and strength, under elastic loads such that neither body undergoes permanent change in shape, with wear-in-cycles to produce surface compatibility, and generally at low to moderate temperatures. Metal-forming operations, on the other hand, involve a hard, nondeforming tool interacting with a soft workpiece at pressures sufficient to cause plastic flow in the weaker material. Only a single pass is involved as the tool shapes the piece, and the workpiece is often at highly elevated temperature. Figure 13-2 shows the change in frictional resistance with variation in contact pressure. At light, elastic loads, friction is proportional to the pressure normal to the interface, the proportionality constant (often denoted as μ) being known as the coefficient of friction. At high pressures, friction becomes independent of contact pressure and is more dependent on the strength of the weaker material. This figure, however, is for unlubricated contact. The presence of a lubricant and variation in type and amount will alter friction behavior. Nevertheless, it is sufficient to note that friction, lubrication, and wear data obtained under conditions typical of mechanical components will be of questionable value when applied to metalworking operations.

Likewise, the importance of wear is different. Since the workpiece only passes over the tooling once, any wear experienced by the workpiece is usually not objectionable. In fact, the shiny, fresh-metal surface produced by wear is often desired by customers. Manufacturers failing to produce enough wear, thereby retaining some of the original dull finish, may actually be accused of selling old or substandard products. Wear on the tooling, however, is quite the reverse. Tooling is expensive and it is expected to shape many workpieces. Wear here generally means that the dimensions of the workpiece will change. Tolerance control is lost and at some point the tools will have to be replaced. Other consequences of tool wear include: increased frictional resistance (increased required power and decreased process efficiency); poor surface finish of the product; and loss of production during tool changeover.

Lubrication is of immense importance in metal forming. Lubricants are se-

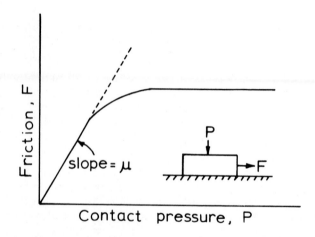

FIGURE 13-2 Effect of contact pressure on frictional resistance.

lected to reduce friction and suppress tool wear. Other considerations in selecting a metalworking lubricant include: its ability to act as a thermal barrier, keeping heat in the workpiece and away from the tooling; its ability to act as a coolant and remove heat from the tools; its ability to retard corrosion if left on the formed product; ease of application and removal; lack of toxicity, odor, and flammability; reactivity or lack of reactivity with material surfaces; thermal stability; adaptability over a useful range of pressure, temperature and velocity; surface wetting characteristics; cost; availability; and its ability to flow or thin and still function. In addition, the behavior of a given lubricant will change with variation in the interface conditions. The exact response will depend on such factors as: the surface finish of both surfaces, area of contact, load, speed, temperature, and amount of lubricant.

In view of the important aspects of lubricants, it is amazing how little science has been applied to their development and selection. Most useful lubricants have been developed on the basis of art and experience. Many alternatives exist, often with proprietary additives or unreported chemistries marketed under exotic trade names. Selection is often a wasteful, hit-or-miss, proposition. Moreover, when a problem is encountered in a process, lubricant variation is often the easiest and least expensive process variable that can be altered.

Nevertheless, the benefits of friction and lubrication control can be great. For example, if one can achieve full fluid separation between a tool and workpiece, the required deformation forces may reduce 30 to 40% and tool wear becomes almost nonexistent. Considerable effort, therefore, has been directed to the measurement of friction under general metalworking conditions and for the various metalforming processes. Hopefully, a science base can be established that will enable the optimum utilization of lubricants in metalworking.

Temperature concerns. The importance of temperature in metalworking is every bit as great as lubrication. The role of temperature in altering material properties and behavior has already been discussed in Chapter 2. In general, an increase in temperature brings about a decrease in material strength, an increase in ductility, and a decrease in the rate of strain hardening—all effects that would tend to promote ease in deformation.

Forming processes tend to be classified as hot working, cold working, or warm working based on the material being formed and the temperature of forming. *Hot working* occurs under conditions of temperature and strain rate such that recrystallization occurs simultaneously with deformation. Generally, the temperature of deformation must exceed 0.6 times the melting point of the workpiece material in degrees Kelvin. *Cold working* is deformation under conditions where the recovery processes are not effective and generally requires working temperatures less than 0.3 times the workpiece melting temperature. *Warm working* is deformation under the conditions of transition, i.e., working temperature between 0.3 and 0.6 times the melting point.

Definition of hot working. Hot working is defined as the plastic deformation of metals above their recrystallization temperature. Here it is important to note that the recrystallization temperature varies greatly with different materials. Lead and tin are hot-worked at room temperature, steels require temperatures near

1100°C (2000°F), and tungsten is still in a cold or warm working state at 1100°C (2000°F). Hot working does not necessarily imply high absolute temperatures.

As was discussed in Chapter 3, plastic deformation above the recrystallization temperature does not produce strain hardening. Therefore, hot working does not cause any increase in yield strength or hardness, or corresponding decrease in ductility. In addition, Figures 2-28 and 2-29 showed that the yield strength of metals decreases as temperature increases, and the ductility improves. The result is a true stress/true strain curve that is essentially horizontal for strains above the yield point, rather than the rising shape observed below the recrystallization temperature. Thus it is possible to alter the shape of metals drastically by hot working without causing them to fracture and without the necessity for using excessively large forces. In addition, the elevated temperatures promote diffusion that can remove chemical inhomogeneities, pores can be welded shut or reduced in size during deformation, and the metallurgical structure can be altered to improve the final properties. For steels, hot working involves the deformation of face-centered cubic austenite, which is rather weak and ductile compared to the b.c.c. ferrite structure found at lower temperatures.

From a negative viewpoint, the high temperatures may promote undesirable reactions between the metal and its surroundings. Tolerances are poorer due to thermal contractions and possible nonuniform cooling. Metallurgical structure may also be nonuniform, the final grain size depending on reduction, temperature at last deformation, cooling history after deformation, and other factors.

Grain alteration through hot working. When metals solidify, particularly in larger sections as in ingots or continuous-cast strands, coarse dendritic grains form and a certain amount of segregation of impurities occurs. Moreover, undesirable grain shapes, such as the columnar grains so common in ingots, can form as seen in Figure 13-3. Small gas cavities or shrinkage porosity can also form. Also, an as-solidified metal typically has a nonuniform grain structure with rather large grain size.

Reheating of the metal without prior deformation will simply promote grain growth and a concurrent decrease in properties. However, if the metal is deformed sufficiently above the recrystallization temperature, the distorted struc-

FIGURE 13-3 Cross section of a cast copper bar (100 mm in diameter) showing as-cast grain structure.

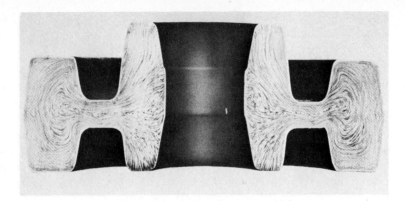

FIGURE 13-4 "Fiber" structure of a hot-formed (forged) transmission gear blank. (*Courtesy Bethlehem Steel Corporation.*)

ture is rapidly eliminated by the formation of new strain-free grains. The metal can then enter into a state of grain growth or be further deformed and recrystallized. The final structure will be that formed by the last recrystallization and subsequent thermal history and will depend on the factors listed previously. Production of a fine, randomly oriented, spherical-shaped grain structure can result in a net increase not only in strength but also in ductility and toughness.

Another improvement that can be obtained from hot working is the reorientation of inclusions or impurity material in the metal. With normal melting and cooling, many impurities tend to locate along grain boundary interfaces and, if unfavorably oriented, can assist a crack in its propagation through a metal. When a piece of metal is plastically deformed, the impurity material often distorts and flows along with the metal. This material, however, does not recrystallize with the base metal and often produces a *fiber structure,* as seen in Figure 13-4. Such a structure clearly has directional properties, being stronger in one direction than in another. Moreover, an impurity originally oriented so as to aid crack movement through the metal is often reoriented into a "crack-arrestor" configuration, perpendicular to crack propagation. Through proper design, extensive use can be made of these results. Figure 13-5 schematically compares a machined thread and a rolled thread in a threaded fastener. By removing potential failure sites where defects intersect the surface, the rolled thread with its fibered structure

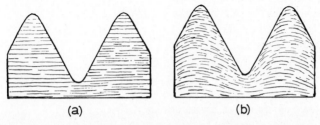

(a)   (b)

FIGURE 13-5 Schematic comparison of the grain flow characteristics in a machined thread (a) and a rolled thread (b). The rolling operation further deforms the axial structure produced by the previous wire- or rod-forming operations, while machining simply cuts through it.

possesses improved strength characteristics. Improper design or deformation, however, can significantly increase the likelihood of failure.

Temperature variation. The success or failure of a hot deformation process often depends on the ability to control thermal conditions. Since over 90% of the energy imparted to a deforming workpiece will be converted into heat, it is possible to produce a temperature rise in the workpiece if deformation is sufficiently rapid. More common, however, is the cooling of the workpiece in its lower temperature environment. Heat is lost through the workpiece surfaces, the bulk of the loss occurring where the workpiece is in contact with lower temperature tooling. As the surfaces cool, nonuniform temperatures are produced within the workpiece. Flow of the hot, weak, interior may well result in cracking of the colder, less ductile, surfaces. It is desirable, therefore, to maintain temperatures as uniform as possible. Heated dies can reduce the rate of heat transfer at the expense of reduced die life. It is not uncommon, therefore, to see die temperatures in the range 320 to 420°C (600 to 800°F) when hot-forming steel. While formers would like to use tooling at temperatures as high as 540 to 650°C (1000 to 1200°F) to produce improved tolerances and permit longer contact times, tool lifetime drops rapidly so as to make these conditions unattractive. When forming complex shaped products, as in hot forging, thin sections cool faster than thick sections and may further complicate flow behavior. In addition, nonuniform cooling from the conditions of working may introduce significant amounts of residual stress in hot-worked products.

Cold working. Plastic deformation of metals below the recrystallization temperature is known as *cold working,* and is generally performed at room temperature. In some cases, however, the working may be done at mildly elevated temperatures to provide increased ductility and reduced strength. From a manufacturing viewpoint, cold working has a number of distinct advantages, and the various cold-working processes have become extremely important. Significant advances in recent years have extended the use of cold forming and the trend appears likely to continue.

When compared to hot working, the advantages of cold working include:

1. No heating is required.
2. Better surface finish is obtained.
3. Superior dimension control.
4. Better reproducibility and interchangeability of parts.
5. Improved strength properties.
6. Directional properties can be imparted.
7. Contamination problems are minimized.

Some disadvantages associated with cold-working processes include:

1. Higher forces are required for deformation.
2. Heavier and more powerful equipment is required.
3. Less ductility is available.
4. Metal surfaces must be clean and scale-free.
5. Strain hardening occurs (may require intermediate anneals).

6. Imparted directional properties may be detrimental.
7. May produce undesirable residual stresses.

If one examines the advantages and disadvantages, it becomes evident that the cold-working processes are particularly suited for large-volume production where the quantity involved can readily justify the cost of the required equipment and tooling. Considerable effort has been devoted to developing and improving cold-forming equipment. In addition, better and more ductile metals and an improved understanding of basic plastic flow have done much to reduce the difficulties experienced in earlier years. To a very large extent, modern mass production has paralleled, and been made possible by, the development of cold-forming processes. Automated, high-quality production enables the manufacture of low-cost metal products. In addition, most cold-working processes eliminate or minimize the production of waste material and the need for subsequent machining. With increasing efforts in conservation and materials recycling, these benefits become quite significant.

Although the cold-forming processes tend to be better suited to large-scale manufacturing, much effort has been devoted to developing methods that enable these processes and their associated equipment to be used economically for quite modest production quantities. A substantial amount of the equipment required is well standardized and not excessively costly.

Relationship of metal properties to cold working.   The suitability of a metal for cold working is determined primarily by its tensile properties, these being directly influenced by its metallurgical structure. Similarly, the cold working of a metal has a direct relationship to the tensile properties of the resulting product. Both of these relationships should be considered by the designer when selecting metals that are to be processed by cold working.

No plastic deformation of a metal can occur until the elastic limit is exceeded. Thus, in Figure 13-6, in order to obtain permanent deformation, the strain must exceed $0–X_1$, the strain associated with the elastic limit, a, on each stress–strain curve. If the strain exceeds $0–X_4$, the metal will rupture. Consequently, from the viewpoint of cold working, two factors are of prime importance: (1) the magnitude of the yield-point stress, which determines the force required to

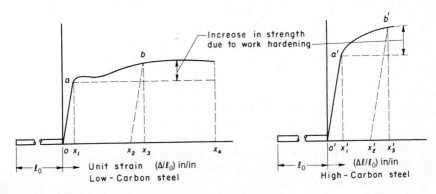

FIGURE 13-6   Relationship of the tensile properties of a metal to its suitability for cold working as shown by its stress-strain diagram.

initiate permanent deformation, and (2) the extent of the strain region from 0 to X_4, which indicates the amount of plastic deformation, or ductility, that is available to be utilized. If considerable deformation must be imparted to a metal without rupture, one having tensile properties similar to those depicted in the left-hand diagram of Figure 13-6 is more desirable than one having properties like those shown in the right-hand diagram. Greater ductility would be available and less force would be required to initiate and continue plastic deformation. A metal having the characteristics of the right-hand diagram would be work-hardened to a greater extent by a given amount of cold working and thus would not be as suitable for most forming operations. On the other hand, it might be more satisfactory for shearing operations and would be easier to machine, as will be discussed in Chapter 18. Cold-working properties are also affected by grain size, too large or too small both producing undesirable results.

Springback is an ever-present phenomenon in cold-working operations that also can be explained with the aid of a stress–strain diagram. When a metal is deformed through the application of a load, part of the resulting total deformation is elastic. For example, if a metal is strained to point X_1 in Figure 13-6 and the load is removed, it will return to its original shape and size because all the deformation is elastic. If, on the other hand, the same metal is strained to X_3, corresponding to point b on the stress–strain curve, the total strain 0–X_3 is made up of two parts, a portion that is elastic and another that is plastic. If the deforming load is removed, the stress reduction will follow line bX_2, and the residual strain will be only 0–X_2. The decrease in strain, $X_3 - X_2$, is known as *elastic springback*. Quite clearly, springback is an important phenomenon in cold working because the deformation must always be carried beyond the desired point by an amount equal to the springback. Moreover, since different materials have different elastic moduli, the amount of springback from a given load will differ from one material to another. Change in material will therefore require changes in the forming process. Springback is a design consideration, and most difficulties can be overcome by proper design procedures.

Preparation of metals for cold working. In order to obtain several of the benefits of cold working, the metal often must receive special treatment prior to processing. First, if better surface finish and dimensional accuracy are to be obtained than those produced by hot working, the starting metal must be free of existing scale to avoid abrasion and damage to the dies or rolls that are used. Scale is removed by *pickling*, in which the metal is dipped in acid and then washed. Second, to assure good dimensional tolerances in cold-worked parts, it often is necessary to start with metal that is uniform in thickness and has a smooth surface. For this reason, sheet metal sometimes is given a light cold rolling prior to the major cold working. This pass also serves to remove the yield-point phenomenon and associated problems of nonuniform deformation and surface irregularities in the product.

A third treatment that may be given to metal prior to cold working is annealing. If the cold working is to involve considerable deformation, it is desirable to have as much ductility available as possible. In many cases, annealing is performed after the workpiece has been partially shaped by cold working. Here annealing restores sufficient ductility to permit the final stages of forming to be done without danger of fracture. The desired grain size can also be obtained by proper control of the annealing process.

Warm forming. Deformation produced at temperatures intermediate to hot and cold forming is known as *warm forming*. Compared to cold forming, warm forming offers the advantages of reduced loads on the tooling and equipment and increased material ductility. For high-carbon steels, it may be possible to eliminate the need to spheroidize anneal the material prior to forming. Also, the favorable as-forged properties may well eliminate subsequent heat treatment operations. Compared to hot forming, the warm regime requires less energy consumption (the decreased energy in heating the workpiece more than offsets the increased deformation energy), produces less scaling and decarburization, and provides better dimensional control, less scrap, and longer tool life (while the tools must exert higher forces, there is less thermal shock and thermal fatigue).

Warm forming, however, is still a developing field and there are several barriers to its growth. Material behavior is less well characterized at these previously little-used temperatures. Lubricants have not been fully developed for operation at these temperatures and pressures. Finally, die design technology is not yet well established for warm working. Nevertheless, the pressures of energy conservation concerns will definitely favor the increased use of warm working.

Isothermal forming. Figure 13-7 shows that some materials, such as titanium, have yield strengths that are strongly dependent on temperature. Within the realm of typical hot-working temperatures, a cooling of as little as 100°C (200°F)

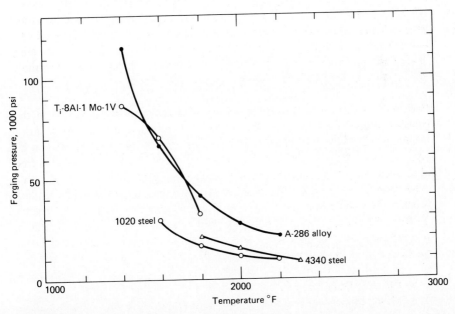

FIGURE 13-7 Variation of material strength (as indicated by pressure required to forge a standard specimen) with temperature. Materials with steep curves may have to be isothermally formed. (*From* A Study of Forging Variables, *ML-TDR-64-95, March 1964; Courtesy Battelle Columbus Laboratories.*)

can produce a doubling in strength. The strength variations, in turn, lead to nonuniform deformation and, often, surface cracking.

To adequately hot deform such materials, the deformation may have to be deformed under isothermal conditions. The dies or tooling must be heated to the workpiece temperature, sacrificing die life for product quality. Deformation speeds must be slowed so that the heat generated by deformation can be removed to maintain uniform and constant temperature. While such methods are indeed costly, they may be the only means of producing a satisfactory product from certain materials.

REVIEW QUESTIONS

1. What are the four basic classes of processes which have been designed to change the shape of engineering materials?
2. Why might large production quantities be necessary to justify metal deformation as a means of manufacture?
3. What is an independent variable?
4. What are some typical independent variables for metal-forming operations?
5. What are dependent variables?
6. What are some typical dependent variables in metal deformation processes?
7. What are some of the ways an engineer can link independent and dependent variables?
8. What are some limitations of an individual operating only on the basis of past experience?
9. What are some limitations on knowledge acquired by experiment?
10. What is a constitutive relation for an engineering material?
11. What changes might occur if the speed of a deformation process is altered?
12. Why are friction and lubrication effects so important in metal deformation operations?
13. How are the friction conditions during metalworking different from the conditions typically observed in mechanical equipment?
14. Why might wear on the workpiece be desirable, whereas wear on the tooling is not?
15. What are some of the features that should be considered when selecting a lubricant for a metalworking operation?
16. What are some of the factors that would influence the performance of a given lubricant under metal-forming conditions?
17. What is hot working?
18. What are some of the advantages of hot working? Disadvantages?
19. Why are heated dies often used in hot forming operations?
20. What is cold working?
21. What are some of the advantages of cold working? Disadvantages?
22. How might stress–strain properties of a metal be useful in assessing the suitability of a metal for cold forming?
23. Why is springback of importance in cold-forming operations?
24. What types of preparations might be required before a metal is cold worked?
25. What is warm forming?
26. What are advantages of warm forming compared to hot forming? Compared to cold forming?
27. What are some of the barriers to the growth of warm working?
28. For what type of materials might isothermal forming be required?

CASE STUDY 13.   The Broken Marine Engine Bearings

The bearings on a small shipboard marine engine have been made of AISI 52100 grade steel that has been austenitized, quenched, and tempered to establish the desired final properties. Performance under normal operating conditions appears adequate. After exposure to a period of subzero temperatures, however, the engines failed. Tear-down revealed rather brittle cracks in the bearings, together with an observed expansion in the bearing dimensions. What would you suspect to be the cause of the failures?

CHAPTER 14

Hot-Working Processes

The shaping of metal by deformation is as old as recorded history. The Bible, in the fourth chapter of Genesis, introduces Tubal-cain and cites his ability as a forger of metal. While we do not know of his equipment, it is well known that forging was commonly practiced before written records. Processes such as rolling and wire drawing were extremely common in the Middle Ages and probably date back much further. In North America, the 1680 Saugus Iron Works near Boston had an operating drop forge, rolling mill, and slitting mill.

While the basic concepts of many forming processes have remained largely unchanged throughout history, the details and equipment have evolved considerably. Manual processes were converted to machine during the industrial revolution. The machinery then became bigger and more powerful. Waterwheel power was replaced by steam and then electricity. Most recently, computer-controlled operations have become increasingly common.

Chapters 14 and 15 will present a survey of metal deformation processes. The division, although somewhat artificial, will be made on the basis of temperature. Processes that are normally performed "hot" will be presented here and processes normally performed "cold" will be deferred to Chapter 15. One should be aware, however, that with increasing emphasis on energy conservation, the growth of "warm working," and new advances in technology, such a temperature classification is often arbitrary. Processes discussed as hot-working processes are often performed cold, and cold-forming processes can often be aided by some degree of heating.

Hot-working processes. The most obvious reason for the popularity of hot working is that it provides an attractive means of forming a desired shape. At elevated temperatures, metals weaken and become more ductile. With continual recrystallization, massive deformation can take place without exhausting material plasticity. In steels, hot forming involves deformation of the weaker austenite structure as opposed to the much stronger, room-temperature, ferrite.

Some of the hot-working processes that are of major importance in modern manufacturing are:

1. Rolling.
2. Forging.
   .a. Smith.
   b. Drop.
   c. Press.
   d. Upset.
   e. Automatic hot.
   f. Roll.
   g. Swaging.
3. Extrusion.
4. Drawing.
5. Spinning.
6. Pipe welding.
   a. Butt.
   b. Lap.
7. Piercing.

Figure 14-1 further emphasizes the significance of these processes by schematically depicting many of the hot-forming operations commonly performed by steel makers and suppliers. Hot-worked products often form the starting material for further processing.

ROLLING

Rolling usually is the first step in converting cast material into finished wrought products. Many finished parts, such as hot-rolled structural shapes, are completed entirely by hot rolling. More often, however, hot-rolled products, such as sheets, plates, bars, and strips, serve as input material for other processing, such as cold forming or machining. From a tonnage viewpoint, hot rolling is predominant among all manufacturing processes, and modern hot-rolling equipment and practices are sufficiently advanced that standardized, uniform-quality products can be produced at low cost. Because the equipment is so massive and costly, however, hot-rolled products normally can be obtained only in standard sizes and shapes for which there is sufficient demand to permit economical production.

The basic rolling processes. Basically hot rolling consists of passing heated metal between rwo rolls that revolve in opposite directions, the space between

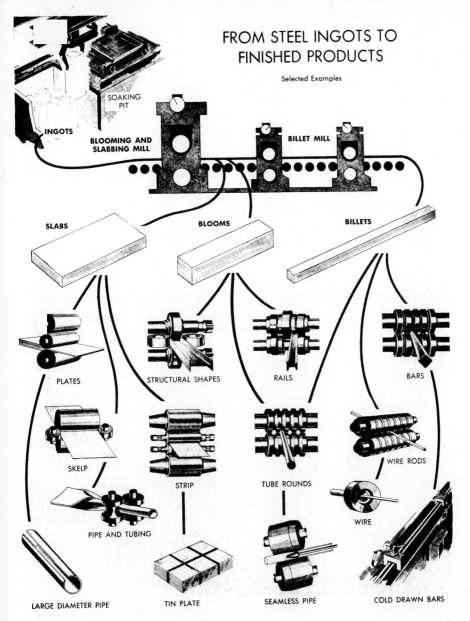

FROM STEEL INGOTS TO FINISHED PRODUCTS

Selected Examples

INGOTS

SOAKING PIT

BLOOMING AND SLABBING MILL

BILLET MILL

SLABS

BLOOMS

BILLETS

PLATES

STRUCTURAL SHAPES

RAILS

BARS

SKELP

STRIP

TUBE ROUNDS

WIRE RODS

PIPE AND TUBING

WIRE

LARGE DIAMETER PIPE

TIN PLATE

SEAMLESS PIPE

COLD DRAWN BARS

FIGURE 14-1 Schematic flow chart for the production of various finished and semi-finished steel shapes. (*Courtesy American Iron and Steel Institute, Washington, D.C.*)

the rolls being somewhat less than the thickness of the entering metal, as depicted in Figure 14-2. Because the rolls rotate with a surface velocity exceeding the speed of the incoming metal, friction along the contact interface acts to propel the metal forward. The metal is squeezed and elongated with a decrease in cross sectional area. The amount of deformation that can be achieved in a single pass between a given pair of rolls depends on the friction conditions along

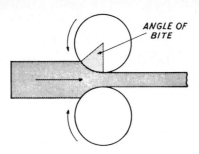

FIGURE 14-2 Schematic representation of the hot-rolling process. (*Courtesy American Iron and Steel Institute, Washington, D.C.*)

the interface. If too much is demanded, the rolls will simply skid over stationary metal. On the other hand, too little deformation per pass results in excessive cost.

Rolling temperatures.   In hot rolling, as in all hot working, it is very important that the metal be heated uniformly throughout to the proper temperature before processing. This usually requires prolonged heating at the desired temperature, a procedure known as *soaking*. If the temperature is not uniform, the subsequent deformation will also be nonuniform. If insufficient soaking has occurred, the hotter exterior flows in preference to the cooler, therefore stronger, interior. If cooling has occurred after removal from soaking or during previous rolling, the colder surface now resists deformation. Cracking and tearing of the metal may result.

Today, much hot rolling is done in integrated mills where the flow of cast ingots or continuous cast material is directed toward certain specific products. Cast ingots are placed in gas- or oil-fired soaking pits as soon as they have solidified and the molds have been stripped. Heat is retained and less energy is required to attain the uniform 1200°C (2200°F) temperature often used for the rolling of carbon steel material. Continuous cast material often goes directly to the rolling stands without additional heating being required.

Hot rolling usually is completed about 50 to 100°C (100 to 200°F) above the recrystallization temperature. Maintenance of such a *finishing temperature* assures the production of a uniform fine grain size and prevents the possibility of unwanted stain hardening.

Rolling mills.   Hot rolling is usually done in stages by a series of rolling mill stands. Cast stock is first rolled into large bars, called *blooms,* usually having a minimum thickness greater than 150 mm (6 inches) and often being square in cross section; or *slabs,* with a distinctly rectangular shape. Blooms, in turn, are reduced in size to form *billets,* and slabs are further rolled into *plate* or *strip*. These products then become the raw material for further hot-working or other forming processes.

Rolling mill stands are available in a variety of roll configurations, as illustrated in Figure 14-3. Early reductions, often called *primary roughing* or *breakdown* passes, usually employ a two-high or three-high configuration with 600 to 1400-mm (24- to 55-inch)-diameter rolls. The three-high mill shown in Figure 14-4 is equipped with an elevator on each side of the stand for raising or lowering the bloom and mechanical manipulators for turning the bloom and shifting it for the various passes as it is rolled back and forth.

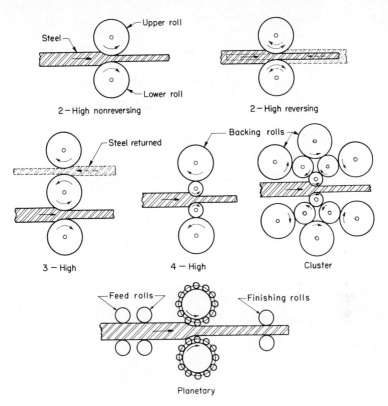

FIGURE 14-3 Roll configurations used in rolling stands.

Smaller-diameter rolls produce less length of contact for a given reduction and therefore require lower loads and less energy for a given change in shape. The smaller cross section, however, provides reduced stiffness, and the rolls are prone to flex elastically under load. Four-high and cluster-roll arrangements use backup rolls to provide the necessary work-roll support. These configurations are used in the hot rolling of wide plate and sheets and in cold rolling, where the deflection in the center of the roll would result in a variation in thickness. Continuous hot-strip mills often involve a roughing train of approximately four four-high mills and a finishing train of six or seven four-high mills.

The planetary mill configuration enables much larger reductions to be performed in a single pass. Each roll consists of a set of planetary rolls carried about a backup or support roll, much like a roller bearing. External drives are required to push the material through the mill and speeds are quite slow. In a typical case, 57.1-mm ($2\frac{1}{4}$-inch) slabs are reduced to 2.5-mm (0.10-inch) sheet in a single operation. Entering speed is only 1.8 meters per minute (6 feet per minute). Production savings relate to the reduction in the required capital investment, floor space, and operating personnel. Speed and productivity are compromised and a small surface reduction usually is required to remove a slight scalloping effect produced by the planetary mill.

In the rolling of nonflat or shaped products, the sets of rolls are grooved to provide a series of openings designed to produce the desired cross section and

FIGURE 14-4 Ingot entering a three-high blooming mill. (*Courtesy Mesta Machine Company.*)

also to provide controlled metal flow. Figure 14-5 shows an example of the roll shapes and pass sequence used in the rolling of an I-beam. In some cases the operator adjusts the roll clearance between the passes, whereas in others the rolls are designed so that the various grooves cut into the rolls provide the proper, decreasing clearance for successive passes without any intermediate adjustment of the rolls.

When the volume of product justifies the investment, finished shapes, such as sheets, bars, plates, and some structurals, may be rolled from billets, blooms, or slabs in a continuous operation on continuous rolling mills. These mills, as shown in Figure 14-6, consist of a series of nonreversing stands through which the metal passes as a continuous piece. The rolls of each successive stand must turn faster than those of the preceding one by a precise amount to accommodate the increased length produced by the reduction in thickness. The metal may leave the final stand at speeds in excess of 110 kilometers per hour (70 miles per hour). In some modern plants continuous rolling mills are fed from continuous casting units so that only a few seconds elapse from the time the metal is solidified until it is rolled into its final shape.

Characteristics, quality, and tolerances of hot-rolled products. Because they are rolled and finished above the recrystallization temperature, hot-rolled products have minimum directional properties and are relatively free of residual stresses. These characteristics, however, often depend on the thickness of the

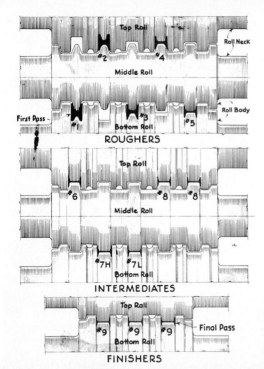

ROUGHERS

INTERMEDIATES

FINISHERS

FIGURE 14-5 Rolling sequence used in producing a structural beam. (*Courtesy American Iron and Steel Institute, Washington, D.C.*)

FIGURE 14-6 A continuous hot strip finishing mill, viewed from the starting end. (*Courtesy Mesta Machine Company.*)

product and the existence of complex sections that may prevent either uniform working in all directions or uniform cooling. Thin sheets show some definite directional characteristics, whereas thicker plate, for example above 20 mm (0.8 inch), usually will have very little. A complex shape, such as an I- or H-beam, will warp substantially if a portion of one flange is cut away because of the residual stress (tension) in the edges.

As a result of the hot plastic working, and the good control that is maintained during processing, hot-rolled products normally are of uniform and dependable quality and considerable reliance can be placed on them. It is quite unusual to find any voids, seams, or laminations when these products are produced by reliable manufacturers.

The surfaces of hot-rolled products are, of course, slightly rough and covered with a tenacious high-temperature oxide known as mill scale. However, with modern procedures, surprisingly smooth surfaces can be obtained.

Dimensional tolerances of hot-rolled products vary with the kind of metal and the size of the product. For most products produced in reasonably large tonnages, the tolerance is from 2 to 5% of the size (height or width).

The variety of hot-rolled products is considerable, despite the fact that only standard sizes and shapes generally are available. Special sizes and shapes can be obtained, but not at reasonable cost unless ordered in considerable volume.

Controlled rolling. As with most deformation processes, rolling is generally viewed as being a means of changing the shape of a material. While heat may be utilized to reduce forces and promote plasticity, thermal processes to produce or control mechanical properties (heat treatments) are usually performed as subsequent operations. *Thermomechanical processing,* of which *controlled rolling* is an example, consists of simultaneously performing both operations so as to produce desired levels of strength and toughness in the as-worked product. The heat for property modification is the same heat used in rolling and subsequent heat treatment is eliminated.

To achieve this goal requires the design of a processing system. Material composition must be selected and maintained. Then a time–temperature–deformation scheme must be developed to achieve the desired objective. Possible goals include: produce fine grain size; control the nature, size and distribution of the transformation products (pearlite, bainite, or martensite in steels); control the reactions producing solid-solution hardening or precipitation hardening; and produce desired levels of toughness. Starting structure (controlled by composition and reheat conditions), deformation details, temperature control during deformation, and cooling from the working conditions must all be controlled.

In view of the benefits of improved product properties and substantial energy savings, it is expected that processes such as controlled rolling will assume increasing significance in the years to come.

FORGING

Forging is the plastic working of metal by means of localized compressive forces exerted by manual or power hammers, presses, or special forging machines. It may be done either hot or cold. However, when it is done cold, special names usually are given to the processes. Consequently, the term "forging" usually implies hot forging done above the recrystallization temperature.

Forging is the oldest known metal working process. From the days when prehistoric peoples discovered that they could heat sponge iron and beat it into a useful implement by hammering with a stone, forging has been an effective method of producing many useful shapes. Modern forging is a development from the ancient art practiced by the armor makers and the immortalized village blacksmith. High-powered hammers and mechanical presses have replaced the strong arm, the hammer, and the anvil, and modern metallurgical knowledge supplements the art and skill of the craftsman in controlling the heating and handling of the metal.

Various forging processes have been developed to provide great flexibility, making it economically possible to forge a single piece or to mass produce thousands of identical parts. The metal may be (1) drawn out, increasing its length and decreasing its cross section; (2) upset, increasing the cross section and decreasing the length; or (3) squeezed in closed impression dies to produce multidirectional flow. As indicated in Table 13-2, the state of stress in the work is primarily uniaxial or multiaxial compression.

The common forging processes are:

1. Open-die hammer or smith forging.
2. Impression-die drop forging.
3. Press forging.
4. Upset forging.
5. Automatic hot forging.
6. Roll forging.
7. Swaging.

Open-die hammer forging. Basically, *open-die hammer forging* is the same type of forging done by the blacksmith of old, but now massive mechanical equipment is used to impart the repeated blows. The metal to be formed is heated throughout to the proper temperature before being placed on the anvil. Gas, oil, or electric furnaces are usually employed, although induction heating has become attractive for many applications. The impact is then delivered by some type of mechanical hammer, the simplest type being the gravity drop or *board hammer*. Here the hammer is attached to the lower end of a hardwood board, which is raised by being gripped between two driven rollers and then released for free-fall. Although some of these are still in use, steam or air hammers, which use pressure to both raise and propel the hammer, are far more common. These give higher striking velocities, more control of striking force, easier automation, and are capable of shaping pieces up to several tons. Computer-controlled hammers can provide predetermined blows of energy for each of the various stages of an operation. Figure 14-7 shows a large double-frame steam hammer. Another style is the open-frame design, which allows more room to manipulate the work and therefore more flexibility. The open-frame design is not as strong as the double frame, however. Open-frame hammers usually range in capacity up to about 2500 kg (5000 pounds) and the double-frame type up to about 12,000 kg (25,000 pounds).

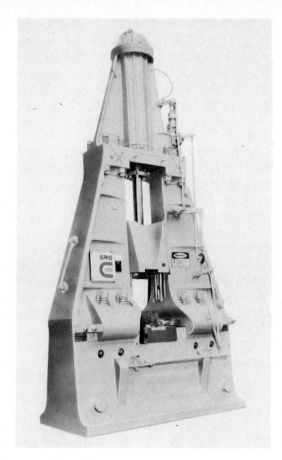

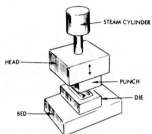

FIGURE 14-7 Double-frame steam drop hammer. (*Courtesy Erie Press Systems, Erie, Pa.*) Inset shows the tooling in schematic.

Open-die forging does not confine the flow of metal, the hammer and anvil often being completely flat. The operator obtains the desired shape by manipulating the workpiece between blows. He may use specially shaped tools or a slightly shaped die between the workpiece and the hammer or anvil to aid in shaping sections (round, concave, or convex), making holes, or performing cutoff operations. Mechanical manipulators are used to hold and manipulate large workpieces, sometimes weighing many tons. Although some finished parts can be made by this technique, it is most often used to preshape metal for some further operation, as in the case of massive parts such as turbine rotors, where the metal is preshaped to minimize subsequent machining. Figure 14-8 shows the formation of a seamless ring and cylindrical shaft by open-die forging.

Impression-die drop forging. The open-die hammer or smith forging is a simple flexible process, but it is not practical for large-scale production because it is slow and the resulting size and shape of the workpiece are dependent on the skill of the operator. *Impression-die* or *closed-die forging* overcomes these difficulties by using shaped dies to control the flow of metal. Figure 14-9 shows a typical set of such dies, one-half of which attaches to the hammer and the other half to the anvil. The heated metal is placed in the lower cavity and struck

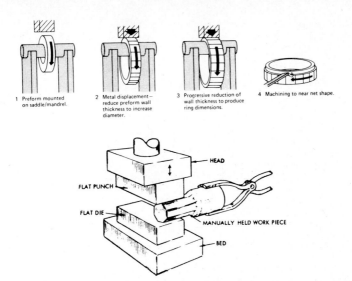

1 Preform mounted on saddle/mandrel.

2 Metal displacement — reduce preform wall thickness to increase diameter.

3 Progressive reduction of wall thickness to produce ring dimensions.

4 Machining to near net shape.

HEAD

FLAT PUNCH

FLAT DIE

MANUALLY HELD WORK PIECE

BED

FIGURE 14-8 *(Top)* Forging of a seamless ring by the open-die method. *(Courtesy Forging Industry Association, Cleveland, Ohio.)* *(Bottom)* Open-die forging of a solid shaft.

one or more blows with the upper die. This hammering causes the metal to flow so as to fill the die cavity. Excess metal is squeezed out between the die faces along the periphery of the cavity to form a *flash*. When forging is completed, the flash is trimmed off by means of a trimming die.

Most drop-forging dies contain several cavities. The first impression usually is an *edging, fullering,* or *bending* impression for roughly distributing the metal in accordance with the requirements of the later impressions. The intermediate impressions are for *blocking* the metal to approximately its final shape. Final shape and size are imparted in the *final impression*. These steps and the shape of a part at the conclusion of each step are shown in Figure 14-9. Because each part produced is shaped in the same die cavities, each is very closely a duplicate of all others, subject to slight die wear.

The restriction to flow that is imposed in certain directions by the shape of the cavity causes the metal to flow in desired directions, and a favorable fiber structure may thus be obtained. In addition, the metal may be placed where it is needed to provide the most favorable section modulus to resist the load stresses. These factors, together with the fine-grain structure and surety of the absence of voids, make it possible to obtain higher strength-to-weight ratios with forgings than with cast or machined parts of the same material.

Board hammers, steam hammers, and air hammers, such as shown in Figure 14-10, are all used in impression die forging. An alternative to the hammer and anvil arrangement is the counterblow or impact machine, such as shown in Figure 14-11. These machines have two horizontal hammers that move together simultaneously and forge the workpiece between them. Because the work is not supported on an anvil, no energy is lost to the machine foundation and the necessity for a heavy base is eliminated. In addition, the machine operates more quietly and with less vibration. In many installations, the operation of the ma-

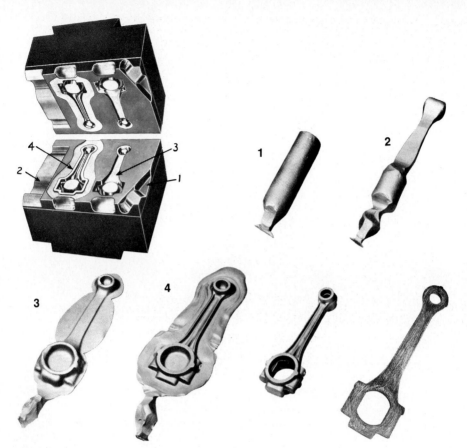

FIGURE 14-9 Impression drop-forging dies and the product resulting from each impression. The flash is trimmed from the finished connecting rod in a separate trimming die. The sectional view shows the grain fiber resulting from the forging process. (*Courtesy Forging Industry Association, Cleveland, Ohio.*)

chine can be almost completely automated. The work can be heated by induction heating and mechanically fed into the machine, forged, and removed.

A recently developed modification of impression die forging utilizes a cast preform that is removed from the mold while hot and forged in a die. After forging, the flash is trimmed in the usual manner. In some cases the four-step process—casting, transfer from the mold to the forging die, forging, and trimming—is completely mechanized. This process is used mostly for nonferrous metals. In another method, preforms have been made by gas-atomizing a stream of molten metal and directing the resulting spray of hot metal particles onto a shaped collector or mold. Final shape and properties are then imparted by forging.

An alternative to conventional impression die forging is *flashless forging*. This is true closed-die forging wherein the metal is deformed in a cavity that allows little or no escape of material. Accurate workpiece sizing is required to assure complete filling of the cavity but not provide any excess. Accurate workpiece positioning is necessary and die design and lubrication must be such as

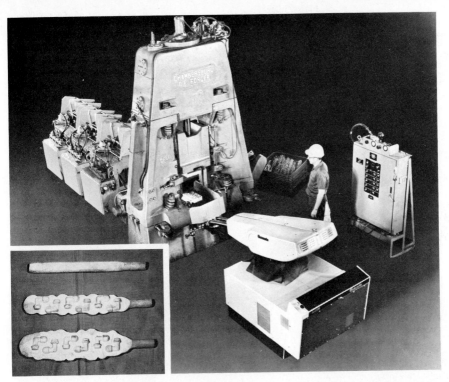

FIGURE 14-10 Fully automated drop-forging system, combining a heating and feeding unit, a pneumatic hammer, and an automatic positioning and handling unit. Inset shows the three states of the forging process. (*Courtesy Chambersburg Engineering Company.*)

to control workpiece flow during the operation. The major advantage is the elimination of the scrap generated in the flash of conventional forgings, which is often in the neighborhood of 20% of the starting material.

Die design factors. There are several important factors that must be kept in mind when designing parts that are to be made by closed-die forging. Several of these relate to the design and maintenance of the forming dies, which are usually made of high alloy or tool steel and are quite costly to construct. Impact resistance, wear resistance, strength at elevated temperature, and the ability to withstand alternating rapid heating and cooling all must be outstanding. Considerable maintenance is often required to assure a smooth and accurate cavity and parting-plane surface. Better and more economical results can be obtained if the following rules are observed in the design of the forging:

1. The parting line of the die should be in a single plane if possible.
2. The parting line should lie in a plane through the center of the forging and not near an upper or lower edge.
3. Adequate draft should be provided—at least 3° for aluminum and 5 to 7° for steel.
4. Generous fillets and radii should be provided.
5. Ribs should be low and wide.

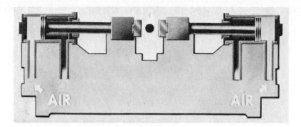

CONVENTIONAL
FORGED DISC WITH
PATHS OF FLOW

DISC FORMED BY
IMPACTER WITH
PATHS OF FLOW

FIGURE 14-11 Automatic impact forging of aluminum turbine blades. Parts are forged between two air-driven rams at the rate of 40 per minute. Upper diagrams show equipment schematic and compare metal flow in both impact and conventional forging. (*Courtesy Chambersburg Engineering Company.*)

6. The various sections should be balanced to avoid extreme differences in metal flow.
7. Full advantage should be taken of fiber flow lines.
8. Dimensional tolerances should not be closer than necessary.

Design details, such as the number of intermediate steps and the shape of each, the amount of excess metal required to assure die filling, and the dimensions of the flash at each step, are often a matter of experience. Each component is a new design entity, and although computer-aided design has made notable advances, good die design is still largely an art.

Good dimensional accuracy is a major reason for using closed-die forging. It must be remembered, however, that dimensions across the parting line are dependent on die wear and maintenance of the parting surfaces. With reasonable precautions, the tolerances shown in Table 14-1 can readily be maintained perpendicular to the parting line. These can be improved by careful practice and die maintenance. Draft angles approaching zero can also be used with some designs, but these are not recommended for general use.

TABLE 14-1 Thickness Tolerances for Steel Drop Forgings

Mass of Forging		Minus		Plus	
kg	lb	mm	in.	mm	in.
0.45	1	0.15	0.006	0.48	0.018
0.91	2	0.20	0.008	0.61	0.024
2.27	5	0.25	0.010	0.76	0.030
4.54	10	0.28	0.011	0.84	0.033
9.07	20	0.33	0.013	0.99	0.039
22.68	50	0.48	0.019	1.45	0.057
45.36	100	0.74	0.029	2.21	0.087

Press forging. In hammer or impact forging, metal flow is a response to the energy in the hammer–workpiece collision. If all the energy can be dissipated through flow of the surface layers of metal and absorption by the press foundation, the interior regions of the workpiece can go undeformed. Therefore, when the forging of large sections is required, *press forging* must be employed. Here the slow squeezing action penetrates throughout the metal and produces a more uniform metal flow. Problems arise, however, because of the long time of die contact with the hot workpiece. If the workpiece surface cools, it becomes stronger and less ductile, and it may crack during forming. Heated dies are often used during press forging operations to minimize this problem.

Forging presses are of two basic types, mechanical and hydraulic, and are usually quite massive. Presses with capacities up to 445 MN (50,000 tons) are currently in operation in the United States. Figure 14-12 shows one of these presses designed and manufactured to enable very large sections of aircraft structures to be forged in a single piece.

Press forgings usually require somewhat less draft than drop forgings and are therefore more accurate dimensionally. In addition, press forgings often can be completed in a single closing of the dies.

Upset forging. *Upset forging* involves increasing the diameter of the end or central portion of a bar of metal by compressing its length. In terms of the number of pieces produced, it is the most widely used of all forging processes, and its use has increased greatly in recent years. Parts are upset-forged both hot and cold on special high speed machines in which the workpiece is moved from station-to-station. Some machines can forge bars up to 250 mm (10 inches) in diameter.

In this type of forging, split dies having several positions or cavities, such as the set shown in Figure 14-13, are commonly used. The split dies move apart slightly for the heated bar to move through them into position. They are then

FIGURE 14-12 A 311-mN (35,000-ton) forging press. Foreground shows a 3.1-meter (121-inch) aluminum part, weighing 119 kg (262 pounds) that was forged on this press. (*Courtesy Wyman-Gordon Company.*)

FIGURE 14-13 Set of upset forging dies having four positions. The product resulting from each position also is shown. (*Courtesy Ajax Manufacturing Company.*)

forced together and a heading tool or ram moves longitudinally against the bar, upsetting it into the die cavity. Separation of the die permits transfer to the next position or removal of the product.

In a modification of the upset-forging process, segments of heated metal can be sheared from the bar as it moves into position in the dies. This permits continuous production of a number of pieces from a single coil or length of feedstock.

Upset-forging machines are used to forge heads on bolts and other fasteners, valves, couplings, and many other small components. The following three rules, illustrated in Figure 14-14, should be followed in designing parts that are to be upset-forged:

1. The limiting length of unsupported metal that can be gathered or upset in one blow without injurious buckling is three times the diameter of the bar.
2. Lengths of stock greater than three times the diameter may be upset successfully provided that the diameter of the die cavity is not more than $1\frac{1}{2}$ times the diameter of the bar.
3. In an upset requiring stock with length more than three times the diameter of the bar and where the diameter of the upset is less than $1\frac{1}{2}$ times the diameter of the bar, the length of unsupported metal beyond the face of the die must not exceed the diameter of the bar.

Figure 14-15 illustrates the variety of parts that can be produced by upsetting and subsequent piercing, trimming, and machining operations.

Automatic hot forging. Several equipment manufacturers are now offering highly automated upset equipment in which mill-length steel bars (typically 8 meters or 24 feet long) are fed in one end at room temperature and hot-forged

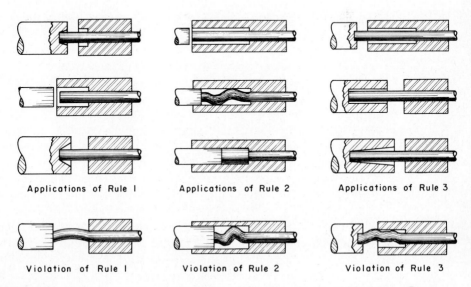

Applications of Rule 1   Applications of Rule 2   Applications of Rule 3

Violation of Rule 1   Violation of Rule 2   Violation of Rule 3

FIGURE 14-14 Rules governing upset forging. (*Courtesy National Machinery Company.*)

FIGURE 14-15 Typical parts made by upsetting and related operations. (*Courtesy National Machinery Company.*)

solid or hollow, round or symmetrical, parts of up to 150-mm (6-inch) diameter and 5-kg (10-pound) weight emerge from the other end at rates of up to 180 parts per minute (86,400 parts per 8-hour shift). Consider the sequence of activities. The input material is the lowest-cost steel bar stock—hot-rolled and air-cooled carbon or alloy steel. The bar is induction heated to 1200 or 1300°C (2200 to 2350°F) in under 60 seconds as it passes through induction heaters. It is descaled by rolls, sheared into individual blanks and transferred through several successive forming stages where it is upset, preformed, final forged, and finally pierced (if necessary). Small parts can be produced at up to 180 parts per minute with rates for larger pieces being near 70 per minute.

The process offers numerous advantages. Low-cost input material and high production speeds have already been cited. Minimum labor is required and, since no flash is produced, material savings may be as much as 20 to 30% over conventional forging. With a consistent finishing temperature near 1040°C (1900°F), an air cool often produces a structure suitable for machining without requiring an anneal or normalizing treatment. Tolerances are generally ± 0.3 mm (± 0.012 inch), surfaces are clean, and draft angles need only be $\frac{1}{2}$ to 1° (as opposed to the conventional 3 to 5°). Tool life is nearly double conventional due to contact times in the order of $\frac{6}{100}$ of a second.

To justify the operation, large quantities of a given product must be required. A single installation of automatic hot-forging equipment may well require an initial investment in excess of $10 million.

Most recently, the technique has been coupled with a cold-forming finishing operation. Preforms are hot-formed at rates of about 180 parts per minute. These are then cold-formed to final shape on another machine at rates of about 90 parts per minute. The benefits include low production cost coupled with the precision, surface finish, and strain hardening of a cold-finished product.

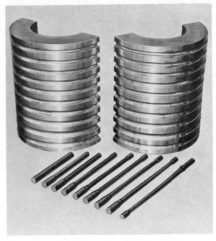

FIGURE 14-16 (*Left*) Roll forging machine. (*Right*) Rolls from a roll forging machine and stages in roll forging a part. (*Courtesy Ajax Manufacturing Company.*)

Roll forging. *Roll forging,* in which round or flat bar stock is reduced in thickness and increased in length, is used to produce such components as axles, tapered levers, and leaf springs. As shown in Figure 14-16, roll forging is done on machines that have two semicylindrical rolls, containing shaped grooves that are slightly eccentric with the axis of rotation. The heated bar is placed between the rolls while they are in the open position. As the rolls turn one half-revolution, the bar is progressively squeezed and rolled out from between them toward the operator. The operator then inserts the forging between another set of smaller grooves and the process is repeated until the desired size and shape are obtained.

Swaging. *Swaging* involves hammering or forcing a tube or rod into a confining die to reduce its diameter, the die often playing the role of the hammer. Repeated blows cause the metal to flow inward and take the form of the die. Figure 14-17 illustrates the application of swaging to close and form the end of a gas cylinder.

FIGURE 14-21. Steps in swaging a tube to form the neck of a cylinder. (*Courtesy United States Steel Corporation.*)

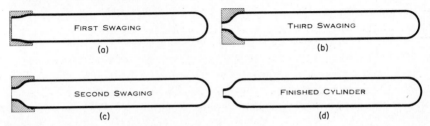

FIGURE 14-17 Steps in swaging a tube to form the neck of a cylinder. (*Courtesy United States Steel Corporation.*)

EXTRUSION

In the *extrusion process,* metal is compressively forced to flow through a suitably shaped die to form a product with reduced cross section. Although extrusion may be performed either hot or cold, hot extrusion is employed for many metals to reduce the forces required, eliminate cold-working effects, and reduce directional properties. Basically, the extrusion process is like squeezing toothpaste out of a tube. In the case of metals, a common arrangement is to have a heated billet placed inside a confining chamber. A ram advances from one end causing the billet to first upset to conform to the confining chamber. As the ram continues to advance, the pressure builds until the material flows plastically through the die, as illustrated in Figure 14-18. The stress state within the material is triaxial compression.

Lead, copper, aluminum, magnesium, and alloys of these metals are commonly extruded, taking advantage of the relatively low yield strengths and extrusion temperatures. Steel is more difficult to extrude. Yield strengths are high and the metal has a tendency to weld to the walls of the die and confining chamber under the conditions of high temperature and pressure. With the development and use of phosphate-based and molten glass lubricants, substantial quantities of hot steel extrusion are now produced. These lubricants adhere to the billet and prevent metal-to-metal contact throughout the process.

As shown in Figure 14-19, almost any cross-sectional shape can be extruded from the nonferrous metals. Size limitations are few because presses are now available that can extrude any shape that can be enclosed within a 750-mm (30-inch) circle. Shapes and sizes are much more limited in the case of steel and the higher-strength metals, but advances are rapidly being made.

Extrusion is an attractive process for numerous reasons. Many shapes can be produced as extrusions that are not possible by rolling, such as ones containing reentrant angles or that are hollow. No draft is required, thereby enabling the

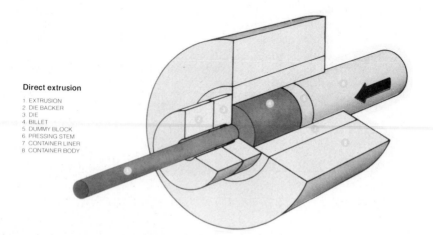

Direct extrusion

1. EXTRUSION
2. DIE BACKER
3. DIE
4. BILLET
5. DUMMY BLOCK
6. PRESSING STEM
7. CONTAINER LINER
8. CONTAINER BODY

FIGURE 14-18 Direct extrusion showing basic equipment components. (*Courtesy Wean United, Inc., Hydraulic Machinery Division.*)

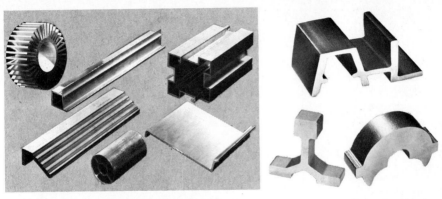

FIGURE 14-19 Typical shapes obtainable by extrusion. (*Left*) Aluminum. (*Courtesy Aluminum Company of America.*) (*Right*) Steel. (*Courtesy Allegheny Ludlum Steel Corporation.*)

saving of metal and weight. Being compressive in nature, the amount of reduction in a single step is limited only by the capacity of the equipment. Billet-to-product area ratios in excess of 100 have become quite common. In addition, extrusion dies are relatively inexpensive—often less than $500—and often only one die is required for a given product. Product changes require only a die change, so small quantities of a desired shape can often be produced economically by extrusion. The major limitation of the process is the requirement that the cross section must be the same for the length of the product being extruded.

The dimensional tolerances of extrusions are very good. For most shapes ±0.003 mm/mm or a minimum of ±0.07 mm is easily attainable. Grain structure is typical of other hot-worked metals, but strong directional properties usually accompany extrusions. Standard product lengths are about 7 to 8 meters (20 to 24 feet). Lengths in excess of 13 meters (40 feet) have been produced.

Extrusion methods. Three basic methods are employed to produce extrusions. Hot extrusion is usually done by either *direct extrusion* or *indirect extrusion*, illustrated in Figure 14-20. Although the indirect configuration reduces friction between the billet and chamber wall, added equipment complexity and restricted length of product favors the direct method. Figure 14-21 shows a typical large extrusion press with the extrusion emerging from the die. The third method, *impact extrusion*, is usually performed cold and will be discussed in Chapter

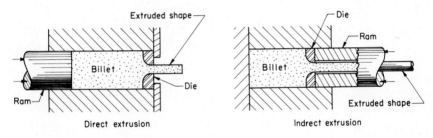

FIGURE 14-20 Direct and indirect extrusion.

FIGURE 14-21 A 125-mN (14,000-ton) extrusion press. Inset shows an extrusion coming from the die. (*Courtesy Aluminum Company of America.*)

15. Extrusion speeds are often rather fast to minimize the cooling of the billet within the metal chamber, but may be restricted by the large amounts of heat generated by the massive deformation being performed.

Extrusion of hollow shapes. Hollow shapes can be extruded by several methods. For tubular products, the stationary or moving mandrel processes of Figure 14-22 are often employed. For more complex internal cavities, a spider mandrel or torpedo die, such as illustrated in Figure 14-23, is used. As the hot metal flows beyond the spider, further reduction between the die and mandrel forces the seams to close and weld back together. Since the metal has never been exposed to contamination, perfect welds result, and the location of the spider

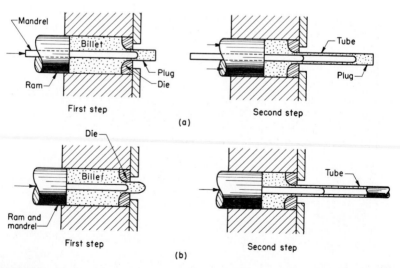

FIGURE 14-22 Two methods of extruding hollow shapes using internal mandrels.

CASTING AND FORMING PROCESSES

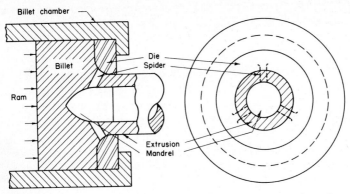

FIGURE 14-23 Extrusion of a hollow shape using a spider mandrel.

seams cannot be found in the product. Obviously, the cost for hollow extrusions is considerably greater than for solid ones, but a wide variety of shapes can be produced that cannot be made by any other process.

Metal flow in extrusion. The flow of metal during extrusion is rather complex, and some care must be exercised to prevent cracks and defects from forming. In direct extrusion, friction restricts motion between the billet and the chamber and forming die. Metal near the center can often pass through the die with little distortion. Metal near the surface undergoes considerable shearing to produce a deformation pattern such as is shown in Figure 14-24. Often flow of metal on the surface cannot keep up with the flow in the center, particularly if the surface regions of the billet have cooled, and surface cracks result. Process control is critical in the areas of design, lubrication, extrusion speed, and temperature uniformity.

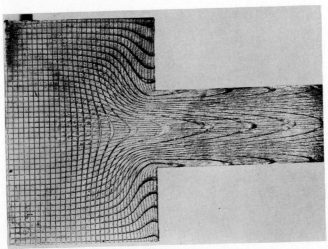

FIGURE 14-24 Grid pattern showing metal flow in direct extrusion. Billet was sectioned and grid pattern engraved prior to extrusion.

HOT DRAWING OF SHEET AND PLATE

Drawing is a plastic forming process in which a flat sheet or plate of metal is formed into a recessed, three-dimensional part with a depth several times the thickness of the metal. As the punch descends into the die (or the die moves over the punch), the metal assumes the configuration of the mating punch and die set. Hot drawing is used for forming relatively thick-walled parts of simple geometry, usually cylindrical. There is often considerable thinning of the metal as it passes through the dies. Cold drawing, on the other hand, utilizes relatively thin metal, changes the thickness very little, or not at all, and produces parts in quite a variety of shapes.

Hot drawing is illustrated schematically in Figure 14-25. As shown, a heated, flat disk of metal is placed on top of a female die. The punch then descends, pushing the metal through the die, converting the flat blank into a cup. If the difference between punch diameter and die opening is less than twice the thickness of the metal being drawn, the cup wall is thinned and elongated during drawing (a process often called *ironing*). The stress condition during deformation, as illustrated in Table 13-2, is primarily biaxial compression and uniaxial tension.

Further reduction in diameter can be obtained by redrawing through a smaller punch and die set as shown in Figure 14-25. This figure also illustrates another possibility wherein the cup is pushed through several dies by a single punch.

Hot drawing is used primarily for forming thick-walled cylindrical components such as oxygen tanks and artillery shells. It can also be used for shaping other parts, where the female die is closed. The male die or punch descends to

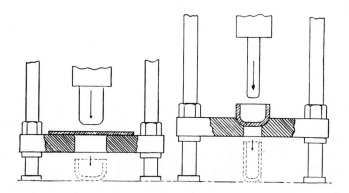

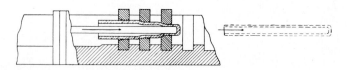

FIGURE 14-25 Methods of cupping or hot drawing by the use of single and multiple dies. (*Courtesy United States Steel Corporation.*)

FIGURE 14-26 Hot drawing of a tank half from 15-mm ($\frac{1}{2}$-inch) plate on a 17.8-mN (2000-ton) press. The dies for this operation weigh 63 500 kg (70 tons). (*Courtesy Lukens Steel Company.*)

form the shape and then retracts to permit removal of the part. Figure 14-26 shows an example of such an operation.

HOT SPINNING

Spinning is the plastic forming of metal parts from a flat rotating disk, the application of localized pressure to one side of the disk forcing the metal to flow against a rotating male form that is held against the opposite side. The basic process is is illustrated in Figure 15-57. Although most spinning is done cold using thin sheets of metal, hot spinning is used to form thicker plates of metal, usually steel, into axisymmetric shapes. Metal up to 150 mm (6 inches) in thickness is routinely spun into dished pressure vessel and tank heads. Thinner plates of hard-to-form metals, such as titanium, are also shaped by hot spinning. The basic theory of spinning is the same whether hot or cold, but, as with other hot processes, hot spinning enables the forming of greater thicknesses of metal with no strain hardening. Simple, hand-held wooden or metal tools are usually employed for cold spinning. In hot spinning, however, heavy rollers are employed with mechanical holding and control mechanisms, as shown in Figure 14-27.

PIPE WELDING

Large quantities of small-diameter steel pipe are produced by two processes that involve hot forming of metal strip and welding of its edges through utilization

FIGURE 14-27 Hot spinning of a large workpiece using a machine equipped with power-assist controls (*Courtesy Spincraft, Inc.*)

of the heat contained in the metal. Both of these processes, *butt welding* and *lap welding* of pipe, utilize steel in the form of *skelp*—long, narrow strips of the desired thickness. Because the skelp has been previously hot rolled and the welding process produces further compressive working and recrystallization, pipe welded by these processes is uniform in quality.

Butt-welded pipe. Figure 14-28 illustrates schematically the *butt-welding process* for making pipe. The skelp is unwound from a continuous coil and is heated to forging temperature as it passes through a furnace. Upon leaving the furnace, it is pulled through forming rolls that shape it into a cylinder. The pressure exerted between the edges of the skelp as it passes through the rolls is sufficient to upset the metal and weld the edges together. Additional sets of rollers size and shape the pipe, after which it is cut to length by "flying shears" as it moves. Normal pipe diameters range from 3 mm ($\frac{1}{8}$ inch) to 75 mm (3 inches). Production rates of a single unit often exceed 10,000 meters per hour (30,000 feet per hour).

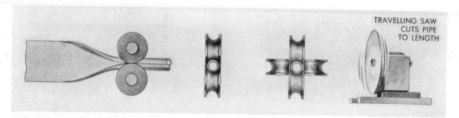

FIGURE 14-28 Method of making butt-welded pipe from continuous skelp. (*Courtesy American Iron and Steel Institute, Washington, D.C.*)

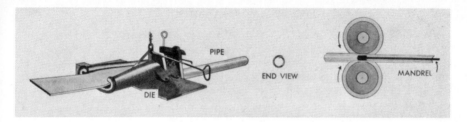

FIGURE 14-29 Method of making lap-welded pipe from skelp. (*Courtesy American Iron and Steel Institute, Washington, D.C.*)

Lap-welded pipe. The *lap-welding process* for making pipe, illustrated in Figure 14-29, differs from butt welding in that the skelp has beveled edges and a mandrel is used in conjunction with a set of rollers to make the weld. The process is used primarily for larger sizes of pipe, from about 50 mm (2 inches) to 400 mm (14 inches) in diameter. Because of the necessity for supporting and removing the mandrel, lengths are limited to about 7 meters (20 feet).

PIERCING

Thick-walled and seamless tubing is made by the *piercing process,* the basic part of which is illustrated in Figure 14-30. A heated, round billet, with its leading end center-punched, is pushed longitudinally in between two large, convex-tapered rolls that revolve in the same direction, their axes being inclined at opposite angles of about 6° from the axis of the billet. The clearance between the rolls is somewhat less than the diameter of the billet. As the billet is caught by the rolls and rotated, their inclination causes the billet to be drawn forward

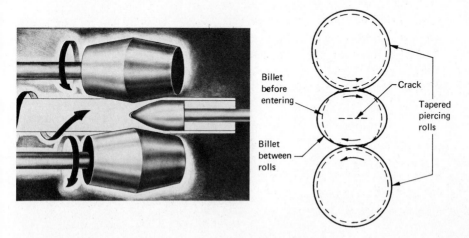

FIGURE 14-30 (*Left*) Principle of the Mannesmann process of producing seamless tubing. (*Courtesy American Brass Company.*) (*Right*) Mechanism of crack formation in the Mannesmann process.

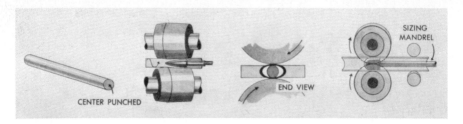

FIGURE 14-31 Schematic diagram of the steps in the production of seamless tubing by the Mannesmann process. (*Courtesy American Iron and Steel Institute, Washington, D.C.*)

into them. The reduced clearance between the rolls forces the rotating billet to deform into an elliptical shape. To rotate with an elliptical cross section, the metal must undergo shear about the major axis, which causes a crack to open. As the crack opens, the billet is forced over a pointed mandrel that enlarges and shapes the opening, forming a seamless tube. After the complete tube has been formed, the piercer point drops off and the mandrel backup bar is removed. The tube is then passed between reeler and sizing rolls, as shown in Figure 14-31, to straighten it and bring it to the desired size. Figure 14-32 shows a thick-walled seamless tube emerging from a Mannesmann mill. When required, the tube may be put through a plug rolling mill, where it is rolled over a larger mandrel to increase its diameter and reduce its wall thickness before it is reeled and sized.

FIGURE 14-32 Pierced tubing emerging from a Mannesmann mill. (*Courtesy Timken Roller Bearing Company.*)

The *Mannesmann-type mill* is used to produce tubing up to ab̶ ̶ (12 inches) in diameter. Larger sizes sometimes are manufactured on piercing mill, which involves the same principle of piercing but uses ̶ ̶ diameter conical disks instead of convex rolls.

REVIEW QUESTIONS

1. Why are hot-rolled products normally available only in standard sizes and shapes?
2. Why is friction between the tool and workpiece not an undesirable feature in hot rolling?
3. Why might small-diameter rolls be desired when rolling thin products?
4. What compensation must be made for the use of small-diameter rolls?
5. What is controlled rolling?
6. What are some of the possible goals of controlled rolling?
7. What is the difference between open-die and impression-die drop forging?
8. Why is open-die forging not practical for large-scale production of identical products?
9. Why are multiple steps often required in the production of an impression-die forging?
10. What is flashless forging? What is its major advantage?
11. What are the advantages of a counterblow or impact forging machine?
12. What are some of the properties desired in forging dies?
13. Why is it necessary to provide a draft angle in an impression forging die?
14. Why must press forging often employ heated dies?
15. What is upset forging?
16. What is automatic hot forging?
17. What are some of the advantages of automatic hot forging?
18. What is roll forging?
19. What development made possible the hot extrusion of steel?
20. What types of billet-to-product area ratios can be achieved in a single extrusion operation?
21. Why might small quantities of a desired shape be economically produced by extrusion?
22. What are the three basic methods used to produce extrusions?
23. How can a spider mandrel or torpedo die produce a seamless tubular product?
24. For what types of products might hot drawing be used instead of cold drawing?
25. What is ironing?
26. What two hot-forming processes can convert steel skelp into pipe?
27. Why is the Mannesmann piercing process not capable of producing extremely long lengths of seamless tubing?

CASE STUDY 14. Material for a Corrosion-Resistant Processing Vessel

You are an engineer employed by the Petro Refining Company and have been assigned to evaluate and make a recommendation regarding two bids which the company has received for a processing vessel. The request for bids stated: "This vessel will be used in refining certain petroleum products that contain weak H_2SO_4. It must operate at approximately 427°C (800°F) for a period of at least 10 years. The corrosion-resistant steel used in its construction must be suitable for these conditions and must have a minimum tensile strength of 483 MPa (70,000 psi) and ductility of at least 15% as measured with a 50.8-mm (2-inch) gage-length specimen."

The Weld-Fab Company's bid is based on AISI type 446 corrosion-resistant steel, whereas that of the Best-Weld Company is for the use of AISI type 302B corrosion-resistant steel. The bid from the Best-Weld Company is approximately 20% higher than that of the Weld-Fab Company, and the company justifies its higher price by stating that type 302B steel is much superior to type 446.

Investigate and give your recommendations and reasons.

CHAPTER 15

Cold-Working Processes

As discussed in Chapter 13, there are a number of advantages to processes in which a metal is deformed "cold." Surface finish and precision are improved markedly. Strain hardening serves to increase the strength of the product and no energy is required to bring the workpiece to forming temperatures.

As a result, numerous cold-working processes have been developed to perform a variety of deformations. The major cold-working operations can be classified basically under the headings of *squeezing, bending, shearing,* and *drawing* as follows:

Squeezing		Bending	
1. Rolling	7. Staking	1. Angle	5. Seaming
2. Swaging	8. Coining	2. Roll	6. Flanging
3. Cold forging	9. Peening	3. Draw and	7. Straightening
4. Extrusion	10. Burnishing	compression	
5. Sizing	11. Die hobbing (hubbing)	4. Roll-forming	
6. Riveting	12. Thread rolling		
	(see Chapter 27)		

Shearing		Drawing	
1. Shearing; slitting	4. Notching; Nibbling	1. Bar and tube drawing	5. Stretch forming
2. Blanking	5. Shaving	2. Wire drawing	6. Shell drawing
3. Piercing; lancing; perforating	6. Trimming	3. Spinning	7. Ironing
	7. Cutoff	4. Embossing	8. Superplastic forming
	8. Dinking		

SQUEEZING PROCESSES

Most of the cold-working squeezing processes have identical hot-working counterparts or are extensions of them. The primary reasons for deforming cold rather than hot are to obtain better dimensional accuracy and surface finish. In many cases the equipment is basically the same, except that it must be more powerful.

Cold rolling. *Cold rolling* accounts for the greatest tonnage of cold-worked products. Sheets, strip, bars, and rods are cold-rolled to obtain products that have smooth surfaces and accurate dimensions. Most cold rolling is done on four-high, cluster, or planetary rolling mills.

Cold-rolled sheet and strip is obtainable in four conditions: *skin-rolled, quarter-hard, half-hard,* and *full-hard*. Skin-rolled metal is given only a $\frac{1}{2}$ to 1 per cent reduction to produce a smooth surface and uniform thickness and to remove the yield-point phenomenon. It is well suited for further cold-working operations, where good ductility is needed. Quarter-hard, half-hard, and full-hard sheet and strip have undergone increasing cold reductions, up to 50%. Consequently, their yield points have been increased, they have definite directional properties, and they have correspondingly decreased ductility. Quarter-hard steel can be bent back on itself across the grain without breaking. Half-hard and full-hard can be bent back 90° and 45°, respectively, about a radius equal to the material thickness.

Swaging. *Swaging* is a process for reducing the diameter, tapering, or pointing round bars or tubes by external hammering. A useful extension of the process involves the formation of internal cavities. Here, a shaped mandrel is inserted inside a tube, and the tube is then collapsed around it by swaging.

Cold swaging is done by means of a rotary machine, as shown in Figures 15-1 and 15-2. Rotation of the spindle within the cage causes the backer blocks to alternately move apart and then be forced inward to pass beneath the rollers. The dies then open and close from various angles around the workpiece. The operator inserts the bar or tube between the dies and gradually pushes it inward until the desired length of material has been swaged. As the diameter is reduced, the product is elongated.

An important modification of rotary swaging utilizes the principle depicted in Figure 15-3. An open- or closed-end tubular workpiece is placed over a shaped mandrel and inserted between the rotating swaging dies. As the swaging dies

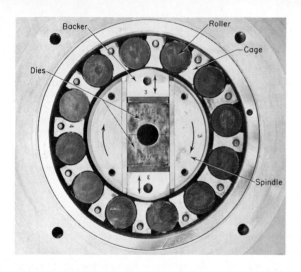

FIGURE 15-1 Basic components and motions in a rotary swaging machine. (*Courtesy Torrington Company.*)

reciprocate and rotate, they force the interior of the workpiece to conform to the shape of the mandrel, thereby imparting an accurate internal shape. The operation requires only a few seconds, and the accuracy and surface finish are excellent. When a long product is desired, the workpiece can be fed over a stationary mandrel, enabling a short, relatively inexpensive mandrel to be used. The process is particularly well suited for making internal gears and splines, recesses, and sockets. Figure 15-4 shows some examples of parts formed by this process.

Cold forging. Extremely large quantities of products are made by *cold forging,* in which the metal is squeezed into a die cavity that imparts the desired shape. *Cold heading,* illustrated schematically in Figure 15-5, is used for making enlarged sections on the ends of a piece of rod or wire, such as the heads on

FIGURE 15-2 Tube being swaged in a rotary swaging machine. (*Courtesy Torrington Company.*)

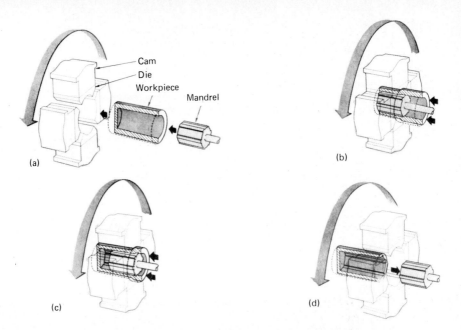

Cam
Die
Workpiece
Mandrel

(a)

(b)

(c)

(d)

FIGURE 15-3 Principle of forming internal details with a mandrel during swaging. (*Courtesy Cincinnati Milacron, Inc.*)

bolts, nails, rivets, and other fasteners. Two variations of the process are used. In the sequence illustrated, a piece of rod is cut off in the first step and is then transferred to a holder–ejector assembly for subsequent operations. Upsetting is then done in one or more strokes of the heading punches. If two or more blows are required, the heading punches rotate into position between strokes. When

FIGURE 15-4 Typical parts swaged with internal details. (*Courtesy Cincinnati Milacron, Inc.*)

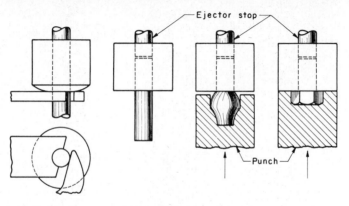

FIGURE 15-5 Steps in a cold-heading operation.

heading is completed, the ejector stop advances to remove the work from the holding die.

In the other variation, a continuous rod is fed forward, clamped, and the head formed. The rod is further advanced, cut to length, and the cycle repeats. This procedure is used for making nails where the point is formed by the cutoff operation.

Enlarged sections at locations other than the ends of rods can be made by the *upsetting* process illustrated in Figure 15-6. In this procedure, ejector pins must be provided in both the punch and the die.

Through the use of various types of dies, and by combining cold heading, upsetting, extrusion, bending, coining, roll threading, and knurling, a wide variety of relatively complex parts can be made to close tolerances very rapidly. Figure 14-15 presented a display of typical products. This is a chipless manufacturing process producing parts which would otherwise be machined from bar or hot forgings. Thus material waste is substantially reduced.

Figure 15-7 compares the manufacture of a spark-plug body machined from hexagonal bar stock to the manufacture by cold forming. Material is saved, machining time and cost are reduced, and the product is stronger, as illustrated by the flow lines in Figure 15-8. Often these features are sufficient to pay for the cost of the forming equipment. While generally associated with small parts of nonferrous metal, cold forming is also used extensively on steel, most recently for parts up to 45 kg (100 pounds) in weight and 180 mm (7 inches) in diameter.

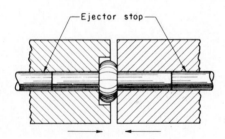

FIGURE 15-6 Method of upsetting the center portion of a rod. Both dies grip the stock during upsetting.

FIGURE 15-7 Manufacturing of a spark-plug body. (*Left*) By machining from hexagonal bar stock. (*Right*) By cold forming. (*Courtesy National Machinery Co.*)

Extrusion. Great advances have been made in cold *extrusion* during recent years, and in combining cold extrusion with cold heading. Figure 15-9 illustrates the basic principles of *forward* and *backward* types of cold extrusion using open and closed dies. This process is often called *impact extrusion* and was first used only with the low-strength, ductile metals such as lead, tin, and aluminum for producing such items as: collapsible tubes for toothpaste, medications, and so forth; small "cans" such as are used for shielding in electronics and electrical apparatus; and larger cans for food and beverages. In recent years, cold extrusion

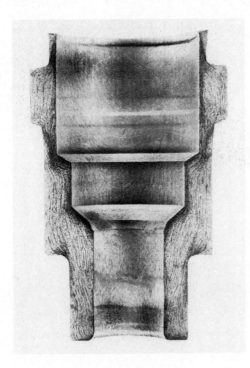

FIGURE 15-8 Section of cold formed spark-plug body of Figure 15-7, etched to reveal flow lines. Cold formed structure produces an 18% increase in strength over machined product. (*Courtesy National Machinery Co.*)

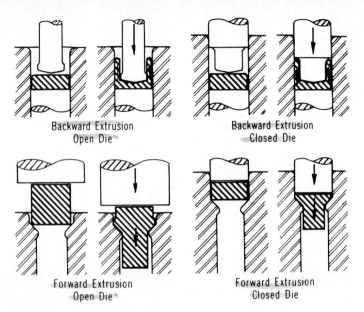

FIGURE 15-9 Methods of cold extrusion.

has been used for forming mild steel parts, often being combined with cold heading. When cold heading alone is used, there is a definite limit to the ratio of the head and stock diameters, as discussed in connection with Figure 14-13. By combining extrusion and cold heading, this difficulty can be easily overcome, as is illustrated by the example shown in Figure 15-10. Not only is considerable metal saved by not machining the parts from large-diameter stock but, in addition to overcoming the limitations of cold heading, the intermediate-size rod used is cheaper than a smaller-diameter rod that might be employed with multiple step heading operations. Figure 15-11 illustrates another cold-forming combination, this time involving two extrusion operations and one upset.

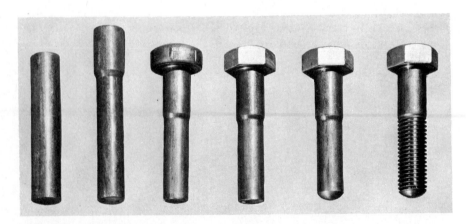

FIGURE 15-10 Steps in cold forming a bolt by extrusion, cold heading, and thread rolling. (*Courtesy National Machinery Co.*)

FIGURE 15-11 Cold-forming sequence involving cutoff, squaring, two extrusions, an upset, and a trimming operation. (*Courtesy National Machinery Co.*)

Another type of cold extrusion, known as *hydrostatic extrusion,* utilizes high fluid pressure to extrude a billet through a die, either into atmospheric pressure or into a lower-pressure chamber. The pressure-to-pressure process, illustrated in Figure 15-12, makes possible the extrusion of relatively brittle materials, such as molybdenum, beryllium, and tungsten. Billet-chamber friction is eliminated, billet-die lubrication is enhanced by the pressure, and the surrounding pressurized atmosphere suppresses crack initiation and growth.

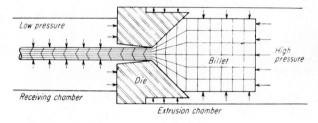

FIGURE 15-12 Hydrostatic extrusion method for the extrusion of relatively brittle materials, using differential hydrostatic pressures. (*Courtesy* American Machinist.)

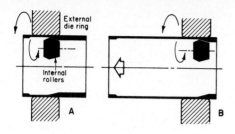

Roll extrusion. Thin-wall cylinders can be produced from thicker-wall material by the *roll-extrusion process* depicted in Figure 15-13. The squeezing action of a rotating roller forces the metal to flow forward between the roller and the external confining ring. Although cylinders from 19 mm (0.75 inch) to 4000 mm (156 inches) in diameter have been made by this procedure, it is most commonly used for those in the range 75 mm (3 inches) to 500 mm (20 inches). An alternative is to use an internal mandrel and an external roller, as in Figure 15-62. Here, however, provision must be made to extract the mandrel.

Sizing. *Sizing* involves squeezing areas of forgings or ductile castings to a desired thickness. It is used principally on bosses and flats, with only enough deformation occurring to bring the region to a desired dimension. By this procedure, designers may make the general dimensional tolerances of a part more liberal, enabling the use of less costly production methods. The few close dimensions are then obtained by one or two simple and inexpensive sizing operations. Sizing usually is done between simple dies in a mechanical-driven press, thereby assuring positive dimensional control.

Riveting. In *riveting,* a head is formed on the shank end of a fastener to provide a permanent method of joining sheets or plates of metal together. Although riveting usually is done hot in structural work, in manufacturing it almost always is done cold. Quite commonly, where there is access to both sides of the work, the method illustrated in Figure 15-14 is used. The shaped punch may be held and advanced by a press or contained in a hand-held pneumatic riveting hammer. When a press is used, the rivet usually is headed by a single squeezing action. Sometimes the heading punch also rotates. Special machines, such as those used in aircraft manufacturing, punch the hole for the rivet, place the rivet in position, and perform the riveting operation in about 1 second.

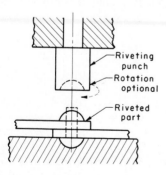

FIGURE 15-14 Method of fastening by riveting.

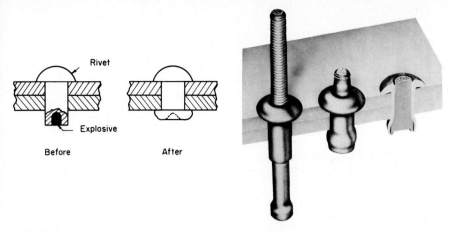

FIGURE 15-15 Rivets for use in "blind" riveting. (*Left*) Explosive type; (*Right*) shank-type pull-up. (*Courtesy Huck Manufacturing Company.*)

It often is desirable to use riveting in situations where there is access to only one side of the assembly. Several types of rivets are available for these applications, two of which are illustrated in Figure 15-15. Both involve cold working. The explosive type is activated by the application of a heated tool to the rivet head, causing the charge to explode and expand the shank into a retaining head. The pull-type or pop-rivet mechanically expands the shank, after which the pull pin is cut off flush with the head.

Staking. *Staking* is a commonly used cold-working method for permanently fastening two parts together where one protrudes through a hole in the other. It is so simple in method and in appearance in the final product that it is often overlooked by designers. As shown in Figure 15-16, a shaped punch is driven into one of the pieces, deforming the metal sufficiently to squeeze it outward against the second piece so that they are tightly locked together. As illustrated, various types of punch patterns may be used. Because the staking punch is simple and the operation is completed with a single stroke of the press, it is a convenient and economical fastening method where permanence is desired and the appearance of the punch mark is not objectionable.

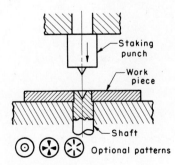

FIGURE 15-16 Fastening by staking.

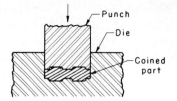

FIGURE 15-17 Coining process.

Coining. *Coining* involves cold working by means of a positive displacement punch while the metal is completely confined within a set of dies. The process, illustrated schematically in Figure 15-17, is used to produce coins, medals, and other products where exact size and fine detail are required in a variable thickness product. Because of the confinement of the metal and the positive displacement of the punch, there is no possibility for excess metal to flow from the die, and very high pressures are required. Pressures as high as 1400 MPa (200 ksi) are often used. Accurate volumetric measurement of the metal put into the die is essential to avoid breakage of the dies or press.

Hobbing. *Hobbing*[1] is a cold-working process that is used to form cavities in various types of dies, such as those used for molding plastics. As shown in Figure 15-18, a male *hob* is made with the contour of the part that ultimately will be formed by the die. After the hob is hardened, it is slowly pressed into an annealed die block by means of a hydraulic press until the desired impression

[1]Hobbing is also a machining process for gear cutting (see Chapter 28).

FIGURE 15-18 Hobbing of a die block in a hydraulic press. Inset shows close-up of the hardened hob and the impression in the die block. The die block is contained in a reinforcing ring. The outer surface of the die block is machined flat to remove bulged metal.

is produced. Flow of metal in the die block can be aided and controlled by machining away some of the metal in the block where large amounts of plastic flow would occur. The die block usually is round during hobbing and is reinforced by a heavy steel ring. When hobbing is completed, the die block is removed from the reinforcing ring, the excess metal is machined away, and the piece is hardened by heat treatment.

Because one hob may be used to form a number of identical cavities in a mold, hobbing is frequently more economical than producing dies by conventional die sinking by machining. The process is also referred to as *hubbing*.

Surface improvement by cold working.   Two cold-working methods are used extensively for improving or altering the surface of metal products. *Peening* involves striking the surface repeated blows by impelled shot or a round-nose tool. The highly localized blows deform and tend to stretch the metal surface. Because the surface deformation is resisted by the metal underneath, the result is a surface layer under residual compressive stresses. This condition is highly favorable to resist cracking under fatigue conditions, such as repeated bending, because the compressive stresses are subtractive from the applied tensile loads. For this reason, shafting, crankshafts, gear teeth, and other cyclic-loaded components are frequently peened.

In most manufacturing, peening is done by means of shot which is impelled by the type of mechanism shown in Figure 15-19. When used after welding, peening is often done by means of manual or pneumatic hammers to avoid distortion and prevent contraction cracking.

Burnishing involves rubbing a smooth hard object under considerable pressure over the minute surface protrusions that are formed on a metal surface during machining or shearing, thereby reducing their depth and sharpness through plastic flow. In one of two major techniques, the edge surfaces of sheet metal stampings are burnished by pushing the stamped parts through a slightly tapered die having its entrance end a little larger than the workpiece and its exit end a slight amount smaller than the workpiece. The rubbing against the sides of the

FIGURE   15-19 Wheelapeening mechanism for impelling shot for peening. Inset shows surface of steel peened with steel shot; 25×. (*Courtesy Wheelabrator-Frye, Inc.*)

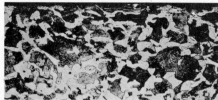
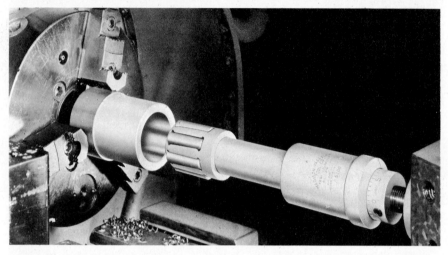

FIGURE 15-20 Tool for roller burnishing. Burnishing rolls are moved outward by means of a taper. (*Top left*) Section of surface before burnishing. (*Top right*) Surface after burnishing. (*Courtesy Madison Industries, Incorporated.*)

die occurs under sufficient pressure to smooth the slightly rough edges that are produced by the blanking of the part (see Figure 15-36).

Roller burnishing is used to improve the size and finish of internal and external cylindrical and conical surfaces after machining by metal cutting. The process is illustrated in Figure 15-20. The hardened rolls of the tool press against the surface and roll the protrusions down into a more nearly flat surface. Being cold-worked and in residual compression, the surface also has better wearing and fatigue properties. See Chapter 39 for discussion on residual stresses.

BENDING

Bending is the plastic deformation of metals about a linear axis with little or no change in the surface area. When two or more bends are made simultaneously with the use of a die, the process is sometimes called *forming*. The two axes involved in forming may be at an angle to each other, but each axis must be linear and independent of the other to be classed as a bending operation. For these cases, only simple bending theory is involved. If the axes about which deformation occurs are not linear, or are not independent, the process is one of *drawing* and not bending.

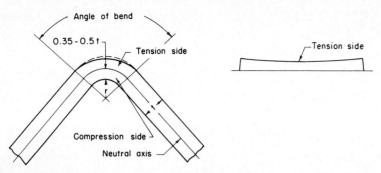

FIGURE 15-21 (*Left*) Nature of a bend in sheet metal. (*Right*) Cross section (exaggerated) of tension side of bent sheet metal or bar, showing variation in thickness due to restraint at edges.

As shown in Figure 15-21, bending causes the metal on the outside of the neutral axis to be stretched, while that on the inside is compressed. Because the yield strength of metals in compression is somewhat higher than the yield strength in tension, the metal on the outer side of the bend yields first, and the neutral axis is not located equidistant between the two surfaces. Instead, the neutral axis tends to be between one-third and one-half the thickness of the metal from the inner surface, depending on the bend radius. Because of the preferred plastic flow of the metal outside the neutral axis, it is thinned somewhat at the bend, being more pronounced in the center of a sheet or bar than at the edges, as shown in Figure 15-21. This can cause difficulties in some applications. There is also a tendency for the metal on the inner side of the neutral axis to be upset plastically as a result of the compressive forces, and to become somewhat wider in the direction parallel with the bend axis. This can be quite substantial and noticeable when thick metals are bent. A further consequence of the condition of combined tension and compression is the tendency of the metal to unbend somewhat after forming (that is, *springback*).

Angle bending. Angle bends up to 150° in sheet metal under about 1.5 mm ($\frac{1}{16}$ inch) in thickness may be made in a *bar folder*, as shown in Figure 15-22. These machines are manually operated and usually are less than $2\frac{1}{2}$ meters (8 feet) long. After the sheet of metal is inserted under the folding leaf and positioned, raising the handle first actuates a cam that causes the blade to clamp the sheet and further motion of the handle then bends the metal to the desired angle.

Bends in heavier sheet metal, and more complex bends in thinner sheets, are made on a *press brake*, such as shown in Figure 15-23. These are mechanically or hydraulically driven presses having a long, narrow bed and relatively slow, short, adjustable strokes. The metal is bent between interchangeable dies that are attached to the bed and the ram. As illustrated in Figure 15-24, different dies can be used to produce many types of bends. The metal can be fed inward between successive strokes to produce various types of repeated bends, such as corrugations. Figure 15-25 illustrates how a complex bend can be formed progressively by repeated strokes and the use of more than one die. Seaming, embossing, punching, and other operations can also be done on press brakes by

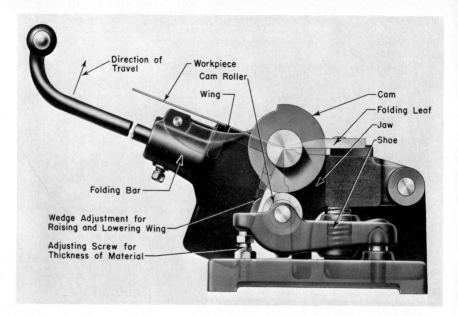

FIGURE 15-22 Phantom section of a bar folder. (*Courtesy Niagara Machine and Tool Works.*)

FIGURE 15-23 (*Left*) Modern press brake with CNC gauging system. (*Courtesy Di-Acro Division, Houdaille Industries, Inc.*) (*Right*) Close-up view of press brake dies. (*Courtesy Cincinnati Incorporated.*)

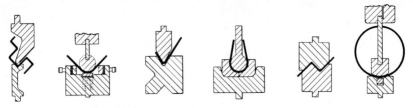

FIGURE 15-24 Several types of dies used on press brakes for forming angles and rounds. (*Courtesy Cincinnati Incorporated.*)

using suitable dies, but they can usually be done more efficiently on other types of equipment when the volume is sufficient to justify their use.

Design for bending. Several factors must be considered when designing parts that are to be made by bending. Of primary importance is the minimum radius that can be bent successfully without metal cracking. This, of course, is related to the ductility of the metal. It has been shown that the ratio of the minimum bend radius R to the thickness of the metal t, for a wide range of metals, can be related to the percent reduction in the area by the curve shown in Figure 15-26. As can be noted, it is seldom feasible to call for a minimum radius of less than the thickness of the metal. Bends should be designed with the largest-possible radii. This permits the designer to select from a much wider variety of materials.

If the metal has been cold-worked previously, or has marked directional properties, this will have considerable effect on its bending properties. Whenever possible, it is wise to make the bend axis normal to the direction of previous

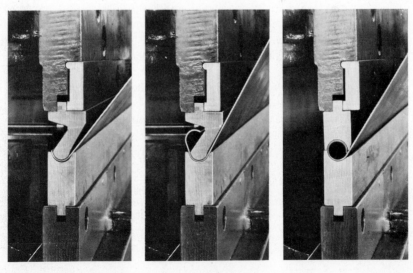

FIGURE 15-25 Dies and steps used in forming a roll bead on a press brake. (*Courtesy Cincinnati Incorporated.*)

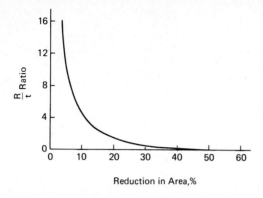

FIGURE 15-26 Curve for relating minimum bend radius (relative to thickness) to the metal ductility as measured by reduction in area in a uniaxial tensile test.

rolling. If two perpendicular bend axes are involved, the metal should be oriented so they are at 45° to the rolling direction, if at all possible.

A second matter of concern to the designer is that of determining the length of a flat blank that will produce a bent part of given dimensions. The fact that the neutral axis is not at the center line of the metal makes it necessary to make some adjustments that are functions of the stock thickness and bend radius. The method shown in Figure 15-27 has been found to give satisfactory results for determining blank length.

A third important design factor is the length of the minimum leg that can be bent successfully. In most cases the leg should be at least $1\frac{1}{2}$ times the metal thickness plus the bend radius.

Whenever possible, the tolerance on bent parts should not be less than 0.8 mm ($\frac{1}{32}$ inch). Ninety-degree bends should not be specified without first determining whether the bending method will permit a full right angle to be obtained. Multibend parts should be designed with most bends of the same radius to reduce setup and tooling costs. Consideration should also be given to providing regions for adequate clamping or handling during manufacture. Bending near the edge of a material will distort the edge. If an undistorted edge is required, additional material must be allowed and a trimming operation performed after bending.

Roll bending. Plates, heavy sheets, and rolled shapes can be bent to a desired curvature on forming rolls of the type shown in Figure 15-28. These usually have three rolls in the form of a pyramid, with the two lower rolls being driven and the upper roll adjustable to control the degree of curvature. Where the rolls are supported by a frame on each end, one of these supports can be swung clear

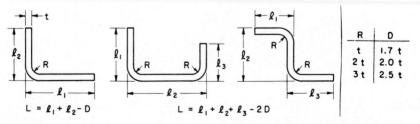

FIGURE 15-27 Method for determining blank length for bending operations.

FIGURE 15-28 Cold-roll bending of structural shapes. (*Courtesy Buffalo Forge Company.*)

to permit removal of a closed shape from the rolls. Bending rolls are available in a wide range of sizes, some being capable of bending plate up to 150 mm (6 inches) thick.

Drawing bending and compression bending. Many modern bending machines utilize a clamp and pressure tool to produce bending about a form block. Draw bending, illustrated in Figure 15-29, is perhaps the most versatile and accurate means of bending. The workpiece is clamped against a bending form and the entire assembly rotates to draw the workpiece across a pressure tool. In compression bending, also shown in Figure 15-29, the bending form remains stationary and the pressure tool traverses to make the bend.

Cold roll-forming. *Cold roll-forming* of flat strip metal into complex sections has been highly developed in recent years. This process, depicted in Figures 15-30 and 15-31, involves the progressive bending of metal strip as it passes through a series of forming rolls. A wide variety of moldings, channeling, and

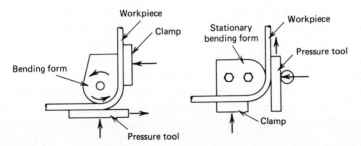

FIGURE 15-29 (*Left*) Draw bending, in which the form block rotates. (*Right*) Compression bending in which a moving tool compresses the workpiece against a stationary form.

FIGURE 15-30 Schematic representation of the cold roll-forming process. Inset shows typical shapes formed by this process. (*Courtesy Van Huffel Tube Corporation.*)

other shapes can be formed on machines that produce up to 3000 meters (10,000 feet) of product per day. By changing the rolls, a single machine can be adapted to the production of many different shapes. Since changeover, setup, and adjustment time may take several hours, it usually is not economical to use the process for less than about 3000 meters (10,000 feet) of product. When tubes or pipe are desired, a resistance welding unit is often combined with roll-forming equipment.

Seaming. *Seaming* is used to join ends of sheet metal to form containers such as cans, pails and drums. Figure 15-32 shows some of the most common types of seams that are used. The seams are formed by a series of small rollers on seaming machines that range from small hand-operated types to large automatic units capable of producing hundreds of seams per minute in the mass production of cans.

Flanging. *Flanges* can be rolled on sheet metal in essentially the same manner as seaming is done. In many cases, however, the forming of flanges and seams involves drawing, since localized bending occurs on a curved axis.

Straightening. *Straightening* or *flattening* has as its objective the opposite of bending and often is done before other cold-forming operations to assure that flat or straight material is available. Two different techniques are quite common.

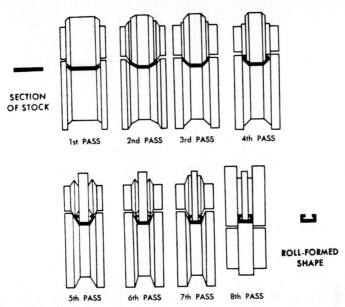

SECTION
OF STOCK

1st PASS   2nd PASS   3rd PASS   4th PASS

5th PASS   6th PASS   7th PASS   8th PASS

ROLL-FORMED
SHAPE

FIGURE 15-31 Eight-roll sequence for forming a box channel. (*Courtesy The Aluminum Association, New York.*)

Roll straightening or *roller leveling,* illustrated in Figure 15-33, involves a series of reverse bends. The rod, sheet, or wire is passed through a series of rolls having decreased offsets from a straight line. These bend the metal back and forth in all directions, stressing it slightly beyond its previous elastic limit and thereby removing all previous permanent set.

Sheet may also be straightened by a process called *stretcher leveling.* As shown in Figure 15-34, the sheets are grabbed mechanically at each end and stretched slightly beyond the elastic limit to remove previous stresses and thus produce the desired flatness.

SHEARING

Shearing is the mechanical cutting of materials in sheet or plate form without the formation of chips or use of burning or melting. When the two cutting blades are straight, the process is called shearing. Processes, in which the shearing blades are in the form of the curved edges of punches and dies, are called by

FIGURE 15-32 Various types of seams used on sheet metal.

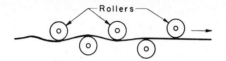

FIGURE 15-33 Method of straightening rod or sheet by passing through a set of straightening rolls. For rod, another set of rolls provides straightening in the other dimension.

other names, such as *blanking, piercing, notching, shaving,* and *trimming.* These all are basically shearing operations, however.

The basic shearing process is illustrated in Figure 15-35. As the punch (upper blade) descends against the workpiece, as shown in the left-hand diagram, the metal is deformed plastically into the die (lower blade). Because the clearance between the punch and the die is only 5 to 10% of the thickness of the metal being cut, the deformation is severely localized. The punch penetrates into the metal, and correspondingly the opposite surface the metal bulges slightly and flows into the die. When the localized penetration reaches about 15 to 60% of the thickness of the metal, depending upon material ductility and strength, the applied stress exceeds the shear strength and the metal suddenly shears or ruptures through the remainder of its thickness. These two stages of the shearing process can often be seen on the edges of sheared parts and are clearly visible in Figure 15-36.

Because of the normal inhomogeneities in a metal and the lack of uniform clearance between the shearing blades, the final shearing does not occur uniformly. Fracture and tearing start at the weakest points and proceed progressively and intermittently to the next-stronger points, producing a somewhat rough sheared edge. If the punch and die or shearing blades have proper clearance and are maintained in good condition, sheared edges may be produced that are often sufficiently smooth for use without further finishing. The quality of sheared

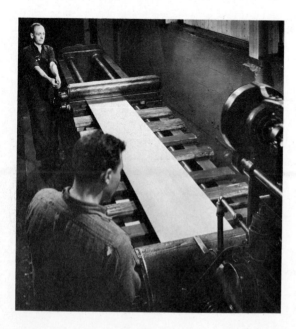

FIGURE 15-34 Straightening sheets of brass by stretching, a process sometimes called stretcher leveling. (*Courtesy Scovill Manufacturing Company.*)

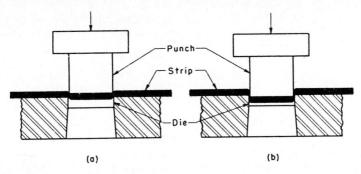

FIGURE 15-35 Basic mechanism of shearing by means of a punch and die.

edges can be further improved if the strip stock is clamped firmly against the die from above, the clearance between the punch and die is reduced to a minimum, and the movement of the piece through the die (and thus the shearing) is controlled by resistance from a plunger acting upward against the workpiece under pressure, such as exerted by a rubber die cushion. These controls cause shearing to take place uniformly around the entire edge rather than randomly at the weakest points.

Research into the effect of superimposed pressure on the shearing process showed that as pressure increased, the relative amount of smooth edge increased. Above a certain pressure, a 100% smooth edge could be obtained. Figure 15-37 schematically presents the production extension of this work, designed to economically produce smooth and square sheared edges. The process is known as *fineblanking*. A V-shaped protrusion is incorporated into the holddown or pressure plate lying slightly external to the contour of the cut. The holddown is first pressed into the plate being sheared and the protrusion places the region to be cut into a localized state of compression. When the punch starts its action, the compressed metal is held tightly against it throughout shearing and a virtually

FIGURE 15-36 Conventionally sheared surface showing the two stages of fracture. (*Courtesy American Feintool Inc.*)

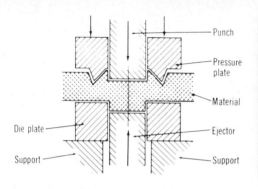

FIGURE 15-37 (*Left*) Method of obtaining a smooth edge in shearing by using a shaped pressure plate to put the metal into localized compression. (*Courtesy* Metal Progress.) (*Right*) Stock skeleton after shearing, showing the compression indentation. (*Courtesy Clark Metal Products Company.*)

smooth and square edge results, as in Figure 15-38. Usually, less clearance between the punch and die is used, and a controlled upward-acting plunger, described previously is employed.

A similar, but somewhat simpler, technique using the same principle is illustrated in Figure 15-39. Incoming bar stock is pressed against the closed end of a feed hole, putting the stock in compression and thereby permitting the shearing of burr-free slugs for use in further processing.

When sheets of metal are to be sheared on a straight line, foot or power-operated *squaring shears* may be used, such as those illustrated in Figure 15-40. As the ram descends, a clamping bar or a set of clamping fingers presses the sheet of metal against the machine table to hold it in position. The moving blade then comes down across the fixed blade and shears the metal. On larger shears the moving blade is usually set at an angle or "rocks" as it descends, to make the cut progressively from one end to the other. This action reduces the amount of cutting force required, although the total energy expended is the same.

Slitting is the shearing process used to cut rolls of sheet metal into several rolls of narrower width. Here, the shearing blades are in the form of circumferential mating grooves on cylindrical rolls, the ribs on one roll mating with

FIGURE 15-38 Fine-blanked surface for the same component shown in Figure 15-36. (*Courtesy American Feintool, Inc.*)

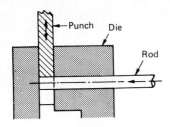

FIGURE 15-39 Method for smooth shearing rod by putting it into compression during shearing.

the grooves in the other. The process is continuous and can be done rapidly and economically. Moreover, because the distance between adjacent sets of shearing edges is fixed, the width of the slit strips is very accurate and constant, more so than can be obtained by alternative procedures.

Piercing and blanking. *Piercing* and *blanking* are shearing operations wherein the shearing blades take the form of closed, curved lines on the edges of a punch and die. They are basically the same cutting action, the difference being primarily one of definition, as shown in Figure 15-41. In blanking, the piece punched out is the desired workpiece, and major burrs or undesirable features should be left on the strip. In piercing, the piece punched out is the scrap and the remainder of the strip becomes the desired workpiece. Piercing and blanking are usually done by some type of mechanical press.

Several variations of piercing and blanking are used and have come to acquire specific names. *Lancing* is a piercing operation that may take the form of a slit in the metal or an actual hole as shown in Figure 15-42. The purpose of lancing is to permit adjacent metal to flow more readily in subsequent forming operations. In the case illustrated, the lancing makes it easier to form the grooves, which were shaped before the ashtray was blanked from the strip of stock and drawn to final shape.

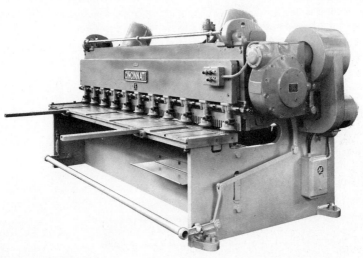

FIGURE 15-40 A 3-meter (10-foot) power shear for 6.4-mm ($\frac{1}{4}$-inch) steel. (*Courtesy Cincinnati Incorporated.*)

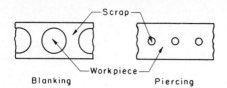

FIGURE 15-41 Difference between piercing and blanking.

Perforating consists of piercing a large number of closely spaced holes.

Notching is essentially the same as piercing except that the edge of the sheet of metal forms a portion of the periphery of the piece that is punched out. It is used to form notches of any desired shape along the edge of a sheet.

Nibbling is a variation of notching in which a special machine makes a series of overlapping notches, each further into the sheet of metal. As can be seen in Figure 15-43, the already sheared edge forms one end of the notch being cut. By repeating the procedure, any desired shape can be cut from sheets of metal up to about 6.5 mm ($\frac{1}{4}$ inch) in thickness. Nibbling machines are simple, inexpensive, and versatile. Moreover, by starting the operation from a punched or drilled hole, the interior of a sheet can be cut away.

Shaving is a finishing operation in which a very small amount of metal is sheared away around the edge of a blanked part. Its primary use is to obtain greater dimensional accuracy, but it also may be used to obtain a square or smoother edge. Because only a very small amount of metal is removed, the punches and dies may be made with very little clearance. Parts, such as small gears, can be shaved to tolerances of 0.025 mm (0.001 inch) after blanking.

Trimming is used to remove the excess metal that remains after a drawing, forging, or casting operation. It is essentially the same as blanking.

A *cutoff* operation is one in which a stamping is removed from a strip of stock by means of a punch and die. The cutoff punch and die cut across the entire width of the strip. Frequently, an irregularly shaped cutoff operation may simultaneously give the workpiece all or part of the desired shape.

Dinking is a modified shearing operation that is used to blank shapes from low-strength materials, primarily rubber, fiber, and cloth. The procedure is illustrated in Figure 15-44. The shank of the die can be struck with a hammer or mallet, or it can be operated by a press of some type.

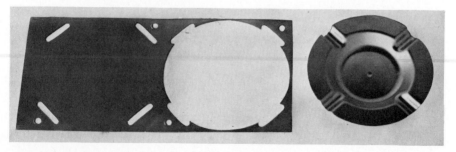

FIGURE 15-42 Steps in making an ashtray. (*Left to right*) Piercing and lancing, blanking, and the final formed ashtray.

FIGURE 15-43 Operations being performed on a nibbling machine. (*Courtesy Tech-Pacific.*)

Piercing and blanking dies. The basic components of piercing and blanking die sets, shown in Figure 15-45, are a *punch,* a *die,* and a *stripper plate.* Theoretically, the punch should be of dimensions such that it would just fit within the die with a uniform clearance approaching zero and, on its downward stroke, it should not enter into the die. In most practice, however, the clearance is from 5 to 12% of the stock thickness, with 5 to 7% being common, and the punch enters the die by a small amount. In operation, the punch and die should be maintained in alignment so that a uniform clearance is obtained around the entire periphery. The die is usually attached to the bolster plate of the press, which, in turn, is attached to the main press frame. The punch is attached to the movable ram, moving in and out of the die with each stroke of the press. Frequently, the punch and die are mounted on a *punch holder* and *die shoe* to form a die set such as is shown in Figure 15-46. The holder and shoe are permanently aligned and guided by two or more guide pins. Once a punch and die are correctly aligned and fastened to the die set, the unit can be inserted directly into a press without having to check the alignment, thereby reducing

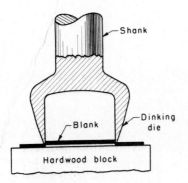

FIGURE 15-44 Dinking process.

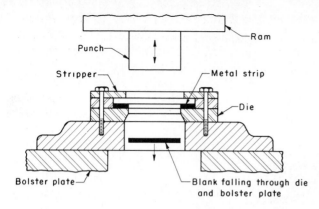

FIGURE 15-45 Basic components in piercing and blanking dies.

setup time. When a given punch and die are no longer needed, they may be removed and new one attached to the shoe and holder assembly.

The punch face may be ground normal to the axis of motion, or it may have a slight angle, referred to by the term *shear*. Shear reduces the maximum cutting force required because all the periphery of a cut is not made at the same time. The amount of shear is limited to the thickness of the metal and frequently is less. One-half shear reduces the cutting load by about 25% and full shear by 50%. The total energy required is not changed, however.

The function of the stripper plate is to prevent the material from riding upward on the punch as it moves upward. It is located a sufficient distance above the die so that the sheet metal can easily slide between it and the die. The hole in the stripper plate is larger than the punch so there will be no friction between them.

In most cases, the punch holder of the die set is attached to the ram of the press so that the punch is raised, as well as lowered, by the press ram. On smaller die sets, a modern practice is for the punch and punch holder to be raised by springs in the die set. These small die-set units are known as *subpress dies*. The press ram simply contacts the top of the punch holder and forces it downward, the springs providing the return. This construction makes the set self-contained so that it can be put into and removed from a press quite rapidly

FIGURE 15-46 Typical die set having two guideposts. (*Courtesy Danly Machine Specialties, Inc.*)

and thereby reduce setup time. Numberous varieties of standardized, self-contained die sets have been developed that can be combined in various patterns on the bed of a press to produce large parts that would otherwise require large and costly die sets. One such setup is shown in Figure 15-47.

Punches and dies usually are made of nondeforming, or air-hardening, tool steel so they can be hardened after machining without danger of warpage. Beyond a depth of about 3 mm ($\frac{1}{8}$ inch) from the face, the die is usually provided with angular clearance or back relief to reduce friction between the part and the die and to permit the part to fall freely from the die after being sheared. The 3-mm land provides adequate strength and sufficient metal so that the die can be resharpened by grinding a few hundredths of a millimeter from its face.

Dies may be made in a single piece, which results in a basically simple, but costly, die, or made in sections that are fastened together on the die shoe. The latter procedure usually simplifies making the die and repairing it in case of damage, because only the broken piece must be replaced. Many standardized punch and die parts are available from which complex die sets can often be constructed at greatly reduced cost. Figure 15-48 shows such a die set. Substantial savings can often be obtained if designers would determine what standard die components are available and modify the design of parts that are to be pierced and blanked so that such components can be utilized. An added advantage is that when the die set is no longer needed, the parts can be disassembled and used to construct another die set.

Another procedure that can be used to cut metal up to about 13 mm ($\frac{1}{2}$ inch) thick is to use "steel-rule" dies, as illustrated in Figure 15-49. Here the die is made from hardened, relatively thin steel strips that are mounted and supported on edge. A die plate takes the place of the conventional punch and holds the strips. Neoprene rubber pads take the place of the usual stripper plate, pushing

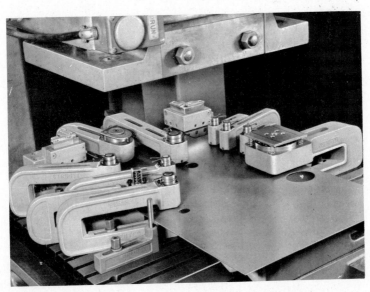

FIGURE 15-47 Typical setup for piercing and blanking using self-contained subpress tooling units. (*Courtesy Strippit Division, Houdaille Industries, Inc.*)

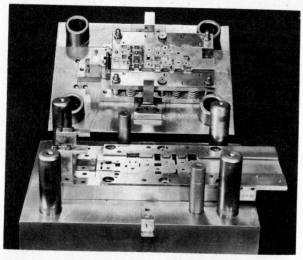

FIGURE 15-48 Progressive piercing, forming, and cutoff die set built up mostly from standard components. Part produced is shown below. (*Courtesy Oak Manufacturing Company.*)

the blank out from between the die strips. Such die sets are usually much less expensive to construct than solid dies and are thereby more feasible for small quantities of parts.

Only piercing or blanking can be done in the simple type of punch and die illustrated in Figure 15-45. Many parts require both piercing and blanking, and these operations can be combined in the two types of dies shown in Figures 15-50 and 15-51. Their operation may be understood by considering the steps required to form a simple, flat washer by piercing and blanking.

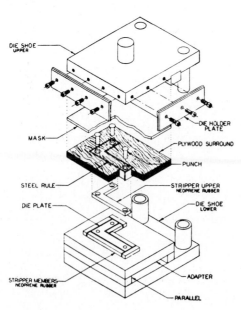

DIE SHOE UPPER

DIE HOLDER PLATE

MASK

PLYWOOD SURROUND

PUNCH

STEEL RULE

STRIPPER UPPER NEOPRENE RUBBER

DIE PLATE

DIE SHOE LOWER

STRIPPER MEMBERS NEOPRENE RUBBER

ADAPTER

PARALLEL

FIGURE 15-49 Construction of a steel-rule die set. (*Courtesy J. J. Raphael.*)

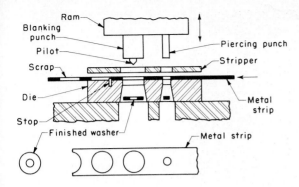

Ram
Blanking punch
Pilot
Scrap
Die
Stop
Finished washer

Piercing punch
Stripper
Metal strip
Metal strip

FIGURE 15-50 Progressive piercing and blanking die for making a washer.

The *progressive die set,* such as depicted in Figure 15-50, is the simpler of the two types, Basically it consists of two or more sets of punches and dies mounted in tandem. The strip stock is fed into the first die where a hole is pierced when the ram closes on the first stroke. When the ram raises, the stock is moved over into position under the blanking punch. Positioning is accomplished automatically by a stop mechanism that engages a previously punched hole. As the ram descends on the second stroke, the pilot on the bottom of the blanking punch enters the hole that was pierced on the previous stroke to ensure accurate alignment. Further descent of the punch blanks the completed washer from the strip of stock and, at the same time, the first punch pierces the hole for the next washer. Another part is completed with each successive stroke of the press.

Progressive dies can be used for many variations of piercing, blanking, forming, lancing, drawing, and so forth. (Examples are shown in Figures 15-42 and 15-43.) They have the advantage of being fairly simple to construct and are economical to repair because a broken punch or die does not necessitate the replacement of the entire set. However, if highly accurate alignment of the various operations is required, they are not as satisfactory as compound dies.

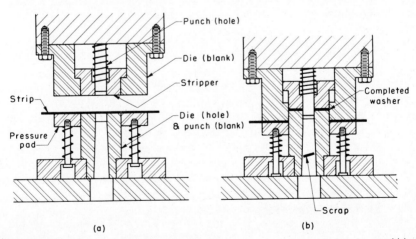

Punch (hole)
Die (blank)
Stripper
Strip
Die (hole) & punch (blank)
Pressure pad
Completed washer
Scrap

(a)   (b)

FIGURE 15-51 Method of making a simple washer in a compound piercing and blanking die. Part is blanked (a) and subsequently pierced (b). Blanking punch contains the die for piercing.

In *Compound dies,* such as shown in Figure 15-51, piercing and blanking, or other combinations, occur progressively within a single stroke of the ram while the strip of stock remains in one position. Dies of this type are more accurate, but they usually are more expensive to construct and are more subject to breakage.

Numerically controlled, turret-type punch presses, such as the one shown in Figure 38-33, are widely used for punching large numbers of holes in sheet metal components. In these, a variety, often up to 60, of piercing punches and dies are contained in coordinated turrets and can quickly be rotated into operation as desired. The workpiece is moved into position through *X–Y* movements of the worktable.

Designing for piercing and blanking.   The construction, operation, and maintenance of piercing and blanking dies can be greatly facilitated if designers of the parts to be fabricated keep a few simple rules in mind:

1. If solid dies are to be used, blank corners should be true radii whenever possible. Square corners are preferred for sectional dies.
2. The width of any projection or slot should be at least $1\frac{1}{2}$ times the metal thickness and never less than $2\frac{1}{2}$ mm ($\frac{3}{32}$ inch).
3. Diameters of pierced holes should not be less than the thickness of the metal or a minimum of 0.65 mm (0.025 inch). Smaller holes can be made, but with difficulty.
4. The minimum distance between holes, or between a hole and the edge of stock, should be at least equal to the metal thickness.
5. Keep tolerances as great as possible. Tolerances below about ± 0.08 mm (± 0.003 inch) mean that shaving will be required.
6. Arrange the pattern of parts on the strip stock to minimize scrap.

DRAWING

Cold drawing is a term that can refer to two somewhat different operations. If the stock is in the form of sheet metal, cold drawing is the forming of parts wherein plastic flow occurs over a curved axis. This is one of the most important of all cold-working operations because a wide range of parts, from small cups to large automobile body panels can be drawn in a few seconds each. Cold drawing is similar to hot drawing, but the higher deformation forces, thinner metal, limited ductility, and closer dimensional tolerances create some distinctive problems.

If the stock is wire, rod, or tubing, "cold drawing" refers to the process of reducing the cross section of the material by pulling it through a die, a tensile equivalent to extrusion.

Bar and tube drawing.   One of the simplest cold-drawing operations, *bar* or *rod drawing,* is illustrated in Figure 15-52. One end of a bar is reduced or pointed, inserted through a die of somewhat smaller cross section than the

FIGURE 15-52 Cold drawing of rods on a chain-driven multiple-die draw bench. (*Courtesy Scovill Manufacturing Company.*) Inset shows schematic of the operation producing finite lengths of straight rod or tube. (*Courtesy Wean United, Inc.*)

original bar, grasped by grips and pulled in tension, drawing the remainder of the bar through the die. The bars reduce in section, elongate, and strain-harden. The reduction in area per pass is usually 20 to 50% to avoid wire fracture, with several steps often being required to obtain a desired product. Intermediate annealing may be necessary to restore ductility and enable further working.

Tube drawing, depicted in Figure 15-53 is used to produce seamless tubing and is essentially the same as the hot drawing of tubes discussed in Chapter 14. Because it is drawn cold from descaled tubing, the product has smoother surfaces, thinner walls, and more accurate dimensions than can be obtained by hot drawing. Mandrels are used for tubes from about 12.5 mm ($\frac{1}{2}$ inch) to 250 mm (10 inches) in diameter. Heavy-walled tubes and those less than 12.5 mm ($\frac{1}{2}$ inch) in diameter are often drawn without a mandrel in a process known as *tube sinking*. Precise control of the inner diameter is sacrificed for process simplicity and the ability to produce extremely long lengths of products. If a controlled internal diameter must be produced in a long-length product, the manufacturer may replace the internal mandrel with a "floating plug" as shown in Figure 15-54. If properly designed, the plug assumes a stable position and sizes the interior while the external die shapes the outside of the tubing.

Wire drawing. *Wire drawing* is essentially the same as bar drawing except that it involves smaller diameters and is generally done as a continuous process

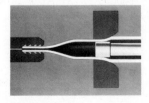

FIGURE 15-53 Method of cold-drawing smaller tubing from larger tubing. The die sets outer dimension while the mandrel sizes the inner diameter. (*Courtesy Copperweld Tubing Group.*)

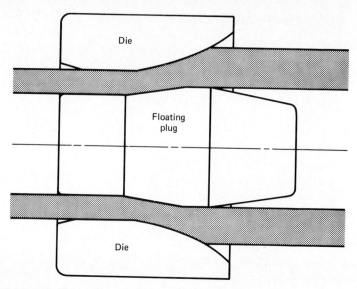

FIGURE 15-54 Tube drawing with a floating plug.

through a succession of drawing dies. Large coils of hot-rolled material roughly 9.5 mm ($\frac{3}{8}$ inch) in diameter are first descaled either by mechanical flexing or acid pickling and rinsing. The cleaned product is then further processed by immersion in a lime bath or other procedure to provide neutralization of remaining acid, corrosion protection, and a carrier for surface lubricant. One end of the coil is pointed, fed through a die, and the drawing process begins. Dies have the configuration shown in Figure 15-55 and usually are made of tungsten carbide or polycrystalline, man-made, diamond. Single crystal diamond dies are often used for drawing very fine wire. Lubrication boxes often precede the individual dies to assure coating of the material.

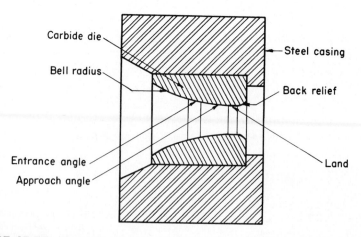

FIGURE 15-55 Cross section through a typical carbide die showing characteristic regions of the contour.

Small-diameter wire usually is drawn on tandem machines of the type shown in Figure 15-56, which contain 3 to 12 dies, each held in water-cooled die blocks. The reduction in each die is controlled so that each station uses about the same power. Speed is controlled at each station to avoid bunch-up or excessive tension in the wire. If extensive deformation is required, intermediate anneals must be performed between the various stages of drawing. Wires of a variety of tempers can be produced by controlling the placement of the last anneal in the process cycle. Full-soft wire is annealed in a controlled-atmosphere furnace after the final drawing.

Special shapes can also be produced by cold drawing of bar stock. Precise dimensions can be produced in complex cross sections with little or no material waste or finish machining. Steels, copper alloys, and aluminum alloys have all been formed in this manner.

Spinning.   *Spinning* is a rather fascinating cold-working operation in which a rotating disk of sheet metal is drawn over a male form by applying localized pressure with a simple, round-ended wooden or metal tool or a small roller. The basic process is illustrated in Figure 15-57.

The form, or *chuck,* is rotated on a rapidly rotating spindle, often in a simple type of lathe. The disk of metal is centered and then held against the small end

FIGURE 15-56   Wire being drawn on a machine having eight die blocks. (*Courtesy Vaughn Machinery Company.*)

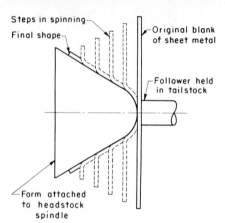

Steps in spinning
Final shape
Original blank of sheet metal
Follower held in tailstock
Form attached to headstock spindle

FIGURE 15-57 Progressive stages of forming a sheet of metal by spinning.

of the form by a follower attached to the tailstock of the lathe. As the disk and form rotate, the operator applies localized pressure against the metal, causing it to flow against the form, as shown in Figure 15-58. Because the final diameter of the formed part is less than that of the initial disk, thus shortening the circumference, the operator must stretch the metal radially a corresponding amount to avoid circumferential buckling. (This is a shrink-forming operation, as explained later and illustrated in Figure 15-67.) Considerable skill is required on the part of the operator. Usually there is some thinning of the metal, but it can be kept to a very small percentage by a skilled worker.

Inasmuch as the metal is not pulled across it under pressure, the form block can often be made of hardwood; it is only essential that it have a smooth surface because any roughness will show in the finished part. Thus, the tooling cost is low, making the process economical for small quantities. The process, however, is also used in many continuous-production applications such as for making lamp reflectors, cooking utensils, bowls, and the bells of some musical instru-

FIGURE 15-58 Two stages in spinning a metal reflector. (*Courtesy Spincraft Inc.*)

FIGURE 15-59 Machine spinning of heavy sheet metal. (*Courtesy Cincinnati Mila-cron Inc.*)

ments. If considerable numbers of identical parts are to be spun, a metal form block can be used.

Spinning is best suited for shapes that can be withdrawn directly from a one-piece chuck, but other shapes can be spun successfully by using multisection or offset chucks. Such shapes have also been spun successfully on form blocks made by freezing water, which is melted out after spinning is completed.

To retain the economic advantages of spinning but not require so much skill from the operator, spinning machines such as that in Figure 15-59 have been developed. The action of the spinning rollers is controlled automatically, being programmed for each particular part. Each successive part then undergoes the same deformation sequence, resulting in better consistency and fewer rejects. Such machines also make possible the spinning of thicker metal. In some cases numerical control is used. The type of machine shown in Figure 14-26 can also be used. On it, the spinning rolls are manipulated by hand through a power-assist system similar to that used on power steering on automobiles.

Shear forming. *Shear forming* or *flow turning,* illustrated in Figure 15-60, is a simplified variation of the spinning process in which the distance of each element in the blank from the axis of rotation remains constant. Because of this fact, the metal flow is entirely in shear and no radial stretch has to take place to compensate for the circumferential shrinkage that occurs in ordinary spinning. As a consequence, conical, hemispherical, and cylindrical shapes, and modifications of them, can be spun more readily by this process, particularly by mechanized means, than by normal spinning. Wall thickness will vary with the angle of the region. As shown in Figure 15-60, for conical parts, the relationship between the wall thickness of the final part and that of the blank is $t_c = t_b \sin \alpha$. If α is less than 30°, it may be necessary to complete the forming in two

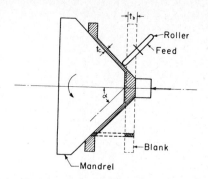

FIGURE 15-60 Basic shear-forming process.

stages with an intermediate anneal between. Reductions in wall thickness as high as 8:1 are possible, but the limit more generally is about 5:1, or an 80% reduction.

Conical shapes usually are shear-formed by the *direct process* depicted in Figure 15-60. They can also be made, however, by the *reverse process,* illustrated in Figure 15-61. By varying the direction of feed of the forming rollers in the reverse process, it is possible to form convex or concave parts without the necessity of having a matching mandrel.

Shear-formed cylinders can be made by either the direct or indirect processes, as shown in Figure 15-62. The reverse process has the advantage that a cylinder can be formed that is longer than the mandrel. Such a reverse flow induced by a roller and mandrel is also known as roll extrusion.

When long, thin-walled cylinders are desired, a ''flo-reform'' process, illustrated in Figure 15-63, can also be used. The process is a combination of shear forming and conventional spinning, the first steps involving no change in the diameter of portions of the workpiece, while such a change does occur in the latter steps. Figure 15-64 shows the type of equipment commonly used in shear forming and a blank being formed into a cone.

Stretch forming. *Stretch forming,* the principle of which is illustrated in Figure 15-65, was developed by the aircraft industry to permit certain sheet-metal parts, particularly large ones, to be formed economically in small quantities. As shown, only a single male form block is required. The sheet of metal is gripped

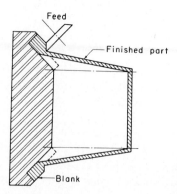

FIGURE 15-61 Forming a conical part by reverse shear spinning.

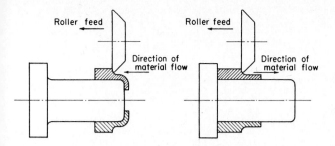

FIGURE 15-62 Shear forming a cylinder by the direct process (*left*) and the reverse process (*right*).

by two or more sets of jaws that stretch it and wrap it around the form block as the latter raises upward. Various combinations of stretching, wrapping, and upward motion of the block are employed, depending on the shape of the part. Figure 15-66 shows a typical part being formed by this process.

Through proper control of the stretching, most or all of the compressive stresses that accompany normal bending and forming are eliminated. Consequently, there is very little springback, and the workpiece conforms closely to the shape of the form block. Because the form block is almost completely in compression, it can be made of wood, Kirksite, or sprayed or laminated plastics. The process is used in making cowlings, wing tips, scoops, and large aircraft panels out of aluminum or stainless steel, and large automobile and truck panels from low-carbon steel. *Stretch-wrap forming* is another name for the process.

Stretch-draw forming is essentially the same process except that mating male and female dies shape the stretched metal.

Shell or deep drawing. The drawing of closed cylindrical or rectangular containers, or a variation of these shapes, with a depth frequently greater than the

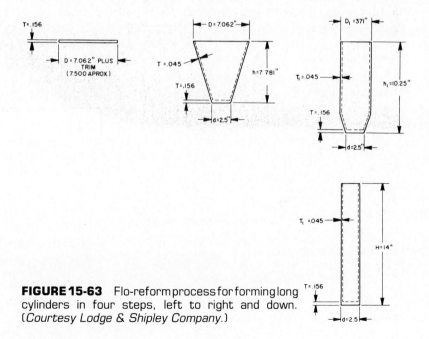

FIGURE 15-63 Flo-reform process for forming long cylinders in four steps, left to right and down. (*Courtesy Lodge & Shipley Company.*)

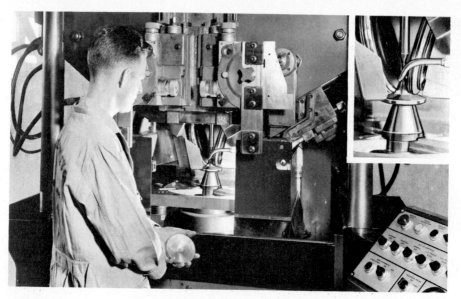

FIGURE 15-64 Cone being shear formed in a Floturn machine. Note blank in operator's hand and action of rollers, shown in inset. (*Courtesy Lodge & Shipley Company.*)

narrower dimension of their opening, is one of the most important and widely used manufacturing processes. Because the process had its earliest uses in manufacturing artillery shells and cartridge cases, it is sometimes called *shell drawing*. When the depth of the drawn part is less than the diameter, or minimum surface dimension of the blank, the process is considered to be *shallow drawing*. If the depth is greater than the diameter, it is considered to be *deep drawing*.

Basically there are two types of drawing, as illustrated in Figure 15-67, both involving multiaxial or curved-axis bending. As noted in Table 13-2, the stress conditions are complex. In *shrink forming,* there is circumferential compression during drawing resulting from the decrease of diameter d_1 to d_1', and the metal tends to thicken. Because the metal is thin, there is a tendency to relieve the circumferential compression by buckling or wrinkling. This tendency must be prevented by compressing the sheet between flat surfaces during forming, forcing a controlled compensating flow of metal as the shell is drawn. In *stretch forming,* on the other hand, there is a thinning of the metal as a result of the circumfer-

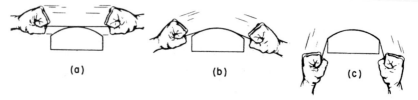

(a)   (b)   (c)

FIGURE 15-65 Schematic representation of the motions and steps involved in stretch-wrap forming.

FIGURE 15-66 Large aircraft part being formed by stretch-wrap forming. (*Courtesy Hufford Machine Works.*)

ential stretching that must occur in order for the diameter d_2 to increase to d_2'. This can lead to tearing of the metal. In many cases drawn parts contain regions of both shrink and stretch forming. This, of course, presents complex problems, and designers should recognize that drawn parts with regions of large shrink or stretch will cost more than those that avoid the associated difficulties.

The design of complex parts that are to be drawn has been aided considerably by computer techniques, but is far from being completely and successfully solved. Consequently, such design still involves a mix of science, experience, empirical data, and actual experimentation. The body of known information is quite substantial, however, and is being used with outstanding results.

To avoid wrinkling and variations in thickness, the flow of metal must be controlled. This is usually accomplished by some type of pressure ring or pad. In single-action presses, as shown in Figure 15-68, where there is only one movement of the slide, spring or air pressure is used to control the flow of metal between the upper die and the pressure ring. When double-action presses are used, with two or more independent rams, as illustrated in Figure 15-69, the force applied to the pressure ring can be controlled independent of the position

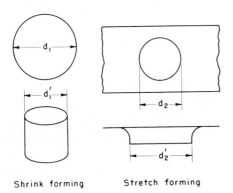

Shrink forming   Stretch forming

FIGURE 15-67 Two basic types of flow during drawing.

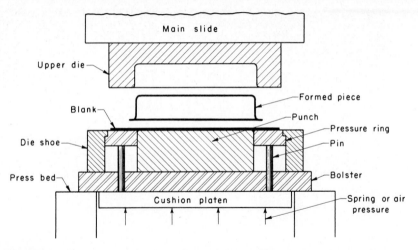

FIGURE 15-68 Method of deep drawing on a single-action press.

of the main slide. This permits the pressure to be varied as needed during the drawing operation. For this reason, double-action presses are usually used for drawing more complex parts, whereas single-action presses are often satisfactory for the more simple types of operations.

Because the flow of metal is often not uniform throughout the workpiece, many drawn parts have to be trimmed, after forming, to remove the excess or undesired metal. Figure 15-70 shows a part before and after trimming. Obviously, such trimming adds to the cost, because it must either be done by hand-guided operations or by use of a separate and special trimming die. In many

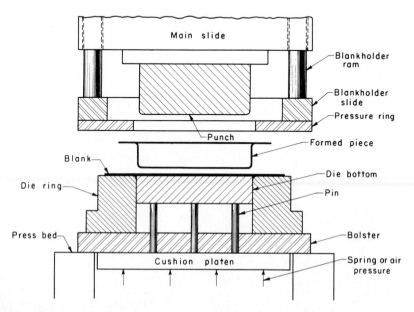

FIGURE 15-69 Drawing dies for use on a double-action press.

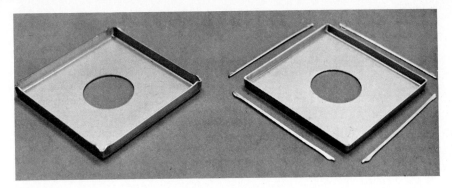

FIGURE 15-70 Pierced, blanked, and formed part before and after trimming.

cases trimming after drawing can be avoided by using a blank that has been cut to a special shape before drawing. Figure 15-71 illustrates this procedure. This approach of course, requires a special-shaped blanking die, but often the drawing blank is produced by blanking and making the blanking die to the desired shape may involve no extra cost. The choice of method is often dependent on the complexity of the part and the quantity to be produced.

When dies for blanking and drawing are made of steel, they are quite expensive. Consequently, numerous procedures and processes have been developed to permit these operations to be done with less expensive tooling, or to permit more extensive deformation to be done with fewer sets of dies or less in-process heat treating. Even though most of these processes have certain limitations as to the shapes or types of metal that can be formed, they are very useful where they can be applied. Some of them will now be discussed.

Forming with rubber or fluid pressure. Several methods of forming employ rubber or fluid pressure to obtain the desired deformation, and thereby eliminate either the male or female member of the die set. The *Guerin process,* depicted in Figure 15-72, utilizes the phenomenon that rubber of the proper consistency, *when totally confined,* acts as a fluid and will transmit pressure uniformly in all directions. Blanks of sheet metal are placed on top of form blocks, which usually are made of wood. The upper ram, which contains a pad of rubber 200 to 250

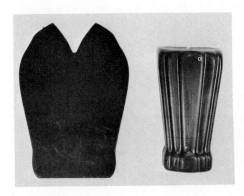

FIGURE 15-71 Irregularly shaped blank and finished drawn stove leg. No trimming is necessary.

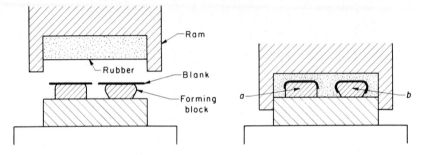

FIGURE 15-72 Guerin process for forming sheet metal.

mm (8 to 10 inches) thick in a steel container, then descends. The rubber pad is confined and transmits force to the metal, causing it to bend to the desired shape. Since no female die is used and form blocks replace the male die, die cost is quite low. Process flexibility is quite high (different shapes can even be formed at the same time), wear on material and tooling is low, and workpiece surface quality is easily maintained. When reentrant sections are formed, as in *b* in Figure 15-72, it must be possible to slide the parts lengthwise from the form blocks or disassemble the form block from within the product.

Guerin forming was developed in the aircraft industry, where small numbers of duplicate parts often must be formed, thus favoring low tooling cost. It can be used for aluminum up to about 3 mm ($\frac{1}{8}$ inch) thick and 1.6 mm ($\frac{1}{16}$-inch) stainless steel. Magnesium can also be formed if it is heated and heated form blocks are used.

Most forming done by the Guerin process is multiple-axis bending, but some shallow drawing can be done. It can also be used for piercing and blanking thin gages of aluminum, as illustrated in Figure 15-73. For this purpose, the blanking blocks, shaped the same as the desired workpiece, have a face, or edge, made of hardened steel. Round-edge supporting blocks are spaced a short distance from the shearing (blanking) blocks to support the scrap skeleton and permit the metal to bend away from the shearing edges.

The *Hydroform process* or "rubber bag forming" replaces the rubber pad with a flexible diaphragm backed by controlled hydraulic pressure. Deeper parts can be formed with truly uniform fluid pressure, as illustrated in Figure 15-74.

In *bulging,* oil or rubber is used for applying an internal bulging force to expand a metal blank or tube outward against a female mold or die, and thereby

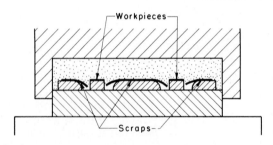

FIGURE 15-73 Method of blanking sheet metal by the Guerin process.

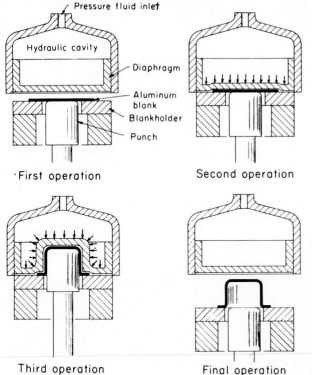

First operation

Second operation

Third operation

Final operation

FIGURE 15-74 Hydroform process showing (1) blank in place, no pressure in cavity; (2) press closed and cavity pressurized; (3) ram advanced with cavity maintaining fluid pressure; and (4) pressure released and ram retracted. (*Courtesy Aluminum Association, New York.*)

eliminates the necessity for a complicated, multiple-piece male die member. Split female dies are often used to facilitate product removal. Complex equipment is used in the fluid technique to provide the necessary seals, yet enable easy input and removal. For less complicated shapes, rubber can replace the internal fluid, as illustrated in Figure 15-75.

Drawing on a drop hammer. Where small quantities of shallow-drawn parts are required, they can often be made most economically through the use of Kirksite dies and a drop hammer. The Kirksite dies can be cast, thus avoiding the expense of costly machined steel dies, and they can be melted down and cast into other shapes when no longer needed. Often, a stack of shims, made

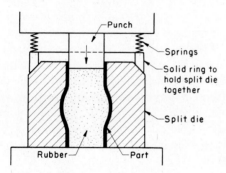

FIGURE 15-75 Method of bulging using rubber tooling.

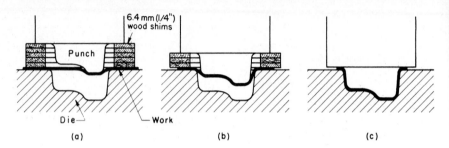

FIGURE 15-76 Method of deep drawing on a drop hammer using Kirksite dies and a book of shims.

of thin plywood, is placed on top of the sheet of metal to permit the ram to descend only partway, as shown in Figure 15-76. These shims further act as a pressure pad to restrict and control the flow of metal and thus inhibit wrinkling. One shim is withdrawn after each stroke of the ram, permitting the ram to fall further and deepen the draw. Wrinkles that form can be hammered out between strokes by means of a hand mallet.

Drawing on a drop hammer is considerably cruder than drawing with steel dies, but it often is the most economical method for small quantities. It is most suitable for aluminum alloys, but thin carbon and stainless sheets can also be drawn successfully.

High-energy-rate forming. A number of methods have been developed for forming metals through the release and application of large amounts of energy in a very short time interval. These processes are called *high-energy-rate forming processes,* commonly designated HERF. Many metals tend to deform more readily under the ultrarapid load application rates used in these processes, a phenomenon apparently related to the relative rates of load application and the movement of dislocations through the metal. As a consequence, HERF makes it possible to form large workpieces and difficult-to-form metals with less-expensive equipment and tooling than would otherwise be required.

Another advantage of HERF is that there is less difficulty due to springback. This is probably associated with two factors: (1) high compressive stresses are set up in the metal when it is forced against the die, and (2) some slight elastic deformation of the die occurs under the high pressure, resulting in the workpiece being slightly overformed and thereby giving the appearance of no springback.

The high energy-release rates are obtained by five methods: (1) underwater explosions, (2) underwater spark discharge (electrohydraulic techniques), (3) pneumatic-mechanical means, (4) internal combustion of gaseous mixtures, and (5) the use of rapidly formed magnetic fields (electromagnetic techniques).

Figure 15-77 illustrates three commonly used procedures involving the use of *explosive charges.* Although these procedures can be used for a wide range of parts they are particularly suitable for large parts of thick material as in the 3-meter (10-foot)-diameter elliptical dome of Figure 15-78. Only a tank of water in the ground with about 2 meters (6 feet) of water above the workpiece is required, along with a female die that can be made of inexpensive materials such as wood, plastic, or Kirksite.

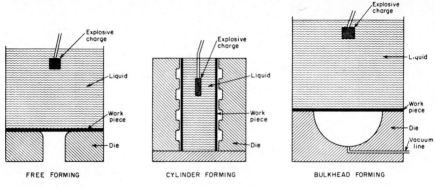

FIGURE 15-77 Three methods of high-energy-rate forming with explosive charges. (*Courtesy* Materials Engineering.)

The *spark-discharge method,* shown in Figure 15-79, uses the energy of an electrical discharge to shape the metal. Electrical energy is stored in large capacitor banks and then released in a controlled discharge, either between two electrodes or across an exploding bridgewire submerged in a medium. High-energy shockwaves propagate from the discharge and deform the metal. The initiating wire can be preshaped, and shock-wave reflectors can be used to adapt the process to a variety of components. The space between the blank and the

FIGURE 15-78 Elliptical dome 3 meters (10 feet) in diameter being removed from explosive forming die. (*Courtesy NASA.*)

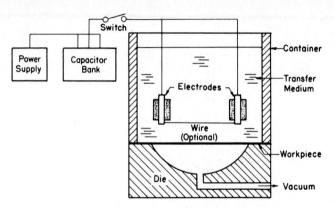

FIGURE 15-79 Components in the spark-discharge method of forming.

die is usually evacuated before the discharge occurs, removing the possibility of puckering due to entrapped air.

The spark-discharge methods are most often used for bulging operations in small parts, such as those in Figure 15-80, but parts up to 1.3 meters (50 inches) in diameter can be formed. Compared to explosive forming, the discharge techniques are much easier and safer, do not require as large tanks, and do not have to be performed in remote areas.

To make the HERF techniques adaptable for rapid use within a plant, the *pneumatic-mechanical* and *internal-combustion* presses were developed. In the pneumatic-mechanical presses one portion of the forming die is attached to the stationary bolster of the press bed and the other to a piston rod. Low-pressure gas acts on the entire area of the piston in a cylinder, holding it up against a seal such that only a small area on the other side of the piston is exposed to

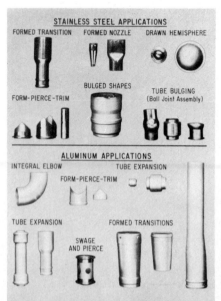

FIGURE 15-80 Some typical operations performed by spark-discharge techniques. (*Courtesy General Dynamics, Fort Worth Division.*)

FIGURE 15-81 Typical parts manufactured by the pneumatic-mechanical HERF technique. (*Courtesy Interstate Drop Forge Inc.*)

high-pressure gas. The small and large areas are balanced so that when the pressure of the high-pressure gas is increased above a certain value, the seal is broken, exposing the total area of the piston to high pressure. The piston is driven downward very rapidly, bringing the dies together. Figure 15-81 shows some typical parts made by this method.

Basically, the internal combustion presses operate on the same principle as an automobile piston and cylinder. A gaseous mixture is exploded within a cylinder, causing the piston to be driven downward very rapidly, one of the die members being attached to its lower end. This type of press can produce ram velocities up to 15 meters per second (50 feet per second) and up to 60 strokes per minute. Either single or repeated strokes can be obtained.

Electromagnetic forming is based on the principle that the electromagnetic field of an induced current always opposes the electromagnetic field of the inducing current. A capacitor is discharged through a coiled conductor that is either within or surrounding a cylinder or adjacent to a flat sheet of metal that is to be formed. This induces a current in the workpiece, causing it to be repelled from the coil and to be deformed against a die or mating workpiece. The process is very rapid and is useful for expanding or contracting tubing to various shapes, coining, forming, swaging, and permanently assembling component parts. Figure 15-82 shows some typical components.

Ironing. *Ironing* is the name given to the process of thinning the walls of a drawn cylinder by passing it between a punch and a die where the separation is less than the original wall thickness. The walls are elongated and thinned while

FIGURE 15-82 Two parts formed by the Magneform process. Both are approximately 125 mm (5 inches) in diameter. (*Courtesy General Atomic Division of General Dynamics Corporation.*)

the base remains unchanged, as shown in the schematic of Figure 15-83. The most common example of an ironed product is the thin-walled all-aluminum beverage can.

Embossing. *Embossing,* shown in Figure 15-84, is a method for producing lettering or other designs in thin sheet metal. Basically, it is a very shallow drawing opeation, usually in open dies, with the depth of the draw being from one to three times the thickness of the metal.

Superplastic sheet forming. The commercial development of superplastic metals (materials with over 100% elongation at selected strain rates and temperatures) has made possible the economical production of complex-shaped products, often in limited production quantities. Deep or complex shapes can now be made as one-piece, single-operation pressings, rather than multistep conventional pressings or multipiece assemblies. Moreover, the required forces are sufficiently low that the techniques are often adaptations of processes used in forming thermoplastics (see Chapter 9) and tooling is relatively inexpensive. Thermoforming, in which a vacuum or pneumatic pressure forces the sheet to conform to a heated male or female die, is quite popular, as are varieties of closed-die forming. Precision is excellent and fine details or textures can be reproduced. The major

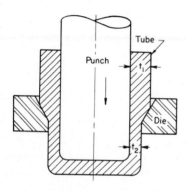

FIGURE 15-83 Ironing process in schematic.

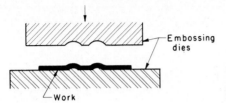

FIGURE 15-84 Embossing.

disadvantage, however, is the inherently low forming rate necessary to maintain superplastic behavior. Typical cycle times may be 1 to 5 minutes per part, rather than the several seconds typical of conventional presswork.

Design aids for sheet metal forming. A majority of sheet metal failures related to forming are the result of either excessive thinning or fracture, both being the result of extensive deformation in a given region. A fast and economical means of evaluating deformation severity in formed parts is by means of *strain analysis* and *forming limit diagrams*. A pattern or grid, such as in Figure 15-85, is placed on the surface of the sheet by means of scribing, printing, or etching. The sheet is then deformed and the distorted pattern is measured and evaluated. Regions where the area has expanded are locations of substantial thinning; contractions in area correspond to sheet thickening and may be locations of buckling or wrinkles. By measuring both the major strain and minor strain at given areas, the values can be plotted on a forming-limit diagram, as shown in Figure 15-85. Strains falling above the limit line indicate areas of probable fracture. Undesirable deformation characteristics should then be changed by methods such as lubricant modification, die design changes, or variation in clamping or hold-down pressure. The method can also be used to orient parts relative to the prior rolling direction, evaluate various lubricants as to their effectiveness, or assist in the design of dies for complex-shaped products.

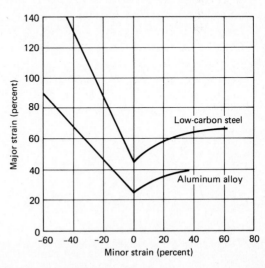

FIGURE 15-85 (*Left*) Typical pattern for sheet metal deformation analysis. (*Right*) Forming limit diagram used to determine whether a metal can be shaped without risk of fracture.

PRESSES

Classification of presses. Many types of presses have been developed to perform the variety of cold-working operations. When selecting a press for a given application, consideration should be given to the capacity required, whether the power source should be *hydraulic* or *mechanical,* and the method of transmitting power to the ram (the type of drive). The table below indicates the various types available and Figure 15-86 illustrates the more important drive mechanisms. In general, mechanical drives provide faster action and more positive displacements than do hydraulic, whereas greater forces can be obtained more readily with hydraulic drives, with more flexibility.

Foot	Mechanical	Hydraulic
Kick presses	Crank	Single-slide
	Single	Multiple-slide
	Double	
	Eccentric.	
	Cam	
	Knuckle joint	
	Toggle	
	Screw	
	Rack and pinion	

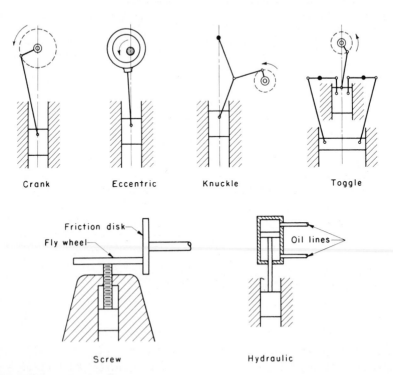

Crank   Eccentric   Knuckle   Toggle

Friction disk

Fly wheel

Oil lines

Screw   Hydraulic

FIGURE 15-86 Schematic representation of the various drive mechanisms for presses.

Foot-operated presses, commonly called *kick presses,* are used only for very light work. *Crank-driven presses* are the most common type because of their simplicity. They are used for most piercing and blanking operations and for simple drawing. Double-crank presses provide a method of actuating blank holders of operating multiple-action dies. *Eccentric* or *cam drives* are used where only a short ram stroke is required. Cam action can provide a dwell at the bottom of the stroke and is sometimes used to actuate the blank holding ring in deep-drawing processes. *Knuckle-joint drives* provide a very high mechanical advantage along with fast action. They are often used in coining, sizing, and Guerin forming. *Toggle mechanisms* are used principally in drawing presses to actuate the blank holder, because they provide two sets of motions.

Hydraulic presses are available in many varieties and sizes. Because almost unlimited capacities can be provided, most large drawing presses are of this type. By using several hydraulic cylinders, programmed loads can be applied to the ram and any desired force and timing can be applied independently to the blank holder. Although most hydraulic presses tend to be relatively slow, types are available that provide up to 600 strokes per minute for high-speed blanking operations.

Press frame types. Another matter of importance in selecting a press is the type of frame, because this often imposes limitations on the size and type of work that can be done. The following is a classification according to frame type:

Arch	Gap	Straight-Sided
Crank or eccentric	Foot	Many variations,
Percussion	Bench	but all with
	Vertical	straight-sided frames
	Inclinable	
	Inclinable	
	Open back	
	Horn	
	Turret	

Arch-frame presses, having their frames in the shape of an arch, are seldom used today, except with screw drives for coining operations.

Gap-frame presses, having a **C**-shaped frame as shown in Figure 15-87 provide good clearance for the dies and permit large stock to be fed into the press. They are made in a wide variety of sizes, capacity ranging from 8.9 to over 2200 *kN* (1 to 250 tons). Often, they can be inclined to permit the completed work to drop out of the back side of the press, as in Figure 15-87. Also included in that press is a sliding bolster plate, a common feature that permits a second die to be set up on the press while another is in operation. Die changeover then requires only a few minutes to unclamp the punch segment of one die set from the ram, move the second die set into position, clamp the upper segment of the new set of the press ram, and start a new operation.

Bench presses are small, inclinable, gap-frame presses, 8.9 to 71.2 kN (1 to 8 tons) in capacity, that are made to be mounted on a bench. *Open-back presses* are gap-frame presses that are not inclinable. *Horn presses* are upright, open-back presses that have a heavy cylindrical shaft or "horn" in place of the usual

FIGURE 15-87 Inclinable gap-frame press with sliding bolster to accommodate two die sets for rapid change of setup. (*Courtesy Niagara Machine & Tool Works.*)

bed. This permits curved or cylindrical workpieces to be placed over the horn for such operations as seaming, punching, riveting, and so forth, as shown in Figure 15-88. On some types both a horn and a bed are provided, with provision for swinging the horn aside when not needed.

Turret presses utilize a modified gap-frame construction but have upper and lower turrets that carry a number of punches and dies. The two turrets are geared together so any desired set can quickly be rotated into position. Another type uses a single turret on which subpress die sets are mounted.

Straight-sided presses are available in a wide variety of sizes and designs. This type of frame is used for most hydraulic presses and for larger-sized and specialized mechanical-driven presses. A typical example is shown in Figure 15-89.

FIGURE 15-88 Making a seam on a horn press. (*Courtesy Niagara Machine & Tool Works.*)

FIGURE 15-89 An 1800-kN (200-ton) straight-side press. (*Courtesy Rousselle Corportion.*)

Special press types. A number of special types of presses are available, designed for doing special types of operations. Two interesting examples are the *transfer press* and the *multislide press*. The transfer press, illustrated in Figure 15-90, has a single, long slide with the provision for mounting a number of die sets side by side. Each die set can be adjusted individually. Stock is fed auto-

FIGURE 15-90 Transfer-type press. Inset shows parts at several of the eight stations. (*Courtesy Verson Allsteel Press Company.*)

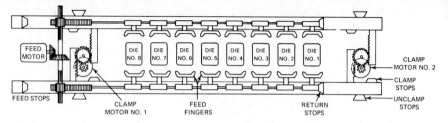

FIGURE 15-91 Transfer mechanism used in transfer presses. (*Courtesy Verson Allsteel Press Company.*)

matically to the first die station. After the completion of each ram stroke, the part is automatically and progressively transferred to the next die station by the mechanism shown in Figure 15-91. Such presses have high production rates, up to 1500 parts per hour, and are also very flexible. Figure 15-92 illustrates the variety of operations that may be sequentially performed in the production of stamped and drawn sheet metal parts.

The multislide machine, shown in Figures 15-93 and 15-94, contains a series of bending or forming slides that move horizontally. As indicated on the schematic diagram, these slides are driven by cams on four shafts that are located on the four sides of the machine, the shafts being driven by means of miter gears. Coiled strip stock is fed into the machine automatically and is progressively pierced, notched, bent, and cut off at the various slide stations in the manner indicated in Figure 15-95. Strip stock up to about 75 mm (3 inches) wide and 2.5 mm ($\frac{3}{32}$ inch) thick and wire up to about 3 mm ($\frac{1}{8}$ inch) in diameter are commonly processed. Parts such as hinges, links, clips, razor blades, and the like are processed at very high rates on such machines.

Press feeding devices. Although hand feeding is still used in many press operations, improved operator safety and increased productivity have motivated

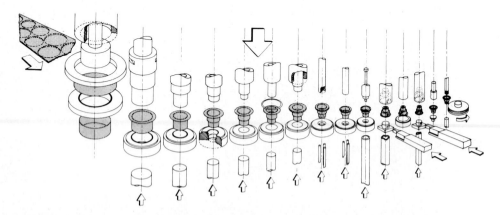

FIGURE 15-92 Possible operations which may be performed during the production of stamped and drawn parts on a transfer press. (*Courtesy U.S. Baird Corporation, Stratford, Conn.*)

FIGURE 15-93 Multislide machine viewed from "back" side. (*Courtesy U.S. Baird Corporation, Stratford, Conn.*)

a strong shift to feeding by some type of mechanical device. *Dial-feed mechanisms,* such as the one shown in Figure 15-96 are often employed. The operator places the workpieces in the front holes of the dial, and the dial then indexes with each stroke of the ram to feed parts into proper position between the punch and die. When continuous strip stock is used, it can be fed automatically into

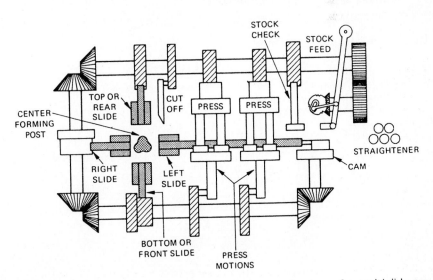

FIGURE 15-94 Schematic diagram of the operating mechanism of a multislide machine. (*Courtesy U.S. Baird Corporation, Stratford, Conn.*)

FIGURE 15-95 Example of the piercing, blanking, and forming of a part on a multislide machine. (*Courtesy U.S. Baird Corporation, Stratford, Conn.*)

a press by press-driven roll feeds mounted on the side of the press. Lightweight parts can be fed into presses by suction-cup mechanisms, vibratory-bed feeders, or similar mechanisms. Robots such as the ones shown in Chapter 38 are being used increasingly to place large parts into presses and remove them after processing. The technology and equipment exist to replace manual feeding in most cases if such a transition is desired.

FIGURE 15-96 Dial feed device being used on a punch press. (*Courtesy E. W. Bliss Company.*)

REVIEW QUESTIONS

1. What are some of the advantages of cold deformation?
2. Why might a sheet or strip be given a skin-rolled reduction pass?
3. What are some of the products that can be produced by swaging?
4. How can cold forging be used to substantially reduce material waste?
5. What are some other advantages that can be obtained by using cold forging?
6. If a product combines a large-diameter head and small-diameter shank, how can the processes of extrusion and cold heading be combined to save material?
7. What are some advantages of hydrostatic extrusion?
8. What types of products are produced by roll extrusion?
9. When might sizing operations be used to provide economical savings?
10. What types of rivets might be used when only one side of the joint is accessible?
11. What is coining?
12. How can peening improve the surface characteristics of a metal?
13. What types of products can benefit from peening?
14. What is burnishing?
15. Why does a metal thin somewhat when bent?
16. Why does springback occur during bending?
17. What is a press brake?

18. How is minimum bend radius related to ductility?
19. What design benefit can be obtained by specifying large bend radii?
20. How should the bend radius be related to the rolling direction of cold-rolled strip?
21. What is the difference between draw bending and compression bending?
22. Why must long lengths be required to justify cold roll-forming?
23. What are two methods to straighten rod or sheet?
24. Why are sheared edges generally not smooth?
25. What process can overcome the tearing segment of metal shearing?
26. What is the major difference between squaring and slitting?
27. What is the common feature of piercing and blanking?
28. What are the three basic components of piercing and blanking die sets?
29. Why might a slight angle be ground on the face of a piercing or blanking punch?
30. What is the major advantage of subpress die sets?
31. What is the advantage of making a shearing die in segments?
32. What is a progressive die set?
33. What two different processes can be classified as cold drawing?
34. Why might tube drawing be performed cold rather than hot?
35. What is the difference between tube drawing and tube sinking?
36. What benefit can be obtained by tube drawing with a "floating plug"?
37. What generally forms the starting material for wire drawing?
38. Why does spinning not require hardened-tool-steel form blocks?
39. What type of products are typically manufactured by stretch forming?
40. What types of defects or failures are associated with shrink forming? Stretch forming?
41. What is the major advantage of the Guerin process or hydroforming?
42. Why might less springback be observed in high-energy-rate forming?
43. What methods have been used to obtain the high-energy-release rates?
44. What common product is often produced by a wall-ironing process?
45. What advantages can be obtained by using the superplastic sheet-forming method?
46. What is the major limitation of superplastic forming?
47. How is strain analysis performed in a sheet metal product?
48. What are the two classes of press frame design most commonly observed today?
49. What benefits can be obtained by using press feeding devices?

CASE STUDY 15.   The Bent Propeller

The propeller of a moderately large pleasure boat has been cast from a nickel–aluminum bronze alloy (82% Cu, 9% Al, 4% Ni, 4% Fe, 1% Mn). It is 330 mm (13 inches) in diameter with three 10-pitch blades and is designed for both fresh- and saltwater usage. One of the blades has struck a rock and is badly bent. A replacement propeller is quite expensive and cannot be obtained for several weeks. The owner, therefore, wishes to have the damaged propeller repaired.

Can it simply be hammered back into shape? Would you recommend any additional processing either before or after the repair? Explain your recommendations.

PART III

MACHINING PROCESSES

CHAPTER 16

Measurement and Inspection

Manufacturing processes create shapes in specific sizes. For the purpose of interchangeable parts, the sizes are often standardized. Thus, 60-watt light bulbs are made with bases to the same standard size. Manufacturers of light sockets make the recepticals to accommodate that bulb. Likewise, you cannot buy a 67-watt bulb because that is not a standard wattage. In order to make interchangeable products economical to the desired level of quality and reliability, precise specifications are proposed in the design, accurate measurements of the product sizes and shapes are made during or after manufacturing, and a means is provided to feed back this measurement information to the processes for the purpose of controlling the quality. Generally speaking, the difficulty and cost to make the product both increase rapidly as one tries to manufacture parts with greater precision and accuracy.

Large-scale manufacturing based on the principles of standardization of sizes and interchangeable parts became common practice in the early 1900s. Size control must be built into the machine tools and the workholding devices through the precision manufacture of these machines and tooling. Then the output of the machines must be carefully checked (1) to determine the capability of specific machines, and (2) to control and maintain the quality of the product. When a designer specifies the dimensions and tolerances of a part, he quite often does so to enhance the function of the product but he is also determining the machines and processes needed to make the part. Quite often it will be necessary for the design engineer to alter the design to improve the ease or cost of either manufacture (producibility) or assembly of the product and he should always be prepared to do this provided that he is not sacrificing functionality, product reliability, or performance.

Attributes versus variables.   The examination of the product either during or after manufacture, either manually or automatically, falls under the providence of *inspection*. Inspection of items or products can basically be done two ways:

1. *By attributes:* uses gages to determine if product is good or bad; a yes or no, go or not-go decision.
2. *By variables:* uses calibrated instruments to determine how good or how bad the product is compared to desired size.

In an automobile a speedometer and an oil pressure gage are examples of variable types of measuring instruments, while an oil pressure light is an example of an attributes type of gage. As typical with an attributes type of gage, you do not know *what* the pressure actually is if the light goes on, only that it is not good.

Measurement is the generally accepted industrial term for the performance of inspection by variables. *Gaging* (or gauging) is the term for determining whether the dimension or characteristic is larger or smaller than the established standard or range of acceptability. Variable types of inspection generally take more time and are more expensive than attributes, but they give more information as the magnitude of the characteristic is known in some standard units of measurement.

STANDARDS OF MEASUREMENT

The four master measures on which all others depend are *length, time, mass,* and *temperature*. These four, along with the *ampere* and the *candela,* provide the basis for all other units of measurement as shown in Figure 16-1. Most mechanical measurements involve combinations of units of mass, length, and time. Thus, the newton, a unit of force, is derived from Newton's second law of motion ($F = ma$) and is defined as the force that gives an acceleration of 1 meter per second per second to a mass of 1 kilogram. Figure 16-1 is in SI units.

Linear measurements.   When man first sought a unit of length, he adopted parts of his body, mainly his hands, arms, or feet. Such tools were not very satisfactory since they were not universally standard in size. Satisfactory measurement and gaging must be based on a reliable, and preferably universal, standard or standards. These have not always existed. For example, although the musket parts made in Eli Whitney's shop were interchangeable, they were not interchangeable with parts made by another contemporary gun maker *from the same drawings,* because the two gunsmiths *had different foot rulers.* Today the entire industrialized world has adopted the *international meter* as the standard of linear measurement. In 1960, it was officially redefined as being 1,650,763.73 wavelengths of the orange-red light given off by electrically excited krypton 86, a rare gas from the atmosphere. Both the United States and English inches have

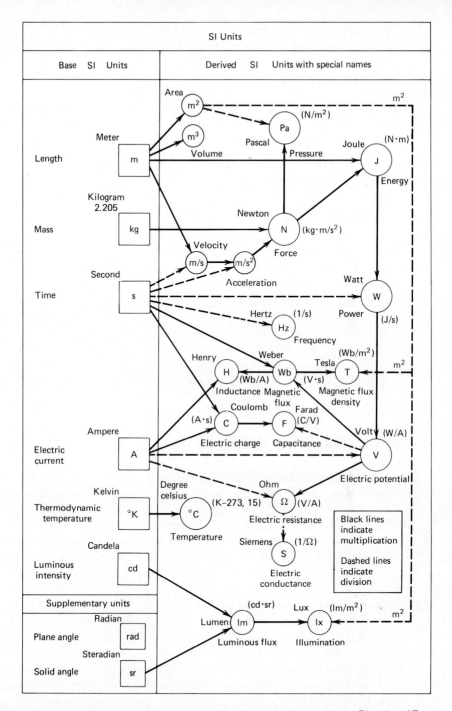

FIGURE 16-1 Relationship of secondary physical quantities to basic SI units. (*From* NBS Technical Note 938, *National Bureau of Standards, Washington, D.C.*)

been officially defined as 2.54 centimeters. Thus the United States inch is 41,929.399 wavelengths of the orange-red light from krypton 86.

Although *officially* the United States is committed to conversion to the metric (SI) system of measurement, which utilizes millimeters for virtually all linear measurements in manufacturing, the English system of feet and inches still is being utilized by most manufacturing plants and its use probably will continue for some time.[1]

Length standards in industry. *Gage blocks* provide industry with linear standards of high accuracy that are necessary for everyday use in manufacturing plants. These are small, steel blocks, usually rectangular in cross section, having two very flat and parallel surfaces that are certain specified distances apart. Such gage blocks were first conceived by Carl E. Johansson in Sweden just prior to 1900. By 1911, he was able to produce sets of such blocks on a very limited scale, and they came into limited but significant use during World War I. Shortly after the war, Henry Ford recognized the importance of having such gage blocks generally available. He arranged for Johansson to come to the United States and, through facilities provided by the Ford Motor Company, methods were devised for large-scale production of gage block sets. As a result, since about 1940, gage block sets of excellent quality have been produced by a number of companies in this country and abroad.

Gage blocks usually are made of alloy steel, hardened and carefully heat-treated (*seasoned*) to relieve internal stresses and minimize subsequent dimensional change. Some are made entirely of carbides, such as chromium or tungsten carbide, to provide extra wear resistance. The measuring surfaces of each block are ground to approximately the required dimension and then lapped to reduce the block to the final dimension and to produce a very flat and smooth surface.

Gage blocks commonly are made to meet the U.S. Federal Specifications GGG-G-15a and 15b, having Grades 1, 2, and 3 with accuracies as follows:

Grade	15a (inches)	15b (millimeters)
1 (laboratory)	±0.000,002	±0.000 05
2 (precision)	+0.000,004 −0.000,002	+0.000 10 −0.000 05
3 (working)	+0.000,008 −0.000,004	+0.000 20 −0.000 10

Blocks up to 1 inch in length have absolute accuracies as stated, whereas the tolerances are per inch of length for blocks larger than 1 inch. Some companies supply blocks in AA quality, which corresponds to Grade 1, and A+ quality, to meet Grade 2.

---

[1]One obvious objection to the exclusive use of millimeters is the magnitude of the numbers that result—for example, 60 inches being 1524 mm. In the aircraft industry numbers above 25 000 are not uncommon.

heat treated   stress relieved hardened steel

FIGURE 16-2   Metric gage block set, containing 88 blocks. (*Courtesy DoALL Company.*)

Metrology = measuring
$20°C = 68°F$

Grade 1 (*Laboratory-Grade*) blocks are used for checking and calibrating other grades of gage blocks. Grade 2 (*Precision-Grade*) blocks are used for checking master gages and Grade 3 blocks. Grade 3 (B or *Working-Grade*) blocks are used to check routine measuring devices, such as micrometers, or in actual gaging operations.

The dimensions of individual blocks are established by light beam interferometric methods whereby it is possible to calibrate these blocks routinely with an uncertainty as low as 1 part per million.

Gage blocks usually come in sets containing various numbers of blocks of various sizes, such as those shown in Figure 16-2. By wringing the blocks together in various combinations, as shown in Figure 16-3, any desired dimen-

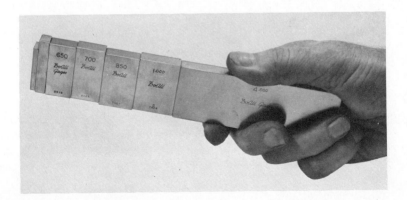

FIGURE 16-3   Seven gage blocks wrung together to build up a desired dimension. (*Courtesy DoALL Company.*)

sion can be obtained. As examples, one manufacturer's 81-block English set and 88-block metric set have the following blocks:

English			Metric		
Series	Number of Blocks	Range (in.)	Series	Number of Blocks	Range (mm)
0.0001 in.	9	0.1001–0.1009		2	0.5 and 1.0005
0.001 in.	49	0.101–0.149	0.001 mm	9	1.001–1.009
0.050 in.	19	0.050–0.950	0.01 mm	49	1.01–1.49
1.000 in.	4	1.000–4.000	0.5 mm	18	1.0–9.5
			10 mm	10	10–120

The 81 blocks in the English system can be combined in over 120,000 combinations in increments of 0.0001 inch from 0.1001 to over 25 inches. When gage blocks are wrung together (slid together firmly), they adhere with considerable force and should not be left in contact for extended periods of time.

Several types of gage blocks are available: fractional series, thin series, angle series, and so on, so that standards of high accuracy can be obtained to fill almost any need. In addition, various auxiliary clamping, scribing, and base block attachments are available that make it possible to form very accurate gaging devices, such as shown in Figure 16-4.

A very useful variation of gage blocks is the device shown in Figure 16-5. It consists of a column of permanently wrung, 1-inch gage blocks arranged in a staggered pattern so that the entire column can be raised and lowered by means

FIGURE 16-4 Wrung-together gage blocks in a special holder and used with a dial gage to form an accurate comparator. (*Courtesy DoALL Company.*)

FIGURE 16-5 Digi-check height gage. (*Courtesy L. S. Starrett Company.*)

of an accurate, calibrated micrometer screw. Direct digital reading of the height of the block steps above the base is shown to 0.001-inch increments and on the micrometer dial to 0.0001-inch increments. Models also are available having direct electronic digital readout to 0.0001-inch increments. Such units are very useful in conjunction with deviation-type gages, which will be discussed later, in making accurate vertical measurements.

Standard measuring temperature. Because all the commonly used metals are affected dimensionally by temperature, a standard measuring temperature of 20°C (68°F) has been adopted for precision-measuring work. All gage blocks, gages, and other precision-measuring instruments are calibrated at this temperature. Consequently, when measurements are to be made to accuracies greater than 0.0025 mm (0.0001 inch), the work should be done in a room in which the temperature is controlled at standard. Although it is true that to some extent both the workpiece and the measuring or gaging device *may* be affected to about the same extent by temperature variations, one should not rely on this. Measurements to even 0.0025 mm (0.0001 inch) should not be relied on if the temperature is very far from 20°C (68°F).

The fit of mating parts. The accuracy that must be specified and achieved in the manufacture of a given part very often is determined by the manner in which it must function with respect to other parts. If it is necessary only that one part always fit inside a second, it makes little difference if the dimension of the smaller part varies considerably as long as its maximum dimension is always smaller than the inside dimension of the part into which it must go. However, if the smaller part is to rotate smoothly within the larger part at high speeds with a minimum of vibration, as in a high-speed ball bearing, considerably more attention must be given to the specification and control of the dimensions of the parts. Thus *function* controls the specification of dimensions that determine the manner in which parts fit together.

Two factors, allowance and tolerance, must be specified to obtain the desired fit between mating parts. *Allowance* is the intentional, desired difference between the dimensions of two mating parts. It is the difference between the dimension of the largest interior-fitting part (shaft) and that of the smallest exterior-fitting part (hole). It thus determines the condition of *tightest* fit between mating parts. Allowance may be specified so that either *clearance* or *interference* exists between the mating parts. With clearance fits, the largest shaft is smaller than the smallest hole, whereas with interference fits, the hole is smaller than the shaft.

Tolerance is an undesirable, but permissible, deviation from a desired dimension in recognition that no part can be made *exactly* to a specified dimension, except by chance, and that such is neither necessary nor economical. Consequently, it is necessary to permit the actual dimension to deviate slightly from the desired, theoretical dimension and to control the degree of deviation such that satisfactory functioning of the mating parts still will be assured.

Tolerance can be specified in three ways: bilateral, unilateral, and limits. *Bilateral* tolerance is specified as a plus or minus deviation from the nominal size, such as 2.000 ± 0.002 in. More modern practice uses the *unilateral* system, where the deviation is in one direction from the basic size, such as

$$2.000 \text{ in.} \begin{array}{c} +0.004 \text{ in.} \\ -0.000 \text{ in.} \end{array} \quad \text{or} \quad 50.8 \text{ mm} \begin{array}{c} +0.1 \text{ mm} \\ -0.0 \text{ mm} \end{array}$$

In the first case, that of bilateral tolerance, the dimension of the part could vary between 1.998 and 2.002 inches, a total tolerance of 0.004 inch. For the example of unilateral tolerance, the dimension could vary between 2.000 and 2.004 inches, again a tolerance of 0.004 inch. Obviously, in order to obtain the same maximum and minimum dimensions with the two systems, different basic sizes must be used. The maximum and minimum dimensions that result from the application of the designated tolerance are called *limit dimensions* or *limits*.

There can be no rigid rules as to the amount of clearance that should be provided between mating parts; the decision must be made by the designer in consideration of how he wants them to function. The American National Standards Institute, Inc. (ANSI) has established eight classes of fits that serve as a useful guide in specifying the allowance and tolerance for typical applications, and that permit the amount of allowance and tolerance to be determined merely by specifying a particular class of fit. These classes are as follows:

Class 1: Loose fit—large allowance. Accuracy is not essential.

Class 2: Free fit—liberal allowance. For running fits where speeds are over 600 rpm and pressures are 4.1 MPa (600 psi) or over.

Class 3: Medium fit—medium allowance. For running fits under 600 rpm and pressures less than 4.1 MPa (600 psi) and for sliding fits.

Class 4: Snug fit—zero allowance. No movement under load is intended and no shaking is wanted. The tightest fit that can be assembled by hand.

Class 5: Wringing fit—zero to negative allowance. Assemblies are selective and not interchangeable.

Class 6: Tight fit—slight negative allowance. An interference fit for parts that must not come apart in service and are not to be disassembled, or disassembled only seldom. Light pressure is required for assembly. Not to be used to withstand other than very light loads.

Class 7: Medium force fit—an interference fit requiring considerable pressure to assemble; ordinarily assembled by heating the external member or cooling the internal member to provide expansion or shrinkage. Used for fastening wheels, crank disks, and the like to shafting. The tightest fit that should be used on cast iron external members.

Class 8: Heavy force and shrink fits—considerable negative allowance. Used for permanent shrink fits on steel members.

The allowances and tolerances that are associated with the ANSI classes of fits are determined according to the theoretical relationship shown in Table 16-1. The actual resulting dimensional values for a wide range of basic sizes can be found in tabulations in drafting and machine-design books.

In the ANSI system, the hole size is always considered basic, because the majority of holes are produced through the use of standard-size drills and reamers. The internal, or shaft, member can be made to any one dimension as readily as to another. The allowance and tolerances are applied to the basic hole size to determine the limit dimensions of the mating parts. For example, for a basic hole size of 50.8 mm (2 inches) and a Class 3 fit, the dimensions would be:

Allowance	0.036 mm (0.0014 in.)
Tolerance	0.025 mm (0.0010 in.)
Hole	
Maximum	50.825 mm (2.0010 in.)
Minimum	50.800 mm (2.0000 in.)
Shaft	
Maximum	50.764 mm (1.9986 in.)
Minimum	50.739 mm (1.9976 in.)

TABLE 16-1 ANSI Recommended Allowances and Tolerances

Class of Fit	Allowance	Average Interference	Hole Tolerance	Shaft Tolerance
1	$0.0025 \sqrt[3]{d^2}$		$+0.0025 \sqrt[3]{d}$	$-0.0025 \sqrt[3]{d}$
2	$0.0014 \sqrt[3]{d^2}$		$+0.0013 \sqrt[3]{d}$	$-0.0013 \sqrt[3]{d}$
3	$0.0009 \sqrt[3]{d^2}$		$+0.0008 \sqrt[3]{d}$	$-0.0008 \sqrt[3]{d}$
4	0		$+0.0006 \sqrt[3]{d}$	$-0.0004 \sqrt[3]{d}$
5		0	$+0.0006 \sqrt[3]{d}$	$+0.0004 \sqrt[3]{d}$
6		$0.00025d$	$+0.0006 \sqrt[3]{d}$	$+0.0006 \sqrt[3]{d}$
7		$0.0005d$	$+0.0006 \sqrt[3]{d}$	$+0.0006 \sqrt[3]{d}$
8		$0.001d$	$+0.0006 \sqrt[3]{d}$	$+0.0006 \sqrt[3]{d}$

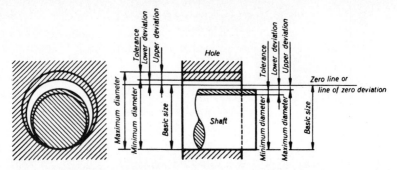

FIGURE 16-6 Basic size, deviation, and tolerance in the ISO system. (*By permission from ISO Recommendation R286-1962, System of Limits and Fits, copyright 1962, American National Standards Institute, New York.*)

It should be noted that for both clearance and interference fits, the permissible tolerances tend to result in a looser fit.

The *ISO System of Limits and Fits*,[2] widely used in a number of leading metric countries, is considerably more complex than the ANSI system just discussed. In this system, each part has a *basic size*. Each limit of size of a part, high and low, is defined by its *deviation* from the basic size, the magnitude and sign being obtained by subtracting the basic size from the limit in question. The difference between the two limits of size of a part is called the *tolerance*, an absolute amount without sign. Figure 16-6 illustrates these definitions.

There are three classes of fits: (1) *clearance fits*, (2) *transition fits* (the assembly may have either clearance or interference), and (3) *interference fits*.

Either a *shaft-basis system* or a *hole-basis system* may be used, as illustrated in Figure 16-7. For any given basic size, a range of tolerances and deviations may be specified with respect to the line of zero deviation, called the *zero line*. The tolerance is a function of the basic size and is designated by a number symbol, called the *grade*—thus the *tolerance grade*. The *position* of the tolerance with respect to the zero line—also a function of the basic size—is indicated by a letter symbol (or two letters), a capital letter for holes and a lowercase

---

[2]This system and the necessary tables for its application are contained in ISO publication R286-1962, *System of Limits and Fits*.

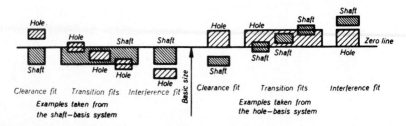

FIGURE 16-7 "Shaft-basis" and "hole-basis" systems for specifying fits in the ISO system. (*By permission from ISO Recommendation R286-1962, System of Limits and Fits, copyright 1962, American National Standards Institute, New York.*)

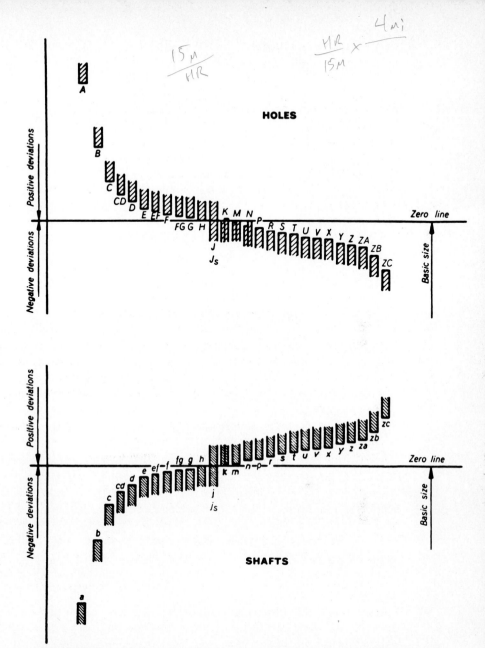

FIGURE 16-8 Positions of the various tolerance zones for a given diameter in the ISO system. *(By permission from ISO Recommendation R286-1962, System of Limits of Fits, copyright 1962, American National Standards Institute, New York.)*

letter for shafts, as illustrated in Figure 16-8.[3] Thus the specification for a hole and shaft having a basic size of 45 mm might be 45 H8/g7.

Eighteen standard grades of tolerances are provided, called IT 01, IT 0, IT 1–16, providing numerical values for each nominal diameter, in arbitrary steps

[3]It will be recognized that the "position" in the ISO system essentially provides "allowance" of the ANSI system.

up to 500 mm (for example 0–3, 3–6, 6–10, . . . , 400–500 mm). The value of the tolerance unit, i, for grades 5–16 is

$$i = 0.45 \sqrt[3]{D} + 0.001D$$

where i is in micrometers and D in millimeters.

Standard shaft and hole deviations similarly are provided by sets of formulas. However, for practical application, both tolerances and deviations are provided in three sets of rather complex tables. Additional tables give the values for basic sizes above 500 mm and for "Commonly Used Shafts and Holes" in two categories—"General Purpose" and "Fine Mechanisms and Horology."

Inasmuch as the ANSI recommended allowances and tolerances, listed in Table 16-1, may be expressed either in inches or millimeters, one may use this relatively simple system or the more complex ISO system in dimensioning metric drawings.

INSPECTION METHODS FOR MEASUREMENT

The world of metrology, even limited to dimensional measurements, is much too great to be covered here, so the rest of the chapter will concentrate on linear dimensions and the associated equipment. Clearly, the distinctions between production inspection and gage calibration have become less significant as requirements for precision machining become more prevalent. This trend toward tighter tolerances and greater reliability in measuring these tolerances has greatly enhanced the need for precision measurement methods.

Table 16-2 gives a summary of inspection methods, listing four basic kinds of devices one can use: air, optical (light and electron), electronic, and mechanical. This table is followed by a discussion of the factors which should be considered in selecting measurement equipment. Most companies, as a minimum, will have gage block comparators, optical flats, perhaps a supermicrometer, a linear measuring machine, and an optical projection comparator along with a set of AA gage blocks and a laser interferometer. In addition, some equipment for measuring surface finish or surface profiles will be important.

Factors in selecting inspection equipment. Greater quality and reliability in consumer and producer goods require greater precision and accuracy in the processes and better, faster inspection. The inspection equipment can be built into the process and is often computer-aided, thus opening the door for feedback sensory data from the process or its output to the computer control of the machine. In addition to this in-process inspection, raw materials need to be inspected along with finished goods. In general, there are six factors which should be considered when selecting equiment for an inspection job by measurement techniques

1. *The-rule-of-10.* The measurement device (or working gage) should be 10 times more precise than the tolerance to be measured. The rule actually

Technique	Typical Accuracy	Major Applications	Comments
Air	0.5 to 10 microinches or 2 to 3% of scale range	Gaging holes and shafts using a calibrated difference in air pressure or airflow, with magnifications of 20,000 to 40,000 to 1; also used for machine control, sorting, and classifying	High precision and flexibility; can measure out-of-round, taper, concentricity, camber, squareness, parallelism, and clearance between mating parts; noncontact principle good for delicate parts
Optical Light energy	0.5 to 2 microinches or better with laser interferometry 0.5 to 1 second of arc in autocollimation	Interferometry; checking flatness and size of gage blocks, find surface flaws, measure spherical shapes, flatness of surface plates, accuracy of rotary index tables; includes all light microscopes and optical comparators	Largest variety of measuring equipment; autocollimators are used for making precision angular measurements; lasers are used to make precision in-process measurements
Electron energy	100 Angstroms (Å)	Precision measurement in scanning electron microscopes of microelectronic circuits and other small precision parts	Part size restricted by vacuum chamber size; electron beam can be used for processing and part testing of electronic circuits
Electronic	0.5 to 10 microinches	Widely used for machine control, on-line inspection, sorting, and classification; OD's, ID's, height, surface, and geometrical relationships, profile tracing for roundness, surface roughness, contours, etc.; most devices are comparators with movement of stylus or spindle producing an electronic signal which is amplified electronically; commonly connected to microprocessors and minicomputers for process adjustment	Electronic gages come in many forms but usually have sensory head or detector combined with amplifier; capable of high magnification with resolution limited by size or geometry of sensory head; readouts commonly have multiple magnification steps; solid-state electronics make these devices small, portable, stable, and extremely flexible, with extremely fast response time
Mechanical	1 to 10 microinches	Large variety of external and internal measurements using gage blocks, dial indicators, micrometers, calipers, and the like; commonly used for bench comparators for gage calibration work	Moderate cost and ease of use make many of these devices the workhorses of the shop floor; highly dependent on workers' skill, and often subject to problems of wear at the contact points and friction in linkages

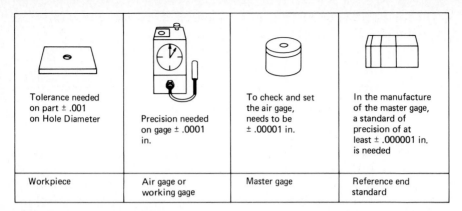

| Workpiece | Air gage or working gage | Master gage | Reference end standard |

Tolerance needed on part ± .001 on Hole Diameter | Precision needed on gage ± .0001 in. | To check and set the air gage, needs to be ± .00001 in. | In the manufacture of the master gage, a standard of precision of at least ± .000001 in. is needed

FIGURE 16-9 The Rule of 10 states that for reliable measurement each successive step in the inspection sequence should have 10 times the *precision* of the preceding step.

applies to all stages in the inspection sequence, as shown in Figure 16-9. The master gage should be precise to $\frac{1}{10}$ of that of the inspection device. The reference standard used to check the master gage should be 10 times more precise than the master gage. The application of this rule greatly reduces the probability of rejecting good parts or accepting bad components and performing additional work on them.

2. *Linearity*. This factor refers to the calibration accuracy of the device over its full working range. Is it linear? What is its degree of nonlinearity? Where does it become nonlinear and what is, therefore, its real linear working region?

3. *Repeat accuracy*. How repeatable is the device in taking the same reading over and over on a given standard?

4. *Stability*. How well does this device retain its calibration over a period of time? This is also called *drift*. As devices become more accurate, they often lose stability and become more sensitive to small changes in temperature and humidity.

5. *Magnification*. The amplification of the output portion of the device over the actual input dimension. The more accurate the device, the greater must be its magnification factor so that the required measurement can be read out (or observed) and compared to the desired standard. Magnification is often confused with resolution, but they are not the same thing.

6. *Resolution*. This is sometimes called sensitivity and refers to the smallest dimensional input that the device can detect or distinguish. The greater resolution of the device, the smaller will be the things it can resolve and the greater will be the magnification required to expand these measurements up to the point where they can be observed by the naked eye.

Some other factors of importance in selecting inspection devices include: the type of measurement information desired; the range or the span of sizes the device can handle versus the size and geometry of the workpieces; the environment; the cost of the device; and the cost to install and use the device. The latter factor depends on the speed of measurement, the degree to which it can be automated, and the functional life of the device in service.

MEASURING INSTRUMENTS

Because of the great importance of measuring in manufacturing, a variety of instruments are available that permit measurements to be made routinely, ranging in accuracy from 0.5 to 0.0003 mm and $\frac{1}{64}$ to 0.00001 inch. The ease of use, precision, and accuracy of making such measurements can be affected by (1) least count of the subdivisions on the instrument, (2) line matching, (3) parallax in reading the instrument, (4) elastic deformation of the instrument and workpiece, and (5) temperature effects. Some instruments are more subject to these factors than others. In addition, the skill of the person making the measurement, and in overcoming these factors, is very important, particularly in using some instruments. The inclusion of digital readout systems in measuring instruments, as will be discussed later, can lessen or eliminate the effect of some of these factors.

Linear measuring instruments. Linear measuring instruments are of two types: direct reading and indirect reading. *Direct-reading instruments* contain a line-graduated scale so that the size of the object being measured can be read directly on this scale. *Indirect-reading instruments* do not contain line graduations and are used to transfer the size of the dimension being measured to a direct-reading scale, thus obtaining the desired size information in an indirect manner.

The simplest, and most common, direct-reading linear measuring instrument is the machinist's rule, shown in Figure 16-10. Metric rules usually have two sets of line graduations on each side—½- and 1-mm divisions—whereas English

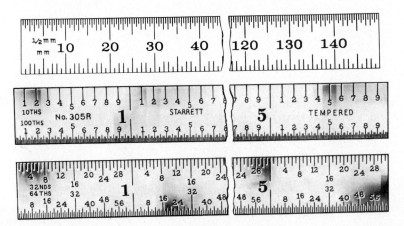

FIGURE 16-10 Machinist's rules. (*Top*) Metric. (*Center and bottom*) Inch graduations; 10ths and 100ths on one side, 32nd and 64ths on opposite side. (*Courtesy L. S. Starrett Company.*)

rules have four sets—$\frac{1}{16}$-, $\frac{1}{32}$-, $\frac{1}{64}$-, and $\frac{1}{100}$-inch divisions. Other combinations can be obtained in each type.

The machinist's rule is an end- or line-matching device; an end and a line, or two lines, must be aligned with the extremities of the object or distance being measured in order to obtain the desired reading. Thus the accuracy of the resulting reading is a function of the alignment and the magnitude of the smallest scale division. Such scales ordinarily are not used for accuracies greater than $\frac{1}{2}$ mm, $\frac{1}{64}$ inch, or 0.01 inch.

Several attachments can be added to a machinist's rule to extend its usefulness. Shown in Figure 16-11, the *square head* can be used as a miter or trisquare or to hold the rule in an upright position on a flat surface for making height measurements. It also contains a small bubble-type level so that it can be used by itself as a level. The *bevel protractor* permits the measurement or layout of angles. The *center head* permits the center of cylindrical work to be determined.

The *rule depth gage,* shown in Figure 16-12, consists of a special head that slides on a small rule and thus permits the depth of holes or shoulders below a given surface to be measured. The slide can be clamped at any desired position on the rule.

The *vernier caliper,* illustrated in Figure 16-13, is an end-measuring instrument, available in various sizes, that can be used to make both outside and inside measurements to theoretical accuracies of 0.01 mm or 0.001 inch. End-measuring instruments are more accurate and somewhat easier to use than line-matching types because their jaws are placed against either end of the object being measured, so that any difficulty in aligning edges or lines is avoided. However, the difficulty in obtaining uniform contact pressure, or "feel," between the legs of the instrument and the object being measured remains.

A major feature of the vernier caliper is the auxiliary scale, shown in Figure 16-13. The main English scale is divided into inches and tenths, with each tenth being subdivided into four divisions of 0.025 inch each. The 25 divisions on

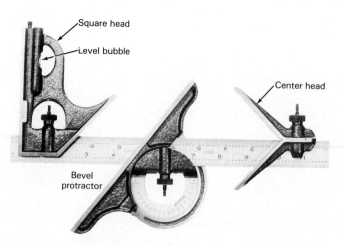

Square head

Level bubble

Center head

Bevel
protractor

FIGURE 16-11 Combination set. *(Courtesy MTI Corporation.)*

MACHINING PROCESSES

FIGURE 16-12 Rule depth gage being used to measure the depth of a groove from a surface. (*Courtesy Brown & Sharpe Mfg. Co.*)

Direct reading instrument = ruler

indirect reading instrument = stick; piece of string

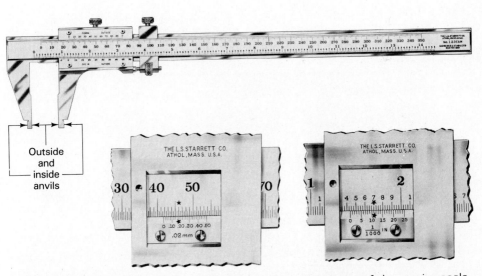

Outside and inside anvils

FIGURE 16-13 Vernier caliper. Insets show enlarged diagrams of the vernier scale. (*Left*) Metric; (*Right*) English. (*Courtesy L. S. Starrett Company.*)

the vernier plate are equal to the length of 24 divisions on the main scale. Thus each division on the vernier scale is ¹⁄₂₅ of (24 × 0.025 inch) = 0.024 inch. This is ¹⁄₁₀₀₀ inch less than each division on the main scale. Thus if the zero readings in each scale were in line, the first lines in each scale would be ¹⁄₁₀₀₀ inch apart, the second lines ²⁄₁₀₀₀ inch apart, and so on. The main English scale, as shown in the enlarged portion of Figure 16-13, is first read in inches and thousandths—1.425 inches. To this is added the reading of the vernier scale—the point at which a line on this scale coincides exactly with a line of the main scale—in this case 11, indicating that the zero line on the vernier scale and the next line to the left of it on the main scale are 0.011 inch apart. Thus the total reading is 1.425 inches plus 0.011 inch, or 1.436 inches. Similarly, the metric reading is 41.5 mm plus 0.18 mm, or 41.68 mm.

Vernier calipers are versatile but are rather slow to use. Consequently, they are not used as extensively as other instruments that have a more limited range but are easier to use.

The *vernier height gage,* shown in Figure 16-14, consists of a rule attached to a special base at one end and having a sliding head containing a vernier scale so that readings can be made to 0.01 mm or 0.001 inch. A beveled point, or other types of contact device, can be attached to the sliding head for making height measurements. It also is used in layout work to mark or transfer a desired height to a workpiece.

The *micrometer caliper,* more commonly called a *micrometer,* is one of the most widely used measuring devices. Until recently the type shown in Figure 16-15 was virtually standard. It consists of a fixed anvil and a movable spindle. By rotating the thimble on the end of the caliper, the spindle is moved away from the anvil by means of an accurate screw thread. On English types, this thread has a lead of 0.025 inch, and one revolution of the thimble moves the spindle this distance. The barrel, or sleeve, is calibrated in 0.025-inch divisions,

FIGURE 16-14 Measuring the height of the bottom of a cup above the base by means of a vernier height gage and special attachment. (*Courtesy L. S. Starrett Company.*)

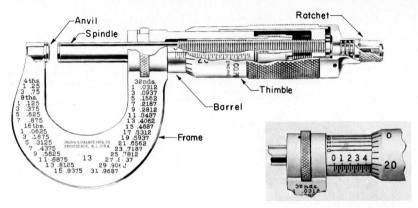

FIGURE 16-15 Cutaway view of a micrometer caliper, measuring in inches. Inset shows the method of reading the scales. (*Courtesy Brown & Sharpe Mfg. Co.*)

with each tenth of an inch being numbered. The circumference at the edge of the thimble is graduated into 25 divisions, each representing 0.001 inch.

A major difficulty of this type of micrometer is in making the reading of the dimension shown on the instrument. To read the instrument, the division on the thimble that coincides with the longitudinal line in the barrel is added to the largest reading exposed on the barrel. For example, on the English-type shown in Figure 16-15, the largest exposed number on the barrel is 4, representing 0.400 inch. Two additional lines exposed on the barrel represent 2 × 0.025 or 0.050 inch. The 20th line on the thimble nearly coincides with the longitudinal line on the barrel, representing 0.020 inch. Thus the total reading is: 0.400 inch + 0.050 inch + 0.020 inch = 0.470 inch. The micrometer shown also has vernier graduations on the barrel that enable it to be read to 0.0001 inch. Thus a more exact reading is 0.4697 inch. However, owing to the lack of pressure control, micrometers can seldom be relied on for accuracy beyond 0.0005 inch, and such vernier scales are not used extensively. On metric micrometers the graduations on the barrel and thimble usually are 0.5 mm and 0.01 mm, respectively (see Figure 16-16).

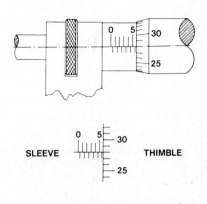

Reading 5.78 mm

FIGURE 16-16 Method of reading scales of a metric micrometer. Reading on sleeve 5.5 mm. Reading on thimble 0.28 mm. Total reading 5.78 mm. (*Courtesy L. S. Starrett Company.*)

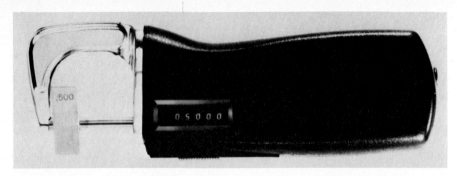

FIGURE 16-17 Electronic micrometer caliper, with LED digital display, which can read in either English or metric units. (*Courtesy Quality Measurement Systems, Inc.*)

Many errors have resulted from the ordinary micrometer being misread, the error being ±0.025 inch or ±0.5 mm. Consequently, several direct-reading types have been developed. Figure 16-17 shows an electronic type that provides constant spindle force, metric or English units at the flick of a switch, and an LED-crystal digital display of the measurement.

Although micrometer calipers are quite easy to use and can be read quickly, the range of each instrument is limited to about 1 inch. Thus a number of micrometers of various sizes are required to cover a wide range of dimensions. Being an end-measuring device, it has the advantages of this type of instrument. Its greatest limitation to accuracy is control of the pressure between the anvil, spindle, and the piece being measured. Most micrometers are equipped with a ratchet or friction device, as shown in Figure 16-15, by means of which the thimble can be turned. This controls the pressure to a fair degree if used carefully. Calipers that do not have this device may be sprung several thousandths by applying excess torque to the thimble. Micrometer calipers usually should not be relied on for measurements of greater accuracy than 0.01 mm or 0.001 inch, except the electronic type shown in Figure 16-17.

Micrometer calipers are available with a variety of specially shaped anvils and/or spindles, such as points, balls, and disks, for measuring special shapes.

Two types of micrometers are available for inside measurements. The caliper type is shown in Figure 16-18 and the plain-rod type in Figure 16-19.

The micrometer principle also is incorporated in the *micrometer depth gage,* shown in Figure 16-20.

The *supermicrometer* is typical of a number of instruments that are designed to make linear measurements directly to 0.0001 inch by incorporating special features that overcome several of the factors mentioned at the beginning of this section. It is made with very heavy members to avoid any possibility of distortion. Although it is an end-measuring instrument, the pressure between the anvil and spindle and the workpieces is maintained constant by a pressure spring which pushes against the anvil. The spindle is tightened against the workpiece until an indicator on the tailstock shows that a selected pressure has been obtained. The spindle screw is very accurate and is actuated by a large dial, that, in conjunction with a vernier, reads directly to 0.0001 inch. An electronic meter permits the instrument to be used as a comparator and to measure to 0.00002

MACHINING PROCESSES

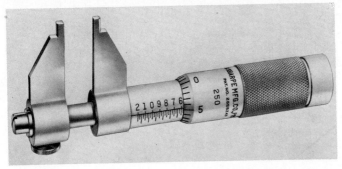

FIGURE 16-18 Caliper-type inside micrometer. (*Courtesy Brown & Sharpe Mfg. Co.*)

inch. Such instruments are used both for inspection and toolroom checking and for production inspection work. Some are available with the anvil connected to an electronic pressure-indicating device for even more precise control of the measuring pressure (called standard measuring machines), they are capable of measuring directly to 0.00001 inch when used in a controlled-temperature room.

The toolmakers' microscope, shown in Figure 16-21, is a versatile instrument that measures by optical means with no pressure being involved. It thus is very useful for making accurate measurements on small or delicate parts. The base, on which the special microscope is mounted, has a table that can be moved in two mutually perpendicular, horizontal directions by means of accurate micrometer screws that can be read to 0.0001 inch, or, if so equipped, by means of the digital readout. Parts to be measured are mounted on the table, the microscope is focused, and one end of the desired part brought into coincidence with the crossline in the microscope. The reading is then noted and the table moved until the other extremity of the part is in coincidence with the crossline. From the final reading, the desired measurement can be determined. In addition to a wide variety of linear measurements, accurate angular measurements can also be made by means of a special protractor eyepiece that can be put into the microscope.

FIGURE 16-19 Inside micrometer being used to measure an inside dimension. This type uses interchangeable rods. (*Courtesy Brown & Sharpe Mfg. Co.*)

FIGURE 16-20 Measuring the depth of a shoulder using a micrometer depth gage. (*Courtesy Brown & Sharpe Mfg. Co.*)

The *optical projector* or *comparator,* shown in Figure 16-22, is a larger optical device on which both linear and angular measurements can be made. As with the toolmakers' microscope, the part to be measured is mounted on a table that can be moved in two directions by accurate micrometer screws. By means of an optical system the image of the part is projected on a screen, magnified from 5 to more than 100 times. Measurements can be made directly, either by means of the micrometer dials or on the magnified image on the screen by means of an accurate rule. A very common use for this type of instrument is the checking of parts, such as dies, screws, and so forth, against a template that is drawn to an enlarged scale, placed on the screen, and compared with the projected image of the part.

Accurate measurement of distances greater than a few inches was very difficult until the development of laser interferometry. Instruments based on this principle permit accuracies of $\pm 0.000\,25$ mm ($\pm .00001$ inch) to be achieved routinely,

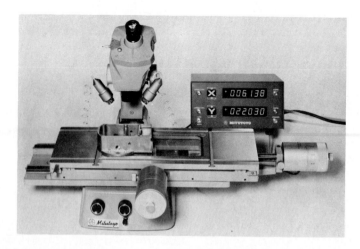

FIGURE 16-21 Toolmaker's microscope having digital readout system for *X* and *Y* table movements. (*Courtesy MTI Corporation.*)

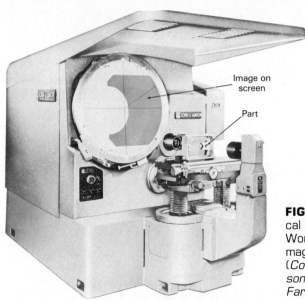

Image on
screen

Part

FIGURE 16-22 Optical contour comparator. Workpiece on table is magnified 20 × on screen. (*Courtesy Jones & Lamson Division of Waterbury Farrel.*)

and substantially greater accuracy is possible. Such equipment is particularly useful in checking the movement of machine tool tables, aligning and checking large assembly jigs, and making measurements of intricate machined parts, such as tire-tread molds. Figure 16-23 shows one type which employs the Doppler effect of a split laser beam, with the reflecting target attached to one end of the distance to be measured. Another type utilizes the principle depicted in Figure 16-24, wherein the laser beam acts as a probe and is reflected by the surface that is a variable distance from the beam source and the electro-optical image-reading system. An accurate measuring screw moves the source-reading head so as to center the reflected image and thus measure the variation in distance. Both types of laser systems incorporate a digital readout.

A number of the commonly used measuring instruments are of the indirect or transfer-measuring type. The most common are inside and outside calipers and dividers. These are shown in Figure 16-25. To obtain a measurement with these instruments they first are adjusted so that their legs just contact the desired portions of the object to be measured. Consequently, a considerable amount of care and experience is required to get the proper "feel" of the caliper legs against the work—a matter of contact pressure. The calipers or dividers are then held against a rule and the distance between the points of the instrument is read from the scale. Although such instruments are versatile, their accuracy of measurement is limited.

Telescoping gages are used for making indirect, internal measurements. *Plain telescoping gages,* such as shown in Figure 16-26, are the most common type. These consist of a knurled handle and a head composed of three sections. The two outer sections telescope inside each other against spring pressure and can be locked in any position by means of a knurled screw on the end of the handle. Their outer ends are rounded to a radius that is less than half the minimum length to which the head will telescope. To use such a gage, the plungers are

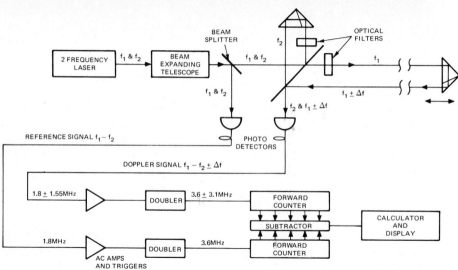

FIGURE 16-23 (*Top*) Calibrating the table movement of a machine tool by means of a laser interferometer. (*Bottom*) Schematic diagram of the components of a laser interferometer. (*Courtesy Hewlett-Packard*).

pushed inward until the telescoping head will fit within the internal dimension that is to be measured. The plungers are then locked and the gage withdrawn. A micrometer caliper commonly is used to measure the distance between the extremities of the gage head.

A *small-hole gage*, shown in Figure 16-27, is a special type of telescoping gage for measuring the ID of holes too small for a standard telescoping gage.

Angle-measuring instruments. Accurate angle measurements usually are more difficult to make than linear measurements. Angles are measured in *degrees—*

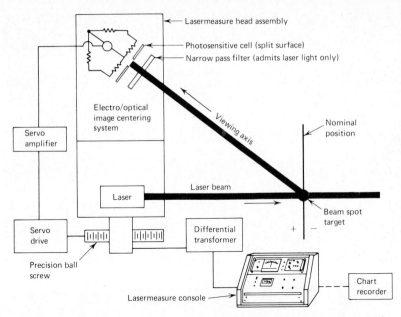

FIGURE 16-24 Arrangement for direct linear measurement by means of a laser beam.

$^{1}/_{360}$ *part of a circle—and decimal subdivisions thereof (or in minutes and seconds of arc).*[4]

The *bevel protractor,* illustrated in Figure 16-28, is the most general angle-measuring instrument. The two movable blades are brought into contact with the sides of the angular part, and the angle can be read on the vernier scale to five minutes of arc. A clamping device is provided to lock the blades in any desired position so that the instrument can be used for both direct measurement and layout work.

As indicated previously, the angle attachment on the combination set also can be used to measure angles in a manner similar to the bevel protractor, but usually with somewhat less accuracy.

The toolmakers' microscope is very satisfactory for making angle measurements, but its use is restricted to small parts. The accuracy obtainable is 5 minutes of arc. Similarly, angles can be measured on the contour projector.

When very accurate angle measurements are required, a *sine bar* may be employed if the physical conditions will permit. This device, as illustrated in Figure 16-29, consists of an accurately ground bar on which two accurately ground pins, of the same diameter, are mounted an exact distance apart. The distances used are usually either 5 or 10 inches, and the resulting instrument is called a 5- or 10-inch sine bar.[5] Measurements are made by using the principle

---

[4]The SI system calls for measurement of plane angles in radians, but degrees are permissible and, for several reasons, it appears likely that the use of degrees will continue to predominate in manufacturing, but with minutes and seconds of arc possibly being replaced by decimal portions of a degree.

[5]Sine bars also are available with millimeter dimensions.

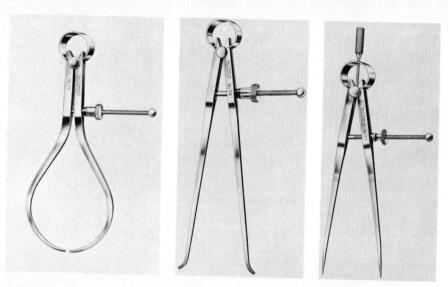

FIGURE 16-25 *(Left to right)* Outside calipers, inside calipers, and dividers.

that the sine of a given angle is the ratio of the opposite side of the right triangle to the hypotenuse.

The object being measured is attached to the sine bar and the inclination of the assembly raised until the top surface is exactly parallel with the surface plate. If a stack of gage blocks is used to elevate one end of the sine bar, as shown, the height of the stack directly determines the difference in height of the two pins. The difference in height of the pins also can be determined by a dial gage or some other type of gage. The difference in elevation is then equal to either five or ten times the sine of the angle being measured, depending on whether a 5- or 10-inch bar is being used. Tabulated values of the angles corresponding to any measured elevation difference for 5- or 10-inch sine bars are

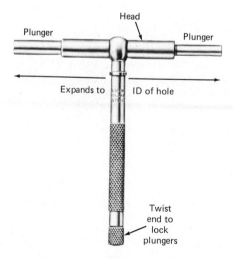

FIGURE 16-26 Telescoping gage used for making internal measurements. (*Courtesy L. S. Starrett Company.*)

FIGURE 16-27 Small-hole gage. (*Courtesy L. S. Starrett Company.*)

available in various handbooks. Several types of sine bars are available to suit various requirements.

Accurate measurements of angle to 1 second of arc can be made by means of *angle gage blocks*. These come in sets of 16 blocks that can be assembled in desired combinations. Angle measurements also can be made to $\pm 0.001°$ on rotary indexing tables having suitable numerical control.[6]

GAGES

In manufacturing, particularly in mass production, it may not be necessary to know the exact dimensions of a part, only that it is within previously established limits. Limits can be determined more easily than specific dimensions by the

[6]See Chapter 38.

FIGURE 16-28 Measuring an angle with a bevel protractor. (*Courtesy Brown & Sharpe Mfg. Co.*)

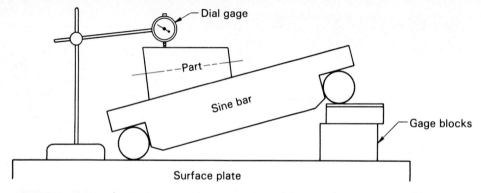

FIGURE 16-29 Setup to measure an angle on a part using a sine bar. The dial indicator is used to determine when the part surface is parallel to the plate.

use of *attribute-type* instruments called gages, either manually or mechanically. They may be either *fixed type* or *deviation type* and are used for both linear and angular dimensions.

Fixed-type gages.   *Fixed-type gages* are designed to gage only one dimension and indicate whether it is larger or smaller than a previously established standard. They do not determine how much larger or smaller the measured dimension is than the standard. Because such gages fulfill a simple and limited function, they are relatively inexpensive and are easy to use.

Gages of this type ordinarily are made of hardened steel of proper composition and heat treated to produce dimensional stability. Hardness is essential so that wear will be minimized and accuracy maintained. Because steels of high hardness tend to be dimensionally unstable, some fixed gages are made of softer steel with a hard chrome plating on the surface to provide surface hardness. Chrome plating also can be used for reclaiming some worn gages. Where gages are to be subjected to great use, they may be made of tungsten carbide at the wear points.

Plug gages are one of the most common types of fixed gages. As shown in Figure 16-30, these are accurately ground cylinders, held in a handle, that are used to gage internal dimensions such as holes. The gaging element of a *plain* plug gage has a single diameter. To control the minimum and maximum limits of a given hole, two plug gages are required. The smaller, or "go," gage controls the minimum because it must go (slide) into any hole that is large enough to meet the required minimum. The larger, "not-go," gage controls the maximum dimension because it must not go into any hole that is not larger than the maximum permissible size. As shown in Figure 16-30, the go and not-go plugs often are fastened into the two ends of a single handle for convenience in use. The not-go plug usually is much shorter than the go plug; it is subjected to little wear because it seldom slides into any holes. *Step-type* go, not-go gages, also shown in Figure 16-30, have the go and not-go diameters on a single plug, the go portion being the outer end. In gaging an acceptable part, the go portion should slide into the hole but the not-go section should not enter. In using a

FIGURE 16-30 (*Top*) Plain plug gage having go member on one end and not-go member on the other. (*Bottom*) Plug gage with stepped go and not-go member. (*Courtesy Bendix Corporation, Automation and Measurement Division.*)

plug or any other type of fixed gage, the gage should never be forced into, or onto, the part being measured.

Plug-type gages also are made for gaging shapes other than cylindrical holes. Three common types are *taper plug gages, thread plug gages,* and *spline gages.* Taper plug gages gage both the angle of the taper and its size. Any deviation from the correct angle is indicated by looseness between the plug and the tapered hole. Size is indicated by the depth to which the plug fits into the hole—the correct depth being denoted by a mark on the plug. Thread plug gages come in go and not-go types. The go gage must screw into the threaded holes while the not-go gage must not enter.

Ring gages are used to gage shafts or other external round members. These also are made in go and not-go types as shown in Figure 16-31. Go ring gages have plain knurled exteriors, whereas not-go ring gages have a circumferential groove in the knurling so that they can easily be distinguished. *Ring thread gages* are made to be slightly adjustable because it is almost impossible to make them exactly to the desired size. Thus they are adjusted to exact, final size after the final grinding and polishing have been completed.

Snap gages are the most common type of fixed gage for measuring external dimensions. As shown in Figure 16-32, they have a rigid, U-shaped frame on which are two or three gaging surfaces, usually made of hardened steel or tungsten carbide. In the adjustable type shown, one gaging surface is fixed while the other(s) may be adjusted over a small range and locked at the desired po-

FIGURE 16-31 Go and not-go ring gages, (*Courtesy Bendix Corporation, Automation and Measurement Division.*)

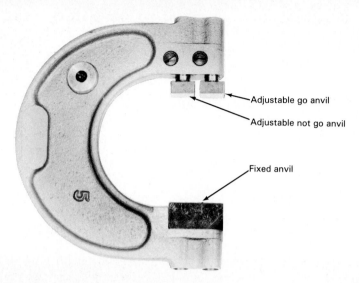

FIGURE 16-32 Adjustable go and not-go snap gage. (*Courtesy Bendix Corporation, Automation and Measurement Division.*)

sition(s). Because, in most cases, one wishes to control both the maximum and minimum dimensions, the *progressive* or *step-type* snap gage, shown in Figure 16-32, is used most frequently. These have one fixed anvil and two adjustable surfaces to form the outer go and inner not-go openings, thus eliminating the use of separate go and not-go gages.

Snap gages are available in several types and a wide range of sizes. The gaging surfaces may be round or rectangular. They are set to the desired dimensions with the aid of gage blocks.

Many types of special gages are avaialble or can be constructed for special applications. The *flush-pin gage,* illustrated in Figure 16-33, is an example for gaging the depth of a shoulder. The main section is placed on the higher of the two surfaces with the movable step pin resting on the lower surface. If the depth between the two surfaces is sufficient but not too great, the top of the pin, but not the lower step, will be slightly above the top surface of the gage body. If the depth is too great, the top of the pin will be below the surface. Similarly, if the depth is not great enough, the lower step on the top of the pin will be above the surface of the gage body. By running a finger, or fingernail, across the top of the pin, its position with respect to the surface of the gage body can readily be determined.

Several types of *form gages* are available for use in checking the *profile* of

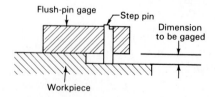

FIGURE 16-33 Flush-pin gage being used to check height of step.

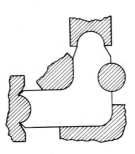

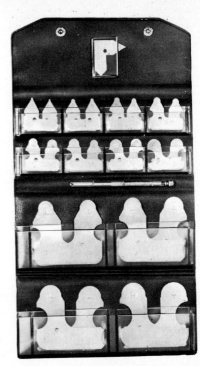

FIGURE 16-34 Set of radius gages, showing how they are used. (*Courtesy MTI Corporation.*)

various objects. Two of the most common types are *radius gages,* shown in Figure 16-34, and *screw-thread pitch gages,* shown in Figure 16-35.

Deviation-type gages. A large amount of gaging, and some measurement, is done through the use of *deviation-type gages,* which determine the amount by which a measured part deviates, plus or minus, from a standard dimension to which the instrument has been set. In most cases the deviation is indicated directly in units of measurement, but in some cases, notably in production inspection, the gage shows only whether the deviation is within a permissible range. Such gages employ mechanical, electrical, or fluidic amplification techniques so that very small linear deviations can be detected. Most are quite rugged, and they are available in a variety of amplifications and sizes. As a consequence, it is quite easy to obtain a suitable gaging device of this type for use in almost any kind of gaging or measuring situation.

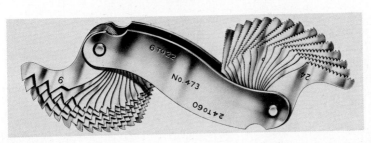

FIGURE 16-35 Thread pitch gages. (*Courtesy L. S. Starrett Company.*)

Dial indicators, such as shown in Figure 16-4, are a simple and widely used form of deviation-type gage. Movement of the gaging spindle is amplified mechanically through a rack and pinion and a gear train and indicated by a pointer on a graduated dial. Most dial indicators have a spindle travel equal to about 2½ revolutions of the indicating pointer and read in either 0.02- or 0.002-mm or 0.001- or 0.0001-inch units. The dial can be rotated by means of the knurled bezel ring to align the zero point with any position of the pointer. The indicator often is mounted on an adjustable arm to permit its being brought into proper relationship with the work. It is important that the axis of the spindle be aligned exactly with the dimension being gaged if accuracy is to be achieved.

Dial indicators should be checked occasionally to assure that accuracy has not been lost through wear in the gear train. Also, it should be remembered that the pressure of the spindle on the work varies due to spring pressure as the spindle moves into the gage. This normally will cause no difficulty unless the spindles are used on soft or flexible parts.

Three types of *deviation comparators* are widely used for routine gaging and inspection. All are set to zero by means of standard gages or gage blocks and indicate the plus or minus deviation of the part being gaged. The mechanical type, shown in Figure 16-36, is quite rugged and readily portable. Depending on the magnification, each division on the dial may be from 0.002 to 0.0002 mm (0.001 to 0.0001 inch), with a total range of about ±0.25 mm (±0.010 inch).

In comparators employing electronic magnification, as shown in Figure 16-37, the gaging head is small, readily portable, and can be mounted in many ways. The end of the sensing lever is shaped so as to automatically compensate for misalignment in the measuring plane up to about ±15°. The indicator may employ either a pointer and graduated scale or a digital readout. Accuracies up to ±0.0003 mm (0.00001 inch) are available, and several ranges usually can be selected by merely turning a knob.

Linear variable differential transformers (LVDT) are used as sensory elements in many electronic gages, usually with a solid-state diode display or in automatic

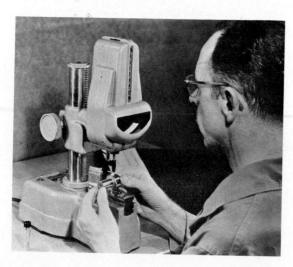

FIGURE 16-36 Visual comparator.

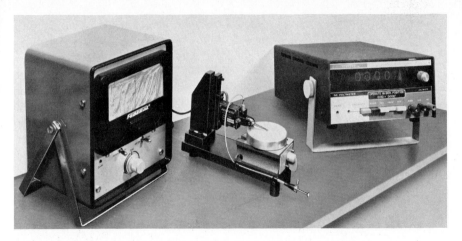

FIGURE 16-37 Electronic-magnification gage being used to gage a computer memory core 0.76 mm (0.030 inch) in diameter and 0.15 mm (0.006 inch) thick. (*Courtesy Federal Products Corporation.*)

inspection setups (see Figure 16-43). These devices can frequently be combined into multiple units for simultaneous gaging of several dimensions. Ranges and resolutions down to 0.013 and 0.000 25 mm (0.0005 and 0.00001 inch), respectively, are available.

Air gages have special characteristics that make them especially suitable for gaging holes or internal dimensions of various shapes. A typical gage of this type is shown in Figure 16-38. These gages indicate the clearance between the

FIGURE 16-38 Gaging the diameter of a hole with an air gage. (*Courtesy Bendix Corporation, Automation and Measurement Division.*)

gaging head and the hole by measuring either the volume of air that escapes or the pressure drop resulting from the air flow. The gage is calibrated directly in 0.02-mm or 0.0001-inch divisions. Air gages have an advantage over mechanical or electronic gages for this purpose in that they not only detect linear size deviations but also out-of-round conditions. Also they are subject to very little wear because the gaging member always is slightly smaller than the hole and the air flow minimizes rubbing. Special types of air gages can be used for external gaging.

Measurement by lightwave interference. The phenomenon of lightwave interference can be used for making very precise measurements, such as in calibrating gage blocks. Three pieces of equipment are required. The first is an *optical flat*. These are quartz or special glass disks, from 50 to 250 mm (2 to 10 inches) in diameter and about 12 to 25 mm (½ to 1 inch) thick, whose surfaces are very nearly true planes and nearly parallel. These flats can be obtained with the surfaces within 0.000 03 mm (0.000001 inch) of true flatness. It is not essential that both surfaces be accurate or that they be exactly parallel, but one must be certain that only the accurate surface is used in making measurements.

The second item of equipment required is a *toolmaker's flat*. These are similar to optical flats but are made of steel and usually have only one surface that is accurate.

The third requirement is a *monochromatic light source*, emitting light of a single wavelength. Selenium, helium, or cadmium sources are commonly used.

Lightwave interference bands are created by the phenomenon illustrated in Figure 16-39. A portion of the light rays coming from the source are reflected from the bottom surface of the optical flat while the remaining portions pass through the flat and are reflected from the surface being tested. If the optical flat and the work surface are separated by a very small angle, at certain intervals along the surface, the distance between the flat and the work surface will be such that the extra distance the light reflected from the work surface must travel will cause it to be 180° out of phase with the portion reflected from the lower surface of the optical flat. As a result, the two portions cancel each other and cause dark lines or interference bands. At *A* in Figure 16-39, the light ray striking at this point is thrown out of phase 180° upon being reflected from the work surface, and it also travels a distance of ½ wavelength in going each direction

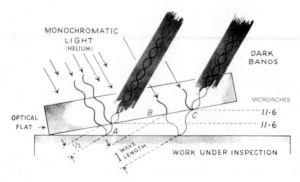

FIGURE 16-39 Explanation of the method of accurate measurement by light-wave interference phenomenon. (*Courtesy DoALL Company.*)

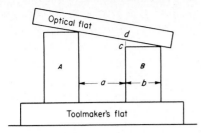

FIGURE 16-40 Method of calibrating gage blocks by light-wave interference.

across the gap. It thus is 180° out of phase with the portion of the ray that is reflected from the lower surface of the optical flat. This produces a dark band at *A*. At *B* the distance between the flat and the work surface is ¾ wavelength, so the 1½ wavelengths of extra distance, plus the 180° phase change upon reflection, puts two reflected rays in phase and produces a bright band. Again at *C* a dark band is produced. Because the difference in the distances between the optical flat and the work surface at *A* and *C* is ½ wavelength, each dark band indicates a change of ½ wavelength in the elevation of the work surface. If a monochromatic light source having a wavelength of 23.2 microinches (0.589 μm) is used, each interference band represents 11.6 microinches (0.295 μm).

Figure 16-40 illustrates how lightwave interference is used for calibrating gage blocks. The block to be calibrated is placed at *B* and a calibrated block at *A*. Distances *a* and *b* must be known but do not have to be measured with great accuracy. By counting the number of interference bands showing on the surface of gage block *B*, the distance *c–d* can be determined and then, by simple geometry, the difference in the heights of the two blocks can be computed. The same method is applicable for making precise measurements of other objects by comparing them with a known gage block.

Light interference also makes it possible to determine easily whether a surface is exactly flat.

SURFACE ROUGHNESS MEASUREMENT

The machining processes to be discussed in later chapters in this part of the book generate a wide variety of surface patterns. **Lay** is the term used to designate the direction of the predominate surface pattern produced by the machining process. In addition, certain other terms and symbols have been developed and standardized for specifying the surface quality. The most important terms are surface roughness, waviness, and lay, which are illustrated in Figure 16-41. *Roughness* refers to the finely spaced surface irregularities. It results from machining operations in the case of machined surfaces. *Waviness* is surface irregularity of greater spacing than occurs in roughness. It may be the result of warping, vibration, or the work deflecting during machining.

Roughness is measured by the heights of the irregularities with respect to an average line, as illustrated in Figure 16-42. These measurements usually are expressed in micrometers or microinches. In most cases the arithmetical average

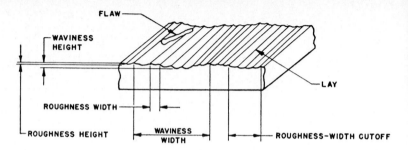

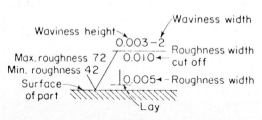

Waviness width

Waviness height

0.003 – 2

Max. roughness 72 — 0.010 ◄--- Roughness width cut off

Min. roughness 42

Surface — 0.005 ◄-- Roughness width
of part

Lay

Lay symbols

= Parallel to the boundary line of the nominal suface

⊥ Perpendicular to the boundary line of the nominal surface

X Angular in both directions to the boundary line of the nominal surface

M Multidirectional

C Approximately circular relative to the center

R Approximately radial relative to the center of the nominal surface

FIGURE 16-41 (*Top*) Terminology used in specifying and measuring surface quality. (*Left*) Symbols used on drawings by designers with definitions of lay symbols. Quite often, only the desired surface roughness will be specified.

(AA) is used. In terms of the measurements indicated in Figure 16-42, this would be as follows:

$$\frac{y_1 + y_2 + y_3 + y_4 + y_5 + y_6 + y_7 + \ldots + y_n}{n}$$

where y_n is a vertical distance from the center line and n is the total number of vertical measurements taken within a specified cutoff distance. Occasionally the *root-mean-square* (rms) value is used. This is defined as

$$\text{rms} = \sqrt{\frac{\Sigma\, y^2}{n}}$$

A variety of instruments are available for measuring surface roughness and surface profiles. The majority of these devices employ a diamond stylus which is moved at a constant rate across the surface, perpendicular to the lay pattern.

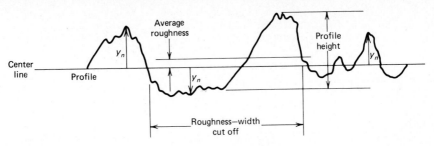

FIGURE 16-42 Schematic of surface profile as produced by a stylus device showing some typical y_n values with respect to the center line.

The rise and fall of the stylus is detected electronically (often using a LVDT device), amplified and recorded on a strip chart, or processed electronically to produce AA or rms readings for a meter (see Figure 16-43). The unit containing the stylus and driving motor may be hand-held or supported on the workpiece or other supporting surface.

The instrument shown in Figure 16-44 is capable of making a series of parallel, offset traces on the surface, thereby providing a two-dimensional profile map as shown in the bottom of the figure. Areas of from 0.13×0.13 mm (0.005×0.005 inch) up to 50.8×50.8 mm (2×2 inches), depending on the magnification selected, can be profiled.

The resolution of these devices is determined by the radius of diameter of the tip of the stylus. When the magnitude of the geometric features begins to approach the magnitude of the tip of the stylus, great caution should be used in interpreting the output from these devices. As a case in point, Figure 16-45 shows a scanning electron micrograph of a face-milled surface, upon which has been superimposed (photographically) a scanning electron micrograph of the tip of a diamond stylus (tip radius of 0.0005 inch). Both micrographs have the same final magnification. Surface flaws of the same general size as the roughness created by the machining process are difficult to resolve with the stylus-type device where both these features are about the same size as the stylus tip.

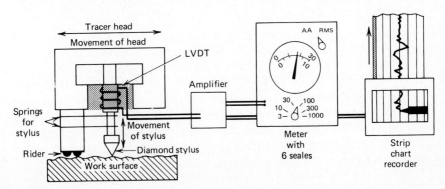

FIGURE 16-43 Schematic of stylus profile device for measuring surface roughness and surface profile with two readout devices shown: a meter for AA or rms values and a strip chart recorder for surface profile.

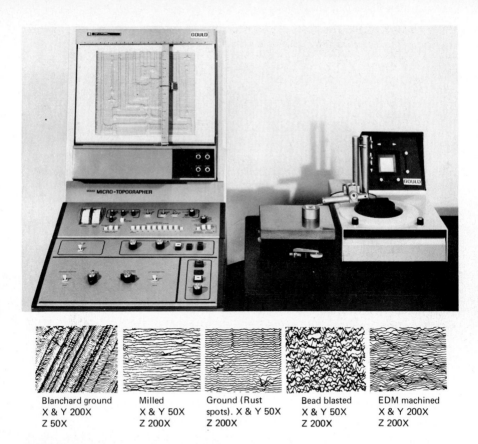

| Blanchard ground X & Y 200X Z 50X | Milled X & Y 50X Z 200X | Ground (Rust spots). X & Y 50X Z 200X | Bead blasted X & Y 50X Z 200X | EDM machined X & Y 200X Z 200X |

FIGURE 16-44 (*Top*) Micro-topographer, used to measure and depict surface roughness and character. (*Bottom*) Some typical surface-roughness profiles. (*Courtesy Gould Inc., Measurement Systems Division.*)

Another problem with these devices is that they produce a reading (a line on the chart) where the stylus tip is not touching the surface. This is demonstrated in Figure 16-46, which shows the "S" from the word TRUST from a U.S. dime. The SEM micrograph was made after the topographical map of Figure 16-46(b) had been made. Both figures are at about the same magnification.

The tracks produced by the stylus tip are easily seen in the micrograph. Notice the difference between the features shown in the micrograph and the trace, indicating that the stylus tip was not in contact with the surface many times during its passage over the surface (left no track in the surface) and yet the trace itself is continuous. The fact that the stylus actually leaves shallow grooves in the surface during its passage is well documented in the literature.

Surface roughnesses that typically are produced by various manufacturing processes are indicated in Figure 16-47. For the designer's use, sets of *surface-finish blocks,* such as are shown in Figure 16-48, are very useful. It is difficult to visualize what a surface having a given microinch roughness looks like, inasmuch as the same value of roughness may reflect different surface characteristics when produced by different processes.

FIGURE 16-45 Typical machined steel surface as created by face milling and examined in the SEM. A micrograph (same magnification) of a 0.0005-inch stylus tip has been superimposed on the upper right.

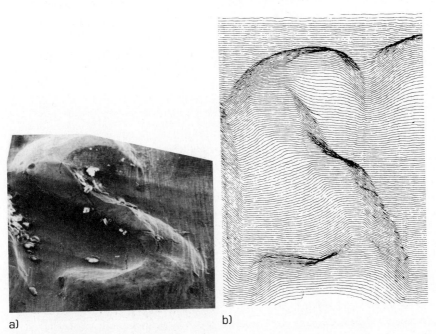

a) b)

FIGURE 16-46 (a) SEM micrograph of a U.S. dime, showing the "S" in the word TRUST after the region has been traced by a stylus-type machine. (b) Topographical map of the "S" region of the word TRUST from a U.S. dime. Compare to Figure 14-46a.

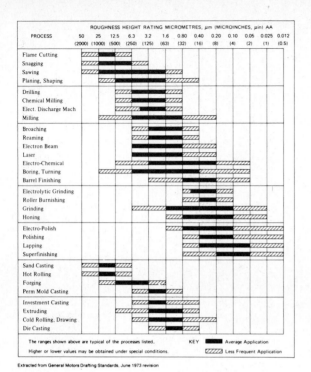

PROCESS	ROUGHNESS HEIGHT RATING MICROMETRES, μm (MICROINCHES, μin) AA												
	50 (2000)	25 (1000)	12.5 (500)	6.3 (250)	3.2 (125)	1.6 (63)	0.80 (32)	0.40 (16)	0.20 (8)	0.10 (4)	0.05 (2)	0.025 (1)	0.012 (0.5)
Flame Cutting													
Snagging													
Sawing													
Planing, Shaping													
Drilling													
Chemical Milling													
Elect. Discharge Mach													
Milling													
Broaching													
Reaming													
Electron Beam													
Laser													
Electro-Chemical													
Boring, Turning													
Barrel Finishing													
Electrolytic Grinding													
Roller Burnishing													
Grinding													
Honing													
Electro-Polish													
Polishing													
Lapping													
Superfinishing													
Sand Casting													
Hot Rolling													
Forging													
Perm Mold Casting													
Investment Casting													
Extruding													
Cold Rolling, Drawing													
Die Casting													

The ranges shown above are typical of the processes listed.

Higher or lower values may be obtained under special conditions.

KEY ■ Average Application ▨ Less Frequent Application

Extracted from General Motors Drafting Standards, June 1973 revision

FIGURE 16-47 Comparison of surface roughness produced by common production processes. (*Courtesy* American Machinist.)

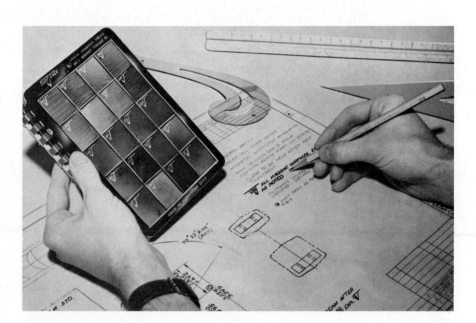

FIGURE 16-48 Set of surface-roughness standards being used in a drafting room. (*Courtesy Surface Checking Gage Co.*)

NONDESTRUCTIVE TESTING AND INSPECTION

Nondestructive testing (NDT) implies that some inspection of the item will be accomplished without destroying the item, so this class of inspection is generally concerned with inspection techniques that can find surface or internal defects without destroying the item. For example, arc welds in a nuclear reactor receive 100% NDT inspection for defects because the cost of a failure here can be great.

While there are many NDT techniques, the primary ones used are the following:

Liquid or dye penetrant testing is used to find discontinuities that are open to the surface of solid and essentially nonporous materials. The procedure is to apply a penetrant liquid to the surface to be inspected, allow time for the penetrant to flow into cracks and gouges by capillary action, and then observe the surface after wiping it clean of the penetrant material. Surface flaws are easily visible. Fluorescent dyes are available when more sensitivity is required. This is one of the most widely used techniques due to its simplicity and low cost.

Magnetic particle testing is a method for locating surface and subsurface discontinuities in ferromagnetic materials. Leakage currents will occur at the discontinuities and surface flaws when the specimen is magnetized. Finely divided magnetic particles, when placed on the test specimen, will then collect at these leakage sites and give a flaw indication. This method is widely used for iron and steel products and complements liquid penetrant inspection for ferromagnetic materials.

Radiographic inspection uses X rays or gamma rays to bombard the part. Differences in part density, thickness, or composition will result in different proportions of the radiation being absorbed. Unabsorbed radiation is recorded on film (or a fluorescent screen; termed fluoroscopy) with the resulting shadow picture having differences in intensity or photographic density. The radiographs of the test parts are compared to standard or perfect parts to detect defects, as interpreted by contrast differences in the film. The process is used extensively on castings and weldments to detect internal flaws in critical units, like high-pressure steam turbines or boiler assemblies. This technique is subject to a host of variables and can be hazardous to the health, so it is best left to experts skilled in its practice and the interpretation of the images.

Neutron radiographic testing uses a neutron source (such as a nuclear reactor or accelerator) to form a radiographic image of the test item. The geometric principles of shadow formation and variation of attenuation with thickness for certain elements gives neutron radiography a high sensitivity for certain elements. Due to the high cost of neutron sources, this method is usually reserved for specialized applications, so not every NDT laboratory will have this capability.

Ultrasonic testing uses beams of high-frequency sound waves to probe the material being inspected. Internal flaws or discontinuities and their location can be determined from analyzing reflected sound waves. This technique is also becoming more popular as the instrumentation improves. It complements radiographic techniques for determining internal flaws.

TABLE 16-3 Summary of Industrial NDT Methods

Method	Representative Applications	Types of Flaws Detected	Comments
1. Acoustic emission	Boilers, building structures, aircraft structures	Growing cracks	Moderate to high cost; used primarily on systems already in the field
2. Eddy current testing	Wire, tubing, bolt and fastener holes, sheet metal, alloy sorting and thickness measurement	Cracks, seams, thickness of plates, tubes, and coatings and variation in alloy composition	Moderate cost; portable, shallow depth penetration; becoming more automated
3. Leak testing	Vacuum systems, gas and liquid storage vessels, piping	Cracks, faulty welds, and other leaks in closed systems	Good sensitivity, various types of instrumentations available, low to moderate cost
4. Liquid or dye penetrant testing	Castings, forgings, weldments	Cracks, gouges, porosity, laps, and seams	Low cost; easy to apply, only works on accessible surfaces
5. Magnetic particle testing	Castings, forgings, and extrusions	Cracks, seams, laps, voids, porosity, and inclusions	Low cost, senses shallow subsurface flaws as well as surface flaws, material must be ferromagnetic
6. Neutron radiography	Inspection of propellant or explosive charge inside ammunition or pyrotechnic devices	Presence, absence, or mislocation of internal components of suitable composition	Expensive; good penetration, high sensitivity to favorable materials
7. Optical holography	Composite structures, tires, turbine blades	Disbonds, delaminations, plastic deformation, subsurface voids	High cost; sensitive, complex test conditions required, interpretation often requires highly skilled personnel
8. Thermal imaging	Laminated structures, composites, electronic circuits	Voids on disbonds in nonmetallics, hot or cold spots in thermally active assemblies	Moderate to high cost; computerized systems give definitive map of surface, often indicating subsurface flaws
9. Ultrasonic testing	Castings, forgings, extrusions, and thickness measurement	Cracks, voids, porosity, delaminations, and thinning of walls	Moderate to high cost; requires good surface preparation; good sensitivity
10. X-ray and gamma radiography	Castings, forgings, weldments, and assemblies	Voids, porosity, inclusions, and cracks	High cost; detects internal flaws on a wide range of materials

Eddy current testing originates from the interaction of a high-frequency electromagnetic field with a metallic specimen. Any flaws or inhomogeneities in the specimen which affect the level of interaction are easily detected. Multifrequency probes and computer-aided interpretation enable eddy current testing to be one of the more sensitive methods for inspection of metallic sheets and tubing.

Acoustic emission occurs in the form of high-frequency stress waves generated by the rapid release of strain energy that occurs within a material during crack growth. This method is used for monitoring structural integrity before failure occurs. Very few applications of this technique are currently being used as a production inspection method; however, a large number of in-service inspections are performed with this method.

Thermal inspection comprises inspection in which heat-sensing devices or substances are used to detect irregular temperatures. A number of contact methods are available; however, there are many limitations when these methods are applied. The noncontact methods, particularly using computer-aided imaging systems, provide a more sensitive test and easier interpretation. This technique is becoming more popular since the instrumentation keeps improving, although it is still relatively expensive.

Optical holography employs the principles of optical interferometry to form a hologram of the specimen, both unstressed and stressed on the same hologram. Any surface displacements occurring due to the stresses show up as fringes or fringe shifts in the hologram. Inspection of tires has been the most frequent application of this NDT technique in manufacturing, although a number of other applications are under development. The high cost of the equipment and level of expertise required to perform the test accurately have hindered its acceptance in industry.

Table 16-3 provides a summary of these NDT methods currently in popular use in the industrial sector. NDT is generally employed where failure of the component will result in serious liability to the manufacturer or harm to the user. Most of these techniques require interpretation of the test results, so only qualified personnel should be employed. It is difficult to discuss resolution for these testing methods (what is the minimum flaw size that can be detected?), because many factors, like flaw orientation, surface preparation, depth of the flaw below surface, and so forth affect the test results. Generally, sensitivity does improve with computer-aided assistance, so many of these techniques are becoming more automated.

REVIEW QUESTIONS

1. What are some of the advantages to the consumer of standardization and of interchangeable parts?
2. Why is it important to interface the manufacturing engineering requirements into the design phase as early as possible?
3. Explain the difference between attributes and variables inspection.
4. Why have so many variables-type devices in autos replaced with attribute-type devices?
5. What are the four basic measures upon which all others depend?
6. What is a pascal and how is it made up of the basic measures?
7. What are the different grades of gage blocks and why do they come in sets?

8. What keeps gage blocks together when they are "wrung together"?
9. What is the difference between tolerance and allowance?
10. What type of fit would describe the following situations?
    (a) The cap of a ball-point pen.
    (b) The lead in a mechanical lead pencil, at the tip.
    (c) A bullet in a barrel of a gun.
11. What items can you think of which are assembled with a medium-force fit?
12. Why might you use a shrink fit rather than welding to join two steel parts? (What does the word "shrink" imply?)
13. What factors should be considered in selecting measurement equipment?
14. How could you determine if your ordinary bathroom scale is linear and has good repeat accuracy?
15. Design and describe a simple experiment which demonstrates the difference between magnification and resolution.
16. What is meant by the statement that usable magnification is limited by the resolution of the device?
17. What is parallax?
18. What are three uses for a combination set?
19. What are the two most likely sources of error in using micrometer calipers?
20. What is the major disadvantage of a micrometer caliper as compared with a vernier caliper? The advantages?
21. What would be the major difficulty in obtaining an accurate measurement with a micrometer depth gage (Figure 16-20) if it were not equipped with a rachet or friction device for turning the thimble?
22. Suppose that you had a 2-foot steel bar in your supermicrometer. Do you think you could detect a length change if the temperature in the room changed 20°F?
23. Why is the toolmarker's microscope particularly useful for making measurements on delicate parts?
24. In what two ways can linear measurements be made using an optical projector?
25. What type of instrument would you select for checking the accuracy of linear movement of a machine tool table through a distance of 50 inches?
26. Upon what principle is a sine bar based?
27. How can the not-go member of a plug gage be easily distinguished from the go member?
28. What is the advantage of a progressive-type snap gage?
29. How does a taper plug gage check both the angle of taper and the size?
30. What is the primary precaution that should be observed in using a dial gage?
31. Why are air gages particularly well suited for gaging the diameter of a hole?
32. Explain the principle of measurement by lightwave interference.
33. In checking a 1-inch-square gage block by means of a helium light source, five dark bands were observed. There was a 2-inch distance between the front edges of the two blocks. What was the difference in height between the two blocks?
34. How does a toolmaker's flat differ from an optical flat?
35. Why may two surfaces that have the same microinch roughness be quite different in appearance?
36. Why are surface-finish blocks often used for specifying surface finish rather than just microinch values?
37. What limits the resolution of a stylus-type surface measuring device in finding profiles?
38. What is the general relationship between surface roughness and tolerance? Between tolerance and cost to produce the surface and/or tolerance?
39. Review the NDT methods and indicate which might be used to inspect the welds on an oil pipeline being welded in the field.

CASE STUDY 16. The Debated Automobile Axle

Figure CS-16(a) shows the heavy wheel cover and left-rear wheel and tire from a heavy passenger car that was badly damaged when it left the road and struck a very large rock on the left-hand side of the highway. The driver of the car claimed that the wheel came off the vehicle while it was proceeding straight along the highway, thereby causing him to lose control and the vehicle to leave the roadway. Examination of the vehicle after the accident revealed that the axle involved had broken just inside the outer axle bearing, with the major portion of the axle (shown in c) remaining inside the axle housing and attached to the differential unit at its inner end. The fracture surface in the axle is shown at b. When a straight edge (x) was clamped to a machined, cylindrical surface on the axle (y), the result is shown at c.

Do you believe the account of the driver of the vehicle, and why?

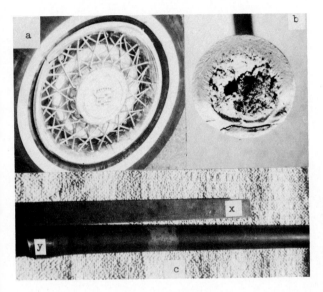

FIGURE CS-16 (a) Wheel, tire, and wheel cover. (b) Fracture surface on broken axle, (c) Broken axle (Y) and straight edge (X) laid side by side.

CHAPTER 17

Process Capability and Quality Control

All manufacturing processes display some level of inherent capability or some inherent uniformity or nature. For example, suppose that we view "shooting at a metal target" as a "process" for putting holes in a piece of metal. I hand you the gun and tell you to take five shots at the bull's-eye. Thus, you are the operator of the process. In order to measure process capability (P.C.)—that is, your ability to consistently hit the bull's-eye—we need to examine the target after you have finished shooting. So process capability is determined by measuring the output of the process. This is true also with quality control (Q.C.). In quality control, we inspect the product (examine the product using the measurement tools described in Chapter 16) to determine whether or not we have accomplished, with the processing, what was specified in the design. Thus, quality control is used to keep tabs on the process by inspection of the output and comparing the findings with what was desired or specified.

P.C. studies use the same sort of statisitical and analytical tools as used in Q.C. except that the results are directed at the machines used in the processing rather than the output or products from the processes. Going back to our example, a P.C. study would be directed toward quantifying the inherent variability in the gun and shooter. A Q.C. program would be looking at the targets to decide if *they were* acceptable (could be sold to a customer) and may be looking at variability as well as other aspects of the acceptability of the product. Thus, the objectives are quite different.

PROCESS CAPABILITY STUDY

When speaking of the "nature of the process," we are referring to the variability or inherent uniformity of the process. Thus, in the target-shooting example, a perfect marksman would be capable of placing five shots right in the middle of the bull's-eye, one right on top of the other. The process displayed no variability! Such performance would be very unusual in a real industrial process. The variability may have assignable causes and may be correctable if we can find and eliminate the cause. That variability to which we can assign no cause and which we cannot eliminate is inherent to the process and is therefore its "nature".

Some examples of causes of assignable variation in processes include multiple machines for the same components, operator blunders, defective materials, or progressive wear of the tools during machining. Sources of inherent variability in the processes include variations in material properties, operator variability, vibrations and chatter, and the wear and the sliding components in the machine, perhaps resulting in sloppier operation of the machine. These kinds of variations, which occur naturally in processes, usually display a random nature, and often can not be eliminated. Sometimes the causes of assignable variation can not be eliminated due to the cost. The question for the manufacturing engineer often is; Can the sources of variability be economically eliminated? Finally, it must be mentioned that almost every process will have multiple causes of variability occurring at the same time. It will be extremely difficult to separate out the effects of the different sources of variability during the analysis.

Accuracy versus precision in processes. It is vitally important that the difference between accuracy and precision be understood, when discussing process capability. Accuracy refers to the ability to hit what is aimed at (the bull's-eye), while precision refers to the repeatability of the process. Suppose that you fire five sets of five shots each at the target. Figure 17-1 shows some of the possible outcomes. In Figure 17-1a, the inspection of the target shows that we have an inherently good process—accurate and precise. In (b) we have precision (repeatability) but not accuracy (poor aim). We could, perhaps improve the process capability by adjusting the sights to account for a steady crosswind. The crosswind would be an assignable cause. In (c), we discover that on the average the process is quite accurate as the " × " is right in the middle of the bull's-eye, but the process has too much scatter. Finally, in (d) we observe a failure to repeat accuracy between samples with respect to time. These four outcomes are typical but not all-inclusive of what might be observed in this process.

Making P.C. studies. The object of the P.C. study is to determine the inherent nature of the process. This means that we must examine the output of the process under normal or what is typically called "hands-off" conditions. The inputs (materials, setups, cycle times, temperature, pressure, operator, etc.) have to be standardized. The process must be allowed to run without tinkering or adjusting while we carefully document the output (i.e., the product or units or components) with respect to (a) time, (b) source, and (c) order of production. A sufficient amount of data has to be taken so that one has confidence in the

a) Accurate and Precise

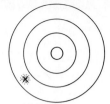

b) Precise, Not accurate

c) Accurate, Not precise

d) Precise within sample
Not precise between samples
Not accurate over all or within sample.

FIGURE 17-1 Accuracy versus precision in process. Dots in targets represent location of shots. Cross represent location of the average position of all shots.

statistical analysis of the data. Be sure that the precision of the measurement system exceeds the process capability by at least one order of magnitude.

In many machining processes in use today, we have replaced the operator with automatic tool positioning capability (see NC or numerical control), which means that we have eliminated some of the variability in the process, thus making the process more repeatable. In the same light, it will be very important in the future for manufacturing engineers to know exactly the process capability of robots they want to use in the workplace.

In summary, then, prior to any data collection, these steps need to be taken:

1. Define "normal" or "hands-off" process conditions; specify machine settings for speed, feed, size of cut, pressures, temperature, fixtures, tool angles, etc.
2. Select a representative operator. This will be critical in processes wherein the operator can dominate the process.
3. Define what will be measured and how it will be measured (the inspection device).
4. Define how much will be measured. A sample of 64 to 100 units is typical.
5. Collect homogeneous amounts of input material to contrast with normal input material. This is critical in those processes which are material dominated.
6. Design data sheets which record date, time, source, order of production, and all process parameters being used (or measured) while the data are being gathered.

Suppose that 70 units have now been manufactured without any adjustment of the process. The units are measured and the data recorded on the data sheet. At this point, it is wise to develop a frequency distribution, or histogram, as shown in Figure 17-2. This histogram shows the raw data and the desired nominal value along with the upper and lower tolerance limits, where LTL = lower

MACHINING PROCESSES

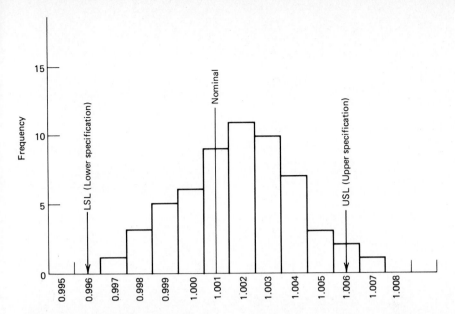

FIGURE 17-2 Histogram of 70 measurements of a parameter. The specification was 1.001 ± 0.005 inch.

tolerance limit and UTL = upper tolerance limit. These are also called the (lower and upper) specifications limits. The statistical data are used to determine the mean and the standard deviation of this distribution.

The mechanics of this statistical analysis are shown in Figure 17-3. The mean of the distribution, designated $\overline{X}'$ (X bar prime) is to be compared with the nominal value. The standard deviation, designated σ' (sigma prime), is compared with the desired tolerance. The purpose of the analysis is to estimate these values, as they are not known. A sample of size 5 was used in this example, so $n = 5$. Fourteen samples were drawn from the process, so $k = 14$. For each sample, the sample mean $\overline{X}$ and sample range R are computed. For large samples, $n > 12$, the standard deviation of each sample is computed rather than the range. Next, the average of the sample averages, $\overline{\overline{X}}$, is computed. This is sometimes called the grand average and it is used to estimate the mean of the process, $\overline{X}'$. The standard deviation of the process, which is a measure of the spread or variability of the process, is estimated from either the average of the sample ranges, $\overline{R}$, or the average of the sample standard deviations, $\overline{\sigma}$, using either $\overline{R}/d_2$ or $\overline{\sigma}/c_2$. The factors d_2 and c_2 depend on the sample size n and are given in Table 17-1. The process capability is defined by $\pm 3\sigma'$ or $6\sigma'$.

Note that a distinction is made between a sample and a population. A sample is of specified, limited size and is drawn from the population. The population is the large source of items, which can include all the items the process will ever produce under the specified conditions. Our calculations assumed that this population was normal or bell shaped. Figure 17-4 shows a typical normal curve and the areas under the curve as defined by the standard deviations. Other distributions, such as binomial, exponential, and Poisson, are possible but in our example, the histogram clearly suggested that this process can best be de-

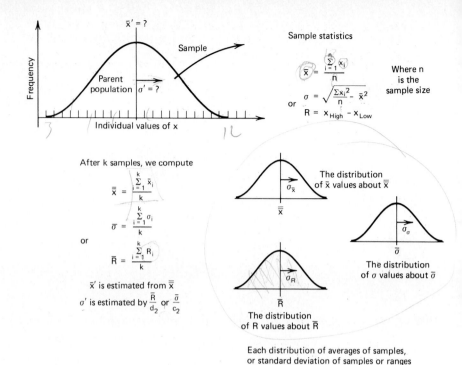

Sample statistics

$$\bar{x} = \frac{\sum\limits_{i=1}^{n} x_i}{n}$$

Where n is the sample size

or

$$\sigma = \sqrt{\frac{\sum x_i^2}{n} - \bar{x}^2}$$

$$R = x_{High} - x_{Low}$$

After k samples, we compute

$$\bar{\bar{x}} = \frac{\sum\limits_{i=1}^{k} \bar{x}_i}{k}$$

$$\bar{\sigma} = \frac{\sum\limits_{i=1}^{k} \sigma_i}{k}$$

or

$$\bar{R} = \frac{\sum\limits_{i=1}^{k} R_i}{k}$$

$\bar{x}'$ is estimated from $\bar{\bar{x}}$

σ' is estimated by $\dfrac{\bar{R}}{d_2}$ or $\dfrac{\bar{\sigma}}{c_2}$

The distribution of $\bar{x}$ values about $\bar{\bar{x}}$

The distribution of σ values about $\bar{\sigma}$

The distribution of R values about $\bar{R}$

Each distribution of averages of samples, or standard deviation of samples or ranges of samples must have its own average and standard deviation.

FIGURE 17-3 Calculations needed to obtain estimates of the mean $(\bar{X}')$ and the standard deviation (σ') of the parent population, as in a process capability study.

fined by a normal probability distribution. Now it remains for the process engineer to combine his knowledge of the process with the results from his analysis in order to draw conclusions about the ability of this process to meet specifications.

What P.C. studies tell about the process. The primary things that P.C. studies tell the process engineer deal with the answers to these questions.

1. Does the process have the ability to meet specification?
2. Is the process well centered with respect to the desired nominal specification?

In finding the answers to these questions, it will be determined what action, if any, is necessary.

In response to the first question, the width of the histogram is compared with the specifications. In addition, the natural specification limits are computed and compared with the upper and lower tolerance limits. Three situations exist.

I. $6\sigma' <$ USL-LSL: process variability less than tolerance spread.
II. $6\sigma' =$ USL-LSL: process variability equal to tolerance spread.
III. $6\sigma' >$ USL-LSL: process variability greater than tolerance spread.

In situation I, the machine is capable of meeting the tolerances applied by the designer. Generally speaking, if machine capability is of the order of $\frac{2}{3}$ to $\frac{3}{4}$

TABLE 17-1 Factors for Estimating the Standard Deviation (σ') of a Parent Population from Sample Data for the Average Range ($\overline{R}$) or the Average Sample Standard Deviation ($\overline{\sigma}$).

Sample Size, or the Number of Observations in Subgroup n	Range Factor, for Estimate from $\overline{R}$, $d_2 \equiv \overline{R}/\sigma'$	Standard Deviation Factor, for Estimate from $\overline{\sigma}$, $c_2 \equiv \overline{\sigma}/\sigma'$	Factor Which Relates σ_R, with σ' (σ_R = Std. Dev. of Range Distribution), $d_3 = \sigma_R/\sigma'$
2	1.128	0.5642	0.8525
3	1.693	0.7236	0.8884
4	2.059	0.7979	0.8798
5	2.326	0.8407	0.8641
6	2.534	0.8686	0.8480
7	2.704	0.8882	0.8330
8	2.847	0.9027	0.8200
9	2.970	0.9139	0.8084
10	3.078	0.9227	0.7970
11	3.173	0.9300	0.7870
12	3.258	0.9359	0.7780
13	3.336	0.9410	For $n > 12$, C_2
14	3.407	0.9453	factors should be
15	3.472	0.9490	used with $\overline{\sigma}$
16	3.532	0.9523	
17	3.588	0.9551	
18	3.640	0.9576	
19	3.689	0.9599	
20	3.735	0.9619	
25	3.931	0.9696	
30	4.086	0.9748	

Source: 1950 Manual on Quality Control of Materials; copyright ASTM, 1916 Race St., Philadelphia, PA, 19103. Adapted, with permission.

of the design tolerance, 100% inspection is not needed, and there is a high probability tnat the process will produce all good work over a long period of time. If the P.C. is on the order of $\frac{1}{2}$ or less of the design tolerance, it may be that the process is "too good". That is, we are making ball bearings when what is called for is marbles. In this case, it may be possible to change processes, or

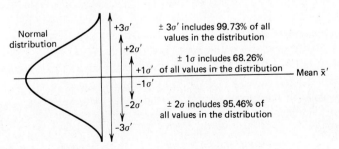

FIGURE 17-4 Distribution of items or values in a normal distribution.

trade off precision in this process for looser tolerances elsewhere, resulting in an overall economic gain.

In situation III, the P.C. was greater than the design tolerance. There are a variety of alternatives here, including:

a. Try to shift this job to another machine with better process capability.
b. Try to get a review of the tolerances to see if they might be loosened.
c. Sort the product, separating the good from the bad. This entails 100% inspection of the product, which may not be a viable economic alternative unless it can be done automatically. The automatic inspection of the product on a 100% basis to ensure perfect quality is called *autonomation* in Japanese industries.
d. Determine whether the precision of the process can be improved by:
   (1) Switching to different machines that do the same process.
   (2) Overhauling existing process.
   (3) Finding and eliminating causes of variability or combinations of the above.

In situation II, the process capability is almost exactly equal to assigned tolerance spread, so if the process is not perfectly centered, defective products will always result. Thus, this situation should be treated like situation III unless the process can be perfectly centered and maintained. Tool wear, which causes the distribution to shift, must be negligible. Then the situation can be treated like that in I, particularly if a small percentage of parts just outside tolerance is acceptable.

The second question deals with the ability of the process to maintain centering so that the average of the distribution comes as close as possible to the desired nominal value. Most processes can be reaimed with proper adjustments and the reasons of poor accuracy are often due to assignable causes, which can be eliminated.

In addition to direct information about the accuracy and precision of the process, P.C. studies can also tell the manufacturing engineer how pilot processes compare to production processes, and vice versa. If the source and time of each product are carefully recorded, information about the instantaneous reproducibility can be found and compared to the repeatability of the process with respect to time (the time-to-time variability). More important, since almost all processes are duplicated, P.C. studies generate information about machine-to-machine variability. Suppose, going back to our target-shooting example, that you used five different guns, all of the same make and type. The results would have been different, just as having five marksmen use the same gun would have resulted in yet another outcome. Thus, P.C. studies generate information about the homogeneity and differences in multiple machines and operators.

It will quite often be the case in such studies that one variable will dominate the process. Target shooting, viewed as a process, is probably "operator" dominated in that the outcome is highly dependent on the skill of the "worker." Those processes which are not well engineered nor highly automated, or where the worker is viewed as "highly skilled," are usually operator dominated. Processes which change or shift uniformly with time but which have good repeatability in the short run are often "machine" dominated. For example, it often happens that machines tend to become more precise (have less variability within a sample) after they have been "broken in" (i.e., the rough contact surfaces have smoothed out due to wear). Other variables that can dominate processes are setup, input components, and even information.

The discussion to this point has assumed that the parent population is normally distributed—has the classic bell-shaped distribution shown in Figure 17-4—wherein the percentages shown are dictated by the number of standard deviations from the central value or mean. This is not necessarily the case in all processes, and the shape of the histogram may reveal the nature of the process to be skewed to the left or right (unsymmetrical), often indicating some natural limit in the process. Drilled holes are this way as the drill tends to make the hole oversize. Filling operations are typically this way. Another possibility is a distribution with two distinct peaks, often caused by two processes being mixed together. The possibilities are endless and will require the engineer to record carefully all the sources of his data in order to track down the factors resulting in loss in precision and accuracy in the processes.

INSPECTION AND QUALITY CONTROL

In virtually all manufacturing, it is extremely important that the dimensions and quality of individual parts be known and maintained. This is of particular importance where large quantities of parts, often made in widely separated plants, must be capable of interchangeable assembly. Otherwise, difficulty may be experienced in subsequent assembly or in service, and costly delays and failures may result. In recent years, defective products resulting in death or injury to the user have resulted in expensive litigation and damage awards against manufacturers. Inspection is that function which checks and controls the quality (dimensions, performance, color, etc.), manually using operators or inspectors or automatically with machines, as discussed previously.

The answer to the economically based question "How much should be inspected" has three answers.

1. *100%*. Inspect all of the product being made. 100% inspection does not ensure perfect quality, owing to the fallibility of inspectors and devices.
2. *Sample*. Inspect some of the product by sampling and make decisions about the quality of the rest based on the sample.
3. *None*. Assume that everything made is acceptable or that the product is inspected by the consumer, who will exchange it if it is defective.

Reasons why all of the product may not be inspected (i.e., why it is sampled) include:

1. Test is destructive.
2. Too much product to inspect all of it.
3. Testing takes too much time or is too complex or too expensive.
4. Not economically feasible to inspect all even though the test is simple, cheap, and quick.

Some characteristics are nondissectible, meaning that they cannot be measured during the manufacturing process because the characteristic does not exist until after a whole series of operations have taken place. The final edge geometry of a razor blade is a good example, as is the yield strength of a rolled bar of steel.

Sampling or looking at some percentage of the whole requires the use of statistical techniques in order to make decisions about the acceptability of the whole based on the quality found in the sample. This is known as *statistical quality control* or SQC.

Statistical quality control.   Looking at some (sampling) and deciding about the behavior of the whole (the parent population) is a common thing in industrial inspection operations. One of the basic statistical quality control (SQC) techniques which employs sampling is *control charts*.

Figure 17-5 shows the basic structure of three charts commonly used for variables types of measurements. Usually, only the $\overline{X}$ chart and the R chart are used unless the sample size is large, and the σ charts are used in place of R charts. The data plotted on these charts are sample values, not individual values, because all sample statistics will tend to be normally distributed about their own mean (see Figure 17-3). Thus $\overline{X}$ values will be normally distributed about X and R values will be normally distributed about $\overline{R}$. Thus, the laws of normality will apply to these charts.

Sampling errors.   It is important to understand that when sampling, two kinds of errors are always possible. Suppose that the process is running perfectly but your sample suggests that something is wrong. You stop the process to make adjustments. This is a type I or α error. Suppose that your process was not running perfectly and was making defective products. However, your sample suggested that nothing was wrong and you did not stop the process and set it right. This is a type II or β error. Both types of errors are possible when you sample. For a given sample size, reducing one type of error will enlarge the other. Increasing the sample size or frequency of sample will reduce the probability of errors but at increased cost of inspection. Ultimately one determines the size of the errors one is willing to accept according to the overall cost of making the errors plus the cost of inspection. If, for example, a type II error is very expensive in terms of product recalls or legal suits against your company, you may be willing to make more type I errors, sample more, or even go to 100% inspection on very critical items to ensure that you are not accepting defective materials as good and passing them on to the customer.

As mentioned earlier, in any continuing manufacturing process, variations from established standards are of two types: (1) *assignable cause variations,* such as those due to malfunctioning equipment or personnel, or to defective material, or due to a worn or broken tool; and (2) normal *chance variations,* resulting from the inherent nonuniformities that exist in materials and in machine motions and operations. Deviations due to assignable causes may vary greatly; their magnitude and occurrence are unpredictable, and one thus wishes to prevent their occurring. On the other hand, if the assignable causes of variation are removed from a given operation, the magnitude and frequency of the chance variations can be predicted with great accuracy. Thus, if one can be assured that

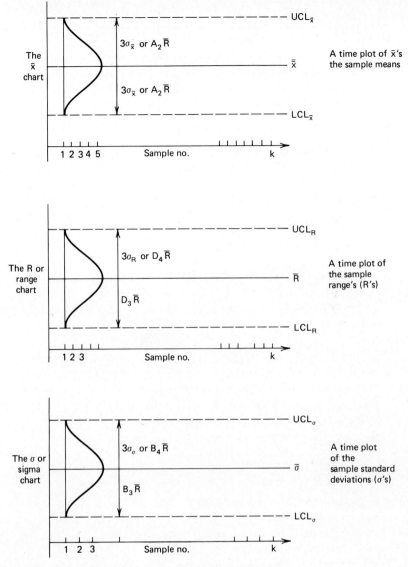

FIGURE 17-5 The basic design of the $\overline{X}$ chart, the R chart, and the σ chart used in SQC. Values for A_2, D_3, D_4 given in Table 17-2.

only chance variations will occur, he knows what the quality of the product will be and manufacturing can proceed with assurance as to the results. By using statistical quality control procedures, one may detect the presence of an assignable-cause variation and remove the cause, often before it causes quality to become unacceptable.

Examine the data in Figure 17-6. This is the frequency distribution of measurements of the diameters of 100 ground pins, which represents the entire population in this case. These pins are supposed to have a nominal diameter of 12.700 mm (0.500 inch) and we observe that this was the size appearing most

TABLE 17-2 Factors for Determining $\overline{X}$ and R Chart Control Limits, 3 Standard Deviations Assumed

Number of Observations in Subgroup n	Factor for $\overline{X}$ Chart A_2	Factors for R Chart	
		Lower Control Limit D_2	Upper Control Limit D_4
2	1.88	0	3.27
3	1.02	0	2.57
4	0.73	0	2.28
5	0.58	0	2.11
6	0.48	0	2.00
7	0.42	0.08	1.92
8	0.37	0.14	1.86
9	0.34	0.18	1.82
10	0.31	0.22	1.78
11	0.29	0.26	1.74
12	0.27	0.28	1.72
13	0.25	0.31	1.69
14	0.24	0.33	1.67
15	0.22	0.35	1.65
16	0.21	0.36	1.64
17	0.20	0.38	1.62
18	0.19	0.39	1.61
19	0.19	0.40	1.60
20	0.18	0.41	1.59

Upper control limit for $\overline{X} = UCL_x = \overline{X}' + A_2\overline{R}$

Lower control limit for $\overline{X} = LCL_x = \overline{X}' - A_2\overline{R}$

Upper control limit for $R = UCL_R = D_4\overline{R}$

Lower control limit for $R = LCL_R = D_3\overline{R}$

Note that since $\sigma_{\bar{x}} = \dfrac{\sigma'}{\sqrt{n}}$, $3\sigma_{\bar{x}} = \dfrac{3\sigma'}{\sqrt{n}} = \dfrac{3\overline{R}}{\sqrt{n}\,d_2} = A_2\overline{R}$ where $A_2 = \dfrac{3}{\sqrt{n}\,d_2}$

Also $UCL_R = \overline{R} + 3\sigma_R = d_2\sigma' + 3d_3\sigma'$ (see Table 17-1).

Thus $UCL_R = (d_2 + 3\sigma_3)\sigma' = \left(1 + \dfrac{3d_3}{d_2}\right)\overline{R} = D_4\overline{R}$

Similarly, $D_3 = \left(1 - \dfrac{3d_3}{d_2}\right)$

Source: 1950 Manual on Quality Control of Materials; copyright ASTM, 1996 Race St., Philadelphia, PA, 19103. Adapted, with permission.

frequently (there were 18 of these). The population is going to be assumed normal and only chance variations are occurring. Even so, we see that we did not find any pins between 0.5003 and 0.5006 inch. The mean for the entire population, μ, is obtained by

$$\mu = \frac{\sum\limits_{i=1}^{n} X_i}{n} = \frac{\sum\limits_{i=1}^{100} X_i}{100} \tag{17-1}$$

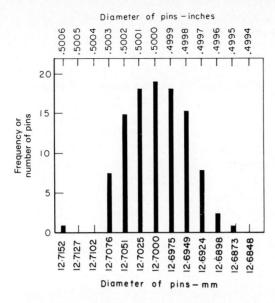

FIGURE 17-6 Frequency distribution of 100 ground pins having a nominal diameter of 12.700 mm (0.5000 inch).

where X is an individual measurement and n the number of items or measurements. It is a measure of the central tendency about which the individual measurements tend to group. The variability of the individual measurements about the average may be indicated by the *standard deviation,* σ, where

$$\sigma = \sqrt{\frac{\sum\limits_{i=1}^{n} (X_i - \mu)^2}{n}} \tag{17-2}$$

The standard deviation is of particular value in that for normal, or chance, distributions 68.26% of all measurement values will lie within $\pm$ 1σ range from the average, 95.46% within $\pm$2σ range, and 99.73% within $\pm$3σ range. Thus, if one knows what the standard deviation is for a population of items, he knows that not over 0.07% of them will fall outside the limits $X \pm 3\sigma$ *as long as only chance variations occur.* For the pin-grinding operation the average diameter, μ, was 12.700 mm (0.5000 inch), and the standard deviation, σ, was 0.0048 mm (0.00019 inch), making 3σ = 0.014 mm (0.0006 inch). The producer could thus set manufacturing limits at

12.714 mm (0.5006 inch)
12.682 mm (0.4994 inch)

with the assurance that fewer than 3 parts in 1000 would fall outside those limits due to chance causes of variation.

Note that these are not the same values as was obtained earlier in the P.C. study. Here we are finding the mean and standard deviation of 100 items. In the P.C. study we were estimating the true mean and standard deviation of the entire population from sample data, which is typically the situation one has to deal with.

Quality control charts. Quality control charts are widely used as aids in maintaining quality and in achieving the objective of detecting trends in quality variation before defective parts are actually produced. These charts are based on the previously discussed concept that if only chance causes of variation are present, the deviations from the specified dimension or attribute will fall within predetermined limits.

In most cases, sampling inspection is used—typical sample sizes of from three to about twelve units being employed. This permits more rapid and more current inspection and also is less costly. Figure 17-7 shows an example of two control charts. The $\bar{X}$ chart tracks the sample averages ($\bar{X}$ values), while the R chart plots the range values (R values). Let us assume that the data shown in Figure 17-6 now represents the results of 20 samples of size 5 taken from a large number of ground pins rather than the entire parent population. The sample data will be used to prepare the control charts shown in Figure 17-7.

The center line of the $\bar{X}$ chart was computed earlier, prior to actual usage of the charts in control work, by $\bar{\bar{X}} = \sum_{i=1}^{k} \bar{X}_i/k$, where $\bar{X}$ was the sample average and k was the number of sample averages ($n = 5$, $k = 20$). The horizontal axis for the charts is time, thus indicating *when* the sample was taken. $\bar{\bar{X}}$ serves as an estimate for $\bar{X}'$, the true center of the process distribution and the center line of the $\bar{X}$ chart. The upper and lower control limits are commonly based on three standard deviations. Thus

$$\text{UCL}_{\bar{X}} = \text{upper control limit } \bar{X} \text{ chart} = \bar{X}' + 3\sigma_{\bar{x}}$$

$$\text{LCL}_{\bar{X}} = \text{lower control limit } \bar{X} \text{ chart} = \bar{X}' - 3\sigma_{\bar{x}}$$

where $\sigma_{\bar{x}} = \dfrac{\sigma'}{\sqrt{n}}$. Recall how $\bar{X}'$ and σ' were estimated when sampling. The standard deviation for the distribution of sample averages, $\sigma_{\bar{x}}$, is determined

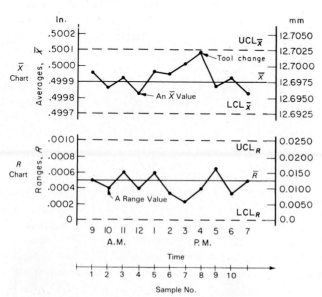

FIGURE 17-7 Statistical quality control charts for 12.700-mm (0.5000-inch)-diameter pins. Note trend in $\bar{X}$ curve before tool change, indicting that the mean was shifting due to tool wear.

directly from the estimate of the standard deviation of the parent population, σ', by dividing σ' by the $\sqrt{n}$. The $\overline{X}$ chart is used to track the central tendency (aim) of the process. In this example, assume that samples are being taken hourly and the average of *each sample* (not individual values) is plotted.

The R chart is used to track the variability or dispersion of the process. A σ chart could also be used. R is computed for each sample $(X_{HIGH} - X_{LOW})$ The value of $\overline{R}$ was determined previously in P.C. study from $\overline{R} = \Sigma_{i=1}^{k} R/k$. $\overline{R}$ represents the average of range values of 20 samples of size n. The range values will be distributed about $\overline{R}$, with standard deviation σ_R. To determine the upper and lower control limits for the charts, the following relationships are used.

$$\mathrm{UCL}_R = \text{upper control limit } R \text{ chart } (\overline{R} + 3\sigma_R) = D_4\overline{R} = 2.115\overline{R}$$

$$\mathrm{LCL}_R = \text{lower control limit } R \text{ chart } (\overline{R} - 3\sigma_R) = D_3\overline{R} = 0\overline{R} = 0$$

where D_4 and D_3 are constants and are given in Table 17-2. For small values of n, the distance between center line $\overline{R}$ and LCL_R is more than $3\sigma_R$, but LCL_R cannot be negative, as negative range values are not allowed, by definition. Hence, $D_3 = 0$ for values of n up to 6.

After control charts have been established and the average and range values plotted for each sample group, the chart acts as a control indicator for the process. If the process is operating under chance cause conditions, the data will appear random (have no trends or pattern). If $\overline{X}$ or R values fall outside the control limits, an assignable cause or change may have occurred and some action should be taken to correct the problem.

As shown in Figure 17-7, trends in the charts often indicate the existence of an assignable cause factor before the process actually produces a point outside the control limit. Thus $\overline{X}$ values for samples 5, 6, 7 and 8 show a trend toward oversize parts. In this grinding operation, the wheel has worn down (become undersize), so now the parts are becoming oversize and corrective action should be taken. (Redress and reset wheel or replace with new wheel.) Note that defective parts can be produced even if the points on the charts are in control. That is, it is possible for something to change in the process, causing defective parts to be made, and the sample point still to be within the control limits. Since no corrective action was suggested by the charts, a type II or β error was made. Subsequent operations will then be performing additional work on products already defective. Thus, it may be that a better economic strategy for the company is to increase the chance of a type I error by using something less than 3σ control limits, which will in turn reduce the probability of a type II error.

With regard to control charts in general, it should be kept in mind that the charts are only capable of indicating to the engineer that something is (or is not) happening and that a certain amount of detective work will be necessary to find out what has occurred to cause a break from the random, normal pattern of sample points on the charts. Keeping careful track of when and where the sample was taken will be very helpful in such investigations.

In summary, we see that the data gathered to develop the P.C. study can be used to prepare the initial control charts from the process after the removal of all assignable causes for variability and proper setting of the process average.

After the charts have been in place for some time, and a large quantity of data obtained from the process output, the P.C. study can be redone to obtain better estimates of the natural spread of the process during actual production.

Total quality control (TQC). In Japan, the attitude toward quality is that it begins with production, with the emphasis on making it right the first time. The responsibility for quality rests with the worker. An attitude of defect prevention and a habit for improvement of quality are fundamental to the JIT system mentioned in Chapter 1. Companies like Toyota have accomplished TQC by extensive education of the workers, giving them the analysis tools they need to find and expose the problems. Workers are encouraged to correct their own errors and 100% inspection (often done automatically) is the rule. Passing defective products on to the next process is not allowed. The goal is perfection. *Quality Circles,* which are now becoming popular in the United States, are just one of the methods used by Japanese industries to achieve perfection. The point, however, is that manufacturing must have primary responsibility for quality, not the quality control department.

REVIEW QUESTIONS

1. Define a process capability study in terms of accuracy or precision.
2. What does the "nature of the process" refer to?
3. Suppose you had a process that was accurate and precise, as shown in Figure 17-1a. What might the target look like if, occasionally while shooting, a sharp gust of wind blew left to right?
4. Review the steps required prior to making a P.C. study of a process.
5. Why don't standard tables exist detailing the natural variability of a given process, like rolling, extruding, or turning?
6. For the items listed below, obtain a quantity of 48. Measure the indicated characteristics and determine the process mean and standard deviation. Use a sample size of 4, so that 12 samples are produced.

Item	Characteristic(s) You Can Measure
Flat washer	Weight, width, diameter of hole, outside diameter
Paper clip	Length, diameter of wire
Coin (penny, dime)	Diameter, thickness at point, weight
Your choice	Your choice

7. Perform a process capability study to determine the P.C. of the process that makes M&M candy. You will need to decide what characteristics you want to measure, (weight, diameter, thickness, etc.), how you will measure it (use rule of 10), and what kind of M&M's you want to inspect (how many bags of M&M's you wish to sample). Take samples of size 4 ($n = 4$). Make a histogram of the individual data and estimate $\bar{X}'$ and σ' as outlined in Figure 17-3. If you decide to measure the weight characteristics, you can check your estimate of $\bar{X}'$ by weighing all the M&M's together and dividing by the total number of M&M's.
8. What are some of the alternatives available to you when you have the situation wherein $6\sigma' > USL - LSL$?

9. Here are some common, everyday processes with which you are familiar. What variable do you think dominates these processes?
   (a) Baking a cake (from scratch; from a cake mix).
   (b) Mowing the lawn.
   (c) Washing dishes in a dishwasher.
10. Explain why holes produced by the process of drilling would have a skewed distribution rather than normal.
11. What are some common consumer items that may not be receiving any final inspection?
12. What are common reasons for sampling inspection rather than 100% inspection?
13. Fill in this table with one of the following statements.
    (a) Type I or α error
    (b) Type II or β error
    (c) No error

		In reality, if we looked at everything the process made, we would know that it had	
		Changed	Not Changed
The sample suggested that the process had:	**Changed**		
	Not Changed		

14. Why, when we sample, can we not avoid making type I and type II errors?
15. Which error can lead to legal action from the consumer for a defective product which caused bodily injury?
16. Define and explain the difference between σ', $\sigma_{\bar{x}}$, and σ_R.

CASE STUDY 17.   The Threaded Ball Studs

Forward Motors Corporation makes about 2,000,000 threaded ball studs each year, as shown in Figure CS-17. They currently are machined from hot-rolled AISI 1030 bar stock of diameter D, heat-treated to 490 MP$_a$ (70,000 psi) ultimate strength, and the ball surface then ground to the required finish. Tolerances on the ball and threaded portion are ±0.05 mm (0.002 inch). Other dimensions are to ±0.64 mm (0.025 inch). Dependable quality is important.

It is desired to reduce the cost of this part and, if possible, reduce the weight. How would you bring about the desired improvement? [D is approximately 38 mm (1½ inches)].

FIGURE CS-17   Threaded ball stud.

CHAPTER 18

Chip-Type Machining Processes (Metal Cutting)

Metal cutting, commonly called *machining,* is the removal of the unwanted metal from a workpiece in the form of chips so as to obtain a finished product of desired size, shape, and finish. U.S. industries annually spend $60 billion to perform metal removal operations because the vast majority of manufactured products require machining at some stage in their production, ranging from relatively rough or nonprecision work, such as cleanup of castings or forgings, to high-precision work involving tolerances of 0.002 mm (0.0001 inch) or less. Thus it undoubtedly is the most important of the basic manufacturing processes.

Over the past 80 years, the process has been the object of considerable research and experimentation, which has led to improved understanding of the nature of both the process itself and the surfaces which are produced by it. While this research effort has led to considerable improvements in machining productivity, the complexity of the process has resulted in poor progress in obtaining a complete theory of chip formation. This is because most theories have ignored the plastic deformation properties of the work material and/or have not been able properly to characterize the interactions at the sliding contact surfaces between the tool and the chip. Metal cutting, an unconstrained deformation process (like tensile testing), is a very large strain plastic deformation process operating at exceptionally high strain rates, which makes it quite unique. The problem is further complicated by tool geometry variations, wide variety of tool materials used in the process, temperature or heat problems, and the great variation in operating conditions of the machines performing this process. In addition, the environment in which the process is performed has a significant influence on the outcome. The objective of this chapter is to put all this in perspective for the practicing engineer.

Basic chip formation processes. There are seven basic chip formation processes: turning, shaping, milling, drilling, sawing, broaching, and abrasive machining (grinding). Most of these processes are shown schematically in Figures 18-1 through 18-7, wherein they are categorized as: single-point tool or multiple-point tool operations. Abrasives or grinding will be treated separately in Chapter 23. For all metal-cutting processes, it is necessary to distinguish between speed, feed, and depth of cut. In general, speed (V) is the primary cutting motion, which relates the velocity of the tool relative to the work. It is generally given in units of surface feet per minute (sfpm), inches per minute (ipm), or meters per minute (m/m) or per second (m/s). Speed (V) is shown in the figures with the heavy dark arrow. Feed (f) is the amount of material removed per revolution or per pass of the tool over the workpiece, so the units are inches/revolution, inches per cycle, inches per minute, or inches per tooth, depending on the process. Feed is shown in the dashed arrows in the figures. The depth of cut reflects the third dimension in these processes and is indicated in the schematics by the letter t. Note that in broaching, for example, the feed is in the direction of depth of cut, so depth of cut becomes width of cut (w) as determined by the width of the tool or the workpiece. In each of the figures, the basic equations for the cutting time per piece (CT) in minutes and the rate of metal removal (MRR) in volume per minute is given. Other main terms used in the equations, and shown in the schematics, are N for revolutions per minute (rpm), L for length of cut, and D for diameter. These basic equations are as fundamental as the processes themselves, so the student should be familiar with them as well as the basic processes. In the main, if one keeps track of the units and visualizes the processes, the equations are, for the most part, straightforward.

The process of milling requires two figures to describe as it takes on different forms depending on the selection of the machine tool and the cutting tool. In milling, like many other multiple tool processes, there are two feeds: the amount of metal an individual tooth removes (called the feed per tooth) and the rate at which the table is translated past the rotating tool (the feed in inches per minute). Table 18-1 summarizes process information for these basic machining processes, noting typical sizes of parts that are manufactured with these processes as well as their typical production rates, tolerances, and surface finishes. Additional specific information on these processes is given in Chapters 19 through 28.

Chip formation in metal cutting. In order to better understand this complex process, the tool geometry is simplified from the three-dimensional (oblique) geometry, which typifies most processes, to a two-dimensional (orthogonal) geometry. The workpiece is a plate as shown in Figure 18-8. This model is sufficient to allow us to consider the behavior of the work material during chip formation; the influence of the most critical elements of the tool geometry (the cutting edge radius and the back rake angle (α)); and the interactions which

TABLE 18-1 Summary of Basic Machining Processes

Applicable Process	Raw Material Form	Size — Maximum	Size — Minimum	Typical Production Rate	Material Choice	Typical Tolerance	Typical Surface Roughness
Shaping	Bar Plate Casing	3 ft × 6 ft	Limited usually by ability to hold part	1–4 parts/hour	Low- to medium-carbon steels and nonferrous metals best; no hardened parts	±0.001–±0.002 in. (larger parts) ±0.0001–±0.0005 in. (small–medium parts)	63–250
Planing	Bar Plate Casting	42 ft wide × 18 ft high × 76 ft long	Parts too large for shaper work	1 part/hour	Low- to medium-carbon steels or nonferrous materials best	±0.001–±0.005 in.	63–125
Gear shaper	Blanks	120-in.-dia. gears 6-in. face width	1 in. dia.	1–60 parts/hour	Any material with good machinability rating	±0.001 in. or better at 200 D.P. to 0.0065 in. at 30 D.P.	63
Turning (engine lathes)	Cylindrical Preforms Castings Forgings	78 in. dia. × 73 in. long	1/64 in. typical	1–10 parts/hour	All ferrous and nonferrous material considered machinable	±0.002 in. on dia. common; ±0.001 in. obtainable	125–250
Turning (turret lathe)	Bar Rod Tube Preforms	36 in. dia. × 93 in. long	1/64 in. dia.	1 part/minute	Any material with good machinability rate	±0.003 in. on dia. where needed; ±0.010 in.	125 average
Turning (automatic screw machine)	Bar Rod	Generally 2 in. dia. × 6 in. long	1/16 in. dia. and less, weight less than 1 ounce	10–30 parts/minute	Any material with good machinability rating	±0.0005 in. possible; ±0.001 to ±0.003 in. common	63 average

Process	Shapes	Size (max)	Size (min)	Production rate	Materials	Tolerance	Surface finish
Turning (Swiss automatic machining)	Rod	Collets adapt to 1/2 in. dia.	Collets adapt to less than 1/2 in.	12–30 parts/minute	Any material with good machinability rating	±0.0002 in. to ±0.001 in. common	63 and better
Boring (vertical)	Casting Preform	98 in. × 72 in.	2 in. × 12 in.	2–20 hrs/pc	All ferrous and nonferrous	±0.0005 in.	90–250
Milling	Bar Plate Rod Tube	4–6 ft long	Limited usually by ability to hold part	1–100 parts/hour	Any material with good machinability rating	±0.0005 in. possible; ±0.001 in. common	63–250
Hobbing (milling gears)	Blanks Preforms Rods	10-ft-dia. gears 14-in. face width	0.100 in. dia.	1 part/minute	Any material with good machinability rating	±0.001 in. or better	63
Drilling	Plate Bar Preforms	3½-in.-dia. drills, (1 in. dia. normal)	0.002 in. drill dia.	2–20 sec/hole after setup	Any unhardened material; Carbides needed for some case-hardened parts	±0.002–±0.010 in. common; ±0.001 in. possible	63–250
Sawing	Bar Plate Sheet	2 in. armor plate (1/2 in. is preferred)	0.010 in. thick	3–30 parts/hour	Any nonhardened material	±0.015 in. possible	250–1000
Broaching	Tube Rod Bar Plate	74 in. long	1 in.	300–400 parts/minute	Any material with good machinability rating	±0.0005–±0.001 in.	32–125
Grinding	Plate Rod Bars	36 in. wide × 7 in. dia.	0.020 in. dia.	1–1000 pieces/hour	Nearly all metalic materials plus many nonmetallic	0.0001 in. and less	16

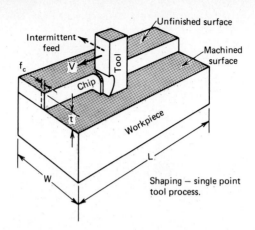

Shaping — single point tool process.

In shaping, the tool (located in RAM) reciprocates over the work with a forward stroke for cutting at velocity (V) and has a quick return velocity of (V_R). The rpm of drive crank (N_S) determines the velocity of the operation.

Stroke ratio, $R_S = \dfrac{\text{Cutting Stroke Angle}}{360 \text{ Degrees}}$

The number of strokes per minute is N_S,

The feed f_c is in inches per stroke,

The length of stroke, ℓ, must be greater than the block length, L, since velocity is position varient. Let $\ell = 2L$. The cut velocity, V, is assumed to be twice the average forward velocity of the ram.

Cutting speed $V = 2 \ell N_S / R_S 12$ (ft/min)

Given a selected cutting speed of V,
$N_S = 12V R_S / 2 \ell$ (strokes/min or rpm)

CT $= W / N_S \cdot f_c$ (minutes)

MRR $= L \cdot W \cdot t / \text{CT}$ (cubic in./min.)

Where W = width of workpiece,
L = length of workpiece,
t = depth of cut.

For a surface of width W, $S = W/f_c$ where S = number of strokes for the job.

Thus CT $= S/N_S = 2 \ell S / 2V R_S$

and MRR $= L \cdot t \cdot N_s \cdot f_c$ in terms of given parameters for the job.

See Chapter 19.

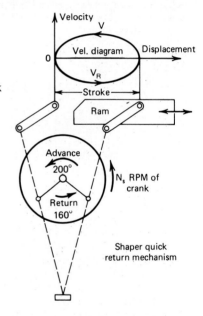

Shaper quick return mechanism

FIGURE 18-1 Basics of the shaping process, including equations for cutting time (CT) and metal-removal rate (MRR).

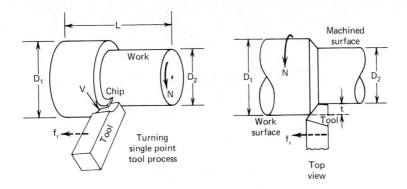

In turning, the primary cutting motion is rotational with the tool feeding parallel to the axis of rotation. The rpm of the rotating workpiece, N, establishes the cutting velocity, V, at the cutting tool according to $V = \pi D N/12$, in surface feet per minute (sfpm). The feed, f_r, is given in inches per revolution (ipr).

The depth of cut, t, is equal to $(D_1 - D_2)/2$ in inches. The length of cut is the distance traveled parallel to the axis, L, plus some overrun to allow the tool to enter and/or exit the cut. Given a selected cutting speed, feed, and depth of cut for a known material being cut with a tool of known cutting tool material, we first need to determine the rpm for the machine.

$N = 12 V/\pi D_1$ using the larger diameter.

The cutting time $= \text{CT} = (L + A)/f_r N$ where A is overrun allowance.

The MRR $= \dfrac{\text{volume removed}}{\text{time}} = \dfrac{\pi D_1^2 L - \pi D_2^2 L}{4L/f_r N}$ (omitting allowance term).

By rearranging and subbing for N, we obtain $12 \dfrac{(D_1^2 - D_2^2)}{4 D_1} f_r V$ which

equals $12 \left(\dfrac{D_1 - D_2}{2}\right)\left(\dfrac{D_1 + D_2}{2D_1}\right) f_r V$ so that

MRR $\cong 12 V f_r t$ cubic inches/min with $\dfrac{D_1 - D_2}{2} = t$ and $\dfrac{D_1 + D_2}{2 D_1} \cong 1$

NOTE THAT THE UNITS FOR THIS APPROXIMATE EQUATION ARE NOT CORRECT. THE APPROXIMATION IS SUFFICIENT EXCEPT FOR VERY LARGE DEPTHS OF CUT.

FIGURE 18-2 Basics of the turning process, including equations for cutting time (CT) and metal-removal rate (MRR).

occur between the tool and freshly generated surfaces of the chip against the rake face and the new surface as rubbed by the flank of the tool.

Basically the chip is formed by a localized shear process which takes place over very narrow regions. This large-strain, high-strain-rate, plastic deformation evolves out of a radial compression zone which travels ahead of the tool as it passes over the workpiece (see Figure 18-9). This radial compression zone has, like all plastic deformations, an elastic compression region which becomes the

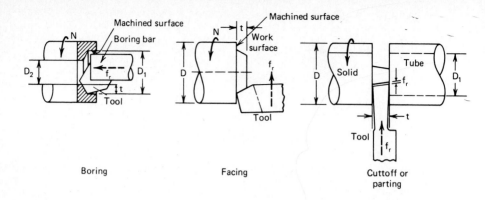

Boring   Facing   Cuttoff or
                                                                          parting

Boring is essentially internal turning while feeding the tool parallel to the rotation axis of the
workpiece.

Given V and f_r, then for a cut of length, L,

The CT $= (L + A)/Nf$, where $N = 12V/\pi D_1$ for $D_1 =$ diameter of bore, where A is overrun
allowance.

$$\text{The MRR} = \frac{(\pi D_1^2 L - \pi D_2^2 L)/4}{L/f_4 N} \cong 12Vf_r t \text{ (omitting allowance term)}$$

In facing, the tool feeds perpendicular to the axis of the rotating workpiece. Because the rpm is
constant, the speed is continually decreasing as the axis is approached. The length, L is $D/2$ or
$(D - D_1)/2$ for a tube.

The cutting time $= \text{CT} = (L + A)/f_r N$ in minutes.

$$\text{The MRR} = \text{Vol/CT} = \frac{\pi D^2 t f_r N}{4L} \cong 6t f_r V \qquad \text{for } L = D/2, \text{ in cubic in./min.}$$

In parting or cutoff work, the tool is fed (plunged) perpendicular to the rotational axis, as it was
in facing. The length of cut for solid bars is $D/2$. For tubes, $L = \dfrac{D - D_1}{2}$

The width of the tool is t in inches.

The equations for CT and MRR are then basically the same as for facing.

In these operations, the speeds and feeds selected are generally less than for turning because of
the large overhang of the tool often needed to complete the cuts. The reduction of the feed (or
depth of cut) reduces the forces operating on the tools and the reduction of the speed usually
reduces the probability of chatter and vibration.

FIGURE 18-3 Basics of boring, facing, and cutoff process. See Figure 18-2 as
reference.

plastic compression region as the field boundary is crossed. The plastic compression generates dense dislocation tangles and networks in annealed metals. When
this work-hardening progresses to a saturated condition (fully work-hardened),
the material has no option but to shear. The shear process itself is nonhomo-
geneous (discontinuous) in which a series of shear fronts or narrow bands produce what is called a lamellar structure in the chips (see Figure 18-10). *This is*

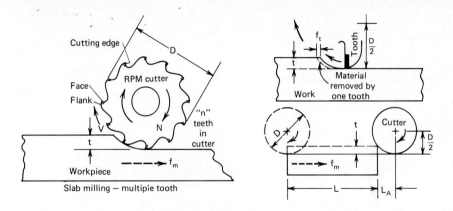

Slab milling — multiple tooth

In milling, the tool rotates (mills) at some rpm (N) while the work feeds past the tool.

The surface cutting speed is established by the cutter of diameter D according to $V = \pi D N/12$ in surface feet per minute where D is in inches.

The depth of cut is t in inches.

The width of cut is the width of the cutter or the work in inches and is given the symbol W.

The length of the cut, L, is the length of the work plus some allowance, L_A, for approach and overtravel.

The feed of the table, f_m, in inches per minute is related to the amount of metal each tooth removes during a revolution (this is called the feed per tooth), f_t, according to $f_m = f_t N n$ where n is the number of teeth in the cutter (teeth/rev.)

Given a selected cutting speed, V, and feed, f_t, for a given work material and tool material combination, we first need to determine the spindle rpm of the machine

$N = 12 V/\pi D$ where D = cutter diameter. Note that the cutting velocity is that which occurs at the cutting edge of the mill.

The table feed is $f_m = f_t n N$ in inches per minute (ipm).

The cutting time = $CT = (L + L_A)/f_m$ in minutes.

The MRR = $\text{Vol}/CT = LWt/CT = Wtf_m$ in^3/min., *ignoring* $\mathbf{L_A}$.

The length of approach = $L_A = \sqrt{\dfrac{D^2}{4} - \left(\dfrac{D}{2} - t\right)^2} = \sqrt{t(D - t)}$ Inches.

See Chapter 22.

FIGURE 18-4 Basics of the milling process (slab milling) as usually performed in a horizontal milling machine, including equations for CT and MRR.

> the fundamental structure which occurs on the microscale in all metals when machined. Through the use of special metallurgical preparations, it can also be observed developing on the side of the work, as shown in Figure 18-11.
> In Figure 18-10, one can see an oxide particle in the surface. This is a hard

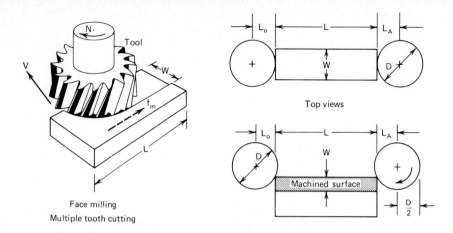

Face milling
Multiple tooth cutting

Top views

In face milling, the tool rotates (face mills) at some rpm (n) while the work feeds past the tool. The surface cutting speed is related to the cutter diameter D and cutter according to

$$V = \pi DN/12 \qquad \text{in surface feet per minute.}$$

The depth of cut is t in inches.

The width of cut is W in inches and may be width of the workpiece or width of the cutter, depending upon the setup.

The length of cut, L in inches is the length of the workpiece, L, plus an allowance for approach for approach, L_A and overtravel, L_O.

The feed of the table, f_m, in inches per minute is related to the amount of metal each tooth removes during a pass over the work and this is called the feed per tooth, f_t, where $f_m = f_t Nn$. The number of teeth in the cutter is n.

Given a selected cutting speed, V, and feed per tooth, f_t, for a given workpiece and cutting tool material combination, we need to determine the spindle rpm and feed rate of the table of the machine.

$N = 12\ V/\pi D$ where D = cutter diameter. Note that the cutting velocity is the tip speed of the cutter teeth.

The table feed is $f_m = f_t\, nN$ in inches per minute (ipm).

The cutting time = $CT = (L + L_A + L_O)/f_m$ in minutes.

The MRR = $\text{Vol}/CT = LWt/CT = Wtf_m$ in inches^3/min. (ignore L_O and L_A).

The length of approach = length of overtravel = $D/2$ inches, usually.

For a setup where the tool does not completely pass over the workpiece,

$$L_O = L_A = \sqrt{W(D - W)} \text{ for } W < D/2 \text{ or}$$

$$L_O = L_A = D/2 \text{ for } W \geqslant D/2.$$

FIGURE 18-5 Basics of the milling process (face and end milling) as performed on a vertical spindle machine, including equations for cutting time and metal-removal rate.

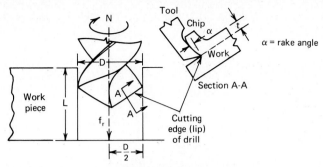

Drilling-multiple edge tool

A conventional drilling process uses a drill of diameter D with two principal cutting edges rotating at rpm of N and feeding axially. The rpm of drill establishes the cutting velocity, V, at the cutter edges of the drill lips according to $V = \pi DN/12$ in surface feet per minute (sfpm).

The feed, f_r, is given in inches per revolution or ipm, $f_r = 2t$. The length of cut in drilling equals the depth of the hole, L, plus an allowance for approach and for the tip of drill, usually $A = D/2$.

Given a selected cutting speed and feed for drilling a hole in a certain metal with a drill of known tool material, determine the rpm of the spindle of the machine.

$N = 12V/\pi D$ where $D = $ drill diameter. Note that the maximum velocity occurs at the extreme ends of the drill lips and the velocity is very small near the center by the chisel end of the drill.

The cutting time $= CT = (L + A)/f_r N$

The MRR $= $ Vol./CT $= (\pi D^2/4)Lf_r N/L$ (omitting allowances)

which reduces to MRR $= (\pi D^2/4)f_r N$ (cubic inches per minute).

An approximate form of MRR $= 3\,DVf_r$ is often used in shop work. Note that the units here are not exactly correct due to the assumption noted above for cutting velocity.

FIGURE 18-6 Basics of the drilling (hole-making) process, including equations for cutting time (CT) and metal-removal rate (MRR).

second-phase particle which acts as a barrier to the shear front dislocations, which cannot penetrate the particle. The dislocations create voids around the particles in the shear front layers. If there are enough particles of the right size and shape, the chip will fracture through the shear zone, forming segmented chips. "Free-machining steels," which have small percentages of hard second-phase particles added to them, use this metallurgical phenomena to break up the chips for easier chip handling.

In Figure 18-10, one can also observe a pattern in the lamella lying perpendicular to the shear fronts. This pattern was produced by the cutting edge of the tool during the previous pass of the tool over the workpiece, subsequently sheared on the next pass. Thus this pattern actually reflects the microgeometry of the cutting edge of the tool.

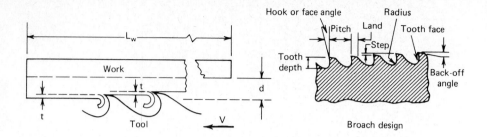

In broaching, the tool (or work) is translated past the work (or tool) with a single stroke of velocity V. The feed is provided by a gradual increase in height of successive teeth. This step height is called the rise per tooth and varies depending on whether the tooth is for roughing (t_r), semifinishing (t_s) or final sizing or finishing (t_f). In a typical broach there are 3 to 5 semifinishing and finishing teeth specified, so we need to estimate the number of roughing teeth needed in order to be able to find the broach length which is needed to estimate the cutting time. Other lengths needed for a typical pull broach, are shown in the sketch below.

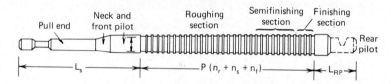

The pitch or distance between teeth is $= .3$ to $.4$ length of cut, where length of cut usually equals L_w.

The number of roughing teeth is $n_r = \dfrac{[d - (n_f t_f) - (n_s t_s)]}{t_r}$ where d is the total amount of metal to be removed and $t_r =$ rise per tooth.

The length of the broach $L_B = (n_r + n_s + n_f) P + L_s + L_{RP}$ for pull broach.

The length of stroke $L = L_B + L_w$ if broach moves past work in inches or $L_w + L_B$ if work moves past broach.

The cutting time, $\text{CT} = L/12V$ where V is the cutting speed, in surface ft. per minute, usually 15–30 sfpm.

The metal removal rate depends upon the number of teeth (roughing) contacting the work.

MRR (per tooth) $= 12\, t_r WV$ in^3/min per roughing tooth.

Maximum number of roughing teeth in contact with part $n = L_w/P$ for broach larger than part.

MRR (for process) $= 12\, t_r\, WVn$ in^3/min where n is usually rounded off to next largest whole number.

FIGURE 18-7 Process basics of broaching, including equations for cutting time and metal removal rate.

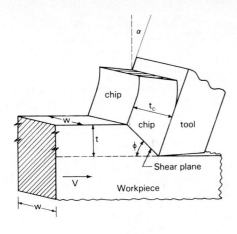

FIGURE 18-8 Schematic of orthogonal machining. The cutting edge of the tool is perpendicular to the direction of motion (V). The back rake angle is α. The shear angle is ϕ.

In Figure 18-11, the narrow shear fronts on the side of the chip are observed to develop at the tool tip and progress toward the free surface, producing ultimately the lamellar structure, seen in Figure 18-10, on the top of the chip. Shear fronts are very narrow (100 to 500 angstroms) compared to the thickness of the lamella (2 to 4 μm) and account for the large strain rates. The micrographs of this figure were made in a scanning electron microscope.

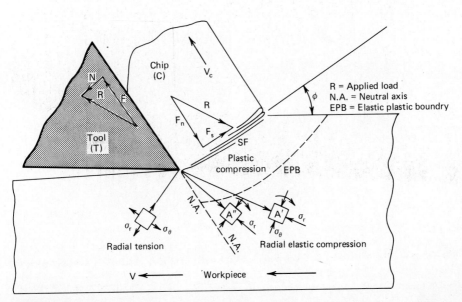

FIGURE 18-9 Machining process produces a radial compression ahead of the shear process. The stress reverses from compression to tension across the neutral axis (N.A.).

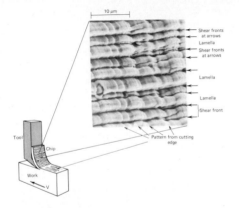

FIGURE 18-10 SEM micrograph of the top of electpolytic tough pitch copper chip, clearly showing the lamella structure produced by the shear process in metal cutting.

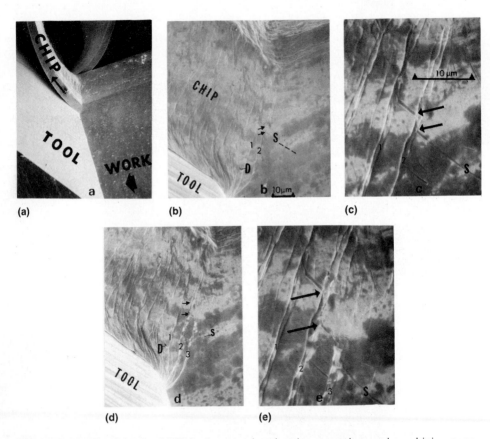

(a) (b) (c)

(d) (e)

FIGURE 18-11 Series of SEM micrographs showing an orthogonal machining operation. (a) Low-magnification view provides orientation. (b) Shear fronts labeled 1 and 2 have progressed toward the free surface, and *D* indicates the location of a defect for later orientation. (c) Central region of (b) is magnified to allow observation of the shear of a surface scratch, labeled *S*; the tool was then advanced slightly so that one additional shear front was generated. (d) New shear front is labeled 3; note the position of the defect *D* with respect to the tool. (e) Additional shear has occurred on shear front 2 and shear front 3 has just sheared the scratch *S*. Note also in (d) the secondary deformation due to the tool/chip interface interaction.

Orthogonal machining is done to test machining mechanics and theory. Orthogonal machining (meaning two forces) can be obtained in practice by

1. Machining a plate as shown in Figure 18-8.
2. End cutting a tube wall in a turning setup.

When the cutting edge and the cutting motion are not perpendicular to each other, we have oblique cutting, which is the common situation described in the preceding section for shaping, drilling, milling, and single-point turning. Since the orthogonal case is more easily modeled, it will be used here to further describe the process.

For the purpose of this modeling activity, we shall assume that the shear process takes place on a single narrow plane rather than a set of shear fronts which actually comprise a narrow shear zone. For modeling purposes, we will also assume that the tool cutting edge is perfectly sharp and no contact is being made between the flank of the tool and the new surface. The workpiece passes the tool with velocity V, which is of course the cutting speed. The depth of cut is t. Ignoring the plastic compression, chips having thickness t_c and velocity V_c are formed by the shear process. The shear process then has velocity V_s and occurs at shear angle ϕ. The tool geometry is given by the back rake angle α and the clearance angle γ. The velocity triangle for V, V_c, and V_s is also shown (Figure 18-12b). The chip makes contact with the rake face of the tool over length l. The plate thickness is called w.

From the geometry, the equation for the chip thickness ratio, r_c, defined as t/t_c can be derived.

$$r_c = \frac{AB \sin \phi}{AB \cos (\phi - \alpha)} \tag{18-1}$$

where AB is the length of the shear plane (see Figure 18-14).

Equation (18-1) may be solved for the shear angle ϕ as a function of the measurable chip thickness ratio by expanding the cosine term and simplifying.

$$\tan \phi = \frac{r_c \cos \alpha}{1 - r_c \sin \alpha} \tag{18-2}$$

There are numerous other ways to measure chip ratios and obtain shear angles both during (dynamically) and after (statically) the cutting process. For example, the ratio of the length of the chip to the length of the cut can be used to determine r_c. Many researchers use the chip compression ratio, which is the reciprocal of r_c, as a parameter. The shear angle can be measured statically by instantaniously interrupting the cut through the use of "quick-stop devices." These devices disengage the cutting tool from the workpiece, while cutting is in progress, leaving the chip attached to the workpiece. Optical and scanning electron microscopy is then used to observe the direction of shear. Figure 1-8 was made this way. High-speed motion pictures have also been used to observe the process at frame rates of as high as 30,000 frames per second. More recently, machining stages have been built that allow the process to be performed inside a scanning electron microscope and recorded on video tapes for high-resolution, high-mag-

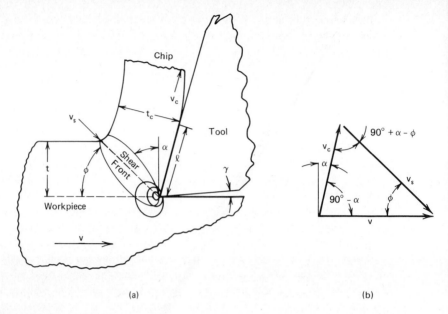

FIGURE 18-12 (a) Schematic of orthogonal machining process. (b), Velocity diagram associated with orthogonal machining.

nification examination of the deformation process. Figure 18-11 was made this way. Sophisticated electronics and slow-motion playback have allowed this technique to be used to measure for the first time the velocity of the shear process during cutting, verifying the existence of the velocity triangle shown in Figure 18-12b. The vector sum of V and V_c equals V_s.

For consistency of volume, we observe that

$$\frac{V_c}{V} = \frac{t}{t_c} = r_c = \frac{\sin \phi}{\cos (\phi - \alpha)} \qquad (18\text{-}3)$$

indicating that the chip ratio (and therefore the shear angle) can be determined dynamically if a reliable means to measure V_c can be found. The ratio of V_s to V is

$$\frac{V_s}{V} = \frac{\cos \alpha}{\cos (\phi - \alpha)} \qquad (18\text{-}4)$$

These velocities are important in power calculations, heat determinations, and vibration analysis associated with chip formation.

During the cutting, the chip undergoes a shear strain ϵ_c of

$$\epsilon_c = \frac{\cos \alpha}{\sin \phi \cos (\phi - \alpha)} \qquad (18\text{-}5)$$

which shows that the shear strain is dependent on the rake angle α and the shear direction, ϕ. Generally speaking, metal cutting strains are quite large compared to other plastic deformation processes, being on the order of 2 to 4 inches per inch.

This large strain occurs however over very narrow regions (the shear fronts) which results in extremely high shear strain rates, $\dot{\epsilon}_c$, which typically are in the range of 10^4 to 10^8 in./in./sec. The strain rate in metal cutting can be estimated by

$$\dot{\epsilon}_c = \frac{V}{t_{SF}} \qquad (18\text{-}6)$$

where t_{SF} is the thickness of the shear front.

It is this combination of large strains and high strain rates operating within a process constrained only by the rake face of the tool which results in great difficulties in theoretical analysis of this process.

Effects of work material properties. As noted previously, the properties of the work material are important in chip formation. High-strength materials require larger forces than do materials of lower strength, causing greater tool and work deflection, increased friction and heat generation and operating temperatures, and requiring greater work input. The structure and composition also influence metal cutting. Hard or abrasive constituents, such as carbides in steel, accelerate tool wear.

Work material ductility is an important factor. Highly ductile materials not only permit extensive plastic deformation of the chip during cutting, which increases work, heat generation, and temperature, but it also results in longer, "continuous" chips that remain in contact longer with the tool face, thus causing more frictional heat. Chips of this type are severely deformed and have a characteristic curl. On the other hand, some materials, such as gray cast iron, lack the ductility necessary for appreciable plastic chip formation. Consequently, the compressed material ahead of the tool fails in a brittle manner along the shear zone, producing small fragments. Such chips are termed *discontinuous* or *segmented* (see Figure 18-13).

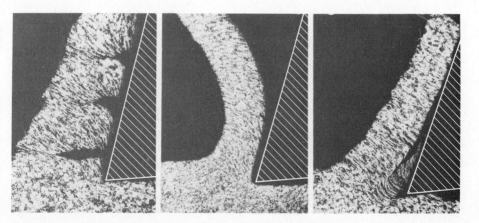

FIGURE 18-13 Three characteristic types of chips. (*Left to right*) Discontinuous, continuous, and continuous with built-up edges with chip samples produced by "quick stop" techniques. (*Courtesy Cincinnati Milacron, Inc.*)

A variation of the continuous chip, often encountered in machining ductile materials, is associated with a "built-up" edge (BUE) on the cutting tool. The local high temperature and extreme pressure in the cutting zone cause the work material to adhere or weld to the cutting edge of the tool forming the built-up edge rather like a dead metal zone in the extrusion process. Although this material protects the cutting edge from wear, it modifies the geometry of the tool. BUE's are not stable and will slough off periodically, adhering to the chip or passing under the tool and remaining on the machined surface. Built-up edge formation often can be eliminated or minimized by reducing the depth of cut, increasing the cutting speed, using positive rake tools, or applying a coolant.

Mechanics of machining. Orthogonal machining has been defined as a two-force system, whereas oblique cutting involves a three-force situation. Consider Figure 18-14, which shows a free-body diagram of a chip which has been separated at the shear plane. It is assumed that the resultant force R acting on the back of the chip is equal and opposite to the resultant force R' acting on the shear plane. The resultant R is composed of the friction force F and the normal force N acting on the tool/chip interface contact area. The resultant force R' is composed of a shear force F_s and a normal F_n, acting on the shear plane area A_s. Since neither of these two sets of forces can usually be measured, a third set is needed which can be measured, using a dynamometer, mounted either in the workholder or the tool holder. Note that this set has resultant R''', which is equal in magnitude to all the other resultant forces in the diagram. The resultant force R''' is composed of a cutting force F_c and a tangential (normal) force F_t. Now it is necessary to express the desired forces (F_s, F_n, F, N) in terms of the dynamometer components, F_c and F_t, and appropriate angles. In order to do this, a circular force diagram is developed in which all six forces are collected in the same force circle. This is shown in Figure 18-15. The only symbol in this figure as yet undefined is β, which is the angle between the normal force

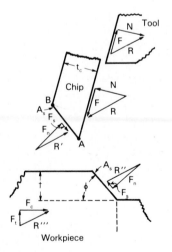

FIGURE 18-14 Free-body diagram of orthogonal chip formation process, showing equilibrium condition between resultant forces R and R'.

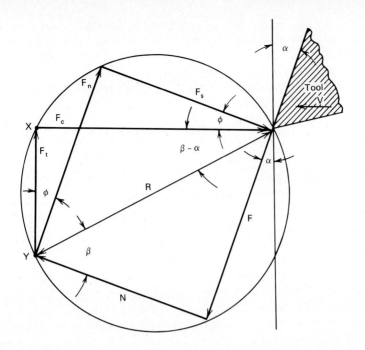

FIGURE 18-15 Circular force diagram used to derive equations for F_s, F_n, F, and N as functions of F_c, F_t, ϕ, α, and β.

N and the resultant R. It is used to describe the friction coefficient μ on the tool/chip interface area, which is defined as F/N, so that

$$\beta = \tan^{-1} \mu = \tan^{-1} \frac{F}{N} \tag{18-7}$$

The friction force F and its normal N can be shown to be

$$F = F_c \sin \alpha + F_t \cos \alpha \quad \text{or} \quad R \sin \beta \tag{18-8}$$

$$N = F_c \cos \alpha - F_t \sin \alpha \quad \text{or} \quad R \cos \beta \tag{18-9}$$

where

$$R = \sqrt{F_c^2 + F_t^2} \tag{18-10}$$

all in pounds. Notice that in the special situation where the back rake angle α is zero, $F = F_c$ and $N = F_t$, so that in this orientation, the friction force and its normal can be directly measured by the dynamometer.

The forces parallel and perpendicular to the shear plane can be shown (from the force circle diagram) to be

$$F_s = F_c \cos \phi - F_t \sin \phi \quad \text{in pounds} \tag{18-11}$$

$$F_n = F_c \sin \phi + F_t \cos \phi \quad \text{in pounds} \tag{18-12}$$

F_s is of particular interest as it is used to compute the shear stress on the shear plane. This shear stress is defined as

$$\tau_s = \frac{F_s}{A_s} \qquad \text{in pounds per square inch (psi)} \qquad (18\text{-}13)$$

where

$$A_s = \frac{tw}{\sin \phi} \qquad \text{(area in inches}^2) \qquad (18\text{-}14)$$

recalling that t was the depth of cut and w was the width of the workpiece. The shear stress is therefore

$$\tau_s = \frac{F_c \sin \phi \cos \phi - F_t \sin^2 \phi}{tw} \qquad \text{psi} \qquad (18\text{-}15)$$

For a given polycrystalline metal, this shear stress is a material constant, not sensitive to variations in cutting parameters, tool material, or the cutting environment. Figure 18-16 gives some typical values for the flow stress for a variety of metals, plotted against hardness.

It is the objective of some metal-cutting researchers to be able to derive (predict) the shear stress τ_s and the shear direction ϕ from dislocation theory, but this has not yet been accomplished. Correlations of the shear stress with

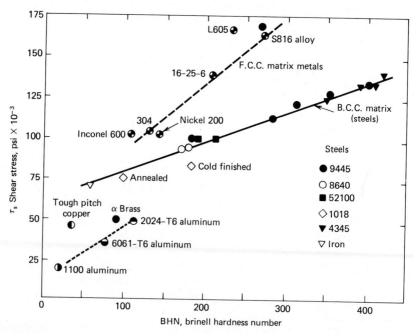

FIGURE 18-16 Shear stress, τ_s, variation with the Brinell hardness number for a group of steels and aerospace alloys. Data of some selected fcc metals are also included. (*Adapted with permission from S. Ramalingham and K. J. Trigger*, Advances in Machine Tool Design and Research, *copyright 1971, Pergamon Press Ltd.*)

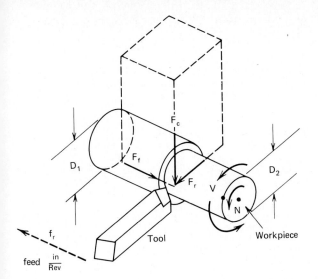

FIGURE 18-17 Three components of measurable forces acting on a single-point turning tool: oblique machining.

metallurgical measures like hardness or dislocation stacking fault energy have been useful in these efforts.

Energy and power in machining. The cutting force system in a conventional, oblique chip formation process, shown schematically in Figure 18-17, is composed of three components.

1. F_c, primary cutting force acting in the direction of the cutting velocity vector. This force is generally the largest force and accounts for 99% of the power required by the process.
2. F_f, feed force acting in the direction of the tool feed. This force is usually about 50% of F_c but accounts for only a small percentage of the power required because feed rates are usually small compared to cutting speeds.
3. F_r, radial or thrust force acting perpendicular to the machined surface. This force is typically about 50% of F_f and contributes very little to power requirements because velocity in the radial direction is negligible.

The energy per unit time, P or power required for cutting is

$$P = F_c V \text{ ft-lb/min} \tag{18-16}$$

The horsepower at the spindle of the machine is therefore

$$\text{HP} = \frac{F_c V}{33,000} \tag{18-17}$$

In metal cutting a very useful parameter is called the unit or specific horsepower HP_s, which is defined as

$$\text{HP}_s = \frac{\text{HP}}{\text{MRR}} \quad (\text{hp/in}^3/\text{min}) \tag{18-18}$$

TABLE 18-2 Values for Specific or Unit Horsepower, HP$_s$, for Various Metals During Metal Removal

	Ferrous Metals and Alloys							High-Temperature Alloys		
	Brinell Hardness Number							Material Classification	Brinell Hardness Number	HP$_s$
AISI	150–175	176–200	201–250	251–300	301–350	351–400		A 286	165	0.82
1010–1025	0.58	0.67	—	—	—	—		A 286	285	0.93
1030–1055	0.58	0.67	0.80	0.96	—	—		Chromoloy	200	0.78
1060–1095	—	—	0.75	0.88	1.0	—		Chromoloy	310	1.18
1112–1120	0.50	—	—	—	—	—		Hastelloy-B	230	1.10
1314–1340	0.42	0.46	0.50	—	—	—		Inco 700	330	1.12
1330–1350	—	0.67	0.75	0.92	1.1	—		Inco 702	230	1.10
2015–2115	0.67	—	—	—	—	—		M-252	230	1.10
2315–2335	0.54	0.58	0.62	0.75	0.92	1.0		M-252	310	1.20
2340–2350	—	0.50	0.58	0.70	0.83	—		TI-150A	340	0.65
2512–2515	0.50	0.58	0.67	0.80	0.92	—		U-500	375	1.10
3115–3130	0.50	0.58	0.70	0.83	1.0	1.0		4340	200	0.78
3160–3450	—	0.50	0.62	0.75	0.87	1.0		4340	340	0.93
4130–4345	—	0.46	0.58	0.70	0.83	1.0				
4615–4820	0.46	0.50	0.58	0.70	0.83	0.87		Nonferrous Metals and Alloys		
5120–5150	0.46	0.50	0.62	0.75	0.87	1.0				HP$_s$
52100	—	0.58	0.67	0.83	1.0	—		Brass Hard		0.83
								Medium		0.50–0.60

6115–6140	0.46	0.54	0.67	0.83	1.0	—
6145–6195	—	0.70	0.83	1.0	1.2	1.3
Plain cast iron	0.30	0.33	0.42	0.50	—	—
Alloy cast iron	0.30	0.42	0.54	—	—	—
Malleable iron	0.42	0.55	—	—	—	—
Cast steel	0.62	0.67	0.80	—	—	—

Soft		0.33
Free machining		0.25–0.41
Bronze		
Hard		1.10–1.20
Medium		0.50–0.60
Soft		0.33–0.40
Copper (pure)		0.90
Aluminum		
2014–T6		0.24
108		0.15
Cast		0.25–0.40
Hard (rolled)		0.33
Monel (rolled)		1.0
Zinc alloy (die cast)		0.25

Other Metals and Alloys

Titanium A-55	0.65–0.76
Titanium C-130	0.81–0.93
S-816	
SAE 302	0.72
SAE 350	1.20
Magnesium alloy	0.39
Nickel	0.89

Source: Data from *Turning Handbook of High-Efficiency Metal Cutting*, courtesy General Electric Company; and from author's sources.

In turning, for example, where MRR $\simeq 12Vf_r d$,

$$HP_s = \frac{F_c}{396,000f_r d} \quad (hp/in^3/min) \quad (18\text{-}19)$$

Thus this term represents the approximate power needed at the spindle to remove a cubic inch of metal per minute. Specific power factors for some common materials are given in Table 18-2. Specific horsepower is related to and correlates well with shear stress for a given metal. Notice the similarity between equations (18-13) and (18-19). The major difference is that unit power is sensitive to material properties (like hardness) as is τ_s, but also rake angle, depth of cut, and feed.

Essentially, the majority of the energy is consumed in shear and tool/chip interface friction.

Thus

$$U = U_s + U_f \quad (18\text{-}20)$$

Where total energy is

$$U = \frac{F_c}{f_r d} \quad (18\text{-}21)$$

and specific shear energy is

$$U_s = \frac{F_s V_s}{f_r dV} = \frac{F_s \cos \alpha}{f_r V \cos (\phi - \alpha)} \quad (psi) \quad (18\text{-}22)$$

Specific friction energy is

$$U_f = \frac{FV_c}{f_r dV} = \frac{Fr_c}{f_r d} \quad (18\text{-}23)$$

with usually 60 to 70% of the total energy going into the shear process.

Specific power can be used in a number of ways. First, it can be used to estimate the motor horsepower required to perform a machining operation for a given material by multiplying HP_s values from the table by the approximate MRR for the process. The motor horsepower is then

$$HP_m = \frac{HP_s \times MRR \times \text{correction factors}}{E} \quad (18\text{-}24)$$

where E is the efficiency of the machine. This factor accounts for the power needed to overcome friction and inertia in the machine and drive moving parts. Correction factors are used to account for variations in cutting speed, feed, and rake angle. There is usually a tool wear correction factor of 1.25 used to account for the fact that dull tools use more power than sharp tools.

The primary cutting force F_c can also be roughly estimated according to

$$F_c \simeq \frac{HP_s \times MRR \times 33,000}{V} \quad \text{pounds} \quad (18\text{-}25)$$

This type of estimate of the major force F_c is useful in analysis of deflection and vibration problems in machining and in the proper design of work-holding devices, as these devices must be able to resist movement and deflection of the part during the process.

In general, increasing the speed, the feed, or the depth of cut will increase the power requirement. Doubling the speed doubles the HP directly. Doubling the feed or the depth of cut doubles the cutting force F_c. In general, increasing the speed does not increase the cutting force F_c, which has always been a puzzle. However, recent research has shown that the periodicity or spacing of the shear fronts remains constant in a material regardless of the cutting speed. Doubling the velocity essentially doubles the number of shear fronts produced in a given amount of time. Put another way, if speed is doubled, chip length is doubled for the same amount of cutting time. For constant shear front spacing (constant lamella size) this means twice as many shear fronts. So energy is doubled. Cutting force F_c, on the other hand, reflects a change in F_s, where $F_s = \tau_s A_s$ and τ_s = shear strength, A_s = shear area. Neither of these two quantities were changed by the change in speed. Thus F_c remains constant. However, speed does have a strong effect on tool life because most all the input energy is

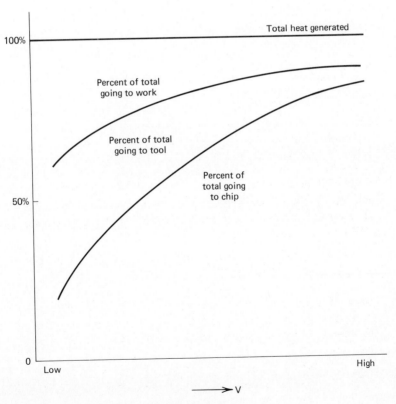

FIGURE 18-18 Distribution of heat generated in machining to the chip, tool, and workpiece. Heat going to the environment is not shown.

converted into heat, which raises the temperature of the chip, the work, and the tool, to the latter's detrement.

Heat and temperature in metal cutting. In metal cutting, the power put into the process (F_cV) is largely converted to heat, elevating the temperatures of the chip, the workpiece, and the tool. These three elements of the process, along with the environment (which includes the cutting fluid), act as the heat sinks. Figure 18-18 shows the distribution of the heat to these three *sinks* as a function of cutting speed.

There are three main *sources* of heat. Listed in order of their heat-generating capacity, they are:

1. The shear zone itself when plastic deformation results in the major heat source. Most of this heat stays in the chip.
2. The tool/chip interface contact region, where additional plastic deformation takes place in the chip and there is considerable heat generated due to sliding friction.
3. The flank of the tool, where the freshly produced workpiece surface rubs the tool.

There have been numerous experimental techniques developed to measure cutting temperatures and some excellent theoretical analysis of this "moving" multiple-heat-source problem. Space does not permit us to explore this problem in depth. Figure 18-19a shows the effect of cutting speed on the tool/chip interface temperature. The rate of wear of the tool at the interface can be shown to be directly related to temperature (see Figure 18-19b). Because cutting forces are concentrated on small areas near the cutting edge, these forces produce large pressures. The tool material must be hard to resist wear and tough to resist cracking and chipping. Tools used in interrupted cutting, like milling, must be able to resist impact loading as well. Tool materials must sustain these properties

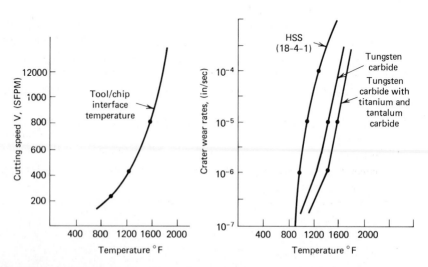

FIGURE 18-19 Typical relationships of temperature to cutting speed and crater wear for machining steels.

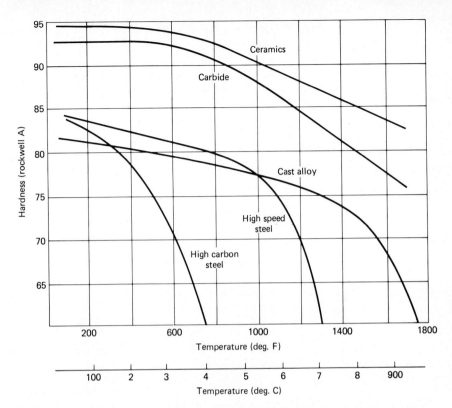

FIGURE 18-20 Hardness of tool materials decreasing with increasing temperature. Some materials display a more rapid drop in hardness above some temperatures. (*From* Metal Cutting Principles, *2nd ed.; courtesy Ingersoll Cutting Tool Company.*)

at elevated temperatures, as shown in Figure 18-20, which relates the hardness of various tool materials to temperature.

The challenge to manufacturers of cutting tools has always been to find materials which satisfy these process needs. Cutting tool materials that do not lose hardness at the high temperatures associated with high speeds are said to have "hot hardness," but obtaining this property usually requires a trade-off in toughness, as hardness and toughness are generally opposing properties.

Cutting tool materials.   In nearly all production machining operations, cutting speed and feed are limited by the capability of the tool material. Chip formation involves high local stresses, sliding friction and abrasion, and considerable heat generation. Speeds and feeds must be kept low enough to provide for a minimum acceptable tool life. If not, the time lost changing tools may outweigh the productivity gains due to increased cutting speed. The materials selected for cutting tools must combine hardness and strength (toughness) with good wear resistance at elevated temperatures. Figure 18-21 presents a chronological rating of cutting tool materials, clearly showing the rapid advances that have occurred in this field in the last two decades. The tool materials are rated by their permissible cutting speed in machining steel materials.

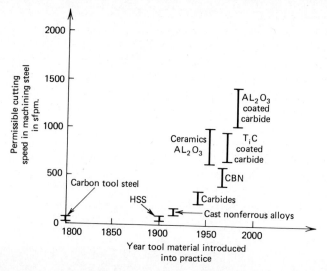

FIGURE 18-21 The acceleration of cutting tool material technology measured by permissible cutting speed (sfpm) for machining steel.

Let us review these materials in their chronological order, beginning with steel-alloyed tools, generally known as tool steels.

Tool steel. Plain carbon steel of 0.90 to 1.30% carbon when hardened and tempered has good hardness and strength, adequate toughness, and can be given a keen cutting edge. However, it loses its hardness at temperatures above 204°C (400°F) because of tempering, and it has largely been replaced by other materials for metal cutting.

High-speed steel. First introduced in 1900, this high-alloy steel is remarkably superior to tool steel in that it retains its cutting ability at temperatures up to 593°C (1100°F), exhibiting "red hardness." Compared with tool steel, it can operate at about double the cutting speed with equal life, resulting in its name *high-speed steel,* often abbreviated HSS. Although several formulations are used, a typical composition is that of the 18–4–1 type (tungsten 18%, chromium 4%, vanadium 1%). Comparable performance can also be obtained by the substitution of approximately 8% molybdenum for the tungsten. High-speed steel still is widely used for drills and many types of general-purpose milling cutters and in single-point tools used in general machining. It has been almost completely replaced for high-production machining by carbides and coated tools.

Cemented carbide. These nonferrous alloys are used for most metal-cutting operations. They are also called sintered carbides because they are manufactured by powder metallurgy techniques. These materials became popular during World War II as they afforded a four- or fivefold increase in cutting speeds. The early versions, which are still widely used, had tungsten carbide as the major constituent, with a cobalt binder in amounts from 3 to 13%. Recent types utilize very fine microparticles dispersed in the carbide structure, improving their toughness and tool life, particularly when subjected to impact, as in making interrupted cuts. Various other carbides, especially titanium, tantalum, and columbium, can be added or substituted for the tungsten. Another type is composed of titanium carbonitrides, titanium-molybdenum transition phases, and a nickel alloy binder, thus requiring no scarce or imported materials.

Carbide tools are extremely hard (90 to $95R_A$) and can be operated at cutting speeds 200 to 500% greater than those used for high-speed steel. Consequently, they have replaced high-speed steel in many metal machining processes. Carbides are not as tough as high-speed steels and sometimes react chemically with iron and steel during cutting. At high cutting speeds, they lose hardness and can plastically deform.

Many carbide tools are made in the form of "throwaway inserts" (see Figure 18-24). They contain from three to eight cutting edges and are held mechanically in a tool holder. When one cutting edge becomes dull, the insert is rotated or turned over to a new edge; when all the edges are dull, the insert is removed from service. It is very important that the carbide inserts be held firmly and backed up well. Table 18-3 shows some of the critical mechanical properties of carbide versus HSS and oxide tools.

Ceramics. A very important addition to the list of cutting-tool materials, ceramics are made of pure aluminum oxide. Very fine particles are formed into cutting tips under a pressure of 267 to 386 MPa (20 to 28 tons/in²) and sintered at about 982°C (1800°F). Unlike the case with ordinary ceramics, sintering occurs without a vitreous phase.

Ceramics usually are in the form of disposable tips. They can be operated at from two to three times the cutting speeds of tungsten carbide, almost completely resist cratering, usually require no coolant, and have about the same tool life at their higher speeds as tungsten carbide does at lower speeds. As shown in Table 18-3, ceramics are usually as hard as carbides but are more brittle (lower bend strength) and therefore require more rigid tool holders and machines in order to take advantage of their capabilities. Their hardness and chemical inertness makes ceramics a good material for finishing.

Diamonds. Diamond is the hardest material known. (Knoop $\simeq$ 7000). Industrial diamonds are now available in the form of polycrystalline compacts which are finding industrial application in machining of aluminum, bronze, and plastics, greatly reducing the cutting forces as compared to carbides. Diamond machining is done at high speeds wih fine feeds for finishing and produces excellent finishes. Single-crystal diamonds, with a cutting-edge radius of 100 angstroms or less, are being used for precision machining of large mirrored surfaces. They

TABLE 18-3   Mechanical Properties of Tool Materials

	Hardness Rockwell A (25°C)	Transverse Rupture (Bend) Strength ($\times 10^{-3}$ psi)	Compressive Strength ($\times 10^{-3}$ psi)	Modulus of Elasticity, E ($\times 10^{-6}$ si)
Cemented carbide	90–95	150–375	600–800	70–100
Tool steel	86	600	600–650	30
Ceramic (oxide)	92–94	100	650	50–60

Source: T. E. Hale et al., *Materials Technology,* Spring 1980, p. 19.

have, for many years, been used to slice biological materials into thin films for viewing in transmission electron microscopes.[1]

Cubic boron nitride. CBN is a man-made tool material, developed by General Electric,[2] and is the hardest material known to man other than diamond. It retains its hardness at elevated temperatures (Knoop 4700 at 20°C, 4000 at 1000°C) and has low chemical reactivity at the tool/chip interface. This material can be used to machine hard aerospace materials like Inconel 718 and René 95 as well as chilled cast iron.

Coated carbide. Coated inserts of carbides are finding more success in many metal cutting applications in recent years. The basic idea is to coat a tough, shock-resistant carbide with a thin, hard, crater-resistant surface layer to provide for a long-lasting, tough tool material. TiC-coated tools were introduced in 1969. These tools have two or three times the wear resistance of the best uncoated tools with the same breakage resistance. This results in a 50 to 100% increase in speed for the same tool life. Most coated inserts cover a broader application range, so fewer grades are needed, resulting in lower inventory costs for the user. Other materials which have found success as coating materials for carbides are titanium nitride (less flank resistance but better crater resistance), hafium nitride (better wear resistance than TiC in steel turning), and aluminum oxide coating for carbides. Al_2O_3-coated carbides permitted the cutting speed to be increased 90% in machining AI1045 steel. Aluminum oxide coatings also demonstrated excellent crater wear resistance by providing a chemical/diffusion reaction barrier at the tool/chip interface.

Coated carbide tools are now more than 10 years old and their acceptance into the metalworking industry has progressed to the place where in the United States about 35% of the carbide tools used in metalworking are coated. Recent improvements in control of thickness, porosity, and the interface metallurgy have resulted in improved coating adhesion and suppression of strength-degrading interface reactions.

Tool Geometry. The geometry of a single-point tool is critical to the performance of the tool during metal removal. There are essentially six angles to be defined. These are shown in Figure 18-22. This is the typical HSS tool used in turning or shaping operations.

The face of the tool over which the chip flows during cutting is a plane of compound slope established by the back rake and side rake angles. For a specific cut, the *true rake* is defined as the inclination of the tool face at the cutting edge as measured in the direction of actual chip flow.

True rake inclination of a cutting tool has a major effect in determining the amount of chip compression and the shear angle. A small rake angle causes high compression, tool forces, and friction, resulting in a thick, highly deformed, hot chip. Increased rake angle reduces the compression, the forces, and the friction, yielding a thinner, less-deformed, and cooler chip. Unfortunately, it is difficult, and usually not possible, to take much advantage of these desirable

---

[1]This process is known as ultramicrotomy and it is one of the few industrial versions of orthogonal machining in common practice.

[2]Called Borazon, it is made in a compact form for tools by a process quite similar to that used for polycrystalline diamonds.

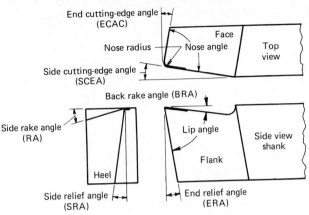

End cutting-edge angle (ECAC)

Nose radius

Nose angle

Face

Top view

Side cutting-edge angle (SCEA)

Back rake angle (BRA)

Side rake angle (RA)

Lip angle

Side view shank

Flank

Heel

Side relief angle (SRA)

End relief angle (ERA)

FIGURE 18-22 Standard terminology to describe the geometry of single-point tools.

effects of larger positive rake angles, since they are offset by the reduced strength of the cutting tool, due to the reduced tool section, and by its greatly reduced capacity to conduct heat away from the cutting edge. An interesting development along these lines will be discussed later.

In order to provide greater strength at the cutting edge and better heat conductivity, zero or negative rake angles commonly are employed on sintered carbide and ceramic cutting tools. These materials tend to be brittle, but their ability to hold their superior hardness at high temperatures makes them preferred for high-speed and continuous machining operations. A negative rake angle increases tool forces to some extent, but this minor disadvantage is offset by the added support to the cutting edge. This is particularly important in making intermittent cuts and in absorbing the impact during the initial engagement of the tool and work.

The introduction of coated tools has spurred the development of improved tool geometries. Specifically, in order to reduce the total energy consumed due to increased cutting speed and to break up the chips into shorter segments, low-force groove geometries have been developed (see Figures 18-23 and 18-24). The effect of these grooves is to effectively increase the rake angle, which increases the shear angle and lowers the cutting force and power. Note that this also means lower cutting temperatures, which helps improve tool life.

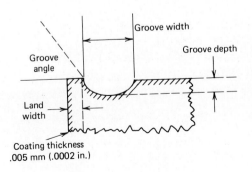

Groove width

Groove depth

Groove angle

Land width

Coating thickness .005 mm (.0002 in.)

FIGURE 18-23 Groove design parameters showing typical low-force groove geometries for coated-carbide insert tools.

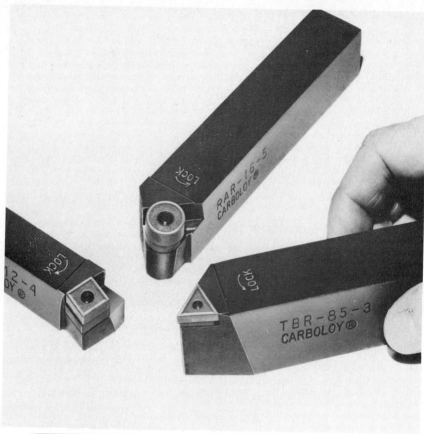

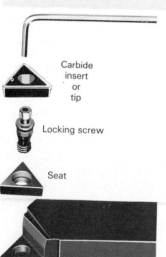

Carbide
insert
or
tip

Locking screw

Seat

FIGURE 18-24 (*Top*) Examples of throwaway carbide cutting tool tips with grooves on rake face. (*Left*) Components of a typical mounting holder. (*Courtesy Metallurgical Products Division, General Electric Company.*)

TABLE 18-4. Representative Machining Conditions for Various Work and Tool–Material Combinations

Lathe Turning Operation Single-Point Tool				Feed: 0.38 mm/rev (0.015 ipr) Depth: 3.18 mm (0.125 in.)	
		Rake Angles (degrees)		Cutting Speed	
Work Material	Tool	Back	Side	m/min	fpm
B1112 steel	HSS	16	22	69	225
	WC	0	3	168	550
	Ceramic	− 5	− 5	427	1400
1020 steel	HSS	16	14	55	180
	WC	0	3	152	500
	Ceramic	− 5	− 5	366	1200
4140 steel	HSS	12	14	40	130
	WC	0	3	91	300
	Ceramic	− 5	− 5	274	900
18–8 steel (stainless)	HSS	8	14	27	90
	WC	4	8	84	275
	Ceramic	− 5	− 5	152	500
Gray cast iron (medium)	HSS	5	12	34	110
	WC	0–4	2–4	69	225
	Ceramic	− 5	− 5	244	800
Brass (free-machining)	HSS	0	0	76	250
	WC	0	4	221	725
Aluminum alloys	HSS	35	15	91	300 plus
	WC	10–20	10–20	122	400 plus
Magnesium alloys	HSS	0	10	91	300 plus
	WC	10	10	213	700 plus
Titanium (turning)	WC	0	5	46	150

As a *chip breaker,* the groove deflects the chip at a sharp angle and causes it to break into short pieces that are easier to remove and are not so likely to become tangled in the machine and possibly cause damage to personnel. This is particularly important on high-speed, mass-production machines.

The shapes of cutting tools as used for various operations and materials are compromises, resulting from experience and research so as to provide good overall performance. Table 18-4 gives representative rake angles and suggested cutting speeds.

In addition to side and back rake, other angles are provided on cutting tools to provide clearance between the tool and the work, thus avoiding rubbing and resulting friction.

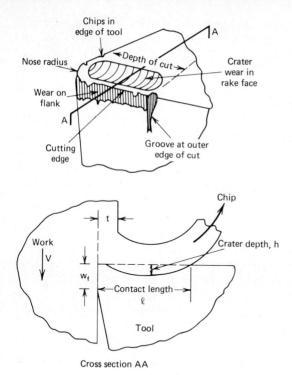

Cross section AA

FIGURE 18-25 Sketch of a worn tool showing various wear elements resulting during oblique cutting.

Tool failure and tool life. In metal cutting, the failure of the cutting tool can be classified into two broad categories, according to the failure mechanisms, which caused the tool to die (or fail).

1. *Slow-death mechanisms:* gradual tool wear on the flank(s) of the tool or on the rake face of tool (called *crater wear*) or both.
2. *Sudden-death mechanisms:* rapid, usually unpredictable and often catastrophic failures resulting from abrupt, premature death of a tool.

Figure 18-25 shows a sketch of a "worn" tool, showing crater and flank wear, along with wear of the tool nose radius and an outer diameter groove. As the tool wears, its geometry changes. This geometry change will influence the cutting forces, the power being consumed, the surface finish obtained, the dimensional accuracy, and even the dynamic stability of the process, as worn tools often chatter in processes usually relatively free of vibration. The actual wear mechanisms active in this high-temperature environment are abrasive, adhesion, diffusion, or chemical interactions. It appears that in metal cutting, any or all of these mechanisms may be operative at a given time in a given process.

The sudden-death mechanisms are more straightforward and are categorized as plastic deformation, brittle fracture, fatigue fracture, or edge chipping. Here again it is impossible to say which mechanism will dominate and result in a tool failure in a particular situation. What can be said is that tools, like people, die (or fail) from a great variety of causes under widely varying conditions, so that tool life should be treated as a random variable, probabilistically, and not as a deterministic quantity.

During machining, the tool is performing in a hostile environment where high contact stresses and high temperatures are commonplace and therefore tool wear is always an unavoidable consequence. In Figure 18-26 four characteristic tool wear curves (average values) are shown for four different cutting speeds, V_1 through V_4, where V_4 is the fastest cutting speed and therefore generates the highest interface temperatures and the fastest wear rates. At lower speeds, such curves often have three general regions as shown in the figure. The central region is a steady-state region (or the region of secondary wear). This is the normal operating region for the tool. Such curves are typical for both flank wear and crater wear. When the amount of wear reaches the value W_f, which is the permissible tool wear on the flank, the tool is said to be "worn out." W_f is typically 0.025 to 0.030 inch. For crater wear, the criteria is related to h, the depth of the crater.

Suppose that the experiment from which the wear curve for V_2 was generated was repeated 15 times, without changing any of the input parameters. The result might look like Figure 18-27, which depicts the variable nature of tool wear and shows why tool wear must be treated as a random variable. In Figure 18-27 the average time is denoted as μ_T and the standard deviation as σ_T, where the wear limit criteria was 0.25 inch. At a given time during the test, 35 minutes, the tool displayed flank wear ranging from 0.13 inch to 0.21 inch.

Other criteria that can be used to define tool death are, in addition to wear limits, the surface finish, failure to conform to size (tolerances), increases in cutting forces and power, or complete failure of the tool. In automated processes, it is very beneficial to be able to monitor the tool wear on-line, so that the tool can be replaced prior to failure wherein defective product may also result. The feed force has been shown to be a good indirect measure of tool wear. That is, as the tool wears and dulls, the feed force increases in proportion to the cutting force.

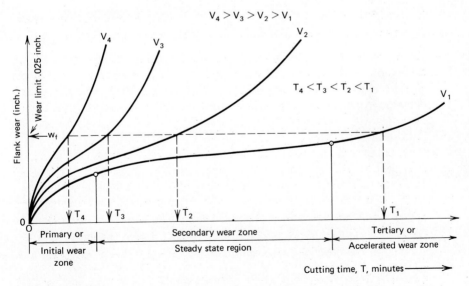

FIGURE 18-26 Typical tool wear curves for flank wear at different velocities.

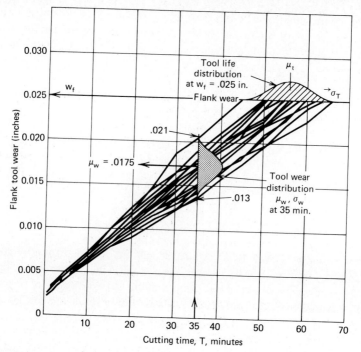

FIGURE 18-27 Tool wear on the flank displays a random nature as does tool life.

Once criteria for failure have been established, tool life is that time elapsed between start and finish of the cut, in minutes. Other ways to express tool life, other than time, include:

1. Volume of metal removed between regrinds or replacement of tool.
2. Number of holes drilled with a given tool.
3. Number of pieces machined per tool.

Taylor's tool life model. F. N. Taylor in 1907, published his now famous "Taylor tool life equation," where tool life (T) was related to cutting speed V and feed (f). These equations had the form

$$T = \frac{\text{constant}}{f_X V_Y}$$

which over the years took the more widely published form

$$VT^n = C \tag{18-26}$$

where n = exponent which depends mostly on tool material but is effected by work material, cutting conditions, and environment and
C = constant which depends on all the input parameters, including feed.

This equation was obtained from emperical log-log plots, like Figure 18-28, which used the values of T (time in minutes) associated with V (cutting speed)

MACHINING PROCESSES

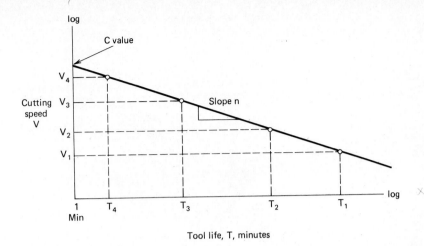

FIGURE 18-28 Construction of Taylor tool life curve from deterministic tool wear plots like those of Figure 18-26.

for a given amount of tool wear, W_f. (See the dashed-line construction in Figure 18-26.) Figure 18-29 shows typical tool life curves for one tool material and three work materials. Notice that all three have about the same slope, n. Typical values for n are 0.14 to 0.16 for HSS, 0.25 for uncoated carbides, 0.30 for TiC inserts and 0.40 for ceramic-coated inserts.

In the main, it takes a great deal of experimental effort to obtain the constants for the Taylor equation, as each combination of tool and work material will have different constants. Note that for a tool life of 1 minute, $C = V$ or the cutting speed that yields about 1 minute of tool life for this tool. A great deal

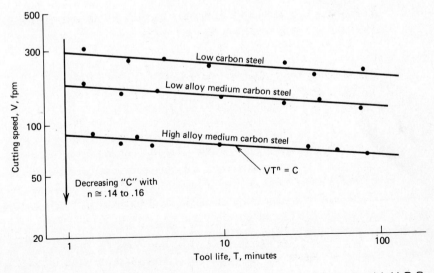

FIGURE 18-29 Log-log tool life plots for three steel work materials cut with H.S.S. tool material.

of research work has gone into developing more sophisticated versions of the Taylor equation wherein constants for other input parameters (typically feed, depth of cut, and work material hardness) are experimentally determined. For example:

$$VT^n f^m d^p = K \qquad (18\text{-}27)$$

where n, m, and p are exponents and K is a constant. Equations of this form are also deterministic.

The problem has been approached probabilistically in the following way. Since $T = g$ (V, feed, materials, etc.), one writes

$$T = \frac{C^{1/n}}{V^{1/n}} = \frac{K}{V^m} \qquad (18\text{-}28)$$

where K is now a random variable which represents the effect of all unmeasured factors and V^m is an input variable.

The sources of tool life variability include factors such as:

1. Variation in work material hardness (from part to part and within a part).
2. Variability in cutting-tool materials, geometry, and preparation.
3. Vibrations in machine tool, including rigidity of work- and tool-holding devices.
4. Changing surface characteristics of workpieces.

The examination of the data from a large number of tool life studies where a variety of steels were machined has shown that regardless of the tool material or process, tool life distributions are usually log normal and typically have a large standard deviation.

Machinability. "Machinability" is a much maligned term which has many different meanings but generally refers to the ease with which a metal is cut. The two principal definitions of the term are entirely different, the first based on material properties and the second based on tool life.

1. Machinability as defined by the ease or difficulty with which the metal can be machined. In this light, specific energy or horsepower or shear stress are used as measures and, in general, the larger the shear stress or specific power values, the more difficult the material is to machine, requiring greater forces and lower speeds. In this definition, the material is the key.
2. Machinability as defined by the relative cutting speed for a given tool life of the tool cutting some material compared to a standard material cut with the same tool material. As shown in Figure 18-30, tool life curves are used to develop machinability ratings. The materials chosen for the standard material was B1112 steel, which has a tool life of 60 minutes at a cutting seed of 100 sfpm. Material "X" then has a 70% rating, which implies that steel "X" has a cutting speed of 70% of B1112 for equal tool lives. Note that this definition assumes that the tool fails when machining "X" by whatever mechanism dominated the tool failure when machining the B1112. There is no guarantee that this will be the case.

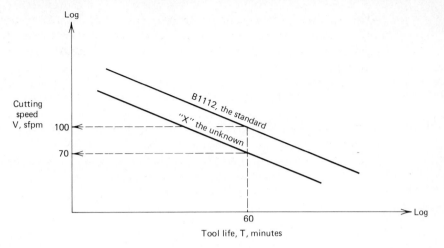

FIGURE 18-30 Machineability ratings defined by deterministic tool life curves.

3. Other definitions of machinability are based on the ease of removal of the chips (chip disposal), the quality of the surface finish of the part itself, the dimensional stability of the process, or the cost to remove a given volume of metal.

Further definitions are being developed based on the probabilistic nature of the tool failure, in which machinability is defined by a tool reliability index. Using such indexes, various tool replacement strategies can be examined and optimum cutting rates obtained. These approaches account for the tool life variability by developing coefficients of variation[3] for common cutting tool/work material combinations. The results to date are very promising. One thing is clear, however, from this sort of research. While many manufacturers of tools have worked at developing materials which have greater tool life at high speed, few have worked to develop tools which have less variability in tool life at all speeds.

The reduction in variability is fundamental to achieving smaller coefficients of variations, which typically are of the order of 0.3 to 0.4. This means that a tool with a 100-minute average tool life has a standard deviation of 30 to 40 minutes and is as likely to fail early as it is to last 100 minutes. In highly automated equipment, where early, unpredicted tool failures are extremely costly, reduction of the tool life variability will pay great benefits in improved productivity and reduced costs.

Cutting fluids. From the day that Taylor demonstrated that a heavy stream of water flowing directly on the cutting process allowed the cutting speeds to be doubled or tripled, cutting fluids have flourished in use and variety, and are employed in virtually every machining process. The cutting fluid acts primarily as a coolant and secondly as a lubricant, reducing the friction effects at the tool/chip interface and the work/flank regions. The cutting fluids also carry away the chips and provide friction reductions (and thereby force) in regions

---

[3]The coefficient of variation is the ratio of the standard deviation divided by the mean of the distribution.

where the bodies of the tools rub against the workpiece. Thus, in processes like drilling, sawing, tapping, and reaming, portions of the tool apart from the cutting edges come in contact with the work and these (sliding friction) contacts greatly increase the power needed to perform the process unless properly lubricated.

The reduction in temperature greatly aids in retaining the hardness of the tool, thereby extending the tool life or permitting increased cutting speed with equal tool life. In addition, the removal of heat from the cutting zone reduces thermal distortion of the work and permits better dimensional control. Coolant effectiveness is closely related to the thermal capacity and conductivity of the fluid used. Water is very effective in this respect but presents a rust hazard to both the work and tools and also is ineffective as a lubricant. Oils offer less effective coolant capacity but do not cause rust and have some lubricant value. In practice, straight cutting oils or emulsion combinations of oil and water or wax and water are frequently used. Various chemicals also can be added to serve as wetting agents or detergents, rust inhibitors, or polarizing agents to promote formation of a protective oil film on the work. The extent to which the flow of a cutting fluid washes the very hot chips away from the cutting area is an important factor in heat removal. Thus the application of a coolant should be copious and with some velocity.

The possibility of a cutting fluid providing lubrication between the chip and the tool face is an attractive one. An effective lubricant can modify the geometry of chip formation so as to yield a thinner, less-deformed, and cooler chip.

Such action further discourages the formation of a built-up edge on the tool and thus promotes improved surface finish. However, the extreme pressure existing between the chip and the tool and the rapid flow of the chip away from the cutting edge make it virtually impossible to maintain a conventional hydro-

TABLE 18-5   Operations and Machines for Machining External Cylindrical Surfaces

Operation	Block Diagram	Most Commonly Used Machines	Machines Less Frequently Used	Machines Seldom Used
Turning		Lathe	Boring mill	Vertical shaper Milling machine
Grinding		Cylindrical grinder		Lathe (with special attachment)
Sawing (of plates)		Contour or band saw	Flame cutting or plasma arc	

MACHINING PROCESSES

TABLE 18-6 Operations and Machines for Machining Flat Surfaces

Operation	Block Diagram	Most Commonly Used Machines	Machines Less Frequently Used	Machines Seldom Used
Shaping		Horizontal shaper	Vertical shaper	
Planing		Planer		
Milling	slab milling	Milling machine		Lathe (with special attachment)
	face milling	Milling machine Machining center		Drill press (light cuts)
Facing		Lathe	Boring mill	
Broaching		Broaching machine		
Grinding		Surface grinder		Lathe (with special attachment)
Sawing		Cutoff saw	Contour saw	

dynamic lubricating film where it would be needed to obtain reduced friction by this method. Consequently, any lubrication that can be provided is associated primarily with the formation of solid chemical compounds of low shear strength on the freshly cut chip face, thereby reducing chip-tool shear forces or friction. For example, carbon tetrachloride is very effective in reducing friction in machining several different metals and yet would hardly be classed as a good lubricant in the usual sense. Chemically active compounds, such as chlorinated or sulfurized oils, can be added to cutting fluids to achieve such a lubrication effect. Extreme-pressure lubricants of this type are especially valuable in severe operations, such as internal threading by a tap where the extensive tool-work contact results in much friction with limited access for a fluid.

In addition to functional effectiveness as coolant and lubricant, cutting fluids should be stable in use and storage, noncorrosive to work and machines, and nontoxic to operating personnel.

Summary. In this chapter the basics of machining have been presented. Chapters 19 through 26 will elaborate on the machine tools which have been developed to perform specific machining operations. Often a machine can perform several of the basic processes and, as indicated in Chapter 1, machining centers are now very much in common use. Tables 18-5 through 18-7 should be studied

TABLE 18-7 Operations and Machines for Machining Internal Cylindrical Surfaces

Operation	Block Diagram	Most Commonly Used Machines	Machines Less Frequently Used	Machines Seldom Used
Drilling		Drill press Machining center Vert. milling machine	Lathe Horizontal boring machine	Horizontal milling machine Boring mill
Boring		Lathe Boring mill Horizontal boring machine Machining center		Milling machine Drill press
Reaming		Lathe Drill press Boring mill Horizontal boring machine Machining center	Milling machine	
Grinding		Cylindrical grinder		Lathe (with special attachment)
Sawing		Contour or band saw		
Broaching		Broaching machine	Arbor Press (keyway broaching)	

carefully before going ahead, as these tables summarize the relationship between the basic processes and the machine tools that can be used to perform these processes. Generally speaking, these machines will be of the A(2) or A(3) level of automation until we begin to discuss machining centers, which are numerical control machines and therefore A(4). For particular machines, you will have to become familiar with new terminology, but in general all will need inputs concerning rpm (given that you selected the cutting speed), feeds, and depths of cut. Note also from these tables that the same process can be performed on two or more different machine tools and there are many ways to produce flat surfaces, internal and external cylindrical surfaces, and special geometries in parts. Generally, the quantity to be made is the driving factor in the selection of processes, as we shall see in later discussions. Machining centers have automatic tool change capability and are usually capable of milling, drilling, boring, reaming, tapping (hole threading), and other minor machining processes.

REVIEW QUESTIONS

1. Why has the metal-cutting process resisted theoretical solution for so many years?
2. What are the variables which must be considered in understanding a machining process?
3. Which of the basic chip formation processes are single point and which are multiple point?
4. Is feed related to speed in all machining operations?
5. Milling has two feeds. What are they and which ones does the operator need to know about?
6. What is the fundamental mechanism of chip formation?
7. What is the main implication of Figure 18-11, given that this series of micrographs was made at very low speed?
8. What is the difference between oblique machining and orthogonal machining?
9. Suppose that you want to shape a block of metal 7 inches wide and 4 inches long, using a shaper as set up in Figure 18-1. You have determined for this metal that the cutting speed should be 25 sfpm, the depth of cut needed here for roughing is 0.25 inch, and the feed will be 0.1 inch per stroke. Determine the approximate crank rpm and then estimate the cutting time and the MRR.
10. In problem 9, why should the block be cut in the 4-inch direction rather than in the 7-inch direction?
11. Could you have saved any time in problem 9 by cutting the block in the 7-inch direction? Redo with $L = 7$ and $W = 4$ inches.
12. If the job described in problem 9 had been done on a planer rather than a shaper, it is likely that the setup would have been made to cut the block in the 7-inch direction. Why?
13. Suppose that you want to do the job described in problem 9 by slab milling. You have selected a 6-inch-diameter cutter with eight carbide insert teeth. The cutting speed will be 500 sfpm and the feed per tooth will be 0.010 ipt. Determine the input parameters for the machine (rpm of arbor and table feed), the CT and MRR. Compare these answers with what you got for shaping the block.
14. How do the magnitude of the strain and strain rate values of metal cutting compare to tensile testing?
15. Why is titanium such a difficult metal to machine?
16. Explain why you get segmented or discontinuous chips when you machine cast iron?
17. Derive equations for F and N using the circular force diagram. (*Hint:* Make a copy of the diagram. Extend a line from point X intersecting force F perpendicularly. Extend a line from point Y intersecting the previous line perpendicularly. Find the angle α made by these constructions.)

18. Derive equations for F_s and F_n using the circular force diagram. (*Hint:* Construct a line through X parallel to vector F_n. Extend vector F_s to intersect this line. Construct a line from X perendicular to F_n. Construct a line through point Y perpendicular to the line through X.)

19. Why is sheer stress so important?

20. Which of the three cutting forces in oblique cutting consumes most of the power? How is that power divided and where does the energy consumed ultimately go?

21. In problem 9, what would you have to know in addition to what has been given in order to estimate the primary cutting force?

22. How is cutting speed related to tool wear?

23. What is the relationship between hardness and temperature in metal cutting tool materials?

24. What is the general strategy behind coated tools?

25. The coefficient of friction between many metals and TiC coatings is lower than many other tool materials. How does this benefit the process?

26. What is meant by the statement that tool life is a random variable?

27. In the figure below are data for cutting speed and tool life. Determine the constants for the Taylor tool life equation (18-26) for these data. What do you think the tool material might have been?

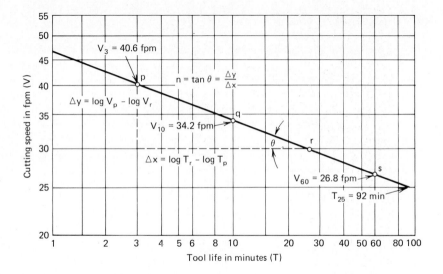

28. How is machinability defined?

29. What are the chief functions of cutting fluids?

CASE STUDY 18. The Broken Surgical Implant

Figure CS-18a shows one portion of a surgical hip-joint implant which failed during the third day the patient, who weighed only 120 pounds, attempted to walk on it. The device fractured through one of the attaching holes (through which no screw had been used) after bending through the angle shown. Figures CS 18(b) and (c) show close-ups of the fracture on the convex and concave side, respectively, and (d) shows an enlarged view of the fracture surfaces. Chemical analysis revealed the composition of the metal to have the following percentages: carbon 0.025; chromium 18.04; nickel 14.17; manganese 1.89; molybdenum 2.26; phosphorus 0.016; silicon 0.4; sulfur 0.11; and titanium 0.008. A duplicate implant, shown at (e), bent through a 30° angle, as shown, without breaking.

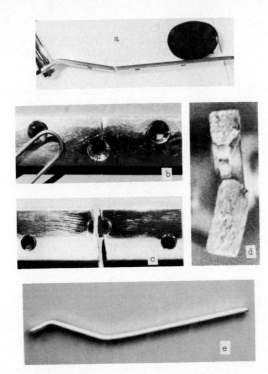

From the evidence given, state your opinion as to the cause of the failure, giving your reasons. Was the device in normal condition at the time it was installed by the surgeon?

Shaping and Planing

From a consideration of the relative motions between the tool and the workpiece, shaping and planing are the simplest of all machining operations, and the machines that do the operations are among the simplest of all machine tools. As indicated in Table 18-6 and Figures 18-1 and 19-1, straight-line cutting motion between a single-point cutting tool and a workpiece is used. A flat surface is generated by the work or the tool being fed at right angles to the cutting motion between successive strokes. In addition to plain flat surfaces, the shapes most commonly produced on the shaper and planer are those illustrated in Figure 19-2. Relatively skilled workers are required to operate shapers and planers, and most of the shapes that can be produced on them also can be made by much more productive processes, such as milling, broaching, or grinding. Consequently, except for certain special types, planers that will do only planing have become obsolete, and shapers are used very little in manufacturing except in tool and die work or in very low volume production. Their basic simplicity and flexibility make them very useful in these situations.

SHAPING AND SHAPERS

Shaping is done on a *shaper*, utilizing a reciprocating, single-point cutting tool that moves in a straight line across the workpiece. To produce a flat horizontal

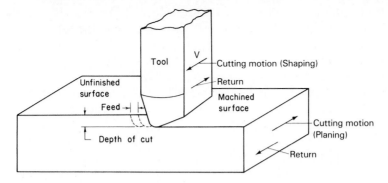

FIGURE 19-1 Basic straight line-relationships of tool motion (speed), feed (inches per stroke), and depth of cut in shaping and planing.

surface, the work is fed across the line of motion of the tool, between strokes. For vertical and angled flat surfaces, the tool usually is fed across the workpiece. Because the tool cuts on only the forward part of the stroke, shaping is a relatively slow process. However, setups on shapers can be made easily and quickly, they are flexible, require no special fixture, and so are very useful where only one or a few identical parts are required.

Shaping usually is confined to machining flat surfaces that do not exceed 610 × 610 mm (24 × 24 inches) in size. External and internal surfaces, either horizontal or inclined, can be produced readily; curved and irregular surfaces also can be made, but with much less ease or by the use of special attachments.

Types of shapers. Shapers usually are classified according to their general design features as follow:

1. Horizontal.
   a. Push-cut.
   b. Pull-cut.
2. Vertical.
   a. Regular or slotters.
   b. Keyseaters.
3. Special.

They are also designated as to the type of drive employed—*mechanical drive* or *hydraulic drive*.

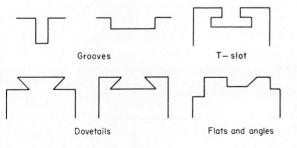

Grooves   T−slot

Dovetails   Flats and angles

FIGURE 19-2 Types of surfaces commonly machined by shaping and planing.

Horizontal push-cut shapers. Most shapers are of the *horizontal push-cut* type (see Figure 19-3), where cutting occurs as the ram *pushes* the tool across the work. The main components of a horizontal push-cut shaper are the *base* and *column,* the *ram,* the *toolhead assembly,* the *driving mechanism,* the *table,* and the *cross-rail* and *feeding mechanism.* The base and column support all the other components, the column containing the ram-driving mechanism. Dovetail ways on the top of the column support the ram, and the cross rail is mounted on vertical ways on the front of the column.

Two types of mechanisms—mechanical and hydraulic—are employed for reciprocating the ram, on which the toolhead assembly is mounted at the forward end. The driving mechanism should provide ample force during the forward, cutting stroke, provide a more rapid return-stroke speed, and permit adjustment of the stroke length and stroke speeds. Figure 18-1 shows the driving mechanism that is used in mechanical-drive shapers. Hydraulic drives provide more uniform

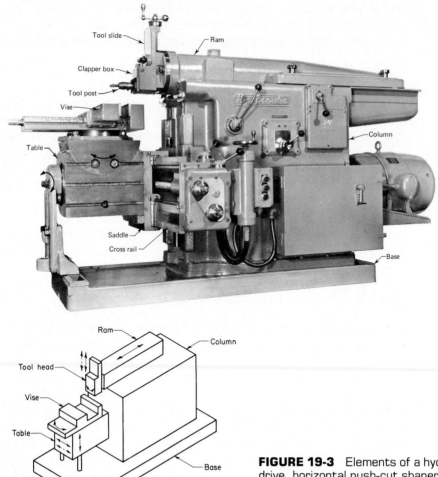

FIGURE 19-3 Elements of a hydraulic-drive, horizontal push-cut shaper with a universal table. (*Courtesy Rockford Machine Tool Company*).

velocity during the cutting stroke and more rapid return-stroke speeds. Mechanical drives are somewhat cheaper and provide absolute limits to the stroke length.

The *ram and toolhead assembly* provides means for adjusting the horizontal position of the ram stroke with respect to the column and for mounting the tool on its forward end. As shown in Figure 19-3 the toolhead assembly has a swivel mounting, which permits the toolslide to be turned about a horizontal axis so that cuts can be made at any desired angle. A graduated scale is provided to facilitate making angular settings. The toolside is carried on the swivel mounting in dovetail ways so that it can be moved a few inches by means of the manually operated lead screw to provide feed or depth of cut, as required, for the tool. A calibrated dial provides a means for controlling this movement.

A two-piece clapper box is mounted on the toolslide. The outer piece is pivoted at its upper end so that the tool, which is mounted on this section of the clapper box by means of a tool post, can lift upward. This permits the tool to clear the work on the return ram stroke. On most shapers the tool slides over the work surface on the backward stroke of the ram, but in some, the tool is lifted mechanically so that it does not slide on the work.

As shown in Figure 19-3 the shaper table is carried on a horizontal cross-rail, which, in turn, is mounted on vertical ways on the front of the shaper column. The cross rail can be raised and lowered by means of an elevating screw, or screws, and hand crank. On some larger shapers the cross rail can be raised and lowered by power. In some machines vertical power feed is provided. The crossrail can be clamped to the column in any desired position to give added rigidity during machining.

A saddle slides on upper and lower ways on the cross rail, and the table is attached to the saddle. A *plain table* is nonadjustable and has T-slots and/or a heavy vise by means of which the work can be held. A *universal table,* shown in Figure 19-3, is provided with a means of rotating the entire table about a horizontal axis parallel with the rail. Such a table is particularly convenient in toolroom work, where inclined flat surfaces must frequently be machined.

In machining horizontal surfaces on a shaper, the depth of cut is set by means of the toolslide, and feed is provided by moving the table along the rail prior to the beginning of each stroke, usually by means of a power mechanism and lead screw. The table remains stationary throughout the cutting stroke. Feed is expressed in *millimeters or inches per stroke.*

To machine a vertical or inclined surface, feed usually is accomplished through manual movement of the tool slide before each cutting stroke. Some larger shapers have vertical power feed for the table so that vertical surfaces can be machined by using this vertical power feed. Also, special attachments can be obtained for providing powered feed to the toolslide, but these are seldom used.

Horizontal draw-cut shapers. In order to eliminate the long, unsupported overhang of the ram that would result when shaping larger surfaces with push-cut shapers, *draw-cut shapers* are available. As illustrated in Figure 19-4 a large overarm supports the outer end of the ram, and cutting is done during the return stroke. Because the cutting forces act inward against the column, it is easier to support the work and avoid vibration and chatter. Such shapers are used primarily in tool and die shops.

FIGURE 19-4 Draw-cut shaper machining a large die block. (*Courtesy Morton Manufacturing Company*).

Vertical shapers and slotters. A vertical shaper, shown in Figure 19-5, has a vertical ram and a round table that can be rotated in a horizontal plane, by either manual or power feed. These machine tools are sometimes called *slotters*. Usually, the ram is pivoted near the top so that it can be swung outward from the column through an arc of about 10°.

Because one circular and two straight-line motions and feeds are available, vertical shapers are very versatile tools and thus find considerable use in die shops. Not only can vertical and inclined flat surfaces be machined, but external and internal cylindrical surfaces can be generated by circular feeding of the table between strokes, or they can be made by turning or boring, respectively, by using a stationary tool and rotating the workpiece. A vertical shaper is partic-

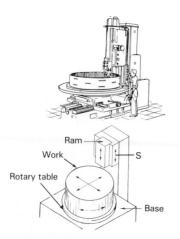

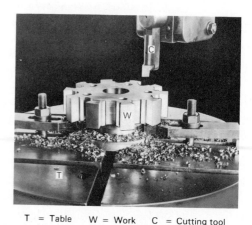

T = Table   W = Work   C = Cutting tool

FIGURE 19-5 (*Left*) Schematic of vertical shaper. (*From* Manufacturing Producibility Handbook; *courtesy General Electric Company.*) (*Right*) Typical job setup in vertical shaper. (*Courtesy Colt Industries.*)

MACHINING PROCESSES

ularly useful in machining curved surfaces, interior surfaces, and arcs by using a stationary tool and rotating the workpiece.

A *keyseater* is a special type of vertical shaper designed and used exclusively for machining keyways on the inside of wheel and gear hubs.

Work setup in shapers. The mounting and holding of the workpiece is of prime importance in all machining operations. The setup not only requires labor, but the machine usually is idle during setup time. If several operations are done on a given machine, or if the workpiece is of an odd shape or is not strong, this usually complicates the setup problem. One of the virtues of a shaper is that only a single, simple cutting action is involved, and the setup usually is very simple. On horizontal push-cut shapers, the work usually is held in a heavy vise mounted on the top side of the table. As shown in Figure 19-6 shaper vises have a very heavy movable jaw, because the vise must often be turned so that the cutting forces are directed against this jaw. Shaper vises can be rotated and clamped about a vertical axis.

In clamping work in a shaper vise, care must be exercised to make sure that it rests solidly against the bottom of the vise, or on parallel bars, so that it will not be deflected by the cutting force, and that it is held securely yet not distorted by the clamping pressure. Figure 19-6 illustrates the use of parallel bars for raising work to the proper height in the vise jaws, and several methods of clamping rough and irregularly shaped work. These latter procedures are required when the workpiece surfaces are rough, or the sides are not parallel, to prevent the clamping action from tilting the work in the vise.

Work that cannot be held conveniently in a vise can be clamped directly to the top or sides of the table, using T-slots in the table.

Shaper tools. Most shaping is done with simple high-speed or carbide-tipped tool bits held in a heavy, forged tool holder, as shown in Figure 20-30. The toolholder is held in a slotted tool post by means of a setscrew. In some cases where a tool of irregular shape is required, as in cutting a T-slot, or where very heavy cuts are to be made, a large, forged tool bit is held directly in the tool post.

Shaper precision. Although shapers are versatile tools, the precision of the work done on them is greatly dependent on the operator. Feed dials on shapers nearly always are graduated in 0.02-mm or 0.001-inch divisions, and work

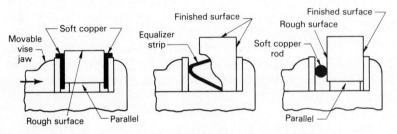

FIGURE 19-6 Methods of clamping workpieces in a shaper vise.

seldom is done to greater precision than this. A tolerance of 0.04 to 0.06 mm (0.002 to 0.003 inch) is desirable on parts that are to be machined on a shaper, because this gives some provision for errors due to clamping, possible looseness or deflection of the table, and deflection of the tool and ram during cutting.

PLANING AND PLANERS

Planing can be used to produce horizontal, vertical, or inclined flat surfaces on workpieces that are too large to be accommodated on shapers. However, as will be pointed out, planing is much less efficient that other basic machining processes that will produce such surfaces; consequently, planing and planers have largely been replaced. However, some understanding of planing and planers is desirable so as to be aware of the possibilities and limitations.

Figure 19-7 shows the basic components and motions of a planer. In most planing, the action is opposite to that of shaping, in that the work is moved past one or more stationary, single-point cutting tools. Because a large and heavy workpiece and table must be reciprocated at relatively low speeds, several tool heads are provided, often with multiple tools in each head, also illustrated in Figure 19-7. In addition, many planers are provided with tool heads arranged so that cuts occur on both directions of the table movement. However, because only single-point cutting tools are used and the cutting speeds are quite low, planers are low in productivity as compared with some other types of machine tools. As a result, they have been replaced, almost entirely, by planer-type milling machines or by similar combination machines which can do both milling and planing. These will be discussed in Chapter 22.

Types of planers. Planers have been made in four basic types. Figure 19-7 depicts the most-common, double-housing type. It has a closed housing structure, spanning the reciprocating worktable, with a crossrail supported at each end on a vertical column and carrying two toolheads. An additional toolhead usually is mounted on each column, so that four tools (or four sets) can cut during each stroke of the table. Obviously, the closed-frame structure of this type of planer limits the size of the work that can be machined. Consequently, open-side planers are available, as indicated in Figure 19-7, which have the crossrail supported on a single column. This provides unrestricted access to one side of the table to permit wider workpieces to be accommodated. Some open-side planers are convertible, in that a second column can be attached to the bed when desired so as to provide added support for the crossrail.

As indicated previously, double-housing and open-side planers seldom are currently made, but two special types of planers probably will continue to have significant, special use. One is the *plate* or *edge planer* designed specifically for planing the edges of plates. The work is stationary, and the tool reciprocates on a side-mounted carriage. The other type is the *pit planer,* shown in Figure 19-8, on which both the work and table are stationary, but the columns and

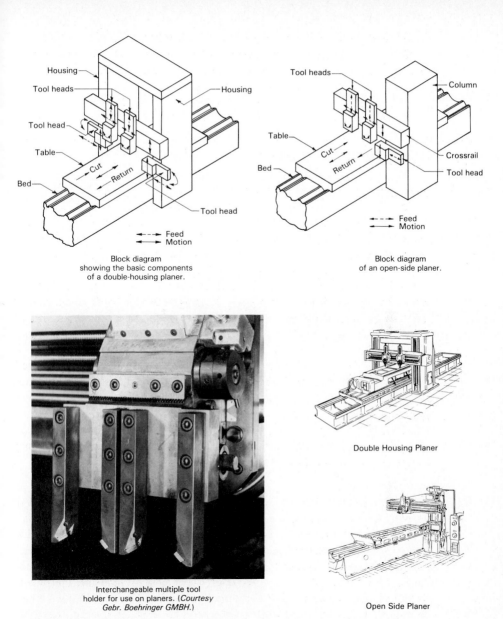

Block diagram showing the basic components of a double-housing planer.

Block diagram of an open-side planer.

Interchangeable multiple tool holder for use on planers. (*Courtesy Gebr. Boehringer GMBH.*)

Double Housing Planer

Open Side Planer

FIGURE 19-7 Elements of double-housing and open-side planers. (*Schematics from Manufacturing Producibility Handbook; courtesy General Electric Company. Photograph, courtesy Gebr. Boehringer GmbH.*)

crossrail move longitudinally on massive rails. These are used only for very large work, where the weight of the workpiece and the required table would make reciprocation difficult and severely limit cutting speeds. The toolheads usually revolve to permit cutting in both directions. They are not common but are very useful for very large work.

FIGURE 19-8 Pit planer. (*Courtesy Consolidated Machine Tool Division, Farrel Birmingham Company, Inc.*)

Work setup on planers. Two factors must be given special consideration in making a work setup on planers. First, because the workpieces are unusually heavy and must be reciprocated, they must be fastened to not only resist the cutting forces but also the high inertia forces that result from the rapid velocity $\left(F = m\dfrac{dv}{dt}\right)$ changes at the ends of the strokes. Special stops often are provided at each end of the workpiece to prevent it from shifting.

The second factor is that considerable time usually is required to set up a workpiece, thus reducing the time the costly machine is available for producing chips. Sometimes special setup plates are used, to which the workpiece is fastened. These plates are designed for quick attachment to the planer table. Another procedure involves a planer having two tables, as shown in Figure 19-9. Work is set up on one table while another workpiece is being machined on the other. Both tables can be fastened together for machining long workpieces. This same table arrangement is available on planer-type milling machines, which will be discussed in Chapter 22.

Planer tools. Because a large percentage of available time is used for setup, and the large workpieces involved usually will permit heavy cutting forces, planing cuts should be as heavy as practicable. Consequently, planer tools usually are quite massive. Usually, the main shank of the tool is made of plain-carbon steel, and a high-speed or carbide tip is clamped or brazed to it. Because the cuts usually are quite long, in planing steel or other ductile materials it is very important that chip breakers be provided to avoid long and dangerous chips.

FIGURE 19-9 Divided-table planer. One workpiece is being set up on a table while another is being machined on a second table. (*Courtesy Cincinnati Shaper Company.*)

Many modern planers have tool holders arranged so that cutting occurs during both directions of the table movement, thereby further increasing the productivity. Others are equipped with hydraulic tool lifters to raise the tool from the work on the return stroke. This is important when using carbide tool tips.

Planer precision and use.   Theoretically, planers have about the same precision as shapers. The feed and other dimension-controlling dials usually are graduated in 0.02-mm or 0.001-inch divisions. However, because larger and heavier workpieces usually are involved, and much longer beds and tables, the working tolerances for planer work should be somewhat greater than for shaping.

REVIEW QUESTIONS

1. What is the basic difference between a shaper and a planer?
2. Why are shapers not well suited for producing curved surfaces?
3. For what type of work are shapers best suited?
4. What is the function of the clapper box on a shaper?
5. What are the advantages and disadvantages of mechanical-drive and hydraulic-drive shapers?
6. Why is a pull-cut shaper more suitable for large work than a push-cut shaper?
7. How is shaper feed expressed?
8. Why is the shaper seldom used for production work?
9. How can angular surfaces be machined on a shaper.
10. Why is it seldom necessary to use carbide cutting tools on a shaper?
11. How many strokes per minute would be required to obtain a cutting speed of 36.6 m (120 feet) per minute on a typical mechanical-drive shaper if a 254-mm (10-inch) stroke is used? Refer to Figure 18-1.
12. How much time would be required to shape a flat surface 254 mm (10 inches) wide and 203 mm (8 inches) long on a hydraulic-drive shaper, using a cutting speed of 45.7 m (150 feet) per minute, a feed of 0.51 mm (0.020 inch) per stroke, and an overrun of 12.7 mm (½ inch) at each end of the cut? Let $N_s = 8V/l$
13. What is the metal-removal rate in question 12 if the depth of cut is 6.35 mm (¼ inch)?

14. If the metal being removed in question 13 is gray cast iron, what will be the power required?

15. A hydraulic-drive shaper has a 5.6-kW (7½-hp) motor, and 75% of the motor output is available at the cutting tool. The specific power for cutting a certain metal is 0.03 W/mm³ (.67 hp/in³/min). What is the maximum depth of cut that can be taken in shaping a surface in this material if the surface is 305 × 305 mm (12 × 12 inches), the feed is 0.64 mm (0.025 inch) per stroke, and the cutting speed is 54.9 m (180 feet) per minute?

16. Why is it difficult to use high cutting speeds on a planer?

17. How does the planer feed differ from shaper feed?

18. What are the basic types of planers?

19. Why are such large tools needed in planers? (Can you estimate F_c from data in problem 15?

20. Why is it so desirable to make more than one cut simultaneously on a planer?

21. What is the basic disadvantage of an open-side planer? What is an inherent disadvantage?

22. Why is a vertical shaper more versatile than an ordinary shaper?

CASE STUDY 19· Aluminum Retainer Rings

The Owlco Corporation has to make 5000 retainer rings, as shown in Figure CS-19. It is essential that the surfaces be smooth with no sharp corners on the circumferential edges. Determine the most economical method for manufacturing these rings.

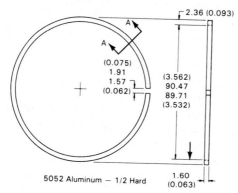

5052 Aluminum — 1/2 Hard

Section A-A

FIGURE CS-19 Aluminum snap-in-retainer ring.

CHAPTER 20

Turning and Boring

Turning provides a widely used means for machining external cylindrical and conical surfaces. As indicated in Figure 20-1, relatively simple work and tool movements are involved in turning a cylindrical surface—the workpiece rotates and a longitudinally fed, single-point tool does the cutting. If the tool is fed at an angle to the axis of rotation, an external conical surface results. This is called taper turning. If the tool is fed at 90° to the axis of rotation, using a tool that is wider than the width of the cut, the operation is called *facing,* and a flat surface is produced. Such a surface can be thought of as a conical surface having a 180° apex angle.

External cylindrical, conical, and irregular surfaces of limited length can also be turned by using a tool having a specific shape and feeding it inward against the work. The shape of the resulting surface is determined by the shape and size of the cutting tool. Such machining is called *form turning*. Obviously, if feeding of the tool continues to the axis of the workpiece, it will be cut in two; this is called *parting* or *cutoff* and a simple, thin tool is used. A similar tool is used for necking or partial cutoff.

Boring is a variation of turning. Essentially it is internal turning, in that a single-point cutting tool produces internal cylindrical or conical surfaces. Consequently, boring can be done on most machine tools that can do turning. However, boring also can be done using a rotating tool with the workpiece remaining stationary. Also, specialized machine tools have been developed that

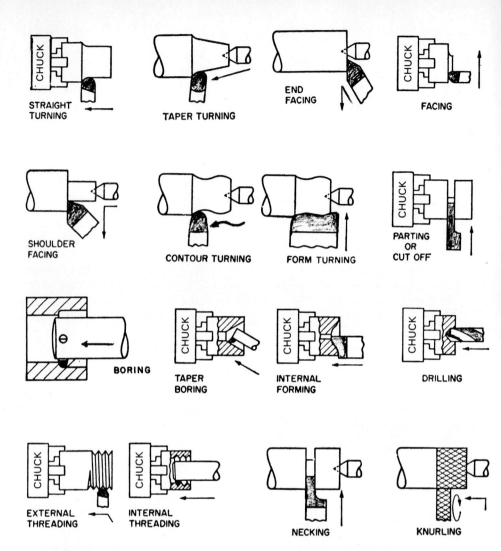

FIGURE 20-1 Turning, facing, boring, and related processes. The arrows indicate the motion of the tool relative to the work.

will do boring, drilling, and reaming but will not do turning. Other operations, like threading and knurling, can be done on machines used for turning. In addition, drilling, reaming, and tapping can be done on the rotation axis of the work.

LATHES

Lathes are machine tools designed primarily to do turning, facing, and boring. As indicated in Tables 18-5 through 18-7, very little turning is done on other types of machine tools, and none can do it with equal facility. Because lathes

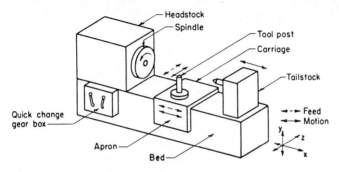

FIGURE 20-2 Block diagram of the basic components of a lathe.

also can do facing, drilling, and reaming, their versatility permits several operations to be done with a single setup of the workpiece. Consequently, more lathes of various types are used in manufacturing than any other machine tool.

Lathes in various forms have existed for more than 2000 years, but modern lathes date from about 1797, when Henry Maudsley developed one with a leadscrew (see Figure 1-13), providing controlled, mechanical feed of the tool. This ingenious Englishman also developed a change-gear system that could connect the motions of the spindle and leadscrew and thus enable threads to be cut.

Lathe construction. The essential components of a lathe are depicted in Figure 20-2. These are the *bed, headstock assembly, tailstock assembly, carriage assembly, quick-change gearbox,* and the *leadscrew* and *feed rod.* A typical modern engine lathe is shown in Figure 20-3.

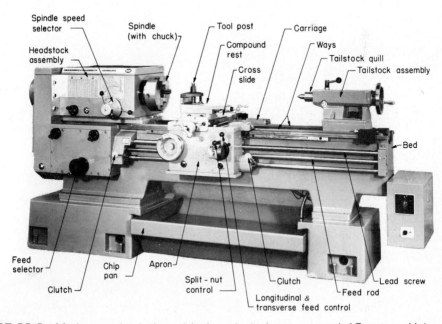

FIGURE 20-3 Modern engine lathe, with the principal parts named. (*Courtesy Heidenreich & Harbeck.*)

The *bed* is the backbone of a lathe. It usually is made of well-normalized or aged gray or nodular cast iron and provides a heavy, rigid frame on which all the other basic components are mounted. Two sets of parallel, longitudinal *ways*, inner and outer, are contained on the bed, usually on the upper side. Some makers use an inverted V-shape for all four ways, whereas others utilize one inverted V and one flat way in one or both sets. They are precision-machined to assure accuracy of alignment. On most modern lathes the ways are surface-hardened to resist wear and abrasion, but precaution should be taken in operating a lathe to assure that the ways are not damaged. Any inaccuracy in them usually means that the accuracy of the entire lathe is destroyed.

The *headstock* is mounted in a fixed position on the inner ways, usually at the left end of the bed. It provides a powered means of rotating the work at various speeds. Essentially, it consists of a hollow spindle, mounted in accurate bearings, and a set of transmission gears—similar to a truck transmission—through which the spindle can be rotated at a number of speeds. Figure 20-4 shows the arrangement of the gears and spindle in a typical lathe. Most lathes provide from eight to 18 speeds, usually in a geometric ratio, and on modern lathes all the speeds can be obtained merely by moving from two to four levers. An increasing trend is to provide a continuously variable speed range through electrical or mechanical drives.

Because the accuracy of a lathe is greatly dependent on the spindle, it is of heavy construction and mounted in heavy bearings, usually preloaded tapered roller or ball types. The spindle has a hole extending through its length, through which long bar stock can be fed. The size of this hole is an important dimension of a lathe because it determines the maximum size of bar stock that can be machined when the material must be fed through the spindle.

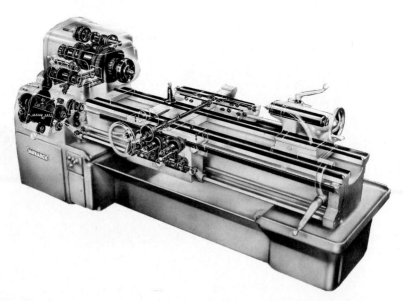

FIGURE 20-4 Phantom view of an engine lathe showing the gear train. (*Courtesy Monarch Machine Tool Company.*)

The inner end of the spindle protrudes from the gearbox and contains a means for mounting various types of chucks, face plates, and dog plates on it.

Small lathes often employ a threaded section to which the chucks are screwed, while most large lathes utilize either *cam-lock* or *key-drive taper* noses. These provide a large-diameter taper that assures the accurate alignment of the chuck, and a mechanism that permits the chuck or face plate to be locked or unlocked in position without the necessity of having to rotate these heavy attachments.

Power is supplied to the spindle by means of an electric motor through a V-belt or silent-chain drive. Most modern lathes have motors of from 5 to 15 horsepower to provide adequate power for carbide and ceramic tools at their high cutting speeds. The tailstock assembly consists, essentially, of three parts. A lower casting fits on the inner ways of the bed and can slide longitudinally thereon, with a means for clamping the entire assembly in any desired location. An upper casting fits on the lower one and can be moved transversely upon it, on some type of keyed ways, to permit aligning the tailstock and headstock spindles. It also provides a method of turning tapers (see Figure 20-48). The third major component of the assembly is the *tailstock quill*. This is a hollow steel cylinder, usually about 51 to 76 mm (2 to 3 inches) in diameter, that can be moved several inches longitudinally in and out of the upper casting by means of a handwheel and screw. The open end of the quill hole terminates in a Morse taper in which a lathe center, or various tools such as drills, can be held. A graduated scale usually is engraved on the outside of the quill to aid in controlling its motion in and out of the upper casting. A locking device permits clamping the quill in any desired position.

The *carriage assembly,* shown in Figure 20-5, provides the means for mounting and moving cutting tools. The *carriage,* a relatively flat H-shaped casting, rests and moves on the outer set of ways on the bed. The *cross slide* is mounted on ways on the transverse bar of the carriage and can be moved by means of a

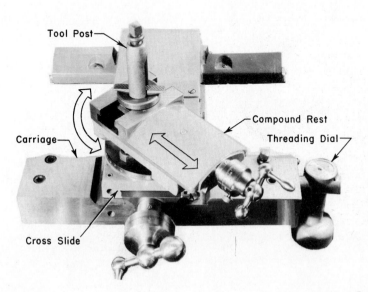

FIGURE 20-5 Lathe carriage assembly. (*Courtesy Sheldon Machine Company, Inc.*)

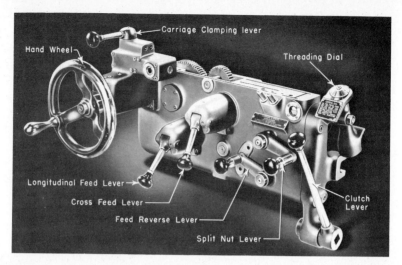

FIGURE 20-6 Front side of a lathe apron. (*Courtesy American Tool Works Company.*)

feed screw that is controlled by a small handwheel and a graduated dial. The cross slide thus provides a means for moving the lathe tool in the direction normal to the axis of rotation of the workpiece.

On most lathes the tool post actually is mounted on a *compound rest*, also shown in Figure 20-5. This consists of a base, which is mounted on the cross slide so that it can be pivoted about a vertical axis, and an upper casting. The upper casting is mounted on ways on this base so that it can be moved back and forth and controlled by means of a short lead screw operated by a handwheel and a calibrated dial.

The *apron*, attached to the front of the carriage, contains the mechanism and controls for providing manual and powered motion for the carriage and powered motion for the cross slide. Figures 20-6 and 20-7 show front and rear views of a typical apron. Manual movement of the carriage along the bed is effected by

FIGURE 20-7 Back side of a lathe apron. (*Courtesy American Tool Works Company.*)

turning a handwheel on the front of the apron, which is geared to a pinion on the back side. This pinion engages a rack that is attached beneath the upper front edge of the bed in an inverted position.

Powered movement of the carriage and cross slide is provided by a rotating *feed rod*, shown in Figure 20-3. The driving of this feed rod will be discussed later.) The feed rod, which contains a keyway throughout most of its length, passes through the two reversing bevel pinions shown in Figure 20-7 and is keyed to them. Either pinion can be brought into mesh with a mating bevel gear by means of the reversing lever on the front of the apron and thus provide "forward" or "reverse" power to the carriage. Suitable clutches connect either the rack pinion or the cross-slide screw to provide longitudinal motion of the carriage or transverse motion of the cross slide.

For cutting threads, a second means of longitudinal drive is provided by a *lead screw*, shown in Figure 20-3. Whereas motion of the carriage when driven by the feed-rod mechanism takes place through a friction clutch in which slippage is possible, motion through the lead screw is by a direct, mechanical connection between the apron and the lead screw. This is achieved by a *split nut*, shown in Figure 20-7, which can be closed around the lead screw by means of a lever on the front of the apron. With the split nut closed, the carriage is moved along the lead screw by direct drive without possibility of slippage.

Modern lathes have *quick-change gearboxes* driven from the spindle by means of suitable gearing and connected to the feed rod and lead screw (see Figure 20-4). Thus, through the quick-change gearbox, the associated gearing, and the lead screw and feed rod, the carriage is connected to the spindle, and the cutting tool can be made to move a specific distance, either longitudinally or transversely, for each revolution of the spindle. Typical lathes may provide, through the feed rod, as many as 48 feeds, ranging from 0.05 to 3 mm (0.002 to 0.118 inch) per revolution of the spindle, and, through the lead screw, leads from 0.28 to 17 mm (1½ to 92 threads per inch). On some small or inexpensive lathes, one or two gears in the gear train between the gearbox and the spindle must be changed to obtain a full range of threads and feeds.

Size designation of lathes.   The size of a lathe is designated by two dimensions. The first is known as the *swing*. This is the maximum diameter of work that can be rotated on a lathe. It is approximately twice the distance between the line connecting the lathe centers and the nearest point on the ways. The second size dimension is the *maximum distance between centers*. The swing thus indicates the maximum workpiece diameter that can be turned in the lathe, while the distance between centers indicates the maximum length of workpiece that can be mounted between centers. The maximum diameter of a workpiece that can be mounted between centers is somewhat less than the swing diameter—usually 127 to 152 mm (5 to 6 inches)—because such a workpiece must clear the carriage assembly, which is above the ways.

Types of lathes.   Lathes used in manufacturing can be classified as speed, engine, toolroom, and special types.

Speed lathes usually have only a headstock, tailstock, and a simple tool post mounted on a light bed. They ordinarily have only three or four speeds and are used primarily for wood turning, polishing, or metal spinning. Spindle speeds up to about 4000 rpm are common.

Engine lathes are the type most frequently used in manufacturing. Figure 20-3 is an example of this type. They are heavy-duty machine tools with all the components described previously and have power drive for all tool movements except on the compound rest. They commonly range in size from 305 to 610 mm (12 to 24 inches) swing and from 610 to 1219 mm (24 to 48 inches) center distances, but swings up to 1270 mm (50 inches) and center distances up to 3658 mm (12 feet) are not uncommon. Most have chip pans and a built-in coolant circulating system. Smaller engine lathes—with swings usually not over 330 mm (13 inches)—also are available in *bench type,* designed for the bed to be mounted on a bench or cabinet.

Toolroom lathes have somewhat greater accuracy and, usually, a wider range of speeds and feeds than ordinary engine lathes. Designed to have greater versatility to meet the requirements of tool and die work, they often have a continuously variable spindle speed range and shorter beds than ordinary engine lathes of comparable swing, since they are generally used for machining relatively small parts. They may be either bench or pedestal type.

Several types of special-purpose lathes are made to accommodate specific types of work. On a *gap-bed lathe,* for example, a section of the bed, adjacent to the headstock, can be removed to permit work of unusually large diameter to be swung. Another example is the *wheel lathe,* which is designed to permit the turning of the journals and wheel treads of railroad-car wheel-and-axle assemblies; a special headstock drives the assembly at a point between the two wheels.

Although engine lathes are versatile and very useful, because of the time required for changing and setting tools and for making measurements on the workpiece, they are not suitable for quantity production. Often the actual chip-production time is less than 30% of the total cycle time. In addition, a skilled machinist is required for all the operations, and such persons are costly and often in short supply. However, much of the operator's time is consumed by simple, repetitious adjustments and in watching chips being made. Consequently, to reduce or eliminate the amount of skilled labor that is required, turret lathes, screw machines, and other types of semiautomatic and automatic lathes have been highly developed and are widely used in manufacturing.

Turret lathes. The basic components of a *turret lathe* are depicted in Figure 20-8. Basically, a longitudinally feedable, multisided turret replaces the compound rest and tool post on the cross slide. The main turret usually has six sides on which tools can be mounted. This turret can be rotated about a vertical axis to bring each tool into operating position, and the entire unit can be moved longitudinally, either manually or by power, to provide feed for the tools. When the turret assembly is moved away from the spindle by means of a capstan wheel (see Figure 20-9), the turret indexes automatically at the end of its movement, thus bringing the next tool into operating position.

The turret on the cross slide can be rotated manually about a vertical axis to bring each of the four tools into operating position. On most machines, the turret

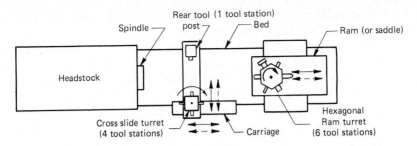

FIGURE 20-8 Block diagram (top view) showing the basic components of a turret lathe.

can be moved transversely, either manually or by power, by means of the cross slide, and longitudinally through power or manual operation of the carriage. In most cases, a fixed tool holder also is added to the back end of the cross slide; this often carries a parting tool.

Through these basic features of a turret lathe, a number of tools can be set up on the machine and then quickly be brought successively into working position so that a complete part can be machined without the necessity for further adjusting, changing tools, or making measurements. A skilled machinist is required for making the setup, but a relatively low-skilled operator thereafter can operate a turret lathe and produce parts with as good accuracy and with as much speed as though the operation were performed by a skilled machinist.

Two basic types of turret lathes are made, differing in the manner in which the main turret is mounted. Figure 20-9 shows a *ram-type turret lathe* on which

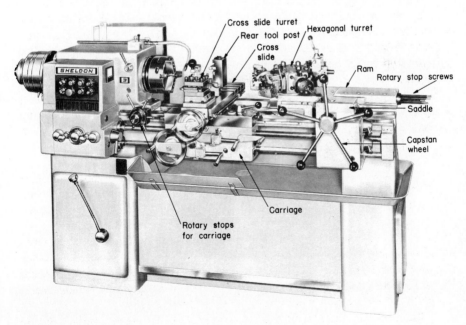

FIGURE 20-9 Ram-type turret lathe. (*Courtesy Sheldon Machine Co., Inc.*)

the main turret is pivoted and carried on a ram that slides back and forth in a saddle. When making a setup, the saddle is moved on the bed ways and clamped in the desired position. Because only the ram and turret must be moved during operation, this type of mounting provides easy and rapid motion.

Ordinarily, the ram and turret are moved up to the cutting position by means of the capstan wheel, and the power feed then is engaged. As the ram is moved toward the headstock, the turret is automatically locked into position so that rigid tool support is obtained. A set of rotary stopscrews, such as those shown in Figure 20-9 control the inward travel of the ram, one stop being provided and set for each face on the turret. The proper stop is brought into operating position automatically when the turret is indexed. A similar set usually is provided to limit movement of the cross slide.

Saddle-type turret lathes provide a more rugged mounting for the main turret than can be obtained by the ram-type mounting. On these lathes, as shown in Figure 20-10, the main turret is mounted directly on the saddle, and the entire saddle and turret assembly reciprocates. Larger turret lathes usually have this type of mounting. However, because the saddle-turret assembly is rather heavy, this type of mounting provides less rapid turret reciprocation. When such lathes are used with heavy tooling for making heavy or multiple cuts, a *pilot arm* attached to the headstock engages a pilot hole attached to one or more faces of the turret to give additional rigidity. Such a device is shown in Figure 20-10.

Turret lathe headstocks have two features not found on ordinary engine lathes. One permits rapid shifting between at least two spindle speeds, with a brake to stop the spindle very rapidly. The second feature is an automatic stock-feeding device for feeding bar stock through the spindle hole.

If the work is to be held in a chuck, some type of air-operated chuck, or special clamping fixture, frequently is employed to reduce the work setup time to a minimum.

Vertical turret lathes. Where chucking-type work is too large and heavy to permit holding it in a vertical chuck and rotating it about a horizontal axis,

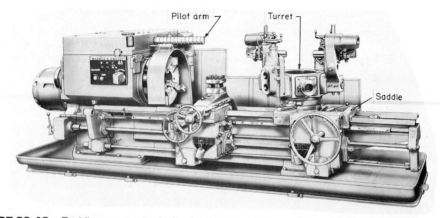

FIGURE 20-10 Saddle-type turret lathe, having a side-hung carriage. (*Courtesy Warner & Swasey Company.*)

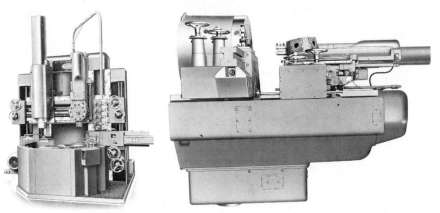

FIGURE 20-11 Vertical turret lathe. (*Left*) Normal front view. (*Right*) Side view, machine turned on its back. (*Courtesy Bullard Company.*)

vertical turret lathes are used. As shown in Figure 20-11 these essentially are regular turret lathes turned on end. These machines resemble certain types of vertical boring mills and frequently are thought of as special types of the latter. Their rotary work tables commonly range from about 610 to 1200 mm (24 to 48 inches) in diameter and are equipped with both removable chuck jaws and T-slots for clamping the work. The cross rail carries one or two five- or six-sided turrets and, often, another smaller, four-sided turret is mounted on one side of the machine on an independent cross slide. Usually, each motion of successive tools can be controlled by means of stops so that duplicate workpieces can be machined with one tooling setup.

Automatic turret lathes. After a turret lathe is tooled, the skill required of the operator is very low, and the motions are simple and repetitive. As a result, several types of automatic turret lathes have been developed which require no operator. A number of machines of this type permit setting the controls for all the machine motions very quickly by means of buttons and knobs on a control panel. A second type, shown in Figure 20-12 has a turret that indexes about a horizontal axis and reciprocates horizontally to provide feed. It has up to three cross slides, and the turret also can be rotated a small distance while making cuts to cut internal grooves. On this machine the operations are controlled by setting trip blocks and pins.

Ordinary turret lathes eliminate the necessity for making tooling setups for each piece machined and minimize machine-controlling time. However, an operator is required to control the machine and to feed the work into machining position. Automatic turret lathes can eliminate these last two functions, but because they usually have provision for manual operation, they are not as productive as *screw machines*, which are lathes designed for completely automatic operation. They originally were designed for machining small parts, such as screws, bolts, bushings, and so on, from bar stock—hence the name "screw machines." Now they are used for producing a wide variety of parts, covering a considerable range of sizes, and are even used for some chucking-type work.

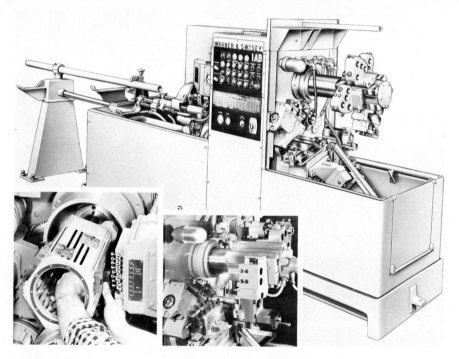

FIGURE 20-12 Automatic single-spindle turret lathe having turret that revolves about a horizontal axis. Insets show setting of control trip blocks and pins and a close-up view of the tooling. (*Courtesy Warner & Swasey Company.*)

Single-spindle screw machines. There are two common types of *single-spindle screw machines*. One, an American development and commonly called the Brown & Sharpe type, is shown in Figure 20-13. The other is of Swiss origin and is referred to as the Swiss type.

As can be seen in Figure 20-13, the *Brown & Sharpe screw machine* is essentially a small automatic turret lathe, designed for bar stock, with the main turret mounted in a vertical plane on a ram. Front and rear toolholders can be mounted on the cross slide. All motions of the turret, cross slide, spindle, chuck, and stock-feed mechanism are controlled by disk cams. These machines usually are equipped with an automatic rod-feeding magazine that feeds a new length of bar stock into the collet as soon as one rod is completely used.

Often Brown & Sharpe-type screw machines are equipped with a transfer, or "picking," attachment. This device swings over and picks up the workpiece from the spindle as it is cut off and carries it to, and holds it in, the position shown while a secondary operation is performed by a small, auxiliary power head. In this manner screwdriver slots are put in screw heads, small flats are milled parallel with the axis of the workpiece, or holes are drilled normal to the axis.

On the *Swiss-type screw machine,* the cutting tools are held and moved in radial slides shown in Figure 20-14. Disk cams move the tools into cutting position and provide feed into the work in a radial direction only; they provide any required longitudinal feed by reciprocating the headstock.

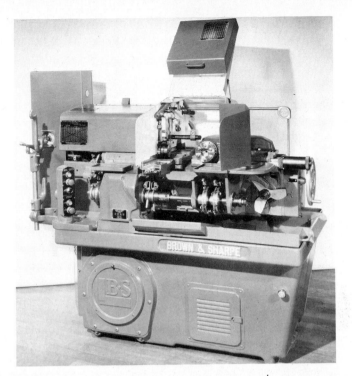

FIGURE 20-13 Brown & Sharpe single-spindle screw machine. (*Courtesy Brown & Sharpe Mfg. Co.*)

FIGURE 20-14 Close-up view of a Swiss-type screw machine, showing the tooling and radial tool sides, actuated by rocker arms. (*Courtesy George Gorton Machine Corporation.*)

Most machining on Swiss-type screw machines is done with single-point cutting tools. Because they are located close to the spindle collet, the workpiece is not subjected to much deflection. Consequently, these machines are particularly well suited for machining very small parts and are used primarily for such work.

Both types of single-spindle screw machines are capable of producing work to close tolerances, the Swiss-type probably being somewhat superior for very small work. Tolerances of 0.005 to 0.013 mm (0.0002 to 0.0005 inch) are not uncommon. The time required for setting the tooling usually is only an hour or two, and one person can tend several machines, once they have been properly tooled. They are highly productive; the entire cycle time frequently is less than ½ minute per piece.

Multiple-spindle screw machines. Although single-spindle screw machines eliminate the need for constant operator attendance, only one or two of the tooling positions are utilized at any given time. Thus the total cycle time per workpiece is the sum of the individual machining and tool-positioning times of the several cutting tools. On *multiple-spindle screw machines,* sufficient spindles, usually four, six, or eight, are provided so that all tools cut simultaneously. Thus the cycle time per piece is equal to the maximum cutting time of a single tool position plus the time required to index the spindles from one position to the next.

The two distinctive features of multiple-spindle screw machines are shown in Figures 20-15 and 20-16. First, the multiple spindles are carried in a rotatable drum that indexes in order to bring each spindle into a different working position.

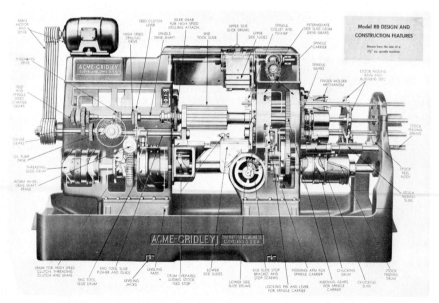

FIGURE 20-15 Mechanism in a six-spindle screw machine. (*Courtesy National Acme Company.*)

MACHINING PROCESSES

FIGURE 20-16 Close-up view of the spindle carrier, spindles, tooling in the end and cross slides, and parts being produced in a six-spindle screw machine. (*Courtesy National Acme Company.*)

Second, a nonrotating tool slide contains the same number of tool holders as there are spindles and thus provides and positions a cutting tool (or tools) for each spindle and imparts feed to these tools by longitudinal reciprocating motion. In addition, most machines have a cross slide at each spindle position so that an additional tool can be fed from the side for facing, grooving, knurling, beveling, and cutoff operations. These slides also are shown in Figure 20-16. All motions are controlled automatically.

With a tool position available on the end tool slide for each spindle (except for a stock-feed stop at one position), when the slide moves forward, these tools cut essentially simultaneously. At the same time, the tools in the cross slides move inward and make their cuts. When the forward cutting motion of the end tool slide is completed, it moves away from the work, accompanied by the outward movement of the radial slides. The spindles are indexed one position, by rotation of the spindle carrier, to position each part for the next operation to be performed. At one spindle position finished pieces are cut off, and the bar stock is fed to correct length for the beginning of the next operation. Thus a piece is completed each time the tool slide moves forward and back.

Multiple-spindle screw machines are made in a considerable range of sizes, determined by the diameter of the stock that can be accommodated in the spindles.

In tooling a multiple-spindle screw machine, it should be remembered that because all cutting operations occur simultaneously, the operating cycle of the main tool slide is determined by the operation that requires the longest time.

Consequently, one attempts to balance all the operations so that each requires the same amount of time. Although this ideal frequently cannot be achieved, careful planning permits it to be approached. Proper sequencing is important, and long operations may be broken up so as to be completed using two or more positions. For example, in the tooling program shown in Figure 20-17 if the ½-inch-diameter hole were drilled before the one having a ⁴⁷⁄₆₄-inch diameter, this operation would require much more time than any of the others. By reversing the sequence, better balance is achieved, a total feed of 0.720 inch being required in one case and 0.750 inch for the other.

Where a simple part is to be made, it sometimes is possible to arrange the tooling on a multiple-spindle automatic so that two pieces are produced in each revolution of the spindle carrier.

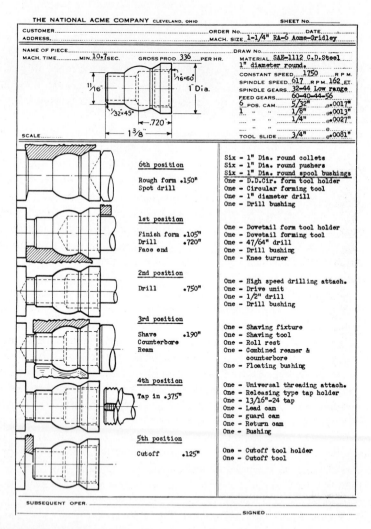

FIGURE 20-17 Tooling sheet for making a part on a six-spindle screw machine. (*Courtesy National Acme Company.*)

The only attention a multiple-spindle screw machine re⬛⬛ bar stock feed rack supplied and to check the finished products ⬛ make sure they are within the desired tolerances. One operator usually ⬛ several machines.

Most multiple-spindle screw machines utilize cams that are composed of specially shaped segments that are bolted onto a drum to control the motions, thereby reducing the need for special cams. Setting these cams and the tooling for a given job may require from 2 to 20 hours. However, once such a machine is placed in operation, the productivity is very great. Often a piece may be completed each 10 seconds. Typically, from 2000 to 5000 parts are required in a lot to justify tooling a multiple-spindle automatic.

The precision of multiple-spindle screw machines is good, but seldom as good as that of single-spindle machines. However, tolerances of from 0.013 to 0.025 mm (0.0005 to 0.001 inch) on the diameter are common.

Multiple-station lathes, available for work that must be held in chucks, are essentially vertical, multiple-spindle screw machines. A number of chuck-equipped spindles are mounted in a rotary indexing table and are indexed successively under a series of vertical rams in which the cutting tools are mounted. One chuck position remains at rest and has no tool ram, so that workpieces can be loaded and unloaded at this position while machining takes place at all the others.

Although screw machines and automatic turret lathes are automatic *types* of lathes, the term *automatic lathe* generally is applied to a group of lathes that are semiautomatic and make simultaneous cuts, but that do not involve the use of the turret or screw-machine principles. The tools are fed to the work and retracted automatically by means of cam-controlled mechanisms. In most cases an operator is required to place the work in and remove it from the machine, so they are not truly automatic. However, in some instances a magazine type of workpiece feeder is employed so that no operator is required. The majority of automatic lathes have only a single spindle, but some specialized multispindle machines are also used.

In a typical *single-spindle automatic lathe,* the cutting tools are held in *tool blocks,* or *slides,* which are power-actuated and controlled to move the tools into and feed them along the work. Sometimes the front block provides only radial motion for facing, form cutting, and cutoff operations. The rear block has both radial and longitudinal motions, which are controlled by a plate cam, as illustrated in Figure 20-18, which shows the motions of the tool blocks and the portion of the metal removed by each tool. On some machines a third, overhead tool block also is provided.

In most cases the work is held between centers, utilizing various types of power-actuated chucks, collets, and tailstocks so that the work-handling time is minimized. In some cases the work is fed into the machine from a hopper-type feeding device and clamped and discharged automatically.

The total machining cycle on an automatic lathe usually is very short—often less than 1 minute. Sometimes a part is put successively into from two to four automatic lathes to complete its machining. They are fairly flexible, so that quite a variety of shapes and sizes can be handled in one lathe by changing the tooling setup.

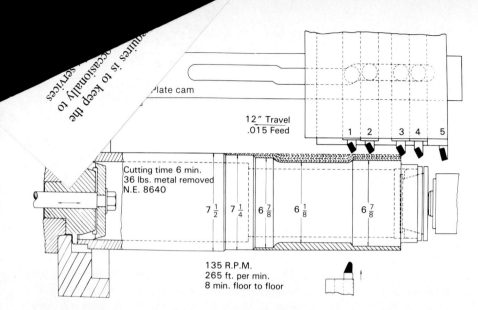

Plate cam

12″ Travel
.015 Feed

1 2 3 4 5

Cutting time 6 min.
36 lbs. metal removed
N.E. 8640

$7\frac{1}{2}$ $7\frac{1}{4}$ $6\frac{7}{8}$ $6\frac{1}{8}$ $6\frac{7}{8}$

135 R.P.M.
265 ft. per min.
8 min. floor to floor

FIGURE 20-18 Movements of the tool blocks for the machining process in a single spindle automatic lathe, with the metal to be removed by each tool indicated. (*Courtesy Gisholt Corporation.*)

Supporting work in lathes. Five methods commonly are used for supporting workpieces in lathes:

1. Held between centers.
2. Held in a chuck.
3. Held in a collet.
4. Mounted on a face plate.
5. Mounted on the carriage.

In the first four of these methods the workpiece is rotated during machining. In the fifth method, which is not used extensively, the tool rotates while the workpiece is fed into the tool.

Lathe centers. Workpieces that are relatively long with respect to their diameters usually are machined between centers. Two *lathe centers* are used, one in the spindle hole and the other in the hole in the tailstock quill. Two types are used. The *plain* or *solid* type, shown in Figure 20-19, is made of hardened steel with a Morse taper on one end so that it will fit into the spindle hole. The other end is ground to a 60° taper; sometimes the tip of this taper is made of tungsten carbide to provide better wear resistance. Before a center is placed in position, the spindle hole should be carefully wiped clean. The presence of

CHICAGO-LATROBE

Taper
end

60°

FIGURE 20-19 Solid or "dead" lathe center. (*Courtesy Chicago-Latrobe Twist Drill Works.*)

TABLE 20-1 Size of Center Holes and Combination Countersink

Diameter of Work		Diameter of Large End of Hole		Diameter of Body of Countersink	
mm	in.	mm	in.	mm	in.
5.0–8.0	3⁄16–5⁄16	3.2	1⁄8	5.15	13⁄64
9.5–25.5	3⁄8–1	4.75	3⁄16	7.95	5⁄16
27.0–51.0	1 1⁄16–2	6.35	1⁄4	7.95	5⁄16
52.5–102.0	2 1⁄16–4	7.95	5⁄16	11.10	7⁄16

foreign material will prevent the center from seating properly and it will not be aligned accurately.

Before a workpiece can be mounted between lathe centers, a 60° center hole must be drilled in each end. This can be done in a drill press, or in a lathe by holding the work in a chuck. A combination center drill and countersink ordinarily is used, taking care that the center hole is deep enough so that it will not be machined away in any facing operation, and yet is not drilled to the full depth of the tapered portion of the center drill. Table 20-1 gives recommended center-hole sizes.

The work and the center at the headstock end rotate together, so no lubricant is needed in the center hole at this end. However, because the center in the tailstock quill does not rotate, adequate lubrication must be provided. This usually is accomplished by putting a mixture of white lead and oil in the center hole before the dead center is tightened in the hole. Failure to provide proper lubrication at all times will result in scoring of the workpiece center hole and the center, and inaccuracy and serious damage may occur.

Proper tightness must be maintained between the centers and the workpiece. The workpiece must rotate freely, yet no looseness should exist. Looseness usually will be manifested in "chattering" of the workpiece during cutting. Tightness of the centers should be checked after cutting for a short time; the resulting heating and thermal expansion of the workpiece will increase the tightness.

Live centers often are used in the tailstock quill. In this type, shown in Figure 20-20, the end that fits into the workpiece is mounted on ball or roller bearings so that it is free to rotate; thus no lubrication of the center hole is required. However, they may not be as accurate as the plain type, so they often are not used for precision work.

A mechanical connection must be provided between the spindle and the workpiece to cause it to rotate. This is accomplished by some type of lathe *dog* and

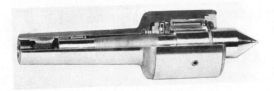

FIGURE 20-20 "Live" type of lathe center. (*Courtesy Motor Tool Manufacturing Company.*)

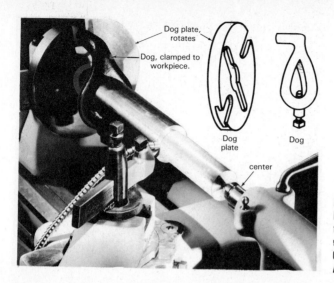

FIGURE 20-21 Work being turned between centers in a lathe, showing the use of a dog and dog plate. (*Courtesy South Bend Lathe.*)

a *dog plate,* as shown in Figure 20-21. The dog is a forging that fits over the end of the workpiece and is clamped to it by means of a setscrew. The tail of the dog enters a slot in the dog plate, which is rigidly attached to the lathe spindle in the same manner as a lathe chuck. If the dog must be attached to work that has a finished surface, a piece of soft metal, such as copper or aluminum, can be placed between the work and the setscrew to avoid marring.

Mandrels. Workpieces that must be machined on both ends, and those that are disklike in shape, are often mounted on mandrels for turning between centers. Three common types of mandrels are shown in Figure 20-22. *Solid mandrels* usually vary from 102 to 305 mm (4 to 12 inches) in length and are accurately ground with a 1:2000 taper (0.006 inch per foot). After the workpiece is drilled and/or bored to fit, it is pressed on the mandrel. The mandrel should be mounted between centers so that the cutting force tends to tighten the work on the mandrel taper. Solid mandrels permit the work to be machined on both ends as well as on the cylindrical surface. They are available in stock sizes but can be made to any desired size.

Gang or *disk mandrels* are used for production-type work, because the workpieces do not have to be pressed on and thus can be put in position and removed more rapidly. However, only the cylindrical surface of the workpiece can be machined when this type of mandrel is used.

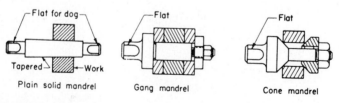

FIGURE 20-22 Three types of mandrels.

Cone mandrels have the advantage that they can be used to center workpieces having a range of hole sizes.

Lathe chucks. Lathe chucks are used to support a wider variety of workpiece shapes and to permit more operations to be performed than can be accomplished when the work is held between centers. Two basic types of chucks are used. These are illustrated in Figure 20-23.

Three-jaw, self-centering chucks are used for work that has a round or hexagonal cross section. The three jaws are moved inward or outward simultaneously by the rotation of a spiral cam, which is operated by means of a special wrench through a bevel gear. If they are not abused, these chucks will provide automatic centering to within about 0.025 mm (0.001 inch). However, they can be damaged through use and will then be considerably less accurate.

Each jaw in a *four-jaw independent chuck* can be moved inward and outward independent of the others by means of a chuck wrench. Thus they can be used to support a wide variety of work shapes. A series of concentric circles, engraved on the chuck face, aid in adjusting the jaws to fit a given workpiece. Four-jaw chucks are heavier and more rugged than the three-jaw type and, because undue pressure on one jaw does not destroy the accuracy of the chuck, they should be used for all heavy work. The jaws on both three- and four-jaw chucks can be reversed to facilitate gripping either the inside or the outside of workpieces.

Combination four-jaw chucks are available in which each jaw can be moved independently or all can be moved simultaneously by means of a spiral cam. Two-jaw chucks are also available. For mass-production work, special chucks often are used in which the jaws are actuated by air or hydraulic pressure, permitting very rapid clamping of the work.

Collets. By the use of *collets,* smooth bar stock or workpieces that have been machined to a given diameter can be held more accurately than with regular

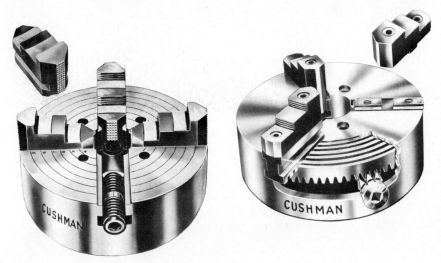

FIGURE 20-23 Lathe chucks. (*Left*) Four-jaw independent. (*Right*) Three-jaw, self-centering. (*Courtesy Cushman Industries, Inc.*)

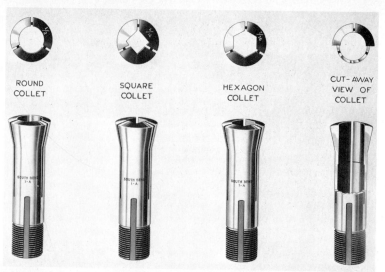

FIGURE 20-24 Several types of lathe collets. (*Courtesy South Bend Lathe.*)

three- or four-jaw chucks. As shown in Figure 20-24, collets are relatively thin tubular steel bushings that are split into three longitudinal segments over about two-thirds of their length. At the split end, the smooth internal surface is shaped to fit the piece of stock that is to be held, and the external surface is a taper that fits within an internal taper of a collet sleeve placed in the spindle hole, as shown in Figure 20-25. When the collet is pulled inward into the spindle, by means of the draw bar that engages threads in its inner end, the action of the two mating tapers squeezes the collet segments together, causing them to grip the workpiece.

As shown in Figure 20-24, collets are made to fit a variety of symmetrical shapes. If the stock surface is smooth and accurate, good collets will provide very accurate centering; maximum runout should be less than 0.013 mm (0.0005 inch). However, the work should be no more than 0.05 mm (0.002 inch) larger

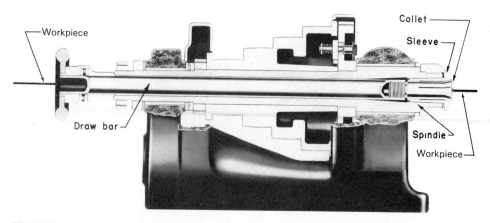

FIGURE 20-25 Method of using a draw-in collet in a lathe spindle. (*Courtesy South Bend Lathe.*)

Face plate

FIGURE 20-26 Boring in a lathe with work (w) held in a special fixture (fix) mounted on a face plate (fp). See Figure 20-35 also.

or 0.13 mm (0.005 inch) smaller than the nominal size of the collet. Consequently, collets are used only on drill-rod, cold-rolled, extruded, or previously machined stock.

Collets which can open automatically and feed bar stock forward to a stop mechanism are commonly used on turret lathes. Another type of collet similar to a Jacobs drill chuck (see Chapter 21) has a greater size range than ordinary collets, so fewer are required.

Face plates. *Face plates* are used to support irregularly shaped work that cannot be gripped easily in chucks or collets. The work can be bolted or clamped directly on the face plate or can be supported on an auxiliary fixture that is attached to the face plate, as shown in Figure 20-26. The latter procedure is time-saving when several identical pieces are to be machined.

Mounting work on the carriage. When no other means is available, boring occasionally is done on a lathe by mounting the work on the carriage, with the boring bar mounted between centers and driven by means of a dog. This procedure is illustrated in Figure 20-27.

Steady and follow rests. If one attempts to turn a long, slender piece between centers, the radial force exerted by the cutting tool, or the weight of the workpiece itself, may cause it to be deflected out of line. *Steady rests* and *follow*

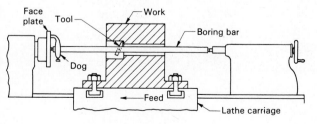

FIGURE 20-27 Method of boring on a lathe with the work mounted on the carriage.

FIGURE 20-28 Cutting a thread on a long, slender workpiece, using a follow rest (*left*) and a steady rest (*right*). (*Courtesy South Bend Lathe.*)

rests, shown in Figure 20-28, provide means for supporting such work between the headstock and the tailstock. The steady rest is clamped to the lathe ways and has three movable fingers that are adjusted to contact the work and align it. A light cut should be taken before adjusting the fingers to provide a smooth contact-surface area.

A steady rest also can be used in place of the tailstock as a means of supporting the end of long pieces, pieces having too large an internal hole to permit using a regular dead center, or work where the end must be open for boring. In such cases the headstock end of the work must be held in a chuck to prevent its moving longitudinally, and tool feed should be toward the headstock.

The follow rest is bolted to the lathe carriage. It has two contact fingers that are adjusted to bear against the workpiece, opposite the cutting tool, so as to prevent the work from being deflected away from the cutting tool by the cutting forces.

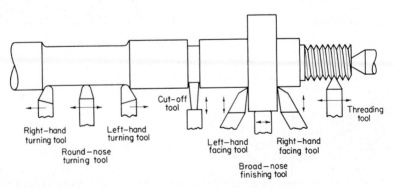

FIGURE 20-29 Shapes and uses of common single-point lathe tools.

Lathe tools. Most lathe operations are done with relatively simple, single-point cutting tools, such as illustrated in Figure 20-29. On right-hand and left-hand turning and facing tools, the cutting takes place on the side of the tool so that the side rake angle is of primary importance and deep cuts can be made. On the round-nose turning tools, cutoff tools, finishing tools, and some threading tools, cutting takes place on or near the end of the tool, so that the back rake is of importance. Such tools are used with relatively light depths of cut.

When using high-speed steel as a tool material, the HSS tool bit is clamped in a forged steel tool holder, which is, in turn, clamped into the tool post (see Figures 20-30 and 20-21).

Because cutting-tool materials are expensive, it is desirable to use as small amounts as possible. At the same time, it is essential that the cutting tool be supported in a strong, rigid manner to minimize deflection and possible vibration. Consequently, lathe tools are supported in various types of heavy, forged tool holders, such as shown in Figure 20-30. The tool bit should be clamped in the tool holder with minimum overhang. Otherwise, tool chatter and poor surface finish may result.

When using carbide, ceramic, or coated carbides for mass production work, throwaway inserts are used, which can be purchased in a great variety of shapes, geometrics (nose radius, tool angles, and groove geometry), and sizes (see Figure 20-31).

Where several different operations on a lathe are performed repeatedly in sequence, the time required for changing and setting tools may constitute as

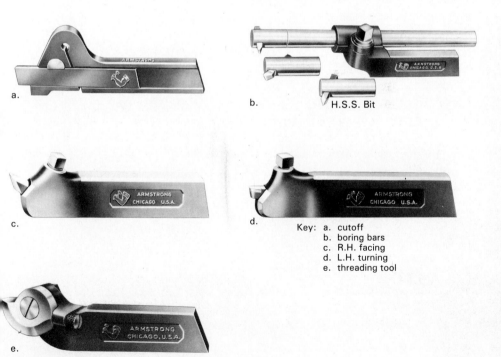

Key: a. cutoff
b. boring bars
c. R.H. facing
d. L.H. turning
e. threading tool

FIGURE 20-30 Common types of forged tool holders. (*Courtesy Armstrong Bros. Tool Company.*)

Insert shape	Available cutting edges	Typical insert holder
Round	4–10 on a side 8–20 total	15° Square insert
80°/100° diamond	4 on a side 8 total	
Square	4 on a side 8 total	0° Triangular insert
Triangle	3 on a side 6 total	
55° diamond	2 on a side 4 total	35° diamond
35° diamond	2 on a side 4 total	5°

FIGURE 20-31 Typical insert shapes, available cutting edges per insert and insert holders. (*Adapted from* Turning Handbook of High Efficiency Metal Cutting; *courtesy General Electric Company.*)

much as 50% of the total time. As a consequence, quick-change tool holders, such as shown in Figure 20-32, are being used increasingly. The individual tools, preset in their holders, can be interchanged in the special tool post in a few seconds. With some systems a second tool may be set in the tool post while a cut is being made with the first tool, and then be brought into proper position by rotating the post.

In lathe work the nose of the tool should be set exactly at the same height as the axis of rotation of the work. However, because any setting below the axis causes the work to tend to "climb" up on the tool, most machinists set their tools a few thousandths of an inch above the axis, except for cutoff, threading, and some facing operations.

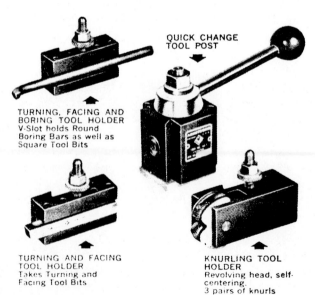

QUICK CHANGE
TOOL POST

TURNING, FACING AND
BORING TOOL HOLDER
V-Slot holds Round
Boring Bars as well as
Square Tool Bits

TURNING AND FACING
TOOL HOLDER
Takes Turning and
Facing Tool Bits

KNURLING TOOL
HOLDER
Revolving head, self-
centering.
3 pairs of knurls

FIGURE 20-32 Quick-change tool post and accompanying tool holders. (*Courtesy Armstrong Bros. Tool Company.*)

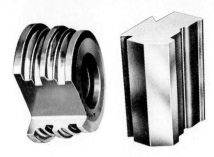

FIGURE 20-33 Circular and block types of form tools. (*Courtesy Speedi Tool Company, Inc.*)

Form tools. Form tools, made by grinding the inverse of the desired work contour on a piece of tool steel, are used to a considerable extent on lathes. For example, a threading tool often is a form tool. Although form tools are relatively expensive to make, they make it possible to machine a fairly complex surface with a single inward feeding of one tool. For mass-production work, adjustable form tools of either flat or rotary types, such as are shown in Figure 20-33, are used. These, of course, are expensive to make but can be resharpened by merely grinding a small amount off the face and then raising or rotating the cutting edge to the correct position.

The use of form tools is limited by the difficulty of grinding adequate rake angles for all points along the cutting edge. A rigid setup is needed to resist the large cutting forces which develop with these tools.

Drill chucks are used on lathes for holding drills, center drills, and reamers mounted in the tailstock quill. Large drills are mounted directly in the quill hole by means of their taper shanks.

Turret-lathe tools. In turret lathes, the work is generally held in collets and the correct amount of bar stock is fed into the machine to make one part. The tools are arranged in sequence at the tool stations with depths of cut all preset. The following factors should be considered when setting up a turret lathe:

1. *Setup time:* time required for setup man to set the tooling and set stops. Standard toolholders and tools should be used as much as possible to minimize setup time.
2. *Workhandling time:* time to load and unload parts and/or stock.
3. *Machine-controlling time:* time required to manipulate the turrets. Can be reduced by combining operations where possible. Dependent on the sequence of operations established by the setup man.
4. *Cutting time:* time during which chips are being produced. Should be as short as economically practical and represent the greatest percentage of the total cycle time as possible.
5. *Cost:* cost of the tool, setup labor cost, lathe operator labor cost, and the number of pieces to be made.

There are essentially 11 tooling stations, as shown in Figure 20-34, with six in the turret, four in the indexable tool post, and one in the rear tool post. The tooling is more rugged in turret lathes because heavy, simultaneous cuts are often made. Tools mounted in the hex turret that are used for turning are often

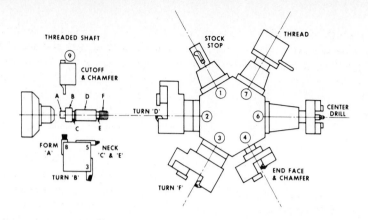

FIGURE 20-34 Turret-lathe tooling for producing the part shown. Numbers in circles indicate the sequence of the operations from 1 to 9. *Note:* 3 is a combined operation.

equipped with pressure rollers set on the opposite side of the rotating workpiece from the tool to counter the cutting forces.

Turret lathes are most economical in producing lots too large for engine lathes but too small for automatic screw machines or automatic lathes. However, in recent years much of this work has been assumed by numerical control lathes, which will be discussed in Chapter 38. Numerical control lathes represent a higher level of automation, A(4), than turret lathes.

LATHE OPERATIONS

Turning. *Turning* constitutes the majority of lathe work. The work usually is held between centers or in a chuck, and a right-hand turning tool is used, so that the cutting forces, resulting from feeding the tool from right to left, tend to force the workpiece against the headstock and thus provide better work support.

If good finish and accurate size are desired, one or more roughing cuts usually are followed by one or more finishing cuts. Roughing cuts may be as heavy as proper chip thickness, tool life, and lathe capacity permit. Large depths of cut and smaller feeds are preferred to the reverse procedure, because fewer cuts are required and less time is lost in reversing the carriage and resetting the tool for the following cut.

On workpieces that have a hard surface, such as castings or hot-rolled materials containing mill scale, the initial roughing cut should be deep enough to penetrate the hard material. Otherwise, the entire cutting edge operates in hard, abrasive material throughout the cut, and the tool will dull rapidly. If the surface is unusually hard, the cutting speed on the first roughing cut should be reduced accordingly.

Finishing cuts are light, usually being less than 0.38 mm (0.015 inch) in depth, with the feed as fine as necessary to give the desired finish. Sometimes a special finishing tool is used, but often the same tool is used for both roughing and finishing cuts. In most cases one finishing cut is all that is required. However, where exceptional accuracy is required, two finishing cuts may be made. If the diameter is controlled manually, it usually is desirable to make a short finishing cut (about 6.4 mm or ¼ inch long) and check the diameter before completing the cut. Because the previous micrometer measurements were made on a rougher surface, some readjustment of the tool setting may be necessary in order to have the final measurement, made on a smoother surface, check exactly.

In turning operations, diameters usually are measured with micrometer calipers, although spring calipers may be used to check roughing cuts or where close accuracy is not required. The method of making length measurements is controlled, primarily, by the shape and accessibility of the surfaces over which measurement must be made. Spring, hermaphrodite, vernier, or micrometer calipers or micrometer depth gages can be used.

Facing.   *Facing* is the producing of a flat surface as the result of the tool being fed across the end of the rotating workpiece. The work may be held in a chuck, on a face plate, or between centers. Unless the work is held on a mandrel, if both ends of the work are to be faced, it must be turned end for end after the first end is completed and the facing operation repeated.

Because most facing operations are performed on surfaces that are away from the headstock, a right-hand tool is used most frequently. The spindle speed should be determined from the largest diameter of the surface to be faced. Facing may be done either from the outside inward or from the center outward. In either case, the point of the tool must be set exactly at the height of the center of rotation. Because the cutting force tends to push the tool away from the work, it usually is desirable to clamp the carriage to the lathe bed during each facing cut to prevent it from moving slightly and thus producing a surface that is not flat.

When facing castings or other materials that have a hard surface, the depth of the first cut should be sufficient to penetrate the hard material to avoid excessive tool wear.

Drilling.   Most *drilling* on lathes is done with the drill held in the tailstock quill and fed against a workpiece that is rotated in a chuck. Drills with taper

FIGURE 20-35   Drilling in a lathe, using a drill chuck. See Figure 20-26 for next operation.

FIGURE 20-36 Drilling in a lathe with the drill held in a drill chuck in the spindle and with the work supported by a drill pad in the tailstock quill. (*Courtesy South Bend Lathe.*)

shanks are mounted directly in the quill hole, whereas those with straight shanks are held in a drill chuck that is mounted in the quill hole, as illustrated in Figure 20-35. Feeding is by hand by means of the handwheel on the outer end of the tailstock assembly.

It also is possible to do drilling on a lathe with the drill mounted and rotated in the spindle while the work remains stationary, supported by a special pad mounted in the tailstock quill. This procedure, illustrated in Figure 20-36, is seldom used but is useful in special cases.

Usual speeds are used for drilling in a lathe. Because feeding is by hand, care must be exercised, particularly in drilling small holes. Coolants should be used where required. In drilling deep holes, the drill should be withdrawn occasionally to clear chips from the hole and to aid in getting coolant to the cutting edges.

Reaming. *Reaming* in a lathe involves no special precautions. Reamers are held in the tailstock quill, taper-shank types being mounted directly and straight-shank types by means of a drill chuck. Rose-chucking reamers usually are used. Fluted-chucking reamers also may be used, but these should be held in some type of holder that will permit the reamer to float.

Parting. *Parting* is the operation by which one section of a workpiece is severed from the remainder by means of a cutoff tool. Because parting tools are

Two forming rolls

FIGURE 20-37 Knurling tool with forming rolls. (*Courtesy Armstrong Bros. Tool Company.*)

quite thin and must have considerable overhang, it is a less accurate and more difficult operation. The tool, shown in Figure 20-31, should be set exactly at the height of the axis of rotation, be kept sharp, have proper clearance angles, and be fed into the workpiece at a proper and uniform rate.

Knurling. *Knurling* produces a regularly shaped, roughened surface on a workpiece. Although knurling also can be done on other machine tools, even on flat surfaces, in most cases it is done on external cylindrical surfaces in some type of lathe. In most cases, knurling is a chipless, cold-forming process, using a tool of the type shown in Figures 20-37 and 20-38. The two hardened rolls are pressed against the rotating workpiece with sufficient force to cause a slight outward and lateral displacement of the metal so as to form the knurling in a raised, diamond pattern. Another type of knurling tool produces the knurled pattern by cutting chips. Because it involves less pressure and thus does not tend to bend the workpiece, it often is preferred for workpieces of small diameter and for use on automatic or semiautomatic machines.

— TOOL OR PIECE CAN BE TURNED.

Boring. *Boring* always involves the enlarging of an existing hole, which may have been made by a drill or be the result of a core in a casting. An equally important, and concurrent, purpose of boring may be to make the hole concentric with the axis of rotation of the workpiece and thus correct any eccentricity that may have resulted from the drill having drifted off the center line. Concentricity is an important attribute of bored holes.

When boring is done in a lathe, the work usually is held in a chuck or on a face plate. Holes may be bored straight, tapered, or to irregular contours. Figure 20-26 shows the relationship of the tool and the workpiece for boring. Many types of boring tools are used. The one shown in Figures 20-26 and 20-30 consists of a conventional left-hand lathe tool held in the end of a round bar that, in turn, is mounted in cantilever fashion in a forged tool holder. Another type is forged from a single piece of tool steel and is held either in the tool post or in a forged tool holder. It is used for boring holes that are too small to permit the entry of the other type of boring bar.

Get notes from Don

on S1-A - 36-6

FIGURE 20-38 Knurling in a lathe, using a forming-type tool, and showing the resulting pattern on the workpiece.

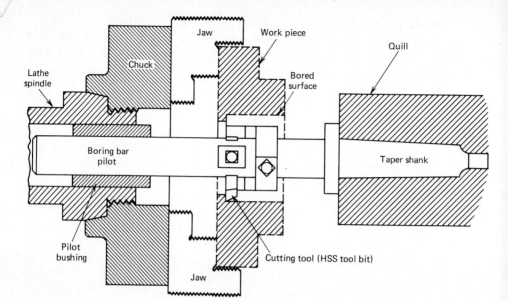

FIGURE 20-39 Pilot boring bar mounted in tailstock of lathe for precision boring large hole in casting. The size of the hole is controlled by the rotation diameter of the cutting tool.

Large holes may be precisoin bored using the lathe setup shown in Figure 20-39, where a pilot bushing is placed in the spindle to mate with the hardened ground pilot of the boring bar. This setup eliminates the cantilever problems common to boring.

In most respects the same principles are used for boring as for turning. However, the tool should be set exactly at the same height as the axis of rotation. Slightly larger end clearance angles sometimes have to be used to prevent the heel of the tool from rubbing on the inner surface of the hole. Because the tool overhangs its support a considerable amount, feeds and depths of cut may have to be somewhat less than for turning to prevent tool vibration and chatter. In some cases the boring bar may be made of tungsten carbide because of its greater stiffness.

There always is a tendency for bored holes to be slightly bell-mouthed because of the tool springing away from the work as it progresses into the hole. This usually can be corrected by repeating the cut with the same tool setting.

Because the rotational relationship between the work and the tool is a simple one and is employed on several types of machine tools, such as lathes, drilling machines, and milling machines, boring very frequently is done on such machines. However, several machine tools have been developed primarily for boring, especially where large workpieces are involved or for large-volume boring of smaller parts. Some of these are also capable of performing other operations, such as milling and turning. Because boring frequently follows drilling, many boring machines also can do drilling, permitting both operations to be done with a single setup of the work.

Vertical boring and turning machines. Figure 20-40 shows the basic elements of a vertical boring and turning machine. It will be noted that these structurally are similar to double-housing planers, except that the table rotates instead of reciprocating. Functionally, a vertical boring machine essentially is the same as a vertical turret lathe, but it usually has two main tool heads instead of a turret. Thus turning, facing, and usually drilling (but not milling) are done on these machines, which often are called *vertical boring mills*.

Vertical boring machines come with tables ranging from about 900 to 12200 mm (3 to 40 feet) in diameter. The toolheads are mounted and provided with both horizontal and vertical feed, so they can be used for boring and for facing cuts. Usually, one or both can also be swiveled about a horizontal axis to permit boring at an angle. Most machines also have a side toolhead, sometimes provided with a four-sided turret. This toolhead has vertical and horizontal feed and is used primarily for turning. Single-point tools customarily are used, and

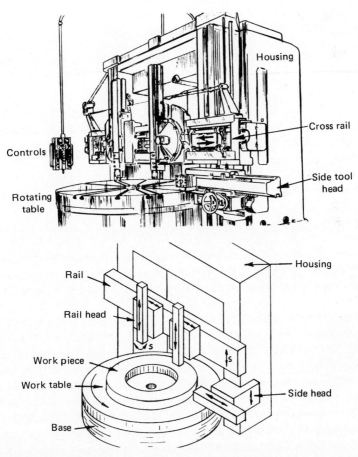

FIGURE 20-40   Sketch and block diagram of vertical boring and turning mill. (Courtesy Giddings and Lewis Machine Tool Co. and *from* Manufacturing Producibility Handbook; *courtesy General Electric Company.*)

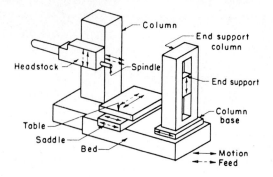

FIGURE 20-41 Block diagram showing the basic components and motions of a horizontal boring, drilling, and milling machine.

turning, facing, and boring, or roughing and finishing cuts can be done simultaneously.

Many modern boring machines are numerically controlled (see Chapter 38). This permits the operator to make tool settings merely by setting dials and also to preset the adjustment for a cut while one is being made. The pressing of a button at the conclusion of a cut causes the tool to move very quickly to the proper position for the next cut. This reduces the amount of machine-controlling time and thereby increases the productivity of such large and costly machines.

Horizontal boring (drilling, and milling) machines. *Horizontal boring machines* are very versatile and thus particularly useful in machining large parts. The basic components of these machines are indicated in Figure 20-41 and an actual machine in use is shown in Figure 20-42. The essential features are as follows:

1. A rotating spindle that can be fed horizontally (tool rotates).
2. A table that can be moved and fed in two directions in a horizontal plane.
3. A headstock that can be moved vertically.
4. An outboard bearing support for a long boring bar.

The spindle is similar to an oversized drilling-machine spindle and will accept both drills and milling cutters. A wide range of speeds is provided, and heavy bearings are incorporated that will absorb thrust in all directions. The spindle also is provided with longitudinal power feed so that drilling and boring can be done through a considerable distance without the table being moved.

Boring on this type of machine is done by means of a rotating single-point tool. The tool can be mounted in either a stub-type bar, held only in the spindle, as shown in Figure 20-44 and Figure 20-45, or in a long line-type bar that has its outer end supported in a bearing on the outboard column, as shown in Figure 20-42. The outboard bearing provides rigid support for the boring bar and permits very accurate work to be done. However, because of the flexibility inherent in a long boring bar and offset tool holder, horizontal boring machines are used primarily for boring holes less than 305 mm (12 inches) in diameter, or for long holes, or for a series of in-line holes. Unless they are very long or unless the shape of the workpiece does not permit, larger holes usually are bored more readily on a vertical boring mill.

FIGURE 20-42 Boring a weldment on horizontal boring, drilling, and milling machine. A line-type boring bar is being used with an outboard bearing support. (*Courtesy Lucas Machine Division, The New Britain Machine Company.*)

Mass-production boring machines. Special boring machines are built for machining specific parts in mass production. In these the workpiece usually remains stationary, and boring is done by one or more rotating boring tools, typically carried in a reciprocating powerhead, such as is shown in Figure 20-43.

In most cases the operation is automatic once the workpiece is placed in the fixture. Such machines usually are very accurate and often are equipped with automatic gaging and sizing controls. (See discussion of transfer lines in Chapter 38.)

With rotating workpieces, the size of the hole is controlled by transverse movement of the tool holder. When boring is done with a rotating tool, size is controlled by changing the offset radius of the cutting-tool tip with respect to the axis of rotation. A general-purpose type of adjustable boring bar is shown in Figure 20-44. The type shown in Figure 20-45 has more precise control and is used on larger-scale manufacturing. Two or more adjustable cutting tools can be built into a single bar, thus permitting more than one diameter to be bored simultaneously. For boring relatively long holes, the type of boring bar shown

FIGURE 20-43 (*Left*) Production-type boring machine, having multiple heads, that completes a part in 51 seconds. (*Above*) Close-up view of one multiple-spindle boring head on a production-type machine. (*Courtesy Heald Machine Company.*)

in Figure 20-46 has a special advantage. As shown, the smaller, forward bit corrects misalignment of the original hole and provides a guide hole for the nose cone. The nose cone then provides good alignment and support for the rear bit, which bores the final hole to size.

Boring machine precision.   Because boring is essentially the same as turning, the precision obtainable is similar except for the fact that the tool support may be less rigid. Thus the precision depends considerably on the rigidity of the tool support. On specialized, production-type boring machines, tolerances are readily held to within 0.013 mm (0.0005 inch) on small diameters, whereas on general-purpose machines tolerances of 0.025 mm (0.001 inch) are typical unless the boring bar overhang becomes excessive.

FIGURE 20-44 Adjustable boring bar, using the offset-radius principle.

FIGURE 20-45 Adjustable boring tool. Extension of the single-point tool from the bar is adjustable, as shown in sectional view. (*Courtesy DeVlieg Machine Co.*)

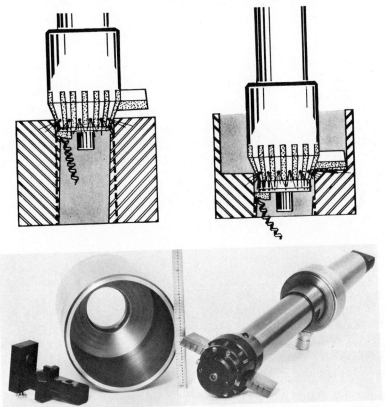

FIGURE 20-46 Boring tool employing a centering tool and conical guide, for boring large holes in a single operation. (*Courtesy Vernon Devices, Inc.*)

Jig borers. *Jig borers* are very precise vertical-type boring machines designed for use in making jigs and fixtures. From the viewpoint of boring operations they contain no unusual features, except that the spindle and spindle bearings are constructed with very high precision. Their unique features are in the design of the worktable controls, which permits very precise movement and control, thus making them especially useful in layout work. These machines will be discussed in Chapter 35. However, the precision of many modern NC machining centers (see Chapter 38) is such that these, to a considerable extent, have taken the place of jig borers.

Turning and boring tapers. The turning and boring of uniform tapers are common lathe operations. Such tapers can be specified either in degrees of included angle between the sides or as the change in diameter per unit of length—millimeters per millimeter or inches per foot.

Three methods are available for turning external tapers on a lathe, and two for boring internal tapers. The simplest is to use the compound rest; this method is suitable for both external and internal tapers. However, because the length of travel of the compound rest is quite limited—seldom over 150 mm (a few inches)—only short tapers can be turned or bored by this method. It is particularly useful for steep tapers. The compound rest is swiveled to the desired angle

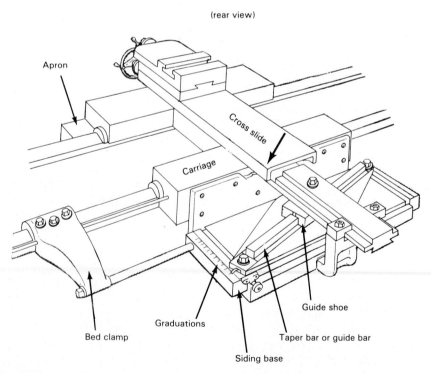

FIGURE 20-47 Taper attachment moves cross slide transversely when carriage moves, but only if the bed clamp is fastened. Taper bar is set for the angle to be machined.

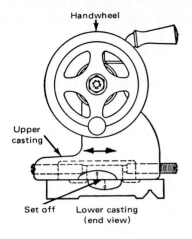

Handwheel

Upper casting

Set off   Lower casting
           (end view)

FIGURE 20-48 Method of turning tapers by setting the tailstock off center.

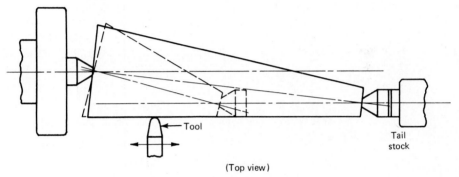

Tool

Tail stock

(Top view)

and locked in position. The compound slide then is fed manually to produce the desired taper. The tool should be set at exactly the height of the axis of rotation of the workpiece in all taper turning and boring.

Because the graduated scale on the base of the compound rest usually is calibrated only to 1° divisions, it is difficult to make the angle setting with accuracy. If accuracy is required, tapers made by this method are checked by means of plug or ring gages, readjusting the setting of the compound rest until the gage fits perfectly. Also, the compound rest cannot be set directly to the correct angle if the taper is dimensioned in millimeters per millimeter or inches per foot.

Both external and internal tapers can be made on a lathe by using a *taper attachment,* such as shown in Figure 20-47. In this device is an *extension* bolted to the rear of the carriage. When the carriage is moved, the cross slide is caused to move transversely.

A raised *guide bar* is pivoted to any desired angle (within its limits). A *guide shoe* slides on the guide bar, so that, when the carriage is moved longitudinally, the guide follows the guide bar and moves the cross slide and tool post transversely to provide the proper tool angle.

Graduations of taper in mm/mm or in./ft are provided at one end (degrees at the other end), so the attachment can be set to the desired taper. While taper attachments provide an excellent and convenient method of cutting tapers, they ordinarily can be used only for tapers of less than 0.5 mm/mm or 6 in./ft.

External tapers also can be turned on workpieces that are mounted between centers by *setting over the tailstock*. This method is illustrated in Figure 20-48. The tailstock is moved out of line with the headstock spindle. The set-off distance from the center line is given by the formula in the figure. This method is limited to small tapers and is seldom used.

When specifying tapers on drawings, the designer or draftsman should remember that it is difficult for the machinist to measure the smaller diameter of a taper accurately if it is the end of a workpiece.

Special lathe accessories.   Several attachments are available that facilitate doing special types of work on lathes. The *milling attachment* is a special vise that attaches to the cross slide to hold work while milling is being done by a cutter that is rotated by the spindle. The work is fed by means of the cross-slide screw.

Tool-post grinders are often used to permit grinding to be done on a lathe. Such an attachment is discussed in Chapter 23.

Duplicating attachments are available that, guided by a template, will automatically control the tool movements for turning irregularly shaped parts. In some cases the first piece, produced in the normal manner, may serve as the template for duplicate parts. To a large extent, duplicating lathes using templates have been replaced by tape- or computer-controlled lathes. These will be discussed in Chapter 38.

REVIEW QUESTIONS

1. What is the tool–work relationship in turning?
2. What different kinds of surfaces can be produced by turning?
3. How does form turning differ from ordinary turning?
4. How does facing differ basically from a cutoff operation?
5. Name six different machining operations that can be done on a lathe.
6. Why is it difficult to make heavy cuts if a form turning tool is complex in shape?
7. What is the ''swing'' of a lathe?
8. Why is a lathe spindle hollow?
9. What functions does a lathe carriage have?
10. How is feed specified on a lathe?
11. What function is provided by the lead screw on a lathe that is not provided by the feed rod?
12. What are four ways for supporting work in a lathe?
13. How is rotation provided to a workpiece that is mounted between centers on a lathe?
14. What will result if work is mounted between centers in a lathe and the centers are not exactly in line?
15. Why is it not advisable to hold hot-rolled steel stock in a collet?
16. How does a steady rest differ from a follow rest?
17. What are the advantages and disadvantages of a four-jaw independent chuck versus a three-jaw chuck?
18. Why should the distance a lathe tool projects from the tool holder be minimized?
19. What occurs if a lathe tool is set below the center line of the workpiece in turning?

20. How can a tapered part be turned on a lathe?
21. Why is it desirable to use a heavy depth and a light feed in turning rather than the opposite?
22. On what diameter is the rpm based for a facing cut, assuming given work and tool materials?
23. Why is it usually necessary to take relatively light cuts when boring on a lathe?
24. What are two basic ways knurling is done?
25. Why are saddle-type turret lathes used much less than ram-type lathes?
26. What important factor must be kept in mind in tooling multiple-spindle screw machines that does not have to be considered in single-spindle machines?
27. At what speed should a 76.2-mm (3-inch)-diameter bar be rotated to provide a cutting speed of 61 meters (2000 feet) per minute?
28. Assume that the workpiece in question 27 is 203.2 mm (8 inches) long and a feed of 0.51 mm (0.020 inch) per revolution is used. How long will a cut across its entire length require?
29. If the depth of cut in question 28 is 4.76 mm ($\frac{3}{16}$ inch), what is the metal removal rate (MRR)?
30. The following data apply for machining a part on a turret lathe and on an engine lathe:

	Engine Lathe	Turret Lathe
Cycle time	20 min	5 min
Labor rate	$5.50/hr	$4.00/hr
Machine rate	$3.50/hr	$5.00/hr

The setup cost and cost for special tooling on the turret lathe would be $30. How many pieces would have to be made to justify using the turret lathe?

31. What would be the effect on the number of pieces required to justify the use of the turret lathe if all labor rates were increased 15%?
32. What are the two objectives of boring?
33. Why does boring assure concentricity between the hole axis and the axis of rotation of the workpiece (for boring tool), whereas drilling does not?
34. Why are vertical boring mills better suited than a lathe for machining large workpieces?
35. What is the principal advantage of a horizontal boring machine over a vertical boring machine for large workpieces?
36. Why is a horizontal boring, drilling, and milling machine such an important tool for machining very large workpieces?
37. A hole 89 mm in diameter is to be drilled and bored through a piece of 1340 steel that is 200 mm long, using a horizontal boring, drilling, and milling machine. High-speed tools will be used. The job will be done by center drilling, drilling with an 18-mm drill, followed by a 76-mm drill, then bored to size in one cut, using a feed of 0.50 mm/rev. Drilling feeds will be 0.25 mm/rev for the smaller drill and 0.64 mm/rev for the larger drill. The center drilling operation requires $\frac{1}{2}$ minute. To set or change any given tool and set the proper machine speed and feed requires 1 minute. Use Figure 39-17 to select cutting speeds, and compute the total time required for doing the job. (Neglect setup time for the workpiece.)
38. In the figure below are three plots of unit production cost ($/unit) versus production volume (Q = build quantity). Note that this plot is made on log-log paper. Cost per unit for a particular process decreases with increased volume. For a particular process there is no minimum cost but rather production volumes within which particular processes are most economical.
(a) For what build quantities is the NC lathe most economical?

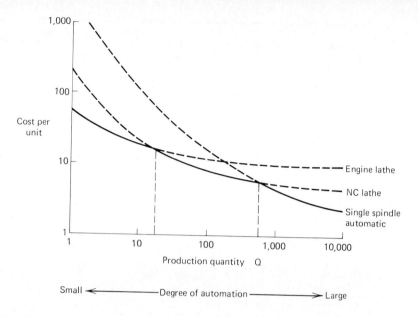

Small ← ————— Degree of automation ————— → Large

(b) What cost per unit does the NC lathe approach as the build quantity becomes very large?

(c) What happens to these plots if you plot them on regular Cartesian coordinates? Try it and comment on what you find.

(d) Many Japanese manufacturers have found innovative ways to eliminate setup time in many of their processes. What is the impact of this on these kinds of plots, on cost per unit economics, and on job shop inventories?

CASE STUDY 20. Machining Economics of Hot-Rolled 8620 Steel Shaft

Listed below are data for the machining of a hot-rolled steel shaft on a lathe using four different tool materials. Triangular insert tooling is being used. The operating cost of the machine tool is $60.00 per hour. It takes 3 minutes to change inserts and 30 seconds to unload a finished part and load a piece of bar stock in this machine. The first column, uncoated inserts, has been completed. Complete the table for the other three tooling alternatives, determining the cost per piece, the percent cost savings, the production rate, and the percent productivity improvement for TiC-coated inserts, Al_2O_3-coated inserts, and Al_2O_3 low-force groove (LFG) inserts. Note that for low-force groove geometries, the insert has only three cutting edges available rather than six and the cost of the insert itself was much higher. The shaft was 6 inches in diameter and the length of cut was about 24 inches.

	Uncoated	TiC-Coated	Al_2O_3-Coated	Al_2O_3 LFG
Cutting speed (surface ft/min)	400	640	1100	1320
Feed (in rev)	0.020	0.022	0.024	0.028
Cutting edges available per insert	6	6	6	3
Cost of an insert ($)	4.80	5.52	6.72	6.72

	Uncoated	TiC-Coated	Al$_2$O$_3$-Coated	Al$_2$O$_3$ LFG
Tool life (pieces/ cutting edge)	40	40	40	40
Tool life (min/ cutting edge)	192	108	60	40
Tool change time per piece (min)	0.075	0.075	0.075	0.075
Nonproductive cost per piece ($)	0.50	0.50	0.50	0.50
Machining cost per piece ($)	4.8	2.7	——	1.00
Machining time per piece (min)	4.8	——	1.5	——
Tool change cost per piece ($)	0.08	0.08	0.08	——
Cutting tool cost per piece ($)	0.02	——	0.03	0.06
Total cost per piece ($)	5.40	——	——	——
Cost savings (%)	0	——	——	70
Production rate (pieces/hr)	11	——	——	——
Improvement in productivity based on pieces/hr (%)	0	——	164	——

Source: Data from T. E. Hale et al., "High Productivity Approaches to Metal Removal," *Materials Technology,* Spring 1980, p. 25.

What do you conclude about the economics of machining in terms of the elements that go to make up the total cost per piece? What is the fallacy of this analysis?

CHAPTER 21

LIP ⟨

LIP CLEARANCE ⟨
12°

Drilling and Reaming

DRILLING

In manufacturing, it is probable that more internal cylindrical surfaces—holes—are produced than any other shape, and a large proportion of these are made by drilling. Consequently, drilling is a very important process. Although drilling, basically, is a relatively simple process, certain aspects of it can cause considerable difficulty. Most drilling is done with a tool having two cutting edges. However, these edges are at the end of a relatively flexible tool, and the cutting action takes place within the workpiece, so that the chips must come out of the hole while the drill is filling a large portion of it. Also, there is friction between the body of the drill and the wall of the workpiece, resulting in heat in addition to that due to chip formation. As a result of these conditions, substantial difficulty can be experienced due to poor heat removal. The counterflow of the chips makes lubrication and cooling difficult. Obtaining desired accuracy and precision often is not easy in drilling.

In recent years, new drill point geometries have resulted in improved hole accuracy, longer life, self-centering action, and increased-feed-rate capabilities. However, virtually 99% of the drills manufactured have the conventional point. It is often left to the user to regrind the drills to fit his application.

Types of drills. The most common types of drills are *twist drills*. These have three basic parts: the *body,* the *point,* and the *shank,* shown in Figure 21-1. The body contains two or more spiral grooves, called *flutes,* in the form of a helix along opposite sides. To reduce the friction between the drill and the hole, each land is reduced in diameter except at the leading edge, leaving a narrow *margin* of full diameter to aid in supporting and guiding the drill and thus aiding in obtaining an accurate hole. The lands terminate in the point, with the leading edge of each land forming a cutting edge.

As shown in Figure 18-6, the principal rake angles behind the cutting edges are formed by the relation of the flute helix angle to the work. This means that the rake angle of a drill varies along the cutting edges (or lips), being 0° close to the point and equal to the helix angle out at the lip. Because the helix angle is built into the drill, the primary rake angle cannot be changed by normal grinding. The helix angle of most drills is 24°, but drills with other helix angles are available. Larger helix angles—often above 30°—are used for materials that can be drilled very rapidly, resulting in a large volume of chips. Helix angles ranging from 0 to 20° are used for soft materials, such as plastics and copper. Straight-flute drills (zero helix and rake angles) also are used for drilling thin sheets of soft materials. It is possible to change the rake angle adjacent to the cutting edge by a special grinding procedure, called *dubbing* (see Figure 9-27).

The cone-shaped point on a drill contains the cutting edges and the various clearance angles. This cone angle affects the direction of flow of the chips across the tool face and into the flute. Obviously, it is very important. The 118° cone angle that is used most often is an arbitrary one that has been found to provide

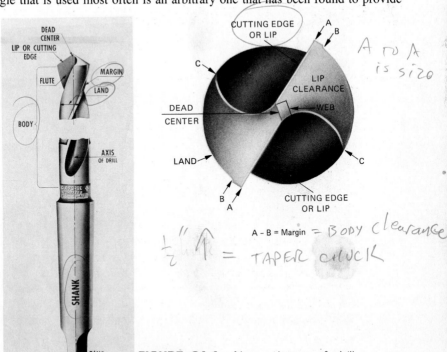

FIGURE 21-1 Nomenclature of drill parts. (*Courtesy Cleveland Twist Drill Company.*)

good cutting conditions and reasonable tool life for mild steel, thus making it suitable for much general-purpose drilling. Smaller cone angles—from 90 to 118°—sometimes are used for drilling more brittle materials, such as gray cast iron and magnesium alloys. Cone angles from 118 to 135° often are used for the more ductile materials, such as alloy steels and aluminum alloys. Cone angles less than 90° frequently are used for drilling plastics. As will be discussed later, several methods of grinding drills have been developed that produce points that are not plain cones.

The flutes serve as channels through which the chips come out of the hole and also to permit coolant to get to the cutting edges. Although most drills have two flutes, some, as shown in Figure 21-2, have three.

As shown in Figure 21-1, the relatively thin *web* between the flutes forms a metal column or backbone. This is an unfortunate feature of a twist drill. If a plain conical point is ground on the drill, the intersection of the web and the cone produces a straight-line *chisel center,* which can be seen in the top views of Figure 21-3. Unfortunately, the chisel point, which also must act as a cutting edge, forms a 56° negative rake angle with the conical surface. Such a large negative rake angle does not cut efficiently, causing excessive deformation of the metal. This results in high thrust forces being required and excessive heat

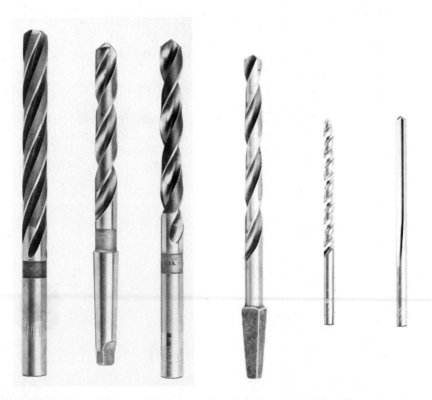

FIGURE 21-2 Types of twist drills and shanks. (*Left to right*) Straight-shank, three-flute core drill; taper-shank; straight-shank; bit-shank; straight-shank, high-helix-angle; straight-shank, straight-flute.

FIGURE 21-3 Cone, lip, and clearance angles for twist drills. (*Courtesy Cleveland Twist Drill, an Acme-Cleveland Company.*)

being developed at the point. This condition is further complicated by the obvious fact that the cutting speed at the drill center is low, approaching zero. As a consequence, drill failure on a normally ground drill occurs both at the center, where the cutting speed is lowest, and at the outer tips of the cutting edges, where the speed is highest. The conventional point also has a tendency to produce a burr on breakthrough.

Another difficulty resulting from a chisel point is that when the rotating, straight-line point comes in contact with the workpiece, it has a distinct tendency to "walk" along the surface, thus moving the drill away from the desired location, unless positive means are provided to prevent this action, usually at extra cost. For these reasons, the conventional point may not be cost effective when used on machining centers or high-speed automatics because additional operations like center drilling, burr removal, and tool change are required, all of which increase total production time and reduce productivity.

Special methods of grinding drill points have been developed to eliminate or minimize the difficulties caused by the chisel point and to obtain better cutting action and tool life (see Figure 21-4). Such methods have had varying degrees of success, and they require special drill-grinding equipment. Another procedure

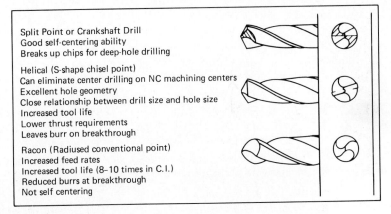

Split Point or Crankshaft Drill
Good self-centering ability
Breaks up chips for deep-hole drilling

Helical (S-shape chisel point)
Can eliminate center drilling on NC machining centers
Excellent hole geometry
Close relationship between drill size and hole size
Increased tool life
Lower thrust requirements
Leaves burr on breakthrough

Racon (Radiused conventional point)
Increased feed rates
Increased tool life (8–10 times in C.I.)
Reduced burrs at breakthrough
Not self centering

FIGURE 21-4 Variations to the conventional drill-point geometry include the split point (crankshaft), helical (spiral point), and radial lip point or Racon point.

is *web thinning,* in which a narrow grinding wheel is used to remove a portion of the web near the point of the drill.

Because of the use of higher cutting speeds, particularly in automatic machines, some type of chip breaker often is incorporated in drills. One procedure is to grind a small groove in the tool face, parallel with and a short distance from the cutting edge. This, of course, requires an additional grinding operation. A more effective procedure is to employ drills that have a special chip-breaker rib as an integral part of the flute. The rib interrupts the flow of the chip, causing an abrupt change in direction, and breaks it into short lengths. Such a drill requires no special grinding.

The original drill point produced by the manufacturer lasts only until the first regrind, then performance and life depend upon the quality of regrind. Overlooking the value of a top quality drill grinder is very costly to companies investing huge sums in NC machining centers which carry labor and burden rates of $50 per hour.

Drill shanks are made in several types. Figure 21-2 shows several shank styles. The two most common types are the straight and the taper. *Straight-shank* drills are used only for sizes up to 12.7 mm (½ inch) and must be held in some type of drill chuck. *Taper shanks* are available on drills from 3.18 to 12.7 mm (⅛ to ½ inch) and are standard on drills above 12.7 mm (½ inch). Morse tapers, having a taper of approximately 1:19 (⅝ inch per foot), are used on taper-shank drills, ranging from a number 1 taper on ⅛-inch drills to a number 6 on a 3½-inch drill.

Taper-shank drills are held in a female taper in the end of the machine tool spindle. If the taper on the drill is smaller than the spindle taper, one or more adapter sleeves are placed in the hole to reduce it to the same size as the taper on the drill. Taper-shank drills have the advantage that the taper assures the drill being accurately centered in the spindle. The tang at the end of the taper shank fits loosely in a slot at the end of the tapered hole in the spindle. Its primary function is to provide a means by which the drill may be loosened for removal by driving a tapered *drift* through a hole in the side of the spindle and against the end of the tang. It also acts as a safety device to prevent the drill from rotating in the spindle hole under heavy loads. However, if the tapers on the drill and in the spindle are in proper condition, no slipping should occur; all driving of the drill is through the friction between the two tapered members.

Deep-hole drills, which contain passages through which coolant is forced to the cutting edges, and which also aids in pushing chips back out of the hole, are used when deep holes are to be drilled. Their special form reduces the tendency of the drill to drift, thus producing a more accurately aligned hole. The one shown in the bottom portion of Figure 21-5 is designed so that the chips are forced back along the central, straight flute, whereas with the type shown in the upper portion of this figure, the chips flow through a hole in the center of the drill. This construction provides better centering support to the 'll and prevents chips from abrading the walls of the hole.

n-flute drills are available that have holes extending throughout the length 'and to permit coolant to be supplied, under pressure, to the point ach cutting edge. These are helpful in providing cooling and also ip removal from the hole when drilling to moderate depths. They

FIGURE 21-5 Drills for drilling deep holes. *(Top)* Central chip-hole type. *(Courtesy Colt Industries, Pratt & Whitney Machine Tool Division.)* *(Bottom)* Single-flute type. *(Courtesy American Heller Corporation.)*

require special fittings through which the coolant can be supplied to the rotating drill and are used primarily on automatic and semiautomatic machines.

Larger holes in thin material may be made with a *trepanning cutter,* shown in Figure 21-6, whereby the main hole is produced by the thin, cylindrical cutter or saw.

Because drills are relatively slender, and also because of the "walking" action of the chisel point, they can be deflected rather easily when starting to drill a hole. Consequently, to assure that a hole is started accurately, a *center drill and countersink,* illustrated in Figure 21-7, is used prior to a regular chisel-point twist drill. The center drill and countersink has a short, straight drill section extending beyond a 60° taper portion. The heavy, short body provides rigidity so that a hole can be started with little possibility of the center drill being deflected. The hole should only be drilled partway up on the tapered section of the center drill (see Figure 21-8). The conical portion of the hole serves to guide the drill being used to make the main hole. Combination center drills are made in four sizes to provide the proper-size starting hole for any drill. If the drill is sufficiently large in diameter, or if it is sufficiently short, satisfactory accuracy often may be obtained without center drilling. Special drill holders are available that permit drills to be held with only a very short length protruding.

Although the use of a center drill aids materially in assuring that a drill starts

FIGURE 21-6 Hole cutter. *(Courtesy Armstrong-Blum Manufacturing Company.)*

Handwritten notes:
1. Prick punch
2. center punch = 3/16
3. go back 1½ size, or under 3/16
4. to accomidate
5. (ream)

FIGURE 21-7 Combination center drill and countersink. (*Courtesy Chicago-Latrobe Twist Drill Works.*)

drilling at the desired location, because of its flexibility the drill may drift off the center line during drilling as the result of nonhomogeneity in the workpiece. Such nonhomogeneities and imperfect drill sharpening also may cause the hole to be oversize. Thus there is little assurance that a drilled hole is accurate as to alignment and size. If accuracy in these respects is desired, it is necessary to follow center drilling and drilling by boring and reaming, as illustrated in Figure 21-8. Boring trues the hole alignment, whereas reaming brings the hole to accurate size and improves the surface finish.

Special *combination drills* are made that can drill two or more diameters, or drill and countersink and/or counterbore, in a single operation. Some of these are illustrated in Figure 21-9. A *step drill* has a single set of flutes and is ground to two or more diameters. *Subland drills* have a separate set of flutes, on a single body, for each diameter or operation; they provide better chip flow, and the cutting edges can be ground to give proper cutting conditions for each operation. Combination drills are expensive and present problems in regrinding but can be economical for production-type operations if they reduce work handling and separate machines and operations.

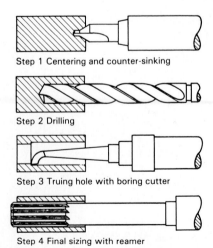

Step 1 Centering and counter-sinking

Step 2 Drilling

Step 3 Truing hole with boring cutter

Step 4 Final sizing with reamer

FIGURE 21-8 Steps required to obtain a hole that is accurate, as to size and aligned on center.

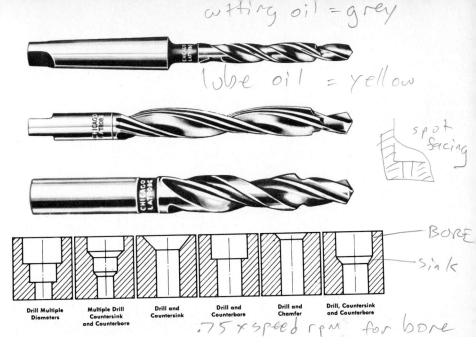

Handwritten annotations:
cutting oil = grey
lube oil = yellow
spot facing
BORE
sink
.75 × speed rpm for bore

| Drill Multiple Diameters | Multiple Drill Countersink and Counterbore | Drill and Countersink | Drill and Counterbore | Drill and Chamfer | Drill, Countersink and Counterbore |

FIGURE 21-9 Special-purpose subland or multicut drills, and some of the operations possible with such drills. (*Courtesy Chicago-Latrobe Twist Drill Works.*)

Spade drills, such as is shown in Figure 21-10, are widely used for making holes 25.4 mm (1 inch) or larger in diameter. Such drills have several advantages: (1) less of the costly cutting-tool material is required because the long supporting bar can be made of ordinary steel; (2) the drill point can be ground with a minimum chisel point; (3) the main body can be made more rigid because

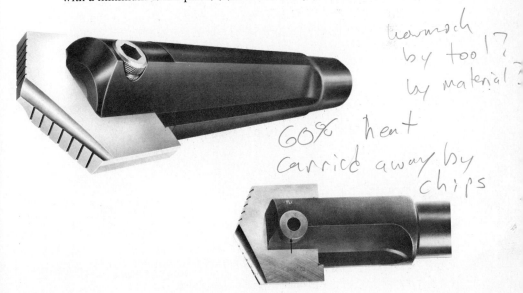

Handwritten annotations:
transmach by tool? by material?
60% heat carried away by chips

FIGURE 21-10 (*Top*) Typical spade drill. (*Bottom*) Spade drill with carbide inserts (arrow) to provide lateral guidance. (*Courtesy Erickson Tool Company.*)

no flutes are required; (4) the cutting blade is easier to sharpen; and (5) the main body can be provided with a central hole through which a fluid can be circulated to aid in cooling and in chip removal.

Spade drills are being used increasingly to machine a shallow locating cone for a subsequent smaller drill and at the same time to provide a small bevel around the hole to facilitate later tapping or assembly operations. Such a bevel also frequently eliminates the need for deburring. This practice is particularly useful on mass-production and numerically controlled machines.

Drill sizes.   Standard drills are available in four size series—the size indicating the diameter of the drill body.

> *Millimeter series:* 0.01- to 0.50-mm increments, according to size, in diameters from 0.015 mm.
> *Numerical series:* No. 80 to No. 1 (0.0135 to 0.228 inch).
> *Lettered series:* A to Z (0.234 to 0.413 inch).
> *Fractional series:* ¹⁄₆₄ to 4 inches (and over) by 64ths.

Drill chucks.   Straight-shank drills must be held in some type of drill chuck. The one shown in Figure 21-11a is adjustable over a considerable size range. The one shown in Figure 21-11b has radial steel fingers that are attached to, and held in position by, synthetic rubber. When the chuck is tightened by means of a chuck key, these fingers are forced inward against the drill. On smaller drill presses, the chuck often is permanently attached to the machine spindle, whereas on larger drilling machines the chucks have a tapered member that fits into the female Morse taper of the machine spindle (see Figure 20-35 for example).

Special types of chucks frequently are used to hold drills in semi- or fully automatic machines. The one shown at the left in Figure 21-12 permits quite a range of drills to be held in a single chuck. The one shown in the middle permits

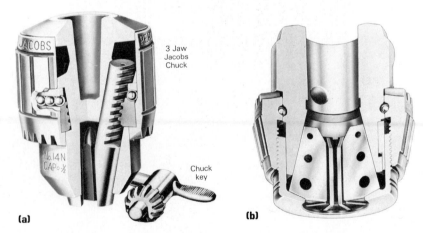

FIGURE 21-11 Two of the most commonly used types of drill chucks. (*Courtesy Jacobs Manufacturing Co.*)

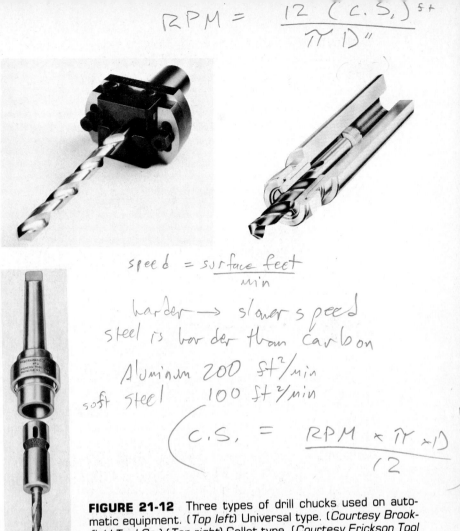

(Handwritten annotations at top and around figure:)

$$RPM = \frac{12 \, (C.S.) \, ft}{\pi \, D \, "}$$

$$speed = \frac{surface \, feet}{min}$$

harder → slower speed
steel is harder than carbon
Aluminum 200 ft^2/min
soft steel   100 ft^2/min

$$C.S. = \frac{RPM \times \pi \times D}{12}$$

FIGURE 21-12 Three types of drill chucks used on automatic equipment. (*Top left*) Universal type. (*Courtesy Brookfield Tool Co.*) (*Top right*) Collet type. (*Courtesy Erickson Tool Company.*) (*Bottom left*) Quick-change drill chuck. (*Courtesy Consolidated Machine Tool Division, Farrel Birmingham Company, Inc.*)

holding the drill with only a controlled amount protruding, so as to reduce flexing and give more accurate centering.

Chucks using chuck keys require the machine spindle be stopped in order to change a drill. To reduce the downtime where drills must be changed frequently, *quick-change chucks* are used. Each drill is fastened in a simple, round collet that can be inserted into, and removed from, the chuck hole while it is turning by merely raising and lowering a ring on the chuck body. By using this type of chuck, center drills, drills, counterbores, reamers, and so on, can be used in quick succession (see Figure 21-12, bottom left).

Drilling speeds and feeds. Cutting speeds specified for drilling are the surface speeds at the outside of the drill. These surface speeds are used to compute the rotational speed of the drill. Also, one should consider whether heat will be conducted away from the cutting edges by conduction through the workpiece and rapid chip flow. In drilling deep holes, or in drilling holes in material that does not conduct heat readily, cutting speeds may have to be reduced unless an ample supply of coolant can be provided at the cutting edges.

Drilling feeds are expressed as *millimeters* or *inches per revolution* (mmpr or ipr). The following are representative:

Diameter of Drill	mmpr	ipr
Less than 3 mm (⅛ in.)	0.03–0.05	0.001–0.002
3 to 6.4 mm (⅛ to ¼ in.)	0.05–0.10	0.002–0.004
6.4 to 12.7 mm (¼ to ½ in.)	0.10–0.18	0.004–0.007
12.7 to 25.4 mm (½ to 1 in.)	0.18–0.38	0.007–0.015
Over 25.4 mm (1 in.)	0.38–0.64	0.015–0.025

If the conditions are unusual, the feed may have to be varied from these suggested values.

Cutting fluids in drilling. For shallow holes, the general rules relating to cutting fluids, as given in Chapter 18, are applicable. Where the hole depth exceeds one diameter, it is desirable to increase the lubricating quality of the fluid because of the rubbing between the drill margins and the wall of the hole. The effectiveness of a cutting fluid as a coolant is quite variable in drilling. While the rapid egress of the chips is a primary factor in heat removal, this action also tends to restrict entry of the cutting fluid. This is of particular importance in drilling materials that have poor heat conductivity.

If the hole depth exceeds two or three diameters, it usually is advantageous to withdraw the drill each time it has drilled about one diameter of depth in order to clear chips from the hole. Some machines are equipped to provide this "pecking" action automatically.

Where cooling is desired, the fluid should be applied copiously. For severe conditions, drills containing coolant holes have considerable advantage; not only is the fluid supplied near the cutting edges, but the flow of the fluid aids in chip removal from the hole. Where feasible, drilling horizontally has distinct advantages over drilling vertically downward.

Drill grinding. Proper grinding of a drill is a more complex and important operation than often is assumed. If satisfactory cutting and hole size are to be achieved, it is essential that the point angle, lip clearance, and lip length be correct. As illustrated in Figure 21-13, incorrect sharpening often results in unbalanced cutting forces at the tip, causing misalignment and oversize holes. Relatively few machinists can sharpen drills correctly and consistently by offhand grinding, even small drills. To obtain correct hole sizes, good drill life, and consistent results, special drill grinders, often computer controlled, should be used that assure the desired lip clearance, cone angle, and balanced lip

Handwritten annotation at top of page:
FEED = inches down / rev

$$\frac{in}{rev} \times \frac{rev}{Min} = \frac{in}{Min}$$

FIGURE 21-13 Effects of improper drill grinding. (*Left*) Angles of two lips are different. (*Right*) Lengths of the lips are not equal. (*Courtesy Cleveland Twist Drill Company.*)

lengths. This is extremely important where drills are used on mass-production or numerically controlled machines. *[handwritten: Pull till chatter, then a little more.]*

Counterboring, countersinking, and spot facing. Drilling often is followed by counterboring, countersinking, or spot facing. As shown in Figure 21-14, each provides a bearing surface at one end of a drilled hole. They usually are done with a special tool having from three to six cutting edges. *[handwritten: if still chatter, not enuff △]*

Counterboring provides an enlarged, cylindrical hole with a flat bottom so that a bolt head, or a nut, will have a smooth bearing surface that is normal to the axis of the hole; the depth may be sufficient so that the entire bolt head or nut will be below the surface of the part (see Figure 21-15). The pilot on the end of the tool fits into the drilled hole and helps to assure concentricity with the original hole. Two or more diameters may be produced in a single counterboring operation. Counterboring also can be done with a single-point tool, although this method ordinarily is used only on large holes and essentially is a boring operation.

Countersinking makes a beveled section at the end of a drilled hole to provide a proper seat for a flat-head screw or rivet. The most common angles are 60, 82, and 90°. Countersinking tools are similar to counterboring tools except that the cutting edges are elements of a cone and they usually do not have a pilot because the bevel of the tool causes them to be self-centering.

Spot facing is done to provide a smooth bearing area on an otherwise rough surface at the opening of a hole and normal to its axis. Machining is limited to the minimum depth that will provide a smooth, uniform surface. Spot faces thus are somewhat easier and more economical to produce than counterbores. They usually are made with a multiedged end-cutting tool that does not have a pilot, although counterboring tools frequently are used.

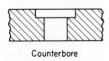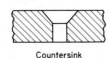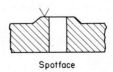

Counterbore   Countersink   Spotface

FIGURE 21-14 Surfaces produced by counterboring, countersinking, and spot facing.

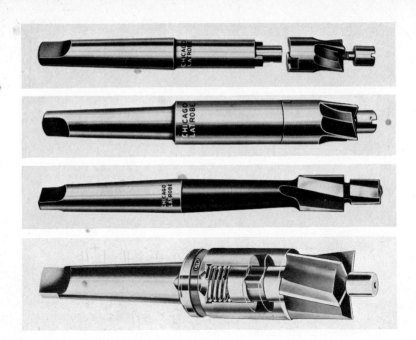

FIGURE 21-15 Counterboring tools. (*Bottom to top*) interchangeable counterbore; solid, taper-shank counterbore with integral pilot; replaceable counterbore and pilot; replaceable counterbore, disassembled. [*Courtesy Ex-Cell-O Corporation, and Chicago-Latrobe Twist Drill Works.*]

REAMING

Reaming is done for two purposes—to bring holes to a more exact size, and to improve the finish of an existing hole by machining a small amount from its surface. Multiedged cutting tools are used, and no special machines are built especially for reaming; it usually is done on the same machine that was employed for drilling the hole that is to be reamed.

In order to obtain proper results, only a minimum amount of material should be left for removal by reaming. As little as 0.13 mm (0.005 inch) is desirable, and in no case should the amount exceed 0.38 mm (0.015 inch). A properly reamed hole should be within 0.03 mm (0.001 inch) of correct size and have a fine finish.

Types of reamers. The principal types of reamers, shown in Figure 21-16 are:

1. Hand reamers.
   a. Straight.
   b. Taper.
2. Machine or chucking reamers.
   a. Rose.
   b. Fluted.

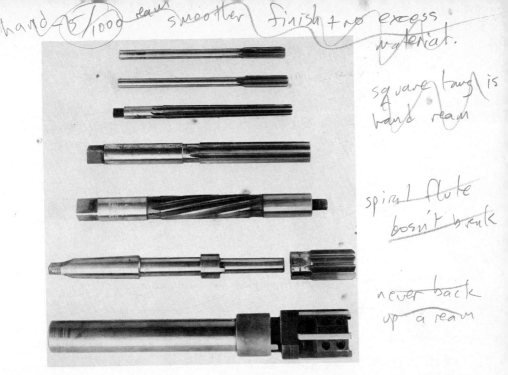

Handwritten annotations:
hand 45/1000 ream smoother finish + no excess material.
square tang is hand ream
spiral flute doesn't break
never back up a ream

FIGURE 21-16 Types of reamers. (*Top to bottom*) Straight-fluted rose reamer; straight-fluted chucking reamer; straight-fluted taper reamer; straight-fluted hand reamer; expansion reamer; shell reamer; adjustable reamer.

Handwritten: ream selection 3/1000 - 5/1000 for hand 3/1000 - 15/1000 for machine

3. Shell reamers.
4. Expansion reamers.
5. Adjustable reamers.

Hand reamers are intended to be turned and fed by hand and to remove only a few thousandths of metal. They have a straight shank with a square tang for a wrench. They can have straight or spiral flutes and be solid or expandable. The teeth have relief along their edges and thus may cut along their entire length. However the reamer is tapered from 0.13 to 0.25 mm (0.005 to 0.010 inch) in the first third of its length to assist in starting it in the hole, and most of the cutting therefore takes place in this portion.

Machine or *chucking reamers* are for use with various machine tools at slow speeds. They have straight or tapered shanks and either straight or spiral flutes. *Rose chucking reamers* are ground cylindrical and have no relief behind the outer edges of the teeth. All cutting is done on the beveled ends of the teeth. *Fluted chucking reamers*, on the other hand, have relief behind the edges of the teeth as well as beveled ends. They thus can cut on all portions of the teeth. Their flutes are relatively short and they are intended for light finishing cuts. For best results they should not be held rigidly but permitted to float and be aligned by the hole.

Shell reamers often are used for sizes over 19 mm (¾ inch) in order to save cutting-tool material. The shell, made of tool steel for smaller sizes and with

Handwritten: 60% of RMP for ream

carbide edges for larger sizes or for mass-production work, is held on an arbor that is made of ordinary steel. One arbor may be used with any number of shells. Only the shell is subject to wear and need be replaced when worn. They may be ground as rose or fluted reamers.

Expansion reamers can be adjusted over a few thousandths of an inch to compensate for wear, or to permit some variation in hole size to be obtained. They are available in both hand and machine types.

Adjustable reamers have cutting edges in the form of blades that are locked in a body. The blades can be adjusted over a considerably greater range than in the case of expansion reamers. This permits adjustment for size and to compensate for regrinding. When the blades become too small from regrinding, they can be replaced. Both tool steel and carbide blades are used.

Taper reamers are used for finishing holes to an exact taper. They may have up to eight straight or spiral flutes. Standard tapers, such as Morse, Jarno, or Brown & Sharpe, come in sets of two. The *roughing reamer* has nicks along the cutting edges to break up the heavy chips that result as a cylindrical hole is cut to a taper. The *finishing reamer* has smooth cutting edges.

MACHINES FOR DRILLING

The basic work and tool motions that are required for drilling—relative rotation between the workpiece and the tool, with relative longitudinal feeding—also occur in a number of other machining operations. Consequently, as indicated in Table 18-7, drilling can be done on a variety of machine tools, such as lathes, milling machines, and boring machines. This chapter will only consider machines that are designed, constructed, and used primarily for drilling. Drilling machines, called drill presses, consist of a *base,* a *column* that supports a *powerhead,* a *spindle,* and a *worktable.* On small machines the base rests on a bench, whereas on larger machines it rests on the floor. The column may be either round or of box-type construction, the latter being used on larger, heavy-duty machines, except in radial types. The powerhead contains an electric motor and means for driving the spindle in rotation at several speeds. On small drilling machines this may be accomplished by shifting a belt on a step-cone pulley, but on larger machines a geared transmission is used. See Figure 21-17 for block diagrams.

The heart of any drilling machine is its spindle. In order to drill satisfactorily, the spindle must rotate accurately and also resist whatever side forces result from the drilling. In virtually all machines the spindle rotates in preloaded ball or tapered-roller bearings. In addition to powered rotation, provision is made so that the spindle can be moved axially to feed the drill into the work. On small machines the spindle is fed by hand, whereas on larger machines power feed is provided. Except on some small bench types, the spindle contains a hole with a Morse taper in its lower end into which taper-shank drills or drill chucks can be inserted.

The worktables on drilling machines may be moved up and down on the column to accommodate work of various sizes. On round-column machines the

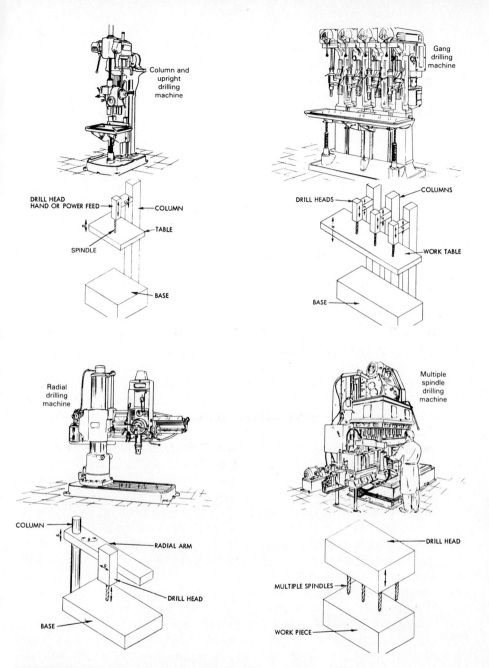

FIGURE 21-17 Four principal types of drilling machines. (*From* Manufacturing Producibility Handbook; *courtesy General Electric Company.*)

table usually can also be rotated out of the way so that workpieces can be mounted directly on the base. On some box-column machines the table is mounted on a subbase so that it can be moved in two directions in a horizontal plane by means of feed screws.

Types of drilling machines. Drilling machines usually are classified in the following manner:

1. Bench.
   a. Plain.
   b. Sensitive.
2. Upright.
   a. Single-spindle.
   b. Turret.
3. Radial.
   a. Plain.
   b. Semiuniversal.
   c. Universal.
4. Gang.
5. Multiple-spindle.
6. Deep-hole.
   a. Vertical.
   b. Horizontal.
7. Transfer.

Bench-type drilling machines. Figure 21-18 shows a typical bench-type drilling machine. The spindle rotates on ball bearings within a nonrotating *quill* that can be moved up and down in the machine head to provide feed to the drill. The vertical motion is imparted by a hand operated capstan wheel through a pinion that meshes with a rack on the quill. A spring raises the quill-and-spindle assembly to the highest position when the hand lever is released. The spindle is driven by means of a step-cone pulley that rides on a splined shaft, thus imparting rotation regardless of the vertical position of the spindle.

Drilling presses of this type can usually drill holes up to 13 mm (½ inch) in diameter. Worktables often contain holes for use in clamping work. The same type of machine can be obtained with a long column so that it can stand on the floor instead of on a bench.

The size of bench and upright drilling machines is designated by *twice* the distance from the center line of the spindle to the nearest point on the column, thus being an indication of the maximum size of the work that can be drilled in the machine. For example, a 380-mm (15-inch) drill press will permit a hole to be drilled at the center of a workpiece 380 mm (15 inches) in diameter.

Sensitive drilling machines are essentially the same as plain bench-type machines except that they usually are smaller, are provided with more accurate spindles and bearings, and operate at higher speeds—up to 30,000 rpm. Very sensitive, hand-operated feeding mechanisms are provided for use in drilling small holes. Such machines are used for tool and die work and for drilling very small holes, often less than 1 millimeter (a few thousandths of an inch) in diameter, where high spindle speeds are necessary to obtain proper cutting speed and sensitive feel in order to provide delicate feeding to avoid the breakage of very small drills.

Upright drilling machines usually have spindle speed ranges from 60 to 3500 rpm and power feed rates, in from four to twelve steps, from about 0.10 to 0.60

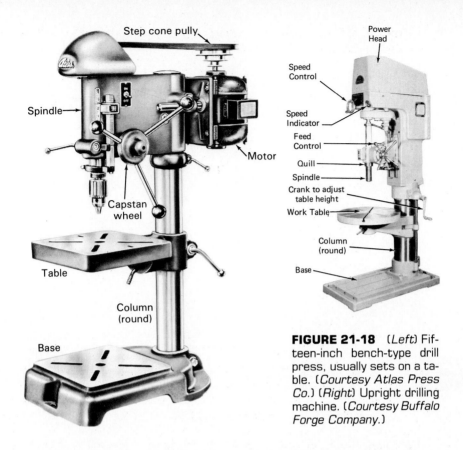

Step cone pully

Spindle

Capstan wheel

Table

Column (round)

Base

Motor

Power Head

Speed Control

Speed Indicator

Feed Control

Quill

Spindle

Crank to adjust table height

Work Table

Column (round)

Base

FIGURE 21-18 (*Left*) Fifteen-inch bench-type drill press, usually sets on a table. (*Courtesy Atlas Press Co.*) (*Right*) Upright drilling machine. (*Courtesy Buffalo Forge Company.*)

mm (0.004 to 0.025 inch) per revolution. Most modern machines use a single-speed motor and a geared transmission to provide the range of speeds and feeds, but some utilize a multispeed motor to obtain some of the spindle speeds. The feed clutch usually is designed so that it disengages automatically when the spindle reaches a preset depth or when it reaches the limits of its travel.

Worktables on most upright drilling machines contain holes and slots for use in clamping work and nearly always have a channel around the edges to collect cutting fluid, when it is used. On box-column machines, the table is mounted on vertical ways on the front of the column and can be raised or lowered by means of a crank-operated elevating screw.

Gang-drilling machines. In mass production, *gang-drilling machines,* shown in Figure 21-17, often are used where several related operations, such as holes of different sizes, reaming, or counterboring, must be done on a single part. These consist essentially of several independent columns, heads, and spindles mounted on a common base and having a single table. The work can be slid into position for the operation at each spindle. They are available with or without power feed. One or several operators may be used.

Turret-type drilling machines. *Turret-type, upright drilling machines,* such as is shown in Figure 21-19, are used where a series of holes of different size,

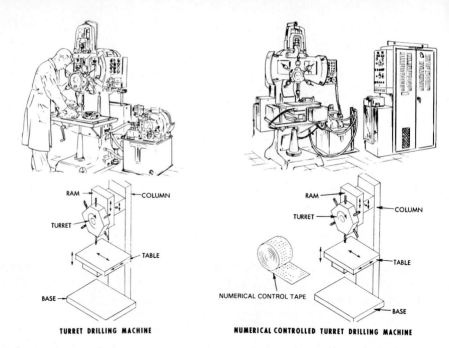

RAM — COLUMN
TURRET
TABLE
BASE
TURRET DRILLING MACHINE

RAM — COLUMN
TURRET
TABLE
NUMERICAL CONTROL TAPE
BASE
NUMERICAL CONTROLLED TURRET DRILLING MACHINE

FIGURE 21-19 Turret drilling machines (*left*) and NC turret drilling machines (*right*) have largely replaced the gang drilling machines in many production facilities. (*From* Manufacturing Producibility Handbook; *courtesy General Electric Company.*)

or a series of operations (such as center drilling, drilling, reaming, and spot facing), must be done repeatedly in succession. After the selected tools are set in the turret, each can quickly be brought into position to be driven by the power spindle merely by rotating the turret, rather than requiring moving and positioning of the workpiece, as with a gang-drilling machine. These machines automatically provide individual feed rates for each spindle and are particularly adaptable for numerical control.

Radial drilling machines. When holes must be drilled at different locations on large workpieces which cannot readily be moved and clamped on an upright drilling machine, *radial drilling machines* are employed. As shown in Figure 21-17, these have a large, heavy, round, vertical column supported on a large base. The column supports a radial arm that can be raised and lowered by power and rotated over the base. The spindle head, with its speed- and feed-changing mechanism, is mounted on the radial arm. It can be moved horizontally to any desired position on the arm. Thus, the spindle can quickly be properly positioned for drilling holes at any desired point on a large workpiece mounted either on the base of the machines or on the floor.

Plain radial drilling machines provide only a vertical spindle motion. On *semiuniversal machines,* the spindle head can be swung about a horizontal axis normal to the arm to permit the drilling of holes at an angle in a vertical plane. On *universal machines,* an additional angular adjustment is provided by rotation of the radial arm about a horizontal axis. This permits holes to be drilled at any desired angle.

MACHINING PROCESSES

Radial drilling machines are designated by the radius of the largest disk in which a center hole can be drilled when the spindle head is at its outermost position. Sizes from 900 to 3650 mm (3 to 12 feet) are available. Usually, the diameter of the column is also given; these range from about 225 to 660 mm (9 to 26 inches). Most radial drilling machines have a wide range of speeds and feeds and include provisions for tapping threads.

Large workpieces usually are fastened directly on the base of radial drilling machines; small pieces can be mounted on a worktable that is attached to the base. Special jigs or fixtures also can be attached to the base to hold the work when multiple pieces are to be drilled.

Most radial drilling machines are equipped with adequately heavy spindle bearings so they can also be used to do boring (see Chapter 20).

Multiple-spindle drilling machines. Where a number of parallel holes must be drilled in a part, *multiple-spindle drilling machines* are used. As shown in Figure 21-20, these are mass-production machines with as many as 50 spindles driven by a single powerhead and fed simultaneously into the work. Figure 21-21 shows the method of driving and positioning the spindles, which permits them to be adjusted over limited, but overlapping, areas so that holes can be drilled at any location within the overall capacity of the head. For example, one machine having 20 spindles can drill holes at any location within a 762-mm (30-inch)-diameter circle. A special drill jig is made for each job to provide accurate guidance for each drill. Although such machines are quite costly, they can readily be converted for use on different jobs where the quantity to be produced will justify the small setup cost and the cost of the jig.

Multiple-spindle drilling machines are available with a wide range of numbers of spindles in a single head, and two or more heads frequently are combined in

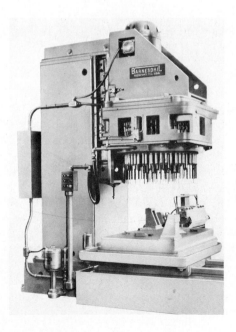

FIGURE 21-20 Multiple-spindle drilling machine equipped with 50 spindles. (*Courtesy Barnes Drill Company.*)

FIGURE 21-21 Multiple-spindle drill head, showing method of driving and positioning the spindles. (*Courtesy Thriftmaster Products Incorporated.*)

a single machine. Often drilling operations are performed simultaneously on two or more sides of a workpiece.

Deep-hole drilling machines. Special machines are used for drilling long (deep) holes, such as are found in rifle barrels, connecting rods, and long spindles. High cutting speeds, very light feeds, a positive and copious flow of cutting fluid to assure rapid chip removal, and adequate support for the long, slender drills are required. In most cases horizontal machines are used. The work is rotated in a chuck with steady rests providing support along its length, as required. The drill does not rotate and is fed into the work. Vertical machines also are available for work that is not very long.

Work holding in drilling. Work that is to be drilled ordinarily is held in a vise or a special jig or fixture. Even in light drilling the work should not be held on the table by hand unless very adequate leverage is available. This is a dangerous practice and can lead to serious accidents, because the drill has a tendency to catch on the workpiece and cause it to rotate. Drilling vises and jigs frequently are made so they can be turned on two faces to permit drilling to be done on two faces of the work with a single clamping.

Work that is too large to be held in a vise can be clamped directly to the machine table, using suitable bolts and clamps and the slots or holes in the table.

REVIEW QUESTIONS

1. What functions are performed by the flutes on a drill?
2. What determines the rake angle of a drill?

MACHINING PROCESSES

3. Basically, what determines what helix angle a drill should have?
4. If an ordinary two-flute twist drill were made to have negative rake angles, describe its appearance as compared with an ordinary positive-rake-angle drill.
5. When a large-diameter hole is to be drilled, why is a small-diameter hole often drilled first?
6. What are the advantages of a Racon point over a conventional point?
7. What can happen when an improperly ground drill is used to drill a hole?
8. Why are most drilled holes oversize with respect to the nominally specified diameter?
9. What are the two primary functions of a combination center drill?
10. What is the function of the margins on a twist drill?
11. What factors tend to cause a drill to ''drift'' off the center line of a hole?
12. For what types of holes are drills having coolant passages in the flutes advantageous?
13. In drilling, the deeper the hole, the greater the torque. Why?
14. Why do cutting fluids for drilling usually have more lubricating qualities than those for most other machining operations?
15. How is feed expressed in drilling?
16. How does a gang-drilling machine differ from a multiple-spindle drilling machine?
17. For what type of work is a quick-change drill chuck advantageous?
18. What may result from holding the workpiece by hand when drilling?
19. What is the rationale behind the operation sequence shown in Figure 21-8?
20. How much time will be required to drill a 25.4-mm (1-inch)-diameter hole through a piece of gray cast iron that is 38 mm (1½ inches) thick, using a high-speed drill? (Values for feed and cutting speed need to be specified by student.)
21. What is the purpose of spot facing?
22. How does the purpose of counterboring differ from that of spot facing?
23. What are the primary purposes of reaming?
24. What are the advantages of shell reamers?
25. A drill that operated satisfactorily for drilling cast iron gave very short life when used for drilling a plastic. Why?
26. What precautionary procedures should be used when drilling a deep, vertical hole in mild steel when using an ordinary twist drill?
27. What is the metal-removal rate when a 1½-inch-diameter hole, 2 inches deep, is drilled in 1020 steel at a cutting speed of 120 fpm with a feed of 0.020 ipr?
28. If the specific horsepower for the steel in question 27 is 0.7, what horsepower would be required?
29. What is the advantage of a spade drill? Is it really a drill?
30. If the specific power of AISI 1020 steel of 0.03 W/mm³, and 75% of the output of the 1.5-kW motor of a drilling machine is available at the tool, what is the maximum feed that can be used in drilling a 51-mm-diameter hole with a HSS drill? (Use the cutting speed suggested, 55m/min.).
31. Let us assume that you are drilling eight holes, equally spaced around in a bolt-hole circle. That is, there would be holes at 12, 3, 6, and 9 o'clock and four more holes equally spaced between them. The diameter of the bolt hole circle is 6 inches. The designer says that the holes must be 45 ± 1° from each other around the circle.
    (a) Compute the tolerance between hole centers.
    (b) Do you think a typical multiple-spindle drill setup (with a jig) could be used to make this bolt circle—using eight drills all at once? Why or why not?

CASE STUDY 21. Break-Even-Point Analysis of Lathe Part

You have received the part drawing for a typical lathe part that will require turning, facing, grooving, boring, and threading as machined from a casting. Unfortunately, you do not yet know what the quantity will be. However, to be prepared, you have developed some cost data for the manufacture of the part by four different lathe processes (see the accompanying table). Complete the table by determining the run cost per batch, the cost per unit at the various quantities, and the total cost per batch.

	Quantity				
	10,000	**1,000**	**100**	**10**	**1**
Cost to produce on six-spindle automatic					
Total cost of batch					
Engineering: 2.5 hr at $40/hr	50.00	50.00	50.00	50.00	50.00
Tooling (cutting tools and work holder)	600.00	600.00	600.00	600.00	600.00
Setup: 8 hr at $15.00/hr	120.00	120.00	120.00	120.00	120.00
Run cost per batch: 50 cents per piece					
Cost each					
Cost to produce on turret lathe					
Total cost of batch					
Engineering: 2 hr at $20/hr	40.00	40.00	40.00	40.00	40.00
Tooling	150.00	150.00	150.00	150.00	150.00
Setup: 4 hr at $12.00/hr	48.00	48.00	48.00	48.00	48.00
Run cost per batch: $8.00 per piece					
Cost each					
Cost to produce on engine lathe					
Total cost of batch					
Engineering: 1 hr at $20/hr	20.00	20.00	20.00	20.00	20.00
Tooling	——	——	——	——	——
Setup: 2 hr at $12.00/hr	24.00	24.00	24.00	24.00	24.00
Run cost per batch: $12 per piece					
Cost each					
Cost to produce on NC lathe					
Total cost of Batch					
Engineering and programming	150.00	150.00	150.00	150.00	150.00
Tooling	100.00	100.00	100.00	100.00	100.00
Setup: 1 hr at $20.00/hr	20.00	20.00	20.00	20.00	20.00
Run cost per batch: $2 per piece					
Cost each					

1. Of the four costs listed for each process, which costs are fixed and which are variable?
2. For which of these costs would you have to be able to estimate the machining time per piece and the cycle time per piece?
3. How would you go about estimating this time and what time elements might be included in the cycle time in addition to machining time?
4. How would you use this estimate (of time) in the cost table?
5. Make a plot of cost (in dollars) versus quantity, with all four methods on one plot.
6. Make a plot of cost per unit versus quantity, again with all four methods on one plot. Find the break-even quantities.
7. Discuss these plots and the break-even quantities that you found.

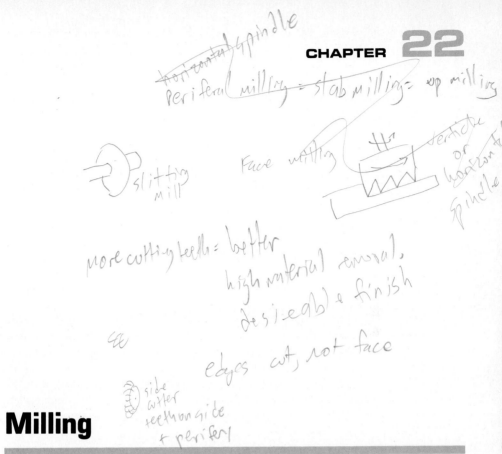

CHAPTER 22

Milling

Milling is a basic machining process by which a surface is generated progressively by the removal of chips from a workpiece as it is fed to a rotating cutter in a direction perpendicular to the axis of the cutter. In some cases the workpiece remains stationary, and the cutter is fed to the work. In nearly all cases, a multiple-tooth cutter is used so that the material removal rate is high. Often the desired surface is obtained in a single pass of the cutter or work and, because very good surface finish can be obtained, milling is particularly well suited, and widely used, for mass-production work. Several types of milling machines are used, ranging from relatively simple and versatile machines that are used for general-purpose machining in job shops and tool-and-die work to highly specialized machines for mass production. Unquestionably, more flat surfaces are produced by milling than by any other machining process.

The tool used in milling is known as a *milling cutter*. It usually consists of a cylindrical body which rotates on its axis and contains equally spaced peripheral teeth that intermittently engage and cut the workpiece. This is called interrupted cutting.

Types of milling operations. Milling operations can be classified into two broad categories, each having many variations. The basic concepts of these two types are indicated in Figures 18-4 and 18-5.

In *peripheral milling* a surface is generated by teeth located on the periphery of the cutter body. The surface is parallel with the axis of rotaton of the cutter. Both flat and formed surfaces can be produced by this method, the cross section of the resulting surface corresponding to the axial contour of the cutter. This process often is called *slab milling*.

In *face milling* the generated surface is at right angles to the cutter axis and is the combined result of the actions of the portions of the teeth located on both the periphery and the face of the cutter. Most of the cutting is done by the peripheral portions of the teeth, with the face portions providing some finishing action. Peripheral milling operations usually are performed on machines having horizontal spindles, whereas face milling is done on both horizontal- and vertical-spindle machines.

The generation of surfaces in milling. In milling, surfaces can be generated by two distinctly different methods, illustrated in Figures 22-1 and 22-2. *Up milling* is the traditional way to mill. Called *conventional* milling, the cutter rotates against the direction of feed of the workpiece. In *climb* or *down milling* the rotation is in the same direction as the feed. The method of chip formation is completely different in the two cases. In up milling the chip is very thin at the beginning, where the tooth first contacts the work, and increases in thickness, becoming a maximum where the tooth leaves the work. The cutter tends to push the work along and lift it upward from the table. This action tends to eliminate any effect of looseness in the feed screw and nut of the milling machine table and results in a smooth cut. However, the action also tends to loosen the work from the clamping device so that greater clamping forces must be employed. In addition, the smoothness of the generated surface depends greatly on the sharpness of the cutting edges.

In down milling, maximum chip thickness occurs close to the point at which the tooth contacts the work. Because the relative motion tends to pull the workpiece into the cutter, all possibility of looseness in the table feed screw must be

smoother

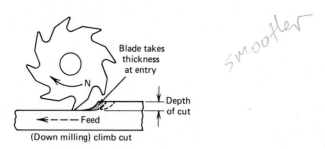

Blade takes thickness at entry

Depth of cut

Feed

(Down milling) climb cut

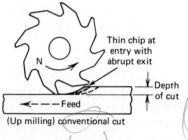

Thin chip at entry with abrupt exit

Depth of cut

Feed

(Up milling) conventional cut

slab

FIGURE 22-1 Comparison of climb and conventional milling with slab mills.

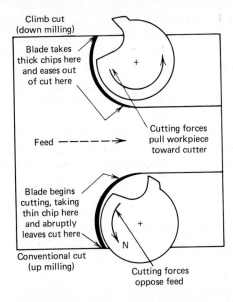

Climb cut
(down milling)

Blade takes
thick chips here
and eases out
of cut here

Feed – – – – ➔

Cutting forces
pull workpiece
toward cutter

Blade begins
cutting, taking
thin chip here
and abruptly
leaves cut here

Conventional cut
(up milling)

Cutting forces
oppose feed

FIGURE 22-2 Comparison of climb and conventional milling with face mills. (*From* Metal Cutting Principles, 2nd ed.; *courtesy Ingersoll Cutting Tool Company.*)

eliminated if down milling is to be used. It should never be attempted on machines that are not designed for this type of milling; virtually all modern milling machines are so equipped. Because the material yields in approximately a tangential direction at the end of the tooth engagement, there is less tendency for the machined surface to show toothmarks than when up milling is used. Another advantage of down milling is that the cutting force tends to hold the work against the machine table, permitting lower clamping forces. However, the fact that the cutter teeth strike against the surface of the work at the beginning of each chip can be a disadvantage if the workpiece has a hard surface, as castings sometimes do. This may cause the teeth to dull rapidly.

Milling is an interrupted cutting process, where entering and leaving the cut subjects the tool to impact loading, cyclic heating, and cycle cutting forces. As shown in Figure 22-3, the cutting force, F_c, builds rapidly as the tool enters the

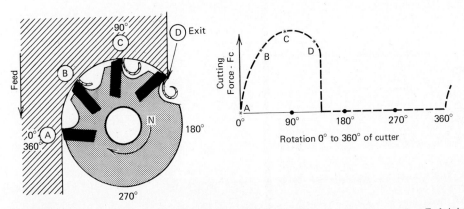

FIGURE 22-3 Conventional face milling (*left*) with cutting force diagram for F_c (*right*), showing the interrupted nature of the process. (*From* Metal Cutting Principles, 2nd ed.; *courtesy Ingersoll Cutting Tool Company.*)

work at Ⓐ and progresses to Ⓑ, peaks as the blade crosses the direction of feed at Ⓒ, decreases to Ⓓ, and then drops to zero abruptly upon exit. The diagram does not indicate the impulse loads caused by impacts. The interrupted cut phenomenan explains in a large part why milling cutter teeth are designed to have small positive or negative rakes, particularly when the tool material is carbide or ceramic. These brittle materials tend to be very strong in compression and negative rake results in the cutting edges being placed in compression by the cutting forces rather than tension. Cutters made from HSS are made with positive rakes, in the main, but must be run at lower speeds. Positive rake tends to lift the workpiece while negative rakes compress the workpiece and allow for heavier cuts to be made.

MILLING CUTTERS

Milling cutters can be classified in several ways. One is to group them into two broad classes, based on tooth relief, as follows:

1. On *profile cutters,* relief is provided on each tooth by grinding a small land back of the cutting edge. The cutting edge may be straight or curved.
2. On *form* or *cam-relieved cutters,* the cross section of each tooth is an eccentric curve behind the cutting edge, thus providing relief. All sections of the eccentric relief, parallel with the cutting edge, have the same contour as the cutting edge. Cutters of this type are sharpened by grinding only the face of the teeth; the contour of the cutting edge thus remains unchanged.

Another useful method of classification is according to the manner of mounting the cutter. *Arbor cutters* have a center hole so they can be mounted on an arbor. *Shank cutters* have either a tapered or straight integral shank. Those with tapered shanks can be mounted directly in the milling machine spindle, whereas straight-shank cutters are held in a chuck. *Facing cutters* usually are bolted to the end of a stub arbor. Common types of milling cutters, classified in this manner, are as follows:

Arbor Cutters	Shank Cutters
Plain	End mills
Side	Solid
Staggered-tooth	Inserted-tooth
Slitting saws	Shell
Angle	Hollow
Inserted-tooth	T-slot
Form	Woodruff key seat
Fly	Fly

Figures 22-4 and 22-5 show several types of arbor-type and shank-type milling cutters, respectively.

Another method of classification applies only to face and end-mill cutters and

FIGURE 22-4 Arbor-type milling cutters. (*Top*) Side, plain, staggered-tooth. (*Center*) Angle, fly. (*Bottom*) Metal-slitting saw, inserted-tooth, form.

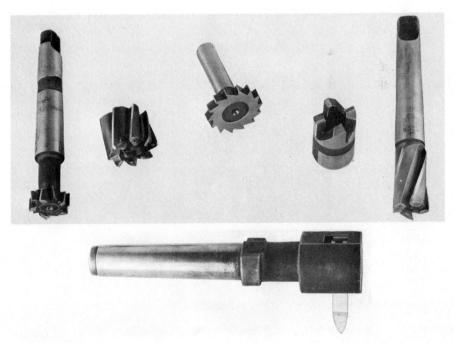

FIGURE 22-5 Shank-type milling cutters. (*Top*) T-slot, shell end mill, Woodruff key seat, hollow end mill, solid end mill. (*Bottom*) Fly cutter.

relates to the direction of rotation. A *right-hand cutter* must rotate counterclockwise when viewed from the front end of the machine spindle. Similarly, a *left-hand cutter* must rotate clockwise. All other cutters can be reversed on the arbor to change them from one hand to the other.

Positive rake angles are used on general-purpose milling cutters, such as in job-shop work, whereas negative rake angles commonly are used on carbide- and ceramic-tipped cutters that are employed in mass-production milling, in order to obtain the greater strength and cooling capacity which they provide. Occasionally dual rake angles are used—a short negative rake face adjacent to the cutting edge, followed by a longer face having a positive rake angle.

Types of milling cutters. *Plain milling cutters* are cylindrical or disk-shaped, have straight or helical teeth on the periphery, and are used for milling flat surfaces. This type of operation is called *plain,* or *slab milling.* As shown in Figure 22-6, each tooth in a helical cutter engages the work gradually, and usually more than one tooth cuts at a given time. This reduces shock and chattering tendencies and promotes a smoother surface. Consequently, this type of cutter usually is preferred over one with straight teeth.

Side milling cutters are similar to plain milling cutters except that the teeth extend radially part way across one or both ends of the cylinder toward the center. The teeth may be either straight or helical. Frequently, these cutters are relatively narrow, being disklike in shape. Two or more side milling cutters often are spaced on an arbor to make simultaneous, parallel cuts, in an operation called *straddle milling.*

Interlocking slotting cutters consist of two cutters similar to side mills, but made to operate as a unit for milling slots. The two cutters are adjusted to the desired width by inserting shims between them.

Staggered-tooth milling cutters are narrow cylindrical cutters having staggered teeth, and with alternate teeth having opposite helix angles. They are ground to

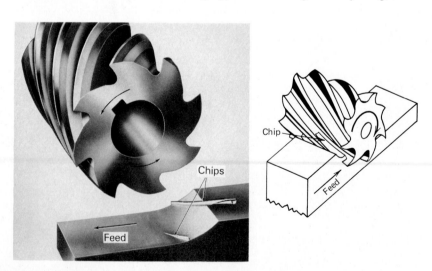

FIGURE 22-6 Manner in which chips are formed progressively by the teeth of a plain helical-tooth milling cutter in up milling. (*Courtesy Cincinnati Milacron, Inc.*)

MACHINING PROCESSES

cut only on the periphery, but each tooth also has chip clearance ground on the protruding side. These cutters have a free cutting action that makes them particularly effective in milling deep slots.

Slitting saws are thin, plain milling cutters, usually from 0.8 to 4.8 mm ($\frac{1}{32}$ to $\frac{3}{16}$ inch) thick, which have their sides slightly "dished" to provide clearance and prevent binding. They usually have more teeth per unit of diameter than ordinary plain milling cutters and are used for milling deep, narrow slots and for cutting-off operations.

Angle milling cutters are made in two types—single-angle and double-angle. *Single-angle cutters* have teeth on the conical surface, usually at an angle of 45 to 60° to the plane face. The teeth may also extend radially on the larger plain face. *Double-angle cutters* have V-shaped teeth, with both conical surfaces at an angle to the end faces, but not necessarily at the same angle. The V-angle usually is 45, 60, or 90°. Angle cutters are used for milling slots of various angles or for milling the edges of workpieces to a desired angle.

Most larger-sized milling cutters are of the *inserted-tooth type*. The cutter body is made of ordinary steel, with the teeth made of high-speed steel, cemented carbide, or ceramic, fastened to the body by various methods. Most commonly, the teeth are throwaway, indexable carbide or ceramic inserts, as shown in Figure 22-7. This type of construction reduces the amount of costly material that is required and can be used for any type of cutter but most often is used with face mills.

Form milling cutters have the teeth ground to a special shape—usually an irregular contour—to produce a surface having a desired transverse contour. They are cam-relieved and are sharpened by grinding only the tooth face, thereby retaining the original contour as long as the plane of the face remains unchanged with respect to the axis of rotation. Convex, concave, corner-rounding, and gear-tooth cutters are common examples.

End mills are shank-type cutters having teeth on the circumferential surface and one end. They thus can be used for facing, profiling, and end milling. The teeth may be either straight or helical, but the latter is more common. Small end mills have straight shanks, whereas taper shanks are used on larger sizes.

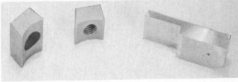

FIGURE 22-7 *(Left)* Inserted-tooth milling cutter, using throwaway tungsten-carbide cutting inserts and locking wedges. *(Above)* Locking wedges and insert support. *(Courtesy Lovejoy Tool Company, Inc.)*

Plain end mills have multiple teeth that extend only about halfway toward the center on the end. They are used in milling slots, profiling, and facing narrow surfaces. *Two-lip mills* have two straight or helical teeth that extend to the center. Thus they may be sunk into material, like a drill, and then fed lengthwise to form a groove.

Shell end mills are solid-type, multiple-tooth cutters, similar to plain end mills but without a shank. The center of the face is recessed to receive a screw head or nut for mounting the cutter on a separate shank or a stub arbor. The back of the cutter contains driving slots that engage collar keys on the shank. This design enables one shank to hold any of several cutters and thus provides great economy for larger-sized end mills. The cutter shown in Figure 22-7 is of this type.

Hollow end mills are tubular in cross section, with teeth only on the end but having internal clearance. They are used primarily on automatic screw machines for sizing cylindrical stock, producing a short, cylindrical surface of accurate diameter.

T-*slot cutters* are integral-shank cutters with teeth on the periphery and *both* sides. They are used for milling the wide groove of a T-slot. In order to use them, the vertical groove must first be made with a slotting mill or an end mill to provide clearance for the shank. Because the T-slot cutter cuts on five surfaces simultaneously, it must be fed with care.

Woodruff keyseat cutters are made for the single purpose of milling the semi-cylindrical seats required in shafts for Woodruff keys. They come in standard sizes corresponding to Woodruff key sizes. Those below 50.8 mm (2 inches) in diameter have integral shanks; the larger sizes may be arbor-mounting.

Occasionally, *fly cutters* may be used as shown in Figure 22-5; these have a single-point cutting tool attached to a special shank, usually with provision for adjusting the effective radius of the cutting tool with respect to the axis of rotation. The cutting edge can be made in any desired shape and, because it is a single-point tool, is very easy to grind. These cutters can be used for face milling and also for boring, and frequently are used where both of these operations need to be done with a single tool at one setup of the work. They are used primarily in experimental and toolroom work.

MILLING MACHINES

Because the milling process is versatile and highly productive, a variety of machines have been developed to employ the milling principle. One type, commonly called *milling machines,* are basic, general-purpose machines that provide a high degree of flexibility. Another type, used exclusively for reproducing parts from templates or patterns, often are called *duplicators.* A third type encompasses *special-purpose machines* that are used in mass-production manufacturing. These will be discussed in Chapter 38. Machines of the fourth type are highly versatile but are designed to do other basic machining operations as well as milling. These commonly are called *machining centers* and will also be discussed in Chapter 38.

Inasmuch as basic milling machines provide an accurate, rugged, rotating spindle, they can also be used for other machining operations, such as drilling and boring. Consequently, because they can do several types of operations and produce several types of surfaces with a single setup of the workpiece, they are among the most important of machine tools. The common types may be classified according to their general design characteristics as follows:

1. Column-and-knee type (general purpose).
   a. Plain.
      (1) Power table feed.
      (2) Hand table feed.
   b. Universal.
   c. Vertical.
   d. Turret-type universal.
2. Bed type (manufacturing).
   a. Simplex.
   b. Duplex.
   c. Triplex.
3. Planer type (large work only).
4. Special.
   a. Rotary table.
   b. Drum type.
   c. Profilers.
   d. Duplicators.

Basic milling-machine construction. Most basic milling machines are of column-and-knee construction, employing the components and motions shown in Figure 22-8. The column, mounted on the base, is the main supporting frame for all the other parts and contains the spindle with its driving mechanism. As indicated, this construction provides controlled motion of the worktable in three mutually perpendicular directions: (1) through the *knee* moving vertically on ways on the front of the column, (2) through the *saddle* moving transversely on ways on the knee, and (3) through the *table* moving longitudinally on ways on the saddle. All these motions can be imparted either by manual or powered means. In most cases, a powered rapid traverse is provided in addition to the regular feed rates for use in setting up work and in returning the table at the end of a cut.

Milling machines having only the three mutually perpendicular table motions just described are called *plain column-and-knee type*. These are available with both horizontal and vertical spindles as shown in Figure 28-8. On the horizontal type, an adjustable over-arm is mounted on the top of the column to provide an outboard bearing support for the end of the cutter arbor, when required. In some vertical-spindle machines the spindle is mounted in a sliding head that can be fed up and down either by power or by hand. Vertical-spindle machines are especially well suited for face- and end-milling operations. They also are very useful for drilling and boring, particularly where holes must be spaced accurately in a horizontal plane, because of the controlled table motion.

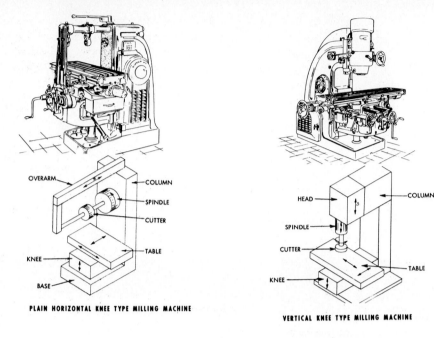

PLAIN HORIZONTAL KNEE TYPE MILLING MACHINE

VERTICAL KNEE TYPE MILLING MACHINE

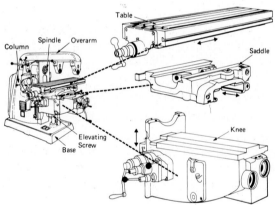

FIGURE 22-8 Major components of a plain column-and-knee-type milling machine, which can have horizontal spindle or vertical spindle. (*Top, from* Manufacturing Producibility Handbook; *courtesy General Electric Company. Bottom, courtesy Cincinnati Milacron, Inc.*)

Universal column-and-knee milling machines. *Universal column-and-knee milling machines* differ from plain column-and-knee machines in that the table is mounted on a housing that can be swiveled in a horizontal plane, thereby increasing its flexibility so as to permit the milling of helices, as found in twist drills, milling cutters, and helical gear teeth. This type of machine is shown in Figure 22-9.

Turret-type milling machines. *Turret-type column-and-knee milling machines,* as shown in Figure 22-10, have dual heads that can be swiveled about

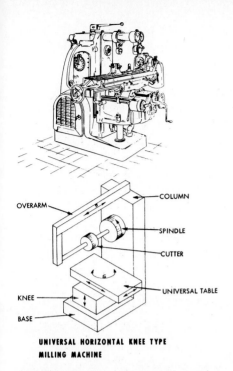

UNIVERSAL HORIZONTAL KNEE TYPE
MILLING MACHINE

FIGURE 22-9 Universal horizontal knee-type milling machine. (*From* Manufacturing Producibility Handbook; *courtesy General Electric Company.*)

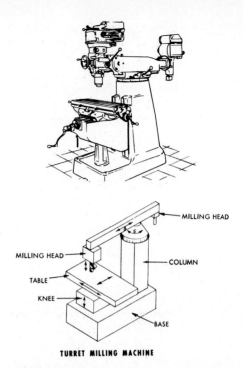

TURRET MILLING MACHINE

FIGURE 22-10 Turret-type milling machine. (*From* Manufacturing Producibility Handbook; *courtesy General Electric Company.*)

$$\left(FEED = IN/REV \right) \times \frac{REV}{MIN} = \left(\frac{in}{min} \right)$$

a horizontal axis on the end of a horizontally adjustable ram. This permits milling to be done horizontally, vertically, or at any angle. This added flexibility is advantageous where a variety of work has to be done, as in tool and die or experimental shops. They are available with either plain or universal tables.

Milling machine size. Milling machine size is designated by numbers from 1 through 6, which are approximate indicators of the longitudinal table travel as follows:

$$\frac{.004 in}{Teeth} \times \frac{\# Teeth}{Rev} = Feed$$

Size:	1	2	3	4	5	6
Table travel { in.	22	28	32	40	50	60
mm	559	711	812	1016	1270	1524

These relationships are not standardized, and a particular machine may vary considerably from these values.

Hand-feed milling machines. *Hand-feed milling machines* are small, simple machines on which the table is fed longitudinally by means of a hand lever which rotates a pinion that engages a rack on the bottom of the table. Such machines are used for light milling of short slots, grooves, and so forth on small parts.

Bed-type milling machines. In production manufacturing operations, ruggedness and the capability of making heavy cuts are of more importance than versatility. *Bed-type milling machines,* such as shown in Figure 22-11, are made for these conditions. The table is mounted directly on the bed and has only longitudinal motion. The spindle head can be moved vertically in order to set up the machine for a given operation. Normally, once the setup is completed, the spindle head is clamped in position and no further motion of it occurs during machining. However, on some machines vertical motion of the spindle occurs during each cycle.

After such milling machines are set up, little skill is required to operate them. This permits the use of semiskilled operators. Some machines of this type are equipped with automatic controls so that the only activity required of the operator is to put the workpiece into a fixture and set the machine into operation. Often a fixture is provided at each end of the table so that one workpiece can be unloaded while another is being machined.

Bed-type milling machines with single spindles sometimes are called *simplex milling machines;* they are made with both horizontal and vertical spindles. Bed-

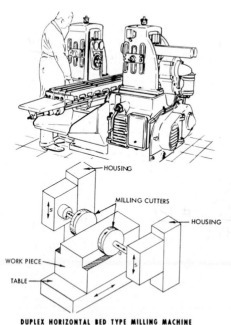

DUPLEX HORIZONTAL BED TYPE MILLING MACHINE

FIGURE 22-11 Duplex horizontal milling machines have two horizontal spindles. (*From* Manufacturing Producibility Handbook; *courtesy General Electric Company.*)

FIGURE 22-12 Triplex milling machine for face milling three surfaces simultaneously. (*Courtesy Cincinnati Milacron, Inc.*)

type machines also are made in *duplex* and *triplex* types, having two and three spindles, respectively. These permit the simultaneous milling of two or three surfaces at a single pass. Figure 22-12 shows a setup of a triplex milling machine for milling engine blocks.

Planer-type milling machines.   As was pointed out in Chapter 18, planers have two serious disadvantages in that they utilize only single-point cutting tools, and a large table and heavy workpiece cannot be reciprocated rapidly. Consequently, they have largely been replaced by *planer-type milling machines,* which, as illustrated in Figure 22-13, utilize several milling heads, which can remove large amounts of metal while permitting the table and workpiece to move quite slowly. Often only a single pass of the workpiece past the cutters is required. Through the use of different types of milling heads and cutters, a wide variety of surfaces can be machined with a single setup of the workpiece. This is a great advantage where heavy workpieces are involved.

Rotary-table milling machines.   Some types of face milling in mass-production manufacturing are often done on *rotary-table milling machines,* such as is shown in Figure 22-14. Roughing and finishing cuts can be made in succession as the workpieces are moved past the several milling cutters while held in fixtures on the rotating table. The operator can load and unload the work without stopping the machine.

FIGURE 22-13   Large planer-type milling machine. Inset shows 90° head being used. (*Courtesy Cosa Corporation.*)

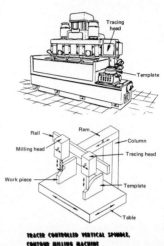

TRACER CONTROLLED VERTICAL SPINDLE,
CONTOUR MILLING MACHINE

FIGURE 22-14 Two-spindle 48-inch rotary-table milling machine being used to rough and finish the surface of automobile timing-gear covers. Eighty covers per hour are machined. (*Courtesy Ingersoll Cutting Tool Company.*)

FIGURE 22-15 Schematics of profiler. Machine on top milling 4 work pieces simultaneously. (*From* Manufacturing Producibility Handbook; *courtesy General Electric Company.*)

Profilers and duplicators. Milling machines that can duplicate external or internal profiles in two dimensions are called *profilers* or tracer-controlled contouring machines. As shown in Figure 22-15, a tracing probe follows a two-dimensional template and, through electronic or air-actuated mechanisms, controls the cutting spindles in two mutually perpendicular directions. The spindles—usually more than one—are set manually in the third dimension.

Duplicators produce forms in three dimensions. A tracing probe follows a three-dimensional master. Often the probe does not actually contact the master, a variation in the length of a spark between the probe and the master controlling the drives to the quill and the table, thereby avoiding wear on the master or possible deflection of the probe. On some machines, the ratio between the movements of the probe and cutter can be varied.

Duplicators are widely used to machine molds and dies and sometimes are called *die-sinking machines*. They are used extensively in the aerospace industry to machine parts from wrought plate or bar stock as substitutes for forgings, where the small number of parts required would make the cost of forging dies uneconomical. Profilers and duplicators have, to a great extent, been replaced by numerically controlled machines in which a punched tape or computer input eliminates the necessity of making a template or master mold.

Accessories for milling machines. The usefulness of ordinary milling machines is greatly extended by employing various accessories.

The *vertical milling attachment,* shown in Figure 22-16, is used on a hori-

FIGURE 22-16 Vertical milling attachment for horizontal milling machine. (*Courtesy Brown & Sharpe Mfg. Co.*)

zontal milling machine to permit vertical milling to be done. Ordinarily, heavy work cannot be done with such an attachment.

The *universal milling attachment,* shown in Figure 22-17, is similar to the vertical attachment but can be swiveled about both the axis of the milling machine spindle and a second, perpendicular axis to permit milling to be done at any angle.

The *slotting attachment* permits adapting a horizontal milling machine to the work of a vertical shaper. Although not used extensively, it is useful for cutting small keyways.

Universal milling attachment

Spindle

Index plate

Crank

FIGURE 22-17 Milling a helical groove using a universal dividing head and a universal milling attachment. (*Courtesy Cincinnati Milacron, Inc.*)

The *universal dividing head* is by far the most widely used milling machine accessory, providing a means for holding and indexing work through any desired arc of rotation. The work may be mounted between centers, as shown in Figure 22-17, or held in a chuck that is mounted in the spindle hole of the dividing head.

Basically, a dividing head is a rugged, accurate, 40:1 worm-gear reduction unit. The spindle is carried in a swivel block so that it can be tilted from about 5° below horizontal to beyond the vertical position.

The spindle of the dividing head is revolved one revolution by turning the input crank 40 turns. An index plate, mounted beneath the crank, contains a number of holes, arranged in concentric circles and equally spaced, with each circle having a different number of holes. A plunger pin on the crank handle can be adjusted to engage the holes of any circle. This permits the crank to be turned an accurate, fractional part of a complete circle as represented by the increment between any two holes of a given circle on the index plate. Utilizing the 40:1 gear ratio and the proper hole circle on the index plate, the spindle can be rotated a precise amount by the applications of either of the following rules.

$$(1) \text{ number of turns of crank } = \frac{40}{\text{cuts per revolution of the workpiece}}$$

$$(2) \text{ holes to be indexed } = \frac{40 \times \text{holes in index circle}}{\text{cuts per revolution of the workpiece}}$$

If the first rule is used, an index circle must be selected that has the proper number of holes to be divisible by the denominator of any resulting fractional portion of a turn of the crank. In using the second rule, the number of holes in the index circle must be such that the numerator of the fraction is an even multiple of the denominator. For example, if 24 cuts are to be taken about the circumference of a workpiece, the number of turns of the crank required would be 1⅔. An index circle having 12, 15, 18, and so on, holes could be used with one full turn plus 8, 10, or 12 additional holes, respectively. Obviously, use of the second rule would give the same answer. Adjustable *sector arms* are provided on the index plate that can be set to a desired number of holes, less than a full turn, so that fractional turns can be made readily without the necessity for counting holes each time.

Dividing heads are made having ratios other than 40:1, so the ratio should be checked before using.

Because each full turn of the crank on a standard dividing head represents 360/40, or 9° of rotation of the spindle, indexing to a fraction of a degree can be obtained. For example, the space between two adjacent holes on a 36-hole circle represents ¼°.

Indexing can be done in three ways. *Plain indexing* is done solely by the use of the 40:1 ratio in the dividing head. In *compound indexing,* the index plate is moved forward or backward a number of hole spaces each time the crank handle is advanced. For *differential indexing* the spindle and the index plate are connected by suitable gearing so that as the spindle is turned, by means of the crank, the index plate is rotated a proportionate amount.

The dividing head also can be connected to the feed screw of the milling-

FIGURE 22-18 Universal vise for use on a milling machine. (*Courtesy Cincinnati Milacron, Inc.*)

machine table by means of gearing. This procedure is used to provide a definite rotation of the workpiece with respect to the longitudinal movement of the table, as in cutting helical gears. This procedure is illustrated in Figure 28-9, the connecting gearing being shown in the latter.

Although T-slots are provided on milling machine tables so that workpieces can be clamped directly to the table, more often various work-holding accessories are utilized. Smaller workpieces usually are held in a vise mounted on the table. A universal vise, shown in Figure 22-18, is particularly useful in tool-and-die work. In mass-production work special fixtures usually are employed, such as those shown in Figure 22-14; these reduce machine-loading time and assure proper clamping. The circular-milling attachment, shown in Figure 22-19, imparts rotary motion to the work, either by manual or power feed, and thus permits cylindrical surfaces to be milled.

FIGURE 22-19 Milling a circular slot using a circular-milling attachment. (*Courtesy Cincinnati Milacron, Inc.*)

Material	Face Mills	Helical Mills	Slotting or Side Mills	End Mills	Saws
Magnesium and aluminum alloys	0.56 (0.022)	0.46 (0.018)	0.33 (0.013)	0.28 (0.011)	0.13 (0.005)
Medium brass	0.36 (0.014)	0.28 (0.011)	0.20 (0.008)	0.18 (0.007)	0.08 (0.003)
Medium cast iron Cast steel Carbon steel Free-machining steel	0.30 (0.012)	0.25 (0.010)	0.18 (0.007)	0.15 (0.006)	0.08 (0.003)

ESTIMATING MILLING TIME

Equations for calculating milling time and metal removal rates were given in Chapter 18, with consideration given to machine capacity and workpiece strength and rigidity. Table 22-1 gives suggested values of feed, in chip thickness per tooth. It must be remembered that it is the chip thickness per tooth that must be used in calculating milling machine feed, and thus milling time (see also Chapter 39).

REVIEW QUESTIONS

1. Why is milling better suited than shaping for producing flat surfaces in mass-production machining?
2. How does face milling differ basically from peripheral milling?
3. Why does down milling tend to produce better surface finish than up milling?
4. Why may down milling dull the cutter more rapidly than up milling when machining castings?
5. What are two common ways of classifying milling cutters?
6. Why do arbor-type slab milling cutters not have to be designated as "right-hand" or "left-hand"?
7. What is the advantage of a helical-tooth cutter over a straight-tooth cutter for slab milling?
8. What is a form cutter and how is it used?
9. Why is a narrow milling cutter not well suited for milling deep slots?
10. Why are helical-tooth cutters preferred over straight-tooth cutters for most slab milling?
11. Explain what steps are required to produce a T-slot by milling.
12. What is the distinctive feature of a fly cutter?
13. Why would a plain column-and-knee milling machine not be suitable for milling the flutes on a large twist drill?
14. Why is a turret-type milling machine often preferred for tool-and-die work?
15. What is the distinctive feature of a universal milling machine?
16. Explain how controlled movements of the work in three mutually perpendicular directions are obtained in column-and-knee-type milling machines.
17. What is a triplex milling machine?

18. Why are bed-type milling machines preferred over column-and-knee types for production milling?

19. How does a duplicator differ from a profiler?

20. Why have planer-type milling machines replaced ordinary planers?

21. What are four common milling machine accessories?

22. What is the basic principle of a universal dividing head?

23. The input end of a universal dividing head can be connected to the feed screw of the milling machine table. For what purpose?

24. What is the purpose of the hole-circle plate on a universal dividing head?

25. Explain how a standard universal dividing head, having hole circles of 21, 24, 27, 30, and 32 holes, would be operated to cut an 18-tooth gear.

26. Explain how table feed (ipm) and spindle rpm is specified or computed for a milling machine.

27. Why must the number of teeth on the cutter be known when calculating milling machine feed and setting it on the machine controls?

28. How long will be required to face mill an AISI 1020 steel surface that is 300 mm long and 127 mm wide, using a 152-mm-diameter, eight-tooth tungsten carbide inserted-tooth cutter? Select values of chip thickness from Table 22-1 and cutting speed from Table 18-4.

29. If the depth of cut is 8 mm, what is the metal-removal rate in question 28?

30. Using data from Table 18-2, estimate the power required for the operation of question 29. Do not forget to consider Figure 22-3.

31. If all the flat surfaces on the bearing block shown in Figure 36-4 must be machined, would you machine them in the same sequence on a shaper as on a milling machine? Explain why.

32. A gray cast iron surface 152 mm wide and 457 mm long may be machined on either a vertical milling machine, using a 203-mm-diameter cutter having 10 inserted HSS teeth, or on a hydraulic shaper with a HSS tool. If milling is used, a chip thickness of 0.25 mm would be employed, and in shaping a feed of 0.38 mm would be used. Setup time for the milling machine would be 30 minutes and for the shaper would be 10 minutes. The time required to put each piece in either machine and to remove it would be 4 minutes. Machine-hour charges for the milling machine and shaper would be $14.50 and $6.50, respectively. Labor cost would be $8.75 per hour in each case. Which machine would be more economical for this job? (Use Table 18-4 for cutting speeds.)

33. In the problem described in question 32, what percentage of the cycle time for the job is consumed in non-metal-removal activities (setup time and part loading/unloading) assuming that 10 parts are to be made?

CASE STUDY 22. HSS Versus Tungsten Carbide

The Quality Machine Works, which does job-shop machining, has received an order to make 40 duplicate pieces, made of AISI 4140 steel, which will require one hour per piece for actual cutting time if an ordinary HSS milling cutter is used. John Young, a new machinist, says the cutting time can be reduced to not over 25 minutes per piece if the company will purchase a suitable tungsten carbide milling cutter. Hans Oldman, the foreman for the milling area, says he does not believe that John's estimate is realistic, and he is not going to spend $450 of the company's money on a carbide cutter that probably would not be used again. The machine-hour rate, including labor, is $30 per hour. John and Hans have come to you, the supervisor of the shop, for a decision on whether or not to buy the cutter, which is readily available from a local supplier.

What are the things you should consider in this situation? Who do you think is right, John or Hans?

CHAPTER 23

Particles on a wheel

Periphery or side used.

accurate size + smooth finish

small amount material.

honing-

Abrasive Machining
Processes

Abrasive machining is the basic process in which chips are formed by very small cutting edges that are integral parts of abrasive particles, usually manmade. Unquestionably, abrasive machining is the oldest of the basic machining processes. Museums abound with examples of utensils, tools, and weapons that ancient man produced by rubbing hard stones against softer materials to abrade away unwanted portions, leaving desired shapes. For centuries, only natural, nonuniform, and relatively ineffective abrasives were available, and abrasive machining was far surpassed in importance and use by more modern, basic machining processes, which were developed around superior cutting materials. However, during this century two developments have changed this situation. First was the development of man-made abrasives. Second, in recent years basic research provided a fundamental understanding of the abrasive machining process. As a consequence, several variations of abrasive machining are among the most important of all the basic machining processes.

The results that can be obtained by abrasive machining range from the finest and smoothest surfaces produced by any machining process, in which very little material is removed, to rough, coarse surfaces that accompany high material-removal rates.

The abrasive particles may be (1) free (see Chapter 26); (2) mounted in resin on a belt (called *coated product*); or (3) close packed into wheels or stones, with abrasives held together by bonding material (called *bonded product*).

TABLE 23-1 Abrasive Machining Processes

Process	Particle Mounting	Features
Grinding	Bonded	Uses wheels, accurate sizing, finishing, low MRR
Abrasive machining[a]	Bonded	High MRR, to obtain desired shapes and approximate sizes
Snagging	Bonded belted	High MRR, rough rapid technique to clean up castings, forgings
Honing	Bonded	"Stones" containing fine abrasives; primarily a hole finishing process
Lapping	Free	Fine particles embedded in soft metal or cloth; primarily a surface-finishing process

[a]The term "abrasive machining" applied to one particular form of the grinding process is unfortunate, because all these processes are doing abrasive machining.

The metal-removal process is basically the same in all three cases but with important differences due to spacing of active grains (grains in contact with work) and the rigidity and degree of fixation of the grains. Table 23-1 describes the primary abrasive processes.

The abrasive machining processes have two unique characteristics. First, because each cutting edge is very small and a number of these edges can cut simultaneously when suitable machines are employed, very fine cuts are possible, and fine surfaces and close dimensional control can be obtained. Second, because extremely hard abrasive particles can be produced, very hard materials, such as hardened steel, glass, carbides, and ceramics, can be machined very readily. As a result, the abrasive machining processes are not only important as manufacturing processes, they are essential. Many of our modern products, such as automobiles, space vehicles, and aircraft, would not be possible without them. To a very large degree they have made possible longer-lived machines and mechanisms that will operate efficiently and dependably at sustained high speeds.

ABRASIVES

An *abrasive* is a hard material that can cut or abrade other substances. As noted previously, certain natural abrasives have existed from the earliest times. Sandstone was used by ancient people to sharpen tools and weapons. Early grinding wheels were cut from slabs of sandstone but, because they were not uniform throughout, they wore unevenly and did not produce consistent results. Emery, a mixture of Al_2O_3 and magnetite, Fe_3O_4, is a natural abrasive used on coated paper and cloth. Corundum (natural Al_2O_3) and diamonds are other naturally occurring abrasive materials. However, the development of artificial abrasives having known, uniform properties has permitted abrasive processes to become a true precision manufacturing process.

Today, the only natural abrasives that have commercial importance are quartz sand, garnets, and diamonds. Quartz sand is used primarily in coated abrasives and in air blasting (which will be discussed later), but artificial abrasives are also making rapid inroads in these applications.

Hardness, the ability to resist penetration, is an important property for an abrasive. Table 23-2 lists the primary abrasives and their approximate Knoop hardness (kg/mm^2).

Two other properties are significant in abrasive grits. *Attrition* or fine, abrasive wear action of the grits results in dulled edges, grit flattening, and wheel glazing. *Friability* refers to the fracture of the grits and is the opposite of toughness. In grinding, it is often important that grits be able to fracture to expose new, sharp edges.

Diamonds are the hardest of all materials. Those that are used for abrasives are either natural, off-color stones (called garnets) that are not suitable for gems, or small, man-made stones that are produced specifically for abrasive purposes. Neither natural nor synthetic diamonds have clear superiority over each other; man-made stones appear to be somewhat more friable and thus tend to cut faster and cooler. They do not perform as satisfactorily in metal-bonded wheels. Diamond abrasive wheels are used extensively for sharpening carbide and ceramic cutting tools. Diamonds also are used for truing and dressing other types of abrasive wheels. Obviously, because of their cost, diamonds are used only when cheaper abrasives will not produce the desired results.

Garnets are used primarily in the form of very finely crushed and graded powders for fine polishing.

Artificial abrasives date from 1891, when E. G. Acheson, while attempting to produce precious gems, discovered how to make *silicon carbide,* SiC. Silicon carbide is made by charging an electric furnace with silica sand, petroleum coke, salt, and sawdust. A temperature of over 2200°C (4000°F) is maintained for several hours, by passing large amounts of current through the charge, and a solid mass of silicon carbide crystals results. After the furnace has cooled, the mass of crystals is removed, crushed, and graded to various desired sizes. As

TABLE 23-2   Knoop Hardness Values for Common Abrasives

Abrasive Material	Year of Discovery	Hardness (Knoop)	Comments
Quartz (SiO$_2$) sand	?	320	Sand blasting
Aluminum oxide	1893	2100	Softer and tougher than silicon carbide; used on steel, iron, brass
Silicon carbide	1891	2400	Used for CI, brass, bronze, aluminum, and stainless
Borazon (cubic boron nitride)	1962	4700	For grinding hard, tough tool steels
Diamond (man-made)	1955	7000	Used to grind tungsten carbide and some die steels

Handwritten annotations:

Boron nitride for machine steel.

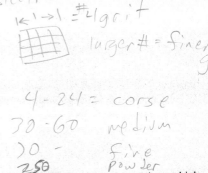

|←1→| #4 grit
larger # = finer grit

4 - 24 = corse
30 - 60 medium
70 - fine
250 powder

FIGURE 23-1 Abrasive grains at high magnification, showing their irregular, sharp cutting edges. (*Courtesy Norton Company.*)

can be seen in Figure 23-1, the resulting crystals, or grains, are irregular in shape, and most of the resulting cutting edges have negative rake.

Silicon carbide crystals are very hard, friable, and rather brittle, and this limits their use. Silicon carbide is sold under the trade names *Carborundum* and *Crystolon*.

Aluminum oxide. Al_2O_3, is the most widely used artificial abrasive. Also produced in an arc furnace, from bauxite, iron filings, and small amounts of coke, it contains aluminum hydroxide, ferric oxide, silica, and some other impurities. The mass of aluminum oxide that is formed is crushed, and the particles are graded to size. Common trade names for aluminum oxide abrasives are *Alundum* and *Aloxite*.

Although aluminum oxide is softer than silicon carbide, it is considerably tougher. Consequently, it is a more general-purpose abrasive.

Another relatively new abrasive is *cubic boron nitride* (CBN). It is harder than either silicon carbide or aluminum oxide and is advantageous for grinding certain hardened tool-and-die steels. It is superior to diamonds, even though it is substantially softer, for grinding steel.

Abrasive grain size and geometry. To assure uniformity of cutting action, abrasive grains are sorted into sizes by mechanical sieving machines. The number of openings per linear inch in a sieve (or screen) through which most of the particles of a particular size can pass determines the grain size.

A 24 grit would pass through a standard screen having 24 openings per inch but would not pass through one having 30 openings per inch. These numbers have since been specified in terms of millimeters and micrometers.[1] Commercial practice commonly designates grain sizes from 4 to 24 inclusive, as *coarse;* 30 to 60, inclusive, as *medium;* and 70 to 600, inclusive, as *fine*. Silicon carbide is obtainable in grit sizes ranging from 2 to 240 and aluminum oxide in sizes from 4 to 240. Sizes from 240 to 600 are designated as *flour* sizes. These are used primarily for lapping, or in fine honing stones (see Figure 23-2).

The grain diameter can be estimated from the screen number (*S*), which

[1]See ANSI B74.12 for details.

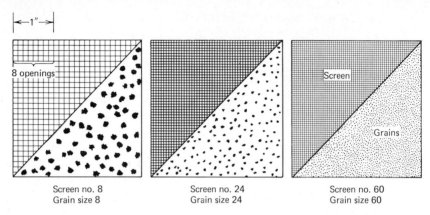

Screen no. 8
Grain size 8

Screen no. 24
Grain size 24

Screen no. 60
Grain size 60

FIGURE 23-2 Typical screens for sifting abrasives into sizes. Screens shown represent 2-inch squares. (*Courtesy Carborundum Company.*)

corresponds to the number of openings per inch. The mean diameter of the grain (*g*) is related to the screen number by $g \simeq 0.7/S$.

Regardless of the size of the grain, only a small percentage (2 to 5%) of the surface of the grain is operative at any one time. That is, the depth of cut for an individual grain (the actual feed per grit) with respect to the grain diameter

FIGURE 23-3 SEM micrograph of a ground steel surface showing a plowed track (T) in the middle and a machined track (M) above. The grit fractured, leaving a portion of the grit in the surface (X), a prow formation (P), and a groove (G) where the fractured portion was pushed further across the surface. The area marked (O) is an oil deposit.

is very small; thus the chips are small. As the grain diameter decreases, the number of active grains per unit area increases. The cuts become finer because grain size is the controlling factor for surface finish (roughness). Of course, the MRR also decreases.

The grain shape is also important, as it determines the tool geometry—that is, the back rake angle and the clearance angle at the cutting edge of the grit. Obviously, there is no specific cutting angle but rather a distribution of angles. Thus a grinding wheel can present to the surface rake angles ranging from +45° to negative 60° or greater. Grits with large negative rake angles or rounded cutting edges do not form chips but rather *plow* a groove in the surface (see Figure 23-3) or just *rub* it. Thus, abrasive machining is a mixture of *cutting, plowing,* and *rubbing* with the percentage of each being highly dependent on the geometry of the grit. As the grits are continuously abraded, fractured, or dislodged from the bond, new grits are exposed and the mixture of cutting, plowing, and rubbing is continuously changing.

GRINDING

Grinding, wherein the abrasives are bonded together into a wheel of some shape, is the most common abrasive machining process. The operation of grinding wheels is greatly affected by the bonding material and the spatial arrangement of the abrasive particles, known as the structure.

Grinding wheel structure.   The spacing of the abrasive particles with respect to each other is called *structure*. Close-packed grains have dense structure, while open structure means widely spaced grains. Open-structure wheels have greater chip clearance but obviously fewer cutting edges per unit area (see Figure 23-4).

In grinding, the chips are small but are formed by the same basic mechanism of compression and shear. See Figure 23-5 for a high-magnification micrograph of some steel chips from a grinding process. The chips often have sufficient heat energy to burn or melt in the atmosphere. Burning chips are the sparks observed during grinding with no cutting fluid. The feeds and depths of cut in

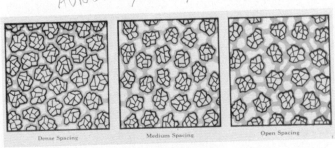

FIGURE 23-4 Meaning of grid wheel "structure." (*Courtesy Carborundum Company.*)

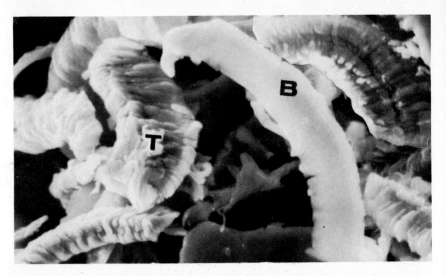

FIGURE 23-5 SEM micrograph of stainless steel chips from a grinding process. The tops (T) of the chips have the typical shear-front-lamella structure while the bottoms (B) of the chips are smooth; 4800×.

grinding are small while the cutting speeds are high, resulting in high specific horsepower numbers. Cutting is obviously more efficient than plowing or rubbing, so grain fracture or grain pullout are natural phenomena used to keep the grains sharp. As the grains become dull, cutting forces increase and there is an increased tendency for the grains to be pulled free from the bonding material. This action can be controlled by varying the strength of the bond, known as the *grade*. Grade thus is a measure of how strongly the grains are held in the wheel. It really is dependent on two factors: the strength of the bonding material, and the amount of the bonding agent connecting the grains. This latter factor is illustrated in Figure 23-6. Because abrasive wheels are usually porous, the grains are connected and held together with ''posts'' of bonding material. If these posts are large in cross section, the force required to break a grain free from the wheel is greater than when the posts are small. If a high dislodging force is required, the bond is said to be *hard*. If only a small force is required, the bond is said to be *soft*. Wheels commonly are referred to as hard or soft, referring to the net

Weak ''Posts''   Medium Strength ''Posts''   Strong ''Posts''

FIGURE 23-6 The grade of a grinding wheel depends on the amount of bonding agent on the holding of abrasive grains in the wheel. (*Courtesy Carborundum Company.*)

strength of the bond resulting from both the strength of the bonding material and its disposition between the grains.

The loss of grains from the wheel means that the wheel is changing size. The grinding ratio or *G* ratio is defined as the cubic inches of stock removed and divided by the cubic inches of wheel lost (see Figure 23-7). In most grinding work the grinding ratio is in the range 20:1 to 80:1. The *G* ratio is a measure of grinding production and reflects the amount of work a wheel can do during its useful life.

A typical vitrified grinding wheel will consist of (volume percentages) 50% abrasive particles, 10% bond, and 40% voids or pores. The manner in which the wheel performs is influenced by:

1. The mean force required to dislodge a grain from the surface (i.e., the grade of the wheel).
2. The void size and distribution (i.e., the structure).
3. The mean spacing of active grains in the wheel surface (i.e., grain size).
4. The properties of the grain (i.e., hardness and friability).
5. The geometry of the cutting edges of the grains (i.e., rake angles and cutting edge radius compared to depth of cut).
6. The process parameters (speeds, feeds, cutting fluids) and type of grinding (surface, cylindrical).

As the wheel is used, there is a tendency for the wheel to become *loaded* (metal chips become lodged in the voids between the grains) and/or glazed (grains abrade, flatten, and polish). Unless the wheel is cleaned and sharpened (i.e., dressed), the wheel will not cut as well and will tend to plow and rub more. Ideally, the dulled grains will cause the cutting forces on the grains to increase, resulting in the grains fracturing or being pulled out of the bond, thus

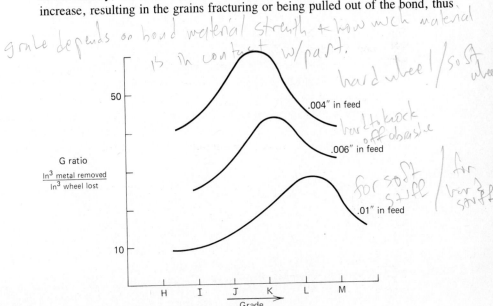

FIGURE 23-7 The *G* ratio varies with wheel grade and feed rate. All other factors held constant.

providing a continuous exposure of sharp cutting edges. Such a continuous action ordinarily will not occur when light feeds and depths of cut are used, as commonly must be employed when grinding to obtain accurate dimensions or fine finishes. However, if considerably heavier cuts are taken, grinding wheels do become self-sharpening and large amounts of metal can be removed very efficiently. The net effects are shown in Figure 23-8. As the downfeed is increased, the required power increases rapidly up to the point at which the wheel becomes completely self sharpening. After this, the required power decreases sharply and later increases more slowly than during the stages of heavier feed. The metal-removal rate increases very markedly, accompanied by increased wheel wear. By operating with the conditions represented by the shaded area, much faster metal removal can be obtained with only slightly greater wheel wear and at much lower cost.

The condition where very rapid metal removal can be achieved by grinding is the one to which some have applied the term *abrasive machining*. The metal-removal rates are comparable to, or exceed, those obtainable by conventional cutter machining, and the size tolerances are comparable. It obviously is just a special type of grinding, using abrasive grains as cutting tools, as do all other types of abrasive machining. Abrasive grinding will produce sufficient localized plastic deformation and heat in the surface so as to develop tensile residual stresses (see Figure 23-9), layers of overtempered martensite (in steels) and even microcracks, as this process is quite abusive. Even conventional grinding should be replaced by procedures which develop lower surface stresses in those applications where service failures due to fatigue or stress corrosion are possible. This is accomplished by employing softer grades of grinding wheels, reducing the grinding speeds and infeed rates, and using chemically active cutting fluids, like highly sulfurized oil or KNO_2 in water. These procedures may require the addition of a variable-speed drive to the grinding machine. Generally, only about 0.005 to 0.010 inch of surface stock needs to be finish-ground in this way, as

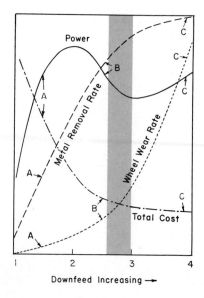

FIGURE 23-8 Relationship between power required, metal-removal rate, wheel wear, and total cost as functions of downfeed. (*Courtesy Norton Company.*)

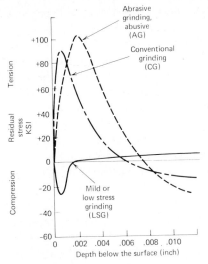

Grinding conditions			
	Abusive AG	Conventional CG	Low stress LSG
Wheel	A46MV	A46KV	A46HV or A60IV
Wheel speed ft/min.	6,000 to 18,000	4,500 to 6,500	2500-3000
Down feed in/pass	.002 to .004	.001 to .003	.0002 to .005
Cross feed in/pass	.040 - .060	.040 - .060	.040-.060
Table speed ft/min	40 to 100	40 to 100	40 to 100
Fluid	Dry	Sol oil (1:20)	Sulfurized oil

FIGURE 23-9 Typical residual stress distributions produced by surface grinding with different conditions. Material 4340 steel. (*From M. Field and W. P. Koster, "Surface Integrity in Grinding," in* New Developments in Grinding, *Carnegie Mellon University Press, Pittsburgh, Pa., 1972, p. 666.*)

the depth of the surface damage due to conventional grinding or abusive grinding is 0.005 to 0.007 inch. High-strength steels, high-temperature nickel, and cobalt-base alloys and titanium alloys are particularly sensitive to surface deformation and cracking problems from grinding. Other post-processing processes, like polishing, honing, and chemical milling plus peening, can be used to remove the deformed layers in critically stressed parts. It is strongly recommended, however, that testing programs be used along with service experience on critical parts before these procedures are employed in production. See discussion of residual stresses in Chapter 39.

Bonding materials for abrasive wheels. Bonding material is a very important factor to be considered in selecting a grinding wheel. It determines the strength of the wheel, thus establishing the maximum speed at which it can safely be operated. It determines the elastic behavior or deflection of the grits in the wheel during grinding. The wheel can be hard or rigid or it can be flexible. Finally, the bond determines the force that is required to dislodge an abrasive particle from the wheel and thus plays a major role in the cutting action. Five types of bonding materials are in common use.

Vitrified bonds are used most extensively. They are composed of clays and other ceramic substances. The abrasive particles are mixed with the wet clays so that each grain is coated. Wheels are formed from the mix, usually by pressing, and then dried. They then are fired in a kiln, which results in the bonding material becoming hard and strong, having properties similar to glass. Vitrified wheels are porous, strong, rigid, and unaffected by oils, water or temperature over the ranges usually encountered. The operating speed range in most cases is from 1676 to 1981 meters (5500 to 6500 feet) per minute, but some wheels now operate at surface speeds up to 4877 meters (16,000 feet) per minute.

Silicate wheels use silicate of soda (waterglass) as the bond, the wheels formed from the mixture being baked at about 260°C (500°F) for a day or more. They are more brittle but not as strong as vitrified wheels, so the abrasive grains are released more readily. Consequently, they operate somewhat cooler than vitrified wheels and are useful in grinding tools where the temperature must be kept to a minimum.

Shellac-bonded wheels are made by mixing the abrasive grains with shellac in a heated mixer, pressing or rolling into the desired shapes, and baking for several hours at about 150°C (300°F). This type of bond is used primarily for strong, thin wheels having some elasticity. They tend to produce a high polish and thus are used in grinding such parts as camshafts and mill rolls.

Rubber bonding is used to produce wheels that can operate at high speeds but must have a considerable degree of flexibility so as to resist side thrust. Rubber, sulfur, and other vulcanizing agents are mixed with the abrasive grains. The mixture then is rolled out into sheets of the desired thickness, and the wheels are cut from these sheets and vulcanized. Rubber-bonded wheels can be operated at speeds up to 4877 meters (16,000 feet) per minute. They commonly are used for snagging work in foundries and for thin cutoff wheels.

Resinoid, or plastic, bonding now is widely used. Because plastics can be compounded to have a wide range of properties, such wheels can be obtained to cover a variety of work conditions. They have, to a considerable extent, replaced shellac and rubber wheels.

Some type of reinforcing frequently is added in rubber- or resinoid-bonded wheels that are to have some degree of flexibility, or are to be used in aplications where they are apt to receive considerable abuse and side loading. Various natural and synthetic fabrics and fibers, glass fibers, and nonferrous wire mesh are used for this purpose.

Dressing grinding wheels. In addition to wearing down and losing shape during grinding, grinding wheels also lose their effectiveness due to *loading* and *glazing*.

Two procedures are employed to expose fresh, sharp cutting edges. One is to *dress* the wheel by using a dressing tool. A steel dressing tool, such as is shown in Figure 23-10 consists of irregularly shaped, hardened steel disks that are free to rotate on an axle. The tool is held at an angle against the rotating grinding wheel and moved across its face. The revolving points of the disks remove the foreign particles that have become lodged between the grains and also remove or fracture dulled grains, thus exposing sharp edges.

Grinding wheels lose their geometry during usage. *Truing* restores the original shape. A single-point diamond dressing tool can be used to *true* the wheel by fracturing abrasive grains to expose new grains and new cutting edges on worn, glazed grains (see Figure 23-10).

Crush dressing consists of forcing a hard steel roll, having the same contour as the part to be ground, against the grinding wheel while it is revolving— usually quite slowly. The crushing action fractures and dislodges some of the abrasive grains, exposing fresh, sharp edges. This procedure usually is employed

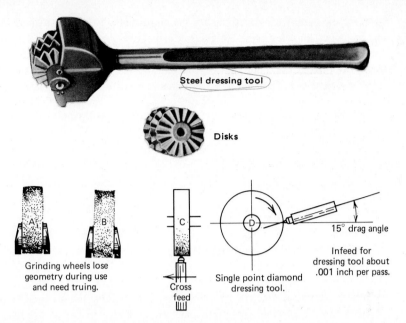

Steel dressing tool

Disks

A B

Grinding wheels lose
geometry during use
and need truing.

Cross
feed

Single point diamond
dressing tool.

15° drag angle

Infeed for
dressing tool about
.001 inch per pass.

FIGURE 23-10 Tools for dressing and truing grinding wheels. Shown below is a single point diamond tool. (*Courtesy Norton Company.*)

to produce and maintain a special contour to the abrasive wheel in form grinding. Crush dressing is a very rapid method of dressing grinding wheels and, because it fractures more of the abrasive grains, results in free cutting and somewhat cooler grinding. The resulting surfaces may be slightly rougher than when diamond dressing is used.

Some machines are equipped so that the wheel can be dressed continuously or intermittently while grinding continues.

Obviously, the self-sharpening tendency of a grinding wheel, determined by the strength or hardness of the bond, is important in grinding-wheel action. A soft-grade wheel loses its grains readily and thus tends to remain sharp at all times. However, it wears rapidly. Conversely, a hard wheel wears less rapidly but is likely to become dull, particularly if used on hard materials. In grinding hard materials, the abrasive grains dull more rapidly than in grinding soft materials. Therefore, a wheel for grinding hard materials should lose its grains rapidly. Thus softer-grade wheels are used for grinding hard materials and harder-grade wheels for soft materials. However, several other factors also affect the apparent hardness or softness of a wheel.

If the speed of the work is held constant, increasing the speed of a grinding wheel makes it act harder; the higher speed tends to dull the grains more rapidly. However, because the path of the grain through the work is decreased, the dislodging force is less. Similarly, increasing the work speed, while maintaining a constant wheel speed, results in a softer action. A larger wheel will act harder than a smaller wheel of the same type and bond because of the larger contact area between the wheel and the work.

The spacing between the abrasive grains also is important in the cutting action.

This spacing must provide room for the chips that are formed. If there is insufficient space, the chips will become wedged between the grains, protrude beyond the periphery of the wheel, and scratch the work surface. Thus, if heavy cuts are to be taken, or if the path of contact is long, greater spacing between the grains is necessary.

Grinding-wheel identification.   Most grinding wheels are identified by a standard marking system that has been established by the American National Standards Institute, Inc. This system is illustrated and explained in Figure 23-11. The first and last symbols in the marking are left to the discretion of the manufacturer.

Grinding-wheel selection.   If optimum results are to be obtained, it is very important that the proper grinding wheel be selected for a given job. Several factors must be considered. Probably the first is the shape and size of the wheel. Obviously, the shape must permit proper contact between the wheel and all of the surface that must be ground. Grinding-wheel shapes have been standardized by the Grinding Wheel Manufacturers' Association. Eight of the most commonly used types are shown in Figure 23-12. Types 1, 2, and 5 are used primarily for grinding external or internal cylindrical surfaces and for plain surface grinding. Type 2 can be mounted for grinding either on the periphery or the side of the wheel. Type 4 is used with tapered safety flanges so that if the wheel is broken in doing rough grinding, such as snagging, these flanges will prevent the pieces of the wheel from flying and causing damage. Type 6, the straight cup, is used primarily for surface grinding but can also be used for certain types of offhand grinding. The flaring-cup type of wheel is used for tool grinding. Dish-type wheels are used for grinding tools and saws.

Straight grinding wheels can be obtained with a variety of standard faces. Some of these are shown in Figure 23-13.

The size of the wheel to be used is determined, primarily, by the spindle speeds available on the grinding machine and the proper cutting speed for the

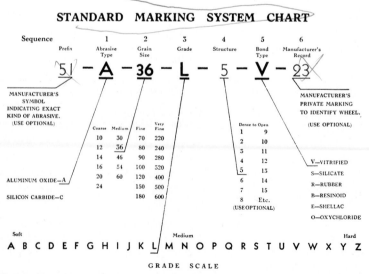

FIGURE 23-11 Standard marking system for grinding wheels.

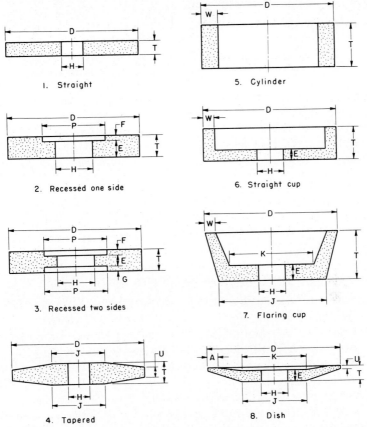

FIGURE 23-12 Common standard grinding-wheel shapes. (*Courtesy Carborundum Company.*)

wheel structure = spacing (tighter space = hard to dog)

wheel, as dictated by the type of bond. For most grinding operations the cutting speed is about 1981 meters (6500 feet) per minute, but different types and grades of bond often justify considerable deviations from this average speed as follows:

Type of Bond	Surface Speed	
	Meters per minute	Feet per minute
Vitrified or silicate		
Soft	1372–1676	4500–5500
Medium	1524–1829	5000–6000
Hard	1676–1981	5500–6500
Organic		
Soft	1829–1981	6000–6500
Medium	2134–2438	7000–8000
Hard	2286–3048	7500–10,000

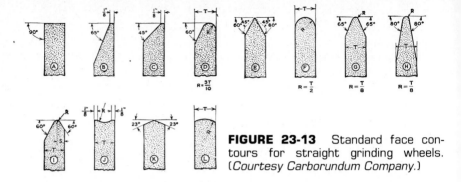

FIGURE 23-13 Standard face contours for straight grinding wheels. (*Courtesy Carborundum Company.*)

These speeds represent the common ranges. For certain types of work using special wheels and machines, as in thread grinding and "abrasive machining," much higher speeds are used.

The operation for which the abrasive wheel is intended will also influence the wheel shape and size. The major use categories are:

1. *Cutting off:* for slicing and slotting parts; use thin wheel, organic bond.
2. *Cylindrical between centers:* grinding OD's between centers.
3. *Cylindrical, centerless:* grinding OD's with work rotated by regulating wheel.
4. *Internal cylindrical:* grinding bores and large holes.
5. *Snagging:* removing large amounts of metal without regard to surface finish or tolerances.
6. *Surface grinding:* where workpiece is flat.
7. *Tool grinding:* for grinding cutting edges on tools like drills, milling cutters, taps, reamers, and single-point, high-speed steel tools.
8. *Off-hand grinding:* where work or the grinding tool is hand-held.

In many cases, the classification of processes coincides with the classification of machines which do the process.

Other factors which will influence the choice of wheel to be selected will include the workpiece material, the amount of stock to be removed, the shape of the workpiece, and the accuracy and surface finish desired.

Workpiece material has the greatest impact on choice of the wheel. Hard, high-strength metals (tool steels, alloy steels) are generally ground with aluminum oxide wheels, although cubic boron nitride wheels are gaining greater acceptance in these materials. Silicon carbide is employed in grinding brittle materials (cast iron and ceramics) as well as softer, low-strength metals like aluminum, brass, copper, and bronze. Diamonds have taken over cutting of tungsten carbides and are used for machining die steel, stainless, and other very hard materials. The truth is that there are so many factors which affect the cutting action, there are no hard and fast rules with regard to abrasive selection.

Selection of grain size is determined by whether course or fine cutting and finish are desired. Course grains take larger bites and cut more rapidly; hard wheels with fine grains leave smaller scratches and thus usually are selected for finishing cuts. Because of their cleaner cutting action, large grains may be used for finishing work if there is a tendency for the work material to load the wheel.

The spacing of the grains (structure) is determined by the problem of chip clearance. With closer spacing, more cutting edges are in contact with the work and a finer finish usually can be obtained. However, when making heavy cuts and in grinding soft materials, too-close grain spacing will not provide adequate clearance for the chips. Also, where the arc of contact is large, greater grain spacing must be used.

As mentioned previously, the grade of the bond should be such as to assure sharp abrasive particles being available for cutting at all times. In general, harder materials call for softer bonds, and vice versa.

Balancing grinding wheels. Because of the high rotative speeds involved, grinding wheels must never be used unless they are in good balance. Not only will slight unbalance produce vibrations that will cause waviness in the work surface, but it may cause a wheel to break, with the probability of serious damage and injury. The wheel should be mounted with proper bushings so that it fits snugly on the spindle of the machine. Rings of blotting paper should be placed between the wheel and the flanges to assure that the clamping pressure is evenly distributed. Most grinding wheels will run in good balance if they are mounted properly and trued. Most machines have provision for compensating for a small amount of wheel unbalance by attaching weights to one mounting flange. Some have provision for semiautomatic balancing with weights that are permanently attached to the machine spindle.

GRIND ON PERIPHERY IF ITS THAT TYPE

Safety in grinding. Because the rotational speeds are quite high, and the strength of grinding wheels usually is much less than the materials on which they may be used or against which they may accidentally be struck, serious accidents occur much too frequently in connection with the use of grinding wheels. Virtually all such accidents could be avoided and are due to one or more of four causes. First, grinding wheels occasionally are operated at unsafe and improper, speeds. All grinding wheels are clearly marked with the maximum rpm at which they should be rotated. They all are tested to considerably above the designated rpm and are safe at the specified speed *unless abused. They should never, under any condition, be operated above the rated speed.* Second, a most common form of abuse, frequently accidental, is dropping the wheel or striking it against a hard object. This can result in a crack, which may not be readily visible, and subsequent failure of the wheel while rotating at high speed. If a wheel is dropped or struck against a hard object, it should be discarded and never used unless tested at above the rated speed in a properly designed test stand. A third common cause of grinding wheel failure is improper use, such as grinding against the side of a wheel that was designed for grinding only on its periphery. The fourth and most common cause of injury from grinding is the absence of a proper safety guard over the wheel and/or over the eyes or face of the operator. The frequency with which people will remove safety guards from grinding equipment, or fail to use safety goggles or face shields, is amazing and inexcusable.

The use of fluids in grinding. Because grinding involves cutting, the selection and use of a cutting fluid is governed by the basic principles discussed in Chapter

18. If a fluid is used, it should be applied in sufficient quantities and in a manner that will assure that the chips are washed away and not trapped between the wheel and the work. This is of particular importance in grinding horizontal surfaces. The use of a fluid can help to prevent fine microcracks that may result from highly localized heating when hardened steels are ground.

Much grinding is done dry—for example, most snagging and off-hand grinding. On some types of material, dry grinding produces a better finish than can be obtained by wet grinding.

Grinding machines. Grinding machines commonly are classified according to the type of surface they produce. The following is such a classification, with further subdivision to indicate characteristic features of different types of machines within each classification:

Type of Machine	Type of Surface	Characteristic Features
External cylindrical	External surface on rotating, usually cylindrical part	Work rotated between centers Centerless Chucking Tool post Crankshaft, cam, etc.
Internal cylindrical	Internal diameters of holes	Chucking Planetary (work stationary) Centerless
Surface conventional	Flat surfaces	Reciprocating table or rotating table Horizontal or vertical spindle
Creep feed	Deep slots, profiles	Reciprocating table Horizontal spindle
Tool grinders	Tool angles and geometries	Universal Special
Other	Special or any of the above	Disk, contour, thread, flexible shaft, swing frame, snag, pedestal, bench

Basically, grinding on all machines is done in two ways. In the first the depth of cut is obtained by *infeed*—moving the wheel into the work, or the work into the wheel. The desired surface is then produced by traversing the wheel across the workpiece, or vice versa. In the second method, known as *plunge-cut* grinding, the basic movement is of the wheel being fed radially into the work while the latter revolves on centers. It is similar to form cutting on a lathe; usually a formed grinding wheel is used.

Grinding machines that are used for precision work have certain important characteristics which permit them to produce parts having close dimensional tolerances. They are constructed very accurately, with heavy, rigid frames to assure permanency of alignment. Rotating parts are accurately balanced to avoid vibration. Spindles are mounted in very accurate bearings, usually of the preloaded ball-bearing type. Controls are provided so that all movements that determine dimensions of the workpiece can be made with accuracy—usually to 0.0025 mm (0.0001 inch).

Another requisite is provision to prevent abrasive dust, resulting from grinding, from entering between moving parts. All ways and bearings must be fully covered or protected by seals. If this is not done, the abrasive dust between moving parts becomes embedded in the softer of the two, causing it to become a "charged" lap. Subsequent movement causes the harder of the two parts to be worn, resulting in permanent loss of accuracy of the machine. Machines that do not have this protection should be located apart from other machine tools, preferable in a separate room with dust control.

These special characteristics add considerably to the cost of these machines. Also, they often must be operated by skilled personnel. Such machines can be operated economically only if they have relatively high use factors. However, production-type grinders are designed to be used by less-skilled labor. Their metal-removal rates are high, and excellent dimensional accuracy and fine surface finish can be obtained very economically on them.

Cylindrical grinding. *Center-type grinding* is commonly used for producing external cylindrical surfaces. Figure 23-14 shows the basic principles and motions of this process. The grinding wheel revolves at an ordinary cutting speed, and the workpiece rotates on centers at a much slower speed, usually from 23 to 38 meters (75 to 125 feet) per minute. The grinding wheel and the workpiece move in opposite directions at their point of contact. The depth of cut is determined by infeed of the wheel or workpiece. Because this motion also determines the finished diameter of the workpiece, accurate control of this movement is required. Provision is made to traverse the workpiece with respect to the wheel, usually by reciprocating the work. However, in very large grinders the wheel is reciprocated due to the massiveness of the work.

A *plain center-type cylindrical grinder* is shown in Figure 23-14. On this type the work is mounted between headstock and tailstock centers. The tailstock center always is dead, and provision usually is made so that the headstock center can be operated either dead or alive. High-precision work usually is ground with a dead headstock center, because this eliminates any possibility of the workpiece running out of round due to any eccentricity in the headstock.

The headstock spindle, mounted in accurate bearings, usually is driven through a belt drive. The workpiece is driven in rotation by means of a face plate and dog.

On this type of machine, the headstock and tailstock are mounted on a table, which can be swiveled approximately 10° about a vertical axis with respect to the table carrier on which it is mounted. This permits either straight cylinders or tapered cylinders, up to about 10° of taper, to be ground.

The table assembly can be reciprocated along the ways on the main frame, either manually or by power. In most cases hydraulic drive is used. The speed of the table movement can be varied, and the length of the movement can be controlled by means of adjustable tripping dogs.

Infeed is provided by movement of the wheelhead at right angles to the longitudinal axis of the table. The spindle is driven by an electric motor that also is mounted on the wheelhead. A flat-belt drive usually is employed to

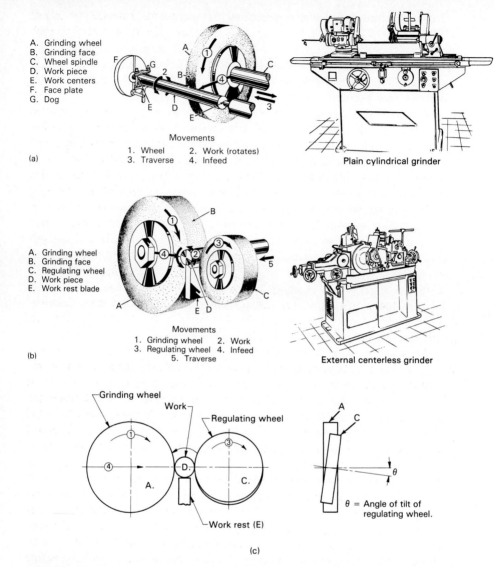

A. Grinding wheel
B. Grinding face
C. Wheel spindle
D. Work piece
E. Work centers
F. Face plate
G. Dog

(a)

Movements
1. Wheel
2. Work (rotates)
3. Traverse
4. Infeed

Plain cylindrical grinder

A. Grinding wheel
B. Grinding face
C. Regulating wheel
D. Work piece
E. Work rest blade

(b)

Movements
1. Grinding wheel
2. Work
3. Regulating wheel
4. Infeed
5. Traverse

External centerless grinder

Grinding wheel
Work
Regulating wheel
A.
D.
C.
Work rest (E)

A
C
θ

θ = Angle of tilt of regulating wheel.

(c)

FIGURE 23-14 External grinding of cylindrical parts. (a) Cylindrical grinding between centers. (b) Centerless grinding. (c) Relationship between the grinding wheel and the workpiece in centerless method. (*Courtesy The Carborundum Company with machine schematics from* Manufacturing Producibility Handbook; *courtesy General Electric Company.*)

connect the driving motor and the spindle because this type of drive minimizes vibrations that might produce chatter marks on the work surface.

The infeed movement usually is controlled manually by some type of vernier drive to provide control to 0.025 mm (0.001 inch) or less. Some machines are equipped with digital readout equipment to show the exact size being produced. Some production-type grinders have devices that automatically infeed the wheel and then retract it when the desired size has been obtained. Such machines usually are equipped with an automatic diamond wheel-truing device that dresses the wheel and resets the measuring element before grinding is started on each piece.

The longitudinal traverse should be about one-fourth to three-fourths of the wheel width for each revolution of the work. For light machines and fine finishes, it should be held to the smaller end of this range.

The depth of cut (infeed) varies with the purpose of the grinding operation and the finish desired. When grinding is done to obtain accurate size, feeds of 0.051 to 0.10 mm (0.002 to 0.004 inch) commonly are used for roughing cuts. For finishing, the feed is reduced to 0.006 to 0.013 mm (0.00025 to 0.0005 inch). The design allowance for grinding should be from 0.13 to 0.25 mm (0.005 to 0.010 inch) on short parts and on parts that are not to be hardened. On long or large parts and on work that is to be hardened, a grinding allowance of from 0.38 to 0.76 mm (0.015 to 0.030 inch) is desirable. When grinding is used primarily for metal removal (so-called abrasive machining), feeds are much higher, 0.51 to 1.02 mm (0.020 to 0.040 inch) being common. Continuous downfeed often is used, rates up to 2.54 mm (0.100 inch) per minute being common.

Plain center-type cylindrical grinders contain systems for storing, filtering, and circulating adequate amounts of grinding fluid. Heavy wheel guards and adequate protection for the ways always are included.

The size of center-type grinders is designated by the *maximum diameter* and *length of work* that can be ground between centers. Plain grinders are obtainable in sizes ranging from 76.2 × 305 mm (3 × 12 inches) up to 610 × 4572 mm (24 × 180 inches) and larger. *Universal center-type grinders* are basically the same as plain center-type grinders, except for two features. First, both the headstock and the wheelhead can be swiveled about vertical axes. This permits the grinding of tapers of all angles and certain other types of work that cannot be done on plain cylindrical grinders. The second feature is that most machines have dual spindles on the swiveling wheelhead, one for external grinding and the other for internal grinding. Either spindle can be brought into use by swiveling or tilting the wheelhead. *Roll grinders* basically are plain center-type machines designed for grinding large, cylindrical mill and calender rolls, which may be up to 1524 mm (60 inches) in diameter. Because of the weight of such workpieces, the wheelhead, instead of the work, reciprocates. Such a machine is shown in Figure 23-15. Often there is provision for grinding a crown, or camber, on the roll to compensate for the deflection that occurs when they are subjected to heavy loads.

Chucking-type grinders. Grinding machines are available in which the workpiece is held in a chuck for grinding both external and internal cylindrical surfaces. *Chucking-type external grinders* are production-type machines, for use in rapid grinding of relatively short parts, such as ball-bearing races (see Figure 23-16a). Both chucks and collets are used for holding the work, dictated by the shape of the workpiece and rapid loading and removal. Frequently, such machines have two head spindles so that work can be removed from one while another piece is being ground in the other.

Chucking-type internal grinding machines employ two basic principles, illus-

FIGURE 23-15 Traveling-head roll grinder (*Courtesy Cincinnati Milacron, Inc.*)

trated in Figure 23-16b. Most commonly, the chuck-held workpiece revolves, and a relatively small, high-speed grinding wheel is rotated on a spindle arranged so that it can be reciprocated in and out of the workpiece. Infeed movement of the wheelhead is normal to the axis of rotation of the work. A machine of this type is shown in Figure 23-17.

On *plain internal grinders* of this type, the workhead can be swiveled so that both straight internal cylinders and beveled holes can be ground. On *universal internal grinders* the workhead not only can be swiveled, but it also is mounted on a cross slide.

On some production-type machines the control of the infeed and traverse of the wheelhead is automatic. The control can be set for the desired size, and the grinding wheel is withdrawn and the machine is shut off automatically when the desired size is achieved. On one type of machine the hole size is automatically gaged after each reciprocation.

Planetary-type internal grinders, illustrated schematically in Figure 23-16b, are used for work that is too large to be rotated conveniently. The revolving grinding wheel also has planetary rotation about an axis that is coincident with the axis of the finished cylinder. The diameter of the ground surface is controlled by adjusting the radius of the planetary rotation. On this type of machine, the work is reciprocated past the wheel.

Tool-post grinders. *Tool-post grinders,* illustrated in Figure 23-18 are used to permit occasional grinding to be done on a lathe. The wheelhead is either a high-speed electric or air motor with the grinding wheel often mounted directly on the motor shaft. The entire mechanism is mounted either on the tool post or on the compound rest. The lathe spindle provides rotation for the workpiece, and the lathe carriage is used to reciprocate the wheelhead.

Although tool-post grinders are versatile and useful, care should be taken to cover the ways of the lathe with a closely woven cloth to provide protection from the abrasive dust that can become entrapped between the moving parts.

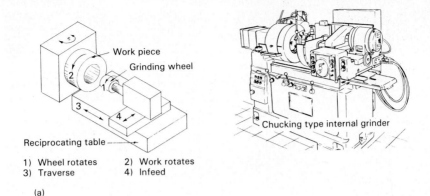

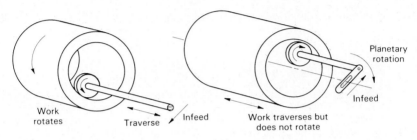

Chucking type internal grinder

1) Wheel rotates   2) Work rotates
3) Traverse   4) Infeed

(a)

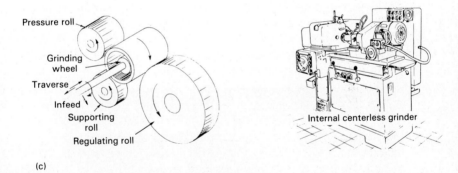

(b) Two basic types of chucking-type internal grinders.

Internal centerless grinder

(c)

FIGURE 23-16 Internal grinding of holes and bores. (a) Internal grinding where work is held in a chuck; in some machines, the work traverses but does not rotate and the grinding wheel is given a planetary motion. (b) Chucking-type internal grinders. (c) Internal centerless grinding. (*Machine schematics from* Manufacturing Producibility Handbook; *courtesy General Electric Company.*)

Centerless grinding. *Centerless grinding* makes it possible to grind both external and internal cylindrical surfaces without the necessity of the workpiece being mounted between centers or in a chuck. This eliminates the requirement of center holes in some workpieces and the necessity for mounting the workpiece, thereby reducing the cycle time.

The principle of *centerless external grinding* is illustrated in Figure 23-14.

FIGURE 23-17 Size-matic chucking-type internal grinder. The machine is set up to grind two inside diameters on a gear. Two truing diamonds are mounted between the work and the grinding wheels. (*Courtesy Herald Machine Company.*)

Two abrasive wheels are used. The larger one operates at regular grinding speeds and does the actual grinding. The smaller wheel is the *regulating* wheel and is mounted at an angle to the plane of the grinding wheel, as shown in Figure 23-14c. Revolving at a much slower surface speed—usually 15 to 61 meters (50 to 200 feet) per minute—it controls the rotation and longitudinal motion of the

FIGURE 23-18 Grinding a bearing seat by means of a tool-post grinder on a lathe. (*Courtesy Dunore Company.*)

MACHINING PROCESSES

workpiece. It usually is a plastic- or rubber-bonded wheel with a fairly wide face.

The workpiece is held against the work-rest blade by the cutting forces exerted by the grinding wheel and rotates at approximately the same surface speed as that of the regulating wheel. This axial feed is calculated approximately by the equation

$$F = dN \sin \theta$$

where F = feed, millimeters or inches per minute
d = diameter of the regulating wheel, millimeters or inches
N = revolutions per minute of the regulating wheel
θ = angle of inclination of the regulating wheel

Four types of external centerless grinding operations can be done. These are illustrated in Figure 23-19. In *thrufeed grinding,* the workpiece is of constant diameter and is fed completely through between the rolls, starting at one end. This is the simplest type and can easily be made automatic.

In *infeed centerless grinding,* the work rest and the regulating wheel are retracted so that the work can be put in position and removed when grinding is completed. When the work is in position, the regulating wheel is fed inward until the desired diameter is obtained. This arrangement permits multiple diameters and curved parts to be ground. When multiple diameters are to be ground, a slight inclination of the regulating wheel holds the work against the end stop. For curved parts, a formed grinding wheel is used in conjunction with a plain regulating wheel to rotate the work about a single axis. For grinding balls, the regulating wheel is grooved and inclined to impart random rotation.

In *endfeed centerless grinding,* both the grinding and regulating wheels are tapered and thus produce tapered workpieces. The stock is fed from one side until it reaches the stop. This arrangement results in exact tapers and size.

Centerless grinding has several important advantages:

1. It is very rapid; infeed centerless grinding is almost continuous.
2. Very little skill is required of the operator.
3. It often can be made automatic.
4. Where the cutting occurs, the work is fully supported by the work rest and the regulating wheel. This permits heavy cuts to be made.
5. Because there is no distortion of the workpiece, accurate size control is easily achieved.
6. Large grinding wheels can be used, thereby minimizing wheel wear.

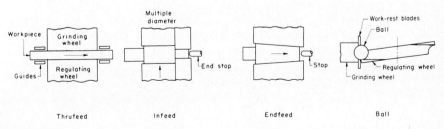

FIGURE 23-19 Four types of external centerless grinding.

Thus centerless grinding is ideally suited to certain types of mass-production operations.

The major disadvantages are as follows:

1. Special machines are required that can do no other type of work.
2. The work must be round—no flats, such as keyways, can be present.
3. Its use on work having more than one diameter is limited.
4. In grinding tubes, there is no guarantee that the OD and ID are concentric.

Centerless internal grinding utilizes the principle illustrated in Figure 23-16c. Three rolls support the workpiece on its outer surface and impart rotation to it; one retracts to permit the work to be placed in position and removed. The grinding wheel traverses into the workpiece. This type of grinding, of course, requires that the external surface of the cylinder be finished accurately before the operation is started, but it assures that the internal and external surfaces will be concentric. The operation is easily mechanized for many applications. Usually, some type of automatic truing device is included on such machines.

Surface grinding. *Surface grinding* is used primarily to grind flat surfaces. However, formed, irregular surfaces can be produced on some types of surface grinders by using a formed wheel.

There are four basic types of surface grinding machines, differing in the movement of their tables and the orientation of the grinding wheel spindles, as follows:

1. Horizontal spindle and reciprocating table.
2. Vertical spindle and reciprocating table.
3. Horizontal spindle and rotary table.
4. Vertical spindle and rotary table.

These are illustrated in Figure 23-20.

Surface grinding machines. The most common type of surface grinding machine has a reciprocating table and horizontal spindle, such as shown in Figure 23-21. Mounted on horizontal ways, the table can be reciprocated longitudinally either by handwheel or by hydraulic power. The wheelhead is given transverse motion at the end of each table motion, again either by handwheel or by hydraulic power feed. Both the longitudinal and transverse motions can be controlled by limit dogs that are easily set. Infeed on such grinders is controlled by a handwheel that lowers the grinding wheel toward the work. Most modern horizontal-spindle surface grinders have the grinding wheel mounted directly on the motor spindle.

The size of such machines is designated by the size of the surface that can be ground. Thus a 203 × 610 mm (8 × 24 inch) surface grinder has a sufficiently large table, and corresponding table and wheel motions, to permit grinding a plane surface 203 mm (8 inches) wide and 610 mm (24 inches) long. A wide range of sizes is available.

In using such machines, the wheel should overtravel the work at both ends of the table reciprocation, so as to prevent the wheel from grinding in one spot while the table is being reversed. The transverse motion should be one-fourth

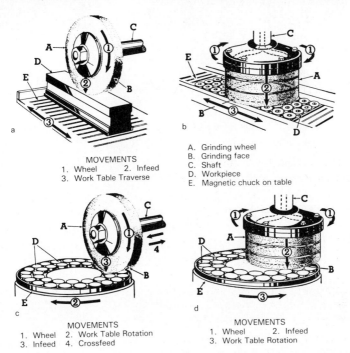

MOVEMENTS
1. Wheel   2. Infeed
3. Work Table Traverse

A. Grinding wheel
B. Grinding face
C. Shaft
D. Workpiece
E. Magnetic chuck on table

MOVEMENTS
1. Wheel   2. Work Table Rotation
3. Infeed   4. Crossfeed

MOVEMENTS
1. Wheel   2. Infeed
3. Work Table Rotation

FIGURE 23-20 Surface grinding. (a) Horizontal surface grinding with reciprocating table. (b) Vertical spindle with reciprocating table. (c) and (d) Both horizontal and vertical spindle machines can have rotary tables. (*Courtesy Carborundum Company.*)

to three-fourths of the wheel width between each stroke. *Vertical-spindle recip-rocating-table surface grinders* differ basically from those with horizontal spindles only in that their spindles are vertical and that the wheel diameter must exceed the width of the surface to be ground. Usually, no traverse motion of either the table or the wheelhead is provided. Such machines can produce very flat surfaces and are used primarily for production-type work.

Rotary-table surface grinders are of two types. Those with horizontal spindles will produce very flat surfaces but, because they are limited in the type of work they will accommodate, they are not used to a great extent. They usually are made in rather small sizes.

Vertical-spindle rotary-table surface grinders are primarily production-type machines. They frequently have two or more grinding heads, so both rough and finish grinding is accomplished in one rotation of the workpiece. The work can be held either on a magnetic chuck or in special fixtures attached to the table.

By using special rotary feeding mechanisms, machines of this type often are made automatic. Parts are dumped on the rotary feeding table and fed automatically onto work-holding devices and moved past the grinding wheels. After they pass the last grinding head, they are automatically removed from the machine.

Creep feed grinding. This is a grinding method, often done in the surface grinding mode, that is markedly different from conventional surface grinding.

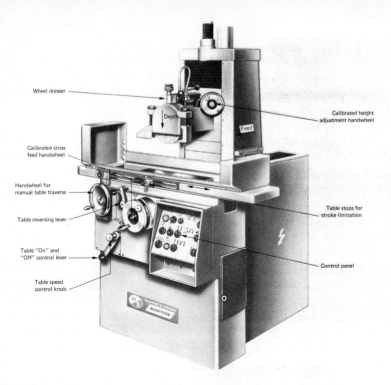

FIGURE 23-21 Reciprocating-table, horizontal-spindle surface grinder. (*Courtesy Warner & Swasey Co., Grinding Machine Division.*)

In contrast to conventional techniques, the depth of cut is increased 1000 to 10,000 times and the work feed is decreased in the same proportion; hence the name "creep feed grinding." The machine tools to perform this type of grinding must be specially designed with high static and dynamic stability, stick-slip free ways, adequate damping, increased horsepower, infinitely variable spindle speed, variable but extremely consistent table feed (especially in the low ranges), high-pressure cooling systems, integrated devices for dressing the grinding wheels, and specially designed (soft with open structure) grinding wheels. The process is mainly being applied to grinding deep slots with straight parallel sides or to grinding complex profiles in difficult-to-grind materials. The process is capable of producing extreme precision at relatively high metal removal rates. Because the process can operate at relatively low surface temperatures, the surface integrity of the metals being ground is good.

Holding work on surface grinders. Workpieces usually are held in a different manner on surface grinders than on other machine tools. To obtain high accuracy, it is desirable to reduce clamping forces and distribute them over the entire area of the workpiece. Also, grinding very frequently is done on quite thin or relatively delicate workpieces, which would be difficult to clamp by normal methods. In addition, there often is the problem of grinding a number of small, duplicate workpieces. Magnetic, electrostatic, and vacuum chucks solve all these problems very satisfactorily.

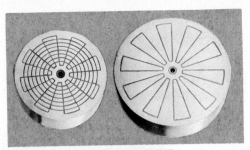

FIGURE 23-22 Two shapes of magnetic chucks. (*Courtesy O. S. Walker Company.*)

Magnetic chucks are used most frequently. Two shapes of electric-powered chucks are shown in Figure 23-22. These use dry-disk rectifiers to provide the necessary direct-current power. Another type of magnetic chuck utilizes permanent magnets. The operation of this type is shown in Figure 23-23. In the operating position the magnetic flux lines pass through the work and thus hold it to the chuck. In the "off" position the top plate short-circuits the flux lines, and no holding force is created between the work and the chuck.

When the cutting forces are not too high, magnetic chucks provide an excellent means of holding workpieces. The holding force is distributed over the entire contact surface of the work, the clamping stresses are low, and there thus is little tendency for the work to be distorted. Consequently, pieces can be held and ground accurately. Also, a number of small pieces can be mounted on a chuck and ground at the same time.

It often is necessary to demagnetize work that has been held on a magnetic chuck. Some electrically powered chucks provide satisfactory demagnetization by reversing the direct current briefly when the power is shut off.

Obviously, magnetic chucks can be used only with ferromagnetic materials. Electrostatic chucks, on the other hand, can be used with any electrically conductive material. Their principle, indicated in Figure 23-24, involves the work being held by mutually attracting electrostatic fields in the chuck and the work-

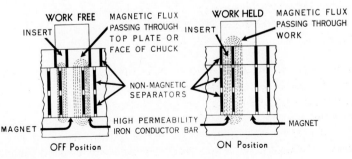

FIGURE 23-23 Principle of operation of permanent-magnet chucks. (*Courtesy Brown & Sharpe Mfg. Co.*)

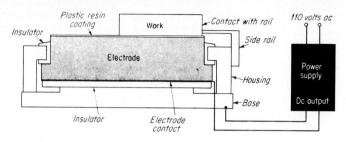

FIGURE 23-24   Principle of electrostatic chuck.

piece. These provide a holding force of up to 21 000 Pa (30 psi). Nonmetal parts usually can be held if they are flashed with a thin metal coating. These chucks have the added advantage of not inducing residual magnetism in the work.

Two types of vacuum chucks are available. In one, illustrated in Figure 23-25, the holes in the work plate are connected to a vacuum pump and can be opened or closed by means of valve screws. The valves are opened in the area on which the work is to rest. The other type has a porous plate on which the work rests. The workpiece and plate are covered with a polyethylene sheet. When the vacuum is turned on, the film forms around the workpiece, covering and sealing the holes not covered by the workpiece and thus producing a seal. The first cut removes the film covering the workpiece. Vacuum chucks have the advantage that they can be used on both nonmetals and metals.

Magnetic, electrostatic, and vacuum chucks have been so satisfactory that they also are used for some light milling and turning operations.

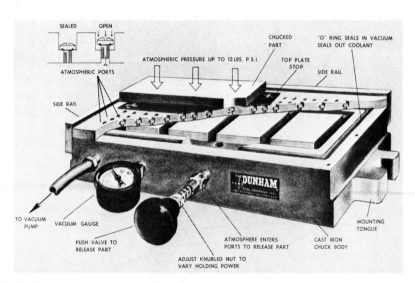

FIGURE 23-25   Cutaway view of a vacuum chuck. (*Courtesy Dunham Tool Company Inc.*)

Disk grinders. *Disk grinders* have relatively large, side mounted abrasive disks. The work is held against one side of the disk for grinding. Both single and double disk grinders are used; in the latter type the work is passed between the two disks and is ground on both sides simultaneously. On these machines, the work is always held and fed automatically. On small, single-disk grinders the work can be held and fed by hand while resting on a supporting table. Although manual disk grinding is nonprecision in nature, fairly flat surfaces can be obtained quite rapidly with little or no tooling cost. On specialized, production-type machines, excellent accuracy can be obtained very economically.

Tool and cutter grinders. Simple, single-point tools often are sharpened by hand on bench or pedestal grinders (*offhand grinding*). More complex tools, such as milling-cutters, reamers, and hobs, and single-point tools for production-type operations, require more complex grinding machines, commonly called *universal tool and cutter grinders.* These machines are similar to small universal cylindrical center-type grinders, but they differ in four important respects:

1. The headstock is not motorized.
2. The headstock can be swiveled about a horizontal as well as a vertical axis.
3. The wheelhead can be raised and lowered and can be swiveled through a 360° rotation about a vertical axis.
4. All table motions are manual, no power feeds being provided.

Specific rake and clearance angles must be created, often repeatedly, on a given tool or on duplicate tools. Tool and cutter grinders have a high degree of flexibility built into them so that the required relationships between the tool and the grinding wheel can be established for almost any type of tool. Although setting up such a grinder is quite complicated and requires a highly skilled worker, after the setup is made for a particular job, the actual grinding is accomplished rather easily. Figure 23-26 shows several typical setups on a tool and cutter grinder.

There are also several specialized types of tool grinders for grinding specific tools. Figure 23-27 shows a surface grinder that is equipped with a projection-type comparator that projects a magnified image of the grinding wheel and work contour on a screen, making it especially useful in grinding form tools.

Snagging. *Snagging* is a type of rough grinding that is done in a foundry to remove fins, gates, risers, and rough spots from castings, preparatory to further machining. The primary objective is to remove substantial amounts of metal rapidly without much regard for accuracy. Pedestal-type or *swing grinders,* shown in Figure 23-28, ordinarily are used. Portable electric or air grinders also are used for this purpose and for miscellaneous grinding in connection with welding.

Mounted wheels and points. *Mounted wheels and points* are small grinding wheels of various shapes that are permanently attached to metal shanks that can be inserted in the chucks of portable, high-speed electric or air motors. They are operated at speeds up to 100,000 rpm, depending on their diameters, and

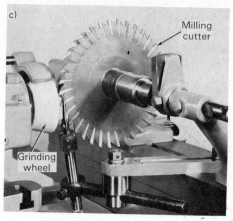

FIGURE 23-26 Three typical setups for grinding single- and multiple-edge tools on a universal tool and cutter grinder. (a) Single-point tool is held in a device that permits all possible angles to be ground. (b) Edges of a large hand reamer are being ground. (c) Milling cutter is sharpened with a cupped Al_2O_3 grinding wheel.

are used primarily for deburring and finishing in mold and die work. Several types are shown in Figure 23-29.

Coated abrasives. *Coated abrasives* are being used increasingly in finishing both metal and nonmetal products. These are made by gluing abrasive grains onto a cloth or paper backing. Synthetic abrasives—Carborundum and Alundum—are used most commonly, but some natural abrasives—sea sand, flint, garnet, and emery—also are employed. Various types of glues are utilized to attach the abrasive grains to the backing, usually compounded to allow the finished product to have some flexibility.

Coated abrasives are available in sheets, rolls, endless belts, and disks of various sizes. Some of the available forms are shown in Figure 23-30. Although the cutting action of coated abrasives basically is the same as with grinding wheels, there is one major difference: they have little tendency to be self-sharpening through dull grains being pulled from the backing to expose sharp particles. Consequently, when the abrasive particles become dull, the belt or other article must be replaced.

Figure 23-31 shows a typical belt grinder such as is used in metal finishing.

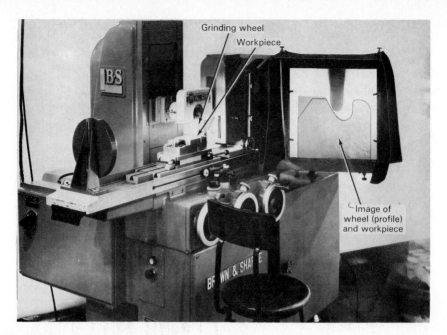

FIGURE 23-27 Micromaster grinder, which combines a projection comparator and a horizontal-spindle grinder to permit grinding work to an enlarged template image. (*Courtesy Brown & Sharpe Mfg. Co.*)

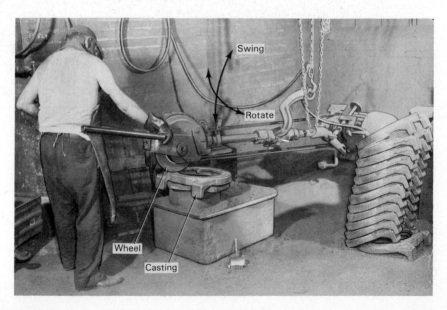

FIGURE 23-28 Hand-operated grinding of castings with a swing grinder. (*Courtesy Norton Company.*)

FIGURE 23-29   Mounted abrasive wheels and points. (*Courtesy Norton Company.*)

HONING

Honing uses fine abrasive stones to remove very small amounts of metal. It is used to size and finish bored holes, removing common errors left by boring (taper, waviness, and tool marks) or to remove the tool marks left by grinding. The amount of metal removed is typically about 0.13 mm (0.005 inch).

Honing stones.   Virtually all honing is done with stones made by bonding together various fine artificial abrasives. *Honing stones* differ from grinding wheels in that additional materials, such as sulfur, resin, or wax, often are added to the bonding agent to modify the cutting action. The abrasive grains range in size from 80 to 600 grit. The stones are equally spaced about the periphery of the tool.

5/1000" or less

2/000" or less = Syper honing

FIGURE 23-30   Various forms of coated abrasives. (*Courtesy Carborundum Company.*)

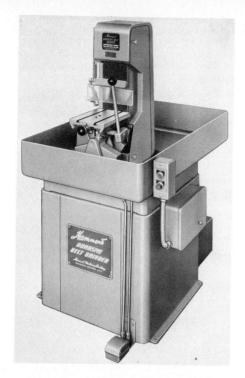

Bored holes are sized + finished by honing

Bore then hone

FIGURE 23-31 Production-type abrasive belt grinder with semiautomatic work table. (*Courtesy Hammond Machinery Builders, Inc.*)

Honing equipment. Although honing occasionally is done by hand, as in finishing the face of a cutting tool, it usually is done with special equipment. Either flat or round surfaces may be honed, but by far the majority of honing is done on internal, cylindrical surfaces, such as automobile cylinder walls. The honing stones usually are held in a honing head, such as is shown in Figure 23-32, with the stones being held against the work with controlled, light pressure. The honing head is not guided externally but, instead, *floats* in the hole, being guided by the work surface.

Surface speeds in honing vary from 15 to 91 meters (50 to 300 feet) per minute. The stones are given a complex motion so as to prevent a single grit from repeating its path over the work surface. For example, in honing internal cylinders a slow rotation is combined with an oscillatory axial motion. For external and flat surfaces, varying oscillatory motions are used. The length of the motions should be such that the stones extend beyond the work surface at the end of each stroke. A cutting fluid is used in virtually all honing operations.

Single- and multiple-spindle honing machines are available in both horizontal and vertical types. Some are equipped with special, sensitive measuring devices that collapse the honing head when the desired size has been reached.

For honing single, small, internal cylindrical surfaces a procedure is often used wherein the workpiece is manually held and reciprocated over a rotating hone.

If the volume of work is sufficient so that the equipment can be fully utilized, honing is a fairly inexpensive process. A complete honing cycle, including loading and unloading the work, often is less than 1 minute. Size control within 0.0076 mm (0.0003 inch) is achieved routinely.

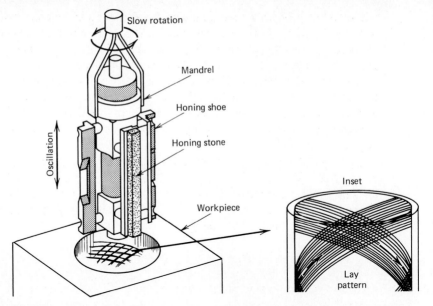

FIGURE 23-32 Schematic of honing head showing the manner in which the stones are held. The machine tool moves the hone in a rotary, oscillatory motion which results in a crosshatched lay pattern on the inside walls of the hole.

Superfinishing is a variation of honing that employes:

1. Very light, controlled pressure 0.07 to 0.28 MPa (10 to 40 psi).
2. Rapid (over 400 per minute), short strokes—less than 6.35 mm (¼ inch).
3. Stroke paths controlled so that a single grit never traverses the same path twice.
4. Copious amounts of low-viscosity lubricant-coolant flooded over the work surface.

This procedure, illustrated in Figures 23-33, results in surfaces of very uniform, repeatable smoothness.

Superfinishing is based on the phenomenon that a lubricant of a given viscosity

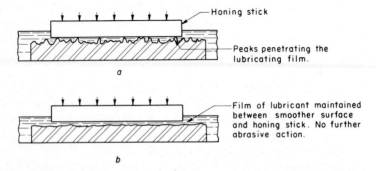

FIGURE 23-33 Manner in which a film of lubricant is established between the work and the abrasive stone in superfinishing as the work becomes smoother.

will establish and maintain a separating, lubricating film between two mating surfaces if their roughness does not exceed a certain value and if a certain critical pressure, holding them together, is not exceeded. Consequently, as the minute peaks on a surface are cut away by the honing stone, applied with a controlled pressure, a certain degree of smoothness is achieved. The controlled-viscosity lubricant establishes a continuous lubricating film between the stone and the workpiece and separates them so that no further cutting action occurs. Thus with a given pressure, lubricant, and honing stone, each workpiece is honed to the same degree of smoothness.

Superfinishing is applied to both cylindrical and plane surfaces. The amount of metal removed usually is less than 0.05 mm (0.002 inch), most of it being the peaks of the surface roughness. The copious amounts of lubricant-coolant maintain the work at a uniform temperature and wash away all abraded metal particles so as to prevent scratching.

LAPPING

Lapping is an abrasive surface-finishing process wherein fine abrasive particles are *charged* (caused to become embedded) into a soft material, called a lap. The material of the lap may range from cloth to cast iron or copper, but it always is softer than the material to be finished, being only a holder for the hard abrasive particles. Lapping is applied to both metals and nonmetals.

As the charged lap is rubbed against a surface, the abrasive particles in the surface of the lap remove small amounts of material from the harder surface. Thus it is the abrasive that does the cutting, and the soft lap is not worn away because the abrasive particles become embedded in its surface instead of moving across it. This action always occurs when two materials rub together in the presence of a fine abrasive—the softer one forms a lap, and the harder one is abraded away.

In lapping, the abrasive usually is carried between the lap and the work surface in some sort of a vehicle, such as grease, oil, or water. The abrasive particles are from 120 grit up to the finest powder sizes. As a result, only very small amounts of metal are removed—usually considerably less than 0.025 mm (0.001 inch). Because it is such a slow metal-removing process, it is used only to remove scratch marks left by grinding or honing, or to obtain very flat or smooth surfaces, such as are required on gage blocks or for liquid-tight seals where high pressures are involved.

Materials of almost any hardness can be lapped. However, it is difficult to lap soft materials because the abrasive tends to become embedded. The most common lap material is fine-grained cast iron. Copper is used quite often and is the common material for lapping diamonds. For lapping hardened metals for metallographic examination, cloth laps are used.

Lapping can be done either by hand or by special machines. In hand lapping the lap is flat, similar to a surface plate. Grooves usually are cut across the surface of a lap to collect the excess abrasive and chips. The work is moved across the surface of the lap, using an irregular, rotary motion, and is turned frequently to obtain a uniform cutting action.

In lapping machines for obtaining flat surfaces, workpieces are placed loosely in holders and are held against the rotating lap by means of floating heads. The holders, rotating slowly, move the workpieces in an irregular path. When two parallel surfaces are to be produced, two laps may be employed, one rotating below and the other above the workpieces.

Various types of lapping machines are available for lapping round surfaces. A special type of centerless lapping machine is used for lapping small, cylindrical parts, such as piston pins and ball-bearing races.

Because the demand for surfaces having only a few micrometers of roughness on hardened materials has become quite common, the use of lapping has increased greatly. However, because it is a very slow method of removing metal, it obviously is costly, compared with other methods, and it should not be specified unless such as surface is absolutely necessary.

REVIEW QUESTIONS

1. What are four machining processes in which abrasive particles are the cutting tools?
2. What is attrition in an abrasive grit?
3. Why is friability an important grit property?
4. What is the relationship between grit size and surface finish?
5. Why is aluminum oxide used more frequently than silicon carbide?
6. Why is CBN superior to silicon carbide as an abrasive in some applications?
7. What materials commonly are used as bonding agents in grinding wheels?
8. Explain what is meant by the grade of a bond in a grinding wheel. Why is it important? How does it differ from structure?
9. What is a hard grinding wheel?
10. Why is a soft abrasive wheel used for grinding hard materials?
11. How does loading differ from glazing?
12. What is meant when we say that grinding is a mixture of processes?
13. What is accomplished in dressing a grinding wheel?
14. How does abrasive machining differ from ordinary grinding?
15. What is a grinding ratio?
16. Basically, what determines the kind of abrasive to use in grinding?
17. Why is grain spacing important in grinding wheels?
18. What are some safety precautions that should be observed in grinding?
19. Why should a cutting fluid be used in copious quantities when doing wet grinding?
20. What is plunge-cut grinding?
21. Why is it not good practice to locate grinding machines among other machine tools?
22. How is the feed of the workpiece controlled in external centerless grinding?
23. What is the purpose of low-stress grinding and how is it done?
24. The number of grains per square inch which actively contact and cut a surface decreases with increasing grain diameter. Why is this so?
25. Is the grain diameter equal to the grain-size number?
26. Why are centerless grinders so popular in industry compared to center-type grinders?
27. How could you vary the through-feed rate in a centerless grinder? Could you do this while the process is in operation?
28. Why can you use vacuum chucks in surface grinding and not in up milling?
29. What is creep feed grinding and how does it differ from conventional surface grinding?
30. Why does a lap not wear, since it is softer than the material being lapped?
31. How do honing stones differ from grinding wheels?
32. What is meant by "charging" a lap?

33. Why is a honing head permitted to float in a hole that is being honed?
34. In what respect does the cutting action of a coated abrasive differ from that of an abrasive wheel?
35. What is crush dressing? Why is it often used in form grinding?

CASE STUDY 23. Bolting Leg on a Casting

Figure CS-23 shows the design of one of four legs on a cast iron casting made by the Hardhat Company. These legs are used to attach this device to the floor. The section drawing to the right shows the typical loading to which the leg is subjected. Currently, the company is drilling the bolt hole and then counterboring the land. Production has experienced some difficulty in machining these legs and sales has recently reported that a substantial number of in-service failures have occurred with these legs. You have obtained a sketch from one of the salesmen showing where one of the legs failed.

1. What machining difficulties would you expect this design to have?
2. What do you think has caused the failures?
3. What do you recommend to solve this problem in the future in terms of materials, design, and manufacture?
4. What do you recommend be done with the units in the field to stop the failures?

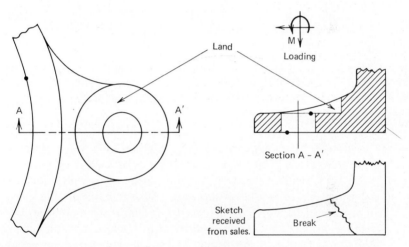

FIGURE CS-23 Bolting leg on cast iron casting and sketch of a broken leg.

CHAPTER 24

Broaching

Broaching is one of the most productive, basic machining processes. As a process, it is similar to shaping, competes economically with milling and boring and is capable of producing precision machined surfaces. The heart of this process lies in the broaching tool wherein roughing, semifinishing, and finishing teeth are combined into one tool. *Broaching* is unique in that it is the only one of the basic machining processes in which the **feed** of the cutting edges into the workpiece, determining the chip thickness, is built into the tool, called a *broach*. The machined surface always is the inverse of the profile of the broach, and, in most cases, it is produced with a single, linear stroke of the tool across the workpiece (or the workpiece across the broach).

As illustrated in Figure 24-1, a broach is composed of a series of single-point cutting edges projecting from a rigid bar, with successive edges protruding farther from the axis of the bar than previous ones. This rise per tooth, known as *step*, determines the depth of cut by each tooth (chip thickness), so that no feeding of the broaching tool is required. The frontal contour of the teeth determines the shape of the resulting machined surface. As the result of these conditions built into the tool, no complex motion of the tool relative to the workpiece is required and the need for highly skilled machine operators is minimized.

Because of the features built into a broach, it is a simple and rapid method of machining. There is a close relationship between the contour of the surface

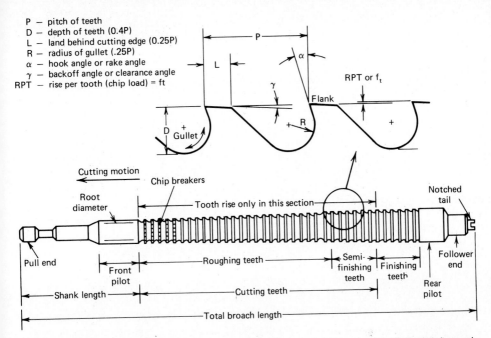

P — pitch of teeth
D — depth of teeth (0.4P)
L — land behind cutting edge (0.25P)
R — radius of gullet (.25P)
α — hook angle or rake angle
γ — backoff angle or clearance angle
RPT — rise per tooth (chip load) = ft

FIGURE 24-1 Basic shape and nomenclature for conventional pull (hole) broach. Note chipbreakers in first section of roughing teeth, which may be extended to more teeth if the cut is heavy or material difficult.

to be produced, the amount of material that must be removed, and the design of the broach. For example, the total depth of the material to be removed cannot exceed the total step provided in the broach, and the step of each tooth must be such as to provide proper chip thickness for the type of material to be machined. Consequently, either a special broach must be made for each job, or the workpiece must be designed so that a standard broach can be used. This means that broaching is particularly well suited for, and is widely used in, mass production, where the volume can easily justify the cost of a rather expensive tool, and for certain simple and standardized shapes, such as keyways, where stock broaches can be used.

Broaching originally was developed for machining internal keyways. However, its obvious advantages quickly led to its development for mass-production machining of various surfaces, such as flat, interior and exterior cylindrical and semicylindrical, and many irregular surfaces. Because there are few limitations as to the contour form that broach teeth may have, there is almost no limitation in the shape of surfaces that can be produced by broaching. The only physical limitations are that there must be no obstruction to interfere with the passage of the entire tool over the surface to be machined and that the workpiece must be strong enough to withstand the forces involved. In internal broaching, a hole must exist in the workpiece into which the broach may enter. Such a hole can be made by drilling, boring, or coring.

Broaching usually produces better accuracy and finish that can be obtained by milling or reaming. Although the relative motion between the broaching tool and the work usually is a single linear one, a rotational motion can be added to permit the broaching of spiral grooves, as in spiral splines or in gun-barrel rifling.

Classification of broaches. Broaches commonly are classified as follows:

Purpose	Motion	Construction	Function
Single	Push	Solid	Roughing
Combination	Pull	Built-up	Sizing
	Stationary		Burnishing

Broach design. Figure 24-1 shows the principal components of a broach and the shape and arrangement of the teeth. Each tooth is essentially a single-edge cutting tool, arranged much like the teeth on a saw except for the step, which determines the depth cut by each tooth. The depth of cut varies from about 0.15 mm (0.006 inch) for roughing teeth in machining free-cutting steel to a minimum of 0.025 mm (0.001 inch) for finishing teeth. The exact amount depends on several factors. Too-large cuts impose undue stresses on the teeth and the work; too-small result in rubbing rather than cutting action. The strength and ductility of the metal being cut are the primary factors.

Where it is desirable for each tooth to take a deep cut, as in broaching castings or forgings that have a hard, abrasive surface layer, *rotor-* or *jump-cut* tooth design, shown in Figure 24-2, may be used. In this design, two or three teeth in succession have the same diameter, or height, but each tooth of the group is notched or cut away so that it cuts only a portion of the circumference or width. This permits deeper, but narrower, cuts by each tooth without increasing the total load per tooth, and also reduces the forces and the power requirements.

Tooth loads and cutting forces also can be reduced by using the *double-cut* construction shown in Figure 24-2. Pairs of teeth have the same size, but the first has extra-wide chip-breaker notches and removes metal over only a part of its width, while the smooth second tooth completes the cut.

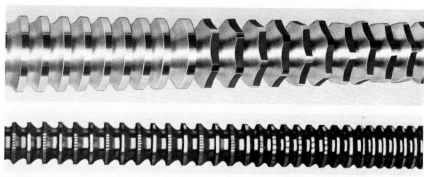

FIGURE 24-2 Special types of broach teeth. (*Top*) Rotor-cut. (*Bottom*) Double-cut with chip-breaker grooves on alternate teeth except at the finishing end. (*Courtesy Colonial Broach & Machine Company.*)

FIGURE 24-3 Progressive surface broach. (*Courtesy Detroit Broach & Machine Company.*)

A third construction for reducing tooth loads utilizes the principle illustrated in Figure 24-3. Employed primarily for broaching wide, flat surfaces, the first few teeth in *progressive* broaches completely machine the center, while succeeding teeth are offset in two groups to complete the remainder of the surface. Rotor, double-cut, and progressive designs require the broach to be made longer than if normal teeth were used, and they thus can be used only on a machine having adequte stroke length.

The faces of the teeth on surface broaches may be either normal to the direction of motion or at an angle of from 5 to 20°. The latter, *shear-cut,* broaches provide smoother cutting action with less tendency to vibrate. Although the majority of surface broaching is done on flat surfaces, other shapes can be broached, as shown in Figure 24-4.

As in saws, the pitch of the teeth and the gullet between them must be sufficient to provide ample room for chip clearance. All chips produced by a given tooth during its passage over the full length of the workpiece must be contained in the space between successive teeth. At the same time, it is desirable to have the pitch sufficiently small so that at least two or three teeth are cutting at all times.

The *hook* determines the primary rake angle and is a function of the material being cut, being 15 to 20° for steel and 6 to 8° for cast iron. *Back-off* or end clearance angles are from 1 to 3° to prevent rubbing.

FIGURE 24-4 Broaching the teeth of a segment gear. (*Courtesy Colonial Broach & Machine Company.*)

Most of the metal removal is done by the *roughing teeth. Semifinishing teeth* provide surface smoothness, whereas *finishing teeth* produce exact size. On a new broach all the finishing teeth usually are made the same size. As the first finishing teeth become worn, those behind continue the sizing function. On some round broaches, *burnishing teeth* are provided for finishing. These have no cutting edges but are button-shaped and from 0.025 to 0.076 mm (0.001 to 0.003 inch) larger than the size of the hole. The resulting rubbing action smooths and sizes the hole. They are used primarily on cast iron and nonferrous metals.

The *pull end* of a broach (Figure 24-1) occurs only on pull broaches and is to provide a means of quickly attaching the broach to the pulling mechanism. The *front pilot* aligns the broach in the hole before it begins to cut, and the *rear pilot* keeps the tool square with the finished hole as it leaves the workpiece. *Shank length* must be sufficient to permit the broach to pass through the workpiece and be attached to the puller before the roughing teeth engage the work. If a broach is to be used on a vertical machine that has a tool-handling mechanism, a *tail* is necessary.

A broach should not be used to remove a greater depth of metal than that for which it is designed—the sum of the steps of all the teeth. In designing workpieces, a minimum of 0.51 mm (0.020 inch) should be provided on surfaces that are to be broached, and about 6.35 mm (¼ inch) is the practical maximum.

Broaching speeds. Broaching speeds are relatively low (25 to 20 sfpm), seldom exceeding 15.24 meters (50 feet) per minute. However, because a surface usually is completed in a single stroke, the productivity is high. A complete cycle usually requires only from 5 to 30 seconds, with most of that time being taken up by the return stroke, broach-handling, and workpiece loading and unloading. Such cutting conditions facilitate cooling and lubrication and result in very slow tool wear. This is an advantage of broaching as a mass-production process because it reduces the necessity for frequent resharpening and prolongs the life of the expensive broaching tool.

For a given cutting speed and material, the force required to pull or push a broach is a function of the tooth width, step, and the number of teeth cutting. Consequently, it is necessary to design or specify a broach within the stroke length and power limitations of the machine on which it is to be used.

Broach materials and construction. Because of the low cutting speeds employed, most broaches are made of alloy or high-speed tool steel, even for some

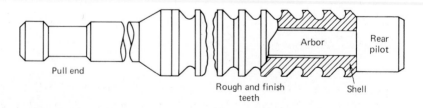

FIGURE 24-5 Shell design. Broach shell sections are mounted on an arbor (for broach sizes over 3 inches in diameter) in this round pull broach.

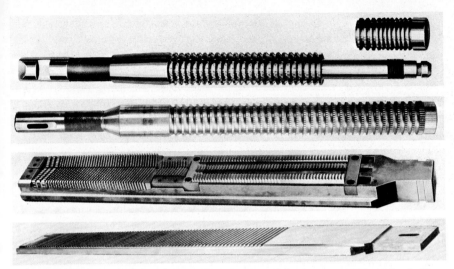

FIGURE 24-6 Types of broach construction. (*Top to bottom*) Round, shell on arbor type, pull broach; Round, solid-type, pull broach; sectional surface broach; solid surface broach; (*Courtesy Ex-Cell-O Corporation and Colonial Broach & Machine Company.*)

mass-production work. Where they are used in continuous mass-production lines, particularly in surface broaching, tungsten carbide teeth may be used, permitting them to be used for long periods of time without resharpening.

Most internal broaches are of solid construction. Quite often, however, they are made of *shells* mounted on an arbor (Figure 24-5). When the broach, or a section of it, is subject to rapid wear, a single shell can be replaced and will be much cheaper than an entire solid broach. Shell construction, however, is more expensive than a solid broach of comparable size.

Small surface broaches may be of solid construction, but larger ones usually are built up from sections, as shown in Figure 24-6. Sectional construction makes the broach easier and cheaper to construct and sharpen. It also often provides some degree of interchangeability of the sections.

Sharpening broaches. Most broaches are resharpened by grinding the hook faces of the teeth. The lands of internal broaches must not be reground because this would change the size of the broach. Lands of flat surface broaches sometimes are ground, in which case all of them must be ground to maintain their proper relationship.

BROACHING MACHINES

Because all the factors that determine the shape of the machined surface and which determine all cutting conditions, except speed, are built into the broaching tool, bro ching machines are relatively simple. Their primary functions are to impart plain reciprocating motion to the broach and to provide a means for automatically handling the broach.

Most broaching machines are driven hydraulically, although mechanical drive is used in a few special types. The major classification relates to whether the motion of the broach is vertical or horizontal as follows:

Vertical	Horizontal	Rotary
Broaching presses (push broaching)	Pull	Special types
Pull down	Surface	
Pull up	Continuous	
Surface		

The choice between vertical and horizontal machines is determined primarily by the length of the stroke required and the available floor space. Vertical machines seldom have greater than 1524 mm (60 inch) strokes, because of height limitations. Horizontal machines can have almost any length of stroke, but they require greater floor space.

Broaching presses. As shown in Figure 24-7, *broaching presses* essentially are arbor presses with a guided ram. They are used with push broaches, have a capacity of from 44.5 to 445 kN (5 to 50 tons), and are used only for internal broaching. The forward guide of the broach is inserted through the hole in the workpiece as it rests on the press table, often in a fixture. As the ram descends, it engages the upper end of the broach and pushes it through the work.

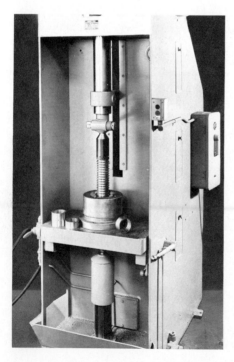

FIGURE 24-7 Broaching internal helical grooves in a broaching press, using a guided ram. (*Courtesy The Oilgear Company.*)

Broaching presses are relatively slow, in comparison with other broaching machines, but they are inexpensive, flexible, and can be used for other types of operations, such as bending and staking.

Vertical pull-down machines. The major components of vertical pull-down machines are a worktable, usually having a spherical-seated workholder, a broach elevator above the table, and a pulling mechanism below the table. As shown in Figure 24-8, when the elevator raises the broach above the table, the work can be placed into position. The elevator then lowers the pilot end of the broach through the hole in the workpiece, where it is engaged by the puller. The elevator then releases the upper end of the broach, and it is pulled through the workpiece. The workpieces are removed from the table, and the broach is raised upward to be engaged by the elevator mechanism. In some cases where machines have two rams, they are arranged so that one broach is being pulled down while the work is being unloaded and the broach raised at the other station.

Vertical pull-up machines. In *vertical pull-up machines* the pulling ram is above the worktable and the broach-handling mechanism below it. The work is placed in position, above the pilot, while the broach is lowered. The handling mechanism then raises the broach until it engages the puller head. As the broach is pulled upward, the work comes to rest against the underside of the table, where it is held until the broach has been pulled through. The work then falls free, often sliding down a chute into a tote bin.

FIGURE 24-8 Broaching a round hole on a vertical pull-down machine having two broaches. (*Courtesy The Oilgear Company.*)

Pull-up machines may have up to eight rams. Because the workpieces need only be placed in the machines, and the broach handling and work removal are automatic, they are highly productive. For certain types of work, automatic feeding can be provided.

Vertical surface-broaching machines.   On *vertical surface-broaching machines* the broaches usually are mounted on guided slides, as shown in Figure 24-9, so as to provide support against lateral thrust. Because there is no need for handling the broach, they are simpler but much heavier than pull- or push-broaching machines. Many have two or more slides so that work can be loaded at one while another part is machined at the other. Inasmuch as there is no handling of the broach, the operating cycle is very short. Usually, slide or rotary-indexing fixtures are used to hold the work, reducing the work-handling time and minimizing the total cycle time.

Horizontal broaching machines.   The primary reason for employing a *horizontal* configuration for pull- and surface-broaching machines is to make possible

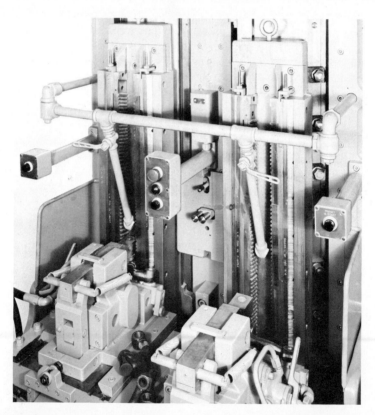

FIGURE 24-9   Close-up view of the fixtures and broaching inserts on a vertical duplex hydrobroaching machine. Work is being broached in the left-hand fixture while a workpiece has just been placed in the right-hand fixture. One fixture moves forward while work is being loaded in the second, retracted fixture. (*Courtesy Cincinnati Milacron, Inc.*)

longer strokes and the use of longer broaches than can conveniently be accommodated in vertical machines. Horizontal pull-broaching machines essentially are vertical machines turned on their side. Where internal surfaces are to be broached, such as holes, the broach must have a diameter-to-length ratio large enough to make it self-supporting without appreciable deflection. Consequently, horizontal machines seldom are used for small holes. In surface broaching, the broach always is supported in guides, so no such limitation is encountered. However, most horizontal broaching machines are quite large. Broaching that requires rotation of the broach, as in rifling and spiral splines, usually is done on horizontal machines.

Horizontal surface-broaching machines vary widely in size and design. As in the vertical type, the broaches are mounted on heavy, ram-driven slides. Some cut in only one direction of the ram stroke, whereas others are provided with two sets of broaches, arranged so that cutting occurs during both motions of the ram. Figure 24-10 shows a typical horizontal surface-broaching machine, such as often is incorporated in production.

Continuous surface-broaching machines. In *continuous surface-broaching machines,* the broaches usually are stationary, and the work is pulled past them on an endless conveyor. Fixtures usually are attached to the conveyor chain so that the workpieces can be placed in them at one end of the machine and removed at the other, sometimes automatically. Such machines are being used increasingly in mass-production lines (see Figure 24-11).

Rotary broaching machines. In *rotary broaching machines,* occasionally used in mass production, the broaches are stationary, and the work is passed beneath,

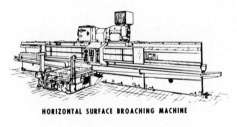

HORIZONTAL SURFACE BROACHING MACHINE

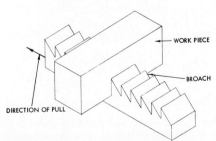

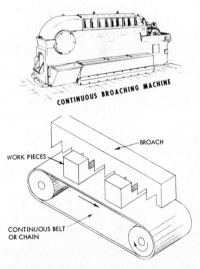

CONTINUOUS BROACHING MACHINE

FIGURE 24-10 Large horizontal surface broaching machine. (*From* Manufacturing Producibility Handbook; *courtesy General Electric Company.*)

FIGURE 24-11 Continuous horizontal, surface broacher. (*From* Manufacturing Producibility Handbook; *courtesy General Electric Company.*)

or between, them while held in fixtures on a rotary table. They have the advantage that there is no lost time due to noncutting, reciprocating strokes.

REVIEW QUESTIONS

1. What is unique about broaching, as compared with the other basic machining processes?
2. Why can a thick saw blade not be used as a broach?
3. Why are broaching machines more simple in a basic design than most other machine tools?
4. Why is broaching particularly well suited for mass production?
5. Explain how internal spiral grooves are produced by broaching.
6. Why is it necessary to relate the design of a broach to the specific workpiece that is to be machined?
7. What two methods can be utilized to reduce the force and power requirements for a particular broaching cut?
8. For a given job, how would a broach having rotor-tooth design compare in length with one having regular, full-width teeth?
9. Why are the pitch and radius between teeth on a broach of importance?
10. Why are broaching speeds usually relatively low, as compared with other machining operations?
11. Why are some broaches made with shell-type construction?
12. Why are most broaches made from alloy or high-speed steel rather than from tungsten carbide?
13. How does a simple broaching press differ from an arbor press?
14. For mass-production operations, why are pull-up broaching machines usually preferred over pull-down machines?
15. Why can continuous broaching machines not be used for broaching holes?
16. The sides of a square, dead-end hole must be machined clear to the bottom. Would it be possible to do this by broaching? Why?
17. The interior, flat surfaces of socket wrenches, which have one "closed" end, often are finished to size by broaching. By examining one of these, determine what design modification was incorporated to make this operation possible.
18. A surface 304.8 mm long is to be machined with a flat, solid broach that has a tooth step of 0.12 mm. What is the minimum cross-sectional area that must be provided in the chip gullet between adjacent teeth?
19. Why is a continuous surface-broaching machine basically more productive than an ordinary horizontal surface-broaching machine?
20. The pitch of the teeth on a simple surface broach can be determined by the equation $P = 0.3 \sqrt{\text{width of cut}}$. If a broach is to remove 6.35 mm of material from a gray iron casting that is 76 mm wide and 450 mm long, and if each tooth has a step of 0.10 mm, what will be the length of the roughing section of the broach?
21. Estimate the approximate maximum horsepower needed to accomplish the operation described in problem 20, at a cutting speed of 10 meters per minute. (*Hint:* First find the HP used per tooth.)
22. Estimate the approximate force acting in the forward direction during cutting for the conditions stated in Problems 20 and 21.

CASE STUDY 24. The Sliding Axle

As a 3-week-old automobile was making a left-hand turn on a city street at about 20 miles per hour, the right-rear axle and wheel assembly came completely off the vehicle. The inner end of the subject axle is shown in Figure CS-24a and b. The axle normally is held in place by a U-shaped clip which fits into the groove near the inner end of the axle and which is bolted to the differential assembly. The clip was designed to resist only outward forces acting on the axle. This clip was found to be broken when the rear axle assembly was examined. The inner end of the axle normally bears against a thrust

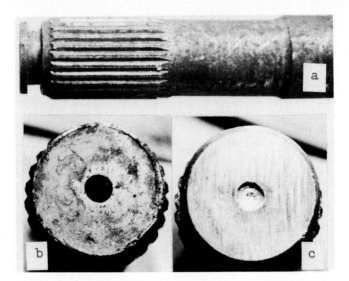

FIGURE CS-24 Photos of right rear automobile axle. a and b show the defective axle, c shows a good axle.

block to provide resistance to inward, lateral thrust. Figure CS-24c shows the inner end of a replacement axle which was known to meet specifications.

From the evidence shown in Figure CS-24, can you explain what caused the difference in appearance of the end surfaces of the two axles? Was the subject axle defective at the time it was assembled, and if so, what was the defect?

CHAPTER 25

Sawing and Filing

SAWING

Sawing is a basic machining process in which chips are produced by a succession of small cutting edges, or *teeth,* arranged in a narrow line on a saw "blade." As shown in Figure 25-1, each tooth forms a chip progressively as it passes through the workpiece, and the chip is contained within the space between two successive teeth until these teeth pass from the work. Because sections of considerable size can be severed from the workpiece with the removal of only a small amount of the material in the form of chips, sawing is probably the most economical of the basic machining processes with respect to the wastage of material and power consumption, and in many instances with respect to labor.

While sawing is an old machining process, during recent years vast improvements have been made in saw blades and the machines that use them and, as a result, the accuracy and precision of the process has been greatly improved. Although it still is true that most sawing is done to sever bar stock and shapes into desired lengths for use in other operations, there are many cases where it is used to produce desired shapes. Frequently, and especially for producing only a few parts, contour sawing may be more economical than any other machining process.

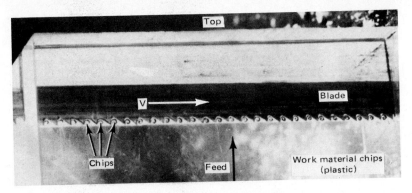

FIGURE 25-1 Formation of chips in sawing. (*Courtesy DoAll Company.*)

Saw blades. Saw blades are made in three basic configurations. The first, commonly called a *hacksaw* blade, is straight, relatively rigid, and of limited length with teeth on one edge. The second is sufficiently flexible so that a long length can be formed into a continuous band with teeth on one edge; these are known as *band-saw* blades. The third form is a rigid disk, having teeth on the periphery; these are *circular saws*.

All saw blades have certain common and basic features. These are (1) material, (2) tooth form, (3) tooth spacing, (4) tooth set, and (5) blade thickness. Small hacksaw blades usually are made entirely of tungsten or molybdenum high-speed steel. Blades for power-operated hacksaws often are made with teeth cut from a strip of high-speed steel that has been electron-beam-welded to the heavy, main portion of the blade, which is made from a tougher and cheaper alloy steel. Band-saw blades frequently are made with this same type of construction, as shown in Figure 25-2, but with the main portion of the blade made of relatively thin, high-tensile-strength alloy steel to provide the required flexibility. Band-saw blades also are available with tungsten carbide teeth.

Three common *tooth forms* are shown in Figure 25-3.

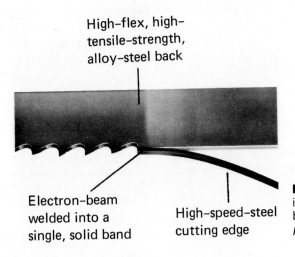

High–flex, high–
tensile–strength,
alloy–steel back

Electron–beam
welded into a
single, solid band

High–speed–steel
cutting edge

FIGURE 25-2 Method of providing HSS teeth on a softer steel band. (*Courtesy DoALL Company.*)

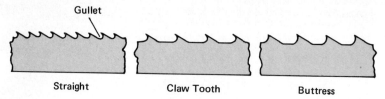

Gullet

Straight   Claw Tooth   Buttress

FIGURE 25-3 Common tooth forms for saw blades.

Tooth spacing is very important in all sawing because it determines three factors. First, it controls the size of the teeth. From a strength viewpoint, large teeth are desirable. Second, tooth spacing determines the space (*gullet*) available to contain the chip that is formed. As shown in Figure 25-1, the chip cannot drop from this space until it emerges from the slot, or *kerf*. The space must be such that there is no crowding of the chip and no tendency for chips to become wedged between the teeth and not drop out. Third, tooth spacing determines how many teeth will bear against the work. This is very important in cutting thin material, such as tubing, as illustrated in Figure 25-4. At least two teeth should be in contact with the work at all times. If the teeth are too coarse, only one tooth rests on the work at a given time, permitting the saw to rock, and the teeth may be stripped from the saw. Hand hacksaw blades can be obtained with tooth spacings from 1.8 to 0.78 mm (14 to 32 teeth per inch). In order to make it easier to start a cut, some hand hacksaw blades are made with a short section at the forward end having teeth of a special form with negative rake angles, as shown in Figure 25-5. Tooth spacings for power hacksaw blades range from 6.4 to 1.4 mm (4 to 18 teeth per inch).

Tooth set, illustrated in Figure 25-6, refers to the manner in which the teeth are offset from the center line in order to make a cut (the kerf) wider than the thickness of the back portions of the blade. This permits the saw to move more freely in the kerf and reduces friction and heating. *Raker-tooth saws* are used in cutting most steel and iron. *Straight-set teeth* are used for sawing brass, copper, and plastics. Saws with *wave-set teeth* are used primarily for cutting thin sheets and thin-walled tubing.

The *blade thickness* of nearly all hand hacksaw blades is 0.64 mm (0.025 inch). Saw blades for power hacksaws vary in thickness from 1.27 to 2.54 mm (0.050 to 0.100 inch).

Hand hacksaw blades come in two standard lengths—254 and 304.8 mm (10 and 12 inches). All are 12.7 mm (½ inch) wide. Blades for power hacksaws

Fine teeth

Coarse teeth
will straddle
work and strip
teeth

FIGURE 25-4 Relationship of tooth size to material thickness in sawing.

FIGURE 25-5 One end of a hacksaw blade having specially shaped starting teeth. (*Courtesy Heller Tool Company.*)

vary in length from 304.8 to 609.6 mm (12 to 24 inches) and in width from 25.4 to 50.8 mm (1 to 2 inches). Wider and thicker blades are desirable for heavy-duty work. The blade should be at least twice as long as the maximum length of cut that is to be made.

Band-saw blades are available in long rolls in straight, raker, wave, or combination sets. In order to reduce the noise from high-speed bandsawing, it is becoming increasingly common to use blades that have more than one pitch, size of teeth, and type of set. One example is shown in Figure 25-7. The most common widths are from 1.6 to 12.7 mm (1/16 to 1/2 inch), although wider blades can be obtained. Blade width is very important in bandsawing because it determines the minimum radius that can be cut. This relationship is illustrated in Figure 25-8. Because wider blades are stronger, as wide a blade as possible should be used. Consequently, cutting small radii requires a narrower and weaker blade and is more time-consuming.

Band-saw blades come in tooth spacings from 12.7 to 0.79 mm (2 to 32 teeth per inch). The *buttress* form, illustrated in Figure 25-3, is often used for tooth spacings of 4.23 mm or greater (6 teeth or less per inch). Other tooth forms are available for special purposes, such as a scallop edge for cutting fabric. Another type is in the form of a tightly wound spiral of metal, forming a blade about 3.2 mm (1/8 inch) in diameter. This provides a continuous cutting edge along the entire spiral so that work can be fed against the blade in all directions.

Circular saws necessarily differ somewhat from straight blade forms. Because they must be relatively large in comparison with the work, only the sizes up to about 457 mm (18 inches) in diameter have teeth that are cut into the disk. Larger saws use either *segment* or *inserted* teeth, illustrated in Figure 25-9. Only the teeth are made of high-speed steel, or tungsten carbide; the remainder of the disk is made of ordinary, less expensive, and tougher steel. Segmental blades

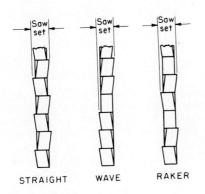

STRAIGHT   WAVE   RAKER

FIGURE 25-6 Basic types of saw-tooth set. Putting "set" in the teeth makes the kerf wider than the body of the blade.

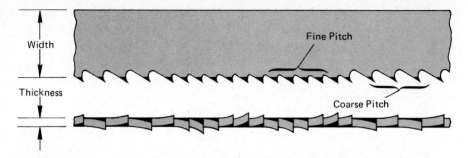

FIGURE 25-7 Use of alternating groups of saw teeth, having either coarse pitch with alternating set or fine pitch with wavy set, in order to reduce noise.

are composed of segments mounted around the periphery of the disk, usually fitted with a tongue and groove and fastened by means of screws or rivets. Each segment contains several teeth. If a single tooth is broken, only one segment need be replaced to restore the saw to operating condition.

Large circular saws, some as large as 1829 mm (72 inches) in diameter, often have inserted teeth. Each tooth is a separate piece of high-speed steel or tungsten carbide and is attached to the central disk by means of wedges or screws, or by brazing.

Figure 25-10 shows a common tooth form used in circular saws, in which successive teeth are beveled on opposite sides to produce a smoother cut. Another method is to bevel both sides of every other tooth. A third procedure has the first tooth beveled on the left side; the second tooth on both sides; the third tooth on the right side; the fourth tooth on the left side, and so forth.

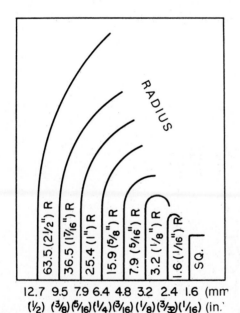

63.5 (2½") R 36.5 (1⁷⁄₁₆") R 25.4 (1") R 15.9 (⁵⁄₈") R 7.9 (⁵⁄₁₆") R 3.2 (⅛") R 1.6 (¹⁄₁₆") R SQ.

RADIUS

12.7 9.5 7.9 6.4 4.8 3.2 2.4 1.6 (mm
(½) (⅜)(⁵⁄₁₆)(¼)(³⁄₁₆) (⅛)(³⁄₃₂)(¹⁄₁₆) (in.'

Saw Blade Width

FIGURE 25-8 Relationship of bandsaw width to the minimum radius that can be cut.

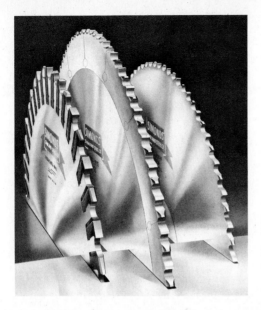

FIGURE 25-9 (*Left to right*) The inserted-tooth, segmental-tooth, and integral-tooth forms of saw constructions. (*Courtesy Simonds Saw and Steel Co.*)

Circular saws for cutting metal are often called *cold saws* to distinguish them from friction-type disk saws that heat the metal to the melting temperature at the point of metal removal. Cold saws cut very rapidly and can produce cuts that are comparable in smoothness and accuracy with surfaces made by slitting saws in a milling machine or by a cutoff tool in a lathe.

Types of sawing machines.   Metal-sawing machines may be classified as follows:

1. Reciprocating saw.
    a. Manual hacksaw.
    b. Power hacksaw.
2. Band saw.
    a. Vertical cutoff.
    b. Horizontal cutoff.

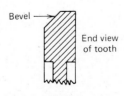

Bevel

End view of tooth

FIGURE 25-10   Method of beveling teeth on alternate sides to produce smoother cutting. (*Courtesy Simonds Saw and Steel Co.*)

c. Combination cutoff and contour.
    d. Friction.
3. Circular saw.
    a. Cold saw.
    b. Steel friction disk.
    c. Abrasive disk.

Power hacksaws. As the name implies, power hacksaws are machines that mechanically reciprocate a large hacksaw blade. As shown in Figure 25-11, they consist of a bed, a work-holding frame, a power mechanism for reciprocating the saw frame, and some type of feeding mechanism. Because of the inherent inefficiency of cutting on only one stroke direction, they have often been replaced by more efficient, horizontal band-sawing machines.

Band-sawing machines. While the earliest metal-cutting band-sawing machines were direct adaptations from wood-cutting band saws, modern machines of this type are much more sophisticated and versatile and have been developed specifically for metal cutting. To a large degree, they were made possible by the development of vastly better and more flexible band-saw blades and simple flash-welding equipment, which can weld the two ends of a strip of band-saw blade together to form a band of any desired length. Three basic types of band-sawing machines are in common use.

Upright, cutoff, band-sawing machines, such as shown in Figure 25-12, are designed primarily for cutoff work on single, stationary workpieces that can be held on a table. On many machines, the blade mechanism can be tilted to about 45°, as shown, to permit cutting at an angle. They usually have automatic power feed of the blade into the work, automatic stops, and provision for supplying coolant.

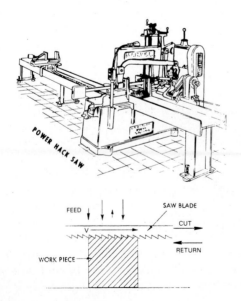

FIGURE 25-11 Power hacksaw with reciprocating blade. (*From Manufacturing Producibility Handbook; courtesy General Electric Company.*)

FIGURE 25-12 Cutting a pipe at a 45° angle on an upright cut-off bandsawing machine. (*Courtesy Armstrong-Blum Mfg. Co.*)

Horizontal, metal-cutting band-sawing machines were developed to combine the flexibility of reciprocating power hacksaws and the continuous cutting action of vertical band saws. These heavy-duty automatic bandsaws feed the saw vertically by a hydraulic mechanism and have automatic stock feed that can be set to feed the stock laterally any desired distance after a cut is completed and automatically clamp it for the next cut. Such machines can be arranged to hold, clamp, and cut several bars of material simultaneously. Smaller and less expensive types have swing-frame construction, with the band-saw head mounted in a pivot on the rear of the machine. Feed is accomplished by gravity through rotation of the head about the pivot point. Because of their continuous cutting action, horizontal band-sawing machines are very efficient (see Figure 25-13).

Combination cutoff and contour band-sawing machines, such as shown in Figure 25-14, can be used not only for cutoff work but also for contour sawing. They are widely used for cutting irregular shapes in connection with making dies and the production of small numbers of parts.

Several features distinguish these machines from regular vertical cutoff band saws. First, the table is pivoted so that it can be tilted to any angle up to 45°. Second, a small flash welder is provided, on the vertical column, to that a straight length of band-saw blade can be welded quickly into a continuous band. A small grinding wheel is located beneath the welder so that the flash can be ground from the weld to provide a smooth joint that will pass through the saw guides. This welding and grinding unit makes it possible to cut internal openings by drilling a hole, inserting one end of the saw blade through the hole, and then butt-welding the two ends together. When the cut is finished, the band is cut apart and removed from the opening. Third, the speed of the saw blade can be varied continuously over a wide range to provide correct operating conditions for any material. Fourth, a method of power feeding the work is provided, sometimes gravity-actuated.

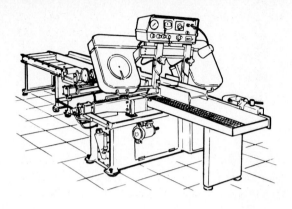

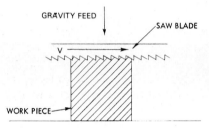

FIGURE 25-13 Horizontal band-sawing machine. (*From* Manufacturing Producibility Handbook; *courtesy General Electric Company.*)

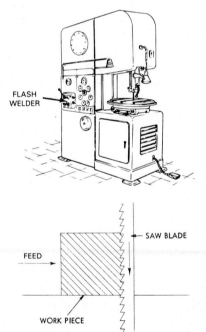

FIGURE 25-14 Combination cutoff and contour band-sawing machine. (*From* Manufacturing Producibility Handbook; *courtesy General Electric Company.*)

Contour-sawing machines are made in a wide range of sizes, the principal size dimension being the throat depth. Sizes from 305 to 1829 mm (12 to 72 inches) are available. The speeds available on most machines range from about 15 to 610 meters (50 to 2000 feet) per minute.

Special band-sawing machines are available with very high speed ranges, up to 4572 meters (15,000 feet) per minute. These are known as *friction band-sawing machines*. Material is not cut by chip formation. Instead, the friction between the rapidly moving saw and the work is sufficient to raise the temperature of the material at the end of the kerf to or just below the melting point where its strength is very low. The saw blade then pulls the molten, or weakened, material out of the kerf. Consequently, the blades do not need to be sharp; they frequently have no teeth, only occasional notches in the blade to aid in removing the metal.

Almost any material, including ceramics, can be cut by friction sawing. Because only a small portion of the blade is in contact with the work for an instant and then is cooled by its passage through the air, it remains cool. Usually the major portion of the work, away from contact with the saw blade, also remains quite cool. Figure 25-15 shows a hardened steel cutter being cut by this method. It also is a very rapid method for trimming the flash from pressed sheet-metal parts.

Circular-blade sawing machines.   Machines employing rotating, circular saw blades are used exclusively for cutoff work. These range from small, simple types, such as shown in Figure 25-16, in which the saw is fed manually, to very large ones having power feed and built-in coolant systems. A common use for the large ones is to cut off hot-rolled shapes as they come from a rolling mill. In some cases friction saws are used for this purpose, having disks up to 1.83 meters (6 feet) in diameter and operating at surface speeds up to 7620 meters (25,000 feet) per minute. Steel sections up to 610 mm (24 inches) can be cut in less than 1 minute by this technique.

Although technically not a sawing operation, cutoff work up to about 152 mm (6 inches) often is done utilizing thin *abrasive* disks. The equipment used

FIGURE 25-15 Cutting a hardened steel cutter by friction sawing. (*Courtesy DoALL Company.*)

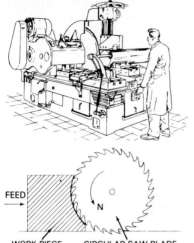

FEED

WORK PIECE   CIRCULAR SAW BLADE

FIGURE 25-16 Rotary circular saw for cutoff operations. (*From* Manufacturing Producibility Handbook; *courtesy General Electric Company.*)

is the same as for sawing. It has the advantage that very hard materials, which would be very difficult to saw, can be cut readily. A thin rubber- or resinoid-bonded abrasive wheel is used. Usually a somewhat smoother surface is produced.

FILING

Basically, the metal-removing action in filing is the same as in sawing, in that chips are removed by cutting teeth that are arranged in succession along the same plane on the surface of a tool, called a *file*. There are two differences in that (1) the chips are very small, so that the cutting action is slow and easily controlled; and (2) the cutting teeth are much wider. Consequently, fine and accurate work can be done.

Types of files. Files are classified according to the following:

1. The type, or *cut*, of the teeth.
2. The degree of coarseness of the teeth.
3. Construction.
   a. Single, solid units for hand use or in die-filing machines.
   b. Band segments, for use in band-filing machines.
   c. Disks, for use in disk-filing machines.

Four types of *cuts* are available. *Single-cut files* have rows of parallel teeth that extend across the entire width of the file at the angle of from 65 to 85°. *Double-cut files* have two series of parallel teeth that extend across the width of the file. One series is cut at an angle of 40 to 45°. The other series is coarser and is cut at an opposite angle that varies from about 10° to 80°. A *vixen-cut file* has a series of parallel, curved teeth, each extending across the file face.

FIGURE 25-17 Four cuts of files. (*Left to right*) Single, double, rasp, and curved (Vixen). (*Courtesy Nicholson File Company.*)

On a *rasp-cut* file, each tooth is short and is raised out of the surface by means of a punch. These four types of cuts are shown in Figure 25-17.

The coarseness of files is designated by the following terms, arranged in order of increasing coarseness: *dead smooth, smooth, second cut, bastard, coarse,* and *rough*. There also is a series of finer Swiss pattern files, designated by numbers from 00 to 8.

Files are available in a number of cross-sectional shapes—*flat, round, square, triangular,* and *half-round*. Flat files can be obtained with no teeth on one or both narrow edges, known as *safe edges*, so as to prevent material from being removed from a surface that is normal to the one being filed.

Most files for hand filing are from 254 to 355 mm (10 to 14 inches) in length and have a pointed *tang* on one end on which a wood or metal handle can be fitted for easy grasping.

Filing machines. Although an experienced machinist can do very accurate work by hand filing, it is a slow and tiresome task. Consequently, three types

FIGURE 25-18 Die-filing machine.

of filing machines have been developed that permit quite accurate results to be obtained much more rapidly and with much less effort. *Die filing machines,* such as is shown in Figure 25-18, hold and reciprocate a file that extends upward through the worktable. The file rides against a roller guide at its upper end, and cutting occurs on the downward stroke, so the cutting force tends to hold the work against the table. The table can be tilted to any desired angle. Such machines operate at from 300 to 500 strokes per minute, and the resulting surface tends to be at a uniform angle with respect to the table. Quite accurate work can be done but, because of the reciprocating action, approximately 50 per cent of the operating time is nonproductive.

Band-filing machines provide continuous cutting action. Most band filing is done on contour band-sawing machines by means of a special band file that is substituted for the usual band-saw blade.

The principle of a band file is shown in Figure 25-19. Rigid, straight file segments, about 3 inches long, are riveted to a flexible steel band near their leading ends. One end of the steel band contains a slot that can be hooked over a pin in the other end to form a continuous band. As the band passes over the drive and idler wheels of the machine, it flexes so that the ends of adjacent file segments move apart. When the band becomes straight, the ends of adjacent segments move together and interlock to form a continuous straight file. Where the file passes through the worktable, it is guided and supported by a grooved guide, which provides the necessary support to resist the pressure of the work against the file.

Band files are available in most of the standard cuts and in several widths and shapes. Operating speeds range from about 15 to 76 meters (50 to 250 feet) per minute.

Although band filing is considerably more rapid than can be done on a die-filing machine, it usually is not quite as accurate. Frequently, band filing may

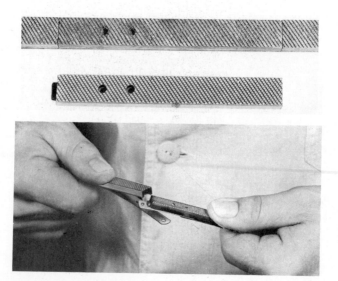

FIGURE 25-19 Band-file segments, and the method of joining the band ends to form a continuous band. (*Courtesy DoALL Company.*)

MACHINING PROCESSES

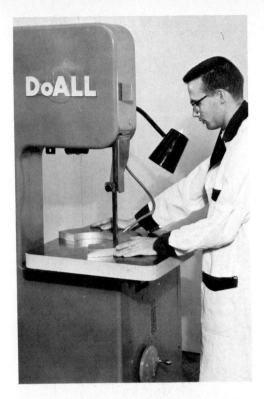

FIGURE 25-20 Typical operation on a band-filing machine. (*Courtesy DoALL Company.*)

be followed by some finish filing on a die-filing machine. Figure 25-20 shows a typical job being done on a band-filing machine.

Some *disk-filing machines* are used, having files in the form of disks, as shown in Figure 25-21. These are even more simple than die-filing machines

FIGURE 25-21 (*Above*) Disk-type filing machine. (*Left*) Some of the available types of disk files. (*Courtesy Jersey Manufacturing Company.*)

and provide continuous cutting action. However, it is difficult to obtain accurate results by their use.

REVIEW QUESTIONS

1. Why is sawing one of the most efficient of the chip-forming processes?
2. Explain why tooth spacing is important in sawing.
3. Why do the specially shaped teeth on the hacksaw blade shown in Figure 25-5 make it easier to start a cut?
4. Why are not all the teeth on a hacksaw blade shaped like the starting teeth shown in Figure 25-5?
5. Explain what is meant by the "set" of the teeth on a saw blade.
6. Why can a band-saw blade not be hardened throughout the entire width of the band?
7. What is the general relationship between the width of a band-saw blade and the radius of a cut that can be made with it?
8. What is the advantage of a spiral-type band-saw blade?
9. What are the advantages and disadvantages of circular saws?
10. Why have band-sawing machines largely replaced those using reciprocating saws?
11. Explain how a square hole can be made on a band-sawing machine.
12. How does friction sawing differ from ordinary band sawing?
13. What is the disadvantage of using gravity to feed a saw in cutting round bar stock?
14. Why does the blade of a friction band saw not melt?
15. To what extent is filing different from sawing?
16. What is a safe edge on a file?
17. Why is a band-filing machine more efficient than a die-filing machine?
18. How does a rasp-cut file differ from other types?
19. In making a 152.4-mm (6 inch) cut in a piece of AISI 1020 CR steel, 25.4 mm (1 inch) thick, the material is fed to a bandsaw blade with teeth having a pitch of 1.27 mm (20 pitch) at the rate of 0.0025 mm (0.0001 inch) per tooth. Estimate the cutting time for the cut.

CASE STUDY 25.   The Jo-Ko Company's Collars

The Jo-Ko Company requires 25,000 of the collars shown in Figure CS-25. They are to be made of 18–8 type stainless steel, and will be purchased from suitable suppliers.

Determine two practicable methods for manufacturing these collars, give an estimate of the relative cost by the two methods, and state the reasons for the difference in cost, if any. (If convenient, contact two companies that would make the collars by the methods you have selected and verify your relative-cost estimate.)

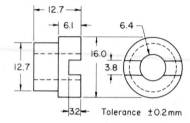

CHAPTER 26

Nontraditional Machining Process

Machining processes that involve compression-shear chip formation have a number of inherently adverse characteristics and limitations. As has been discussed previously, the formation of chips, basically, is an expensive process. Large amounts of energy are utilized in producing an unwanted product—chips. Further expenditure of energy and money is required to remove these chips and to dispose of, or recycle, them. A large amount of energy ends up as undesirable heat that often produces problems of distortion and cooling. The high forces involved create problems in regard to holding the work, and sometimes cause distortion. Undesirable cold working and residual stresses in the workpiece often require further processing to remove the effects. Finally, there are definite limitations in regard to the delicacy of the work that can be done. For example, the production of such parts as the semiconductor "chip," shown in Figure 26-1, which now play such an important role in our economy, would not be possible with any of the chip-making processes. In view of these adverse and limiting characteristics, it is not surprising that in recent years substantial effort has been devoted to developing and perfecting material-removal processes that typically do not involve the formation of chips. Nontraditional machining processes (NTM) is one designation for this diverse family of unconventional processes, which are generally nonmechanical, do not produce chips or a lay pattern in the surface and often involve new energy modes.

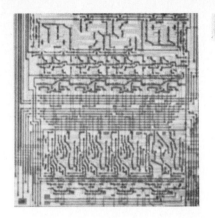

FIGURE 26-1 (*Left*) Enlarged view of one portion of a microprocessor chip. (*Above*) The chip, shown in full size, measures only 5 millimeters on a side and contains over 3000 transistors. (*Courtesy Bell Laboratories.*)

The NTM processes can be divided into four basic categories:

1. *Chemical.* Chemical reaction, sometimes enhanced by electrical or thermal energy, is the dominant mode of material removal.
2. *Electrochemical.* Electrolytic dissolution dominates the material removal process.
3. *Mechanical.* Multipoint cutting or erosion dominates the removal process.
4. *Thermal.* High temperatures in very localized regions to melt and vaporize material dominate the removal process.

Tables 26-1 through 26-4 give a summary of the characteristics for these procedures. Not included in this grouping are the new forming processes, like hot isostatic pressing or HERF, discussed in earlier chapters, which do not remove material. When examining these tables, recall that conventional turning has these typical values: surface finish, 32 to 250μ in. AA; MRR, 100 to 200 in^3/min; HP$_s$, 0.5 to 2 hp/in^3/min; V, 100 to 1000 ft/min; Accuracy $\cong$ .002 inch. One observes that the NTM processes typically have low metal-removal rates compared to machining and very high specific horsepowers. They typically have better accuracy, usually at slow rates of processing, which often results in less subsurface damage than conventional processing. These processes are usually employed when conventional machining or grinding cannot be used, often because the materials are too hard. There are numerous hybrid forms of all these processes, generally developed for special applications. Only the main ones will be described here, due to space limitations.

CHEMICAL MACHINING

Basically, *chemical machining* is the simplest and oldest of the chipless machining processes. It has been used for many years in the production of engraved plates for printing and in making small name plates. However, in its use as a machining process, it is applied to parts ranging from very small electronic circuits, such as shown in Figure 26-1, to very large parts up to 15 meters (50 feet) long.

TABLE 26-1 Summary of Chemical NTM Processes

Process	Typical Surface Finish AA (μ in.)	Typical Metal Removal Rate	Typical Specific Horsepower (hp/in.^{3}/min)	Typical Penetration Rate (ipm) or Cutting Speed (sfpm)	Typical Accuracy (in.)	Comments
Chemical machining	63–250, but can go as low as 8	30 in^3/min	Chemical energy	0.001–0.002 ipm	0.001–0.006; material and process dependent	Most all materials possible; depth of cut limited to 1/2 inch; no burrs; no surface stresses; tooling low cost
Electropolishing	4–32, but can go as low as 2 or 1 or better	Very slow	50–200 amperes per square foot	0.0005–0.0015 ipm	NA[a]; process used to obtain finish	High quality, no stress surface; removes residual stresses; makes corrosion resistant surfaces; may be considered to be an electrochemical process
Photochemical machining (blanking)	63–250, but can go as low as 8	Same as chemical milling	DC power	0.0004–0.0020 ipm	10% of sheet thickness or 0.001–0.002 inch	Limited to thin material; burr-free blanking of brittle material; tooling low cost; used microelectronic
Thermochemical machining (combustion machining)	Burr-free	Minute with rapid cycle time	NA[a]	NA[a]	NA[a]	For burrs and fins on cast or machined parts; deburr steel gears automatically

[a]NA, not applicable for this process.

TABLE 26-2 Summary of Electrochemical NTM Processes

Process	Typical Surface Finish AA (μin.)	Typical Removal Rates (in³/1000 Amp-Min)	Typical Specific Horsepower (hp/in³/min)	Typical Penetration Rate (ipm) or Cutting Speed (sfpm)	Typical Accuracy (in.)	Comments
Electrochemical machining (ECM)	16–63	0.06 in W, Mo 0.16 in CI 0.13 in steel, Al 0.60 in Cu	160	.1 to .5 ipm	0.0005–0.005 ≅ 0.002 in cavities	Stress free metal removal in hard to machine metals; tool design expensive; disposal of chemicals a problem; MRR independent of hardness; deep cuts will have tapered walls
Electrochemical grinding (ECG)[a]	8–32	0.010	High	Cutting rates about same as grinding; wheel speeds, 4000–6000	0.001–0.0005	Special form of ECM; grinding with ECM assist; good for grinding hard conductive materials like tungsten carbide tool bits; no heat damage, burrs, or residual stresses
Electrolytic hole machining (Electrostream)[b]	16–63	NA[c]	NA[c]	0.060–0.120 ipm	≅ 0.001 or 5% of dia. of hole	Special version of ECM for hole drilling small round or shaped holes; multiple-hole drilling; typical holes 0.004 to 0.03 inch in diameter with depth-to-diameter ratio of 50:1

[a]Honing can also be done with EC assistance which can quadruple the MRR over conventional honing and yield 2 μ in. AA finish.
[b]Trademark of General Electric Company.
[c]NA, not applicable for this process.

TABLE 26-3   Summary of Mechanical NTM Processes

Process	Typical Surface Finish AA (μ/in.)	Typical Metal Removal Rate (in^3/min)	Typical Specific Horsepower (hp/in^3/min)	Typical Penetration Rate (ipm) or Cutting Speed (sfpm)	Typical Accuracy (in.)	Comments
Abrasive flow machining	30–300; can go as low as 2	Low	NA[a]	Low	0.001–0.002	Typically used to finish inaccessible integral passages; often used to remove recast layer produced by EDM; used for burr removal; (cannot do blind holes)
Abrasive jet machining	10–50	Very low; fine finishing process, 0.001	NA[a]	Very low	$\cong 0.005$ typical, $\cong 0.002$ possible	Used in heat-sensitive or brittle materials; produces tapered walls in deep cuts
Hydrodynamic machining	Generally 30–100	Depends on material	NA[a]	Depends on material	0.001 possible	Used for soft nonmetallic slitting; no heat-affected zone; produces narrow kerfs (0.001–0.020 inch); high noise levels
Ultrasonic machining	16–63; as low as 8	Slow, 0.05 typical	200	0.02–0.150 ipm	0.001–0.0005	Most effective in hard materials, R_c >40; tool wear and taper limit hole depth to width at 2.5 to 1; tool also wears

[a]NA, not applicable for this process.

TABLE 26-4 Summary of Thermal NTM Processes

Process	Typical Surface Finish AA (μ/in.)	Typical Metal Removal Rate (in^3/min)	Typical Specific Horsepower (hp/in^3/min)	Typical Penetration Rate (ipm) or Cutting Speed (sfpm)	Typical Accuracy (in.)	Comments
Electron beam machining (EBM)	32–250	0.0005 max.; extremely low	10,000	200 sfpm; 6 ipm	0.001–0.0002	Micromachining of thin materials and hole drilling minute holes with 100:1 depth to diameter ratios; work must be placed in vacuum but suitable for automatic control; beam can be used for processing and inspection; used widely in microelectronics.
Laser beam machining (LBM)	32–250	0.0003; extremely low	60,000	4 ipm	0.005–0.0005	Can drill 0.005 to 0.050 inch dia. holes in materials 0.100 inch thick in seconds; same equipment can weld, surface heat treat, engrave, trim, blank, etc.; has heat affected zone and recast layers which may need to be removed.
Electrical discharge machining (EDM)	32–105	0.3	40	0.5 ipm	0.002–0.00015 possible	Oldest of NTM processes; widely used and automated; tools and dies expensive; cuts any conductive material regardless of hardness; delicate, burr free parts possible; always for recast layer
Electrical discharge wire cutting	32–64	0.10–0.3	40	4 ipm	$\cong$ 0.0002	Special form of EDM using traveling wire; cuts straight narrow kerfs in metals 0.001 to 3 inches thick; wire diams. of 0.002 to 0.010 used; N/C machines allow for complex shapes
Plasma beam machining (PBM)	25–500	10	20	50 sfpm; 10 ipm; 120 ipm in steel	0.1–0.02	Clean, rapid cuts and profiles in almost all plates up to 8 inches thick with 5 to 10° taper

In chemical machining, material is removed from selected areas of a workpiece by immersing it in a chemical reagent. Material is removed by microscopic electrochemical cell action, as occurs in corrosion or chemical dissolution of a metal; no external circuit is involved. This controlled chemical dissolution of the workpiece will simultaneously etch all exposed surfaces, which enhances the productivity even though the penetration rates of the etch may be only 0.0005 to 0.0030 inch per minute. The basic process takes many forms: *chemical milling,* for pockets, contours, and overall metal removal, so named because in its earliest use, it replaced milling; *chemical blanking,* for etching through thin sheets; sometimes called *photochemical machining* when photosensitive resists are used for masks, resulting in very detailed parts which in fact could not be blanked by conventional means: *gel milling,* using reagant in gel form; *chemical or electrochemical polishing,* where weak chemical reagants are used, sometimes with remote electric assist, for polishing or deburring; *chemical jet machining,* using a single, chemically active jet.

Chemical milling or blanking. The material removal processing steps in chemical milling and chemical blanking are:

1. *Prepare.* Degrease, clean, rinse, pickle, or preclean to provide for good adhesion for masking material.
2. *Mask.* Coating or covering areas not to be etched.
3. *Etch.* Chemical dissolution by spray or dip followed by rinse.
4. *Remove mask.* Strip or demask, clean and desmut as necessary.
5. *Finish.* Post-treatments and finish inspection.

After cleaning and masking, the part is immersed or sprayed with the proper etchant and is permitted to remain in the reagant until the desired amount of material has been removed. If the immersion technique is employed, the bath is agitated or circulated so as to sweep away the waste products and thereby keep the metal exposed to the reagant. With the spray technique this is not required, so this procedure usually is preferred when the size and shape of the workpiece permit. The major complexity in the process involves providing a maskant on the surface of the workpiece so that removal will occur only on desired areas. This can be accomplished by photographic means through the use of *photosensitive resists,* or achieved by completely covering the surface with a maskant and then manually peeling away the maskant from those areas where metal removal is desired.

Chemical machining with photosensitive resists. Figure 26-2 shows the steps that are involved when chemical machining is done through the use of photosensitive resists. These are as follows:

1. Prepare the "artwork." An accurate drawing of the workpiece is made, usually on polyester drafting film or glass and up to 50 or more times the size of the final part. With such magnifications, an accuracy of 0.64 mm (0.025 inch) in the original drawing will permit 0.013 mm (0.0005 inch) to be achieved in the workpiece. By special procedures, lines of 2 μm can be made.

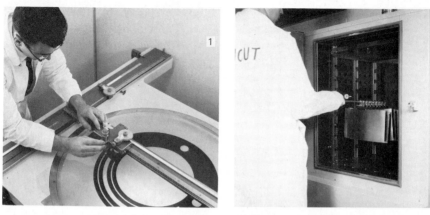

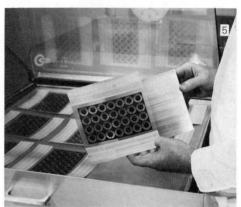

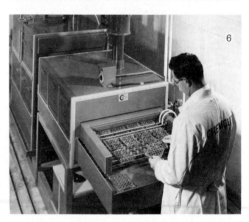

FIGURE 26-2 Basic steps in chemical machining with the use of photoresists. (1) Preparing the artwork. (2) Making reduced-size template from the artwork by photography. (3) Dipping and draining metal sheets in photoresist. (4) Drying photoresist coating in oven. (5) Placing sensitized metal behind template. (6) Chemically machined parts emerging from etching machine. (*Courtesy Chemcut Corporation.*)

2. Reduce the original drawing by photographic means to obtain a negative, master pattern that is exactly the size of the finished part. This reduction may require several steps, using regular industrial photographic equipment.
3. Coat the workpiece with a light-sensitive emulsion, usually by dipping or by spraying. The emulsion, or resist, then is dried, usually in an oven.
4. After the sensitized workpiece has been placed against the negative, usually in a vacuum frame to ensure good contact, it is exposed to blue light, passing through the negative. Mercury-vapor lamps commonly are employed as the light source. Exposure to the light hardens the selected areas of the resist so that it will not be washed away in the subsequent developing.
5. Develop the workpiece. This removes, or dissolves away, the unexposed areas of the resist, thereby exposing the areas of the workpiece that are to be acted upon by the chemical reagant. The final developing step is to rinse away all residual material.
6. Spray the workpiece with (or immerse it in) the reagant.

Chemical machining with the aid of photosensitive resists has been widely used for the production of small, complex parts, such as electronic circuit boards, and very thin parts, such as are shown in Figure 26-3, that are too small to permit their being blanked by ordinary blanking dies. Supplemented by plating, sputtering, and vacuum deposition, which make it possible to deposit metallic films of controlled thickness, it has made possible the tremendous developments of solid-state circuitry, much of it in miniature size.

The use of scribed maskants. Although photosensitive resists are used in the majority of chemical machining operations, there are some cases where the older scribed-and-peeled maskants are employed. These usually are (1) where the workpiece is not flat, (2) where it is very large, and (3) for low-volume work where the several steps required in using photosensitive resists are not economically justified. In this procedure the maskant is applied to the entire surface of the workpiece, either by dipping or spraying. It then is removed from those areas where metal removal is desired, by scribing through the maskant with a

FIGURE 26-3 Typical parts produced by chemical machining. (*Courtesy Chemcut Corporation.*)

knife and peeling away the desired portions. Where volume permits, scribing templates can be used to assist in the scribing. Figure 26-4 shows the use of a scribing template and the stripping of the maskant from the workpiece.

Chemical machining to multiple depths.   If all areas are to be machined to the same depth, only a single masking, or resist application sequence, and immersing are required. Machining to two or more depths, called *step machining,* can be accomplished by removing the maskant from additional areas after the original immersion. Figure 26-5 illustrates the steps required for stepped chemical machining.

Parts having either uniformly or variably tapered cross sections can be produced by chemical machining through the relatively simple procedure of withdrawing them vertically from the bath at controlled rates. In this way, different areas are exposed to the chemical action for differing amounts of time.

Design factors in chemical machining.   When designing parts that are to be made by chemical machining, several unique factors related to the process must be kept in mind. First, dimensional variations can occur through size changes in the art work due to temperature and humidity changes. These usually can be eliminated or controlled by drawing the art work on thicker polyester films or on glass. If very accurate dimensions must be held, the room temperature and humidity should be controlled. The photographic film used in making the master negative also can be affected to some degree by temperature and humidity, but control of handling and processing conditions can eliminate this difficulty.

The second item that must be considered is the *etch factor* or *etch radius,* which describe the undercutting of the maskant. The etchant acts on whatever surface is exposed. Areas that are exposed longer will have more metal removed

FIGURE 26-4   (*Left*) Scribing a maskant on a part to be chemically machined, using a scribing template. (*Right*) Stripping the maskant from a part after it has been scribed. (*Courtesy United States Chemical Milling Corporation.*)

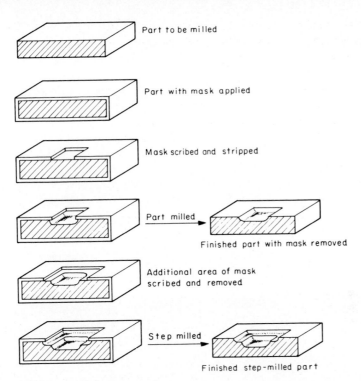

Part to be milled

Part with mask applied

Mask scribed and stripped

Part milled → Finished part with mask removed

Additional area of mask scribed and removed

Step milled → Finished step-milled part

FIGURE 26-5 Steps required to produce a stepped contour by chemical machining.

from them. Consequently, as the depth of etch increases, there is a tendency to undercut or etch under the maskant, as illustrated in Figure 26-6. When the etch depth is only a few hundredths of a millimeter, as often is the case, this causes little or no difficulty. But when the depth is substantial, whether etching from only one or both sides, and when doing chemical blanking, the conditions shown in Figure 26-6 result. In making grooves, the width of the opening in the maskant must be reduced by an amount sufficient to compensate for the etch radius. This radius varies from about one-fourth to three-fourths of the *depth* of the etch (for breakthrough in chemical blanking, $t/2$), depending on the type of material and to some extent on the depth of the etch. Consequently, it is difficult to produce narrow grooves except when the etch depth is quite small.

All allowances for the etch factor must be taken into account in designing the part and the original artwork or scribing template. The values indicated in Figure 26-6 are *minimum* values; it has been found that results will vary between etching machines, and actual etch allowances will have to be somewhat greater and adapted to the specific conditions.

In chemical blanking, with etching occurring from both sides, a sharp edge remains along the line at which breakthrough occurs, as in Figure 26-6c. Such an edge usually is objectionable, so etching ordinarily is continued to produce the straight-wall condition shown in (d).

Etching from both sides, of course, requires the preparation of two maskant patterns and careful registration of them on the two sides of the workpiece.

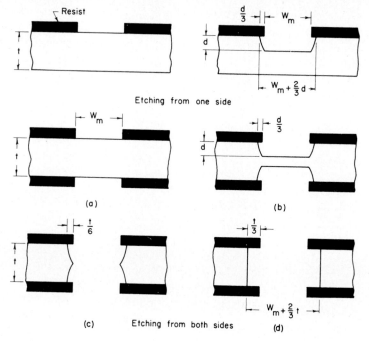

Etching from one side

(a)　(b)

(c)　Etching from both sides　(d)

FIGURE 26-6　Effect of the "etch factor" in chemical machining.

If the bath is not agitated properly the "overhang" condition depicted in Figure 26-7 may result, particularly on deep cuts. Not only is the resulting dimension of the opening incorrect, but a very sharp edge may be produced. Other common defects, like islands and dishing, are also depicted in Figure 26-7. Other common processing problems include selective etching at the grain boundries, producing microcracks, and pitting, caused by selective etching (etching at unequal etch rates) of the workpiece.

Advantages and disadvantages of chemical machining.　Chemical machining has a number of distinct advantages. Except for the preparation of the

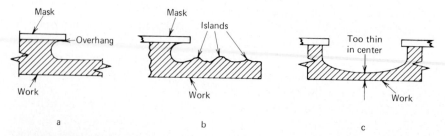

a　b　c

FIGURE 26-7　Typical chemical milling defects. (a) Overhang—deep cuts with improper agitation. (b) Islands—isolated high spots from dirt, residual maskant or work material inhomogeneity. (c) Dishing—thinning in center due to improper agitation or stacking of parts in tank.

artwork and master negative, or a scribing template, the process is relatively simple, does not require highly skilled labor, induces no stresses or cold working in the metal, and can be applied to almost any metal—aluminum, magnesium, titanium, and steel being most common. Large areas can be machined—tanks for parts up to 3.7 × 15.2 meters (12 × 50 feet) are available. Machining can be done on parts of virtually any shape, and thin sections, such as honeycomb, can be machined because there are no mechanical forces involved. Consequently, chemical machining is very useful and economical for weight reduction. Figure 26-8 shows a typical large, thin part that has been chemically milled.

The tolerances obtainable with chemical machining are good—from ± 0.013 mm (± 0.0005 inch) with care on small etch depths to ± 0.076 mm (± 0.003 inch) in routine production involving substantial depths. The surface finish is good, seldom having a roughness greater than 0.0025 mm (0.0001 inch).

In using chemical machining, some disadvantages and limitations should be kept in mind. The metal-removal rate is slow *in terms of unit area exposed,* being about 0.1 to 0.2 kg per minute per square meter exposed (0.2 to 0.04 pound per minute per square foot exposed) in the case of steel. However, where large areas can be exposed at a time, or where the metal is quite thin, the overall removal rate may compare favorably with other metal-removal processes.

The soundness and homogeneity of the metal are very important. Wrought materials should be uniformly heat-treated and stress-relieved prior to processing. Although chemical machining induces no stresses, it may release existing residual stresses in the metal and thus cause warping. Castings can be chemically machined provided they are sound and have uniform grain size. Lack of the

FIGURE 26-8 Inspecting curved parts that have been chemically machined. (*Courtesy North American Aviation, Inc.*)

latter can cause difficulty. Because of the different grain structures that exist near welds, weldments usually are not suitable for chemical machining.

Sophisticated versions of chemical machining are extensively used in the fabrication of integrated microelectronic circuits.

Chemical milling and conventional machining often can be combined advantageously for producing parts, a fact that often is overlooked.

The tolerance in chemical milling increases with the depth of cut and with faster etch rates and varies for different metals, as shown in Figure 26-9. A detailed description of this process has been prepared by Bellows.[1]

Thermochemical machining has been developed for the removal of burrs and fins by exposing the workpiece to hot corrosive gases, for a shorter period of time. The workpiece remains unaffected and relatively cool because of its low surface-to-mass ratio and the short exposure time. Alternatively, fine burrs can be removed quickly by exposing the parts to a suitable chemical spray and at much less cost than if done by hand. *Of course, some smaller amounts of metal*

---

[1]*Chemical Machining: Production with Chemistry,* MDC 82-102, Metcut Research Associates Inc., Machinability Data Center, Cincinnati, Ohio, 1982.

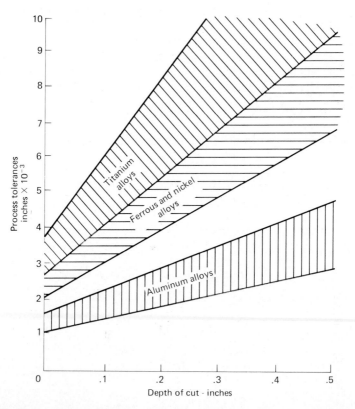

FIGURE 26-9 Chemical milling tolerance bands showing variation with respect to depth of cut and different metals. (*Adapted from* Chemical Machining: Production with Chemistry, *MDC 82-102, Metcut Research Associates Inc., Machinability Data Center, Cincinnati, Ohio, 1982.*

is removed from all exposed surfaces, and this must be permissible if the process is to be used. Consequently, the procedure usually can be used only for removing very small burrs.

The hot gases are formed by detonating explosive mixtures of oxygen, hydrogen, and natural gas in a chamber with the parts. A thermal shock wave vaporizes the burrs found on gears, die castings, valves, and so on, in a few milliseconds. The process has been automated and cycle times of 15 seconds are possible.

ELECTROCHEMICAL MACHINING

Electrochemical machining, commonly designated ECM, removes material by anodic dissolution with a rapidly flowing electrolyte. It is basically a deplating process in which the tool is a cathode and the workpiece is the anode, so both must be electrically conductive. The electrolyte, which can be pumped rapidly through or around the tool, sweeps away the waste product (sludge) and captures it by settling in filters. The shape of the cavity is the mirror image of the tool, which is advanced via a servomechanism which controls the gap (0.003 to 0.030 inch, with 0.010 inch typical) between the electrodes. The tool advances into the work at a constant feed rate which matches the dissolution rate of the electrodes. This is essentially a deplating operation. The electrolytes are highly conductive solutions of inorganic salts, usually NaCl, KCl, $NaNO_3$ (or other proprietary mixtures) and are operated at about 90 to 125°F with flow rates ranging from 50 to 200 feet per second (fps). Tools are usually made of copper or brass and sometimes stainless steel. The process is shown schematically in Figure 26-10.

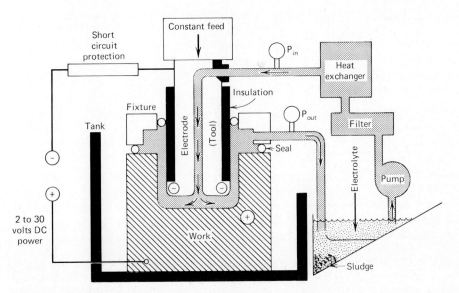

FIGURE 26-10 Schematic diagram of electrochemical machining process (ECM).

As shown in Figure 26-11 the metal-removal rate, in terms of penetration of the tool into the workpiece, is primarily a function of the current density. Consequently, current densities from 2.33 to 3.10 A/mm² (1500 to 2000 amperes per square inch) are used and, in suitable applications, ECM provides metal-removal rates on the order of 0.1 in³/min/1000 A. The cutting rate is solely a function of the ion-exchange rate and is not affected by the hardness or toughness of the work material. Cutting rates up to 2.54 mm (0.1 inch) of depth per minute are obtained routinely in Waspalloy, a very hard metal alloy.

ECM is well suited for mass production of complex shapes in difficult-to-machine but conductive materials. The principal tooling cost is for the preparation of the tool electrode, which can be time consuming and costly, requiring several "cut and try" efforts, except for simple shapes. There is no wear of the tool during actual cutting as the tool is protected cathodically. The process produces a stress-free surface, and the ability to cut the entire cavity simultaneously aides productivity. Process control must be exact to obtain tight tolerances, and the tools must be designed to compensate for variable current densities produced by electrode geometries or electrolyte variations. For example, corners in cavities are automatically rounded due to the concentration of the current density at the edge of the tool.

The ECM process has a number of modified forms. Electrochemical polishing operates essentially the same as ECM but the feed is halted. Lower current densities and slower electrolyte flow rates greatly reduce the metal removal rates so that the surface develops a fine finish, 10 to 12 μin. AA being typical.

Electrochemical hole drilling processes have been developed for drilling very small holes using high voltages and acid electrolytes. The tool is a drawn glass nozzle with an internal electrode. Multiple sets of glass tubes are employed and over 50 holes per stroke can be done. This technique was developed to drill the cooling holes in gas turbine blades. Stress-free holes from 0.004 to 0.030 inch in diameter with 50:1 depth-to-diameter ratios are routinely accomplished in nickel and cobalt alloys. Acid is used so that the dissolved metals go into solution instead of forming a sludge.

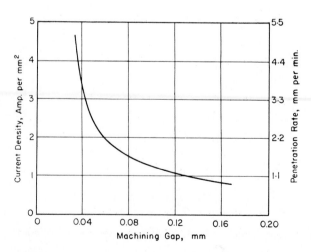

FIGURE 26-11 Relationship of current density, penetration rate, and machining gap in electrochemical machining.

The process can be used to drill shaped holes in difficult-to-machine, conductive metals. Holes up to 24 inches deep with diameters ranging from 0.020 to 0.250 inch are possible. The major differences between this process and the hole drilling process described above are the reduced voltage levels (5 to 10 volts dc) and special electrodes, which are long, straight, acid-resistant tubes coated with an enamel insulation. The acid is pressure fed through the tube and returns via the gap (0.001 to 0.002 inch) between the insulated tube wall and the hole wall.

Electrochemical grinding. *Electrochemical grinding,* commonly designated as ECG, is a variant of electrochemical machining in which the tool electrode is a rotating, metal-bonded, diamond grit grinding wheel. The arrangement of the tool and workpiece is indicated in Figure 26-12. The metal bond of the wheel is the cathode. The diamond particles serve three functions: (1) as insulators to preserve a small gap between the cathode and the work, (2) to wipe away the residue, and (3) to cut chips if the wheel should contact the workpiece, particularly in the event of a power failure. When operated properly, less than 5% of the material is removed by normal chip forming. The process is used for shaping and sharpening carbide cutting tools, which cause high wear rates on expensive diamond wheels in normal grinding. Electrochemical grinding greatly reduces this tool wear. Fragile parts (honeycomb structures), surgical needles, and tips of assembled turbine blades have been ECG-processed successfully. The lack of heat damage, burrs, and residual stresses are very beneficial, particularly when coupled with MRRs that are competitive with conventional grinding but with far less tool wear.

Electrochemical deburring. Limited use is made of the ECM principle for removing burrs from parts. The work is put into a rotating, electrically insulated drum that contains two current-carrying electrodes that are insulated from the drum. Small graphite spheres, added to the electrolyte, receive an inductive charge from the electrodes and thus have sufficient potential gradient across the sphere-to-workpiece gap to cause electrochemical machining to occur as they

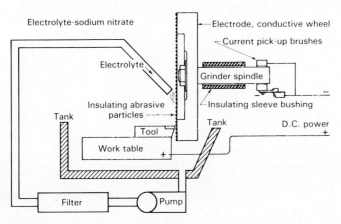

FIGURE 26-12 Equipment setup and electrical circuit for electrochemical grinding.

move randomly over all areas of the workpiece. Because the current density is higher at the protrusions of the burrs than at smooth areas on the workpiece, they are preferentially removed. As in chemical deburring, there is a slight dimensional change throughout the workpiece, in this case due to the general ECM action and to the natural abrasive character of the graphite spheres.

MECHANICAL NTM PROCESSES

Ultrasonic machining employs an ultrasonically vibrating tool to impel the abrasive in a slurry against the workpiece. The tool forms a reverse image in the workpiece as the abrasive-loaded slurry abrades (machines) the material. Boron carbide, aluminum oxide, and silicon carbide are the most commonly used grit material. The process can cut virtually any material but is most effective on materials with hardness greater than $R_c = 40$. Figure 26-13 shows a simple schematic of this process.

Ultrasonic machining uses a transducer to impart high-frequency vibrations to the toolholder. Abrasive particles in the slurry are accelerated to great speed by the vibrating tool and perform the actual cutting. The tool materials are usually brass, carbide, or mild or tool steel and will vary in tool wear depending upon their hardness. Wear ratios of 1:1 to 100:1 (material removed versus tool lost to wear) are possible. The tool must be strong enough to resist fatigue failure.

The cut will be oversize by about twice the size of the abrasive grit being

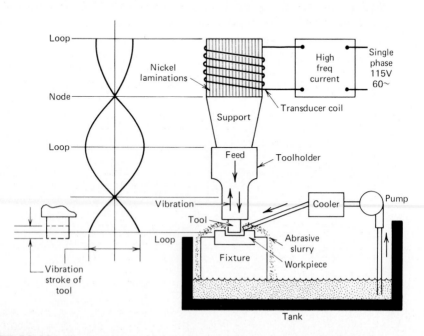

FIGURE 26-13 Ultrasonic machining shown sinking a hole in a workpiece.

used and holes will be tapered, usually limiting the hole depth to a diameter ratio of about 3:1. Surface roughness is controlled by the size of the grit (finer finish with smaller grits). Holes, slots, or shaped cavities can be readily eroded in any material, conductive or nonconductive, metallic, ceramic, or composite.

Ultrasonics have also been employed for coining, lapping, deburring, and broaching. Plastics can be welded using ultrasonics.

Three other mechanical NTM processes are listed in Table 26-3.

Hydrodynamic jet machining uses a high-velocity fluid jet (Mach 2) impinging the workpiece, performing a slitting operation. A long-chain polymer is added to the water of the jet to make the jet coherent (not come out of the mozzle as a mist) under the 10,000- to 60,000-psi nozzle pressure. The jet is typically 0.002 to 0.040 inch in diameter. This process is mainly used to cut soft, non-metalics like acoustic tile, plastics, paperboard, asbestos (with no dust), leather, rubber, and fiberglass. Generally, the kerf of the cut is about 0.001 inch greater than the diameter of the jet and is not sensitive to dwell. Cutting rates vary from 250 fpm for acoustic tile up to 6000 fpm for paper products (see Figure 26-14).

Abrasive jet machining is a finishing process that removes material with the abrasive action of a gas jet (air, nitrogen, or carbon dioxide) loaded with abrasive powder particles (silicon carbide, aluminum oxide, or glass). Jet velocity is of the order of 1000 fps for this hand-held process. One can deburr, polish, etch, scribe, slit, groove, or cut shapes or holes. The process works well in heat-sensitive, brittle, thin, or hard materials.

Abrasive flow machining uses a slurry of abrasive material flowing over or through a part, under pressure, to perform edge finishing, deburring, radiusing, polishing, or minor surface machining. Aluminum oxide, silicon carbide, boron carbide, and diamonds are used as abrasives in grit sizes ranging from No. 8 to No. 700. This process is useful for polishing and deburring inaccessible internal passageways and works for most materials.

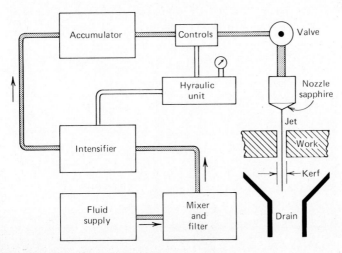

FIGURE 26-14 Schematic diagram of hydrodynamic jet machining. The intensifier elevates the fluid to the desired nozzle pressure while the accumulator smooths out pulses in the fluid jet.

THERMAL PROCESSES

Electrodischarge machining (EDM) cuts metal by discharging electric current stored in a capacitor bank across a thin gap between the tool (cathode) and the workpiece (anode). Literally thousands of sparks per second are generated and each spark produces a tiny crater by vaporization, thus eroding the shape of the tool into the workpiece. The dielectric fluid (kerosene) flushes out the "chips" and confines the spark. Of all the exotic metalworking processes, none has gained greater industry-wide acceptance than EDM. Figure 26-15 shows a schematic of the process.

In EDM, each spark contains a discrete, measured, and controllable amount of energy, so MRR and surface finish can be predicted, while size is carefully controlled. The heat generated by the spark melts the metal, and the impact of the spark causes it to be ejected, possibly vaporized, and recast in the dielectric as spheres. Materials of any hardness can be cut as long as the material can conduct electricity. Any shape that can be cut into the tool can be reproduced in the workpiece. Spark reversals and unflushed chips cause sparking on the tool, producing unwanted tool wear and taper, so tools are usually made in duplicate sets by precision numerical control machining, for production runs. About 80 to 90% of EDM work is in the manufacture of tool and die sets for production casting, forging, stamping, and extrusion.

The absence of almost all mechanical forces makes it possible to machine fragile parts without distortion. In addition, fragile tools, such as wires, can be used. While EDM is slow compared to conventional methods and even ECM and produces a brittle, recast layer on the surfaces, its controllability, versatility, and accuracy usually result in superior design flexibility. Significant cost reduc-

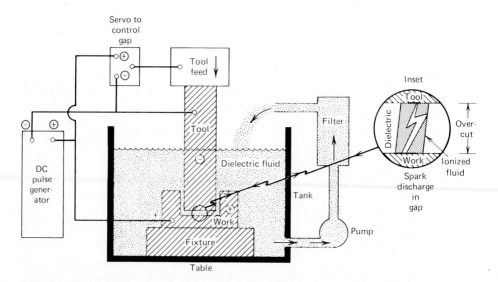

FIGURE 26-15 EDM or spark erosion machining of metal, using high-frequency spark discharges in a dielectric, between the shaped tool (cathode) and the work (anode). The table can make *X-Y* movements.

tions in the manufacture of tools and dies from steels of any hardness and carbides can be achieved.

The surface finish produced by EDM is a matte finish, composed of many small craters. See Figure 26-16, which shows an EDM surface on top of a ground surface. Note the small sphere in the lower right corner attached to the surface. In EDM, surface finish varies widely as a function of the spark frequency, voltage, and current, parameters which, of course, also control the MRR.

The distance between the surface of the electrode and the surface of the workpiece represents the *overcut* and it is equal to the length of the spark, which is essentially constant over all areas of the electrode, regardless of size or shape. Typical overcut values range from 0.0005 to 0.020 inch. Overcut depends on the gap voltage plus the "chip" size, which varies with the amperage. EDM equipment manufacturers publish overcut charts for the different power supplies for their machines, but these values should be used mainly as guides for the tool designer. The dimensions of the electrode (the tool) are basically equal to the desired dimensions of the part less the overcut values.

Many different materials are used for electrode material, with the best choice depending on such things as the application, how easily the electrode material machines, how fast it wears, how fast it cuts, what kind and quality of finish it can produce, what type of power supply is being used, and how much the material costs. Graphite, copper, brass, copper–tungsten, aluminum, 70/30 zinc–tin, and other alloys are used for electrode materials.

The dielectric fluid has four main functions: insulation between tool and work; spark conductor; coolant; and flushing medium. This fluid must ionize to provide a channel for the spark and deionize quickly to become an insulator. Polar compounds, like glycerine–water (90:10) with triethylene oil as an additive, have been shown to improve the MRR and decrease the tool wear when compared with traditional oil cutting fluids, like kerosene.

A special form of EDM is shown in Figure 26-17 wherein the electrode is a

FIGURE 26-16 EDM surface on top of a ground surface in steel. Spherodial nature of debris from the surface in evidence around the craters.

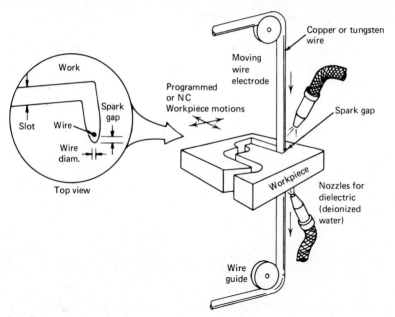

FIGURE 26-17 Schematic diagram of equipment for electrodischarge machining using a moving wire electrode.

continuously moving conductive wire. The tensioned wire of copper, brass, or tungsten is used only once, traveling from a take-off spool to takeup spool while being "guided" to produce a straight narrow kerf in plates up to 3 inches thick. The wire diameter ranges from 0.002 to 0.010 inch with positioning accuracy up to ± 0.0002 inch in machines with NC or tracer control. The dielectric is usually deionized water. This process is widely used for the manufacture of punches, dies, and stripper plates, with modern machines capable of cutting die relief, intricate openings, tight radius contours, and corners routinely.

Advantages and disadvantages of EDM.   Electrodischarge machining is applicable to all materials that are fairly good electrical conductors, thus including metals, alloys, and most carbides. The melting point, hardness, toughness, or brittleness of the material impose no limitations. Thus it provides a relatively simple method for making holes of any desired cross section in materials that are too hard or brittle to be machined by most other methods. There are virtually zero forces between the tool and the workpiece, so that very delicate work can be done. The process leaves no burrs on the edges. Its use has expanded very rapidly, and it not only is used to produce the type of work shown in Figure 26-18, but now it is widely used to produce large body-forming dies in the automotive industry.

On most materials, the process produces a thin, hard recast surface, which may be an advantage or a disadvantage, depending on the use. When the workpiece material is one that tends to be brittle at room temperatures, the surface may contain fine cracks caused by the thermally induced stresses. Consequently,

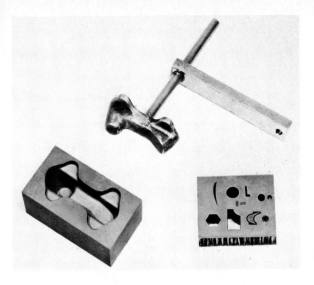

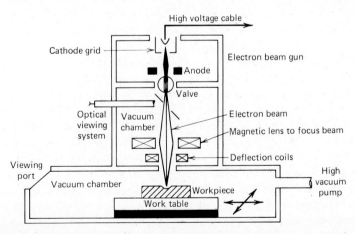

FIGURE 26-18 (*Left*) Die cavity formed by electrodischarge machining. (*Top*) Electrode used for forming the die cavity. (*Right*) Holes formed in honeycomb panel by electrodischarge machining. (*Courtesy Cincinnati Milacron, Inc.*)

some other finishing process often is used subsequent to EDM to remove a thin surface layer.

Electron Beam Machining (EBM) is a thermal NTM process which uses a beam of high-energy electrons focused on the workpiece to melt and vaporize metal. This micromachining process is performed in a vacuum chamber (10^{-5} mm of mercury). Magnetic lenses are used to focus the beam and deflection coils control the position of the beam. The desired beam path can be programmed with a computer to produce any desired pattern in the work. The spot sizes are on the order of 0.0005 to 0.001 inch and holes or narrow slits with depth-to-width ratios of 100:1 can be "machined" with great precision in a short time

FIGURE 26-19 Electron beam machining uses high-energy electron beam (10^9 watts/in²) to melt and vaporize metal in 0.001-inch-diameter spots. (*Reprinted, by permission, from* Nontraditional Machining Guide: 26 Newcomers for Production, *Metcut Research Associates Inc., Machinability Data Center, Cincinnati, Ohio, 1982.*)

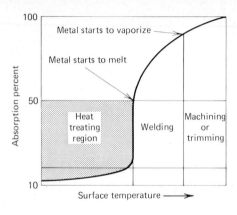

FIGURE 26-20 The power density input and the percent of surface heat absorbed determines whether the laser cuts, welds, or heat treats.

in any material. The interaction of the beam with the surface produces dangerous X-rays, so shielding is necessary. The layer of recast material and the depth of heat damage is very small and for micromachining, processing speeds can exceed that of EDM or ECM. Typical tolerances are about 10% of the hole diameter or slot width. These machines require high voltages (50 to 200 kilovolts) to accelerate the electrons to speeds of 0.5 to 0.8 the speed of light and should be operated by fully trained personnel. See Figure 26-19 for a schematic.

Laser Beam Machining (LBM) is a thermal NTM process which uses a laser to melt and vaporize materials. The beam can be focused down to 0.005 inch in diameter for drilling microholes through materials as thick as 0.100 inch, but hole depth-to-diameter ratios of 10:1 are more typical. High-energy solid-state and gas lasers are needed with the optical characteristics of the workpiece determining what wavelength of light energy should be used. The range is 0.69 μm for ruby lasers up to 10.6 μm for carbon dioxide gas lasers. The small beam divergence, high peak powers (in pulsed lasers), and single frequency provide power densities of the order of 10^5 to 10^9 watts per square inch, allowing holes

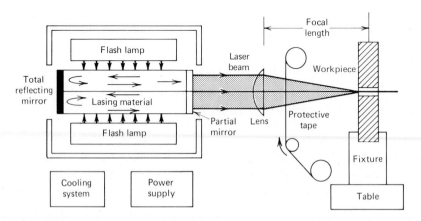

FIGURE 26-21 Schematic diagram of a laser beam machine, a thermal NTM process which can micromachine any material. (*Reprinted, by permission, from* Nontraditional Machining Guide: 26 Newcomers for Production, *Metcut Research Associates Inc., Machinability Data Center, Cincinnati, Ohio, 1982.*)

0.020 inch in diameter to be drilled in milliseconds with accuracy of $\pm$ 0.001 inch. However, this is not a mass metal-removal process. Hole geometry is irregular, and there will be a recast layer and a heat-affected zone which can be detrimental to material properties. The same laser can be used to cut, weld, heat-treat, and trim by varying the power density along with appropriate adjustments in output beam intensity, focus, and duration (see Figure 26-20).

Off-the-shelf laser systems are now available with NC controls and are being used for applications ranging from cigarette paper cutting to drilling microholes in turbine engine blades. Figure 26-21 shows a schematic of laser machining. Protective materials are absolutely necessary when working around laser equipment due to the potential damage to eyesight from either direct or scattered laser light.

REVIEW QUESTIONS

1. What are the four basic types of NTM processes?
2. Give three reasons why chipless machining processes are likely to have greater importance in the future.
3. What is the purpose of a photosensitive resist?
4. What are the six steps in chemical machining using photosensitive resists?
5. Why is it perferable in chemical machining to apply the etchant by spraying instead of by immersion?
6. For what types of parts are photosensitive resists not suitable?
7. Explain how multiple depths can be produced by chemical machining.
8. Would it be feasible to produce a groove 2 mm wide and 3 mm deep by chemical machining?
9. A drawing calls for making a groove 23 mm wide and 3 mm deep by chemical machining. What should be the width of the opening in the maskant?
10. Could an ordinary steel weldment be chemically machined? Why or why not?
11. How would you produce a tapered section by chemical machining?
12. What distinguishes chemical blanking from ordinary chemical machining?
13. Is ECM related to chemical machining?
14. What effect does work-material hardness have on the metal-removal rate in ECM?
15. Explain the basic principle involved in electrochemical deburring.
16. What is the principal cause of tool wear in ECM?
17. Basically, what type of process is electrochemical grinding?
18. Would electrochemical grinding be a suitable process for sharpening ceramic tools? Why or why not? What about using ultrasonics?
19. Upon what factors does the metal-removal rate depend in ECM?
20. Why is the tool insulated in the ECM schematic?
21. Is ultrasonic machining really a chipless process?
22. Where, in the ultrasonic stroke, is the acceleration and deceleration of the tool the greatest? Given that $f = ma$, what does this mean about the force given to the grits in the slurry?
23. What is the nature of the surface obtained by electrodischarge machining?
24. What is the principal advantage of using a moving-wire electrode in electrodischarge machining?
25. What effect would increasing the voltage have on the metal-removal rate in electrodischarge machining? Why?
26. If the metal from which a part is to be made is quite brittle and the part will be subjected to repeated tensile loads, would you select ECM or electrodischarge machining for making it? Why?

27. If you had to make several holes in a large number of duplicate parts, would you prefer ECM, EDM, EBM, or LBM? Why?
28. What process would you recommend to make many small holes in a very hard alloy where the holes will be used for cooling and venting?
29. Why are the specific power values so large for LBM?
30. Explain (using a little physics and metallurgy) why the "chips" in thermal processes are often hollow spheres?

CASE STUDY 26. The Vented Cap Screws

The machine shop at the Hi-Fly Space Laboratories received an order from their engineering department for 10 of the vented cap screws shown in Figure CS-26. The machine shop foreman returned the order, stating that there was no practical way to make these other than by hand. The design engineer insisted that the vent slot, as designed, was essential to assure no pressure buildup around the threads of the screw body in the intended application, and that he knew they could be made because he had seen such cap screws.

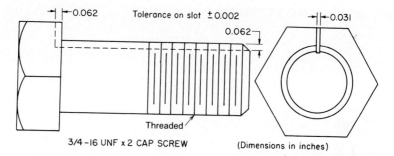

FIGURE CS-26 Design for venting threads of a cap screw.

1. Who was correct?
2. If there is a practicable way to make the slot, how should it be done?
3. How could the slot in the part be redesigned to lower further the cost of manufacture?

Making threads
Thred rolling
Thread casting
Thread cutting
 lathe
 grinding wheel
 tap +, die - ext
  int

CHAPTER 27

Thread Cutting and Forming

Screw threads probably are the most important of all the machine elements. Without them, our present technological society would come to a grinding halt. More are made each year than any other machine element. They range in size from those used in the balance wheels of small watches to more than 254 mm (10 inches) in diameter, used to withstand or transmit tremendous forces. They are made in quantities ranging from one to several million duplicate threads. Their accuracy varies from that of cheap dime-store screws to that of micrometer calipers and lead screws on the most accurate machine tools. Consequently, it is not surprising that several very different procedures have been developed for making screw threads and that the production cost by the various methods varies greatly. Fortunately, some of the most economical methods can provide very accurate results. However, as in the design of most products, the designer can greatly affect the ease and cost of producing specified screw threads; thus reasonable understanding of the various processes is most helpful in permitting the designer to specify and incorporate screw threads into his designs while avoiding needless and excessive cost.

Technically, a screw thread is a ridge of uniform section in the form of a helix on the external or internal surface of a cylinder, or in the form of a conical spiral on the external or internal surface of a frustrum of a cone. These are called *straight* or *tapered* threads, respectively. Tapered threads are used on pipe joints or other applications where liquid-tight joints are required. Straight threads, on the other hand, are used in a wide variety of applications, most commonly on fastening devices, such as bolts, screws, and nuts, and as integral elements on parts that are to be fastened together. But, as mentioned previously, they find very important applications in transmitting controlled motion, as in lead screws and precision measuring equipment.

The basic problem in manufacturing screw threads is how to produce the desired ridge on the workpiece. Three basic methods are used: *cutting, rolling,* and *casting.* Although both external and internal threads can be cast, relatively few are made in this manner, primarily in connection with die casting, investment casting, or the molding of plastics. Today by far the largest number of threads are made by rolling. Both external and internal threads can be made by rolling, but the material must be ductile, and it is a less flexible process than thread cutting, thus it essentially is restricted to standardized and simple parts. Consequently, large numbers of threads still are, and will continue to be, made by cutting, including grinding.

Screw-thread standardization. Starting with Sir Joseph Whitworth in England, in 1841, and William Sellers in the United States, in 1864, a great amount of effort has been devoted to screw-thread standardization. In 1948, representatives of the United States, Canada, and Great Britain adopted the Unified and American Screw Thread Standards, based on the form shown in Figure 27-1.

UNITED NAT. STANDerd
American nat. Standard

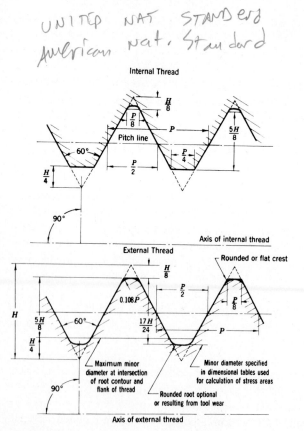

FIGURE 27-1 Unified and American screw-thread form.

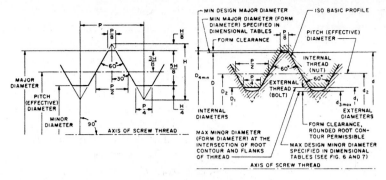

FIGURE 27-2 ISO metric screw-thread forms. (*Left*) Basic profile; (*right*) design dimensions. (*Report B1*, ISO Metric Screw Threads; *courtesy* ASME.)

In 1968 the International Organization for Standards (ISO) recommended the adoption of a set of metric standards, based on the basic thread profile shown in Figure 27-2. It appears likely that both types of threads will continue to be used for some time to come.

Screw-thread nomenclature. The standard nomenclature for screw-thread components is illustrated in Figure 27-3. In both the *Unified* and the *ISO systems*, the crests of external threads may be flat or rounded. The root usually is made rounded to minimize stress concentration at this critical area. The internal thread has a flat crest in order to mate with either a rounded or V-root of the external thread. A small round is used at the root to provide clearance for the flat crest of the external thread.

In the metric system, the pitch always is expressed in millimeters, whereas in the American (Unified) system it is a fraction having as the numerator 1 and as the denominator the number of threads per inch—thus 1/16 pitch being 1/16 of an inch. Consequently, in the Unified system, threads more commonly are expressed in terms of threads per inch rather than by the pitch.

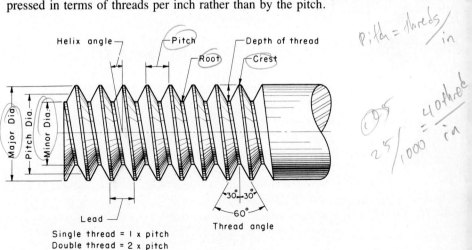

FIGURE 27-3 Standard screw-thread nomenclature.

While all elements of the thread form are based on the *pitch diameter,* screw-thread sizes are expressed in terms of the *outside,* or *major,* diameter and the *pitch* or *number of threads per inch.* In threaded elements, lead refers to the axial advance of the element during one revolution, so lead equals pitch on a single-thread screw.

Types of screw threads. Eleven types, or series, of threads are of commercial importance, several having equivalent series in the metric system and Unified systems:

1. *Coarse-thread series* (UNC and NC). For general use where not subjected to vibration.
2. *Fine-thread series* (UNF and NF). For most automotive and aircraft work.
3. *Extra-fine-thread series* (UNEF and NEF). For use with thin-walled material or where a maximum number of threads are required in a given length.
4. *Eight-thread series* (8UN and 8N). Eight threads per inch for all diameters from 1 through 6 inches. It is used primarily for bolts on pipe flanges and cylinder-head studs where an initial tension must be set up to resist steam or air pressures.
5. *Twelve-thread series* (12 UN and 12N). Twelve threads per inch for diameters from ½ through 6 inches. It is not used extensively.
6. *Sixteen-thread series* (16 UN and 16N). Sixteen threads per inch for diameters from ¾ through 6 inches. It is used for a wide variety of applications that require a fine thread.
7. *American Acme thread.*
8. *Buttress thread.*
9. *Square thread.*
10. *29° Worm thread.* The last four threads, shown in Figure 27-4, are used primarily in transmitting power and motion.
11. *American standard pipe thread.* This thread, shown in Figure 27-4, is the standard tapered thread used on pipe joints in this country. The taper on all pipe threads is ¾ inch per foot.

As has been indicated, the Unified threads are available in a coarse (UNC and NC), fine (UNF and NF), extra-fine (UNEF and NEF), and three "pitch" (8, 12, and 16) series, the number of threads per inch being according to an arbitrary determination based on the major diameter. Similarly, the ISO threads have a "Coarse" and a "Fine" series, with the pitch varying with the major diameter. In addition, they have a "Constant Pitch" series, wherein the pitch (from 6 mm to 0.2 mm) is the same for all diameters. Table 27-1 gives a comparison between some Unified and Metric threads.

The availability of fasteners, particularly nuts, containing plastic inserts to make them self-locking and thus able to resist loosening due to vibration, and the use of special coatings that serve the same purpose, have resulted in less use of finer-thread-series fasteners in mass production. Coarser-thread fasteners are easier to assemble and less subject to cross threading (binding).

Thread classes. In the Unified system, manufacturing tolerances are specified by three classes. Class 1 is for ordnance and other special applications. Class 2

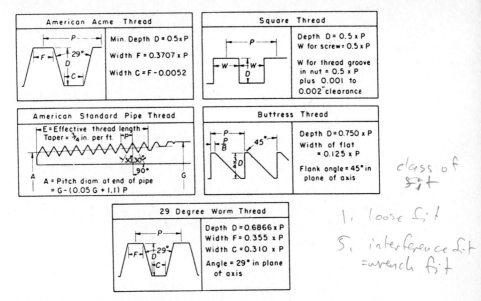

FIGURE 27-4 Special thread forms.

threads are the normal production grade, and Class 3 threads have minimum tolerances where tight fits are required. The letters A and B are added after the class numerals to indicate external and internal threads, respectively.

In the ISO system, tolerances are applied to "positions" and "grades." Tolerance positions denote the limits of pitch and crest diameters, using "e" (large), "g" (small), and "h" (no allowance) for external threads and "G" (small) and "H" (no allowance) for internal threads. The grade is expressed by numerals 3 through 9. Grade 6 is medium quality and for normal length of engagement. Below 6 is fine quality and/or short engagement. Above 6 is coarse quality and/or long length of engagement.

TABLE 27-1   Comparison Between Selected Unified and ISO Threads

	Unified			ISO	
Diameter		Threads per Inch		(Threads per Inch)	
Number; inches	mm	UNC	UNF	Coarse	Fine
#2	2.18	56	64	M2 × 0.4(63.5)	
#4	2.84	40	48	M2 × 0.45 (56.4)	
#8	4.17	32	36	M4 × 0.7 (36.3)	
#10	4.82	24	28	M5 × 0.8 (31.8)	
¼ in.	6.35	20	28	M6 × 1.0 (25.4)	
½ in.	12.7	13	20	M12 × 1.75 (14.5)	M12 × 1.25(20.3)
¾ in.	10.05	10	16	M20 × 2.5 (10.2)	M20 × 1.5 (16.9)
1 in.	25.4	8	14	M24 × 3 (8.47)	M24 × 2 (12.7)

Thread designation. In the Unified system, screw threads are designated by symbols as follows:

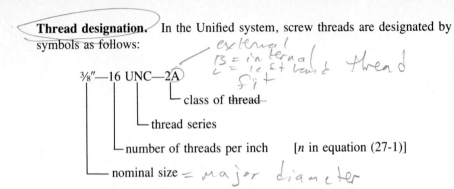

3/8"—16 UNC—2A

— class of thread
— thread series
— number of threads per inch   [*n* in equation (27-1)]
— nominal size = major diameter

handwritten annotations: A = external, B = internal, C = left hand thread, fit

This type of designation applies to right-hand threads. For left-hand threads, the letters LH are added after the thread class symbol.

In the ISO system, threads are designated as follows:

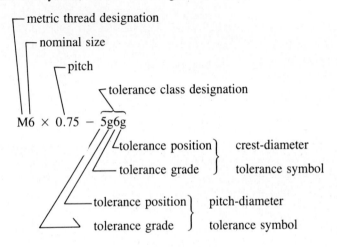

— metric thread designation
— nominal size
— pitch
— tolerance class designation

M6 × 0.75 — 5g6g

— tolerance position ⎫ crest-diameter
— tolerance grade ⎭ tolerance symbol

— tolerance position ⎫ pitch-diameter
— tolerance grade ⎭ tolerance symbol

THREAD CUTTING

Threads can be cut by the following methods:

External	Internal
On an engine lathe	On an engine lathe
With a stock and die (manual)	With a tap and holder (manual, semiautomatic, or automatic)
With an automatic die (turret lathe or screw machine)	With a collapsible tap (turret lathe, screw machine, or special threading machine)
By milling	
By grinding	By milling

Cutting threads on a lathe. Lathes provided the first method for cutting threads by machine. Although most threads now are produced by other methods, lathes still provide the most versatile and fundamentally simple method. Consequently, they often are used for cutting threads on special workpieces where the configuration or nonstandard size does not permit them to be made by less costly methods.

There are two basic requirements for cutting a thread on a lathe. The first is an accurately shaped and mounted tool, because thread cutting is a form-cutting operation; the resulting thread profile is determined by the shape of the tool and its position relative to the workpiece. The second requirement is that the tool must move longitudinally in a specific relationship to the rotation of the workpiece, because this determines the lead of the thread. This requirement is met through the use of the lead screw and the split nut, which provide positive motion of the carriage relative to the rotation of the spindle.

External threads can be cut with the work mounted either between centers, as shown in Figure 27-5, or held in a chuck. For internal threads, the work must be held in a chuck. The cutting tool usually is checked for shape and alignment by means of a thread template, as indicated in Figure 27-6. Figure 27-7 illustrates two methods of feeding the tool into the work. If the tool is fed radially, cutting takes place simultaneously on both sides of the tool. With this true form-cutting procedure, no rake should be ground on the tool, and the top of the tool must be horizontal and be set exactly in line with the axis of rotation

FIGURE 27-5 Cutting a screw thread on a lathe, showing the method of supporting the work and the relationship of the tool to the work. Inset shows face of threading dial. See Figures 20-5 and 20-6 for location. (*Courtesy South Bend Lathe.*)

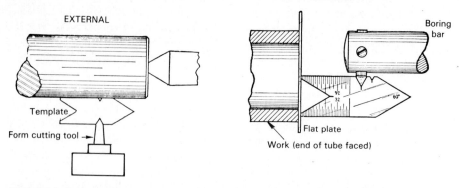

EXTERNAL   INTERNAL

Boring bar

Template

Form cutting tool →

Flat plate

Work (end of tube faced)

FIGURE 27-6 Methods of checking the form and setting of the cutting tool for thread cutting by means of a template. (*Courtesy South Bend Lathe.*)

of the work, as shown in Figure 27-8; otherwise, the resulting thread profile will not be correct. An obvious disadvantage of this method is that the absence of side and back rake will not produce proper cutting except on cast iron or brass. On steel the surface usually will be rough. Consequently, the second method commonly is used, with the compound swiveled 29°. The cutting then occurs primarily on the left-hand edge of the tool, and some side rake can be provided.

Proper speed ratio between the spindle and the lead screw is set by means of the gear-change box. Modern industrial lathes have ranges of ratios available so that nearly all standard threads can be cut merely by setting the proper levers on the quick-change gear box.

To cut a thread, it also is essential that a constant positional relationship be maintained between the workpiece, the cutting tool, and the lead screw. If this is not done, on successive cuts the tool will not be positioned correctly in the thread space. Correct relationship is obtained by means of a *threading dial*, shown in Figure 27-5, which is driven directly by the lead screw through a worm gear. Because the workpiece and the lead screw are directly connected, the threading dial provides a means for establishing the desired positional relationship between the workpiece and the cutting tool.

As shown in Figure 27-5, the threading dial is graduated into an even number

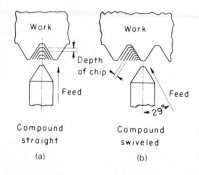

Work   Work

Depth of chip

Feed   Feed

←29°→

Compound straight   Compound swiveled

(a)   (b)

FIGURE 27-7 Two methods of feeding the tool into the work in cutting threads on a lathe. (a) Radial feed, (b) Half thread angle feed.

FIGURE 27-8 Proper relationship of the thread-cutting tool to the workpiece center line. (*Courtesy South Bend Lathe.*)

of major and half divisions. If the split nut is closed in accordance with the following rules, correct positioning of the tool will result:

1. *For even-number threads:* at any line on the dial.
2. *For odd-number threads:* at any numbered line on the dial.
3. *For threads involving ½ numbers:* at any odd-numbered line on the dial.
4. *For ¼ or ⅛ threads:* return to the original starting line on the dial.

To start cutting a thread, the tool usually is fed inward until it just scratches the work, and the cross-slide dial reading is then noted, or set at zero. The split nut is engaged and the tool permitted to run over the desired thread length. When the tool reaches the end of the thread, it is quickly withdrawn by means of the cross-slide control. The split nut is then disengaged and the carriage returned to the starting position, where the tool is clear of the workpiece. At this point the future thread will be indicated by a fine scratch line. This permits the operator to check the thread lead, by means of a scale, or thread gage, to assure that all settings have been made correctly.

Next, the tool is returned to its initial zero depth position by returning the cross slide to the zero setting. By using the compound rest, the tool can be fed inward the proper amount for the first cut. A depth of 0.25 to 0.38 mm (0.010 to 0.015 inch) usually is used for the first cut, and smaller amounts on each successive cut, until the final cut is made with a depth of only 0.025 to 0.076 mm (0.001 to 0.003 inch) to produce a good finish.

Each successive cut is repeated in the manner described. The tool is fed inward to its previous position by means of the cross slide, and additional depth for the next cut provided by the compound rest. When the thread has been cut nearly to its full depth, it is checked for size by means of a mating nut or thread gage. Cutting is continued until a proper fit is obtained.

To cut right-hand threads, the tool is moved from right to left. For left-hand threads the tool must be moved from left to right. Otherwise, the procedure is essentially the same. Internal threads are cut in the same basic manner except that the tool is held in a boring bar. Tapered threads can be cut either with a taper attachment or by setting the tailstock off center. It should be remembered that in cutting a tapered thread the tool must be set normal to the axis of rotation of the workpiece, not the tapered surface.

It is apparent that cutting screw threads on a lathe is a slow, repetitious process and requires considerable skill on the part of the operator. The cutting speeds usually employed are from one-third to one-half of regular speeds to enable the operator to have time to manipulate the controls and to ensure better cutting. Consequently, the total process is quite costly, and it is evident why other methods are used whenever possible.

Cutting threads with dies. Straight and tapered external threads up to about 38 mm (1½ inches) in diameter can be cut quickly by means of threading dies, such as are shown in Figure 27-9. Basically, these are similar to hardened, threaded nuts with several longitudinal grooves that expose multiple cutting edges formed by the intersection of each groove and each thread. The cutting edges at the starting end are beveled to aid in starting the dies on the workpiece. As a consequence, a few threads at the inner end of the workpiece are not cut to full depth. Such threading dies are made of carbon or high-speed tool steel.

The solid-type dies, shown in Figure 27-9a, seldom are used in manufacturing because they have no provision for compensating for wear. The solid-adjustable type, shown in (b), is split and can be adjusted over a small range by means of a screw. This permits some adjustment to compensate for wear or to provide a variation in the fit of the resulting screw thread.

These types of threading dies usually are held in a *stock* and are rotated by hand. In using them, a suitable lubricant is desirable to produce a smoother thread and to prolong the life of the die, since there is a large amount of friction because the die is fed onto the workpiece by the screw action of the die on the threads being, or already, cut.

Self-opening die heads. A major disadvantage of solid-type threading dies is that they must be unscrewed from the workpiece to remove them. They thus are not suitable for use on high-speed, production-type machines. Therefore, *self-opening die heads* are used on turret lathes, screw machines, and special threading machines for cutting external threads.

There are three types of self-opening die heads, all having four sets of adjustable, multiple-point cutters that can be removed for sharpening or for interchanging for different thread sizes. This permits one head to be used for a range of thread sizes (see Figure 27-10). The cutters can be positioned radially or tangentially, resulting in less tool flank contact and friction rubbing. In some self-opening die heads, the cutters are circular, with an interruption in the cir-

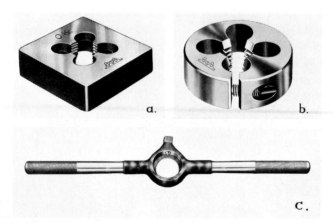

a. b.

c.

FIGURE 27-9 (a) Solid threading die. (b) Solid-adjustable threading die. (c) Threading-die stock. (*Courtesy TRW-Greenfield Tap & Die.*)

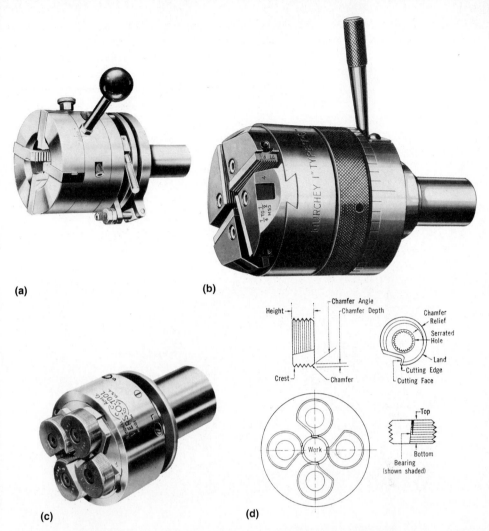

(a)

(b)

(c)

(d)

Chamfer Angle
Chamfer Depth
Height
Crest
Chamfer
Chamfer Relief
Serrated Hole
Land
Cutting Edge
Cutting Face
Work
Top
Bottom
Bearing (shown shaded)

FIGURE 27-10 Self-opening die heads, with (a) radial cutter, (b) tangential cutters, and (c) circular cutters. (d) Terminology of circular chasers and their relation to the work. (*Courtesy Geometric Tool Co., Warner & Swasey Co., National Acme Co., TRW-Greenfield Tap & Die, respectively.*)

cular form to provide an easily sharpened cutting face. The cutters are mounted on the holder at an angle equal to the helix angle of the thread.

As the name implies, the cutters in self-opening die heads are arranged to open automatically when the thread has been cut to the desired length, thereby permitting the die head to quickly be withdrawn from the workpiece. On die heads used on turret lathes, the operator usually must reset the cutters in the closed position before making the next thread. The die heads used on screw machines and automatic threading machines are provided with a mechanism that automatically closes the cutters after the heads are withdrawn.

Cutting threads by means of self-opening die heads frequently is called *thread chasing*. However, some people apply this term to other methods of thread cutting—even to cutting a thread in a lathe.

Thread tapping. The cutting of an internal thread by means of a multiple-point tool is called *thread tapping,* and the tool is called a *tap*. A hole of diameter slightly larger than the minor diameter of the thread must already exist, made by drilling, boring, or casting.

For small holes, solid taps, such as shown in Figure 27-11, usually are used. These are similar to threaded bolts, with from one to four flutes to provide cutting edges. Such taps are made of either carbon or high-speed steel. The flutes can be either straight, helical, spiral, or spiral-pointed.

Hand taps, shown in Figure 27-11, have square shanks and are made in three types, usually in sets. The tapered end of *taper taps* will enter the hole a sufficient distance to help align the tap. In addition, the threads increase grad-

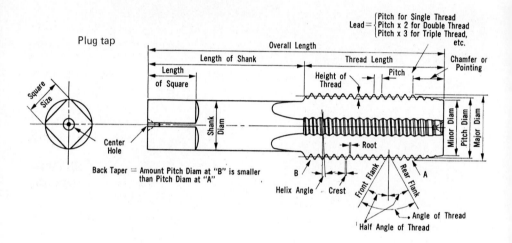

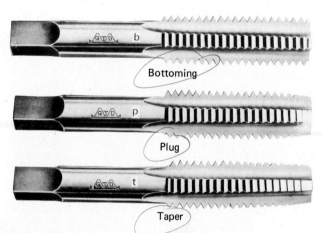

FIGURE 27-11 Terminology for a plug tap with photographs of taper (t), plug (p), and bottoming (b) taps which are used serially in threading holes. (*Courtesy TRW-Greenfield Tap & Die.*)

FIGURE 27-12. Spiral-flute tap. (*Courtesy TRW-Greenfield Tap & Die.*)

ually to full depth, so this type of tap requires less effort to use. However, only a through hole can be threaded completely with a taper tap, because it cuts to full depth only behind the tapered portion. If a blind, or dead-end, hole must be threaded to the bottom, all three types of taps should be used in succession. After the taper tap has the thread started in proper alignment, a *plug tap*, which has only a few tapered threads, to provide gradual cutting of the threads to depth, is used to cut the threads as deep into the hole as its shape will permit. A *bottoming tap*, having no tapered threads, is used to finish the few remaining threads at the bottom of the hole to full depth. Obviously, producing threads to the full depth of a blind hole is time-consuming, and it also frequently results in broken taps and spoiled workpieces. Such configurations usually can be avoided if designers will give reasonable thought to the matter.

Taps operate under very severe conditions, both because of the severe friction involved and the difficulty of chip removal. Also, taps are relatively fragile. Spiral-fluted taps, illustrated in Figure 27-12, provide better removal of chips from a hole—particularly in tapping materials which produce long, curling chips. They also are helpful in tapping holes where the cutting action is interrupted by slots or keyways. In tapping through holes or blind holes that are not tapped to the bottom, the type of tap shown in Figure 27-13 is very useful. These have a spiral point that projects the chips ahead of the tap so that they do not interfere with the cutting action and the flow of cutting fluid into the hole.

Care must be exercised in using taps, particularly in tapping by hand. Proper cutting fluid should be used, and the tap should be reversed a partial turn after each two or three forward turns to assist in clearing the chips.

Solid taps also are used in tapping operations on machine tools, such as lathes, drill presses, and special tapping machines. In tapping on a drill press, a tapping attachment often is used. These devices rotate the tap slowly when the drill press spindle is fed downward against the work. When the tapping is completed and the spindle raised, the tap is automatically driven in the reverse direction at a higher speed to reduce the time required to back the tap out of the hole. Some modern machine tools provide for extremely fast spindle reverse for backing taps out of holes.

When solid taps are used on a screw machine or turret lathe, a special holder is employed in which a pin prevents the tap from turning while it is being fed into the work. As the tap reaches the end of the hole, it pulls the pin away from its stop so that the tap is free to rotate with the work. The rotation of the work is then reversed and the pin again prevents the tap from turning while it is backed out of the hole.

FIGURE 27-13 Spiral-point, straight-flute tap. (*Courtesy TRW-Greenfield Tap & Die.*)

Collapsing taps.   *Collapsing taps* are similar to self-opening die heads in that the cutting elements collapse inward automatically when the thread is completed. This permits withdrawing the tap from the workpiece without the necessity of unscrewing it from the thread. They can be either self-setting, for use on automatic machines, or require manual setting for each cycle. As shown in Figure 27-14, two types are available. The one with radial cutters generally is used for smaller size threads, whereas the type having circular cutters is used for larger-sized threads.

Tap-drill sizes.   In most cases the hole that must be made before an internal thread is tapped is produced by drilling. Consequently, the drill size that is used is very important because it determines the depth of the thread contour and the force required in the tapping operation. In most applications a drill size is selected that will result in the thread having about 75% of full thread depth. This practice makes tapping much easier than if full thread depth were attempted, and the resulting thread is only slightly less strong. Table 27-2 gives the drill sizes used to provide 75% thread depth for several sizes of UNC threads. Full tables of tap-drill sizes for both Unified and Metric threads are readily available in machinists' handbooks and textbooks on drafting practice.

Tapping Cutting Time.   The time to tap a hole is

$$CT = \pi DLn/8V \qquad (27\text{-}1)$$

where D = tap diameter (in.)
   L = depth of tapped hole (in.)
   n = number of threads per inch (tpi)
   V = cutting speed (sfpm)

FIGURE 27-14   Collapsing taps. (*Left*) Tap using circular cutters. (Courtesy Landis Machine Co.) (*Right*) Tap using radial cutters. (*Courtesy Geometric Tool Company Division, TRW-Greenfield Tap & Die.*)

TABLE 27-2 Recommended Tap-Drill Sizes for Standard Screw-Thread Pitches, American National Course-Thread Series

Number or Diameter	Threads per Inch	Outside Diameter of Screw	Tap Drill Sizes	Decimal Equivalent of Drill
6	32	0.138	36	0.1065
8	32	0.164	29	0.1360
10	24	0.190	25	0.1495
12	24	0.216	16	0.1770
1/4	20	0.250	7	0.2010
3/8	16	0.375	5/16	0.3125
1/2	13	0.500	27/64	0.4219
3/4	10	0.750	21/32	0.6562
1	8	1.000	7/8	0.875

Threading and tapping machines. Special machines are available for production threading and tapping. Threading machines usually have one or more spindles on which a self-opening die head is mounted, with suitable means for clamping and feeding the workpiece. A typical machine of this type is shown in Figure 27-15.

Some special tapping machines are similar in construction, with self-collapsing taps substituted for the threading dies. More commonly, tapping machines resemble drill presses, modified to provide spindle feeds both upward and downward, with the speed and feed more rapid on the upward motion.

For threading nuts, the type of machine shown in Figure 27-16 is used, in conjunction with a bent tap as pictured. The bent shank of the tap prevents it from rotating, yet permits the nuts to be rotated onto the top and threaded, then slid continuously up off the shank.

Thread milling. High-accuracy threads, particularly in larger sizes, are often cut by milling. Either a single- or a multiple-form cutter may be used, but the procedures are quite different.

A single-form cutter has a single, annular row of teeth. As shown in Figure 27-17, with the cutter tilted through an angle equal to the helix angle of the thread, it is fed inward radially to full depth while the work is stationary. The workpiece then is rotated slowly, and the cutter simultaneously is moved longitudinally, parallel with the axis of the work (or vice versa), by means of a lead screw, until the thread is completed. The thread can be completed in a single cut, or roughing and finish cuts can be used. This process is used primarily for large-lead or multiple-lead threads.

Some threads can be milled more quickly by using a multiple-form cutter, having several annular rows of teeth that are perpendicular to the cutter axis (the rows having no lead). The cutter must be slightly longer than the thread to be cut. It is set parallel with the axis of the workpiece and fed inward to full-

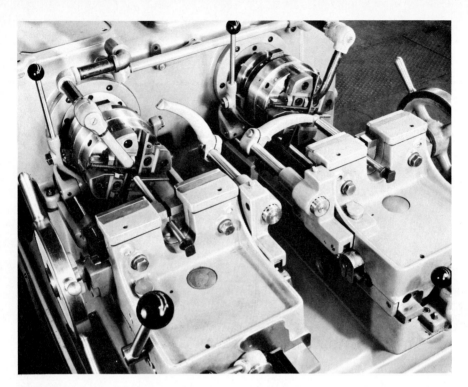

FIGURE 27-15 Two-spindle, automatic threading machine. (*Courtesy Landis Machine Company.*)

thread depth while the work is stationary. The work then is rotated slowly for a little over one revolution, and the rotating cutter is simultaneously moved longitudinally with respect to the workpiece (or vice versa) according to the thread lead. When the work has revolved one revolution, the thread is complete.

This process cannot be used on threads having a helix angle greater than about 3°, because clearance between the sides of the threads and the cutter depends on the cutter diameter being substantially less than that of the workpiece. Thus, although the process is rapid, its use is restricted to threads of substantial diameter and not more than about 51 mm (2 inches) long.

Thread grinding.   Grinding can produce very accurate threads, and it also permits threads to be made on hardened materials. Three basic methods are used. *Center-type grinding with axial feed* is the most common method, being similar to cutting a thread on a lathe; a shaped grinding wheel replaces the single-point tool. Usually, a single-ribbed grinding wheel is employed, but multiple-ribbed wheels are used occasionally. The grinding wheels are shaped by special diamond dressers or by crush dressing and must be inclined to the helix angle of the thread. Wheel speeds are in the high range. Several passes usually are required to complete the thread. Figure 27-18 shows a thread being ground by this procedure.

Center-type infeed thread grinding is similar to multiple-form milling in that a multiple-ribbed wheel, as wide as the length of the desired·thread, is used.

Bent tap

FIGURE 27-16 Section view of automatic nut-tapping machine, using a bent tap. Inset shows a bent tap. (*Courtesy National Machinery Company.*)

FIGURE 27-17 Milling a large thread with a single-form cutter. The cutter can be seen behind the thread. (*Courtesy Lees-Bradner Company.*)

FIGURE 27-18 (*Left*) Grinding a worm, using a form wheel; (*Above*) Close-up view of the abrasive wheel and thread. (*Courtesy Ex-Cell-O Corporation.*)

The wheel is fed inward radially to full thread depth, and the thread blank is then turned through about 1½ turns as the grinding wheel is fed axially a little more than the width of one thread.

Centerless thread grinding, illustrated in Figure 27-19 is used for making headless set screws. The blanks are hopperfed to position *A*. The regulating wheel causes them to traverse the grinding wheel face, from which they emerge at position *B* in completed form. A production rate of 60 to 70 screws of 12.7-mm (½-inch) length per minute is possible.

Thread rolling. As stated previously, virtually all threads that are produced in substantial quantities are made by *rolling*. This is a simple cold-forming operation in which the threads are formed by rolling a thread blank between hardened dies that cause the metal to flow radially into the desired shape. Because no metal is removed in the form of chips, less material is required, resulting in substantial savings. In addition, because of cold working, the threads have greater strength than cut threads, and a smoother, harder, and more wear-resistant surface is obtained. In addition, the process is almost unbelievably fast; the action on an actual machine cannot be observed without slowing the machine or using a stroboscopic light. On large threads, hot rolling is used occasionally.

Thread rolling is done by four basic methods. The simplest of these employs

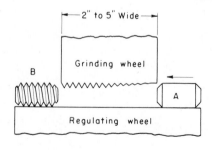

FIGURE 27-19 Principle of centerless thread grinding.

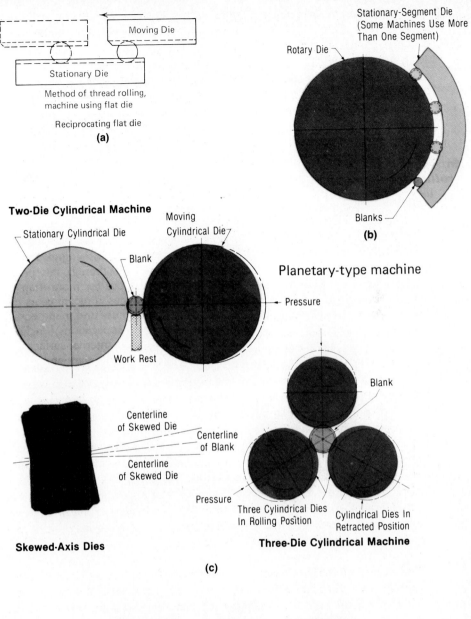

Method of thread rolling,
machine using flat die

Reciprocating flat die

(a)

Stationary-Segment Die
(Some Machines Use More
Than One Segment)

Rotary Die

Blanks

(b)

Two-Die Cylindrical Machine

Stationary Cylindrical Die

Moving
Cylindrical Die

Blank

Planetary-type machine

Pressure

Work Rest

Blank

Centerline
of Skewed Die

Centerline
of Blank

Centerline
of Skewed Die

Pressure

Three Cylindrical Dies
In Rolling Position

Cylindrical Dies In
Retracted Position

Skewed-Axis Dies

Three-Die Cylindrical Machine

(c)

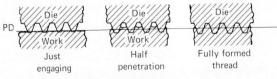

PD

Die

Work

Just
engaging

Die

Work

Half
penetration

Die

Fully formed
thread

Action of die in forming thread

(d)

FIGURE 27-20 Schematic diagrams, (a) Method of rolling threads with a flat die.
(b) Planetary-type machine. (c) Two die and three die cylindrical machines. (d) Action
of the die in forming the threads. (*Courtesy TRW-Greenfield Tap & Die.*)

one fixed and one movable flat rolling die, as illustrated in Figure 27-20a. For Class 2 threads, the normal commercial grade, the diameter of the thread blank is equal to the pitch diameter of the thread. After the blank is placed in position on the stationary die, movement of the moving die causes the blank to be rolled between the two dies and the metal in the blank is displaced to form the threads. As the blank rolls, it moves across the die parallel with its longitudinal axis. Prior to the end of the stroke of the moving die, the blank rolls off the end of the stationary die, its thread being completed.

One obvious characteristic of a rolled thread is that its major diameter always is greater than the diameter of the blank. When an accurate class of fit is desired, the diameter of the blank is made about 0.05 mm (0.002 inch) larger than the thread-pitch diameter. If it is desired to have the body of a bolt larger than the outside diameter of the rolled thread, the blank for the thread is made smaller than the body (see Figure 15-10).

Thread rolling can be done with cylindrical dies. Figure 27-20c illustrates the three-roll method and shows the type of dies commonly employed on turret lathes and screw machines. Two variations are used. In one, the rolls are retracted while the blank is placed in position. They then move inward radially, while rotating, to form the thread. In the more common procedure, used on turret lathes and screw machines, the three rolls are contained in a self-opening die head, similar to the conventional type used for cutting external threads. The die head is fed onto the blank longitudinally and forms the thread progressively as the blank rotates. With this procedure, as in the case of cut threads, the innermost $1\frac{1}{2}$ to 2 threads are not formed to full depth because of the progressive action of the rollers.

The two-roll method, shown in Figure 27-20c, is commonly employed for automatically producing large quantities of externally threaded parts.

Not only is thread rolling very economical, the threads are excellent as to form and strength. The cold working contributes to increased strength, particularly at the critical root areas. There is less likelihood of scratches and tears that can result from machining, which can act as stress raisers.

Large numbers of threads are rolled on thin, tubular products. In this case external and internal rolls are used. The threads on electric lamp bases and sockets are examples of this type of thread.

Cold forming internal threads. Unfortunately, most internal threads cannot be made by rolling; there is insufficient space within the hole to permit the required rolls to be arranged and supported, and the required forces are too high. However, many internal threads, up to about 12.7 mm ($\frac{1}{2}$ inch) in diameter,

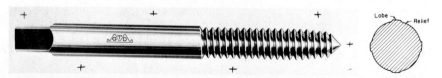

FIGURE 27-21 (*Left*) Fluteless tap for forming internal threads. (*Right*) Cross section of fluteless tap. (*Courtesy Besley-Welles Corporation.*)

FIGURE 27-22 Action of a fluteless tap in forming an internal thread.

are cold formed in holes in ductile metals by means of *fluteless taps*. Such a tap and its special cross section are shown in Figure 27-21. As illustrated in Figure 27-22, the forming action is essentially the same as in rolling external threads. Because of the forming involved and the high friction, the torque required is about double that for cutting taps. Also, the hole diameter must be controlled carefully to obtain full thread depth without excessive torque. However, fluteless taps produce somewhat better accuracy than cutting taps. A lubricating fluid should be used—water-soluble oils being quite effective.

Fluteless taps are especially suitable for forming threads in dead-end holes because no chips are produced. They come in both plug and bottoming types.

REVIEW QUESTIONS

1. How does the pitch diameter differ from the major diameter?
2. For what types of threads are the pitch and the lead the same?
3. Why are pipe threads tapered?
4. By what three basic methods can external threads be produced?
5. Explain the meaning of ¼"-20 UNC-3A.
6. What is meant by the designation M20 × 2.5-6g6g?
7. What are two reasons why fine-series threads are being used less now than in former years?
8. In cutting a thread on a lathe, how is the pitch controlled?
9. Why, when possible, should parts be designed so that any required threads can be made by methods other than cutting on a lathe with a single-point tool?
10. What is the function of a threading dial on a lathe?
11. What controls the lead of a thread when it is cut by a threading die?
12. What is the basic purpose of a self-opening die head?
13. Why is it essential that if threads are to be cut on a turret lathe, they should be designed to a standard diameter, whereas it is not so if they are to be cut on a lathe?
14. What is the reason for using a taper tap before a plug tap in tapping a hole?
15. What difficulties are encountered if full threads are specified to the bottom of a dead-end hole?
16. Why can a fluteless tap not be used for threading a hole in gray cast iron?
17. What provisions should a designer make if she or he desires a dead-end hole to be threaded?
18. What is the major advantage of a spiral-point tap?
19. Why can a fluteless tap not be used for threading to the bottom of a dead-end hole?
20. Is it desirable for a tapping fluid to have lubricating qualities? Why?
21. How does thread milling differ when using single- and double-form cutters?
22. What are the advantages of making threads by grinding?
23. Why has thread rolling become the most commonly used method for making threads?

24. How may one determine whether a thread has been produced by rolling rather than by cutting?

25. How long will it take to tap a ¾ × 2 inch hole, using a tap with 10 tpi at 30 fpm?

CASE STUDY 27.   The Bronze Bolt Mystery

You are employed by the Mountainous Irrigation Company, which had 1000 special bolts of phosphor bronze (10% Sn) made to the design shown in Figure CS-27 for use in its various pumping plants. Within a few weeks after a number of these bolts were installed, the heads broke off several of them, the fracture being along the dashed line shown in Figure CS-26. The broken bolts were replaced, with special precautions being taken to assure that they were not overloaded. Again, after only a few days, several of these replacement bolts broke in the same manner. You have been assigned the task of determining the cause of the failures. Upon examining several unused bolts, you discover that many of them have fine cracks at the intersection of the head and body. Upon checking with the manufacturer, you learn that the bolts were made by ordinary heading procedures.

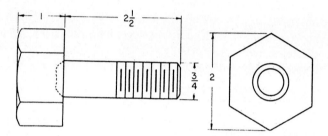

FIGURE CS-27   Special bronze bolt, showing location of fracture (dashed line).

1. Outline the procedure you will use to determine the cause of the cracks.
2. What do you suspect is the cause?
3. How could the difficulty have been avoided?

CHAPTER 28

Gear Manufacturing

Gears transmit power or motion mechanically between parallel, intersecting, or nonintersecting shafts. Although unsung and usually hidden from sight, they are one of the most important mechanical elements in our civilization, possibly even surpassing the wheel, since most wheels would not be turning without power being applied to them through gears. They operate at almost unlimited speeds under a wide variety of conditions. Millions are produced each year in sizes from a few millimeters up to more than 6 meters (20 feet) in diameter. Often the requirements that must be, and routinely are, met in their manufacture are amazingly precise. Consequently, the machines and processes that have been developed for producing them are among the most ingenious we have. In order to understand the functional requirements of these machines and processes, it is helpful first to consider the basic theory of gears and their operation.

Gear theory and terminology. Gears, basically, are modifications of friction disks, as illustrated in Figure 28-1, teeth being added to prevent slipping and to assure that their relative motions are constant. However, it should be noted that the addition of teeth does not change the relative velocities of the disks and shafts; the velocity ratio is determined by the diameters of the disks.

Although wooden teeth or pegs were attached to disks to make gears in ancient times, the teeth of modern gears are produced by making cuts into disks that are sufficiently large to contain the outer portions of the teeth, or by forming

(a) Friction disks   (b) Teeth attached to disks

FIGURE 28-1 Relationship between the transmission of rotation through friction disks and by gear teeth.

processes that cause the metal in the teeth to plastically flow outward from a disk. But the basic concept of a disk remains, with the *pitch circle,* shown in Figure 28-2, corresponding to the diameter of the friction disk. Thus the angular velocity of a gear is determined by the all-important diameter of this imaginary pitch circle, and all design calculations relating to gear performance are based on the pitch-circle diameter or, more simply, the *pitch diameter (PD)*.

For two gears to operate properly, their pitch circles must be tangent to each other. The point at which the two pitch circles are tangent, and at which they intersect the center line connecting their centers of rotation, is called the *pitch point.* The common normal at the point of contact of mating teeth must pass through the pitch point. This condition is illustrated in Figure 28-3.

To minimize friction and wear, and thus increase their life and efficiency, gears are designed to have rolling motion between mating teeth, rather than sliding motion. To achieve this condition, most gears utilize a tooth form that is based on an *involute curve.* This is the curve that is generated by a *point* on a straight line when the line rolls around a *base circle.* A somewhat simpler method of developing an involute curve is that shown in Figure 28-4, by unwinding a tautly held string from a base circle; point A generates an involute curve.

There are four reasons for using the involute form for gear teeth. First, such a tooth form provides the desired pure rolling action. Second, even if a pair of involute gears is operated with the distance between the centers slightly too large

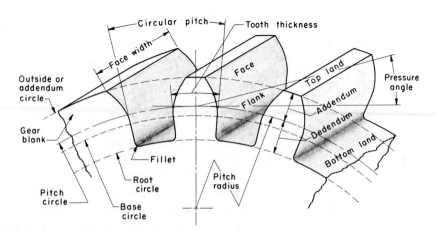

FIGURE 28-2 Gear-tooth nomenclature.

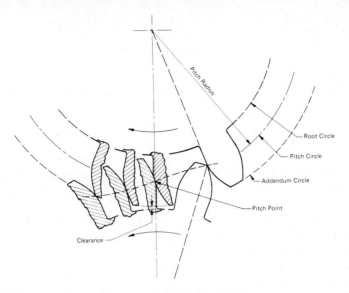

FIGURE 28-3 Action between mating gear teeth.

or too small, the common normal at the point of contact between mating teeth will always pass through the pitch point. Obviously, the theoretical pitch circles in such cases will be increased or decreased slightly. Third, the *line of action,* or *path of contact,* that is, the locus of the points of contact of mating teeth, is a straight line that passes through the pitch point and is tangent to the base circles of the two gears. The fourth very important reason is that a true involute tooth form can be generated by a rack that has straight-sided teeth. This permits a very accurate tooth profile to be obtained through the use of a simple and easily made cutting tool.

The basic size of gear teeth may be expressed in two ways. The common practice, especially in the United States and England, is to express the dimensions as a function of the *diametral pitch (DP),* which is *the number of teeth (N) per unit of pitch diameter (PD);* thus $DP = N/PD$. Dimensionally, DP involves inches in the English system and millimeters in the SI system, and it is a measure of tooth size. The second method for specifying gear tooth size is by means of the *module (M),* defined as *the pitch diameter divided by the number*

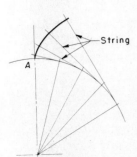

FIGURE 28-4 Method of generating an involute curve by unwinding a string from a cylinder.

of teeth, thus $M = PD/N$. It thus is the reciprocal of diametral pitch and is expressed in inches or millimeters. Any two gears having the same diametral pitch or module will mesh properly if they are mounted so as to have the correct distances and relationship.

The important tooth elements, shown in Figure 28-2, can be specified in terms of the diametral pitch or the module and are as follows:

1. *Addendum:* the radial distance from the pitch circle to the outside diameter.
2. *Dedendum:* the radial distance from the pitch circle to the root circle. It is equal to the addendum plus the *clearance,* which is provided to prevent the outer corner of a tooth from touching against the bottom of the tooth space.
3. *Circular pitch:* the distance between corresponding points of adjacent teeth, measured along the pitch circle. It is numerically equal to π/diametral pitch.
4. *Tooth thickness:* the thickness of a tooth, measured along the pitch circle. When tooth thickness and the corresponding *tooth space* are equal, no *backlash* exists in a pair of mating gears.
5. *Face width:* the length of the gear teeth in an axial plane.
6. *Tooth face:* the mating surface between the pitch circle and the addendum circle.
7. *Tooth flank:* the mating surface between the pitch circle and the root circle.

Four shapes of involute gear teeth are used in this country:

1. 14½° pressure angle, full-depth (used most frequently).
2. 14½° pressure angle, composite (seldom used).
3. 20° pressure angle, full-depth (seldom used).
4. 20° pressure angle, stub-tooth (next most common).

In the 14½° full-depth system the tooth profile outside the base circle is an involute curve. Inward from the base circle the profile is a straight, radial line that is joined with the bottom land by a small fillet. With this system the teeth of the basic rack have straight sides.

The 14½° composite system and the 20° full-depth system provide somewhat stronger teeth. However, with the 20° full-depth system considerable undercutting occurs in the dedendum area, so stub teeth often are used; in these the addendum is shortened by 20 per cent, thus permitting the dedendum to be shortened a similar amount. This results in very strong teeth without undercutting.

Table 28-1 gives the formulas for computing the dimensions of gear teeth in the 14½° full-depth and 20° stub-tooth systems.

Physical requirements of gears. A consideration of gear theory leads to five requirements that must be met in order for gears to operate satisfactorily:

1. The actual tooth profile must be identical to the theoretical profile.
2. Tooth spacing must be uniform and correct.
3. The *actual* and theoretical pitch circles must be coincident and be concentric with the axis of rotation of the gear.
4. The face and flank surfaces must be smooth and sufficiently hard to resist wear and prevent noisy operation.

TABLE 28-1 Standard Dimensions for Involute Gear Teeth

	14½° Full Depth	20° Stub Tooth
Pitch diameter	$\dfrac{N}{DP}$	$\dfrac{N}{DP}$
Addendum	$\dfrac{1}{DP}$	$\dfrac{0.8}{DP}$
Dedendum	$\dfrac{1.157}{DP}$	$\dfrac{1}{DP}$
Outside diameter	$\dfrac{N+2}{DP}$	$\dfrac{N+1.6}{DP}$
Clearance	$\dfrac{0.157}{DP}$	$\dfrac{0.2}{DP}$
Tooth thickness	$\dfrac{1.5708}{DP}$	$\dfrac{1.5708}{DP}$

5. Adequate shafts and bearings must be provided so that desired center-to-center distances are retained under operational loads.

The first four of these requirements are determined by the material selection and manufacturing process. The various methods of manufacture that are used represent attempts to meet these requirements to varying degrees with minimum cost, and their effectiveness must be measured in terms of the extent to which the resulting gears embody these requirements.

The more common types of gears are shown in Figure 28-5. *Spur gears* have straight teeth and are used to connect parallel shafts. They are the most easily made and the cheapest of all types.

The teeth on *helical gears* lie along a helix, the angle of the helix being the angle between the helix and a pitch cylinder element parallel with the gear shaft. Helical gears can connect either parallel or nonparallel, nonintersecting shafts. Such gears are stronger and quieter than spur gears because the contact between mating teeth increases more gradually and more teeth are in contact at a given time. Although they usually are slightly more expensive to make than spur gears, they can be manufactured in several ways and are produced in large numbers.

Helical gears have one disadvantage in that a side thrust is created when they are loaded, which must be absorbed in the bearings. *Herringbone gears* neutralize this side thrust by having, in effect, two helical-gear halves, one having a right-hand and the other a left-hand helix. The *continuous* herring-bone type, shown in Figure 28-5, is rather difficult to machine but is very strong. A modified herringbone type is made by machining a groove, or gap, around the gear blank where the two sets of teeth would come together. This provides a runout space for the cutting tool in making each set of teeth.

A *rack* is a gear with infinite radius, having teeth that lie on a straight line on a plane. The teeth may be normal to the axis of the rack or helical, so as to mate with spur or helical gears, respectively.

A *worm* is similar to a screw. It may have one or more threads, the multiple-

FIGURE 28-5 Several types of gears. (*Top*) Spur gear and rack, worm and worm gear, continuous herringbone gears. (*Center*) Spiral bevel gear; helical gears; crown gear. (*Bottom*) Straight, zerol, and hypoid bevel gears. (*Courtesy Gleason Works*).

thread type being very common. Worms usually are used in conjunction with a *worm gear*. High gear ratios are easily obtainable with this combination. The axes of the worm and worm gear are nonintersecting and usually are at right angles. If the worm has a small helix angle, it cannot be driven by the mating worm gear. This principle frequently is employed to obtain nonreversible drives. Worm gears usually are made with the top land concave, to permit greater area of contact between the worm and the gear. A similar effect can be achieved by using a *conical worm,* in which the helical teeth are cut on a double-conical blank, thus producing a worm that has an hourglass shape.

Bevel gears are used to transmit motion between intersecting shafts. They are conical in form, the teeth being cut on the surface of a truncated cone. Several types of bevel gears are made—the types varying as to whether the teeth are straight or curved, and whether the axes of the mating gears intersect. On *straight-tooth* bevel gears the teeth are straight, and if extended all would pass through a common apex. *Spiral-tooth* bevel gears have teeth that are segments of spirals. Like helical gears, this design provides tooth overlap so that more teeth are engaged at a given time and the engagement is progressive. *Hypoid* bevel gears also have a curved-tooth shape but are designed to operate with nonintersecting axes. They are used in the rear axles of most automobiles so that the drive shaft axis can be below the axis of the axle and thus permit a lower floor height. *Zerol* bevel gears have teeth that are circular arcs, providing somewhat stronger teeth that can be obtained in a comparable straight-tooth gear. They are not used extensively. When a pair of bevel gears are the same size and have their shafts at right angles, they are termed *miter gears*.

A *crown gear* is a special form of bevel gear having a 180° cone apex angle. In effect, it is a disk with the teeth on the side of the disk. It also may be thought of as a rack that has been bent into a circle so that its teeth lie in a plane. The teeth may be straight or curved. On straight-tooth crown gears the teeth are radial. Crown gears seldom are used, but they have the important quality that they will mesh properly with a bevel gear of any cone angle, provided that the bevel gear has the same tooth form and diametral pitch. This important principle is incorporated in the design and operation of two very important types of gear generating machines that will be discussed later.

Most gears are of the external type, the teeth forming the outer periphery of the gear. Internal gears have the teeth on the inside of a solid ring, pointing toward the center of the gear.

GEAR MANUFACTURING

Gears are made in very large numbers by both machining and by cold-roll forming. In addition, significant quantities are made by extrusion, by blanking, and some by powder metallurgy and by a forging process. However, it is only by machining that all types of gears can be made in all sizes, and although roll-formed gears can be made with accuracy sufficient for most applications—even for automobile transmissions—machining still is unsurpassed for gears that must have very high accuracy. Also, roll forming can be used only on ductile metals.

Basic methods for machining gears. Three basic methods are employed for machining gears, each having certain advantages and limitations as to quality, flexibility, and cost.

Form cutting utilizes the principle illustrated in Figure 28-6, the cutter having the same form as the *space* between adjacent teeth. Usually a multiple-tooth form cutter (see Figure 28-10) is used, either rotating about an axis as shown in Figure 28-6 or, occasionally, rotating about an axis that is normal to the axis of the gear blank. However, it is possible to use a single-point cutter, and the

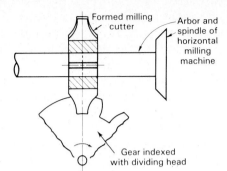

FIGURE 28-6 Basic method of machining a gear by form milling or form cutting.

Formed milling cutter

Arbor and spindle of horizontal milling machine

Gear indexed with dividing head

principle can be employed with a reciprocating tool. The tool is fed radially toward the center of the gear blank, to obtain the desired tooth depth; it then is moved across the tooth face to obtain the required tooth width. When one tooth has been completed, the tool is withdrawn, the gear blank is indexed, and the cutting of the next tooth space is started. It is possible to cut all the tooth spaces simultaneously by using a number of cutting tools equal to the number of teeth in the gear; this is done in one type of gear broaching.

Basically, form cutting is a simple and flexible method of machining gears. The equipment and cutters required are relatively simple, and standard machine tools (milling machines) often are used. However, in most cases the procedure is quite slow, and considerable care is required on the part of the operator, so it usually is employed where only one or a few gears are to be made.

Template machining utilizes a simple, single-point cutting tool that is guided by a template. This principle is illustrated in Figure 28-7. By using a template that is several times larger than the gear tooth that is to be cut, good accuracy can be achieved. However, the equipment is specialized, and the method is seldom used except for making large bevel gears.

Most high-quality gears that are made by machining are made by the *generating process*. This process is based on the principle that any two involute gears,

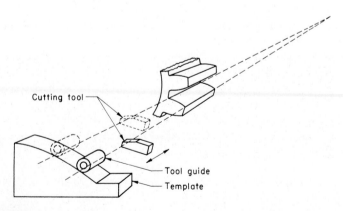

Cutting tool

Tool guide

Template

FIGURE 28-7 Method of machining gear teeth by means of a tool that is guided by a template.

or any gear and a rack, of the same diametral pitch will mesh together properly. Utilizing this principle, if one of the gears (or the rack) is made into a cutter by proper sharpening, it can be used to cut into a mating gear blank and thus generate teeth on the blank.

To carry out the process, the cutter gear and the gear blank must be attached rigidly to their corresponding shafts, and the two shafts must be interconnected by suitable gearing so that the cutter and the blank rotate positively with respect to each other and with the same pitch-line velocities. To start cutting a gear, the cutter gear is reciprocated and is fed radially into the blank between successive strokes. When the desired tooth depth has been obtained, the cutter and blank are then rotated slightly after each cutting stroke. The resulting generating action is indicated schematically in the upper diagram of Figure 28-8 and shown in the cutting of an actual gear tooth in the photographs in the lower portion of the same figure.

Machines for form-cutting gears. In machining gears by the form-cutting process, the form cutter is mounted on the machine spindle, and the gear blank

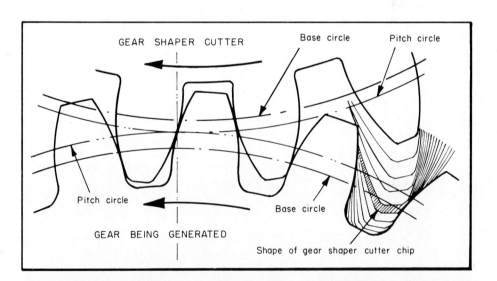

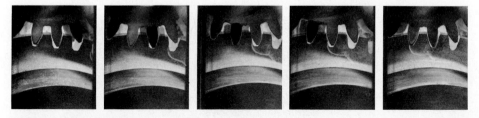

FIGURE 28-8 (*Top*) Generating action of a Fellows gear-shaper cutter. (*Bottom*) Series of photographs showing various stages in generating one tooth in a gear by means of a gear shaper, action taking place from right to left, corresponding to a diagram above. One tooth of the cutter was painted white. (Top *Courtesy Fellows Gear Shaper Co.*)

is mounted on an arbor held between the centers of some type of indexing device. Figure 28-9 shows the arrangement that is employed when, as often is the case, the work is done on a universal milling machine; the cutter is mounted on the spindle, and a dividing head is used to index the gear blank. When a helical gear is to be cut, as in the case shown, the table must be set at an angle equal to the helix angle, and the dividing head is geared to the longitudinal feed screw of the table so that the gear blank will rotate as it moves longitudinally.

Standard cutters usually are employed in form-cutting gears. In the United States, these come in eight sizes for each diametral pitch and will cut gears having the number of teeth indicated in the following tabulation:

Cutter Number	Gear Tooth Range
1	135 teeth to rack
2	55–134
3	35–54
4	26–34
5	21–25
6	17–20
7	14–16
8	12–13

FIGURE 28-9 Form cutting a helical gear on a universal milling machine. The gear train, which connects the table lead screw and the universal dividing head, is shown in the foreground. Inset shows a close-up view of the cutter and the gear blank.

A single cutter will not produce a theoretically perfect tooth profile for all sizes of gears in the range for which it is intended. However, the change in tooth profile over the range covered by each cutter is very slight and for most purposes satisfactory results are achieved. Where greater accuracy is required, half-number cutters (such as 3½) can be obtained. A typical cutter is shown in Figure 28-10. Cutters are available for all common diametral pitches and 14½ and 20° pressure angles.

To cut a gear on a milling machine, the geometric center of the cutter must be exactly aligned vertically with the center line of the index-head spindle. A gear blank of the proper outside diameter is placed on an arbor, which, in turn, is mounted between a foot-stock dead center and the live center in the index head and is connected positively to the latter by a dog. The table of the machine is raised until the cutter just makes contact with the periphery of the gear blank. The vertical feed dial then is set to zero and the table moved horizontally until the cutter clears the blank. The table is then fed upward an amount equal to the desired tooth depth, or a lesser amount if two or more cuts are to be made. The longitudinal power feed of the table then is engaged, and the tooth space is cut in the blank. After one tooth space is cut, the table movement is reversed until the cutter is again clear of the blank. The blank is then indexed to the proper position for cutting the next tooth space. This cutting procedure is repeated until all teeth have been formed.

If the amount of metal that must be removed to form a tooth space is large, roughing cuts may be taken with a *stocking cutter*, shown in Figure 28-11. The stepped sides of the stocking cutter remove most of the metal and leave only a small amount to be removed subsequently by the regular form cutter in a finish cut.

Straight-tooth bevel gears can be form cut on a milling machine, but this seldon is done. Because the tooth profile varies from one end of the tooth to the other, after one cut is taken to form the correct tooth profile at the smaller end, the relationship between the cutter and the blank must be altered. Shaving cuts then are taken on the side of each tooth to form the correct profile throughout the entire tooth length.

FIGURE 28-10 Typical form cutter for machining gear teeth. (*Courtesy Brown & Sharpe Mfg. Co.*)

FIGURE 28-11 Stocking cutter for making roughing cuts in machining gear teeth. (*Courtesy Brown & Sharpe Mfg. Co.*)

Although the form cutting of gears on a milling machine is a flexible process and is suitable for gears that are not to be operated at high speeds, or that need not operate with extreme quietness, the process is slow and requires skilled labor. It obviously is not suitable for quantity production.

Semiautomatic machines are available for making gears by the form-cutting process. Such a machine is shown in Figure 28-12. The procedure utilized is essentially the same as on a milling machine, except that after it is set for depth of cut, indexing, and so on, the various operations are completed automatically. Gears made on such machines are no more accurate than those produced on a milling machine, but the possibility of error is less, and they are much cheaper because of reduced labor requirements. But for large quantities, form cutting is not used.

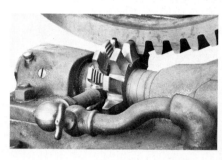

FIGURE 28-12 Cutting a gear on a semiautomatic gear-cutting machine. (*Above*) Cutters making simultaneous roughing and finish cuts. (*Courtesy Brown & Sharpe Mfg. Co.*)

Cutter-gear generating machines. A machine that generates gears by means of a gear-type cutter is shown in Figure 28-13. Figure 28-14 shows details of the cutter design. The gear blank is mounted on a vertical spindle and the cutter on the end of a second, vertical, reciprocating spindle. The two spindles are connected by means of change gears so that the cutter and gear blank revolve with the same pitch-line velocity. Cutting occurs on the down stroke on some machines and on the up stroke on others. At the end of each cutting stroke the spindle carrying the blank retracts slightly to provide clearance between the work and the tool on the return stroke. Because of the reciprocating action of the cutter, these machines commonly are called *gear shapers.*

To start cutting a gear, the cutter starts to feed inward before each cutting stroke, as it and the blank rotate. When the proper depth is reached, the inward feed stops and the cutter and blank continue their rotation until all the teeth have been formed by the generation process.

Either straight- or helical-tooth gears can be cut on gear shapers. To cut helical teeth, both the cutter and the blank are given an oscillating rotational motion during each stroke of the cutter, turning in one direction during the cutting stroke and in the opposite direction during the return stroke. Because the cutting stroke can be adjusted to end at any desired point, gear shapers are particularly useful for cutting cluster gears. Some machines can be equipped with two cutters

FIGURE 28-13 Fellows gear shaper, showing the motions of the gear blank and the cutter. Inset shows a close-up view of the cutter and teeth. (*Courtesy Fellows Gear Shaper Company.*)

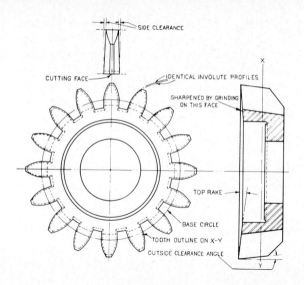

FIGURE 28-14 Details of the cutter used on the Fellows gear shaper (*Courtesy Fellows Gear Shaper Company.*)

to simultaneously cut two gears, often of different diameters. Gear shapers also can be adapted for cutting internal gears.

Two special types of gear shapers have been developed for mass-production purposes. The *rotary gear shaper* essentially is 10 shaper units mounted on a rotating base and having a single drive mechanism. Nine gears are cut simultaneously while a finished gear is removed and a new blank is put in place on the tenth unit. In the *planetary gear shaper,* shown in Figure 28-15, six gear blanks move in planetary motion about a large, central gear cutter. The cutter has no teeth in one portion to provide a space where the gear can be removed and a new blank placed on the empty spindle.

The *Sykes gear-generating machine* is employed primarily for cutting continuous herringbone gears; it is the only machine that will cut gears of this type.

FIGURE 28-15 Cutter and six gear blanks in a planetary gear shaper. (*Courtesy Fellows Gear Shaper Company.*)

Its unique feature is that it employs two cutter gears, mounted a fixed distance apart, and that either the cutters or the gear blank are reciprocated with respect to each other, parallel with the gear-blank axis, as shown in Figure 28-16. One cutter cuts one half of the gear as the reciprocating motion takes place in one direction, and the other cutter cuts the other half on the return stroke. Cutting takes place as each cutter moves toward the center line of the gear. Change gears connect the gear blank and the cutters to provide the necessary continuous rotation to bring about the generating process. In addition, the cutters and blank are given an oscillatory, twisting motion during each reciprocation cycle to provide the desired helix angle. These machines produce gears of excellent quality.

Gear-hobbing machines. As mentioned previously, involute gear teeth can be generated by a cutter that has the form of a rack, having teeth with straight sides. Although the use of a rack as a cutter would have a great advantage—in that the cutter would be simple to make—such a use has two major disadvantages. First, the cutter or the blank would have to reciprocate, with cutting occurring during only one stroke direction. Second, the rack would have to move longitudinally as the blank rotated (or the blank roll, in mesh, on the rack). Unless the rack were very long, or the gear very small, the two would not be in mesh after a few teeth were cut.

A *hob* overcomes the preceding two difficulties. As shown in Figures 28-17 and 28-18, a hob can be thought of, basically, as one long rack tooth that has been wrapped around a cylinder in the form of a helix and gashed at intervals to provide a number of cutting edges. Relief is provided behind each of the helically arranged cutting faces. As shown in Figure 28-18, the cross section of

FIGURE 28-16 Cutting a continuous herring-bone gear on a Sykes gear-cutting machine. (*Top*) Gear blank at left end of the stroke, and (*bottom*) at the right end. Cutters, behind gear, are denoted by arrows. (*Courtesy Western Gear Works.*)

FIGURE 28-17 Typical gear hob. (*Courtesy Barber-Colman Company.*)

each tooth, normal to the helix, is the same that of a rack tooth. A hob also can be thought of as a gashed worm.

The action of a hob in cutting a gear is illustrated Figure 28-18, with the actual hobbing of a small spur gear shown in Figure 28-19. To cut a spur gear, the axis of the hob must be set off from the normal to the rotational axis of the blank by the helix angle of the hob. In cutting helical gears, the hob must be set over an additional amount equal to the helix angle of the gear.

The cutting of a gear by means of a hob is a continuous action. The hob and the blank are connected by means of proper change gearing so they rotate in

FIGURE 28-18 Relationship of the hob and the gear blank in machining a spur gear by hobbing.

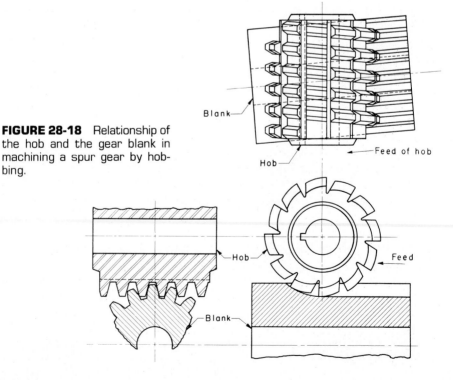

FIGURE 28-19 Setup for hobbing a spur gear, with gear blank mounted on mandrel between centers. (*Courtesy Barber-Colman Company.*)

mesh. To start cutting a gear, the hob is located so as to clear the blank and then moved inward until the proper setting for tooth depth is obtained. The hob is then fed in a direction parallel with the axis of rotation of the blank. As the gear blank rotates, the teeth are generated and the feed of the hob across the face of the blank extends the teeth to the desired tooth face width.

Because hobbing is a continuous action involving a multipoint cutting tool, it is rapid and economical. More gears are cut by this process than by any other. The process produces excellent gears and can also be used for splines and sprockets. Single-, double-, and triple-thread hobs are used. Multiple-thread types increase the production rate but do not produce accuracy as high as single-thread hobs.

Gear-hobbing machines are made in a wide range of sizes. Figure 28-19 shows a small machine while Figure 28-20 shows a hobbing machine cutting a gear over 2.44 meters (8 feet) in diameter. Such machines for cutting accurate, large gears frequently are housed in temperature-controlled rooms, and the temperature of the cutting fluid is controlled to avoid dimensional changes due to variations in temperature.

Bevel-gear generating machines. The machines used to generate the teeth on various types of bevel gears are among the most ingenious of all machine tools. Two basic types utilize the principle that a crown gear will mate properly with any bevel gear having the same diametral pitch and tooth form. One is used for cutting straight-tooth bevel gears, and the other for cutting curved-tooth bevel gears.

The basic principle of these machines is indicated in Figure 28-21. The blank and a connecting gear, having the same cone angle and diametral pitch, are

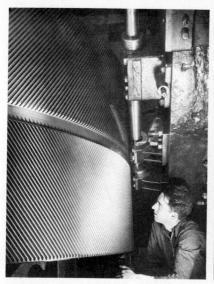

FIGURE 28-20 (*Left*) Close-up showing the hobbing of a large gear. (*Right*) Completed gear. This gear will transmit 22.4 MW (30,000 hp). Allowable error in tooth spacing is 0.0025 mm (0.0001 inch). (*Courtesy Westinghouse Electric and Manufacturing Company.*)

mounted on a common shaft so that the connecting gear meshes with the regular teeth on the crown gear, which also has a reciprocating cutter tooth that meshes with the teeth that it will cut in the gear blank. As the connecting gear rolls on the crown gear, the reciprocating cutter generates a tooth in the gear blank. Obviously, with only one cutter tooth, only a single tooth space would be cut in the gear blank. However, this limitation is overcome in actual machines by two modifications. First, by indexing the connecting gear and blank unit, the process is repeated as often as required to cut all the teeth in the blank. Second,

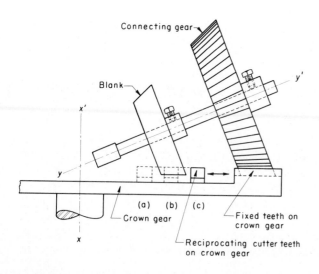

instead of a single, reciprocating cutter, two half-tooth cutters are used—approximating the inner sides of two adjacent teeth, so that they cut both faces of one tooth on the blank.

The application of this modified basic principle in actual machines is shown in Figures 28-22 and 28-23. The cradle acts as a crown gear and carries the two cutters, which reciprocate in slides simultaneously in opposite directions. It can be noted in Figure 28-23 that straight lines through the faces of the cutters, the axis of rotation of the gear blank, and elements of the pitch cone of the gear blank all meet at a common point that is in the plane of the imaginary crown gear and on the axis of the cradle, thus fulfilling the requirements shown in Figure 28-21.

In operation, while the cradle is at the extreme upward position, the gear blank is fed inward toward the cutter until the position for nearly full tooth depth is reached. The cradle then starts to roll downward, and a tooth is generated during the downward roll. When the cradle reaches the full down position, the blank is automatically fed inward a small amount—usually 0.25 to 0.38 mm (0.010 to 0.015 inch)—and a finish cut is made during the upward roll of the cradle. At the completion of the upward roll, the blank is withdrawn, indexed, and moved inward automatically, ready to start the next tooth.

Provision is made so that the spindle on which the gear blank is mounted can be swung through an angle to accommodate bevel gears of different cone angles. This spindle is connected to the cradle by means of suitable internal change gears to provide the required positive rolling motion between the two.

For cutting small straight-tooth bevel gears, the type of machine shown in Figure 28-24 often is used. These employ two revolving, disk-type cutters having interlocking teeth. These cutters reciprocate on the cradle slides as they rotate. Because of their multiple cutting edges and continuous rotary, rather than reciprocating, cutting action, they are much more productive than the other type.

Generating machines for straight-tooth bevel gears often are provided with a mechanism that produces a slight crown on the teeth. Such teeth are slightly

FIGURE 28-22 (*Top*) Schematic diagram showing the roll of the gear blank and cutters during the cutting of one tooth on a bevel-gear generating machine, as seen from the front. (*Bottom*) Photographs showing the same roll action, as seen from the back of the machine. (*Courtesy Heidenreich & Harbeck.*)

FIGURE 28-23 Cutters and gear blank on a Gleason straight-tooth gear-generating machine, showing the relationship between the axis of rotation of the blank and the axis of the cradle, which with the cutter simulates a crown gear. (*Courtesy Gleason Works.*)

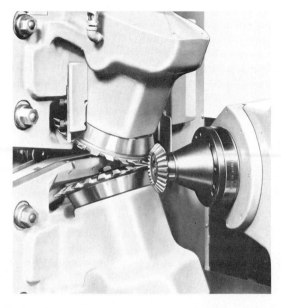

FIGURE 28-24 Bevel-gear generating machine, utilizing two disk-type cutters. (*Courtesy Gleason Works.*)

thicker at the middle than at the ends, to avoid having applied loads concentrated at the tooth ends where they are weakest.

Machines used for generating spiral, Zerol, and hypoid bevel gears employ the same basic crown-gear principle, but a multitooth rotating cutter is used. This cutter has its axis of rotation parallel with the axis of roll of the cradle, as is shown in Figure 28-25. Figure 28-26 shows the relationship of the gear blank to the axis of the theoretical crown gear (the cradle) and of the cutter teeth to the theoretical mating bevel gear.

In cutting a spiral bevel gear, the blank is fed inward toward the cutter to the full tooth depth while the cradle is in the downward position. In this position the cutter clears the blank. As the cradle and the gear blank roll upward together, the cutter engages the blank, starting at the smaller end of the gear, and the teeth are generated. At the same time the cutter progresses across the width of the tooth face. In some cases both sides of a gear tooth are generated simultaneously, whereas in others only one side is cut in a single roll of the gear and cradle.

A special process is sometimes employed to make curved-tooth pairs of bevel gears somewhat more cheaply than can be done by true generation. It produces fairly satisfactory gears where the gear ratio is greater than 3:1 or 4:1. In this process the larger gear is cut on a special machine, not by generation. The smaller, mating gear is then generated to mate properly with the tooth profile of the larger gear. Such gears are known as *Formate gears*.

Cold roll-forming of gears. The manufacture of gears by cold roll-forming has been highly developed and widely adopted in recent years. Currently, mil-

FIGURE 28-25 Cutting a hypoid bevel gear on a Gleason curved-tooth generating machine. (*Courtesy Gleason Works.*)

FIGURE 28-26 Relationship of the gear blank to the axis of the theoretical crown gear (the cradle) and of the cutter teeth to the theoretical mating bevel gear. (*Courtesy Gleason Works.*)

lions of high-quality gears are produced annually by this process; many of the gears in automobile transmissions are made this way. As indicated in Figure 28-27, the process is basically the same as that by which screw threads are roll-formed, except that in most cases the teeth cannot be formed in a single rotation of the forming rolls; the rolls are fed inward gradually during several revolutions.

Because of the metal flow that occurs, the top lands of roll-formed teeth are not smooth and perfect in shape—a depressed line between two slight protrusions often can be seen, as shown encircled in Figure 28-28. However, because the top land plays no part in gear-tooth action, if there is sufficient clearance in the mating gear, this causes no difficulty. Where desired, a light turning cut is used to provide a smooth top land and correct addendum diameter.

The hardened forming rolls are very accurately made, and the roll-formed gear teeth usually have excellent accuracy. In addition, because the severe cold working produces tooth faces that are much smoother and harder than those on ordinary machined gears, they seldom require hardening or further finishing, and they have excellent wear characteristics.

Owing to the rapidity of the process, its ease of being mechanized, the fact that no chips are made and thus less material is needed, and because skilled labor is not required, roll-formed gears are rapidly replacing machined gears whenever the process can be used. Small gears often are made by rolling a length of shaft and then slicing off the individual gear units.

Other gear-making processes. Gears can be made by the various casting processes. *Sand-cast gears* have rough surfaces and are not accurate dimen-

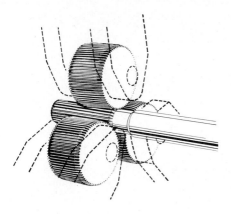

FIGURE 28-27 Method for forming gear teeth and splines by cold forming.

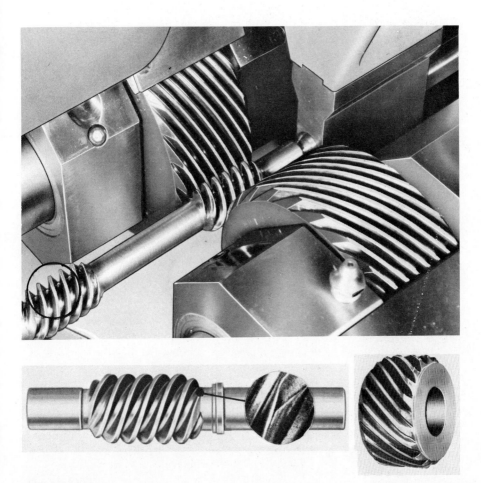

FIGURE 28-28 (*Top*) Worm being rolled by means of rotating rolling tools. (*Bottom left*) Typical worm made by rolling, with enlarged view of end of one tooth. (*Right*) Gear made by rolling. (Courtesy Landis Machine Company.)

sionally. They are used only for services where the gear moves slowly and where noise and inaccuracy of motion can be tolerated. Gears made by *die casting* are fairly accurate and have fair surface finish. They can be used to transmit light loads at moderate speeds. Gears made by *investment casting* may be accurate and have good surface characteristics. They can be made of strong materials to permit their use in transmitting heavy loads. In many instances gears that are to be finished by machining are made from cast blanks, and in some larger gears the teeth can be cast to approximate shape to reduce the amount of machining.

Large quantities of gears are produced by *blanking* in a punch press. The thickness of such gears usually does not exceed about 1.6 mm (¹⁄₁₆ inch). By shaving the gears after they are blanked, excellent accuracy can be achieved. Such gears are used in clocks, watches, meters, and calculating machines.

Excellent gears can be made by *broaching*. However, because of the very high cost of the broach, and the fact that a separate broach must be provided for each size of gear, this method is used relatively little.

High-quality gears, both as to dimensional accuracy and surface quality, can be made by the powder metallurgy process. Usually, this process is employed only for small sizes, ordinarily less than 25 mm (1 inch) in diameter. However, larger and excellent gears are made by forging powder metallurgy preforms. An example is shown in Figure 28-29. As discussed in Chapter 12, this results in a product of much greater density and strength than usually can be obtained by ordinary powder metallurgy methods, and the resulting gears give excellent service at reduced cost. Gears made by this process often require little or no finishing.

Large quantities of plastic gears are made by *plastic molding*. The quality of such gears is only fair, and they are suitable only for light loads. Accurate gears suitable for heavy loads frequently are machined out of laminated plastic materials. When such gears are mated with metal gears, they have the quality of reducing noise.

Quite accurate small-sized gears can be made by the *extrusion* process. Typically, long lengths of rod, having the cross section of the desired gear, are

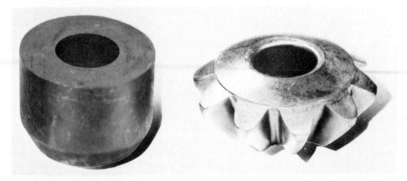

FIGURE 28-29 Powder metallurgy preform and finished forged gear. (*Courtesy Cincinnati Incorporated.*)

extruded. The individual gears are then sliced from this rod. Materials suitable for this process are brass, bronze, aluminum alloys, magnesium alloys, and occasionally, steel.

Flame machining (oxyacetylene cutting) can be used to produce gears that are to be used for slow-moving applications, where accuracy is not required.

A few gears are made by the hot roll-forming process. In this process a cold master gear is pressed into a hot blank as the two are rolled together.

Gear finishing.   In order to operate efficiently and have satisfactory life, gears must have accurate tooth profiles, and the faces of the teeth must be smooth and hard. These qualities are particularly important when gears must operate quietly at high speeds. When they are produced rapidly and economically by most of the processes except cold-roll forming, the tooth profiles may not be as accurate as desired, and the surfaces are somewhat rough and subject to rapid wear. Also, it is difficult to cut gear teeth in a hardened gear blank, so economy dictates that the gear be cut in a relatively soft blank and subsequently be heat treated to obtain greater hardness, if this is required. Such heat treatment usually results in some slight distortion and surface roughness. Although most roll-formed gears have sufficiently accurate profiles, and the tooth faces are adequately smooth and frequently have sufficient hardness, this process is feasible only for relatively small gears. Consequently, a large proportion of high-quality gears are given some type of finishing operation after they have received primary machining or heat treatment. Most of these finishing operations can be done quite economically, because only minute amounts of metal are removed.

Gear shaving is the most commonly used method for gear finishing. The gear is run, at high speed, in contact with a shaving tool, usually of the type shown in Figure 28-30. Such a tool is a very accurate, hardened, and ground gear that contains a number of peripheral gashes, or grooves, thus forming a series of sharp cutting edges on each tooth. The gear and shaving cutter are run in mesh

FIGURE 28-30   (*Left*) Gear being shaved by a rotary shaving cutter. (*Right*) Cutter for rotary gear shaving. (*Courtesy National Broach and Machine Company.*)

with their axes crossed at a small angle, usually about 10°. As they rotate, the gear is reciprocated longitudinally across the shaving tool (or vice versa). During this action, which usually requires less than a minute, very fine chips are removed from the gear-tooth faces, thus eliminating any high spots and producing a very accurate tooth profile.

Rack-type shaving cutters sometimes are used for shaving small gears—the cutter reciprocating lengthwise, causing the gear to roll along it, as it is moved sideways across the cutter and fed inward.

Although shaving cutters are costly, they have a relatively long life because only a very small amount of metal is removed—usually 0.025 to 0.10 mm (0.001 to 0.004 inch). Most gears are not hardened prior to shaving, although it is possible to remove very small amounts of metal from hardened gears, if they are not too hard. However, modern heat-treating equipment makes it possible to harden gears after shaving without harmful effects, so this practice usually is followed.

Some gear-shaving machines produce a slight crown on the gear teeth during shaving.

Gear burnishing is a finishing process that is used to a limited extent on unhardened gears. The unhardened gear is rolled under pressure contact with three hardened, accurately formed, burnishing gears. As a result of this action, any high points on the unhardened gear are plastically deformed so that a smoother surface and more accurate tooth form are achieved. Because the operation is one of localized cold working, some undesirable effects may accrue, such as localized residual stresses and nonuniform surface characteristics.

Grinding is used to obtain very accurate teeth on hardened gears. Two methods are used. One employs a formed grinding wheel that is trued to the exact form of a tooth by means of diamonds mounted on a special holder and guided by a large template. The other method utilizes a straight-sided grinding wheel that simulates one side of a rack tooth. The surface of the gear tooth is ground by the generating principle as the gear rolls past the grinding wheel, in the same manner that it would roll on a rack. This method has the advantage of using a simpler shape of grinding wheel, but it is much slower than form grinding because only one side of a tooth is ground at a time. Grinding produces very accurate gears, but because it is slow and expensive, it is used only on the highest quality, hardened gears.

Lapping also can be used for finishing hardened gears. The gear to be finished is run in contact with one or more cast-iron lapping gears under a flow of very fine abrasive in oil. Because lapping removes only a very small amount of metal, it usually is employed on gears that previously have been shaved and hardened. This combination of processes produces gears that are nearly equal to ground gears in quality, but at considerably lower cost.

Gear inspection.   As with all manufactured products, gears must be checked to determine whether the resulting product meets the design specifications and requirements. Because of their irregular shape and the number of factors that must be measured, such inspection of gears is somewhat difficult. Among the factors to be checked are the linear tooth dimensions—thickness, spacing, depth,

FIGURE 28-31 Using gear-tooth vernier calipers to check the tooth thickness at the pitch circle.

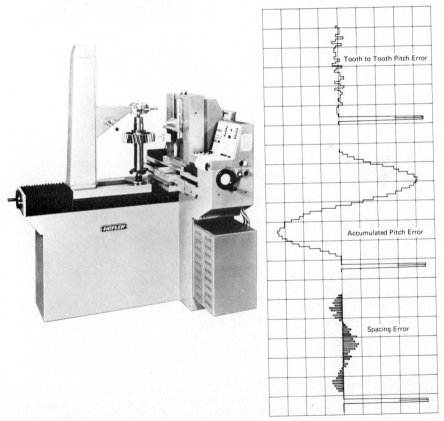

Tooth to Tooth Pitch Error

Accumulated Pitch Error

Spacing Error

FIGURE 28-32 (*Above*) Gear being checked on a special gear-checking machine. (*Right*) Resulting charts. (*Courtesy American Pfauter Corporation.*)

and so on—tooth profile, surface roughness, and noise. Several special devices, most of them automatic or semiautomatic, are used for such inspection.

Gear-tooth vernier calipers can be used to measure the thickness of gear teeth on the pitch circle, as shown in Figure 28-31. However, most inspection is made by special machines, such as shown in Figure 28-32, which in one or a series of operations check several factors, including eccentricity, variations in circular pitch, variations in pressure angle, fillet interference, and lack of continuous action. The gear usually is mounted and moved in contact with a master gear. The movement of the latter is amplified and recorded on moving charts, as shown in Figure 28-32.

Because noise level is important in many applications, not only from the viewpoint of noise pollution but also as an indicator of probable gear life, special equipment for its measurement is quite widely used, sometimes integrated into mass-production assembly lines.

REVIEW QUESTIONS

1. Why can the relative angular velocities of two mating spur gears not be determined by their outside diameters alone?
2. Why is the involute form used for gear teeth?
3. What is the diametral pitch of a gear?
4. What is the relationship between the diametral pitch and the module of a gear?
5. On a sketch of a gear, indicate the pitch circle, addendum circle, dedendum circle, and the circular pitch.
6. What five requirements must be met in order for gears to operate satisfactorily? Which of these are determined by the manufacturing process?
7. What are the advantages of helical gears, compared with spur gears?
8. What is the principal disadvantage of helical gears?
9. A gear that has a pitch diameter of 152.4 mm (6 inches) has a diametral pitch of 4. What number of form cutter would be used in cutting it?
10. What difficulty would be encountered in hobbing a herringbone gear?
11. What is the only type of machine on which full-herringbone gears can be cut?
12. What modification in design is made to enable the major advantage of herringbone gears to be obtained and still permit them to be cut by hobbing?
13. Why are not more gears made by broaching?
14. What is the most important property of a crown gear?
15. What are three basic processes for machining gears?
16. Which basic gear-machining process is utilized in a Fellows gear shaper?
17. Could a helical gear be machined on a plain milling machine? Why?
18. Explain how the gear blank and the machine table are interconnected when a helical gear is machined on a milling machine.
19. What is the relationship between a crown gear and Gleason gear generators?
20. Why is a gear-hobbing machine much more productive than a gear shaper?
21. Why is a gear shaper more likely to be used for machining cluster gears than a hobbing machine?
22. What are the advantages of cold roll-forming for making gears?
23. Assume that 10,000 spur gears, 28.6 mm (1⅛ inch) in diameter and 9.53 mm (⅜ inch) thick are to be made of 70–30 brass. What manufacturing methods would you consider?
24. If only three gears described in question 23 were to be made, what process would you select?
25. Why is cold roll-forming not suitable for making gray cast iron gears?
26. Under what conditions can shaving not be used for finishing gears?

27. What inherent property accrues from cold roll-forming of gears that may result in improved gear life?
28. Why can lapping not be used to finish cast iron gears?
29. What factors usually are checked in inspecting gears?
30. A single thread hob that has a pitch diameter of 76.2 mm is used to cut a gear having 36 teeth. If a cutting speed of 27.4 meters per minute is used, what will be the rpm of the gear blank?
31. If the gear in question 30 has a face width of 76.2 mm, a feed of 1.9 mm per revolution of the workpiece is used, and the approach and overtravel distances of the hob are 38 mm, how much time will be required to hob the gear?

CASE STUDY 28. How Did the Tubing Get Flared?

Figure CS-28 shows an enlarged view of a portion of the inside of a piece of 19.05 mm (0.750-inch)-OD soft steel tubing, which had a wall thickness of 24.67 mm (0.097 inch). This tubing has been assembled into a self-flare hydraulic fitting. If the tubing was assembled properly in the fitting, the inside would be forced against a smooth surface in the fitting, and a bevel would be formed on the inside of the end of the tubing. The joint failed in service, resulting in a fire and very substantial damage to a die-casting plant.

FIGURE CS-28 View of end of piece of hydraulic tubing showing marks on bevel on inside of tubing and variation of remaining wall thickness on end of tubing.

In subsequent litigation, one of the parties alleged that the tubing and fitting had been properly assembled. The other party claimed that someone had beveled the inside of the tubing by some means—probably by some hand-machining process—prior to assembly into the fitting.

Using your knowledge of how such tubing is made, of how metals react when cold-worked and/or machined, and how the resulting metal surface would appear, which claim do you believe was correct? State your reasons.

PART IV

JOINING PROCESSES

CHAPTER 29

Forge, Oxyfuel Gas, and Arc Welding

Welding is a process in which two materials, usually metals, are permanently joined together through localized coalescence, resulting from a suitable combination of temperature, pressure, and metallurgical conditions. Because the combination of temperature and pressure can range from high temperature with no pressure to high pressure with no increased temperature, welding can be accomplished under a very wide variety of conditions, and numerous welding processes have been developed and are used routinely in manufacturing. The average person usually has little concept of the importance of welding as a manufacturing process, yet a large proportion of our metal products would have to be drastically modified, would be considerably more costly, or could not perform as efficiently, if it were not for the use of welding.

To obtain coalescence between two metals, there must be a combination of sufficient proximity and activity between the atoms of the pieces being joined to cause the formation of common metallic crystals. The ideal metallurgical bond requires: (1) perfectly smooth, flat or matching surfaces; (2) clean surfaces, free from oxides, absorbed gases, grease, or other contaminants; (3) no impurities be present within the metals; and (4) that both metals be single crystals with identical crystallographic orientation. Obviously, these conditions are difficult to obtain, even under ideal conditions, and are impossible under normal conditions. Consequently, the various joining methods are designed to overcome or compensate for the inability to achieve ideal conditions. Surface roughness is overcome either by force, causing plastic deformation of the asperities, or by melting the two surfaces, so that fusion occurs. In solid-state welding, contaminated layers can be removed by mechanical or chemical cleaning prior to welding or by causing sufficient metal flow along the interface so that they are squeezed out of the weld. In fusion welding, where a pool of molten metal exists, the contaminants are removed by the use of fluxing agents. If welding is done in a vacuum, either by solid-state or fusion processes, the contaminants are removed much more easily, and coalescence is established with considerable ease. Thus, in outer-space, mating parts may weld under light loads, even when such was not intended.

The various welding processes not only differ considerably in the manner in which temperature and pressure are combined and achieved, they also vary as to the attention that must be given to the cleanliness of the metal surface prior to welding and to possible oxidation or contamination of the metal during welding. If high temperatures are used, most metals are affected more adversely by the surrounding environment and, if actual melting occurs, serious modification of the metal may result. Also, the metallurgical structure and quality of the metal may be affected—usually adversely—as a consequence of the heating and cooling, and these effects should be taken into account.

In summary, in order to obtain satisfactory welds it is desirable to have (1) a satisfactory heat and/or pressure source, (2) a means of protecting or cleaning the metal, and (3) avoidance of, or compensation for, harmful metallurgical effects.

A chart of welding processes. The various welding processes have been defined and classified by the American Welding Society as shown in Figure 29-1 and assigned letter symbols to facilitate their designation. These processes provide a variety of ways of meeting the three requirements stated previously, and make it possible to achieve effective and economical welds in nearly all metals and combinations of metals. The result has been that welding has replaced other types of permanent fastening to such a degree that a large proportion of manufactured products contain one or more welds.

FORGE WELDING

Although little *forge welding* (FOW) is done today, it is the most ancient of the welding processes and a review of it is of both historical and practical use in understanding how and why modern welding processes were developed. The armor makers of ancient times owed their positions of prominence in their societies as much to their ability to join two pieces of steel into a single strong piece as to their ability to harden and temper the metal. Perhaps the master of forge welding was the colorful village blacksmith, who, with his forge, hammer, and anvil, could join pieces of metal to form a variety of products.

The blacksmith used a charcoal forge as his source of heat. Pieces that were to be welded were heated to a forging temperature and the ends scarfed by

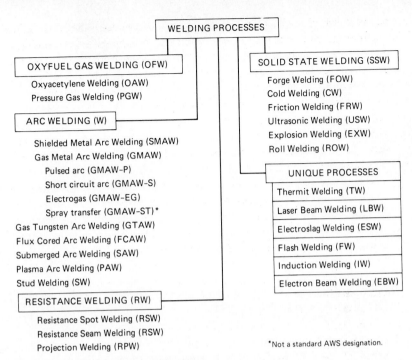

FIGURE 29-1 Chart of welding processes.

In the chart:

WELDING PROCESSES

OXYFUEL GAS WELDING (OFW)
Oxyacetylene Welding (OAW)
Pressure Gas Welding (PGW)

ARC WELDING (W)
Shielded Metal Arc Welding (SMAW)
Gas Metal Arc Welding (GMAW)
Pulsed arc (GMAW-P)
Short circuit arc (GMAW-S)
Electrogas (GMAW-EG)
Spray transfer (GMAW-ST)*
Gas Tungsten Arc Welding (GTAW)
Flux Cored Arc Welding (FCAW)
Submerged Arc Welding (SAW)
Plasma Arc Welding (PAW)
Stud Welding (SW)

RESISTANCE WELDING (RW)
Resistance Spot Welding (RSW)
Resistance Seam Welding (RSW)
Projection Welding (RPW)

SOLID STATE WELDING (SSW)
Forge Welding (FOW)
Cold Welding (CW)
Friction Welding (FRW)
Ultrasonic Welding (USW)
Explosion Welding (EXW)
Roll Welding (ROW)

UNIQUE PROCESSES
Thermit Welding (TW)
Laser Beam Welding (LBW)
Electroslag Welding (ESW)
Flash Welding (FW)
Induction Welding (IW)
Electron Beam Welding (EBW)

*Not a standard AWS designation.

hammering to permit them to be fitted together without undue thickness. The ends were again heated until they were nearly to the proper temperature. They then were withdrawn and dipped into some borax, which acted as a flux. Heating was then continued for a short interval until the blacksmith judged by their color that the workpieces were at the proper temperature for welding. He then withdrew them from the forge, struck them sharply against the anvil, or with his hammer, to knock off the scale and impurities. He then placed the ends of each piece on the anvil and hammered them into the necessary proximity to complete the weld. Thus the competent blacksmith did solid-state welding and was able to produce a welded joint that was as strong as the original metal. However, because of the crudeness of his heat source, the uncertainty of his temperature control, and the difficulty of maintaining metal cleanliness, a great amount of skill was required, and the results were variable. The quality of deformation welds clearly depends upon: the degree of deformation, surface cleanliness, and temperature.

Forge-seam welding. Although forge welding, as done by the blacksmith, is seldom done today, very large amounts of *forge-seam welding* are done in the manufacture of pipe, as was described in Chapter 14. A heated strip of steel is formed into a cylinder and the edges welded together in either a lap or butt joint by the pressure that is exerted as the formed metal is either pulled through a conical welding bell or passed between formed rolls (see Figures 14-28 and 14-29).

Cold welding.   *Cold welding* (CW), a solid-state process, is a unique variation of forge welding, in that no heating is used and only a single blow or pressure application is employed. Coalescence results solely through the rapid application of pressure, so it represents one extreme of the possible temperature–pressure possibilities.

To produce a cold weld the faying surfaces are cleaned, usually by wire brushing, placed in contact, and subjected to localized pressure sufficient to produce about 30 to 50% cold working. This can be done in a punch press or a special tool. The result is as shown in Figure 29-2 (left). Undoubtedly, some heating does occur as a result of the severe cold working of the metal, but the high localized pressure is the primary factor in producing coalescence.

The use of cold welding is generally confined to fastening small parts, such as the electrical connections shown in Figure 29-2 (right). However, *roll bonding,* wherein two or more sheets are joined by passing simultaneously through a rolling mill, can be performed either cold or hot.

OXYFUEL GAS WELDING[1]

Oxyfuel gas welding (OFW) covers a group of welding processes that utilize as the heat source a flame resulting from the burning of a fuel gas and oxygen, mixed in proper proportions. The oxygen usually is supplied in relatively pure form, but may, in rare cases, come from air.

It was the development of a practical torch to burn acetylene and oxygen,

[1]Why the AWS committee came up with this term, when they were clearly trying to imply "gas–oxygen flame welding," is a mystery.

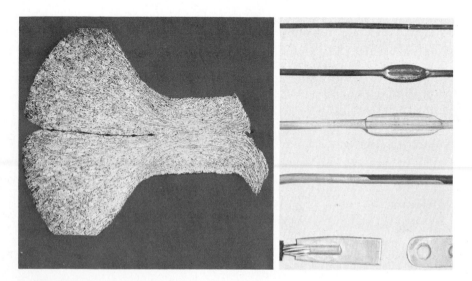

FIGURE 29-2   (*Left*) Section through a cold weld. (*Right*) Small parts joined by cold welding. (*Courtesy Koldweld Corporation.*)

shortly after 1900, that brought welding out of the blacksmith's shop, demonstrated its potential, and started its development as a manufacturing process. However, gas-flame welding has largely been replaced by other welding processes, except for some repair work and a few special applications. Acetylene still is the principal fuel gas employed in the process.

The combustion of oxygen and acetylene (C_2H_2), by means of a welding torch of the type shown in Figure 29-3, produces a temperature of about 3482°C (6300°F) in a two-stage reaction. In the first stage the oxygen and acetylene react in this manner:

$$C_2H_2 + O_2 \rightarrow 2CO + H_2$$

This reactions occurs near the end of the torch tip. The second stage of the reaction—combustion of the CO and H_2—occurs just beyond the first combustion zone in two reactions:

$$2CO + O_2 \rightarrow 2CO_2$$

$$H_2 + \tfrac{1}{2}O_2 \rightarrow H_2O$$

The oxygen for these secondary reactions is obtained from the atmosphere. This two-stage process produces a flame having two distinct zones, with the maximum temperature occurring at the end of the inner cone (see Figure 29-4) where the first stage of combustion is complete. Most welding should be done with the torch held so that this point of maximum temperature is just off the metal being welded. The outer zone of the flame serves to preheat the metal and at the same time provides some shielding from oxidation because some of the oxygen from the surrounding air is consumed in secondary combustion.

As shown in Figure 29-4, three types of flames can be obtained by varying the oxygen-to-acetylene ratio. If the ratio is about 1:1 to 1.15:1, all reactions

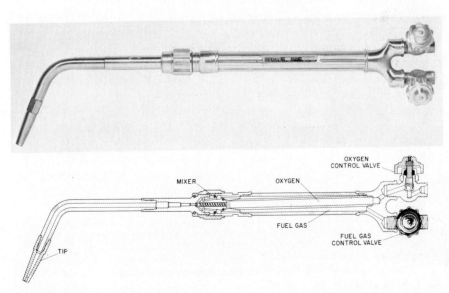

FIGURE 29-3 Oxyacetylene welding torch. (*Courtesy Victor Equipment Company.*)

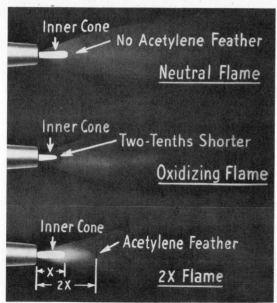

are carried to completion and a *neutral flame* is produced. Most welding is done with a neutral flame, since such a flame has a minimum chemical effect on most heated metals.

A higher ratio produces an *oxidizing flame,* quite similar in appearance, but possessing an excess of oxygen. Such flames are used only when welding copper and copper alloys and as a decarburizing flame for steels, the excess oxygen reacting with carbon in the steel.

Excess fuel produces a flame that is *carburizing.* The excess fuel decomposes to carbon and oxygen, and the flame temperature is not as great. Flames with a slight excess of fuel are reducing flames. No carburization occurs, but the metal is well protected from oxidation. Flames of this type are used in welding Monel, low-carbon steels, and some alloy steels, and in applying some common hard facing materials. While discussed in terms of acetylene, similar flame variation can be obtained with other fuel gases, such as hydrogen, methane, or propane.

For welding purposes, acetylene most often is obtained in portable storage tanks holding up to 8.5 cubic meters (300 ft³) at 1.72-MPa (250-psi) pressure. Because acetylene is not safe when stored as a gas at over 0.1 MPa (15 psi), it is commonly dissolved in acetone. The storage cylinders are filled with a porous filler, such as balsa-wood chips and infusorial earth. Acetone fills the voids in the filler material and serves as a medium for dissolving the acetylene.

When large quantities of acetylene are needed, acetylene generators can be employed. These operate on the principle of dropping pieces of calcium carbide into water to generate the acetylene.

Stabilized methylacetylene propadiene, known best under the trade name MAPP gas, has become a competitor to acetylene particularly where portability is important. While flame temperature is slightly lower, it is more dense, thus providing more energy for a given volume, and it can be stored safely in ordinary pressure tanks. Oxygen for gas-flame welding almost always is obtained from pressure tanks.

The pressures used in gas-flame welding ordinarily vary from 6.9 to 103.4 kPa (1 to 15 psi), pressure regulators being employed to reduce and maintain the desired values. Because mixtures of acetylene and oxygen or air are highly explosive, precautions must be taken to avoid mixing the gases improperly or by accident. All acetylene fittings have left-hand threads, whereas those for oxygen are equipped with right-hand threads, thus preventing improper connections from being made.

The tip size (or orifice diameter) of the torch can be altered to control the shape of the inner cone, flow rate of gases, and the size of the material that can be welded. Larger tips permit greater flow of gases, resulting in greater heat input without the high gas velocities that would blow the molten metal from the weld puddle. Thicker metal requires larger torch tips, and larger torch tips operate with higher gas pressure.

Uses, advantages, and limitations.   Almost all oxyfuel gas welding is *fusion* welding; the metals being joined are simply melted at the point where welding occurs, and no pressure is involved. Because a slight gap usually exists between the pieces being joined, filler material usually must be added in the form of a wire or rod that is melted in the flame or in the pool of weld metal. Welding rods come in standard sizes, with diameters from 1.6 to 9.5 mm ($\frac{1}{16}$ to $\frac{3}{8}$ inch) and lengths from 0.6 to 0.9 meter (24 to 36 inches). They are available in standard grades to provide specified minimum tensile strengths or in compositions to match the base metal. Fluxes can be added as a powder, or the welding rod can be dipped in a flux paste or precoated.

Good-quality welds can be obtained by the OFW process if proper technique and care are used. Control of the temperature of the work is easily accomplished. However, exposure of the heated and molten metal to various gases in the flame and the atmosphere, without effective shielding, makes it difficult to prevent contamination. In addition, because the heat source is not concentrated, considerable areas of the metal are heated and distortion is likely to occur. Flame welding processes have largely been replaced by shielded-arc and inert gas metal arc welding processes.

Pressure gas welding.   *Pressure gas welding* (PGW) is a process used to make butt joints between such objects as pipe and railroad rails. The ends are heated with a gas flame to a temperature below the melting point and then forced together under considerable pressure. The process thus is a type of solid-state welding. Figure 29-5 shows this process being used to join sections of pipe.

ARC WELDING

Almost from the time electricity became a commercial reality, it was recognized that an electric arc between two electrodes was a concentrated heat source ap-

FIGURE 29-5 Pressure welding of pipe. (*Courtesy Linde Division, Union Carbide Corporation.*)

proaching 3871°C (7000°F) in temperature. As early as 1881, various persons attempted to use an arc between a carbon electrode and metal workpieces as the heat source for fusion welding, using the basic circuit shown in Figure 29-6. As in gas-flame welding, filler metal was added in the form of a metallic wire. Later, the bare metal wire was used as the electrode; it melted in the arc and thus automatically supplied the needed filler metal. However, the results were very uncertain. Because of the instability of the arc, a great amount of skill was required to maintain it, and contamination and oxidation of the weld metal resulted from its exposure to the atmosphere at such high temperatures. Furthermore, there was little or no understanding of the metallurgical effects and requirements. Consequently, although the great potential of the process was recognized, very little use was made of the process until after World War I. About 1920, shielded metal electrodes were developed, which provided a stable arc, shielding from the atmosphere, and some fluxing action for the molten pool of metal—thus solving the major problems related to arc welding. The use of the process has expanded very rapidly ever since. Today, a considerable variety of arc-welding processes are available.

All arc-welding processes employ essentially the same basic circuit depicted in Figure 29-6, except that alternating current is used at least as much as direct current. When direct current is used, if the work is made positive (the anode of the circuit) and the electrode is made negative, *straight polarity* (spdc) is said

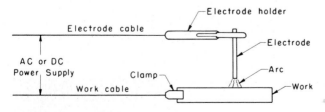

FIGURE 29-6 Basic arc-welding circuit.

to be employed. When the work is negative and the electrode is positive, the polarity is *reversed* (rpdc). When bare electrodes are used, greater heat is liberated at the anode. Certain shielded electrodes, however, change the heat conditions and are used with reverse polarity.

All arc welding is done with metal electrodes. In one type the electrode is consumed and thus supplies needed filler metal to fill in the voids in the joint and thus speed the welding process. In this case the electrode has a melting point below the temperature of the arc. Small droplets are melted from the end of the electrode and pass to the parent metal. The size of these droplets varies greatly and the mechanism of the transfer varies with different types of electrodes and processes. Figure 29-7 depicts metal transfer by the globular, spray, and short-circuit modes. As the electrode melts, the arc length and the resistance of the arc path vary. This requires that the electrode be moved toward or away from the work to maintain the arc and satisfactory welding conditions.

No ordinary manual arc welding is done today with bare electrodes; shielded (covered) electrodes always are used. A large amount of automatic and semiautomatic arc welding is done in which the electrode is a continuous, bare metal wire. However, this is always in conjunction with a separate shielding and arc-stabilizing medium and automatic feed-controlling devices that maintain the proper arc length.

In the other type of metal arc welding, the electrode is made of tungsten, which is not consumed by the arc except by relatively slow vaporization. Here, a separate filler wire must be used to supply the needed metal.

In summary, the various arc welding processes require selection or specification of the welding voltage, welding current, arc polarity (straight polarity, reversed polarity, or alternating current), arc length, welding speed (how fast the electrode is moved across the workpiece), arc atmosphere, electrode or filler material, and flux. The filler metal is selected to match the base metal with respect to properties and/or alloy content (composition).

Shielded metal arc welding. *Shielded metal arc welding* (SMAW) uses electrodes that consist of metal wire, usually from 1.59 to 9.53 mm ($\frac{1}{16}$ to $\frac{3}{8}$ inch) in diameter, upon which is extruded a coating containing chemical components

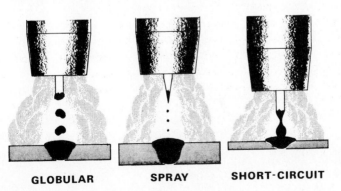

GLOBULAR　　　　SPRAY　　　SHORT-CIRCUIT

FIGURE 29-7 Basic modes of metal transfer during arc welding. (*Courtesy Republic Steel Corporation.*)

that add a number of desirable characteristics, including all or a number of the following:

1. Provide a protective atmosphere.
2. Stabilize the arc.
3. Act as a flux to remove impurities from the molten metal.
4. Provide a protective slag to accumulate impurities, prevent oxidation, and slow down the cooling of the weld metal.
5. Reduce weld-metal spatter and increase the efficiency of deposition.
6. Add alloying elements.
7. Affect arc penetration.
8. Influence the shape of the weld bead.
9. Add additional filler metal.

Coated electrodes are classified on the basis of the tensile strength of the deposited weld metal, the welding position in which they may be used, the type of current and polarity (if direct current), and the type of covering. A four- or five-digit system of designation is used, as indicated in Figure 29-8. As an example, type E7016 is a low-alloy steel electrode that will provide a deposit having a minimum tensile strength of 70,000 psi in the non-stress-relieved condition; it can be used in all positions, with either alternating current or reverse-polarity direct current, and it has a low-hydrogen-type coating.

In general, the cellulosic coatings contain about 50% SiO_2; 15% TiO_2; small amounts of FeO, MgO, and Na_2O; and about 30% volatile matter. The titania coatings have about 30% SiO_2, 50% TiO_2; small amounts of FeO, MgO, Na_2O, and Al_2O_3; and about 5% volatile material. The low-hydrogen coatings have various compositions designed to eliminate dissolved hydrogen in the deposited weld metal and thus prevent microcracking. To be effective they must be baked just prior to use to assure the removal of all moisture from the coating.

All electrodes are marked with colors in accordance with a standard established by the National Electrical Manufacturers Association so they can be read-

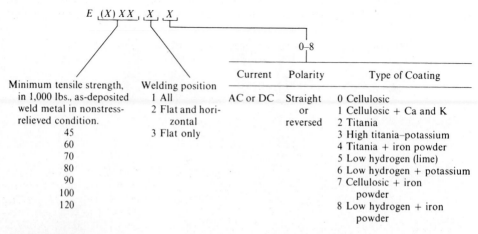

FIGURE 29-8 Designation system for arc-welding electrodes.

ily identified as to type. Electrode selection consists of determining: electrode coating, coating thickness, electrode composition, and the electrode diameter.

As the coating on the electrode melts and vaporizes, it forms a protective atmosphere that stabilizes the arc and protects the molten and hot metal from contamination. Fluxing constituents unite with any impurities in the molten metal and float them to the surface to be entrapped in the slag coating that forms over the weld. This slag coating protects the cooling metal from oxidation and slows down the cooling rate to prevent hardening. The slag is easily chipped from the weld when it has cooled. Figure 29-9 depicts the way in which metal is deposited from a shielded electrode.

Electrodes having iron powder in the coating are used extensively in production-type welding, since they significantly increase the amount of metal that can be deposited with a given size electrode wire and current. Other electrodes possess coatings designed to melt more slowly than the filler wire, such that if the electrode is dragged along the work, the center wire will be recessed by the proper arc length. These are called *contact* or *drag electrodes*.

Carbon steels, alloy steels, stainless steels, and cast irons are commonly welded by this process. Reverse-polarity dc is used to obtain deep penetration, with alternate modes being employed when welding thin sheet. Metal transfer is either globular or short circuit and arc temperatures are rather low (5000°C or 9000°F). Typical welding voltages are 15 to 45 volts with currents between 10 and 500 amps.

In order to provide electrical contact to the center, filler-metal wire, most SMAW electrodes are finite-length "sticks." Length is limited since the current must be supplied near the arc or the electrode will overheat and ruin the coating. Although some techniques have been developed to provide continuous shielded metal arc welding, the general trend is to use alternative methods for heavy-use production manufacturing. Stick electrodes, however, are still used in many job-shop or repair welding operations.

Gas tungsten arc welding. *Gas tungsten arc welding* (GTAW)[2] was one of the first major developments away from the use of ordinary shielded electrodes. Originally developed for welding magnesium, it employs a tungsten electrode, held in a special holder through which an inert gas is supplied with sufficient

---

[2]This process formerly was known as TIG welding—for Tungsten, Inert-Gas.

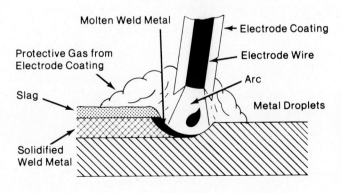

Molten Weld Metal
← Electrode Coating
Protective Gas from Electrode Coating
Electrode Wire
Arc
Slag
Metal Droplets
Solidified Weld Metal

FIGURE 29-9 Schematic diagram of shielded metal arc welding (SMAW). (*Courtesy American Iron and Steel Institute, Washington, D.C.*)

flow to form an inert shield around the arc and the molten pool of metal, thereby shielding them from the atmosphere. Argon or helium, or a mixture of them, is used as the inert shielding medium. Because the tungsten electrode is not consumed at arc temperatures in these inert gases, the arc length remains constant so that the arc is stable and easy to maintain. The tungsten electrodes often are treated with thoria or zirconium to provide better current-carrying and electron-emission characteristics. A high-frequency, high-voltage current usually is superimposed on the regular ac or dc welding current to make it easier to start and maintain the arc. Figure 29-10 shows a typical, water-cooled, GTAW torch.

If filler metal is required, it must be supplied by a separate wire, as indicated in Figure 29-11, and is generally identical to the metal being welded. In applications where a close fit exists, no filler metal may be needed. Where higher deposition rates are desired, a separate circuit is applied to electrically preheat the filler wire. As shown in Figure 29-12, the deposition rate can be several times that with a cold wire, and it can be further increased by oscillating the filler wire from side to side when making a weld pass. The hot-wire process is not practical with copper or aluminum, however, because of the low resistivities of the filler wire.

Gas tungsten arc welding produces very clean welds, and no special cleaning or slag removal is required because no flux is employed. With skillful operators, welds often can be made that are scarcely visible. However, the surfaces to be welded must be clean and free of oil, grease, paint, or rust, because the inert gas does not provide any cleaning or fluxing action.

All alloys can be welded by this method. Maximum penetration is obtained with straight polarity dc conditions, although ac may be specified to break up surface oxides. Reverse polarity tends to melt the tungsten electrode. Weld voltage is typically 20 to 40 volts and weld current varies from less than 125 amperes for rpdc to 1000 amperes for spdc.

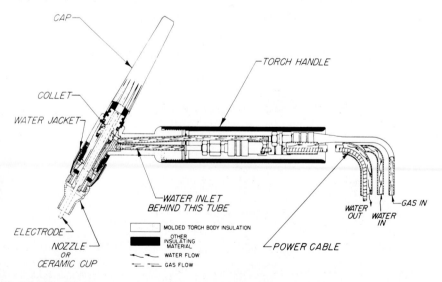

FIGURE 29-10 Welding torch used in nonconsumable metal-electrode, inert-gas (GTAW) welding. (*Courtesy Linde Division, Union Carbide Corporation.*)

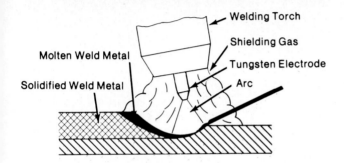

FIGURE 29-11 Schematic diagram of gas tungsten arc welding (GTAW). (*Courtesy American Iron and Steel Institute, Washington, D.C.*)

Gas tungsten arc spot welding. A variation of gas tungsten arc welding is employed for making spot welds between two pieces of metal without the necessity of having access to both sides of the joint. The basic procedure is shown in Figure 29-13. A modified and vented inert-gas, tungsten arc gun and nozzle are used, with the nozzle pressed against one of the two pieces of the joint. The workpieces must be sufficiently rigid to sustain the pressure that must be applied to one side to hold them in reasonably good contact. The arc between the tungsten electrode and the upper workpiece provides the necessary heat, and an inert gas, usually argon or helium, flows through the nozzle and provides a shielding atmosphere. Automatic controls move the electrode to make momentary contact with the workpiece to start the arc, and then withdraw and hold it at a correct distance to maintain the arc. The duration of the arc is timed automatically so that the two workpieces are heated sufficiently to form a spot weld under the pressure of the ''gun'' nozzle. The depth and size of the weld nugget are controlled by the amperage, time, and type of shielding gas.

Because access to only one side of the work is required, this type of spot welding has an advantage over resistance spot welding in certain applications, as in fastening relatively thin sheet metal to a heavier framework, illustrated in Figure 29-14.

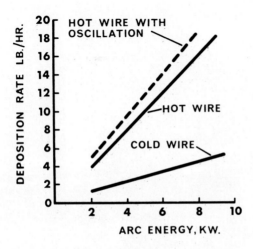

FIGURE 29-12 Comparison of the metal deposition rates with hot and cold filler wire in GTAW welding. (*Courtesy Welding Journal.*)

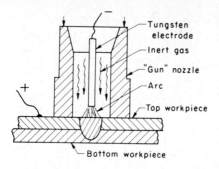

FIGURE 29-13 Schematic diagram of the method of making spot welds by the inert-gas-shielded tungsten arc process.

Gas metal arc welding. *Gas metal arc welding* (GMAW)[3] was a logical outgrowth of gas tungsten arc welding, differing in that the arc is maintained between an automatically fed, consumable wire electrode and the workpiece. It thereby automatically provides the additional filler as depicted in Figure 29-15. Figure 29-16 shows a schematic of GMAW equipment showing the electrical circuits and mechanisms for water cooling and flow of the shielding gas. Although argon and helium, or mixtures of them, can be used for welding virtually any metal, they are used primarily for welding nonferrous metals. For welding steel, some O_2 or CO_2 usually is added to improve the arc stability and to reduce weld spatter. The cheaper CO_2 alone can be used for welding steel, provided that a deoxidizing electrode wire is employed.

The shielding gases have considerable effect on the nature of the metal transfer from the electrode to the work and also affect the heat transfer behavior, penetration, and tendency for undercutting (weld pool below the level of the base

[3]Formerly called MIG (metal, inert-gas) welding.

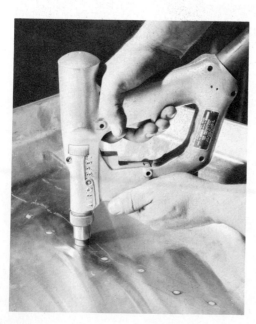

FIGURE 29-14 Making a spot weld by the inert-gas-shielded tungsten arc process. (*Courtesy Air Reduction Company, Inc.*)

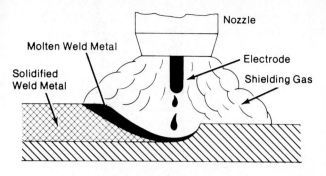

FIGURE 29-15 Schematic diagram of gas metal arc welding (GMAW). (*Courtesy American Iron and Steel Institute, Washington, D.C.*)

metal). Several types of electronic controls, which alter the wave form of the current, make it possible to vary the mechanism of metal transfer—by drops, spray, or short-circuiting drops. Some of these variations in the basic process are: *pulsed arc welding* (GMAW-P), *short circuiting arc welding* (GMAW-S), and *spray transfer welding* (GMAW-ST). *Buried arc welding* (GMAW-B) is another variation in which carbon dioxide-rich gas is used and the arc is buried in its own crater.

Gas metal arc welding is fast and economical because there is no frequent changing of electrodes, as with stick-type electrodes. In addition, there is no slag formed over the weld, the process often can be automated and, if done manually, the welding head is relatively light and compact, as shown in Figure 29-17. A reverse-polarity dc arc is quite commonly used because of its deep

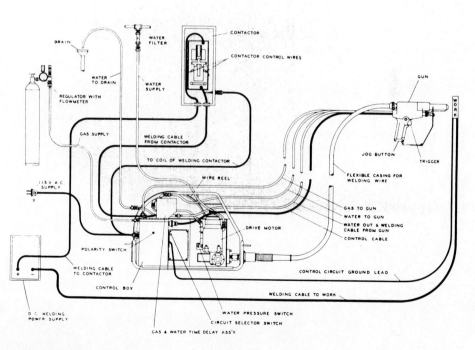

FIGURE 29-16 Schematic diagram of equipment for gas metal arc welding (GMAW). (*Courtesy Air Products and Chemicals, Inc.*)

FIGURE 29-17 Welding with a manually held GMAW welding gun. (*Courtesy Air Products and Chemicals, Inc.*)

penetration, spray transfer, and smooth welds with good profile. Process variables include: type of current, current magnitude, shielding gas, type of metal transfer, electrode diameter, electrode composition, electrode stickout (extension beyond the gun), welding speed, welding voltage, and arc length. Almost all metals and alloys can be welded by this process. A number of robots are now available to perform GMAW. The computer electronics of these robots, however, must be shielded from the high-frequency interference from the welding process.

Flux-cored arc welding. Flux-cored arc welding (FCAW) utilizes a continuous, hollow, electrode wire filled with a granular flux, as shown in Figure 29-18. It can be viewed as an adaptation of the shielded-metal arc process wherein the electrode is now continuous and less bulky, since no binder is required to hold the flux onto the electrode. A protective atmosphere is provided by vaporized flux and further protection results from the slag overlayer.

For enhanced weld properties, an externally supplied shielding gas, generally CO_2, may be employed—essentially creating a gas metal arc process with a flux-cored electrode. It is used primarily for welding ferrous material, almost always with a rpdc power source.

Submerged arc welding. In *submerged arc welding* (SAW), as shown in Figure 29-19, the arc is maintained beneath a blanket of granular flux. Either ac or dc current can be used as the power source. The flux is deposited just ahead of the electrode, which is in the form of coiled wire—copper-coated to provide good electrical contact. Because the arc is completely submerged in the flux, only a few small flames are visible. The granular flux provides excellent

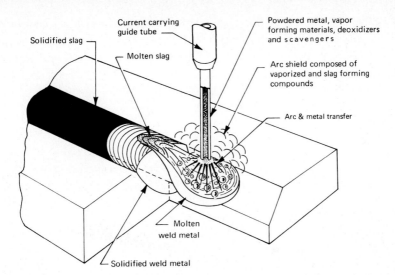

FIGURE 29-18 Schematic representation of the flux-cored arc welding process (FCAW). (*Courtesy The American Welding Society, New York.*)

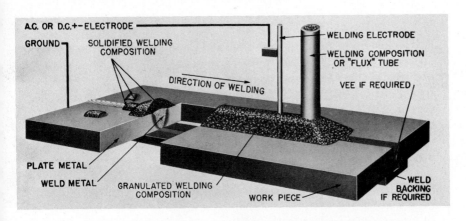

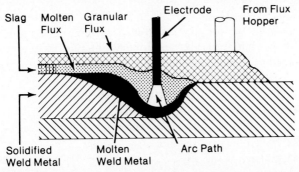

FIGURE 29-19 (*Top*) Basic features of the submerged arc welding process (SAW). (*Courtesy Linde Division, Union Carbide Corporation.*); (*Bottom*) Cut-away schematic of submerged arc welding. (*Courtesy American Iron and Steel Institute, Washington, D.C.*)

shielding of the molten metal and, because the pool of molten metal is relatively large, good fluxing action occurs, so as to remove impurities. Consequently, very high quality welds are obtainable. A portion of the flux is melted and solidifies into a glasslike covering over the weld. This, with the flux that is not melted, provides a thermal coating that slows down the cooling of the weld area and thus helps to produce a soft, ductile weld. The solidified flux cracks loose from the weld on cooling and is easily removed. Surplus unmelted flux is recovered by a vacuum system.

Submerged arc welding is most suitable for making flat butt or fillet welds in low-carbon steel (< 0.3%C). With some pre-heat and post-heat precautions, medium carbon and alloy steels and some copper alloys can be welded. The process is not suitable for high-carbon tool steels, cast irons, or aluminum, magnesium, titanium, lead, or zinc.

High welding speeds, high deposition rates, deep penetration, and high cleanliness (due to the flux action) are characteristics of submerged arc welds. Welding speeds of 760 mm (30 inches) per minute in 25-mm (1-inch)-thick steel plate or 300 mm (12 inches) per minute in 40-mm ($1\frac{1}{2}$-inch) plate are common. Single-pass welds 40 mm ($1\frac{1}{2}$ inches) deep can be made and almost any thickness of base metal can be joined. Because the metal is deposited in fewer passes than with alternative processes, there is less possibility of entrapped slag or voids, and weld quality is further enhanced. When higher deposition rates are desired, multiple electrode wires may be employed. This technique is widely used in large-volume welding, as in the building of ships or the manufacture of large-diameter steel pipe or tanks.

Limitations to the process include: the need for extensive flux handling, possible contamination of the flux by moisture (leading to porosity), the large volume of slag that must be removed, the restriction to flat welding due to the slag, the high heat inputs promoting grain coarsening, and the slow cooling rate, which permits segregation and possible hot cracking. Also, chemical control is important since the electrode material often comprises over 70% of the molten weld region.

The electrodes are classified by composition and are available in diameters from 1.1 mm (0.045 inch) to 9.5 mm ($\frac{3}{8}$ inch). Larger electrodes carry higher currents and provide for rapid deposition rates, but penetration is shallower. The wire may be solid alloy material, plain steel (with alloy additions coming from the flux), or tubular wire with an alloy element core. The fluxes are classified according to the properties of the weld metal and are designed to have low melting temperatures, good fluidity at high temperatures, and be brittle after cooling.

Submerged arc equipment may be semiautomatic, with the operator controlling the speed, or fully automatic. They may be manual, portable (wherein the welder transverses over a stationary workpiece), or stationary (wherein the workpiece passes under the arc). Figure 29-20 shows a manual unit and Figure 29-21 shows a high-volume production setup for welding large steel pipe.

In a modification of the process, often called *bulk welding,* iron powder is deposited in the prepared gap between the plates to be joined (ahead of the flux and on top of a backing strip) as a means of increasing deposition rate. A single

FIGURE 29-20 Manual submerged arc welding. (*Courtesy Lincoln Electric Company.*)

pass can apply 50 kg/hr (110 lb/hr) of weld metal, equivalent to seven or eight conventional submerged arc passes.

By using the arrangement shown in Figure 29-22, vertical welds can be made by the submerged arc process. Stationary, copper side molds are employed, and a stationary, consumable wire guide is used. The consumable wire guide, coated with flux, melts as it enters the molten flux pool, thereby replacing the flux that solidifies at the copper–weld interface. Good-quality welds up to 102 mm (4

FIGURE 29-21 Production setup for welding large pipe by the submerged arc process. (*Courtesy Linde Division, Union Carbide Corporation.*)

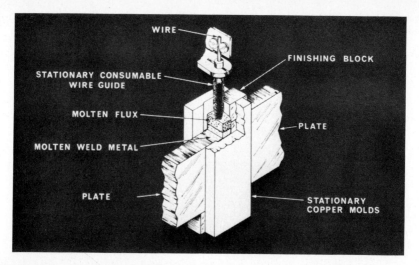

FIGURE 29-22 Setup for making vertical welds by the submerged arc process. (*Courtesy Welding Journal.*)

inches) thick can be made by this procedure, but for plate thicker than about 51 mm (2 inches), the electroslag process, which will be discussed later, is usually more economical.

Plasma arc welding. In plasma arc welding (PAW), the arc is struck between a nonconsumable electrode and either the welding gun or the workpiece, as in Figure 32-10. The arc serves to heat an orifice gas, which is then constricted through a nozzle and directed at the workpiece. This technique provides greater energy concentration, fast welding speeds, deep penetration, a narrow heat-affected zone, reduced distortion, less use of filler metal, higher temperatures, and a process that is insensitive to arc length. Nearly all metals and alloys can be welded by the plasma arc technique.

With a low-pressure plasma, the metal is simply "melted in" and a filler material is often used. At higher pressures a "keyhole" effect occurs, where the plasma arc forms a hole completely through the sheet (up to 6 mm, or $\frac{1}{4}$ inch thick), surrounded by molten metal. As the arc travels, the liquid metal fills in the keyhole. If the pressure is further increased, the molten metal is expelled from the region and the process becomes plasma cutting (see Chapter 32).

Many plasma welding torches employ a small nontransferred arc within the torch, which heats the orifice gas, ionizing it and thereby forming a conductive path for the main transferred arc. This permits instant ignition of a low-current arc, which can be lower in magnitude, more stable, and more readily controlled than in an ordinary plasma torch. Separate dc power supplies are used for the pilot and main arcs. An inert shielding gas usually is supplied through an outer cup surrounding the torch.

Stud welding. *Stud welding* (SW) is an arc-welding process by which fasteners can be welded into place in lieu of riveting or drilling and tapping. A special gun is used, such as is shown in Figure 29-23, into which the stud is inserted. A dc arc is established between the end of the stud and the workpiece until a small amount of metal is melted. They are then brought together under light pressure and allowed to solidify. Automatic equipment controls the establishing of the arc, its duration, and the application of pressure to the stud.

A wide variety of studs is available, as shown in Figure 29-24. The recessed end of the stud is filled with flux. A ceramic ferrule, which is placed over the end of the stud before it is positioned in the gun, is an important factor in the process, acting to concentrate the arc heat and to protect the metal from the atmosphere. It also confines the molten or plastic metal to the weld area and shapes it around the base of the stud. After the weld is complete, the ferrule is broken from the stud. Burnoff or melting of the stud reduces its length, a factor which should be compensated for in design.

The process requires almost no skill on the part of the operator; once the stud and ferrule are placed in the gun chuck and the gun is positioned on the work, all the operator has to do is pull the trigger. The remainder of the cycle is automatic, consuming less than 1 second. It thus is well suited to, and widely used in, manufacturing, eliminating the necessity for drilling and tapping many holes. Special production-type stud-welding machines are available that can form over 1000 welds per hour.

Advantages and disadvantages of arc welding. Because of its great flexibility and the variety of processes that are available, arc welding is an extremely useful, versatile, and widely used process. However, except for gas-shielded

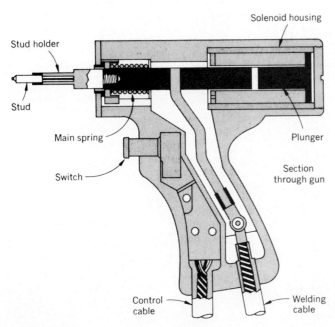

FIGURE 29-23 Schematic diagram of a stud welding gun. (*Courtesy* American Machinist.)

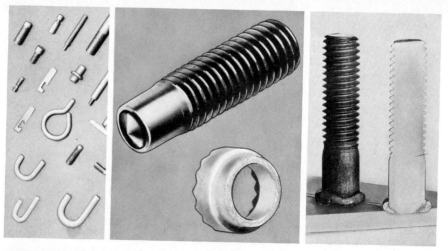

FIGURE 29-24 (*Left*) Some of the types of available studs for stud welding. (*Center*) Stud and ceramic ferrule. (*Right*) Stud after welding, and a section through a welded stud. (*Courtesy Nelson Stud Welding Co.*)

tungsten arc spot welding, stud welding, and, to some degree, submerged arc welding, the various arc welding processes have one disadvantage—the quality of the weld depends ultimately on the skill and integrity of the worker who does the welding. While automation and robotics are helping to reduce this problem, proper training, selection, and supervision of personnel are still of great importance. Numerous nondestructive inspection techniques have been developed to assure weld quality.

Power sources for arc welding. Arc welding requires a large amount of current that does not change in magnitude as the voltage varies over a considerable range. The load voltage usually is from 30 to 40 volts, although the actual voltage across the arc ordinarily varies from about 12 to 30 volts, depending primarily on the arc length. Both dc and ac sources are available, generally having "drooping voltage" characteristics, as shown in Figure 29-25, and ca-

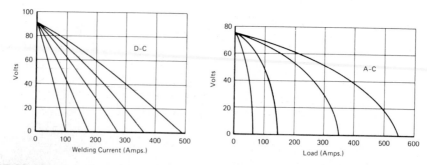

FIGURE 29-25 Drooping-voltage characteristics of typical arc-welding power sources. (*Left*) Direct current; (*right*) alternating current.

pacities from 150 to 1000 amperes. These characteristics assure that the current does not vary greatly as the voltage fluctuates over the usual operating range.

In earlier years most direct current for welding was provided by motor-generator sets. However, today, solid-state transformer-rectifier machines, such as shown in Figure 29-26, are the most generally used power source. Using a three-phase power supply, these usually can provide both ac and dc output. For field work where electric power is not available, gasoline-engine-driven dc generators are used.

If only ac welding is to be done, relatively simple transformer-type machines are available, such as shown in Figure 29-27. Usually, these are single-phase devices having low power factors, but when several machines are to be operated, as in production-type situations, they may be connected to the several phases of a three-phase supply and thus help to balance the load. Usually, they have internal capacitors to improve the power factor.

FUSION WELD TYPES AND JOINTS

There are four basic types of fusion welds, as illustrated in Figure 29-28. *Bead welds* require no edge preparation. However, because the weld is made on a flat surface and the penetration thus is limited, they are suitable only for joining thin sheets of metal, for building up surfaces, or for applying hard facing metals.

Groove welds are used where full-thickness strength is sought on thicker materials. These require some type of edge preparation to make a groove between the abutting edges. V, double V, U, and J configurations are most common, usually produced by oxyacetylene flame cutting. The type of groove configuration depends primarily on the thickness of the work, the welding process to be employed, and the position of the work, the primary consideration being to obtain a sound weld throughout the full thickness with a minimum deposit of weld metal. The weld may be made in either a single pass or by multiple-pass procedures, depending on the thickness of the material and the welding process used. As shown in Figure 29-29, special consumable insert rings, or strips, often are used to assist in obtaining proper spacing between the mating edges and to aid in assuring proper quality in the root pass. These are especially useful in pipeline welding, particularly under field conditions and where the welding must be done from only one side of the work.

Fillet welds are used for tee, lap, and corner joints. The size of fillet welds is measured by the leg of the largest 45° right triangle that can be inscribed within the contour of the weld cross section. This is shown in Figure 29-30, which also indicates the proper shape for fillet welds to avoid excess metal and to reduce stress concentration. Fillet welds require no special edge preparation. They may be continuous or made intermittently, spaces being left between short lengths of weld.

Plug welds are used to attach one part on top of another, replacing rivets or bolts. A hole is made in the top plate, and welding is started at the bottom of this hole. They offer substantial saving in weight as compared with riveting or bolting.

FIGURE 29-26 Rectifier-type dc and ac welding machine, 300 amperes capacity, with cover removed to show interior. (*Courtesy Lincoln Electric Company.*)

Figure 29-31 shows the five basic types of joints that can be made through the use of bead, groove, and fillet welds, and Figure 29-32 shows several methods to make these joints. In selecting the type of weld joint to be used, the primary consideration should be the type of loading that will be applied. Too

FIGURE 29-27 Ac welding machine being used on a typical welding job. (*Courtesy Lincoln Electric Company.*)

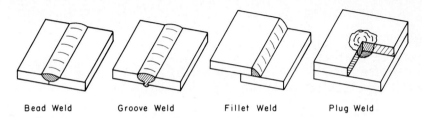

Bead Weld   Groove Weld   Fillet Weld   Plug Weld

FIGURE 29-28 Four basic types of fusion welds.

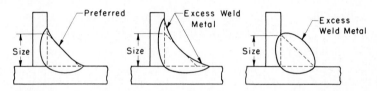

Insert in place.   Insert tack-welded.   Insert consumed.   Completed weld.

FIGURE 29-29 Use of a consumable backup insert in making arc welds. (*Courtesy Arcos Corporation.*)

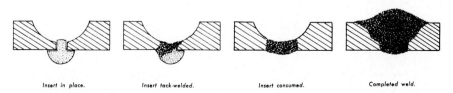

FIGURE 29-30 Preferred shape of fillet welds and the method of measuring the size of a fillet weld.

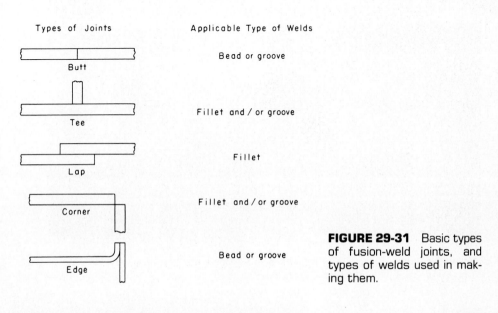

Types of Joints   Applicable Type of Welds

Butt   Bead or groove

Tee   Fillet and / or groove

Lap   Fillet

Corner   Fillet and / or groove

Edge   Bead or groove

FIGURE 29-31 Basic types of fusion-weld joints, and types of welds used in making them.

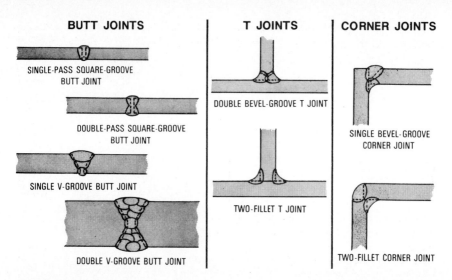

BUTT JOINTS

SINGLE-PASS SQUARE-GROOVE BUTT JOINT

DOUBLE-PASS SQUARE-GROOVE BUTT JOINT

SINGLE V-GROOVE BUTT JOINT

DOUBLE V-GROOVE BUTT JOINT

T JOINTS

DOUBLE BEVEL-GROOVE T JOINT

TWO-FILLET T JOINT

CORNER JOINTS

SINGLE BEVEL-GROOVE CORNER JOINT

TWO-FILLET CORNER JOINT

FIGURE 29-32 Various weld procedures used to form several common joints. (*Courtesy Republic Steel Corporation.*)

frequently this basic fact is neglected, and a large proportion of what are erroneously called "welding failures" are the result of such oversight. Cost and accessibility for welding are important, but secondary, factors in joint selection. Cost is affected by the required edge preparation, the amount of weld metal that must be deposited, the type of process and equipment that must be used, and the speed and ease with which the welding can be accomplished. Accessibility obviously will have considerable influence on several of these factors.

In production welding, extensive use is made of welding jigs or fixtures (also called positioners) into which the work is clamped. These make it possible for

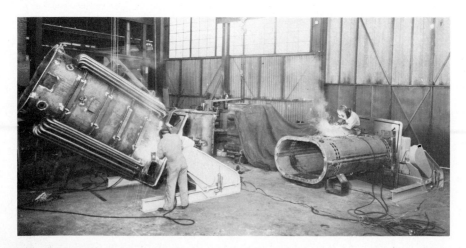

FIGURE 29-33 Transformer cases being welded while held on a welding positioner. (*Courtesy Panjiris Weldment Company.*)

the work to be manipulated and positioned to permit the welding to be done in the most favorable position. Figure 29-33 shows a large weldment being fabricated on a positioner of moderate size. Special positioners are employed which are capable of holding large sections of ships weighing many tons.

REVIEW QUESTIONS

1. What is welding?
2. What conditions are required for an "ideal metallurgical bond?"
3. What are some of the problems that may occur when high temperatures are used in welding?
4. What factors influence the quality of deformation welds?
5. What are the two stages of the combustion of oxygen and acetylene?
6. What is the location of the maximum temperature in an oxyacetylene flame?
7. What are the three types of flames in oxyacetylene welding?
8. What are some of the attractive features of MAPP gas?
9. What problems may occur because the heat source is not concentrated in oxyfuel gas welding?
10. What sort of problems plagued the early attempts to develop arc welding?
11. What are the three basic types of current and circuit conditions used in arc welding?
12. What are the three modes of metal transfer that can occur in arc welding?
13. What are some of the variables that must be specified in arc welding processes?
14. What are some of the roles of the electrode coatings in shielded metal arc welding?
15. What is the role of the slag coating that forms in shielded metal arc welding?
16. What problems exist in developing continuous shielded metal arc welding?
17. What are two of the techniques that may be used to produce higher deposition rates in gas tungsten arc welding?
18. What are some of the advantages of gas tungsten arc welding?
19. What is the major difference between the GMAW and GTAW processes?
20. Why might gas metal arc welding be preferred over the shielded metal arc technique for production welding?
21. What advantages can be obtained by placing the flux in the core of the electrode?
22. What are some of the functions of the flux in submerged arc welding?
23. What are some of the significant production characteristics of submerged arc welding?
24. Why can submerged arc welding only be performed in certain restricted positions?
25. What characteristics are desirable in a submerged arc welding flux?
26. What is bulk welding? What is its advantage?
27. What is the "keyhole effect" in plasma arc welding of thin sheet?
28. What is stud welding?
29. What functions are performed by the ceramic ferrule in stud welding?
30. What are the four basic types of fusion welds?
31. What factors influence the type of groove welds?
32. What is the role of positioners in welding?
33. A weldment, made from AISI 1025 steel, is being welded with E6012 electrodes. Some difficulty is being experienced with cracking in the weld beads and in the heat-affected zones. What possible corrective measures can you suggest?
34. A base for a special machine tool will weigh 635 kg (1400 pounds) if made as a gray iron casting. Pattern cost would be $450. and the foundry has quoted a price of $1.32 per kilogram ($0.60 per pound) for making the casting. If the part is made as a weldment, it will require 363 kg (800 pounds) of steel costing $0.31 per kilogram ($0.14 per pound). Cutting, edge preparation, and setup time will require 30 hours at a rate of $10.00 per hour for labor and overhead. Welding time will be 55 hours at an hourly rate of $9.50. Ninety-one kilograms (200 pounds) of electrode will be required, costing $0.37 per kilogram ($0.17 per pound).

(a) Which method of fabrication will be more economical if only one part is required?

(b) What number of parts would be required for welding and casting to break even?

(c) Since the base of the machine tool will now be made of steel rather than cast iron, what special property of the cast iron will be lost and why must this be considered in deciding whether to make this design change regardless of economics?

CASE STUDY 29. Cargo-container Racks

On container ships, the cargo containers are held in racks on deck, stacked as many as six high. On one ship, the racks were designed and constructed of four vertical 203-mm (8-inch) steel H sections, attached to transverse 254-mm (10-inch) box sections by means of 19-mm ($\frac{3}{4}$-inch) fillet welds along the outside edges of the flanges, as shown schematically in Figure CS-29. On the ship's second voyage, during a Pacific storm, two of the racks failed at the attaching point of the vertical members, and six containers of cargo were lost overboard.

Was there a basic deficiency in the design? If so, what was it, and how could it have been corrected?

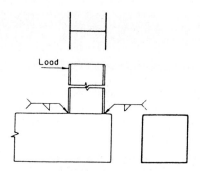

FIGURE CS-29 Method of attaching vertical legs to horizontal member of a cargo container rack.

CHAPTER 30

Resistance Welding

RESISTANCE-WELDING THEORY

In resistance welding, both heat and pressure are utilized in producing coalescence. The heat is the consequence of the electrical resistances of the workpieces and the interface between them. A certain amount of pressure is applied initially to hold the workpieces in contact, thereby controlling the electrical resistance at the interface, and is increased when the proper temperature is attained to facilitate the coalescence. Because of the pressure utilized, coalescence occurs at a lower temperature than with oxyfuel gas or arc welding. Consequently, in modern resistance welding, intentional melting of the metal need not occur. Thus these can be solid-state processes, although they are not officially classified as such by the American Welding Society.

In some of the resistance welding processes, additional pressure can be applied immediately after coalescence is achieved to provide a certain amount of forging action, with some accompanying grain refinement. Also, some additional heating can be induced, as a part of the process, to provide tempering and/or stress relief. Usually, the required temperature can be obtained and the weld completed in a few seconds or less. Consequently, resistance welding is a very rapid and economical process, extremely well suited to manufacturing.

Heating.   Heat for resistance welding is obtained by passing a large electrical current through the workpieces for a short interval, utilizing the basic relationship

$$H = I^2RT \tag{30-1}$$

where H is the heat, I the current, R the electrical resistance of the circuit, and T the time or duration of the current flow. In most cases, alternating current is used. As indicated in Figure 30-1, the work is a part of the electrical circuit. It is important to note that the total resistance of the assembly between the electrodes consists of three parts: (1) the resistance of the workpieces, (2) the contact resistance between the electrodes and the work, and (3) the resistance between the surfaces to be joined, known as faying surfaces. Because it is desired to have the maximum temperature occur at the point where the weld is to be made, it is essential to keep resistances (1) and (2) as low as possible with respect to resistance (3). Obviously, with materials having low electrical resistance, such as aluminum and copper, this condition is difficult to achieve, and they require much larger currents and more attention to the interface conditions than when welding steel.

The resistance of the workpieces is determined by the type and thickness of the metal. It usually is much less than the other two resistances because of the larger area involved and the relatively high electrical conductivity of most metals. The resistance between the work and the electrodes is minimized by using electrode materials that are excellent electrical conductors, by controlling the shape and size of the electrodes, and by using proper pressure between the work and the electrodes. Because any change in the pressure between the work and the electrodes also tends to change the pressure between the faying surfaces, only limited control of the electrode-to-work resistance can be obtained in this manner.

The resistance between the faying surfaces is a function of (1) the quality of the surfaces; (2) the presence of nonconductive scale, dirt, or other contaminants; (3) the pressure; and (4) the contact area. These factors must be controlled to obtain uniform results.

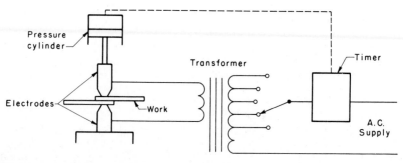

FIGURE 30-1   Fundamental resistance-welding circuit.

As indicated in Figure 30-2, the objective is to bring the faying surfaces simultaneously to the proper elevated temperature while keeping the remaining material and the electrodes at much lower temperatures. The electrodes usually are water-cooled to keep their temperature low and to aid in maintaining them in proper condition. When metals of different thickness or of different conductivities are to be welded, they can be brought to the proper welding temperature simultaneously by using a larger electrode or one having higher conductivity against the thicker, or high-resistance, material.

Pressure. Because pressure in resistance welding affects the contact resistance, welds can be made at lower temperatures with forging action. The control of both the magnitude and timing of the pressure is very important. If too little pressure is used, the contact resistance is high and surface burning and pitting of the electrodes may result. On the other hand, if too high pressure is employed, molten or softened metal may be squirted or squeezed from between the faying surfaces, or the work may be indented by the electrodes. Ideally, a moderate pressure should be applied prior to and during the passage of the welding current, to establish proper contact resistance. It should then be increased considerably just as the proper welding heat is attained, to complete the coalescence and forge the weld to produce a fine grain structure.

On small, foot-operated machines only a single spring-controlled pressure is used. On larger, production-type welders, the pressures usually are applied through air or hydraulic cylinders that are controlled and timed automatically.

Current control. With the surface conditions held constant and the pressure controlled, the temperature in resistance welding is regulated by controlling the magnitude and timing of the welding current. Very precise and sophisticated controls are available for this purpose.

The current usually is obtained from a "step-down" transformer. On small machines the magnitude is controlled through taps on the primary of the transformer or by an autotransformer that varies the primary voltage supplied to the main transformer. On larger machines several methods are used. In *phase-shift control* the magnitude and wave shape of the primary current are altered. With *slope control* the current is permitted to rise gradually to full magnitude in from about 3 to 25 cycles.

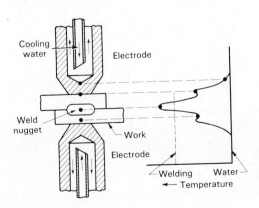

FIGURE 30-2 Desired temperature distribution across the electrodes and the workpiece in lap resistance welding.

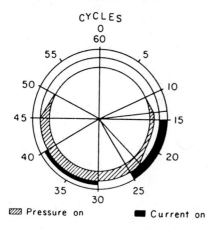

CYCLES

FIGURE 30-3 Typical current and pressure cycle for spot welding. Cycle includes forging and postheating.

▨ Pressure on   ■ Current on

In large, production-type welders, not only is the magnitude of the current controlled, but the extent of the current flow and the application of the current and pressure are carefully programmed. Figure 30-3 shows a relatively simple current and pressure cycle for spot welding that includes forging and postheating.

Power supply. The magnitude of the current required for resistance welding (up to 100,000 amps) is so great that special types of circuits are employed in most machines to reduce the load on the power lines. Ordinary single-phase circuits are used only in smaller machines. Larger machines employ three-phase circuits. In one type, three-phase power is rectified and the energy stored in a special transformer. When the dc flow through the transformer is interrupted, the collapse of the field provides the required heavy current flow in the secondary circuit. Many modern resistance welders utilize dc welding current, obtained through solid-state rectification of three-phase power. Such machines reduce the current demand per phase, give a balanced load, and produce excellent welds.

RESISTANCE WELDING PROCESSES

Resistance spot welding. *Resistance spot welding* (RSW) is the simplest and most widely used of this type of welding. As shown in Figure 30-4, the over-lapping work is positioned between water-cooled electrodes, which have reduced areas at the tips to produce welds that usually are from 1.6 to 12.7 mm ($\frac{1}{16}$ to $\frac{1}{2}$ inch) in diameter. After the electrodes are closed on the work, a controlled cycle of pressure application and current flow occurs, producing a weld at the interface. The electrodes then open, and the work is removed.

A satisfactory spot weld, such as is shown in Figure 30-5, consists of a *nugget* of coalesced metal formed between the faying surfaces. There should be no, or only a very slight, indentation of the metal under the electrodes. The strength of the weld should be such that in a tensile or tear test the weld will remain intact and failure will occur in the heat-affected zone surrounding the nugget, as illustrated in Figure 30-6. If proper current density and timing, electrode

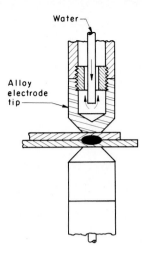

Water

Alloy
electrode
tip

FIGURE 30-4 Arrangement of electrodes and the work in spot welding.

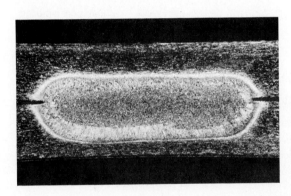

FIGURE 30-5 Spot-weld nugget between two sheets of 1.3-mm (0.051-inch) aluminum alloy. The radius of the upper electrode was greater than that of the lower electrode. (*Courtesy Lockheed Aircraft Corporation.*)

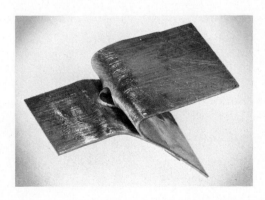

FIGURE 30-6 Tear test of a satisfactory spot weld, showing how failure occurs outside the weld.

pressure and shape, and surface conditions are maintained, sound spot welds can be obtained with excellent consistency.

Spot-welding machines. Three general types of spot-welding machines are available. For light-production-type work where complex current-pressure cycles are not required, the simple *rocker-arm* type, shown in Figure 30-7, is often used. On these machines the lower electrode arm is stationary, and the upper electrode, mounted on a pivoted arm, is brought down into contact with the work by means of a spring-loaded foot pedal. On larger machines and on machines for larger-volume work, this motion is obtained through an air cylinder or an electric motor. Machines of this type are available with throat depths up to about 1220 mm (48 inches) and transformer capacities up to 50 kVa. They are used primarily on steel.

Most large spot welders, and those used for great production rates, are of the *press type,* such as is shown in Figure 30-8. In these the movable electrode has a straight-line motion, provided by an air or hydraulic cylinder. Such machines are adaptable to any type of pressure-controlled cycle that may be desired. Capacities up to 500 kVa and 1524-mm (60-inch) throat depth are common. Special-purpose spot welders of this type, employing multiple welding heads, are widely used in mass-production industries. Some, such as is shown in Figure 30-9, can make up to 200 spot welds in 60 seconds.

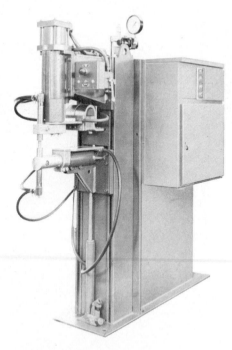

FIGURE 30-7 Foot-operated, rocker-arm, spot-welding machine. (*Courtesy Sciaky Bros., Inc.*)

FIGURE 30-8 Single-phase, air-operated, press-type resistance welder with microprocessor control. (*Courtesy Sciaky Bros., Inc.*)

The application of spot welding is greatly extended through the use of *portable spot welding guns,* such as shown in Figure 30-10. Each gun is connected to the power supply and control units by flexible air hoses and electrical cables and by water hoses for cooling, where required. Such equipment permits bringing the welding unit to the work, thus greatly extending the use of spot welding in applications where work is too large to be moved to a welding machine.

Quite frequently, portable welding guns are installed on industrial robots, which can be programmed to position the gun to the desired (three-dimensional) location and produce spot welds automatically, without the need for an operator. Such a procedure is illustrated in Figure 38-38.

Spot-weldable metals.   One of the great advantages of spot welding is the fact that virtually all the commercial metals can be spot-welded, and most of them can be spot-welded to each other. In only a few cases do the welds tend to be brittle. Table 30-1 shows combinations of metals that can be spot-welded satisfactorily.

Although the majority of spot welding is done on wrought sheet, other forms of metal also can be spot-welded. Sheets can be spot-welded to rolled shapes and steel castings, and some types of die castings can be welded without difficulty. Except for aluminum, most metals require no special preparation, except

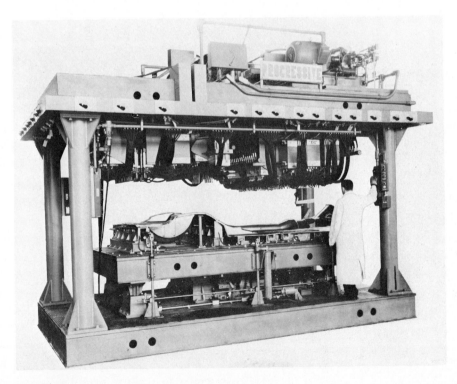

FIGURE 30-9 Large press-type spot welder which has 50 transformers and makes 200 spot welds on an automobile underbody in less than 6 seconds. (*Courtesy Progressive Welder Sales Company.*)

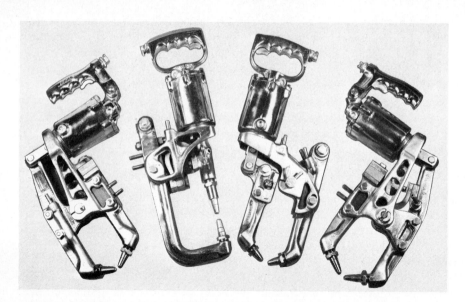

FIGURE 30-10 Typical portable spot-welding guns. (*Courtesy Progressive Machinery Corporation.*)

TABLE 30-1 Metal Combinations That Can Be Spot-Welded

Metal	Aluminum	Brass	Copper	Galvanized Iron	Iron (Wrought)	Monel	Nickel	Nickel Silver	Nichrome	Steel	Tin Plate	Zinc
Aluminum	×										×	×
Brass		×	×	×	×	×	×	×	×	×	×	×
Copper		×	×	×	×	×	×	×	×	×	×	×
Galvanized iron		×	×	×	×	×	×	×	×	×	×	
Iron (wrought)		×	×	×	×	×	×	×	×	×	×	
Monel		×	×	×	×	×	×	×	×	×	×	
Nickel		×	×	×	×	×	×	×	×	×	×	
Nickel silver		×	×	×	×	×	×	×	×	×	×	
Nichrome		×	×	×	×	×	×	×	×	×	×	
Steel		×	×	×	×	×	×	×	×	×	×	
Tin plate	×	×	×	×	×	×	×	×	×	×	×	
Zinc	×	×	×									×

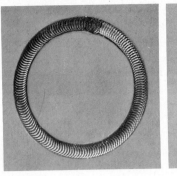

FIGURE 30-11 Seam welds made with overlapping spots of varied spacing. (*Courtesy Taylor-Winfield Corporation.*)

to be sure that the surface is free of corrosion and is not badly pitted. For best results, aluminum and magnesium should be cleaned immediately prior to welding by mechanical or simple chemical means. Metals that have high electrical conductivity require clean surfaces to assure that the electrode-to-metal resistance is low enough for adequate temperature to be developed in the metal itself. Silver and copper are difficult to weld because of their high heat conductivity. However, many copper alloys are readily resistance-welded. Water cooling adjacent to the spot-weld area can be used to assure that adequate welding temperature is obtained only in the desired spot area of such materials.

The practical limit of thicknesses that can be spot-welded by ordinary processes is about 3.18 mm ($\frac{1}{8}$ inch), where each piece is the same thickness. However, a thin piece can readily be welded to another piece that is much thicker than 3.18 mm. Spot welding of two 12.7-mm ($\frac{1}{2}$-inch) steel plates has been done satisfactorily as a replacement for riveting.

Resistance seam welding. *Resistance seam welds* (RSEW) are made by two distinctly different processes. In most cases, where the weld is between two sheets of metal, the seam actually is a series of overlapping spot welds, as shown in Figure 30-11. The basic equipment is the same as for spot welds, except that two rotating disks are used as electrodes, arranged as shown in Figure 30-12. In the roll-spot weld process, the metal passes between the electrodes

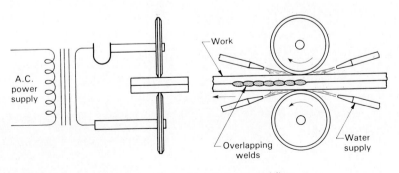

FIGURE 30-12 Schematic representation of seam welding.

and timed pulses of current pass through it to form the overlapping, elliptical welds. The timing of the welds and the movement of the work must be adjusted so that the workpieces do not get too hot. External cooling of the work, by air or water, often is used. Continuous seam welding is also possible with this equipment, by applying a continuous current through the rotating electrodes. A typical seam-welding machine is shown in Figure 30-13.

This type of seam welding is used primarily for the production of liquid- or pressure-tight vessels, such as gasoline tanks, automobile mufflers, or heat exchangers.

The second type of resistance seam welding is being used increasingly to make butt welds between metal plates. In this process the electrical resistance of the abutting metal(s) is utilized, but a high-frequency current is employed—up to 450 kHz—that confines the heating adjacent to the surfaces to be joined. (An alternative method employs high-frequency induction heating.) As they reach the welding temperature, the heated surfaces are pressed together by passing through pressure rolls. The most extensive use of this process is in making

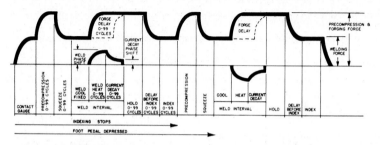

VARIABLE ELECTRODE FORCE — INTERMITTENT DRIVE — SINGLE IMPULSE WITH CURRENT DECAY FOR WELDING LIGHT ALLOYS

FIGURE 30-13 (*Left*) 125-kVa seam welder. (*Top*) Typical cycle of electrode force and current used in spot and seam welding. (*Courtesy Sciaky Bros., Inc.*)

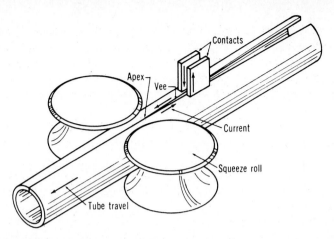

FIGURE 30-14 Method of making a seam weld in pipe by means of a high-frequency current as the heat source.

pipe, using the arrangement illustrated in Figure 30-14. However, it is also being used for making structural shapes from flat stock, as shown in Figure 30-15. Material from 0.13 mm (0.005 inch) to more than 19.0 mm ($\frac{3}{4}$ inch) in thickness can be welded at speeds up to 82 meters (250 feet) per minute. The combination of high-frequency current and high welding speed produces a very narrow heat-affected zone. Almost any type of metal can be welded, including dissimilar metals. The process is particularly attractive for high-conductivity metals like aluminum and copper.

Projection welding. Two disadvantages of spot welding are that electrode maintenace is a considerable problem and, usually, only one spot weld is made at a time. If more strength, or attachment, is required than can be provided by one spot weld, several such welds must be made. *Projection welding* (RPW)

FIGURE 30-15 Fabricating an I-beam from three plates by means of two simultaneous high-frequency resistance welds. (*Courtesy AMF Thermatool, Inc.*)

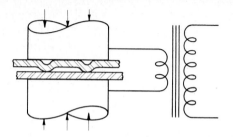

FIGURE 30-16 Principle of projection welding.

provides a means of overcoming both of these disadvantages and thus is particularly well adapted to mass production.

The principle of projection welding is illustrated in Figure 30-16. A *dimple* is embossed on one of the workpieces at the location where a weld is desired. The workpieces then are placed between plain, large-area electrodes in the projection-welding machine, and pressure and current are applied, as in spot welding. Because the contact area between the work and the electrodes is much greater than the area on the end of the dimple, nearly all the resistance of the circuit is at the dimple, and virtually all the heating occurs at the point where the weld is desired. As the metal becomes plastic, due to heating, the pressure causes the dimple to flatten as the weld is formed, and the workpieces are forced tightly together.

Several projection welds can be made at one time by having several dimples between the electrodes, the number being limited only by the capacity of the machine to provide the required current and pressure. Another important advantage is that the dimples, or projections, can be made almost any shape—such as round, oval, or circular—to produce welds of desired shapes as required for design purposes. Many spot-welding machines are convertible to projection welding by changing the electrodes.

Because the projections are formed on the work in punch presses, they often can be formed concurrently with other blanking or forming operations at virtually no additional cost. They should be shaped so that the weld forms outward from the center of each projection. Bolts and nuts are often attached to other metal parts by projection welding. Such bolts and nuts are available with the projections formed on them, ready for welding.

Advantages and disadvantages of resistance welding. Resistance welding processes have a number of distinct advantages that account for their wide use, particularly in mass production:

1. They are very rapid.
2. The equipment is semiautomatic or fully automated.
3. They conserve material; no filler metal is required.
4. Skilled operators are not required.
5. Dissimilar metals can easily be joined.
6. A high degree of reliability and reproducibility can be achieved.

Resistance welding, of course, has some disadvantages, the principal ones being:

1. The equipment has a high initial cost.
2. There are limitations to the types of joints that can be made (mostly lap joints).

3. Skilled maintenance personnel are required to service the control equipment.
4. For some materials the surfaces must receive special preparation prior to welding.

Resistance welding is one of the most common methods of high-volume joining. However, due to rapid heat inputs, short welding times, and rapid quenching by both the base metal and the electrodes, cooling rates are extremely high in spot and seam welds. Martensite readily forms in steels containing more than 0.15% carbon, and here, a post-weld heating is generally required to temper the weld.

REVIEW QUESTIONS

1. Why is it important to have the resistance between the faying surfaces larger than other resistances in the welding assembly?
2. What factors can control or alter the resistance between the faying surfaces?
3. What design features can be altered to permit the joining of different thickness or different conductivity metals?
4. What can happen if too little pressure is applied? Too much?
5. What are the three general types of spot-welding equipment?
6. How can industrial robots be utilized in spot welding?
7. What is the difference between roll-spot welding and continuous seam welding?
8. What is the benefit of high-frequency current as a heat source for seam welding?
9. What benefits can be gained by making I-beams in the manner of Figure 30-15 as opposed to conventional rolling?
10. What two disadvantages of spot welding can be overcome by projection welding?
11. Why do the dimples in projection welding tend to flatten during the welding process?
12. What difficulties might be encountered when spot welding medium- or high-carbon steel?
13. What restriction in joint design do we typically have in resistance welding?

CASE STUDY 30. The Broken Trailer Hitch

Figure CS-30 shows three views of a broken trailer hitch, which was fabricated by welding. Two of the views show broken sections. The manufacturer claimed that the hitch had been overloaded, whereas the owner of the hitch claimed that it was defectively manufactured. Who was correct? Cite your reasons for your conclusions.

FIGURE CS-30 Broken welded trailer hitch.

CHAPTER 31

Miscellaneous Welding and Related Processes

As indicated in Figure 29-1, there are a number of very useful welding processes that utilize heat sources other than an oxyfuel gas flame, electrical resistance, or an electric arc. Although some are quite old, others are among the newest of the welding processes.

THERMIT WELDING

Thermit welding (TW) is an old process which has been replaced by alternate methods for most applications. However, it still is very effective and is used extensively for joining thick sections of steel, particularly where the contour varies, as in joining railroad rails or steel castings. It also is very useful in repairing large, broken or cracked, steel castings.

Heating and coalescence are produced by superheated molten metal and slag, obtained from the reaction between a metal oxide and aluminum. In addition to furnishing heat, the molten metal also supplies any required filler metal.

The thermit process utilizes a mechanical mixture of about one part of finely divided aluminum and three parts of iron oxide. When the mixture is ignited, it reacts according to the chemical equation

$$8Al + 3Fe_3O_4 = 9Fe + 4Al_2O_3 + heat$$

producing a temperature of over 2760°C (5000°F) in about 30 seconds. The ignition temperature of about 1150°C (2100°F) is supplied by a magnesium fuse.

The essential steps in welding by the thermit process are shown in Figure 31-1. The sections to be welded must be prepared to provide clearance between them. Wax is used to fill in the gap and is built up to form the desired shape of the weld, riser, and runner system—similar to the procedure of investment

FIGURE 31-1 Steps in repairing a large casting by thermit welding. (*Top left*) Broken casting. (*Center left*) Crack gouged out ready for wax. (*Bottom left*) Wax pattern in place. (*Top right*) Thermit powder being melted and flowing into completed mold. (*Bottom right*) Welding completed, gates and risers removed, and ready for finishing. (*Courtesy United Chromium Division, Metal & Thermit Corporation.*)

casting. A box is placed around the work and rammed with a material similar to molding sand to form a mold.

When the mold is completed, a heating torch is used to dry it and to melt the wax. Heating is continued until the faces of the work are at a red heat. One or more crucibles, filled with the thermit mixture, are set atop the mold, holes in their bottoms connecting with the runner system in the mold. The mixture in the crucibles is ignited, and the molten metal flows out of the crucible, filling the mold and at the same time supplying sufficient heat to raise the surfaces of the workpieces to a high-enough temperature to produce coalescence. After the deposited metal has cooled, the mold is removed. The weld then can be ground to final shape if desired.

ELECTROSLAG WELDING

Electroslag welding (ESW), illustrated in Figure 31-2, is a very effective process for welding thick sections of steelplate. Heat is derived from the passage of current through a liquid slag where resistance heating raises the temperature up to 1760°C (3200°F). There is no arc involved—so the process is entirely different from submerged arc welding—and the electrical resistance of the metal being welded plays no part in producing heat. Instead, the molten slag melts the edges of the pieces that are being joined, as well as the continuously fed electrodes which supply the filler metal. Multiple electrodes are often used to provide an adequate feed of metal. Typically, there is about 65 mm ($2\frac{1}{2}$ inches) depth of molten slag, which also serves to protect and cleanse the molten metal, and 12 to 20 mm ($\frac{1}{2}$ to $\frac{3}{4}$ inch) of molten metal. These liquids are confined in the gap between the plates being welded by means of sliding, water-cooled plates.

Obviously, the best conditions for maintaining a deep slag bath exist in ver-

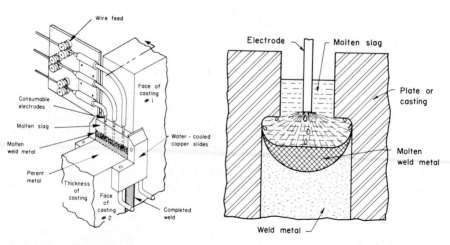

FIGURE 31-2 (*Left*) Arrangement of equipment and work for making a vertical weld by the electroslag process. (*Right*) Section through workpieces and weld during the making of an electroslag weld.

tical joints, so the process is used most frequently for this type of work. However, it also is used successfully for making circumferential joints in large pipe, using special curved slag-holder plates and rotating the pipe to maintain the area where welding is occurring in a vertical position.

Because very large amounts of weld metal and heat can be supplied, electroslag welding is the best of all the welding processes for making welds in thick plates. Thickness of plate varies from 13 mm to 900 mm ($\frac{1}{2}$ to 36 inches) and the length of the weld is almost unlimited. Edge preparation is minimal, requiring only squared edges separated by 25 to 35 mm (1 to $1\frac{1}{2}$ inches). Applications have included building construction, machine manufacture, heavy pressure vessels, and the joining of large castings and forgings.

Control of the solidification is vitally important to obtaining a good weld, since a coarse columnar grain structure tends to result. Associated cracking tendencies can be suppressed by promoting a wide, shallow pool of molten metal through control of current, voltage, slag depth, number of electrodes, and electrode extension. A large heat-affected zone and extensive grain growth are also common for the process. The long thermal cycle, however, minimizes residual stresses, distortion and cracking in the heat-affected zone. If good fracture resistance is desired, subsequent heat treatment will be required.

ELECTRON BEAM WELDING

Electron beam welding (EBW) is a fusion welding process in which heating results from the impingement of a beam of high-velocity electrons on the metal to be welded. Originally developed for obtaining ultra-high-purity welds in reactive and refractory metals, its unique qualities have led to its substantial use in numerous applications.

The electron optical system employed is shown in Figure 31-3. The electrons must be generated and focused in a very high vacuum, and the welding usually is performed at a pressure of from 0.13 to 133 mPa. However, the process can be adopted to weld in pressures of from 0.13 to 13 Pa or even at atmospheric pressure, but the penetration of the beam and the depth-to-width ratio are reduced as the pressure increases.

A high-voltage current heats the tungsten filament to about 2204°C (4000°F), causing it to emit high-velocity electrons. By means of a control grid, accelerating anode, and focusing coils, the electrons are converted into a concentrated beam and focused onto the workpiece in a spot from 0.8 to 3.2 mm ($\frac{1}{32}$ to $\frac{1}{8}$ inch) in diameter. The work is enclosed and moved under the electron gun in the vacuum chamber. Under these conditions, the vacuum assures degassification and decontamination of the molten weld metal, and very high quality welds can be obtained. However, the size of the vacuum chamber required naturally imposes serious limitations on the size of the workpiece that can be accommodated and productivity. As a consequence, electron-beam welding machines have been developed that permit the workpiece to remain outside the vacuum chamber. In these machines the electron beam emerges through a small orifice in the vacuum chamber to strike the adjacent workpiece. High-capacity vacuum

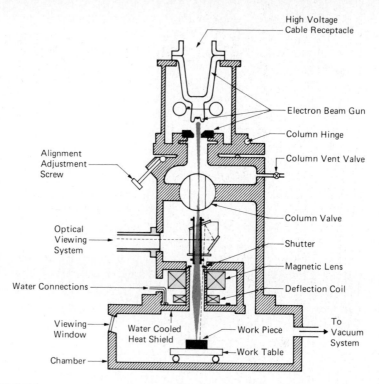

FIGURE 31-3 Schematic diagram of the electron beam welding process. (*Courtesy* American Machinist.)

pumps take care of the leakage through the orifice. Although these machines have some of the advantages of the total-vacuum types, because they do not operate at as low pressure they do not have as great penetrating power and they produce considerably wider welds.

In general, two ranges of voltages are employed in electron-beam welding. High-voltage equipment employs 50 to 100 kilovolts and provides a smaller spot size and greater penetration than does the lower-voltage type, which uses from 10 to 30 kilovolts. However, the high-voltage units, with their high electron velocities, emit considerable quantities of harmful X-rays and thus require expensive shielding and indirect viewing systems for observing the work. The X-rays produced by the low-voltage types are sufficiently soft to be absorbed by the walls of the vacuum chamber. They are less critical in adjustment and the work can be observed directly through viewing ports.

Materials that are difficult to weld by other processes, such as zirconium, beryllium, and tungsten, can be welded successfully by electron-beam welding, but the weld configuration should be simple and preferably flat. As shown in Figure 31-4, very narrow welds can be obtained as well as remarkable penetrations. The high power and heat concentrations can produce fusion zones with depth-to-width ratios of 25:1, with low heat input, low distortion, and a very narrow heat-affected zone. Heat sensitive materials can be welded without damage to the base metal. High welding speeds are common; no filler metal is

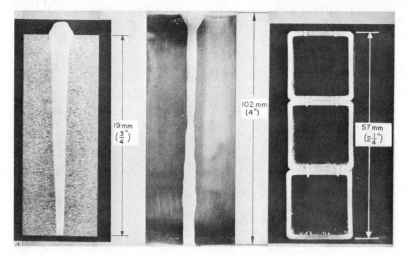

FIGURE 31-4 (*Left to right*) Electron beam welds in 7079 aluminum, thick stainless steel, and a multiple-tier weld in stainless steel tubing. (*Courtesy Hamilton Standard Division of United Aircraft Corp.*)

required; the process can be performed in all positions; and preheat or post-heat is generally unnecessary. However, the equipment is quite expensive and extensive joint preparation is required. Furthermore, the vacuum chamber tends to limit production rate and the size of piece that can be welded.

The process is best employed where extremely high quality welds are required or where other processes will not produce the desired results. Nevertheless, its unique capabilities have resulted in its routine use in a number of applications, particularly in the automotive and aerospace industries.

LASER BEAM WELDING

The heat source in *laser beam welding* (LBW) is a focused laser beam, usually providing power intensities in excess of 10 kilowatts per square centimeter, but with low heat input—0.1 to 10 joules. The high-intensity beam produces a very thin column of vaporized metal, extending into the base metal. The column of vaporized metal is surrounded by a liquid pool, which moves along as welding progresses, resulting in welds having depth-to-width ratios greater than 4:1. Laser beam welds are most effective for simple fusion welds without filler metal, but filler metal can be added.

Deep-penetration welds produced by lasers are similar to electron-beam welds, but LBW has several additional advantages:

1. A vacuum environment is not required.
2. No X-rays are generated.
3. The laser beam is easily shaped and directed with reflective optics.
4. Because only a light beam is involved, there does not need to be any physical

FIGURE 31-5 Typical welds made by laser welding. (*Courtesy Linde Division, Union Carbide Corporation.*)

contact between the workpieces and the welding equipment, and the beam will pass through transparent materials, permitting welds to be made inside transparent containers.

Because laser beams are concentrated, laser welds are small—usually less than 0.025 mm (0.001 inch). They thus are very useful in the electronics industry in connecting leads on small electronic components and in integrated circuitry. Lap, butt, tee, and cross-wire configurations are used. It also is possible to weld lead wires having polyurethane insulation without removing the insulation. The laser evaporates the insulation and completes the weld. Figure 31-5 shows examples of laser welds.

The equipment required for laser beam welding is high in cost but usually is designed for use by semiskilled workers. Most industrial lasers are of the CO_2 variety. In general, however, lasers are not very efficient and tend to consume large amounts of power. In addition, reflected or scattered laser beams can be dangerous to human eyes, even at great distances from the welding site. Eye protection is a must.

FLASH WELDING

In *flash welding* (FW) two pieces are mounted in a machine and lightly touched together. An electric current may be passed through the joint to provide preheat (optional), after which the pieces are withdrawn slightly. A flashing action (arc) then occurs which melts the interface and expels the liquid and oxides. The pieces are then forced together under high pressure to upset the joint and form the weld. The force is maintained until solidification is complete, after which the product is removed from the machine and the upset portion is machined away. Figure 31-6 provides a schematic of the process.

The flashing action must be long enough to provide heat for melting and to lower the metal strength to allow for plastic deformation. Sufficient upsetting

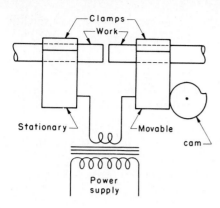

FIGURE 31-6 Principle of the flash-welding process.

should occur that all impure metal is squeezed out into the *flash* and only sound metal exists in the weld. Figure 31-7 shows parts before and after flash welding.

Flash welding can be employed for butt welding solid or tubular metals. It is widely used in manufacturing such products as tubular metal furniture, metal windows, and pipe. Except for very small sections, the equipment required is rather large and expensive, but excellent welds can be made at high production rates, making the process well suited to mass production. Although the resulting flash usually must be removed (may be a problem inside a tube), in most cases no surface preparation is required.

Percussion welding is a similar method where heating is obtained by an arc produced by a rapid discharge of electrical energy and is followed by a rapid application of force to expel the metal and produce the joint. Here the arc duration is only 1 to 10 milliseconds. While the heat is intense, it is also highly concentrated. The adjacent heat-affected zone is quite small, so the process is attractive in applications where adjacent components may be heat sensitive, as in the electronics industry.

FRICTION WELDING OR INERTIA WELDING

The heat for *friction welding* (FRW)—sometimes called *inertia welding*—is the result of mechanical friction between two abutting pieces of metal that are held

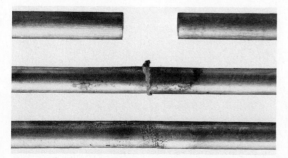

FIGURE 31-7 Flash-welded parts. (*Top*) Before welding; (*center*) welded, showing flash; (*bottom*) flash ground smooth.

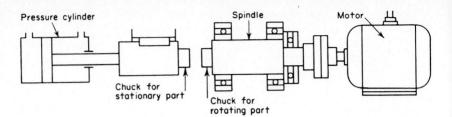

Chuck for
stationary part
Chuck for
rotating part

FIGURE 31-8 Arrangement of equipment for inertia (friction) welding. (*Courtesy Materials Engineering.*)

together while one rotates and the other is held stationary. Two basic procedures are used. In one, as illustrated in Figure 31-8, the moving part is held and rotated in a motor-driven collet while the stationary part is pressed against it with sufficient pressure so that the friction quickly generates enough heat to raise the abutting surfaces to welding temperature. As soon as the welding temperature is reached, rotation is stopped, and the pressure is maintained or increased until the weld is completed.

In the second (inertia) process, as indicated in Figure 31-9, the one workpiece is gripped in a rotating flywheel, with the kinetic energy of the flywheel being converted into heat by pressing the two workpieces together when the flywheel has attained the desired velocity. This method results in extremely repeatable conditions (consistent welds) and can be automated. Figure 31-10 shows an example of the equipment used and the relationship between surface velocity, torque, and upset.

In both procedures the total cycle time for a weld usually is less than 25 seconds, whereas the actual time for heating and welding is about 2 seconds.

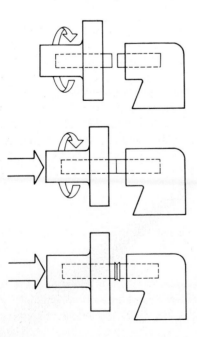

FIGURE 31-9 Schematic representation of the three steps in inertia welding.

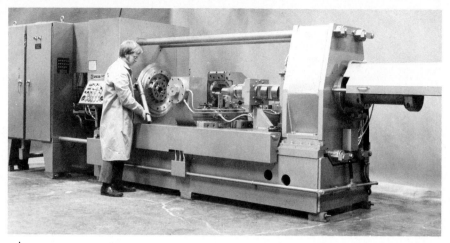

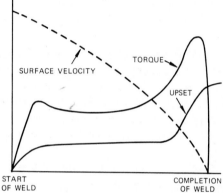

SURFACE VELOCITY

TORQUE

UPSET

START
OF WELD

COMPLETION
OF WELD

FIGURE 31-10 (*Top*) Inertia-type friction-welding machine and welded part. (*Bottom*) Relationship between surface velocity, torque, upset, and time in friction welding. (*Courtesy Production Technology, Inc.*)

No material is melted. Because of the very short period of heating—and thus the lack of time for the heat to flow away from the joint—the weld and heat-affected zones are very narrow. Surface impurities are displaced radially into a small flash that can be removed after welding if desired. Because virtually all the energy used is converted into heat, the process is very efficient, and it can be used to join many metals or combinations of dissimilar metals. Since grain size is refined due to hot working, the strength of the weld is almost the same as the base metal. Nevertheless, the process is restricted to joining round bars and tubes of the same size, or to joining bars and tubes to flat surfaces. The ends of the workpieces must be cut true and fairly smooth. Figure 31-11 shows some typical applications.

ULTRASONIC WELDING

Ultrasonic welding (USW) is a solid-state process wherein coalescence is produced by localized application of very high frequency (10,000 to 200,000 cps) vibratory energy to the workpieces as they are held together under pressure. The

FIGURE 31-11 Some typical friction welded parts. (*Left*) Impeller made by joining a chrome-moly steel shaft to a nickel–steel casting. (*Center*) Stud plate with two mild steel studs joined to a square plate. (*Right*) tube component where a turned segment is joined to medium-carbon steel tubing. (*Courtesy Newcor Bay City, Div. of Newcor, Inc.*)

basic components of the process are shown in Figure 31-12. Although there is some increase in temperature at the faying surfaces, it always is far below the melting points of the materials. It appears that the rapid reversals of stress along the bond interface play an important role in facilitating coalescence by breaking up and dispersing the mating-surface films.

The ultrasonic transducer used is essentially the same as is employed in ultrasonic machining and shown in Figure 26-13. It is coupled to a force-sensitive system that contains a welding tip on one end. The pieces to be welded are placed between this tip and a reflecting anvil, thereby concentrating the vibratory energy within the work. Either stationary tips, for spot welds, or rotating disks, for seam welds, can be used.

Ultrasonic welding is restricted to joining thin materials—sheet, foil, and wire; the maximum thickness is about 2.5 mm (0.1 inch) for aluminum and 1.0

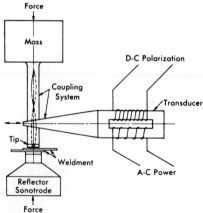

FIGURE 31-12 Schematic diagram of the equipment used in ultrasonic welding.

TABLE 31-1 Metal Combinations Weldable by Ultrasonic Welding

Metal	Aluminum	Copper	Germanium	Gold	Molybdenum	Nickel	Platinum	Silicon	Steel	Zirconium
Aluminum	×	×	×	×	×	×	×	×	×	×
Copper		×		×		×	×		×	×
Germanium			×	×		×	×	×		
Gold				×		×	×	×		
Molybdenum					×	×			×	×
Nickel						×	×		×	×
Platinum							×		×	
Silicon										
Steel									×	×
Zirconium										×

mm (0.04 inch) for harder metals. However, as indicated in Table 31-1, it is particularly valuable in that numerous dissimilar metals can readily be joined by the process. Because the temperatures involved are low, and no current flow or arcing is involved intermetallic compounds seldom are formed, and there is no contamination of surrounding areas. The equipment is simple and reliable, and only moderate skill is required of the operator. Typical applications include: joining the dissimilar metals of bimetallics, making microcircuit electrical contacts, welding refractory or reactive metals, bonding ultra-thin metal, and encapsulating explosives or chemicals.

DIFFUSION WELDING

Diffusion welding (DFW) occurs in the solid state when properly prepared surfaces are maintained in contact under proper conditions of pressure and time at elevated temperature. In contrast to deformation welding methods, deformation here is limited and the principal bonding factor is diffusion. Under low pressure and elevated temperature, a well-prepared interface is essentially a planar grain boundary with voids and impurities. Additional time then allows for void shrinkage and grain-boundary migration to complete the joint.

The process is controllable through control of surface condition and preparation, time, temperature, pressure, and the possible use of intermediate materials or interlayers. The latter, being of a dissimilar metal, can serve to promote diffusion, or minimize undesirable intermetallic compounds. Some form a temporary liquid eutectic which significantly speeds up the rate of interdiffusion.

EXPLOSION WELDING

Explosion welding (EXW) is used primarily for bonding sheets of corrosion-resistant metals to heavier plates of base metals (*cladding*), particularly where large areas are involved. An explosive material, usually in the form of a sheet, is placed on top of the two layers of metal and detonated progressively. A compressive stress wave, on the order of thousands of megapascals, progresses across the surface of the plates, so that a small open angle is formed between the two colliding surfaces. Surface films are liquefied or scarfed off the surfaces and jetted out of the interface, leaving clean surfaces which coalesce under the high pressure. The result is a cold weld having a wavy configuration at the interface.

SURFACING

Surfacing is the deposition of a layer of metal or other material of one composition upon the surface or edge of a base metal of a different composition. The usual objectives are to obtain improved resistance to wear, abrasion, or chemical reactions. Gas-flame, arc, or plasma-arc methods are used. The process often is called *hard facing,* because the deposited surfaces usually are harder than the base metal. However, this is not always true. In some cases a softer metal, such as bronze, is applied to a harder base metal.

Surfacing materials. The materials most commonly used for surfacing include (1) carbon and low-alloy steels; (2) high-alloy steels and irons; (3) cobalt-base alloys; (4) nickel-base alloys, such as Monel, Nichrome, and Hastelloy; (5) copper-base alloys; (6) stainless steels; and (7) ceramic and refractory carbides, oxides, borides, silicides, and so on.

FIGURE 31-13 Applying a hard-surfacing layer to a cylindrical drum by the submerged arc process. (*Courtesy Linde Division, Union Carbide Corporation.*)

Surfacing methods and application. Surfacing materials can be deposited by nearly all of the gas-flame or arc welding methods, including oxy-fuel gas, shielded metal-arc, gas metal-arc, gas tungsten-arc, submerged arc, and plasma-arc. Arc welding is frequently used for the deposition of high-melting-point alloys. Submerged-arc welding is useful when large areas are to be surfaced or a large amount of surfacing material is to be added, as in Figure 31-13. The plasma-arc process (see Figure 32-10) further extends capabilities because of its extreme temperatures. To obtain true fusion of the surfacing material, a transferred arc is used and the surfacing material is injected in the form of a powder. If a nontransferred arc is used, only a mechanical bond is produced. This is a form of metallizing, which will be discussed next. Laser hard-facing has also been performed.

METALLIZING

In metallizing, surfacing materials are melted and atomized in a special torch, or gun, and sprayed onto a base material. The process sometimes is called metal spraying. Figure 31-14 illustrates an oxyacetylene metal-spraying gun. As a wire of the surfacing metal is fed automatically through the center of the flame, it melts and is simultaneously atomized by a stream of compressed air and blown onto the base material. Almost any kind of metal that can be made into a wire can be deposited by this procedure.

Another type of gun utilizes materials in the form of powders, which are blown through the flame. This has the advantage that not only metals but also cermets, oxides, and carbides can be sprayed.

Most sophisticated is the plasma spray process, illustrated in Figure 31-15. Since plasma temperatures can reach 16,650°C (30,000°F), it can be used to spray materials with melting points up to 3300°C (6000°F). Metals, alloys, ceramics, carbides, cermets, intermetallics, and plastic-base metal powders have all been successfully deposited.

The metallizing guns can be either hand-held or machine-mounted and me-

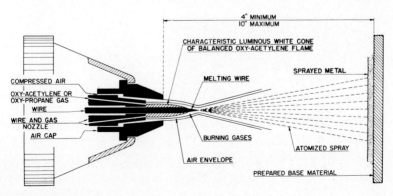

FIGURE 31-14 Schematic diagram of an oxyacetylene metal-spraying gun. (*Courtesy Metallizing Engineering Company, Inc.*)

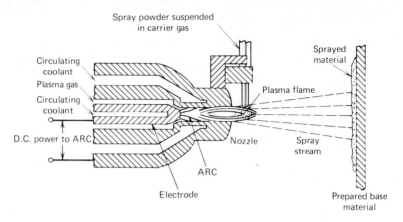

Spray powder suspended
in carrier gas

Circulating
coolant

Plasma gas

Circulating
coolant

D.C. power to ARC

Sprayed
material

Plasma flame

Nozzle

Spray
stream

ARC

Electrode

Spray
stream

Prepared base
material

FIGURE 31-15 Schematic of a plasma-spray gun. (*Courtesy METCO Inc., Westbury, NY, 11590.*)

chanically driven as shown in Figure 31-16. A stand-off distance of 150 to 250 mm (6 to 10 inches) is usually maintained between the spray nozzle and the workpiece.

Surface preparation.   Because the bond obtained between the deposited and base materials by metallizing is purely mechanical, it is essential that the base be prepared properly so that good mechanical interlocking can be obtained. With any method of surface preparation, the base surface must be clean and free of oil.

The most common method of surface preparation is grit blasting. For this purpose the grit should be sharp enough to produce a really rough surface. On cylindrical surfaces that can be rotated in a lathe, an effective method is to turn very rough threads and then roll the crests over slightly with a knurling tool. A modification that can be used on flat surfaces is to cut a series of parallel grooves, using a rounded grooving tool, and then roll over the lands between the grooves,

FIGURE 31-16 Spraying martensitic stainless steel to improve the wear resistance of an integral piston rod, rotating the workpiece in a lathe. Gun shown uses metal wire (other types use powder feed). (*Courtesy Wall Colmonoy Corporation.*)

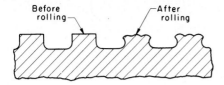

FIGURE 31-17 Method of preparing surfaces for metal spraying by machining grooves and rolling the edges.

as indicated in Figure 31-17. Figure 31-18 shows a comparison of the surfaces obtained by grit blasting and rough machining. If the sprayed surface is to be machined, the base should be prepared by either rough machining or grooving so as to provide maximum interlocking.

Characteristics of sprayed metals.   The properties of sprayed metals are what would be expected from the fact that they are broken into fine, molten particles, mixed with air, and then cooled rapidly upon hitting the base material. As a result, such coatings contain particles of oxidized metal and are harder, more porous, and more brittle than in the wrought state. In general, they have many of the characteristics of cast metals. However, these properties make sprayed metals particularly well suited to resist abrasion and wear and to serve as bearing surfaces. The characteristic porosity retains lubricants, which adds to their quality as lubricated wearing surfaces.

Although sprayed coatings have only about 85 to 90% the density and about one-third to one-half the strength of wrought metals, the electrical conductivity is nearly as good.

Applications of metal spraying.   Metal spraying has a number of important applications.

1. *Protective coatings.* Zinc and aluminum are sprayed on steel and iron to provide corrosion resistance.
2. *Building up worn surfaces.* Worn parts can be salvaged by adding metal to desired areas by spraying.

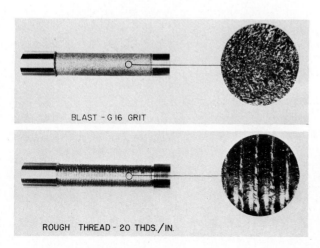

BLAST - G I6 GRIT

ROUGH   THREAD - 2O THDS./IN.

FIGURE   31-18 Surfaces obtained by two methods of preparation for metal spraying. (*Courtesy Metallizing Engineering Company, Inc.*)

3. *Hard surfacing.* Although metal spraying is not to be compared with hard surfacing by depositing weld metals, it is useful where thin coatings are adequate.
4. *Applying coatings of expensive metals.* Metal spraying provides a simple method for applying thin coatings of noble metals to small areas where conventional plating would not be economical.
5. *Electrical conductivity.* Because metal can be sprayed on almost any surface, metal or nonmetal, it provides a simple means of supplying a conductive surface on a poor or nonconductor. Copper or silver frequently is sprayed on glass or plastics for this purpose.
6. *Reflecting surfaces.* Aluminum, sprayed on the back of glass by a special fusion process, makes an excellent reflecting surface that is used in traffic markers and for similar applications.
7. *Decorative effects.* One of the earliest and still important uses of metal spraying was to obtain decorative effects. Because sprayed metal can be treated in a variety of ways, such as buffed, wire-brushed, or left in the as-sprayed condition, it is used in both products and architectural works as a decorative device.

FLAME STRAIGHTENING

Basically, *flame straightening* is the creation of controlled, localized upsetting in order to straighten warped or buckled plates. The theory of the process is illustrated in Figure 31-19. If a straight piece of metal is heated in a localized area, as indicated by the curved path in the upper diagram, the material adjacent to *b* will be upset as it tries to expand against the restraining cool metal. When the upset portion cools, it will contract and the piece will be shorter on edge *b* and bend, as shown in the lower diagram.

If the same procedure is used on a bent or warped piece of material, using the heated path shown in the lower diagram of Figure 31-19, the piece may be straightened because of the upsetting and subsequent contraction of the material near side *a'*. This procedure has been used to straighten various structures that have had members bent through accident.

A similar process can be used to flatten metal plates that have become dished due to buckling. In this case spots about 51 mm (2 inches) in diameter are heated quickly to upsetting temperature so that the surrounding metal remains relatively

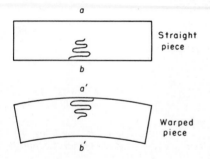

Straight piece

Warped piece

FIGURE 31-19 Theory of flame straightening.

cool. Cool water then is sprayed on the plate and the contraction of the upset spot straightens the buckle over a considerable area. To remove large buckles it is necessary to repeat the process on several spots.

Flame straightening cannot be used on thin material, however. The metal adjacent to the heated area must have sufficient rigidity to resist transferring the buckle from one area to another.

WELDING OF PLASTICS

Plastics of the thermoplastic type can be welded successfully using either a hot-gas torch of the type shown in Figure 31-20 or an electrically heated tool. In the hot-gas torch, a gas—usually air—is heated by an electrical coil as it passes through the torch. Electrical tools are similar to an electric soldering iron and are moved in contact with the material until the desired temperature is achieved. The heating is localized, at from 246 to 357°C (475 to 675°F), until the plastic softens. Some pressure then is applied to produce coalescence.

V-groove butt or fillet welds usually are employed. Because the plastic cannot be made to flow as in fusion welding of metals, some filler material usually has to be added. This is done by using a plastic filler rod that is heated simultaneously with the workpieces and then pressed or stretched into the joint, thus supplying some of the pressure needed to complete coalescence. This procedure is shown in Figure 31-20.

Heated-tool welding is also employed for making lap-seam welds in flexible plastic sheets. Pressure is applied by a roller, or other pressure device, after the material has been heated.

Butt welds sometimes are made in plastic pipe and rods by friction welding.

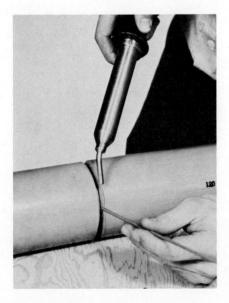

FIGURE 31-20 Using a hot-gas torch to make a weld in plastic pipe.

REVIEW QUESTIONS

1. What is the source of heat and filler metal in thermit welding?
2. How does electroslag welding differ from either arc or resistance methods?
3. What are the various functions of the slag in electroslag welding?
4. What type of edge preparation is required for electroslag welding?
5. What compromises are made when electron beam welds are made on parts that remain outside the vacuum system?
6. What are the drawbacks of high-voltage electron beam welding equipment?
7. What are some of the attractive features of electron beam welding?
8. What advantages does laser beam welding have over the electron beam method?
9. Describe the weld sequence of flash welding.
10. Why is it important that sufficient upsetting occur in flash welding?
11. What is the difference between the two basic procedures of friction welding?
12. What are the geometrical limitations of the friction welding process?
13. What is the procedure by which a joint is produced by ultrasonic welding?
14. What geometrical restriction or limitation exists for ultrasonic welding?
15. What conditions are required for diffusion welding?
16. What type of joint (size and shape) would be appropriate for explosion welding?
17. What are some of the reasons why a surface deposit might be desired on a base metal?
18. What are the distinct advantages of plasma spray metallizing?
19. Why is surface preparation so important in metallizing?
20. How can metal spraying be used to salvage worn parts?
21. Why can flame straightening not be performed on thin material?
22. Why can thermosetting plastics not be welded by the heat-type methods?

CASE STUDY 31. Shaft and Circular Cam

The Cab-Con Corporation makes the part shown in Figure CS.31 in large quantities. (Only the essential dimensions are shown.) Currently, it is machining the part from AISI 1120 bar stock that is 76.2 mm (3 inches) in diameter and then carburizing the outer periphery of the disk cam, which must have a hardness of at least $55R_C$, although the depth of the hardness need not exceed 0.15 mm (0.006 inch). Obviously, the machining cost and the wastage of material are excessive.

The chief engineer has assigned you to devise a more economical procedure for producing this part. How would you propose to produce the part? (Assume that the volume is sufficient to justify any required new equipment, or that some operations can be subcontracted, if that would be more economical.)

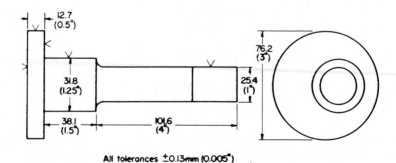

All tolerances ±0.13mm (0.005")

FIGURE CS-31 Shaft with eccentric circular cam.

CHAPTER 32

Torch and Arc Cutting

For many years, metal sheets and plates have been cut by means of oxyfuel torches and electric arc equipment. Developed originally for use in salvage and repair work, then later for preparing plates for welding, these processes are now widely used for cutting metal sheets and plates into desired shapes for assembly and other processing operations. In recent years the development of laser and electron beam equipment has made possible the cutting of both metals and nonmetals, with cutting speeds up to 25.4 meters (1000 inches) per minute (see Chaper 26). Accuracies up to 0.254 mm (0.010 inch) are readily attainable, and speeds of 1.27 meters (50 inches) per minute are common. Figure 32-1 shows the commonly used torch and arc cutting processes with their AWS designations.

OXYGEN TORCH CUTTING

Oxyfuel gas cutting. By far the majority of metal cutting is done by *oxyfuel gas cutting* (OFC). In a few cases, primarily where the metal is nonferrous, the metal is merely melted by means of the flame of the oxyfuel gas torch and blown away to form a gap, or kerf in the metal. However, where ferrous metal is being

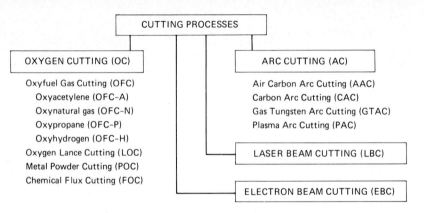

FIGURE 32-1 Common cutting processes and their AWS designations.

cut, the process is one of rapid oxidation (burning) of iron at high temperatures according to the chemical equation

$$3Fe + 2O_2 = Fe_3O_4 + heat$$

Because this reaction does not occur until steel is at approximately 870°C (1600°F), an oxyfuel flame is first used to raise the metal to the temperature at which burning will start. Then a stream of pure oxygen is added to the torch, or the oxygen content of the oxyfuel mixture is increased, to oxidize the iron. The liquid iron oxide is then expelled from the joint by the kinetic energy of the oxygen–gas stream.

Theoretically, no further heat is required, but in most cases some additional heat must be supplied to compensate for losses to the atmosphere and the surrounding metal and thus to assure that the reaction progresses, particularly in the desired direction. Under some ideal conditions, where the area of burning is sufficiently confined so as to conserve the heat of combustion, no supplementary heating is required, and a supply of oxygen through a small pipe will keep the cut progressing. This is known as *oxygen lance cutting* (LOC), and a temperature of about 1204°C (2200°F) has to be achieved in order for this procedure to be effective.

Fuel gases for oxyfuel gas cutting. Acetylene is by far the most common fuel used in oxyfuel gas cutting; thus the process often is called oxyacetylene cutting (OFC-A). The type of torch commonly used is shown in Figure 32-2. The tip contains a circular array of small holes through which the oxygen–acetylene mixture is supplied for the heating flame. A larger hole in the center supplies a stream of oxygen, controlled by a lever valve. The rapid flow of the cutting oxygen not only produces rapid oxidation but also blows the oxides from the cut. If the torch is adjusted and manipulated properly, a smooth cut results, as shown in the top example of Figure 32-3. As shown, quality cutting requires selection of a proper preheat condition, oxygen flow rate, and cutting speed. Oxygen purities over 99.5% are required for most efficient cutting.

The cutting torch can be manipulated manually. However, in most manufacturing operations, the traverse of the desired path is controlled by mechanical

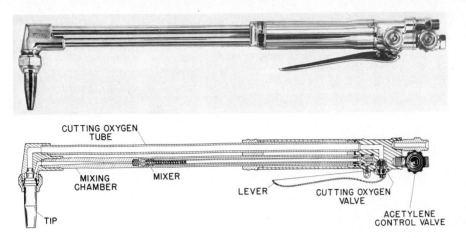

CUTTING OXYGEN
TUBE

MIXING
CHAMBER

MIXER

LEVER

CUTTING OXYGEN
VALVE

ACETYLENE
CONTROL VALVE

TIP

FIGURE 32-2 Oxyacetylene cutting torch. (*Courtesy Victor Equipment Company.*)

or programmable means. Figure 32-4 shows a portable, electrically driven carriage that is commonly used. For straight cuts, the device travels along a section of portable track. In addition, it can be adapted to cut circles of various radii. Where duplicate or more complex shapes are required, a template-controlled flame cutting machine, as shown in Figure 32-5, is widely used. More recently, computer numerical controlled (CNC) cutting machines have become quite popular. All of these types produce desired shapes with remarkable accuracy. Accuracies of ±0.38 mm (0.015 inch) are possible, but accuracies of 0.76 to 1.0 mm (0.030 to 0.040 inch) are more common.

Fuel gases other than acetylene also are used for oxyfuel gas cutting, the most common being natural gas (OFC-N) and propane (OFC-P). Their use is a matter of economics due to special availability. For certain special work, hydrogen may be used (OFC-H).

In preparing plate edges for welding, two or three simultaneous cuts often are made, as shown in Figure 32-6.

Stack cutting. In order to cut a stack of thin sheets of steel successfully, two precautions must be observed. First, the sheets should be flat and smooth and be free of scale. Second, they must be clamped together tightly so there are no gaps between them that could interrupt uniform oxidation and permit slag and molten metal to be entrapped. *Stack cutting* is a useful technique where a modest quantity of duplicate parts is required but is insufficient in number to justify the construction of a blanking die. Obviously, the accuracy obtainable is not as good as can be obtained by blanking.

Metal powder cutting and chemical flux cutting. Hard-to-cut materials can often be processed by modified torch techniques. *Metal powder cutting* (POC) uses an injection of iron powder into the flame to raise the cutting temperature. *Chemical flux cutting* (FOC) adds a fine stream of special flux to the cutting oxygen to increase the fluidity of the high-melting-point oxides. Both methods

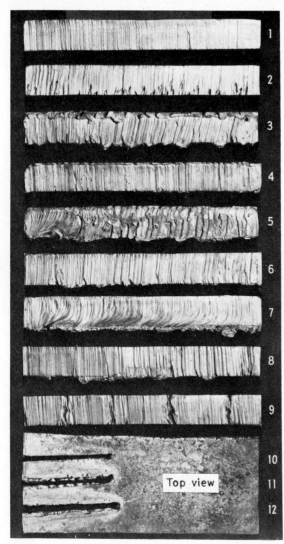

(1) Correct Procedure
Compare this cut in 1-in. plate with those below. The edge is square, the drag lines are vertical and not too pronounced.

(2) Preheat Flames Too Small
They are only about ⅛ in. long. Result: cutting speed was too slow, causing bad gouging effect at bottom.

(3) Preheat Flames Too Long
They are about ½ in. long. Result: surface has melted over, cut edge is irregular, and there is too much adhering slag.

(4) Oxygen Pressure Too Low
Result: top edge has melted over because of too slow cutting speed.

(5) Oxygen Pressure Too High
Nozzle size also too small. Result: entire control of the cut has been lost.

(6) Cutting Speed Too Slow
Result: irregularities of drag lines are emphasized.

(7) Cutting Speed Too High
Result: a pronounced rake to the drag lines and irregularities on the cut edge.

(8) Blowpipe Travel Unsteady
Result: the cut edge is wavy and irregular.

(9) Lost Cut Not Properly Restarted
Result: bad gouges where cut was restarted.

(10) Good Kerf
Compare this good kerf (viewed from the top of the plate) with those below.

(11) Too Much Preheat
Nozzle also is too close to plate. Result: bad melting of the top edges.

(12) Too Little Preheat
Flames also are too far from the plate. Result: heat spread has opened up kerf at top. Kerf is tapered and too wide.

FIGURE 32-3 Appearance of edges of metal cut properly and improperly by the oxyacetylene process. (*Courtesy Linde Division, Union Carbide Corporation.*)

have been replaced to a considerable extent by plasma arc torches (PAC), which will be discussed later.

Underwater torch cutting. Steel can be cut underwater by use of a special torch that supplies a flow of compressed air to provide the secondary oxygen for the oxyacetylene flame and to keep the water away from the zone where the burning of the metal occurs. Such a torch, shown in Figure 32-7, contains an auxiliary skirt, surrounding the main tip, through which the compressed air flows. The torch is either ignited in the usual manner before descent or by an electric spark device after being submerged.

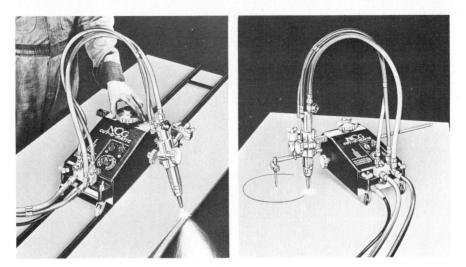

FIGURE 32-4 Oxyacetylene cutting with the torch carried by a small portable, electrically driven carriage. (*Courtesy National Cylinder Gas Company.*)

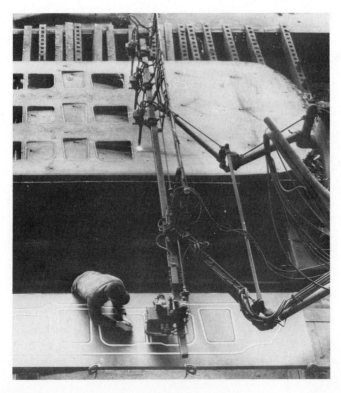

FIGURE 32-5 Three parts being cut simultaneously from heavy steel plate, using a tracer-pantograph machine that is guided by the template at the bottom. (*Courtesy Linde Division, Union Carbide Corporation.*)

FIGURE 32-6 Plate edge being prepared by three simultaneous cuts with oxyacetylene cutting torches. (*Courtesy Linde Division, Union Carbide Corporation.*)

Acetylene gas usually is utilized for depths up to about 7.6 m (25 feet). For greater depths hydrogen is used, because the pressure involved is too great for safe operation with acetylene.

ARC CUTTING

Virtually all metals can be cut by electric arc procedures, wherein the metal is melted by the intense heat of the arc and then permitted, or forced, to flow from the kerf. Most processes are adaptations of the previously discussed arc welding procedures of Chapter 29.

Carbon-arc (CAC) and shielded metal-arc cutting (SMAC). These methods utilize the arc to melt metal, which is then removed from the cut by gravity or

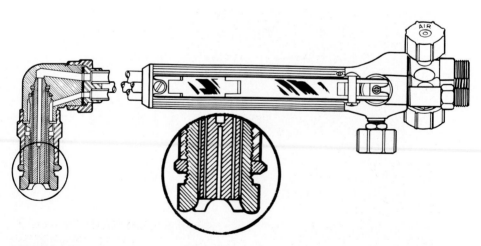

FIGURE 32-7 Underwater cutting torch. Note extra set of gas openings in the nozzle to provide an air bubble and the extra control valve. (*Courtesy Bastian-Blessing Company.*)

the force of the arc. Use is limited largely to small shops, garages, and homes where equipment investment is small.

Oxygen-arc cutting (AOC).   In this process, an electric arc and a stream of oxygen are employed to make the cut. The electrode is a coated ferrous metal tube, the tube serving as the conductor for maintaining the arc while the bore of the tube conveys and directs oxygen at the area of incandescence. In easily oxidizable metals, such as steel, the arc simply preheats the base metal, which is oxidized, liquefied, and expelled by the oxygen stream.

Air carbon-arc cutting (AAC).   Here an arc is maintained between a carbon electrode and the workpiece, and high-velocity jets of air are directed at the molten metal from holes in the electrode holder, as shown in Figure 32-8. While there is some oxidation, the primary function of the airstream is to blow molten material from the cut and the process can be employed on metals which do not readily oxidize. The process is particularly effective for cutting cast iron and for gouging grooves in steel plates preparatory to welding. Speeds up to 610 mm (24 inches) per minute are readily attained. For cutting stainless steel and nonferrous metals, plasma arc cutting is more efficient. The disadvantages of arc air cutting are that it is quite noisy, and the hot metal particles tend to be blown out over a substantial area.

Gas metal-arc cutting (GMAC).   If the wire feed rate and other variables of MIG welding are adjusted so that the electrode penetrates completely through the plate, then cutting rather than welding will occur. Wire feed rate controls the quality of the cut and voltage determines the width of the kerf.

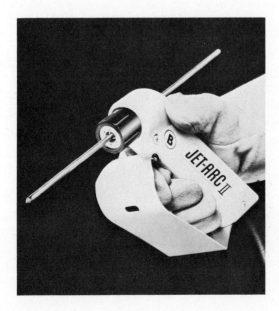

FIGURE 32-8 Gun used in the arc air process. Note air holes in holder surrounding electrode. (*Courtesy Jackson Products.*)

Gas tungsten-arc cutting (GTAC). The same basic circuit and shielding gas is used as in gas tungsten-arc welding, with a high-velocity jet of gas passing through the same nozzle as the arc serving to expel the molten metal. The process is quite useful for making holes—up to 9.5 mm ($\frac{3}{8}$ inch)—in sheet metal, as shown in Figure 32-9.

Plasma arc cutting (PAC). The torches used in *plasma arc cutting* produce the highest temperature available from any practicable source. They thus are very useful for cutting metals, particularly nonferrous and stainless types that cannot be cut by the usual rapid oxidation induced by ordinary flame torches. Two types of torches are used, shown in Figure 32-10. Both are arranged so that the arc column is constricted within a small-diameter nozzle through which inert gas is directed. Because the arc fills a substantial part of the nozzle opening, most of the gas must flow through the arc and, as a consequence, is heated to a very high temperature, forming a plasma.

With the nontransferred-type torch, the arc column is completed within the nozzle, and a temperature of about 16,649°C (30,000°F) is obtainable. With the transferred-type torch, wherein the arc column is between the electrode and the workpiece, the temperatures obtainable are estimated to be up to 33,316°C (60,000°F). Obviously, such high temperatures provide a means of very rapid cutting of any material by melting and blowing it from the cut. Cutting speeds up to 7620 mm (300 inches) per minute have been obtained in 6.35 mm ($\frac{1}{4}$ inch) aluminum, and 2500 mm (100 inches) per minute in 12.7-mm ($\frac{1}{2}$-inch) steel. The combination of extremely high temperature and jetlike action of the plasma produces narrow kerfs and remarkably smooth surfaces—nearly as smooth as can be obtained by sawing. Integration with CNC machines provides fast, clean and accurate cutting. Transferred-type torches are usually used for cutting metals, whereas the nontransferred type must be used for nonmetals.

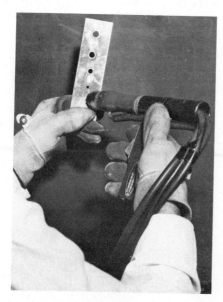

FIGURE 32-9 Making holes in sheet metal by the inert-gas arc process. (*Courtesy Hobart Brothers Company.*)

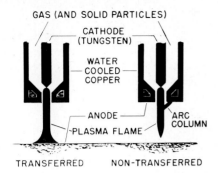

GAS (AND SOLID PARTICLES)

CATHODE
(TUNGSTEN)

WATER
COOLED
COPPER

ANODE

ARC
COLUMN

PLASMA FLAME

TRANSFERRED NON-TRANSFERRED

FIGURE 32-10 Principle of arc plasma torches. (*Courtesy Linde Division, Union Carbide Corporation.*)

Argon, helium, nitrogen, and mixtures of argon and hydrogen are used. Mixtures of 65 to 80% argon and 20 to 35% hydrogen are very common.

LASER BEAM CUTTING

Laser beam cutting (LBC) uses the intense heat from a laser beam to melt and/or evaporate the material being cut. Any known material can be cut by this process. For some nonmetallic materials the mechanism is purely evaporation, but for many metals a gas may be supplied, either inert to blow away the molten metal and provide a smooth, clean kerf or oxygen to speed the process through oxidation. The temperature achieved may be in excess of 11,093°C (20,000°F), and cutting speeds of the order of 25.4 meters (1000 inches) per minute are not uncommon in nonmetals and 508 mm (20 inches) per minute in tough steels. As shown in Figure 32-11, very accurate cuts can be made. The kerf and heat-affected zone are narrower than with any other thermal cutting process, and it is quite easily adapted to CNC control units. No post-cut finishing is required in most applications, even though the process does produce a thin, recast surface. [See Chapter 26 for laser beam machining (LBM).]

METALLURGICAL AND HEAT CONSIDERATIONS

Flame and arc cutting obviously involve high temperatures, often localized. Consequently, it is always possible that the use of these processes could have harmful metallurgical and other heat effects, and such possibilities should not be overlooked. Fortunately, in most cases little or no difficulty is experienced, but in others, definite steps should be taken to avoid or overcome the harmful effects.

In low-carbon steel, below 0.25% carbon, oxyacetylene cutting usually causes no serious metallurgical effects. Although there often is some minor hardening in a thin zone near the cut, and a small amount of grain growth, these effects usually will be eliminated if any subsequent welding of the cut edges is done. However, in steels of higher carbon content, these effects can be quite serious,

FIGURE 32-11 Cutting sheet metal with a plasma torch. (*Courtesy GTE Sylvania.*)

and preheating and/or postheating may be required. For alloy steels, additional consideration should be given to the effects of the various alloy elements. Chromium, molybdenum and tungsten can be detrimental to cutting.

From a heat-effect viewpoint, arc air cutting is about the same as arc welding. If welding follows the use of this process, its heat effects will replace those left by the cutting process, and no special precautions are required. However, if no subsequent welding is to be done, consideration should be given to whether the resulting heat effects will be damaging in view of the load stresses that are anticipated.

Plasma arc and laser cutting are so rapid, and the heat is so localized, that the heat-affected zone usually is less than 2.38 mm ($\frac{1}{32}$ inch), and the original hardness of the metal exists beyond 1.59 mm ($\frac{1}{16}$ inch) from the cut.

All these processes leave some residual stresses, with the cut surface in tension. Except in the case of thin sheet, gas or arc cutting usually will not produce warping in themselves. However, if subsequent machining removes only a portion of the total cut surface, or not all of the depth in residual tension, the resulting unbalancing of stresses may produce warping. Thus, if subsequent machining is to be done, it may be necessary to remove all cut surfaces to a substantial depth in order to achieve dimensional stability. Machining cuts should be sufficiently deep to get below the hardened surface in one pass and avoid dulling tools.

All flame- or arc-cut edges are rough, to varying degrees, and thus contain geometrical notches that can act as stress raisers and thus reduce the endurance strength. Consequently, if such edges are to be subjected to high or repeated tensile stressing, the cut surface and the heat-affected zone should be machined away or, as a minimum, given a stress-relief heat treatment.

REVIEW QUESTIONS

1. How does the torch cutting of ferrous metal differ from the cutting of nonferrous metal?
2. Why might continuously cast steel strands be effectively cut by oxygen lance cutting when they emerge from the casting operation?

3. What are some of the means by which a cutting torch can be mechanically manipulated?
4. What two precautions must be observed in order to cut a stack of thin sheets successfully?
5. What additional feature must be added to permit torch cutting under water?
6. What is the only attractive feature of carbon-arc and shielded metal-arc cutting?
7. Why might a metal such as stainless steel be difficult to cut by oxygen-arc cutting?
8. How can a gas metal-arc welding unit be adapted for cutting?
9. Why are plasma arc processes attractive for cutting metals that do not readily oxidize?
10. Why must nontransferred type plasma torches be used to cut nonmetals?
11. Why can laser beam cutting be performed at such high cutting rates?
12. Why is there little concern regarding heat-affected zones on cut edges that will subsequently be welded?
13. Under what conditions might the residual stresses induced by cutting become objectionable?
14. Why might it be wise to finish machine the cut edge of a highly stressed machine part?

CASE STUDY 32. The Broken Axle

Figure CS-32 shows the outboard fracture surface of a broken rear axle of an automobile. The axle broke just outside an axle bearing. Significant features found were the relatively smooth gouge (a), which extended across the wheel-mounting hub into the axle, and several straight, smooth scratches (b), which extended from the edge of the gouge across the face of the hub. Investigation revealed that the axle bearing had been replaced a few months prior to the breaking of the axle, having been done at a small, local garage.

What do you believe caused the axle failure, and what had taken place to bring about the failure?

FIGURE CS-32 Fracture surface of broken automobile axle.

CHAPTER **33**

Brazing, Soldering, Adhesive Bonding, and Mechanical Fastening

There are many joining or assembly operations where welding may not be the best choice. Perhaps the heat of welding is objectionable or the materials possess poor weldability or the economics of alternative methods are superior. In such cases, low-temperature joining methods may be employed. These include: brazing, soldering, adhesive joining, and the use of mechanical fasteners. In brazing and soldering, a low-melting-point metal is melted, drawn into the space between two solid surfaces by capillary action, and allowed to solidify. Adhesive bonding utilizes a polymerizable resin which fills the space between the surfaces to be joined. Variations in surface finish are more tolerable, since capillary action is not required and oxides may not be a problem since the adhesive may actually adhere better to a tight oxide layer. Mechanical fasteners span a wide spectrum, including rivets, bolts, screws, nails, and others. While some may be permanent, others offer the advantage of easy disassembly for service or replacement of components.

BRAZING

Brazing is the joining of metals through the use of heat and a filler metal whose melting temperature is above 450°C (840°F)[1] but below the melting point of the metals being joined. It differs from welding in the following ways:

1. The *composition* of the brazing alloy is significantly different from the base metal.
2. The *strength* of the brazing alloy is substantially lower than the base metal.
3. The *melting point* of the brazing alloy is lower than that of the base metal, so the base metal is not melted.
4. Bonding requires *capillary action*.

Because of these differences, brazing has several distinct advantages:

1. Virtually all metals can be joined by some type of brazing metals. The process is ideally suited for dissimilar metals, such as joining nonferrous to ferrous, or metals with widely different melting points.
2. Since less heating is required than for welding, the process can be performed quickly and economically.
3. The lower temperatures reduce problems associated with heat-affected zones, warping, or distortion and thinner and more complex assemblies can be joined successfully.
4. Brazing is highly adaptable to automation and performs well when mass-producing delicate assemblies. A strong permanent joint is formed.

The major disadvantage of brazing is the fact that reheating can cause inadvertent melting of the braze metal, causing it to run, thus weakening or destroying the joint. Too often this occurs when people apply heat to brazed parts in attempting to repair or straighten such devices as bicycles or motorcycles. Such a consequence, of course, is not a defect of brazing, but it can lead to most unfortunate results. Consequently, if brazing is specified for use in products that later might be subjected to such abuse, adequate warning should be given to those who will use the product.

The nature and strength of brazed joints. Just as in welding, brazing forms a strong, metallurgical bond at the interfaces. The bonding is enhanced by clean surfaces, proper clearance, good wetting, and good fluidity. Resulting strength can be very high, certainly higher than the strength of the brazing alloy and even higher than the strength of the base metal. Attainment of this strength, however, requires optimum processing and design.

Bond strength is a strong function of the clearance between the parts to be joined. There must be sufficient clearance so that the braze metal will wet the joint and flow into it, but, beyond this, strength decreases rapidly as a function of joint thickness, to that of the braze metal itself. Proper clearance varies considerably, depending primarily on the type of braze metal. Copper requires virtually no clearance when heated in a hydrogen environment. Silver–alloy

---

[1]The temperature is an arbitrary one, set to distinguish brazing from soldering.

brazing metals require about 0.04 to 0.05 mm (0.0015 to 0.002 inch) and 60–40 brass uses a clearance of about 0.50 to 0.75 mm (0.02 to 0.03 inch) when brazing iron and copper.

Wettability is a strong function of the surface tensions between the braze metal and base alloy. Generally, wettability is good when the two metals will form intermediate diffused alloys. Sometimes, this property can be improved, as in tin-plating steel, so that it will accept a lead–tin solder.

Fluidity is a measure of the flow characteristics of the molten braze metal and is a function of the metal, its temperature, surface cleanliness, and clearance.

Brazing metals. The most commonly used brazing metals are copper and copper alloys, silver and silver alloys and aluminum alloys. Table 33-1 lists some of the most frequently used brazing metals and their usages.

Copper is used only for brazing steel and other high-melting-point alloys, such as high-speed steel and tungsten carbide. Its use is confined almost exclusively to furnace heating in a protective hydrogen atmosphere in which the

TABLE 33-1   Commonly Used Brazing Metals and Their Uses

Braze Metal	Composition	Brazing Process	Base Metals
Brazing brass	60% Cu, 40% Zn	Torch Furnace Dip Flow	Steel, copper, high-copper alloys, nickel, nickel alloys, stainless steel
Manganese bronze	58.5% Cu, 1% Sn, 1% Fe, 0.25% Mn, 39.5% Zn	Torch	Steel, copper, high-copper alloys, nickel, nickel alloys, stainless steel
Nickel silver	18% Ni, 55–65% Cu, 27–17% Zn	Torch Induction	Steel, nickel, nickel alloys
Copper silicon	1.5% Si, 0.25% Mn, 98.25% Cu, 1.5% Si, 1.00% Zn, 97.5% Cu	Torch	Steel
Silver alloys (no phosphorus)	5–80% Ag, 15–52% Cu, balance Zn + Sn + Cd	Torch Furnace Induction Resistance Dip	Steel, copper, copper alloys, nickel, nickel alloys, stainless steel
Silver alloys (with phosphorus)	15% Ag, 5% P, 80% Cu	Torch Furnace Induction Resistance Dip	Copper, copper alloys
Copper phosphorus	93% Cu, 7% P	Torch Furnace Induction Resistance	Copper, copper alloys

copper is extremely fluid and requires no flux. Copper brazing is used extensively for assemblies composed of low-carbon steel stampings, screw-machine parts, and tubing, such as are common in mass-produced products.

Copper alloys were the earliest brazing materials. Today, copper–zinc alloys are used extensively for brazing steel, cast irons, and copper. Copper–phosphorus alloys are useful for fluxless brazing of copper since the phosphorus can reduce the copper oxide film. Manganese bronzes are also used for brazing purposes.

Pure silver is used for brazing titanium. *Silver solders,* silver–copper–zinc alloys, are used in joining steels, copper, brass, and nickel. Although these brazing alloys are expensive, such a small amount is required that cost per joint is low. The silver alloys are also used in brazing stainless steel. However, because the brazing temperatures are in the range of carbide precipitation, only stabilized or low-carbon stainless steels should be brazed with these alloys if continued corrosion resistance is desired.

Aluminum–silicon alloys, containing about 6 to 12 per cent silicon, are used for brazing aluminum and aluminum alloys. By using a braze metal that is not greatly unlike the base metal, the possibility of galvanic corrosion is reduced. However, because these brazing alloys have melting points of about 610°C (1130°F), when the melting temperature of commonly brazed aluminum alloys, such as 3003, is around 669°C (1290°F), control of the temperature used in brazing is quite critical. In brazing aluminum, proper fluxing action, surface cleaning, and/or use of a controlled-atmosphere or vacuum environment must be utilized to assure adequate flow of the braze metal and yet avoid damage to the base metal.

A commonly used procedure in connection with brazing aluminum is to use sheets that have one or both surfaces coated with the brazing alloy to a thickness of about 10% of the total sheet thickness. These "brazing sheets" have sufficient coating to form adequate fillets, and joints are made merely by coating the joint area with suitable flux followed by heating.

A nickel–chromium–iron–boron brazing alloy is frequently used to braze heat-resistant alloys which are to be used at high temperatures. Although the service temperature is above the initial melting point of the brazing alloy, the boron diffuses from the braze metal into the base metal, raising the melting point of the braze alloy to above the service temperature.

Cast iron is not readily wettable because of the graphite. Consequently, before cast iron can be brazed, the graphite must be removed by etching.

Fluxes.   *Fluxes* play a very important part in brazing by (1) dissolving oxides that may be on the surface prior to heating, (2) preventing the formation of oxides during heating, and (3) lowering the surface tension of the molten brazing metal and thus promoting its flow into the joint.

One of the primary factors affecting the quality and uniformity of brazed joints is cleanliness. Although fluxes will dissolve modest amounts of oxides, *they are not cleaners*. Before a flux is applied, dirt, particularly oil, should be removed from the surfaces that are to be brazed. The less the flux has to do prior to heating, the more effective it will be during heating.

Borax has been a commonly used brazing flux. *Fused* borax should be used because the water in ordinary borax causes bubbling when the flux is heated. Alcohol can be mixed with fused borax to form a paste.

Many modern fluxes are available that have lower melting temperatures than borax and are somewhat more effective in removing oxidation. Fluxes should be selected with reference to the base metal. Paste fluxes usually are utilized for furnace, induction, and dip brazing and either paste or powdered fluxes are used for torch brazing. In furnace, induction, and dip brazing the flux ordinarily is brushed onto the surfaces. In torch brazing it frequently is applied by dipping the heated end of the filler wire into the flux.

Fluxes for aluminum usually are mixtures of metallic halide salts, the base typically being sodium and potassium chlorides, comprising from 15 to 85% of the flux. Activators, such as fluorides or lithium compounds, are added. These fluxes do *not* dissolve the surface oxide film on aluminum.

Most brazing fluxes are corrosive, and the residue should be removed from the work after the brazing is completed. This is particularly important in the case of aluminum. Considerable effort has been devoted to developing fluxless procedures for brazing aluminum, as will be discussed later.

Applying brazing metal. Brazing metal is applied to joints in three ways. The oldest, and a common method used in torch brazing, uses the brazing metal in the form of a rod or wire. When the joint has been heated to a sufficient temperature so that the base metal will melt the brazing wire or rod, the wire or rod is then melted by the torch and capillary attraction causes it to flow into the joint. Although the base metal should be hot enough to melt the braze metal, and assure its remaining molten and flowing into the joint, the actual melting should be done with the torch.

Obviously, this method of braze metal application requires considerable labor, and care is necessary to assure that it has flowed to the inner portions of the joint because it always is applied from the outside. To avoid these difficulties, the braze metal often is applied to the joint before heating, in the form of wire or shims. In cases where it can be done, rings or shims of braze metal are fitted into internal grooves in the joint before the parts are assembled, as shown in Figure 33-1. When this procedure is employed, the parts usually must be held together by press fits, riveting, staking, tack welding, or a jig, to assure their proper alignment. In such preloaded joints, care must be exercised to assure that the filler metal is not pulled away from the intended surface by the capillary action of another surface with which it may be in contact. Capillary action always will pull the molten braze metal into the smallest clearance, whether or not such was intended.

Another precaution that must be observed is that the flow of filler metal not be cut off by the absence of required clearances or by no provision for the escape of trapped air. Also, fillets or grooves within the joint may act as reservoirs and trap the metal.

Special brazing jigs or fixtures often are used to hold the parts that are to be brazed in proper relationship during heating, especially in the case of complex assemblies. When these are used, it usually is necessary to provide springs that will compensate for expansion, particularly when two or more dissimilar metals are being joined. Figure 33-2 shows an excellent example of this procedure.

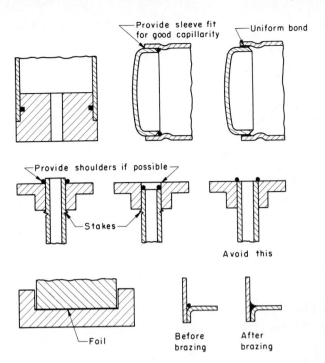

Provide sleeve fit for good capillarity

Uniform bond

Provide shoulders if possible

Stakes

Avoid this

Foil

Before brazing

After brazing

FIGURE 33-1 Methods of applying braze metal in sheet or wire form to assure proper flow into the joint.

Heating methods used in brazing. A common source of heat for brazing is a gas-flame torch. In this *torch-brazing procedure*, oxyacetylene, oxyhydrogen, or other gas-flame sources can be used. Most repair brazing is done in this manner because of its flexibility and simplicity, but the process also is widely used in production brazing, as illustrated in Figure 33-3. Its major drawbacks are the difficulty in obtaining uniform heating, proper control of the temperature,

FIGURE 33-2 Brazing fixture, with provision for expansion caused by brazing heat. (*Courtesy Aluminum Company of America.*)

FIGURE 33-3 Manual torch brazing of large heat exchangers using preplaced rings of brazing alloy. (*Reprinted with permission of Handy & Harman, from* The Brazing Book.)

and the requirement of costly skilled labor. In production-type torch brazing, specially shaped torches often are used to speed the heating and to aid in reducing the amount of skill required.

Large amounts of brazing are done in controlled-atmosphere or vacuum furnaces. In such *furnace brazing,* the brazing metal must be preloaded into the work. If the work is not of such a nature that its preassembly will hold the parts in proper alignment and with adequate pressure, brazing jigs or fixtures must be used. Assemblies that are to be brazed usually can be designed so that such jigs or fixtures will not be needed; often a light press fit will suffice. Figure 33-4 shows a number of typical furnace-brazed assemblies.

Because excellent control of brazing temperatures can be obtained and no skilled labor is required, furnace brazing is particularly well suited for mass production. Either box- or continuous-type furnaces can be used, the latter being more suitable for mass-production work. If the furnace atmosphere can reduce the oxide film, flux may not be required. Reactive metals are often furnace brazed in a vacuum.

In *salt-bath brazing* the parts are heated by dipping in a bath of molten salt that is maintained at a temperature slightly above the melting point of the brazing metal. This method has three major advantages: (1) the work heats very rapidly because it is completely in contact with the heating medium, (2) the salt bath acts as a protective medium to prevent oxidation, and (3) thin pieces can easily be attached to thicker pieces without danger of overheating because the temperature of the salt bath is below the melting point of the parent metal. This latter feature makes this process well suited for brazing aluminum.

It is essential that the parts be held in jigs or fixtures (or that they be prefas-

FIGURE 33-4 Furnace-brazed assemblies. (*Courtesy Pacific Metals Company.*)

tened in some manner) and that the brazing metal be preloaded into the work. Also, to assure that the bath remains at the desired temperature, the volume must be rather large, depending on the weight and quantity of the assemblies that are to be brazed.

In *dip brazing*, the assemblies are immersed in a bath of molten brazing metal. The bath thus provides both the required heat and the braze metal for the joint. Because the braze metal usually will coat the entire work, it is wasteful of braze metal and is used primarily only for small parts, such as wire.

Induction brazing utilizes high-frequency induction currents for heating. The process has the following advantages, which account for its extensive use:

1. Heating is very rapid—usually only a few seconds being required for the complete cycle.
2. The operation can easily be made semiautomatic, so that semiskilled labor may be used.
3. Heating can be confined to the localized area of the joint, because of the shape of the heating coils and the short heating time. This reduces scale, discoloration, and distortion.
4. Uniform results are readily obtained.
5. By making new, and relatively simple, heating coils, a wide variety of work can be done with a single power unit.

The high-frequency power-supply units are available in both small and large capacities at very modest costs. The only other special equipment required for adapting induction brazing to a given job is a simple heating coil to fit around the joint to provide heating at the desired area. These coils are formed of copper

FIGURE 33-5 Typical induction brazing operation. (*Courtesy Lepel High Frequency Laboratories, Inc.*)

tubing through which cooling water is carried. Although the filler material can be added to the joint manually after it is heated, the usual practice is to use preloaded joints to speed the operation and obtain more uniform joints. Figure 33-5 shows a typical induction-brazing operation.

Induction brazing is so rapid that it can often be used to braze parts with high surface finishes, such as silver plating, without affecting the finish.

Some *resistance brazing* is done, in which the parts to be joined are held under pressure between two electrodes, similar to spot welding, and an electrical current is passed through them. However, unlike resistance welding, most of the resistance is provided by the electrodes, which are made of carbon or graphite. Most of the heating of the metal thus is by conduction from the hot electrodes. This process is used primarily in the electrical equipment manufacturing industry for brazing conductors, cable connections, and so on. Regular resistance welders and their timing equipment often are adapted for resistance brazing.

Flux removal. Although not all brazing fluxes are corrosive, most of them are. Consequently, flux residues usually must be completely removed. Most of the commonly used fluxes are soluble in hot water, so their removal is not difficult. Immersion in a tank of hot water for a few minutes will give satisfactory results, provided that the water is kept really hot. Usually, it is better to remove flux residue while the flux is still hot. Blasting with sand or grit is also an effective method of flux removal, but this procedure cannot be utilized if the surface finish must be maintained. Such drastic treatment seldom is necessary.

Fluxless brazing. Obviously, both the application of brazing flux and the removal of flux residues involve significant costs, particularly where complex joints and assemblies are involved. Consequently, a large amount of work has been devoted to the development of procedures whereby flux is not required, particularly for brazing aluminum. This work has been spurred by the obvious advantages of aluminum as a lightweight and excellent heat conductor for use in heat-transfer applications such as radiators in automobiles, where weight reduction is of increasing importance.

The brazing of aluminum is complicated by the refractory oxide surface film, its low melting point, and its high galvanic potential. However, the successful fluxless brazing of aluminum has been achieved by employing rather complicated vacuum-furnace techniques, utilizing vacuums up to 0.0013 Pa (1×10^{-5} torr). Often a "getter" metal is employed to aid in absorbing the small amount of oxygen, nitrogen, and other occluded gases that remain in the "vacuum," or that may be evolved from the aluminum being brazed. The aluminum must be carefully degreased prior to brazing, and the design of the joint is quite critical. Sharp V-edge joints appear to give the best results.

Some success has been achieved in the fluxless induction brazing of aluminum in air, where an aluminum braze metal having about 7% silicon and 2.5% magnesium is used. The resulting magnesium vapor apparently reduces some of the oxide on the surface of the aluminum and thus permits the braze metal to flow and cover the aluminum surface.

Design of brazed joints. Three types of brazed joints are used: *butt, scarf,* and *lap* or *shear*. These, together with some examples of good and poor design details, are shown in Figure 33-6. Because the basic strength of a brazed joint is generally somewhat less than that of the parent metals, desired strength must be obtained by utilizing sufficient joint area. Often, some type of lap joint is employed when maximum strength is required. If joints are made very carefully, a lap of 1 or $1\frac{1}{4}$ times the thickness of the metal can develop strength as great as that of the parent metal. However, for joints that are to be made in routine production, it is best to use a lap equal to three times the material thickness. Full electrical conductivity usually can be obtained with a lap about $1\frac{1}{2}$ times the material thickness.

If maximum joint strength is desired, it is important to have some pressure applied to the parts during heating and until the braze metal has cooled sufficiently to attain most of its strength. In many cases, the needed pressure can be obtained automatically through proper joint selection and design.

In designing joints that are to be brazed, one must make sure that no gases are trapped within the joint. Trapped gas may prevent the filler metal from flowing throughout the joint, owing to pressure developed during heating.

Braze welding. Braze welding differs from straight brazing in that capillary action is not used to distribute the filler metal. Here the molten filler is deposited by gravity as in oxyacetylene gas welding. Because relatively low temperatures are required and warping is minimized, it is very effective for repair of steel parts and ferrous castings. It is also attractive for joining certain cast irons since the low heat does not alter the graphite shape and the process does not require

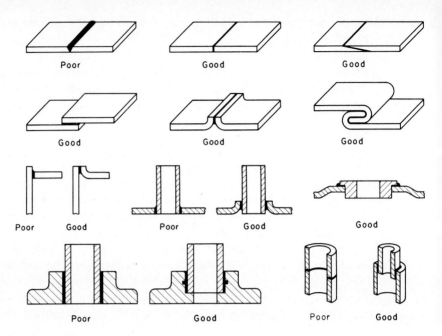

FIGURE 33-6 Examples of good and bad joint design for brazing.

good wetting characteristics. Strength is determined by the braze metal used, and considerable buildup of the braze metal may be required if full strength is required in the repaired part.

Virtually all braze welding is done with an oxyacetylene torch. The surfaces should be "tinned" with a thin coating of the brazing metal before the remainder of the filler metal is added. Figure 33-7 shows an example of a casting being repaired by braze welding.

FIGURE 33-7 Large casting being repaired by braze welding. Four hours were required for preparation of the joint and 3 hours for welding. Twenty pounds of braze metal were used. (*Courtesy Anaconda Brass Company.*)

SOLDERING

By definition, *soldering* is a brazing-type operation where the filler metal has a melting temperature below 450°C (840°F). Bond strength is relatively low, the joining being the result of adhesion between the solder and the parent metal.

Solder metals.   Most solders are alloys of lead and tin with the addition of a very small amount of antimony—usually less than 0.5%. The three most commonly used alloys contain 60, 50, and 40% tin, and all melt below 241°C (465°F). Because tin is expensive, those having higher proportions of tin are used only where the high fluidity is required. For *wiped* joints and for filling dents and seams, as in automobile body work where no strength is required, solder containing only 20 or 30% tin is used.

Other soldering alloys have been developed for various purposes. Tin–antimony alloys are useful in electrical applications. Bismuth alloys have very low melting points. Aluminum is often soldered with tin–zinc, cadmium–zinc, or aluminum–zinc alloys. Lead–silver or cadmium–silver alloys can be used for high-temperature service and indium–tin alloys are useful when joining metal to glass.

Soldering fluxes.   As in brazing, soldering requires that the metal must be clean. Fluxes are used for this purpose, but it is essential that all dirt, oil, and grease be removed before the flux is applied. Soldering fluxes are not intended to, and will not, remove any appreciable amount of contamination.

Soldering fluxes are classified as *corrosive* or *noncorrosive*. A common noncorrosive flux is rosin[1] in alcohol. This is suitable for copper and brass and for tin-, cadmium-, or silver-plated surfaces, if the surfaces are clean. Aniline phosphate is a more active noncorrosive flux, but it has limited use because it gives off toxic gases. In addition to being suitable for copper and brass, it can be used on aluminum, zinc, steel, and nickel.

The two most commonly used corrosive-type fluxes are muriatic acid and a mixture of zinc and ammonium chlorides. Acid fluxes are very active but are highly corrosive. Chloride fluxes are effective on aluminum, copper, brass, bronze, steel, and nickel if no oil is on the surface.

Heating for soldering.   Although any method of heating that is suitable for brazing can be used for soldering, furnace and salt-bath heating are seldom used. Dip soldering is used extensively for soldering wire ends, particularly in electronics work, for automobile radiators, and for tinning. Induction heating is used extensively where large numbers of identical parts are to be soldered. However, a large amount of soldering still is done with electric soldering irons or guns. The principal requisites of these are that they have sufficient heat capacity and that the surface that is held against the work be flat and well tinned so as to assure good heat transfer. For low melting point solders, infrared heat sources may be employed.

As in brazing, the joints can be preloaded with solder, or the filler metal can

---

[1]Rosin left after distilling turpentine.

be supplied from a wire. The method of heating usually determines which procedure is used.

Design and strength of soldered joints. Soldered joints seldom will develop shear strengths in excess of 1.72 MPa (250 psi). Consequently, if appreciable strength is required, soldered joints should not be used or some type of mechanical joint, such as a rolled-seam lock joint, should be made prior to soldering. Butt joints should never be used, and designs where peeling action is possible should be avoided.

In making soldered joints, the parts must be held firmly so that no movement can occur until the solder has cooled well below the solidification temperature. Otherwise, the joint will be full of cracks and have very little strength.

Flux removal. The flux usually must be removed from soldered joints, either to prevent corrosion or for the sake of appearance. Flux removal usually is easily accomplished if the type of solvent used in the flux is known. Water-soluble fluxes can be removed with hot water and a brush. Alcohol will remove most rosin fluxes. When the flux contains a grease, as in most paste fluxes, a grease solvent can be used, followed by a hot-water rinse.

ADHESIVE BONDING

Tremendous advances have been made in recent years in the development, use, and reliability of adhesive bonding. Its use, even in such critical applications as automobiles and aircraft, is increasing rapidly. The adhesives used are thermoplastic and thermosetting resins, several artificial elastomers, and some ceramics. Both metals and nonmetals can be joined satisfactorily.

Adhesive materials and properties. Structural adhesives are actually composite systems with several components and may be available as liquids, pastes, solids, pellets, cartridges, tapes, or films. Commonly used adhesives include:

1. *Epoxies.* The bonding system here includes a resin and curing agent, additives, and a curing operation. One-component epoxies require elevated temperature curing; two-component epoxies generally cure at room temperature.
2. *Cyanoacrylates.* These are liquid monomers that polymerize when spread into a thin film between surfaces, thereby providing a one-component system that cures at room temperature. Most of the consumer-oriented "super glues" are of this variety.
3. *Anaerobics.* These one-component, room-temperature-curing, polyester acrylics remain liquid when exposed to air, but cure rapidly when removed from oxygen (as in a joint to be bonded). Rubber-modified anaerobics remove odor, flammability, and toxicity, while speeding the curing operation. They can bond almost anything, including oily surfaces.
4. *Acrylics.* Most of these involve systems where a catalyst primer is first

applied to the surfaces to be bonded. Then, when the adhesive is applied, it acts like an anaerobic, curing to a strong thermoset bond at room temperature, or faster if heated.

5. *Urethanes or polyurethanes.* Both one-part thermoplastic and two-part thermosetting systems are available.
6. *Silicones.* While generally used for their sealing characteristics, these materials can form low-strength structural joints.
7. *High-temperature adhesives.* Where strength must be retained at temperatures over 250°C (500°F), structural adhesives such as epoxy phenolics, modified silicones or phenolics, polyamides, or ceramics are used.
8. *Hot melts.* While generally not considered to be structural adhesives, these materials are being used increasingly to transmit loads between pieces. These are thermoplastic materials which are solid at room temperature, but change to a liquid when heated. They are generally applied as heated liquids and form a bond as the molten adhesive cools.

Table 33-2 lists some of the popular structural adhesives with their service and curing temperatures and expected strengths.

Joint design and preparation.   Adhesive-bonded joints often are classified as either continuous surface or core-to-face. In *continuous-surface bonds,* both of the adhered surface areas are relatively large and are of the same size and shape. *Core-to-face bonds* have one adhered area that is very small compared with the other, as in bonding lightweight honeycomb core structures to the face sheets. In designing a bonded joint, consideration must be given to the types of stress to which it will be subjected. As shown in Figure 33-8, these are tension, shear, cleavage, and peel. Because, as noted in Table 32-2, most of the adhesives are much weaker in peel and cleavage, it is important that as much of the stress as possible be shear or tension. In this way, all of the bonded area will share the load equally.

As noted in Table 33-2, the lap shear strengths of some of the common adhesives range from 13.8 to 41.4 MPa (2000 to 6000 psi) at room temperature and their tensile strengths from 4.14 to 8.28 MPa (600 to 1200 psi). It thus is evident why joints should be designed to utilize the superior lap shear strengths whenever possible. Figure 33-9 shows some of the commonly used types of joints. Additional strength can be obtained by increasing the bond area.

To obtain satisfactory and consistent quality in adhesive-bonded joints, it is essential that the surfaces be prepared properly, that a standard procedure be established, and that adequate and frequent checks be made to assure that it is being followed. A four-step procedure customarily is used:

1. *Cleaning.* All contaminants and grease *must* be removed.
2. *Etching.* The surface must be made chemically receptive to the adhesive primer and provide maximum wetting characteristics.
3. *Rinsing.*
4. *Drying.*

A low-viscosity primer may be applied, in one or more coats, by spraying or

TABLE 33-2 Some Common Adhesives with Their Preparation, Service Temperatures, and Strengths

Chemical Type	Cure Temperature [°C (°F)]	Service Temperature [°C (°F)]	Lap Shear Strength[a] [MPa at °C (psi at °F)]	Peel Strength at Room Temperature [N/cm (lb/in.)]
Butyral-phenolic	135–177 (275–350)	−51–79 (−60–175)	17.2(2500) at RT 6.9 at 79 (1000 at 175)	17.5 (10)
Epoxy (room-temperature cure)	16–32 (60–90)	−51–82 (−60–180)	17.2(2500) at RT 10.3 at 82 (1500 at 180)	7.0 (4)
Epoxy (elevated-temperature cure)	93–177 (200–350)	−51–177 (−60–350)	17.2(2500) at RT 10.3 at 177 (1500 at 350)	8.8 (5)
Epoxy-nylon	121–177 (250–350)	−251–82 (−420–180)	41.4(6000) at RT 13.8 at 82 (2000 at 180)	122.6 (70)
Epoxy-phenolic	121–177 (250–350)	−251–260) (−420–500)	17.2(2500) at RT 6.9 at 79 (1000 at 175)	17.5 (10)
Neoprene-phenolic	135–177 (275–350)	−51–82 (−60–180)	13.8(2000) at RT 6.9 at 82 (1000 at 180)	26.3 (15)
Nitrile-phenolic	135–177 (275–350)	−51–121 (−60–250)	27.6(4000) at RT 13.8 at 121 (2000 at 250)	105.1 (60)
Polyimide	288–343 (550–650)	−251–538 (−420–1000)	17.2(2500) at RT 6.9 at 538 (1000 at 1000)	5.3 (3)
Urethane	24–121 (75–250)	−251–79 (−420–175)	17.2(2500) at RT 6.9 at 79 (1000 at 175)	87.6 (50)

[a]RT, room temperature.

brushing. After the primer has dried, the adhesive is applied, usually in liquid or paste form. If the adhesive contains a solvent, most of it must be removed before the joint is closed.

When elevated temperatures are required for curing thermosetting adhesives, heat lamps, ovens, heated-platen presses, or autoclaves are employed, depending on the conditions.

For added strength, systems have been developed to combine structural adhesives and spot welding.

Advantages and disadvantages of adhesive bonding. Adhesive bonding has a number of obvious advantages. Almost any material or combination of materials can be joined. For most adhesives, the curing temperatures are quite low,

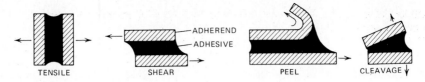

FIGURE 33-8 Types of stresses in adhesive-bonded joints.

seldom exceeding 177°C (350°F), and a substantial number that cure at room temperature, or only slightly above, will provide adequate strength for many applications. Very thin and quite delicate materials, such as foils, can be joined to each other or to heavier sections. Because continuous bonding can be obtained throughout a joint, good load distribution and fatigue resistance can be obtained. Similarly, because of the large amount of contact area that usually can be obtained, the total joint strength compares favorably with that resulting from other methods of attachment. Smooth contours are obtainable, and no holes have to be made, as with riveting or bolting. The adhesive can provide thermal and electrical insulation, act as a vibration damper, and provide protection against galvanic action where dissimilar metals are joined. Surface preparation may be reduced, since bonding can occur with an oxide film in place and rough surfaces are beneficial because of the increased contact area. Cost savings can result from simplified machining and assembly, reduced finishing requirements, elimination of mechanical fasteners, and the absence of highly skilled labor. Adhesives are generally inexpensive and weigh less than fasteners.

The major disadvantages of adhesive bonding are: (1) most of the adhesives are not stable above 177°C (350°F), although a few are quite good up to 260°C (500°F); (2) it is difficult to determine the quality of an adhesive-bonded joint by nondestructive means, although some methods give fairly good results for some types of joints; (3) surface preparation, adhesive preparation, and curing procedures are quite critical if good and consistent results are to be obtained; (4) life expectancy is hard to predict; (5) assembly time is often greater than for

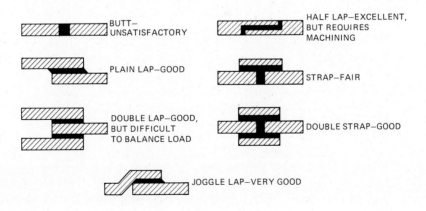

alternative methods; and (6) some adhesives contain objectionable chemicals or solvents. Nevertheless, the extensive and successful use of adhesive bonding is proof that these factors can be overcome if adequate quality-control procedures are adopted and followed.

While the unit strengths that can be obtained by adhesive bonding are relatively low, this is not a major disadvantage or limitation in most cases, since sufficient areas usually can be utilized if joints are designed properly.

MECHANICAL FASTENING

Often assembly in manufacturing can best be performed by some type of mechanical fastening. These may be in the form of integral fasteners; discrete fasteners such as bolts, screws, and rivets; or shrink fits. Here the joint is formed by mechanical interlocking or interference and no fusion or bonding of the parts is required.

Integral fasteners are formed areas of the component that interfere or interlock with other areas of the assembly and are commonly employed with sheet metal products. Examples include lanced or shear-formed tabs, extruded hole flanges, embossed protrusions, edge seams, and crimps. The common beverage can includes several of these types: an edge seam to join the top to the can body and an embossed protrusion which is subsequently fitted to attach the pull-ring.

Discrete fasteners are separate pieces whose primary function is to join the components and include such items as bolts, screws, nuts (with accessory washers, etc.), rivets, quick-release fasteners, staples, and wire stitches. Over 150 billion such fasteners are consumed annually, 27 billion alone by the automotive industry. A typical railroad box car requires 1200 mechanical fasteners; a numerically controlled turret lathe, 1700; and a standard telephone, over 70. The variety here is so immense that the major challenge is to select an appropriate fastener for the task at hand, and, if possible, an optimum fastener.

Shrink and expansion fits form the third class of mechanical joining. Here a dimensional change is introduced to one or both of the parts by heating or cooling (heating one part only, heating one and cooling the other, or cooling one), assembly is effected, and a strong interference fit is established when temperature uniformity resumes. Joint strength is exceptionally high. In addition, such a procedure can be used to produce a prestressed condition in a weak, low-cost material and enable it to replace a more costly, stronger one. Similarly, a corrosion-resistant cladding or lining can be imparted to a less costly bulk material. *Press fits* produce essentially the same results with mechanical force producing the assembly, not differential temperatures.

Of the three forms of mechanical fastening, discrete fasteners are, by far, the most utilized.

Problems with fasteners. When a product has fastened joints, the fasteners become a vulnerable site for failure. The cause of such failures generally relates to one of four areas: (1) the design of the fastener and the manufacturing techniques used to make it, (2) the material from which the fastener is made, (3) joint design, or (4) the means and details of installation. Fasteners may have insufficient strength or corrosion resistance, or may be subject to hydrogen embrittlement. Installation may have provided too much or too little preload. The joint surfaces may not be flat or parallel and the area under the head may be insufficient to bear the load. Vibration loosening and fatigue are common causes of failure.

Nearly all fastener failures can be avoided by proper design and fastener selection. Consideration should be given to the operating environment, required strength, and magnitude and frequency of vibration. The need for weight savings and the possible desire for disassembly will influence decisions. Standard fasteners should be used whenever possible and as little variety as necessary should be specified.

Fastener design should consider a shank-to-head fillet, rolled threads, and corrosion-resistant coatings. Joint design should seek to avoid such features as offsets and oversized holes. Proper tightening should be assured.

REVIEW QUESTIONS

1. How does brazing differ from welding?
2. Why might brazing be used for joining thinner material than welding?
3. How does the strength of a brazed joint depend on the clearance between the pieces?
4. What are some of the more commonly used brazing metals?
5. What cautions should be exercised when using silver solder to braze stainless steels?
6. Why might it be desirable to use a braze metal of composition similar to the base metal?
7. What are the three functions of a flux in brazing?
8. What are some of the drawbacks of the torch-brazing procedure?
9. Why might a controlled atmosphere or a vacuum be employed in a furnace brazing operation?
10. What are some of the attractive features of induction brazing?
11. Why is it usually important to remove all flux residues?
12. What is the major advantage of fluxless brazing?
13. Why is it important to vent gases from a brazing joint?
14. How does braze welding differ from straight brazing?
15. How does soldering differ from brazing?
16. What alloy system contains the majority of solders?
17. Why should soldering not be employed where appreciable strength is desired in a joint?
18. What are some of the available forms of structural adhesives?
19. How do anaerobic (absence of air) adhesives work?
20. Why is it best for adhesive joints to be designed to be in pure shear or pure tension?
21. What are some of the attractive features of adhesive bonding?
22. What are the three major types of mechanical fastening?
23. How do press fits differ from shrink or expansion fits?
24. What are some common reasons for fastener failures?

CASE STUDY 33.   The Industrial Disposal Impeller

The impeller of a large industrial disposal unit is made of a circular segment of hot-rolled steel plate. Pieces of tungsten carbide are brazed into recesses in the plate with a copper–base brazing alloy as shown in Figure CS-33. In service, the impeller rotates at approximately 700 rpm, shredding the refuse (paper, bottles, chemicals, etc.) to a pulp,

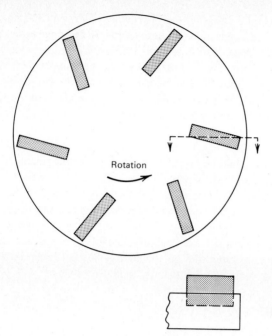

FIGURE CS-33 Schematic of impeller design.

eliminating all identification and permitting discharge into a refuse dump. After 5 weeks of service, the impeller must be removed. Several pieces of tungsten carbide have broken free and are chewing up the unit. Large craters have formed in the steel plate adjacent to the carbide inserts.

What do you suspect was the cause of the problem? How would you alter the design, materials, or fabrication to prevent its recurrence?

CHAPTER 34

Potential Problems in Welding and Cutting

It has been observed that the majority of engineering students enter industry without so much as a basic understanding of the problems inherent in welding and frequently encountered in its use. Welding is a unique process and should only be used when proper consideration is given to its particular characteristics and requirements. Joint design is critical to the successful application of welding. Process selection is a complex procedure due to the large number of available processes, the large number of possible joint configurations, and the numerous parameters that must be specified for each operation. Heating, melting, and resolidification can cause numerous problems, including drastic modification of material properties. Metal properties may further change due to dilution of the filler by base material, vaporization of various elements, or a variety of gas–metal reactions.

Common types of weld defects include: cracks of a variety of types, cavities (both gas and shrinkage), inclusions (slag, flux, and oxides), incomplete fusion between the weld and base metals, incomplete penetration (insufficient weld depth), unacceptable weld shape or contour, arc strikes, spatter, undesirable metallurgical changes (aging, grain growth, or transformations), and excessive distortion.

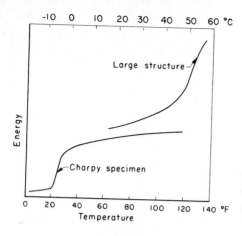

FIGURE 34-1 Effect of size on the energy-absorbing ability (*transition temperature*) of a steel specimen.

DESIGN CONSIDERATIONS

Welding is a unique process that cannot be substituted directly for another method of fastening without proper consideration being given to its peculiar characteristics and requirements. Unfortunately, welding is so easy and convenient to use that these facts are often overlooked—and the cause of many so-called welding failures can be traced to such negligence. Proper design must be employed.

One very important factor that must be kept in mind in the use of welding is that it produces monolithic, or one-piece, structures. When two pieces are welded together, they become one piece. This can cause complications if not properly taken into account. For example, a crack in one piece of a multipiece structure may not be serious, because it seldom will progress beyond the single piece in which it occurs. However, when a large structure, such as a welded ship, a pipeline, storage tank, or a pressure vessel, consists of only one piece, a crack that starts in a single plate or weld may progress for a great distance and cause complete failure.[1] Obviously, such a situation is not the fault of welding but, rather, a failure to take that simple fact into account when designing the structure.

Another factor that relates to design is that a given material in small pieces may not behave as it does in a large piece, as is often produced by welding. The importance of this fact is illustrated vividly in Figure 34-1, which shows the relationship between energy absorption and temperature for the same steel when tested in a Charpy impact specimen and in a large, welded structure. In a Charpy bar the material exhibited ductile behavior and good energy absorption at temperatures down to −4°C (25°F), but when welded into a large structure it exhibited ductile behavior down to only 43°C (110°F). Thus the notch-ductility

---

[1]Propagation velocities up to 1524 meters (5000 feet) per second and distances of 304.8 meters (1000 feet) have been recorded.

characteristics of steel to be used in large welded structures may be of great importance, and ordinary specifications for steel frequently do not control this quality. More than one welded structure has failed because the designer did not take this fact into account.

Another common error is to make structures too rigid by welding, thereby restricting their ability to redistribute high stresses and avoid failure. Considerable thought may be required to design structures and joints that permit sufficient flexibility, but the multitude of successful welded structures attest to the fact that it can be done.

Accessibility of the joint for welding, welding position, component matchup, and the specific nature of the joint are all elements of design considerations.

HEAT EFFECTS

Welding metallurgy. Heating and cooling are essential and integral components of all welding processes (except cold welding) and tend to produce metallurgical changes that usually are not desired. In fusion welding, the heating is sufficient to produce some melting, which is then followed by rapid cooling. The thermal effects tend to be most pronounced for this type of welding, but also exist to a lesser degree in other types of welding, where the heating–cooling cycle is less severe. However, if these thermal effects are considered properly, adverse results can be avoided, and excellent service performance can be obtained through the use of welding. If they are overlooked, the results can be disastrous—the failure being that of the designer or fabricator, not of the welding process.

Because such a wide range of metals is welded and a variety of processes is used, welding metallurgy is an extensive subject. However, a few basic considerations will enable anyone who has a reasonable working knowledge of metallurgy to understand the changes in microstructure that occur. In fusion welding a pool of molten metal, either from the parent plate or from both the parent plate and an electrode or filler rod, is created. This pool is contained in a metal mold, formed by the parent plate, and is generally very small compared to the surrounding metal. Thus the process, metallurgically, is one of *casting* a small amount of molten metal into a metal mold, and the resulting metallurgical and strength characteristics can be anticipated and explained on this basis.

Figure 34-2 shows a typical microstructure that results from a fusion weld. In the center of the weld is a zone that is made up primarily of weld metal that has solidified from the molten state. Actually, it is a mixture of parent metal and electrode or filler metal, the ratio depending on the welding process used, the type of joint, and the edge preparation. This zone is cast metal with a microstructure reflecting the cooling rate in the weld, and, as such, it cannot be expected to have the same properties as the *wrought* parent metal. It can achieve equal properties only through the addition of filler metal which, in the *as-cast* condition, has equal or superior properties. Thus one should select filler rods or electrodes which, in the *as-deposited* condition, have properties that equal, or

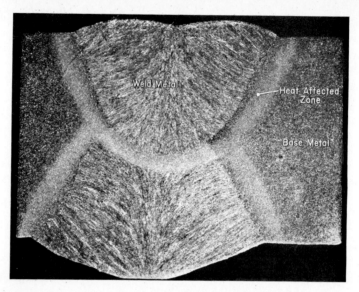

FIGURE 34-2 Grain structure and zones in a fusion weld.

are superior to, those of the parent metal. This requirement is the basis for several AWS specifications for electrodes and filler rods.

The grain structure in the weld metal zone may be fine or coarse and equiaxed or dendritic, depending on the type and volume of weld metal and the cooling rate. Most electrode and filler rod compositions tend to produce fine, equiaxed grains, but the volume of weld metal and the cooling rate can easily defeat these objectives.

Thus, fusion weldments are prone to all of the problems and defects associated with metal casting, such as gas porosity, inclusions, blowholes, cracks, and shrinkage. Because the amount of metal that is molten is usually small relative to the total mass of the workpiece, rapid solidification and rapid cooling of the solidified metal are quite common. Associated with this may be the inability to expel dissolved gases, chemical segregation, grain-size variation, grain shape problems, and orientation effects.

Adjacent to the weld metal is the ever-present, and generally undesirable, *heat-affected zone* (HAZ). In this region, the parent metal is not melted, but is subjected to elevated temperatures for a brief period of time. Since the temperature and its duration varies widely with location, welding might be more appropriately described as "a casting and an abnormal, widely varying, heat treatment." The adjacent, unmelted metal often experiences enough heat to bring about structure and property modifications, such as phase transformations, grain growth, precipitation, embrittlement, or even cracking. The variation in thermal history produces a variety of microstructures and a range of properties. In plain carbon steel, the structures may range from regions of hard martensite to rather weak, coarse pearlite.

The heat-affected zone no longer possesses the desirable properties of the parent material, and since it was not molten, it does not have the selected properties of the weld metal. Consequently, it is often the weakest area in a weld *in the as-welded condition*. Except where there are obvious defects in the weld deposit, most welding failures originate in the heat-affected zone. Outside the heat-affected zone is the base metal that is not affected by the welding process.

It is apparent that the structure and properties of a weld are complex and varied. Associated problems, however, can be reduced or eliminated in several ways. First, consideration should be given to the thermal characteristics of the various processes. Table 34-1 classifies some of the more common welding processes as to their rate of heat input. Processes with low rates of heat input (slow heating) tend to produce a high total heat content in the metal, slow cooling rates, and a large heat-affected zone. High-heat-input processes have low total heats, fast cooling rates, and small heat-affected zones. The size of the heat-affected zone will also increase with increased starting temperature, decreased welding speed, increased thermal conductivity of the base metal, and a decrease in base metal thickness. Weld geometry is also important; fillet welds produce smaller heat-affected zones than butt welds.

When the results are unacceptable the entire piece may be heat treated after welding. Much of the variation can be reduced or eliminated but the results are limited to structures that can be produced by heat treatment. In addition, numerous problems may occur in producing controlled heating and cooling in the often large, complex-shaped structure typically produced by welding.

Another procedure that can reduce the variation in microstructure, particularly the sharpness of the variation, is to preheat the base metal adjacent to the weld just prior to welding. For plain carbon steels, a temperature of 93 to 204°C (200 to 400°F) usually is adequate. This procedure reduces the cooling rate of the weld deposit and the immediately adjacent metal in the heat-affected zone, producing a more gradual change in microstructures and thereby eliminating a metallurgical stress raiser.

TABLE 34-1 Rates of Heat Input for Common Welding Processes

Low heat input	High rate of heat input
Oxyfuel welding	Plasma-arc welding
Electroslag welding	Electron beam welding
Flash welding	Laser welding
Moderate heat input	Spot and seam resistance welding
Shielded metal-arc welding	Percussion welding
Flux cored-arc welding	
Gas metal-arc welding	
Submerged-arc welding	
Gas tungsten-arc welding	

If the carbon content of plain carbon steels is greater than about 0.3 percent, the cooling ranges encountered in normal welding are sufficient to cause hardening and loss of ductility, unless precautionary measures are taken. This also is true for many alloy steels, because of their high hardenability. Thus special pre- and post-welding heat cycles must be used when such steels are welded. It is for this reason that the various weldable, low-alloy steels have gained so much acceptance, since they usually can be welded without the necessity for preheating or postheating.

Where little or no melting occurs and there is considerable pressure applied to the heated metal, as in forge or resistance welding, the weld may retain some of the characteristics of wrought material.

The discussion thus far has largely considered the metallurgical effects related to the welding of steel. The effects of heating and cooling accompanying the welding of other metals will, of course, depend on the transformations and changes that can occur in such metals as they are subjected to the heating and cooling cycles that accompany the welding process.

Thermal stresses.   Another effect of the heating and cooling that accompanies welding is the introduction of residual stresses. (See Chapter 39 for further discussion.) In welding, these may be of two types and are most pronounced in fusion welding, where maximum heating occurs. Their effect may be observed in the form of dimensional changes, distortion, or even cracking.

Residual welding stresses are the result of restraint to thermal expansion and contraction *offered by the pieces being welded.* They may exist independent of whether the pieces are attached to other portions of the structure or are restrained in any manner. The way in which these residual stresses occur may be explained by means of Figure 34-3. As the weld is made, the liquid region conforms to the "mold" shape, and the adjacent metal becomes hot and undergoes solid expansion. Expansion perpendicular to the weldment can be absorbed by flow of the molten pool, but expansions parallel to the weld line tend to be restrained by the cooler, stronger parent metal in region C. This resistance can be sufficient to induce plastic deformation of the heat-affected zone. The material is *upset,* accommodating the increase in volume by becoming thicker instead of longer.

After the weld metal solidifies, the weld pool and adjacent heat-affected region cool and contract. This contraction is also resisted by the cooler surrounding

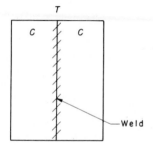

FIGURE 34-3 Conditions that cause residual welding stresses.

material. The cooling metal wants to contract but is restrained in a "stretched" condition, known as residual tension (region T). Likewise, the forces exerted by this contracting region on the adjacent metal bring about a "squeezed" state known as residual compression. While the net force is zero, in keeping with the basic laws of physics and mechanics, the localized forces can be substantial.

Thermal contractions occur both parallel (longitudinal) and perpendicular (transverse) to the weld. However, the transverse stresses may be relieved by movement of the base material. That is, the thermal contraction simply results in the welded assembly being shorter when cool than when it was welded. If the base metal is restrained from movement, however, significant thermal stresses can be induced and these are known as *reaction stresses*. The magnitude of the reaction stresses is primarily an inverse function of the length of material between the weld joint and point of maximum rigidity. It can never exceed the yield strength of the parent metal and generally is less than the magnitude of the stresses parallel to the weld, seldom exceeding 10.3 MPa (15,000 psi) in steel.

The longitudinal residual stresses are a maximum in the weld metal, being slightly above the yield strength of the parent plate, depending on the thickness of the metal and the amount and type of weld metal. For example, in a butt-fusion weld of 25.4-mm (1-inch) mild-steel plate, having a yield strength of 221 MPa (32,000 psi), the longitudinal residual stresses had a maximum value of 331 MPa (48,000 psi). These stresses can be greater than the yield strength of the plate because the weld metal, in the as-deposited condition, has a yield strength of about 359 MPa (52,000 psi), being designed to be stronger than the plate. The resulting weld, being a mixture of weld metal and plate metal, has properties between those of the pure weld metal and the wrought plate metal. The longitudinal residual stresses decrease rapidly away from the weld and become compression stresses of low magnitude—usually not exceeding 34.5 MPa (5000 psi).

Figure 34-4 depicts the typical distribution of residual stresses in a butt-fusion weld.

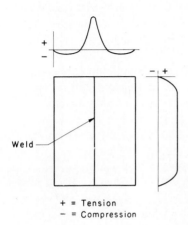

+ = Tension
− = Compression

FIGURE 34-4 Typical pattern of residual stresses in a butt weld.

Effects of thermal stresses. The effects of thermal stresses resulting from welding have been widely misunderstood. The most obvious result is that they can cause distortion, or warping as the material seeks to reduce the unbalanced stresses.

No fixed rules can be given for avoiding warping because the possible conditions that cause it are widely varied. Several procedures may be employed. In general, a welding sequence should be used that permits the plates to have as much freedom as possible during the making of each weld. If possible, it is beneficial to weld toward the point of greatest freedom. Alternatively, the plates may be prepositioned so that the resulting distortion leaves them in the desired shape. Another common procedure is to completely restrain the plates during welding, thereby forcing some plastic flow in the plates and/or the cooling weld metal. This procedure is utilized most readily on small weldments. Still another procedure is to balance the resulting thermal stresses by depositing the weld metal in predetermined patterns and areas, often in short lengths. Warping frequently can be minimized by the use of *peening,* which imparts a residual compressive surface stress to counteract the residual tensile stress of the weld. Each pass of multipass welds, except the first and last passes, is hammered with a peening tool to cause a small amount of localized plastic flow and thus induce the needed movement.

As for residual stresses, there is no substantial evidence that they have a harmful effect on the strength performance of weldments *except in the presence of notches or in very rigid structures* where no plastic flow can occur. These are two conditions that should not exist in weldments *if they are properly designed and if proper workmanship is employed.* Unfortunately, welding makes it easy to inadvertently join heavy sections of steel and produce rigid configurations that will not permit the small amounts of elastic or plastic movement required to reduce peak or concentrated stresses. Too often, geometric notches, such as sharp, interior corners, are incorporated in welded structures. Proper design can eliminate both rigidity and geometric notches. Other harmful notches, such as gas pockets, rough beads, porosity, and arc "strikes," can serve as initiation sites for weld failures, but these, too, can be avoided by proper welding procedures, good workmanship, and adequate supervision and inspection.

Residual stresses can cause additional warpage when weldments are machined, so as to unbalance the stress equilibrium. Consequently, weldments that are to undergo appreciable machining are frequently given a stress-relief heat treatment prior to machining.

The reaction stresses constitute a system of stresses that are superimposed on the applied loads to the system. They, too, can cause distortion. However, their most frequent effect is their tendency to cause cracking during or immediately following welding as the weld cools. This is particularly likely when welds are made under conditions where there is great restraint to the normal shrinkage that occurs transverse to the length of the weld. Often, where a multipass weld is being made, such a crack may occur in one of the early beads, when there is insufficient metal to withstand the shrinkage stresses. Such a condition may be serious if the crack goes undetected and is not chipped out and repaired or is not completely rewelded as subsequent beads are deposited. When a weld must be made under restrained conditions, it is good practice to preheat the adjacent metal for a few inches on either side of the weld, as this will often eliminate this type of cracking.

Certain welding codes require some types of weldments to be stress-relieved by heat treatment prior to use. In many cases it appears that the improved performance that results from such treatment is due as much to the improvement in microstructure that results as it is to the reduction of residual stresses.

Summary. By now it should be obvious that control of heating and cooling should be considered when designing and producing weldments if the potential benefits of the process are to be obtained and the harmful side effects avoided. Nevertheless, there are numerous examples, in the form of failures in welded components, which show that designers and fabricators of welded structures often fail to give these factors proper consideration.

Flame and arc cutting also involve localized heating and subsequent cooling, and thus often experience similar problems. In general, however, the problems there are less severe and less extensive.

Welding is an excellent fabricating process but, if satisfactory results are to be obtained, it cannot be used thoughtlessly, any more than can the rivets not be headed in a riveted connection or the nuts be left off bolts in a bolted connection. Before welding is adopted and used in manufacturing a product or structure, one should be sure that it has been designed properly for welding.

REVIEW QUESTIONS

1. What are some common types of weld defects?
2. Why is it important to consider welded products as monolithic structures?
3. How might the properties of a large piece of metal differ from those of a small piece?
4. How might excessive rigidity actually be a liability in a welded structure?
5. Describe the cast segment of a fusion weld.
6. On what basis are electrode or filler rod materials often selected?
7. What types of defects may occur in a weldment due to casting problems or rapid cooling after solidification?
8. Why do properties vary widely in most welding heat-affected zones?
9. What types of structure and property modifications can occur in heat-affected zones?
10. Why do most welding failures originate in the heat-affected zone?
11. Describe the cooling rates and size of heat-affected zone for processes with low rates of heat input.
12. What difficulties or limitations might be encountered in heat treating products after welding?
13. What is the purpose of pre- and post-welding heat cycles?
14. What causes residual welding stresses?
15. What are some of the observed effects of thermally induced stresses in welding?
16. Under what conditions do residual stresses exert a harmful effect on the performance of weldments?
17. Why might it be desirable to stress relieve a weldment prior to machining?

CASE STUDY 34. The Cracked Propeller

A ship propeller, identical to the one in Case Study 15, is again damaged by impact with a rock. This time, however, the damage is a crack at the base of one of the blades, as shown in Figure CS-34. Since the crack does not penetrate to the hub, a repair is suggested using a welding or brazing technique.

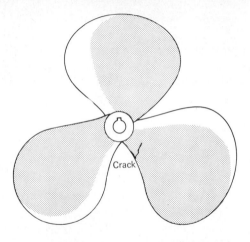

FIGURE CS-34 Ship propeller showing location of a crack at the base of one of the blades.

How would you suggest that such a repair be made? Provide details and justification for your solution. What, if any, sacrifice in quality or performance will accompany such a repair?

PART V

PROCESSES AND TECHNIQUES RELATED TO MANUFACTURING

CHAPTER 35

Layout

After the configuration of a part is set forth on a drawing, the dimensions that determine the size and location of its various geometrical surfaces must be established on the workpiece. The process of establishing these dimensions is called *layout*. It probably is most evident in the cases of machining and pattern making, but it also is involved in pressworking, welding, and other processes.

In most cases, layout involves transferring to a workpiece dimensions that are shown on a drawing. However, in some cases no drawings may be involved; the dimensions are only in someone's mind. Increasingly in modern manufacturing, the dimensions may be transferred from a person's mind to a punch-tape or a computer and then from the tape or computer memory to the machine tool. These procedures will be discussed in Chapter 38.

Designers and draftsmen should be aware of their relationship to layout. A drawing that is designed and dimensioned so that it can easily be converted into reality can have a marked effect on reducing costs and, indirectly, quality. Unfortunately, sufficient examples exist to keep alive the belief among many shop people that the designers' motto is: "We design it; let the shop find a way to make it."

When only one or a few workpieces are involved, layout usually is accomplished manually, either directly on the workpiece or through the use of the controls available on the machine tools that are used. Digital controls make this latter procedure easier. When larger quantities are to be made, layout often is accomplished through the use of jigs and dies. These will be discussed in Chapter 36, but layout is an extremely important factor in their construction. If numerically controlled machines are used in manufacturing, even more attention may have to be given to the layout dimensions so that the required operations and sequences are compatible with the machine capabilities.

In order to understand the problems associated with layout, consider the simple part depicted in Figure 35-1. All the required finished surfaces can be produced without difficulty. Dimensions *e*, *f*, *g*, and *h* can be obtained by direct surface-to-surface measurements, using ordinary measuring instruments. However, before the two holes can be machined, dimensions *a*, *b*, *c*, and *d* have to be established accurately on the workpiece. These are called *location dimensions* because they determine the location of geometrical shapes—in this case holes—with respect to another geometrical shape, the rectangular solid. Two methods for accomplishing this layout—manual and machine—will now be discussed.

MANUAL LAYOUT METHODS

The required dimensions *a*, *b*, *c*, and *d* can be established with a fair degree of accuracy by simple manual methods. In one method, surfaces *A* and *B* would be coated with a substance that could easily be scratched with a fine scribing point. If only fair accuracy were required, chalk rubbed on the surface might be satisfactory. However, because the chalk is likely to be rubbed from the surface in handling the workpiece, a more permanent coating usually is used. Several commercial colored lacquers are available that can be brushed or sprayed on the surface to form a very thin coating that will remain until it is removed with a lacquer thinner. These lacquers dry quickly, and a scriber will readily scratch through the coating, leaving a fine, easily seen mark.

The next step usually would be to set the workpiece on a surface plate or on a pair of parallel bars, as shown in Figure 35-2. For accurate work, a vernier height gage with a scribing attachment would be set to the desired dimension *c*, or *d*, and slid along the surface plate with the scribing point in contact with

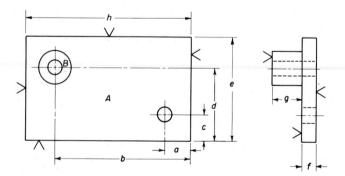

FIGURE 35-1   A part, showing dimensions that must be established on the workpiece in order to locate two holes.

PROCESSES AND TECHNIQUES
RELATED TO MANUFACTURING

FIGURE 35-2 Scribing layout lines on a workpiece by means of a vernier height gage and a pair of dividers.

the coated work surface, as shown in Figure 35-2, thus scribing a fine line at the desired distance from the edge that was resting on the surface plate or parallel bars. By thus using a vernier height gage, or a height-gage attachment with gage blocks, lines can be located to an accuracy of about ±0.02 mm (0.001 inch) if care is exercised. If less accuracy is required, dividers can be used to establish the desired line, as also shown in Figure 35-2. If even less accuracy is required, a combination rule and square head may be used (see Figure 16-11), with the square head held against one finished side of the workpiece.

Dimensions *a* and *b* could be established in a similar manner by standing the block on end.

Unfortunately, although the centers of the two holes can be established within 0.02 mm (0.001 inch) by the procedure described, this does not assure that the centers of the actual holes will be located this accurately. Consideration of the operations that are required to produce the holes makes it clear that such probably would not be true. To produce the holes to accurate size would require the following steps:

1. Center-punch the intersections of the layout lines.
2. Center-drill at the center-punched locations.
3. Drill the holes to approximate size. *pilot drilling*
4. Ream the drilled holes to final size.

Several factors in these steps tend to reduce the accuracy of the location of the centers. First, it is difficult to locate and hold the point of the center punch exactly at the intersection of the two fine, scribed layout lines. Second, when the center punch is struck, local surface irregularities and the nonhomogeneity of the metal in the workpiece may tend to cause the center punch to drift slightly from the desired location. It is true that a skilled worker can, after the first light striking of the center punch, by additional blows correct this drifting to some extent, but one cannot expect to have the final center-punch mark located more accurately than about 0.05 to 0.08 mm (0.002 to 0.003 inch).

Third, the center-drilling operation will usually decrease the accuracy of location to some extent. Again it is difficult to locate the mark from the center punch exactly under the center of the center drill. Also, the center drill may not be perfectly accurate in all respects, and it may wobble slightly.

The drilling operation may add to the inaccuracy. The drill may not start drilling exactly in the center of the center-drilled hole, and it may be deflected slightly as it drills into the workpiece. Thus, although the layout lines might be accurate to 0.02 mm (0.001 inch), the actual center of the hole may easily be off by as much as 0.13 mm (0.005 inch). Further, there is little assurance that the axis of the drilled hole is exactly normal to the surface on which the work was resting when the drilling was done, as a drill is slightly flexible and it can be deflected from the desired axis.

Similar difficulties may be encountered in producing other geometric surfaces at desired locations. Thus it is apparent that the problem of translating specified dimensions from a drawing into actual locations of component parts of a finished product is not only an important one but also a complicated one, and it is a matter that is of direct concern to the designer.

Toolmakers buttons. *Toolmaker's buttons* are a means for considerably improving the manual layout for holes. The essential features of these devices are shown in Figure 35-3. They are small, hollow, accurately ground cylinders, about 12.7 mm ($\frac{1}{2}$ inch) high and available in several diameters; the 12.7 mm ($\frac{1}{2}$ inch) size is most common.

To use toolmaker's buttons, the approximate location of the center of the desired hole is established by ordinary layout methods. The location does not have to be established exactly, but it should be within about 0.25 mm (0.010 inch). This location is center-punched, drilled, and tapped for a No. 5-40 screw. A toolmaker's button then is attached to the work by means of a No. 5-40 machine screw and a small washer, the screw being tightened sufficiently to hold the button firm, yet loose enough to permit it to be moved slightly under the washer by light tapping. Then, by using a dial gage or other suitable indi-

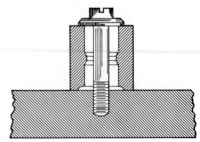

Sectional View of Button Applied

FIGURE 35-3 Toolmaker's button mounted on a workpiece, and sectional view showing the method of attachment.

cating device, the button is adjusted until its *axis* is exactly coincident with the axis of the desired hole. The attaching screw then is tightened to hold the button firmly in this position.

With the button properly located, the work is mounted in a lathe or other suitable machine tool, and located, by means of a dial or wiggler gage, so that the axis of the toolmaker's button is coincident with the axis of rotation of the work (or drill or boring tool if a rotating tool is to be used). The toolmaker's button then is removed. The hole is center-drilled and drilled somewhat under-size. Next, it is bored to approximate size. Because the axis of the bored hole is coincident and properly aligned with the axis of the toolmaker's button, it is exactly at the desired location, and all inaccuracies that might have existed in the drilled hole are eliminated. Such a hole often is finished to exact size by reaming. Although holes may be located to within ± 0.013 mm (0.0005 inch) by means of toolmaker's buttons, it is obvious that the procedure is very labo-rious. It is also evident why machine layout methods, which require less skilled labor, are more economical and are used whenever possible.

MACHINE LAYOUT METHODS

By making use of the controlled movements of their worktables, layout can be accomplished on several machine tools in far less time and with greater accuracy than by hand methods. Referring to the part shown in Figure 35-1, if one of the longer edge surfaces were machined on a vertical milling machine, dimension *e* could readily be established merely by moving the machine table a distance equal to dimension *e* plus the diameter of the cutter. With the usual table controls the dimension could quickly be established to an accuracy of 0.02 mm (0.001 inch). On most machines equipped with digital readout systems, even greater accuracy could be obtained. Similarly, after completion of the lower edge (as pictured), dimension *c* could be established by moving the table a distance equal to *c* plus the radius of the milling cutter.

It is apparent that layout by means of the controlled motions available on a machine tool has a number of advantages over manual layout. It is both quicker and more accurate. Direct measurement is not required. As a consequence, vertical and horizontal milling machines, horizontal drilling, boring, and milling machines, and vertical drilling machines with controlled-movement tables are widely used in this manner. Since such machines tools usually can do several operations, such as milling, drilling, boring, and reaming, they are very flexible and useful because the necessary layout and all required machining operations often can be accomplished with a single setup.

Digital readout equipment. Digital readout devices, such as shown in Figure 35-4 and referred to previously, have markedly improved the controlling of table movements on machine tools, and have made layout on machines equipped with these devices much more rapid and accurate. These devices, usually electro-mechanical or optical-electrical, automatically indicate the position of a work-table in either two or three axis directions. Some operate through the table feed

FIGURE 35-4 Jig borer equipped with digital readout indicators, being used for boring a hole in a fixture. (*Courtesy Bendix Corporation, Automation and Measurement Division.*)

screws; others are connected between the table and the machine bed. Most systems have a "floating zero," meaning that the zero can be set for any position of the table merely by pushing a button or setting a switch. Accuracies and indications of 0.02 and 0.002 mm (0.001 and 0.0001 inch) are most common, although others can be obtained. Some are available that will read in either millimeters or inches by merely throwing a switch.

Such equipment can be used in several ways. One is to control all table movements from a single zero location. Referring to Figure 35-1, for example, the distances c and d, the locations of the two holes on one axis, would be obtained by moving the table the distance c and d, respectively, with the lower edge of the workpiece being at zero.

The use of digital readout equipment not only provides better accuracy and reduces errors in doing layout on a machine tool, but it also speeds the total machining process. The machinist can move the table, and thus the work, any desired distance more quickly without the necessity of counting the turns of a crank or reading graduated dials. He can know at all times the exact location of the machine spindle axis relative to some zero point on the workpiece. Most new machine tools are equipped with digital readout equipment, and many older machines are being retrofitted with such equipment.

Layout on jig borers. Where extremely accurate layout is required, as in making jigs, fixtures, and dies, special machine tools, called *jig borers*, are often used. Essentially, these are very accurately made vertical drilling machines that incorporate four special features:

1. Provision for accurately locating the worktable with respect to the cutting tool in at least two directions in a plane.
2. Oversized spindle, bearings, and quill so that light boring and grinding can be done, as well as drilling.
3. High accuracy in the movement of the critical parts, such as the rotation and feed of the spindle.
4. Provision to avoid the harmful effects of temperature changes.

Because the first three of these features are now routinely incorporated in modern numerically controlled vertical milling machines and machining centers, permitting accuracies of ± 0.0025 mm (0.0001 inch) to be obtained readily, such machines are now used for much of the work that formerly was done only on jig borers. However, if greater accuracy is required, or if a considerable quantity of such high precision work must be done, jig borers generally are used, their use being restricted to high-precision jobs, usually not on large, heavy workpieces. In this manner there is less danger of the accuracy of the machine being lost through abuse.

In jig borers, accurate positioning of the worktable, in two mutually perpendicular directions, is achieved in several ways. One is the use of gage-block-quality end-measuring rods and a dial gage. Another is the use of very accurate lead screws. Most now utilize digital readout equipment that measures directly between the worktable and the machine bed. This method is employed in the jig borer shown in Figure 35-4 and provides rapid location to 0.0013 mm (0.00005 inch). Such a system does not require the table traversing screws to be made with extreme accuracy, and any wear or temperature effects in the screws do not affect the accuracy of the machine.

The jig borer shown in Figure 35-5 is equipped with both digital readout and NC tape control. It is arranged so that a program may be prepared in the usual manner, or one can be prepared automatically by machining one part, using the digital readout system. The tape or program so prepared then can be used to control the machine automatically in machining subsequent, duplicate parts. This feature is particularly valuable where one may want to exactly duplicate a very accurate part at some later time.

Jig borers are built with the finest possible precision. Particular attention is given to the design of the spindle and spindle bearings to assure accurate rotation and feed. The castings used are carefully annealed so that they will not warp with age. In several makes certain critical castings are made of Invar or other materials that have low-temperature coefficients, so that changes in temperature will not reduce the ultimate accuracy. Handwheels and other controls are designed to minimize the effect of the body heat of the operator.

When high accuracy is essential, it is desirable to locate jig borers in temperature-controlled rooms. In addition, some care may be necessary to provide coolant at a uniform temperature so that the work can be maintained at a fairly uniform temperature.

Although light milling can be done on some jig borers, such use is not rec-

FIGURE 35-5 Jig borer equipped with digital tape control. (*Courtesy Société Genevoise D'Instruments de Physique.*)

ommended. They primarily are for center drilling, drilling, and boring. Some are equipped so that light grinding can be done, and similar machines are made specifically for jig grinding.

Layout on jig borers can be speeded materially if drawings of parts that are to be machined are dimensioned properly. In general, base-line, or reference-line, dimensioning should be used, as illustrated in Figure 35-6. The method shown, of enclosing the dimensions that the jig borer operator must use in boxes, is a good one; the other dimensions shown are for use in checking the work. This type of dimensioning is particularly helpful when digital readout equipment is used. However, digital readout equipment that has provision for a floating zero permits point-to-point dimensioning to be employed with less difficulty than would be experienced otherwise.

Because layout accuracy within 0.0025 mm (0.0001 inch) is readily obtainable on jig borers, in former years they often were used to produce small quantities of duplicate parts where extreme accuracy was required. At present, this seldom is done, because numerically and computer-numerically controlled milling machines and machining centers routinely do such work at less cost.

Layout by photographic methods. Extensive use is made of photographic methods of layout, particularly in making large templates and in transferring location dimensions to large workpieces, such as are encountered in the aircraft industry. An accurate drawing is prepared, or a very accurate layout is made on a special type of plastic sheet. It then is photographed onto a sensitized glass plate, using a special template camera. After the glass plate has been developed,

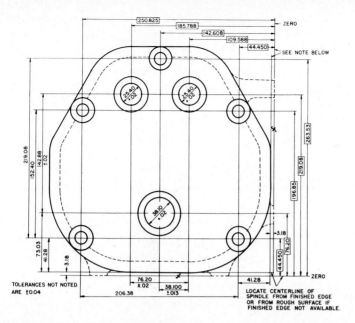

FIGURE 35-6 Metric drawing dimensioned for use with a jig borer.

it can be used for projecting the layout image, full size, onto a sheet of metal that has been sensitized with a photographic emulsion. The exposed metal sheet then is developed in a large tank, thereby having the layout on it. In some cases the exposed and developed metal sheet is machined, according to the layout, and used as a template for layout of additional workpieces.

REVIEW QUESTIONS

1. What does layout involve?
2. In Figure 35-1, what do the **V**-shaped marks indicate?
3. Define location dimensions.
4. If you were making a large number of parts from plate stock, would you use layout methods to establish the location of critical geometrical points on the part?
5. What are some factors which tend to reduce the accuracy and the precision of the location of the centers of holes?
6. What advantages does machine layout have over manual layout?
7. What is a jig borer and what special features does it have which facilitate extremely accurate and precise layout?
8. In Figure 35-6, there are two holes dimensioned 25.40. What is the nominal distance between these two holes and the tolerance?
9. What does "floating zero" refer to in connection with a digital readout system?
10. Manual and machine layout is seldon done these days. Why?

CASE STUDY 35. The Steering Shaft Connection
Figure CS-35 shows the design of a connection that was incorporated in the steering shaft of an internationally known sports car. A failure occurred, resulting in a death and a large damage claim and award.

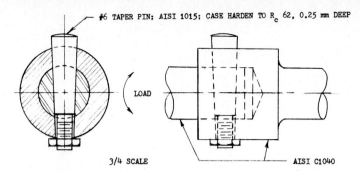

FIGURE CS-35 Method of coupling two sections of steering column shaft in an automobile.

1. Analyze the design and determine its inherent deficiencies.
2. What manufacturing difficulties would you anticipate with this design, and which might led to failure?
3. How would you modify the design to provide proper performance?

CHAPTER 36

Jigs and Fixtures

In previous chapters attention has been directed repeatedly to the manner in which workpieces are mounted and held in the various machines. In Chapter 35 the very important subject of layout was discussed, with special emphasis on the time and skill required to achieve accurate results. In this chapter jigs and fixtures will be considered as important production tools or adjuncts, with primary attention being directed toward their essential characteristics, the manner in which they accomplish layout and/or work holding, their relationship to the machine tools and manufacturing processes, and the economics of their use. Only minimal attention will be given to the details of their design. Just as the engineer who is concerned with manufacturing must understand the functioning, capabilities, and proper utilization of machine tools but almost never designs them, similarly he or she should have a thorough understanding of the basic principles of jigs and fixtures so as to utilize them effectively. However, in most cases, he can leave the design of their details to the tool designer.

Definitions. To understand and utilize jigs and fixtures effectively, it is important to define them clearly. Unfortunately, and particularly in earlier years, the terms "jigs" and "fixtures" sometimes are interchanged. This is not too surprising in that a jig quite often also performs the function of a fixture; the reverse is never true.

In defining jigs and fixtures, and in avoiding confusing the two, it is helpful to consider the subject of dimensioning as used in drafting practice. Dimensions are of two types—*size* and *location*. Size dimensions denote the size of geometrical shapes—cylinders, cubes, parallelopipeds, etc.—of which objects are composed. Location dimensions, on the other hand, determine the position or location of these geometrical shapes *with respect to each other*. Thus, in Figure 35-1, *a, b, c,* and *d* are location dimensions while *e, g,* and *h* are size dimensions. With location dimensions in mind, one can precisely define a jig as follows: *a jig is a special device which, through built-in features, determines location dimensions that are produced by machining or fastening operations.* The key requirement of a jig is that it determines a location dimension. Thus jigs automatically accomplish layout.

In establishing location dimensions, jigs may do a number of other things. They frequently guide tools, as in drill jigs, and thus determine the location of a component geometrical shape. However, they do not always guide tools. In the case of welding jigs component parts are held in a desired relationship with respect to each other while an unguided tool accomplishes the fastening. Thus the guiding of a tool is not a necessary requirement of a jig.

Similarly, jigs usually hold the work that is to be machined, fastened, or assembled. However, this is not always true because in certain cases the work actually supports the jig. Thus, although a jig *may* incidentally perform other functions, the basic requirement is that, through qualities that are built into it, certain critical dimensions of the workpiece are determined.

A *fixture is a special device that holds work during machining or assembly operations.* The key characteristic is that it is a *special* work-holding device, designed and built for a particular part or shape. A general-purpose device, such as a vise or clamp, is not a fixture. Thus a fixture has as its specific objective the facilitating of *setup,* or making holding easier.

Because many jigs hold the work while determining critical dimensions, it is apparent that in such cases they meet all the requirements of a fixture. But it is equally evident that fixtures never determine dimensions of parts which they hold—a basic requirement of a jig.

From their definition and function, it follows that jigs are associated with operations, whereas fixtures most commonly are related to specific machine tools. Thus the most common jigs are drilling jigs, reaming jigs, welding jigs, and assembly jigs. In these uses they often are not fastened to the machine table, being free to be moved so as to permit the proper registering of the work and the tool. Fixtures, on the other hand, most frequently are attached to a machine tool or table. Consequently, they are associated, in name, with the particular tool with which they are used. Because their function of holding is a broad one, there are many types of fixtures, such as milling fixtures, broaching fixtures, lathe fixtures, grinding fixtures, assembly fixtures, and plating fixtures.

Some basic factors in jig and fixture design.   In order to assure their optimum functioning, the following six factors should be considered in designing jigs and fixtures:

1. Clamping of the work.
2. Support of the work against any forces that are imposed by the process.

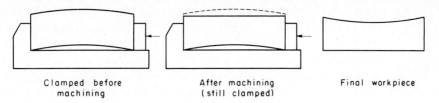

Clamped before
machining

After machining
(still clamped)

Final workpiece

FIGURE 36-1 Exaggerated illustration of the manner in which too-large clamping forces can affect the final dimensions of a workpiece.

3. Location of the work to provide required dimensional control.
4. Guiding of the tool, when required.
5. Provision for chips, if they are made.
6. Rapid and easy operation.

Items 3 and 4 apply only to jigs.

Clamping of the work is closely related to support of the work. However, there are certain principles that relate specifically to clamping. First, clamping *stresses* should be kept low. Any clamping, of course, induces some stresses that tend to cause some distortion of the workpiece, usually elastic. If this distortion is measurable, it will cause some inaccuracy in final dimensions, as illustrated in an exaggerated manner in Figure 36-1. The obvious solution is to spread the clamping forces over a sufficient area to reduce the stresses to a magnitude that will not produce appreciable distortion.

Second, the clamping forces should direct the work against the points of location and work support. This principle is illustrated in Figure 36-2. It is not safe to assume that if there is no *designed* clamping-force component either upward or downward, the work will be held properly and will not tend to rise from the supports. Clamped surfaces often have some irregularities that may produce force components in an undesired direction. Consequently, clamping forces should be applied in directions that will assure that the work will remain in the desired position.

An important corollary to the second principle is that, whenever possible, jigs and fixtures should be designed so that the forces induced by the cutting process act to hold the workpiece in position against the supports. These forces are predictable, and proper utilization of them can materially aid in reducing the magnitude of the required clamping stresses.

The third principle is that as many operations as practicable should be performed with each clamping of the workpiece. This principle has both physical and economic aspects. Because some stresses result from each clamping, with

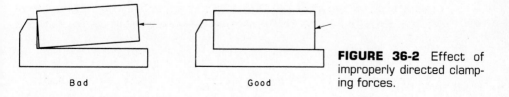

Bad

Good

FIGURE 36-2 Effect of improperly directed clamping forces.

the possibility of accompanying distortion, greater accuracy is probable if as many operations as practicable are performed with each clamping. Thus a jig or fixture should be designed to serve for several operations when possible. From the economic viewpoint it is obvious that if the number of jigs or fixtures is reduced, a smaller investment of capital usually will be required and fewer man-hours will be necessary for handling the workpiece in putting it in and removing it from the jigs or fixtures.

The locating of a workpiece in a jig or fixture requires adherence to the "three-two-one" principle—at least three locating points, or stops, are required to locate an object in the first plane, at least two points in a second plane to locate it in that plane, and at least one point to locate it in a third plane, assuming that the planes are not parallel and preferably are perpendicular to each other.

In addition to locating the work properly, the stops or work-supporting areas must be arranged so as to provide adequate support against the forces imposed on the workpiece by the cutting tool. Such consideration of the cutting forces should be a routine step in the design of all jigs and fixtures that are used in connection with machining operations. Figure 36-3 illustrates this concept. As shown in Figure 36-3a, whenever possible the cutting force should always act against a fixed portion of the jig or fixture and not against a movable section. This aids materially in permitting lower clamping stresses to be utilized. Figure 36-3b illustrates the principle of keeping the points of clamping as nearly as possible in line with the action forces of the cutting tool so as to reduce their

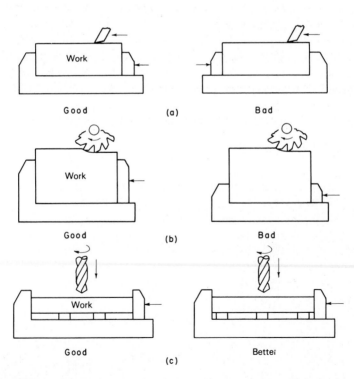

FIGURE 36-3 Proper work support to resist the forces imposed by cutting tools. In (c), three buttons form triangle for work to rest on.

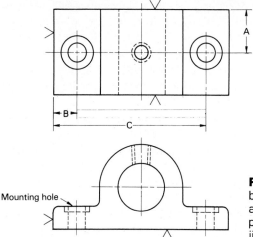

FIGURE 36-4 Drawing of a bearing block. The surfaces marked with a Vee are reference surfaces and are finished prior to insertion of part into the box drill jig.

tendency to pull the work from the clamping jaws. Compliance with this principle results both in lower clamping stresses and less massive clamping devices. The location points should be as far apart as possible but positioned so as to not allow the cutting forces to distort the work. If the principle illustrated in Figure 36-3c is not followed, the action forces of the cutting tool may distort the work, with resulting inaccuracy, or broken tools. Compliance with these three principles also aids materially in reducing the possibility of workpiece vibration and tool chatter.

The principles of work location and tool guidance are illustrated in Figures 36-4 and 36-5. The two mounting holes in the base of the bearing block are located and drilled with the use of the drill jig shown in Figure 36-5. The dimensions *A, B,* and *C,* shown in Figure 36-4, are determined by the jig. There also is one other location dimension that must be controlled. This is the implied

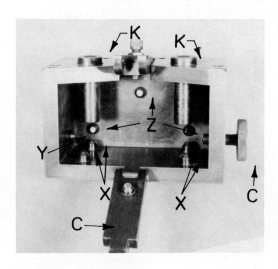

FIGURE 36-5 Box-type drill jig for drilling the mounting holes in the bearing block shown in Figure 36-4. C designates clamps.

requirement that the axes of the mounting holes be at right angles to the bottom surface of the block.

The way in which these dimensions are determined in the finished block can be seen by reference to Figures 36-5 and 36-6. The bearing block rests in the jig on four buttons marked x in Figure 36-5. These buttons, made of hardened steel, are set into the bottom plate of the jig and are accurately ground so that their surfaces are in a single plane. The left-hand end of the block is held against another button y in Figure 36-5. This locating button is built into the jig so that its surface is at right angles to the plane of the x buttons. When the block is placed in the jig, its rear surface rests against three more buttons marked z. These buttons are located and ground so that their surfaces lie in a plane that is at right angles to the planes of both the x and y buttons.

The use of four buttons in the bottom of this jig (x buttons) appears not to adhere to the three–two–one principle stated previously. However, although only two x buttons would have been required for complete location, the use of only

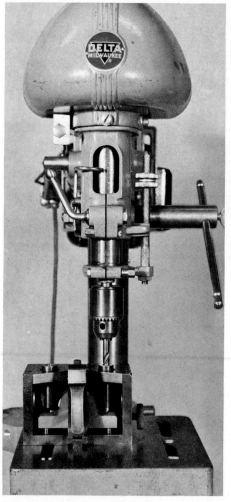

FIGURE 36-6 (*Left*) Drilling the mounting holes in a bearing block, using the drill jig shown in Figure 36-5. (*Right*) Close-up view, showing the block in the jig, resting on the location buttons, and the drill being guided by the drill bushing.

PROCESSES AND TECHNIQUES RELATED TO MANUFACTURING

two buttons would not have provided adequate support adjacent to the holes that were to be drilled. This would have violated one of the fundamental design principles. Thus the three–two–one principle is a *minimum* concept, and often must be exceeded.

To assure that the mounting holes are drilled in their proper locations, the drill must be located and then guided during the drilling process. This is accomplished by the two drill bushings, marked K in Figure 36-5. Such drill bushings are accurately made of hardened steel with their inner and outer cylindrical surfaces concentric. The inner diameter is made just enough larger than the drill that is to be used—usually 0.013 to 0.025 mm (0.0005 to 0.001 inch)—so that the drill can turn freely but not drift appreciably. The bushings are mounted in the upper plate of the jig by means of a press fit and positioned so that their axes are exactly perpendicular to the plane of the x buttons, at a distance A from the z buttons and at distances B and C, respectively, from the plane of the y button. Note that the bushings are sufficiently long that the drill is guided close to the surface where it will start drilling. Consequently, when the bearing block is properly placed and clamped in the jig (see Figure 36-6), the drill will be located and guided by the bushings so that the critical dimensions on the workpiece will be correct.

When jigs or fixtures are used in connection with chip-making operations, adequate provision must be made for the chips that are produced and for their easy removal. This is essential for several reasons. First, if chips become packed around the tool, heat will not be carried away, and tool life can be decreased or the tool actually broken. Figure 36-7 illustrates how insufficient clearance between the end of a drill bushing and the workpiece can prevent the chips from escaping, whereas too much clearance may not provide adequate drill guidance and can result in broken drills.

A second reason why chips must be considered relates to preventing them from interfering with the proper seating of the work in the jig or fixture. This is illustrated in Figure 36-8. Even though chips and dirt always have to be cleared from the locating and supporting surfaces by a worker or by automatic means, such as an air blast, the design details should be such that chips and other debris will not readily adhere to, or be caught in or on, the locating surfaces, corners, or overhanging elements and thereby prevent the work from seating properly. Such a condition results in distortion, high clamping stresses, and incorrect finished dimensions.

Another very important factor in the design and selecting of jigs and fixtures is the ease and rapidity with which they can be utilized by the operating per-

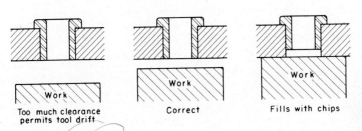

FIGURE 36-7 Importance of proper chip clearance.

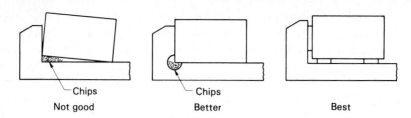

Chips
Not good

Chips
Better

Best

FIGURE 36-8 Methods of providing chip clearance to assure proper seating of the work.

sonnel. Jigs and fixtures are used only when considerable quantities of production are involved, and their primary purpose is to increase the productivity of workers and machines. While work is being put into or being taken out of jigs and fixtures they, and usually also the machines with which they are used, are nonproductive. Although a small amount of time saved is of some importance when a machine of low productivity is involved, the same amount of saved time is of much greater importance when a machine of higher productivity is being utilized.

This important relationship is shown in Table 36-1. In this table the effect of reducing setup time by 1 minute is shown for two conditions—one where the machining time is 10 minutes and the other where it is only 5 minutes. In the first case, A, a skilled operator is used, who receives $6.00 per hour, and the cost for the machine is $3.50 per hour, which includes all overhead and interest costs. In the second case, B, a less-skilled operator, paid $4.00 per hour, can be employed because the machine is more nearly automatic, but this results in the cost for the machine being $6.00 per hour. The result of decreasing the setup time by a constant amount are very evident. Although the cost saving in each case happens to be the same ($0.16 per unit), the percentage savings and the percentage increase in productivity are much greater in the case of the more productive machine. This is typical of what can happen when machine-hour costs and productivity are high and emphasizes the importance of tooling adjuncts that reduce setup time when costly machine tools are utilized. Such tools are increasingly common.

TABLE 36-1 Effects of Reducing Setup Time by a Constant Amount for Two Production Cycles

Case	Setup Time (min)	Machining Time (min)	Total Cycle Time (min)	Cost per Unit	Cost Saving per Unit (%)	Productivity Increase (%)
A						
Labor: $6/hr	3	10	13	$2.06	=7.8	8.3
Machine: $3.50/hr	2	10	12	$1.90		
B						
Labor: $4.00/hr	3	5	8	$1.33	=12.0	14.1
Machine: $6/hr	2	5	7	$1.17		

There are several ways in which jigs and fixtures can be made easier and more rapid to use. An important one relates to the method in which the work is clamped. Some methods can be operated much more readily than others. For example, in the drill jig shown in Figure 36-5, a knurled clamping screw is used to hold the block against the buttons at the end of the jig. To clamp or unclamp the block in this direction requires several motions. On the other hand, a cam latch is used to close the jig and hold the block against both the bottom and rear locating buttons. This type of latch can be operated with a single motion.

Ease of operation of jigs and fixtures, as well as the controls on machine tools, not only directly increases the productivity of such equipment, it also does so indirectly through better satisfied workers and better treatment of the tools.

Transfer of skill. The construction of a jig, with the required locating surfaces, clamps, drill bushings, locating buttons, and so on, requires careful layout and machining. This requires the services of a skilled toolmaker and the use of precision tools, such as a jig borer. In the making of the jig, the highly skilled toolmaker transfers some of his skill to the jig, so that through the use of the jig a less skillful person can produce a workpiece with as high a degree of accuracy as if it had been made by the more highly skilled worker. This relationship is expressed by the equation

$$\frac{\text{total skill required}}{\text{for the job}} = \frac{\text{skill built into}}{\text{the jig}} + \frac{\text{skill required from}}{\text{the worker}}$$

This relationship is an extremely important one in manufacturing. The greater the skill that is built into a jig or fixture, the less is the amount required of the worker who uses the tooling in producing a product. Similarly, the more of the crtical setup that is built into a fixture or jig, the less is the time required to set up each piece in a machine tool. Consequently, well-designed jigs and fixtures not only reduce the requirement for highly skilled workers in production operations, they also increase the productivity of both workers and machines and thereby the availability of goods for society. This is especially important when very costly machines are utilized, because nonproductive time is reduced.

JIGS

Type of jigs. Because jigs are designed to facilitate certain processes, they are made in several basic forms and carry names that are descriptive of their general configurations or predominant features. Several of these are illustrated in Figure 36-9.

A *plate jig* is one of the simplest types, consisting only of a plate that contains the drill bushings, and a simple means of clamping the work in the jig, or the jig to the work. In the later case, where the jig is clamped to the work, it sometimes is called a *clamp-on jig.* Such jigs frequently are used on large parts, where it is necessary to drill one or more holes that must be spaced accurately

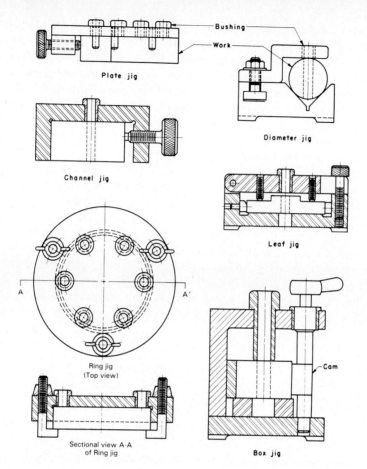

FIGURE 36-9 Common types of jigs.

with respect to each other, or to a corner of the part, but that need not have an exact relationship with other portions of the work.

Channel jigs also are simple and derive their name from the cross-sectional shape of the main member. They can be used only with parts having fairly simple shapes.

Ring jigs are used only for drilling round parts, such as pipe flanges. The clamping must be sufficient to prevent the part from rotating in the jig.

Diameter jigs provide a means of locating a drilled hole exactly on a diameter of a cylindrical or spherical piece.

Leaf jigs derive their name from the hinged leaf or cover that can be swung open to permit the workpiece to be inserted and then closed to clamp the work in position. Drill bushings may be located in the leaf as well as in the body of the jig so as to permit locating and drilling holes on more than one side of the workpiece. Such jigs, which require turning to permit drilling from more than one side, are called *rollover jigs*.

Box jigs are very common, deriving their name from their boxlike construc-

PROCESSES AND TECHNIQUES
RELATED TO MANUFACTURING

tion. They have five fixed sides and a hinged cover or leaf, or, as shown in Figure 36-9, a cam that locks the workpiece in place. Usually, the drill bushings are located in the fixed sides to assure retention of their accuracy. The fixed sides of the box usually are fastened by means of dowel pins and screws so that they can be taken apart and reassembled without loss of accuracy. Because of their more complex construction, box jigs are costly, but their inherent accuracy and strength make them justified where there is sufficient volume of production. They have two obvious disadvantages: (1) it usually is more difficult to put work into them than into more simple types, and (2) there is a greater tendency for chips to accumulate within them. Figure 36-5 shows a box-type jig.

Because jigs must be constructed very accurately and be made sufficiently rugged so as to maintain their accuracy despite the abuse to which they inevitably are subjected in use, they are expensive. Consequently, several methods have been devised to aid in lowering the cost of making jigs. One method involves the use of simple, standardized plate and clamping mechanisms, called *universal jigs,* such as shown in Figure 36-10. These can easily be equipped with suitable locating buttons and drill bushings to construct a jig for a particular job. Such universal jigs are available in a variety of configurations and sizes and, because

FIGURE 36-10 (*Top*) Two types of universal jigs, manual (left) and power-actuated (right). (*Left*) Completed jig, made from unit shown at upper right. (*Courtesy Cleveland Universal Jig Division, The Industrial Machine Company.*)

they can be produced in quantities, their cost is relatively low. However, the variety of work that can be accommodated by such jigs obviously is limited.

A second method of reducing the cost of constructing jigs and fixtures is the use of standardized components, such as drill bushings, locating buttons, and clamping devices. As shown in Figure 36-11, a wide variety of such components is available.

Because *assembly jigs* usually must provide for the introduction of several component parts and the use of some type of fastening equipment, such as welding or riveting, they commonly are of the open-frame type. Such jigs are widely used in the automobile and aircraft industries. A very large jig of this type is shown in Figure 36-12.

Since fixtures are for the sole purpose of facilitating setup and clamping, they tend to be simpler than jigs and more open in design, to permit easy placement and removal of the workpiece. However, those details in jigs that involve support and clamping apply in the same manner to fixtures. In most instances, fixtures are attached to a machine table or to a workbench, so some attention must be given to this matter. If there is some possibility that the fixture may need to be used on more than one machine, the method of attachment should be designed so as to permit this to be done.

Because fixtures are less complex than jigs, they are much less expensive to construct and thus can be economically justified for smaller production quantities. Although most fixtures are made specifically for given jobs, in many cases satisfactory results can be obtained by simple modification of standard devices.

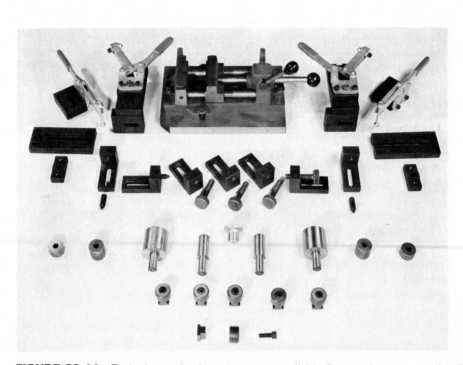

FIGURE 36-11 Typical standard components available for use in constructing jigs and fixtures. (*Courtesy Brown & Sharpe Mfg. Co.*)

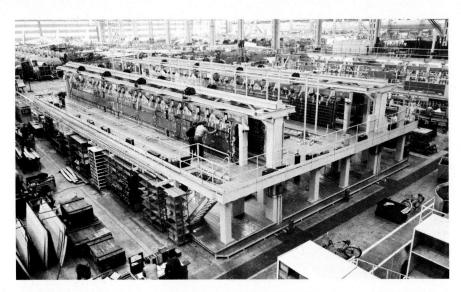

FIGURE 36-12 Large assembly jig used in an aircraft factory. This jig is constructed mainly of reinforced concrete. (*Courtesy Boeing Company.*)

An example is the addition of special jaws to a standard vise, as shown in Figure 36-13. Frequently, minor modifications can be made to standard clamping components, such as those shown in Figure 36-11. By exercising a little ingenuity, very effective fixtures often can be constructed at very modest cost and thus make their use economical for quite small production volumes, with accompanying substantial gains in productivity.

ECONOMIC JUSTIFICATION OF JIGS AND FIXTURES

As has been discussed previously, jigs and fixtures are expensive, even when designed and constructed by using standard components. Obviously, their cost is a part of the total cost of production, and one must determine whether they can be justified economically by the savings in labor and machine cost that will result from their use.

In order to determine the economic justification of any special tooling, the following factors must be considered:

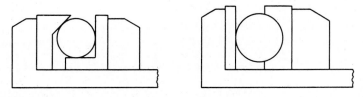

FIGURE 36-13 Use of special adaptors for vise jaws.

1. The cost of the special tooling.
2. Interest or profit charges on the tooling cost.
3. The savings in production labor cost resulting from the use of the tooling.
4. The savings in production machine cost due to increased productivity.
5. The number of units that will be produced using the tooling.

The economic relationship between these factors can be expressed in the following manner:

savings per piece (exclusive of tooling costs)

total cost per piece without tooling	+	total cost per piece using tooling (exclusive of tooling costs)	+	tooling cost per piece
labor cost per piece without tooling	machine and overhead cost per piece without tooling	labor cost per piece with tooling	machine and overhead cost per piece with tooling	cost of tooling	interest on tooling cost

$$[(R)(t) + (R_m)(t)] \quad - \quad [(R_t)(t_t) + (R_m)(t_t)] \quad \geq \quad \frac{C_t + (C_t/2)(n)(i)}{N}$$

(36-1)

where R = labor rate per hour, without tooling
R_t = labor rate per hour, using tooling
t = hours per piece, without tooling
t_t = hours per piece, using tooling
R_m = machine cost per hour, including all overhead
C_t = cost of the special tooling
n = number of years tooling will be used
i = interest or profit rate invested capital is worth
N = number of pieces that will be produced with the tooling

This equation can be expressed in a simpler form:

$$(R + R_m)t - (R_t + R_m)t_t \geq \frac{C_t}{N}\left(1 + \frac{n \times i}{2}\right)$$

(36-2)

This equation assumes straight-line depreciation and computes interest on the average amount of capital invested throughout the life of the tooling.[1] Where the time over which the tooling is to be used is less than 1 year, companies often do not include an interest cost. If this factor is neglected, the right-hand term of equation (36-2) reduces to C_t/N.

Equations (36-1) and (36-2) assume that the material cost will be the same regardless of whether special tooling is used. This is not always true. Although

[1]For the use of the slightly more complex concept of "average interest" or the use of more accurate sinking fund depreciation, see E. P. DeGarmo, J. R. Canada, and W. G. Sullivan, *Engineering Economy,* 6th ed., Macmillan Publishing Co., Inc., New York, 1979.

these equations are not completely accurate for all cases, they are satisfactory for determining tooling justification in most cases, because the life of such tooling seldom exceeds 5 years, and more frequently does not exceed 2 years. Furthermore, the savings usually should exceed the tooling costs by a substantial margin before special tooling is adopted.

The following example illustrates the use of equation (36-2) to determine tooling justification. In drilling a series of holes on a radial drill, the use of a drill jig will reduce the time from $\frac{1}{2}$ hour per piece to 12 minutes. If a jig is not used, an A-grade machinist must be used, whose hourly rate is $8.00. If the jig is used, the job can be done by a B-grade machinist, whose rate is $6.50 per hour. The hourly rate for the radial drill is $8.75.

The cost of making the jig would include $250 for design, $75 for material, and 50 hours of toolmaker's labor, which is charged at the rate of $12 per hour to include all machine and overhead costs in the toolmaking department.

Investment capital is worth 16% to the company. It is estimated that the jig would last 3 years and that it would be used for the production of 300 parts over this period. Is the jig justified? How many parts would have to be produced with the jig for it to "break even"?

To use equation (36-2), the cost of the jig, C_t, must be determined.

$$C_t = \$250 + \$75 + \$12 \times 50 = \$925$$

Substituting the values given in equation (36-2) yields

$$(\$8.00 + \$8.75)0.5 - (\$6.50 + \$8.75)0.2 \geq \frac{\$925}{300}\left(\frac{3 \times 0.16}{2}\right)$$
$$\$5.33 > \$3.82$$

Thus, use of the jig is justified. By omitting the value 300 in the solution above and solving for N, it is found to be 215+ pieces, so that at least 216 pieces would have to be produced with the jig for it break even or pay out.

It should be noted that equations (36-1) and (36-2) assume that the production time of the machines that would be made available by the use of the special tooling can be used for other operations. If this is not the case, there being no other use for the machines, the cost analysis should be altered to take this important fact into account. Otherwise, the tooling justification may be substantially in error.

The concepts of group technology and NC machines, discussed in Chapter 38, may eliminate the need for designing and building a new jig or fixture every time a new part is designed.

REVIEW QUESTIONS

1. What distinguishes a jig from a fixture?
2. Jigs have been called "automatic layout devices." Explain.
3. An early treatise defined a jig as "a device that holds the work and guides a tool." Why was this definition incorrect?
4. Why would an ordinary vise not be considered to be a fixture?
5. What six basic factors should be considered in designing jigs and fixtures?

6. What difficulties can result from not keeping clamping stresses low in designing jigs and fixtures?

7. Explain the three-two-one concept related to jig design.

8. Which of the six basic design principles relating to jigs and fixtures would most likely be in conflict with the 3-2-1 location concept?

9. Why would you want a drill bushing to be readily removable?

10. What are two reasons for not having drill bushings actually touching the workpiece?

11. Why does the use of down milling often make it easier to design a milling fixture than if up milling were used?

12. Explain what is meant by the expression, "a jig is a skill-transfer device?"

13. A large assembly jig for an airplane-wing component gave difficulty when it rested on four points of support but was satisfactory when only three supporting points were used. Why?

14. Explain why the use of a given fixture may not be economical when used with one machine tool but may be economical when used in conjunction with another machine tool.

15. What are roll-over jigs and what advantages do they offer?

16. Using the following values, determine the number of pieces that would have to be made to justify the use of a jig costing $3000:

$$R = \$5.75 \qquad t_t = 1\tfrac{1}{4} \qquad R_m = \$4.50$$
$$R_t = \$4.50 \qquad t = 2\tfrac{1}{4} \qquad n = 3$$
$$i = 10\%$$

17. In order to drill 24 holes in a large, curved plate, an air frame manufacturer mounted the plates vertically in front of a robot and put a drill in the hand of the robot (see Figure 38-38). Then they programmed the robot to drill the holes. Later they found they had to build a plate jig for this job because the robot could not locate the holes within specifications. What might you have tried here instead of going to a jig?

18. What design errors can you spot in the jigs in Figure 36-9?

CASE STUDY 36. Overhead Crane Installation

In order to install an overhead crane (Figure CS-36) in one bay of an assembly plant, brackets for the rails of the crane are to be mounted on eight columns, four on each side of the bay area facing each other. The rails for the crane will span four columns. Each bracket on each column will need six holes in a circular pattern. The holes must be

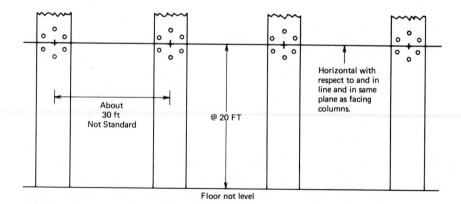

FIGURE CS-36 Hole pattern locations on four of the eight columns. Other four columns face these four.

accurately spaced within 5 minutes of the arc of each other. The axis of the holes must be parallel and normal to the face of the columns. The center of the bolt-hole circle must be at a height of at least 20 feet from the floor, but the centers of all the eight bolt-hole circles must be on the same parallel plane so that the rails for the crane are level and parallel with each other. Four of the columns along the wall have their faces flush with the wall surface so that mechanical clamping or attachments cannot be used. The building code will permit no welding of anything to these columns.

1. How would you proceed to get the bolt holes located in the right position on the beams?
2. How would you get the hole patterns located properly with respect to each other on all eight beams?
3. List the equipment you will need.
4. Make a sketch of any special tool you recommend.

CHAPTER 37

Decorative and Protective Surface Treatments

A large proportion of all manufactured products must be given some type of decorative or protective surface treatments before they can be sold or used. Handling and the various manufacturing processes leave scratches, burrs, fins, fine pores, or other blemishes which detract from their appearance or present possible hazards to users. Many commonly used materials, such as most irons and steels, do not inherently possess the colors that customers want in products—especially in large-volume, consumer goods. Materials often are not adequately resistant to the environments in which they will be used. As materials become more scarce and more costly, there is a pressing need to substitute for them basically inferior materials with modified surfaces. As a consequence, after achieving their desired shape, most manufactured products require one or more additional operations to clean, protect, or color them.

These important decorative and protective surface treatments add to the cost of manufactured products. Further, as with other manufacturing processes, there is often a definite relationship between design and these finishing processes. In recent years, a large amount of attention has been devoted to developing finishing processes and equipment that enable them to be accomplished successfully in mass quantities at low cost. Consequently, designers should be knowledgeable about them. Through proper coordination of design and the shape-producing processes, occasionally the need for finishing operations can be eliminated, and invariably finishing costs can be reduced or better results achieved. In a few cases, processes that were developed for finishing have been adapted for producing desired shapes.

CLEANING AND SMOOTHING PROCESSES

The first step in finishing usually is cleaning and/or smoothing. Mechanical, chemical, and electrochemical methods are used.

Abrasive cleaning. One of the most common steps preliminary to the application of decorative or protective surface treatments is the removal of sand or scale from metal parts. Sand often adheres to certain types of castings, and scale often results when metal has been processed at elevated temperatures. In some cases, sand may be removable by simple vibratory shaking, but often some type of *abrasive cleaning* is employed to remove such foreign materials. Some type of abrasive, usually sand or steel grit or shot, is impelled against the surface to be cleaned. Fine glass shot may be used for some materials.

In *shot blasting,* a high-velocity air blast is used as the impelling agent. Air pressures from 0.4 to 0.69 MPa (60 to 100 psi) are used for ferrous metals and from 0.07 to 0.4 MPa (10 to 60 psi) for nonferrous metals. A nozzle with about a 9.5-mm ($\frac{3}{8}$-inch) diameter is often used. For large or only a few parts, the blast may be directed by hand, as shown in Figure 37-1. Quantities of small parts usually are moved past sets of stationary nozzles inside a hood. When done manually, protective clothing and breathing equipment must be provided and precautions taken to prevent the resulting dust from being spread. Ordinarily, a separate, well-vented room or booth is used with suitable collection equipment so as to avoid air pollution.

FIGURE 37-1 Shot blasting in a special cleaning room. (*Courtesy Norton Company.*)

Equipment that impels the abrasive particles by mechanical means is being used increasingly. This type of equipment employs the centrifugal principle illustrated in Figure 15-19. It is more economical in the use of energy, and less difficulty is experienced in avoiding pollution.

If sand is used, it should be clean, sharp-edged, silica sand. Steel grit will clean much more rapidly than sand and causes much less dust. However, sand has a lower cost and somewhat greater flexibility.

Obviously, abrasive cleaning is effective only if the abrasive can reach all the areas of the surface that must be cleaned, which may be difficult if the surface is very complex. Edges are rounded considerably during abrasive cleaning and, therefore, it cannot be used if sharp edges and corners must be maintained. If the nozzles are directed manually, the labor cost tends to become high.

Tumbling. Tumbling is a widely used method for mechanical cleaning. The parts are placed in a special barrel or drum until it is nearly full. Occasionally, the loaded barrel is rotated without the addition of any abrasive agents. However, in most cases metal slugs or jacks, or some abrasive, such as sand, granite chips, slag, or aluminum oxide pellets, are added. The rotation of the barrel causes the parts to be carried upward and then to tumble and roll over each other as they slide downward, as depicted in Figure 37-2. This produces a cutting action that usually will remove fins, flashes, scale, and sand. The process can be used only on parts that are sufficiently rugged to withstand the tumbling action. However, by a suitable selection of abrasives, fillers, barrel size and speeds, and careful packing of the barrel, an amazing range of parts can be tumbled successfully. Delicate parts should not shift loosely during tumbling. In some cases such parts must be attached to racks within the barrel so that they will not strike each other.

Tumbling is an inexpensive cleaning method. Various shapes of slug materials are used. Several shapes often are mixed in a given load so that some will reach into all sections and corners to be cleaned. Tumbling usually is done dry, but it can also be done wet. Obviously, tumbled parts will have rounded edges and corners. The equipment can be arranged so that loading and unloading are accomplished readily and so that the slug material is separated from the workpieces by falling through suitable grid tables.

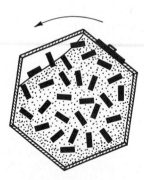

FIGURE 37-2 Principle of tumbling.

Wire brushing. A high-speed, rotary wire brush sometimes is used to clean surfaces; it also does a minor amount of smoothing. Wire brushing can be done by hand application of the workpiece to the brush, but more commonly automatic machines are used in which the parts are moved past a series of rotating brushes, similar to the procedure shown in Figure 37-8.

Wire brushing normally removes very little metal, except fine, sharp high spots. It produces a surface composed of fairly uniform, fine scratches that, for some purposes, is satisfactory as a final finish, or can easily be removed by barrel finishing or buffing.

Belt sanding. A simple and common method for obtaining smooth surfaces is by *belt sanding*. The workpieces are held against a moving, abrasive belt until the desired degree of finish is obtained. A series of belts of varying degrees of fineness can be used. If flat surfaces are desired, the belt passes over a flat table at the point where the work is held against it.

Belt sanding, as used for finishing, cannot be considered a sizing operation, because only sufficient sanding is done to remove the sharp, high spots and thus produce a smoother surface. The resulting surface is composed of very fine scratches, their fineness depending on the grit of the belt.

Although greatly improved sanding belts now are available and fairly smooth surfaces can be obtained, belt sanding is a hand operation and therfore is quite slow and costly in labor. Consequently, it should not be used except where more economical methods cannot be utilized. It has the further disadvantage of not being effective where recesses or interior corners are involved. As a result, it is used primarily for delicate parts or where small quantities are involved.

Barrel finishing. Since specially shaped, artificial abrasive pellets became available, about 1946, great progress has taken place in *barrel finishing*. The advent of vibratory finishing equipment during recent years has given further impetus to this method of finishing.

In this smoothing process, the workpieces and abrasive pellets are placed in a container and caused to move relative to each other, either by the container rotating, similar to barrel tumbling, or by the container vibrating through short strokes at from 900 to 3600 cycles per minute. The general actions involved are indicated in Figure 37-3.

Although the results obtained are essentially the same, vibratory finishing tends to be somewhat faster because the entire mass of workpieces and abrasive pellets are in constant, relative motion. Vibratory finishing tends to be superior to barrel finishing for smoothing and deburring interior surfaces.

Although some cleaning may occur incidentally in barrel finishing the primary objective is to "cut down" the surface through the movement of the abrasive pellets across the surface of the workpiece. To assure this action, the work surface, the ratio of the work to abrasive, the speed of rotation or vibration, and the shape of the abrasive pellets relative to the workpiece configuration are very important. As shown in Figures 37-4 and 37-5, a considerable variety of pellet sizes and shapes is available to permit selection of shapes that will slide across all the required surfaces and not become lodged in restricted areas. Water and

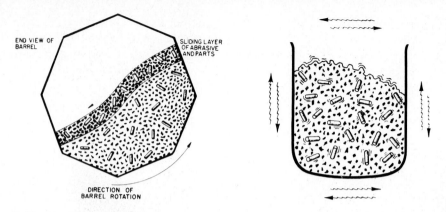

FIGURE 37-3 Comparison of actions in rotary and vibratory finishing. (*Courtesy Norton Company.*)

other compounds, such as dilute acids or soaps, frequently are added. Several natural abrasives, including slag, cinders, sharp sand, and granite chips, still are used, but in most cases the artificial abrasives have replaced these because of their greater uniformity of shape and cutting action.

Parts that are to be barrel or vibratory finished must be of such shape that they will not lock together, thus preventing them from moving freely. Some type of filler often is added to act as a carrying agent for the abrasive and to provide the necessary bulk to keep the workpieces separated from each other. Scrap punchings or mineral matter can be used in wet finishing, and in dry finishing, hardwood sawdust or leather scraps sometimes are employed. Heavy or intricate parts can be finished in a barrel by being racked in special fixtures to prevent them from becoming nicked. When sharp edges must be maintained, or when certain areas are not to be smoothed, masking can be employed. Obviously, it is preferable for such restrictions to be avoided through design.

By using the proper abrasive, a wide range of finishes can be obtained, the best being virtually free of visible scratches. However, unusually deep scratches should be removed by wire brushing prior to barrel finishing if a uniform surface

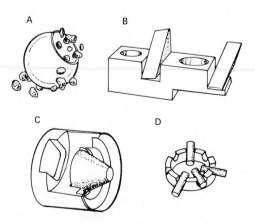

FIGURE 37-4 Some shapes of abrasives used for finishing various shapes of workpieces.

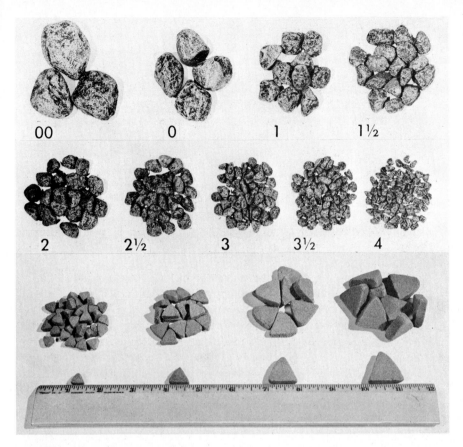

FIGURE 37-5 Various types and shapes of artificial pellets used for rotary and vibratory finishing. (*Courtesy Norton Company.*)

finished is desired. Figure 37-6 shows several examples of parts before and after barrel finishing. Finishing time varies from 10 minutes for soft, nonferrous parts to 2 or more hours for steel parts. Although they are batch processes, barrel and vibratory finishing are quite simple and economical. Sometimes the parts may be put through more than one barrel, using increasingly fine abrasivies.

Figure 37-7 shows an example of a vibratory finishing machine with an automatic system for handling the work and the abrasive.

Buffing. *Buffing* is a polishing operation in which the workpiece is brought in contact with a revolving cloth buffing wheel that has been charged with a very fine abrasive, such as polishing rouge. Obviously, buffing is closely related to lapping in that the cloth buffing wheel is a carrying vehicle for the abrasive. The abrasive removes minute amounts of metal from the workpiece, thus eliminating fine scratch marks and producing a very smooth surface. When softer metals are buffed, there is some indication that a small amount of metal flow may occur that helps to reduce high spots and produce a high polish.

Buffing wheels are made of disks of linen, cotton, broadcloth, or canvas that

FIGURE 37-6 Examples of typical parts before and after barrel finishing. (*Courtesy Norton Company.*)

are made more or less firm by the amount of stiching used to fasten the layers of cloth together. Buffing wheels for very soft polishing, or which can be used to polish into interior corners, may have no stiching, the cloth layers being kept in proper position by the centrifugal force resulting from the rotation of the wheel. Various types of polishing rouges are available, most of them being primarily ferric oxide in some type of binder.

Buffing should be used only to remove very fine scratches, or to remove oxide or similar coatings that may be on the work surface. If it is done by manually holding the work against the rotating buffing wheel, it is quite expensive because

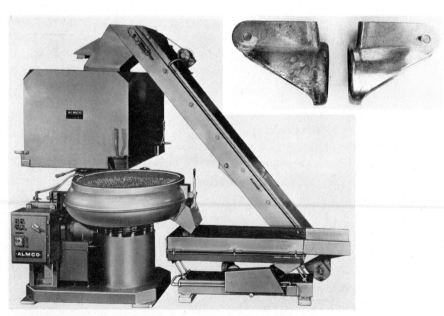

FIGURE 37-7 Vibratory-type finishing machine and close-up view of part before and after finishing. (*Courtesy Queen Products Division, King-Seeley Thermos Co.*)

of the labor cost. However, semiautomatic buffing machines are available, such as is shown in Figure 37-8 in which the workpieces are held in fixtures on a rotating, circular worktable and moved past a series of individually driven buffing wheels. The buffing wheels can be adjusted to desired positions so as to buff different portions of specific workpieces. If the workpieces are not too complex, very good results can be obtained quite economically with such equipment. Obviously, part design plays an important role where automatic buffing is to be employed.

Barrel burnishing. In many cases, results comparable with those obtained by buffing can be obtained by barrel burnishing. In this procedure, balls, shot, or round-ended pins are added to the work in a rotating barrel. No cutting action is involved. Instead, the slug material produces peening and rubbing actions, reducing the minute irregularities and producing an even surface. Burnishing will not remove visible scratches or pits, so in most cases the parts should first be rolled or vibrated with a fine abrasive. However, the resulting surface is smooth, uniform, and free of porosity.

Barrel burnishing is normally done wet, using water to which some lubricating or cleaning agent has been added, such as soap or cream of tartar. Because the rubbing action between the work and the shot material is very important, the barrel should not be loaded more than half full with work and shot, and the ratio of shot to work should be about 2 volumes to 1. The ratio should be such that the workpieces do not rub against each other. The speed of rotation of the barrel should be adjusted so that the workpieces will not be thrown out of the mass as they reach the top position and roll down the inclined surfaces.

It usually is necessary to use several shapes of shot material to assure that it can come in contact with inside corners and other recesses that must be rubbed. Balls from 3.2 to 6.4 mm ($\frac{1}{8}$ to $\frac{1}{4}$ inch) in diameter, pins, jacks, and ball cones commonly are used.

Parts that cannot be permitted to bump against each other can be burnished

FIGURE 37-8 Parts being buffed on an automatic machine. (*Courtesy Murray-Way Corporation.*)

successfully by fastening them in racks inside the burnishing barrel. The shot material is then added, and burnishing is carried out in the usual manner.

Electropolishing. *Electropolishing,* the reverse of electroplating, sometimes is used for polishing metal parts. The workpiece is made the anode in an electrolyte with a cathode added to complete the electrical circuit. In the resulting deplating, material is removed most rapidly from raised, rough spots, producing a very smooth, polished surface. It usually is not economical to remove more than about 0.025 mm (0.001 inch) of material, so the process is used primarily to produce mirrorlike surfaces, and the initial surface must be quite smooth. A final finish of less than 0.05 μm (2 microinches) can be obtained if the initial roughness does not exceed 0.18 to 0.020 μm (7 to 8 microinches) rms.

Electropolishing was originated for polishing metallurgical specimens and later was adapted for polishing stainless steel sheets and parts. It is particularly useful for polishing irregular shapes that would be difficult to buff. For best results the metal should be fine grain and free of surface defects exceeding the coarseness of a 180-grit scratch.

CHEMICAL CLEANING

At some stage in the finishing of virtually all metal products it is necessary to employ *chemical cleaning* to remove oil, dirt, scale, or other foreign material that may adhere to the surface so that subsequent painting or plating can be done successfully. One or more of three cleaning processes is used.

Alkaline cleaning. *Alkaline cleaning* employs such agents as sodium metasilicate or caustic soda with some type of soap to aid in emulsification. Wetting agents often are added to assist in obtaining thorough cleaning. The cleaning action is by emulsification of the oils and greases. Thus the solution must penetrate any dirt that covers them. It also is necessary to thoroughly rinse the cleansing solution from the work surface so as not to leave any residue.

The cleansing bath must be controlled to maintain a constant and proper pH value. Too high as well as too low pH levels may produce poor results.

Cleaning with emulsifiable solvents. *Cleaning with emulsifiable solvents* is done by combining an organic solvent with a hydrocarbon-soluble emulsifying agent such as sulfonated castor oil with water, or blending a soap and an organic solvent such as kerosene with a small amount of water. Cresylic acid or some other blending agent is used.

The work is dipped in the solvent solution and then rinsed once or twice. If the work is to be electroplated, it should have a subsequent treatment in an alkaline cleaner to remove any organic matter that remains on the metal. As a preparation for painting, solvent cleaning and rinsing usually are adequate. Solvent cleaning is used extensively for metals such as aluminum, lead, and zinc, which are chemically active and might be attacked by alkaline cleaners.

Vapor degreasing. *Vapor degreasing* is widely used to remove oil from ferrous parts and from such metals as aluminum and zinc alloys, which would be attacked by alkaline cleaners. A nonflammable solvent, such as trichlorethylene, is heated to its boiling point, and the parts to be cleaned are hung in its vapors. The vapor condenses on the work and washes off the grease and oil. Excess vapor is condensed by cooling coils in the top of the vapor chamber. Although grease and oil from the work are washed off into the liquid solvent, causing the bath to become dirty, because they are only slightly volatile at the boiling temperatures of the solvent, the vapor remains relatively clean at all times and so continues to clean effectively.

It must be remembered that vapor degreasing is effective only if the vapor condenses on the work. Thus the work must remain relatively cool. This offers no difficulty except in the case of thin sheets that contain considerable amounts of oil and do not have sufficient heat capacity to remain cool and thus condense enough vapor to bring about satisfactory cleaning.

Vapor degreasing has the advantages of being rapid and of having almost no visible effect on the surface. Its major disadvantages are that vapor alone does not remove solid dirt, and it frequently must be followed by alkaline cleaning to remove remaining organic matter. If the surface has substantial solid dirt in addition to oil, this may be removed by passing the work through a boiling liquid, thereby removing most of the dirt and some of the oil, then through cold liquid to cool the work, and finally through hot vapor to remove the remaining oil.

Pickling. *Pickling* involves dipping metal parts in dilute acid solutions to remove the oxides and dirt that are left on the surface by various processing operations. The most commonly used pickling solution is a 10% sulfuric acid bath at temperatures from 66 to 85°C (150 to 185°F). Muriatic acid also is used, either cold or hot. When used cold, pickling baths have approximately equal parts of acid and water. At temperatures ranging from 38 to 60°C (100 to 150°F), more dilute solutions are used.

It is very important that parts be thoroughly cleaned prior to pickling; the pickling solution will not act as a cleaner, and any dirt or oil on the surface will result in an uneven removal of the oxides. Alkaline cleaning is usually employed for this purpose. Pickling *inhibitors,* which decrease the attack of the acid on the metal but do not interfere with the action of the acid on the oxides, frequently are added to the pickling bath.

After the parts are removed from a pickling bath, they should be rinsed thoroughly, to remove all traces of acid, and then dipped in a bath that is slightly alkaline to prevent subsequent rusting. Where it will not interfere with further processing, a dip in cold milk of lime often is used for this purpose. Parts should not be overpickled, as this may result in roughening the surface.

Ultrasonic cleaning. *Ultrasonic cleaning* is used extensively where very high quality cleaning is required for relatively small parts. In this process the parts are suspended, or placed in wire baskets, in a cleaning bath, such as Freon, that contains an ultrasonic transducer operating at a frequency that causes cavitation in the liquid. Excellent results can be obtained in from 60 to 200 seconds in most cases. It usually is best to remove gross dirt, grease, and oil before doing ultrasonic cleaning.

FINISHES

Paints are by far the most widely used finishes on manufactured products, and a great variety is available to meet a wide range of requirements. Today, most paints and enamels are synthetic organic compounds that dry by polymerization or by a combination of polymerization and adsorption of oxygen. Water is frequently the carrying vehicle for the pigments. Moderate amounts of heat can be used to accelerate the drying, but many synthetic paints and enamels will dry in less than 1 hour without the use of heat. The older, oil-base paints and enamels require too much drying time for use in mass production and thus seldom are used.

Table 37-1 lists the more commonly used organic finishes and their important characteristics. *Nitrocellulose lacquers,* although very fast drying and capable of producing very beautiful finishes, are not sufficiently durable for most commercial applications. The *alkyds* are general-purpose paints but do not have sufficient durability for hard service conditions. The *acrylic enamels* are widely used for automobile finishes. *Silicones* and *fluoropolymers* are specialty finishes; their high cost is justified only where their special properties are important.

Asphaltic paints, which are solutions of asphalt in some type of solvent, such as benzine or toluol, are still used extensively, especially in the electrical in-

TABLE 37-1 Commonly Used Organic Finishes and Their Qualities

Material	Durability (Scale of 1–10)	Relative Cost (Scale of 1–10)	Characteristics
Nitrocellulose lacquers	1	2	Fast drying; low durability
Epoxy esters	1	2	Good chemical resistance
Alkyd-amine	2	1	Versatile; low adhesion
Acrylic lacquers	4	1.7	Good color retention; low adhesion
Acrylic enamels	4	1.3	Good color retention; tough; high baking temperature
Vinyl solutions	4	2	Flexible; good chemical resistance; low solids
Silicones	4–7	5	Good gloss retention; low flexibility
Fluoropolymers	10	10	Excellent durability; difficult to apply

dustry, where resistance to corrosion is required but appearance is not of prime importance.

Paint application. In manufacturing, almost all painting is done by one of four methods: *dipping, hand spraying, automatic spraying,* or *electrocoating.* In most cases at least two coats of paint are required. The first, or prime, coat serves primarily to (1) assure adhesion, (2) provide a leveling effect by filling in minor porosity and other surface blemishes, and (3) improve corrosion resistance and thus prevent later coatings from being dislodged in service. These properties are less obtainable in the more highly pigmented paints that are used for final coats because of their better color and appearance. In using multiple coats, one must be sure that the carrying vehicles in the final coats do not unduly soften the previous coats.

Paint application by *dipping* is used extensively. The parts are either dipped manually into the paint or are passed down into the paint while on a conveyor. Obviously, all of the workpiece is coated, and thus it is a very simple and generally economical technique where all surfaces require painting. Consequently, it is used for prime coats and for small parts where the loss of paint due to overspray would be excessive if ordinary spray painting were used. On the other hand, the unnecessary amount of paint used can make the process uneconomical if only some surfaces actually require painting or in cases where very thin, uniform coatings of some of the modern primers are adequate, particularly on large objects such as automobile bodies. Other difficulties with dipping are the tendency of the paint to run, thus producing a wavy surface, and the final drop of paint that usually is left at the lowest drip point. It also is essential that the paint in the dip tanks be kept stirred at all times and be of uniform viscosity.

Spray painting is probably the most widely used painting process, owing to its versatility and economy in the use of paint. The paint is atomized by three methods: air, mechanical pressure, or electrostatically. Either manual or automatic application is used. When hand spraying is used, either air or mechanical atomization is employed, and the spray of paint is directed against the work by means of a hand-manipulated gun. The worker must exercise considerable skill in obtaining proper coverage without allowing the paint to ''run'' or ''drape'' downward. Consequently, only a very thin film can be deposited at one time— usually not over 0.025 mm (0.001 inch)—if conventional methods are used. As a result, several coats usually must be applied with some intervening time for drying. Somewhat thicker coatings can be applied in one operation by using a *hot spray* method. In this procedure, the paint is sprayed while hot.

Obviously, spray painting by hand is costly from the viewpoint of labor expense and therefore is replaced by automatic methods whenever practicable. The simplest type of automatic equipment consists of a chain conveyor on which the parts are moved past a series of spray heads. However, if regular spray heads are used, results are often not satisfactory. A large amount of the paint may be wasted and it is difficult to get uniform coverage. Robots have been successfully used in spray painting. The robot is programmed to spray in a pattern and to turn on and off during part coverage. This application removes the human being from an unpleasant, even unhealthy environment.

Good results are obtained from either manual or automatic spray painting by using the electrostatic principle. In the air process, the spray gun atomizes the paint, giving the particles an electrostatic charge and considerable velocity. The atomized particles are attracted to, and deposited on, the work, which is grounded electrically. Under proper control, painting efficiencies of from 75 to 95% can be obtained.

In a second, airless electrostatic method, the paint is fed onto the interior of a rapidly rotating cone or disk that is one electrode of a high-potential electrostatic circuit. The rotation of the cone or disk causes the paint to flow outward to the edge by centrifugal force. As the thin film of paint reaches the edge and then is spun off, the particles are charged electrostatically and atomized without the need for any air pressure. With the workpiece being the other electrode of the circuit, the paint is transferred, as in the previously described method. The primary advantages of this method, which is illustrated in Figure 37-9, are that because no air pressure is used for atomization, there is less spray loss, much less extensive provision must be made to take care of fumes, and there is a higher efficienty in the use of paint—running as high as 99%.

With automatic spray-painting systems it often is necessary to do some touch-up work manually where the coverage is not completely uniform. There also is a tendency for the paint to be drawn to the nearest edge or surface, making it difficult to get paint into deep recesses. Of course, the workpiece must be electrically conductive.

Electrocoating is the most recent basic development in paint application. It permits the economy of ordinary dip painting to be achieved, but overcomes its disadvantges while permitting thinner and more uniform coatings and superior

FIGURE 37-9 Electrostatically finishing aluminum extrusions, using two reciprocating disks. (*Courtesy Ransburg Corporation.*)

coverage in interior recesses. The principle of the process is shown in Figure 37-10. The paint particles, in a water solvent, are given an electrostatic charge by applying a dc voltage between the tank (cathode) and the workpiece (anode). As the workpiece enters and passes through the tank, the paint particles are attracted to, and deposited on, it in a uniform, thin coating from 0.02 to 0.038 mm (0.0008 to 0.0015 inch) thick. When the coating reaches the desired thickness, determined by controlling the conditions, no more paint is deposited. The water in the deposited film is drawn away by electroosmosis, leaving a coating that is composed of more than 90% resins and pigments. The workpiece is removed from the dip tank, rinsed by a water spray, and baked for about 25 minutes at about 190°C (375°F).

Electrocoating is especially suitable for applying the prime coat to complex metal structures, such as automobile bodies, where good corrosion resistance is necessary. The flow of paint to hard-to-reach areas can be improved by placing electrodes in the workpiece at strategic locations. In addition, because the solvent is water, there is no fire danger as exists when large-area tanks are employed with regular dipping primers. As shown in Figure 37-11, electrocoating is readily adapted to conveyor-line production.

A new development is the application of paint in powder form by an electrostatic spray process. Several coats, such as primer and finish, can be applied and then followed by a single baking, instead of requiring baking after each coat as in conventional spray processes.

Drying.   Most paints and enamels used in manufacturing require from 2 to 24 hours to dry at normal room temperatures. This obviously is not practical. They can be dried satisfactorily in from 20 minutes to 1 hour at temperatures of from 135 to 232°C (275 to 450°F). Consequently, some drying at elevated temperatures usually is done, using either a baking oven or, more often, a tunnel or panel of infrared heat lamps. The latter involve relatively low investment, not much floor space, and are very flexible.

Although drying at elevated temperatures can be accomplished without difficulty on metal parts, this is not the case with wooden products. The temperatures are high enough to expand the gases, moisture, and sap that are in the wood, even though it is quite dry. These are forced to the surface after the paint

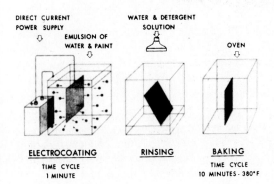

FIGURE 37-10   Principle of the electrocoating method of painting. (*Courtesy* Materials Engineering.)

FIGURE 37-11 Automobile bodies entering an electrocoating paint tank. (*Courtesy Ford Motor Company.*)

has started to harden and form small bubbles that roughen the surface or, if they break, leave small holes in the painted surface.

Hot-dip coatings. Large quantities of metal parts are given corrosion-resistant coatings by being dipped into certain molten metals. Those most commonly used are zinc, tin, and an alloy of lead and tin.

Hot-dip galvanizing is the most widely used method of providing steel with a protective coating. After the parts, or sheets, have been cleaned, they are fluxed by dipping them into a solution of zinc chloride and hydrochloric acid. They then are dipped into a molten zinc bath. The resulting zinc coating is complex, consisting of a layer of $FeZn_2$ at the metal surface, an intermediate layer of $FeZn_7$, and an outer layer of pure zinc. Hot-dip galvanizing gives a good degree of corrosion resistance.

The coating thickness should be controlled; coatings that are too thick crack and peel. A wide variety of "spangle" patterns can be obtained by proper processing. When galvanizing is done properly, considerable subsequent bending and forming can be done without damaging the coating. However, rimmed steel should not be galvanized.

Tin "plating" also can be done by hot dipping. After the steel has been cleaned, it is dipped into the molten tin, the surface of which is covered with a layer of zinc chloride. In this manner the work passes through the zinc chloride before entering the molten tin. As the work leaves the tin bath it passes through rollers that are immersed in palm oil, thus removing the excess tin. However, most tin plate now is produced by an electrolytic process that gives more uniform coating with less tin being required.

Terne coating is similar to hot-dip coating, but an alloy of 15 to 20% tin and the remainder lead is used in place of pure tin. This process is thus cheaper than tin coating and provides satisfactory corrosion resistance for some purposes.

Phosphate coatings. Two phosphate coating processes are used extensively to provide corrosion resistance, usually to steel. In these processes the surface of the metal is converted into an insoluble crystalline phosphate by treatment with a dilute acid phosphate solution.

Parkerizing produces a fairly corrosion-resistant coating from 0.004 to 0.008 mm (0.00015 to 0.0003 inch). The treatment requires about 45 minutes and provides quite good corrosion resistance for parts that are to be kept painted. *Bonderizing* is similar to Parkerizing, but its primary purpose is not to give corrosion resistance but to form a surface to which paint will adhere tightly. The coating is thinner than that obtained by Parkerizing, but it reduces the activity of the metal surface so that corrosion at the paint–metal interface is retarded. As a result, if the paint coat is scratched, there is less likelihood of rust starting and progressing and thereby causing the paint adjacent to the scratch to loosen.

Blackening. Many steel parts are treated to produce a black, lustrous surface that will be resistant to rusting when handled. Such coatings usually are obtained by converting the surface into black iron oxide. One method is to heat the parts in a closed box of spent carburizing compound at 649°C (1200°F) for about $1\frac{1}{2}$ hours and then quench them in oil. Another method consists of immersing the parts in special blackening salts at 149°C (300°F) for about 15 minutes. A third method is to heat the parts in a rotary retort furnace to about 399°C (750°F). A small quantity of linseed or fish oil is then added. After a few minutes the parts are removed from the furnace, spread out, and allowed to cool. When they have cooled they are dipped into an oil that retards rust.

A *gun-metal finish* is obtained by heating the parts in a retort with a small amount of charred bone to 399°C (750°F). When the parts are oxidized, they are allowed to cool to about 343°C (650°F). A mixture of bone and some carbonic oil is then added and the heating continued for several hours. The work is then removed from the furnace and dipped in sperm oil.

ELECTROLYTIC FINISHES

Large quantities of both metal and plastic parts are electroplated to provide corrosion or wear resistance, improved appearance (such as color or luster), or an increase in dimensions. Plating is applied to virtually all base metals–copper, brass, nickel–brass, aluminum, steel, and zinc-base die castings—and also to plastics. In order to plate plastics, they first must be coated with some electrically conductive material.

The most common plating metals are tin, cadmium, chromium, copper, gold, platinum, silver, and zinc. Except for the making of tin plate for the container industry, chromium is by far the most common metal plated as the surface layer. However, in most cases a thin layer of both copper and nickel is deposited beneath the chromium. Gold, silver, and platinum are very important plating metals in the jewelry and electronics industries.

All electroplating processes essentially are the same, although the methods may vary somewhat in details. The basic process is indicated in Figure 37-12. The parts to be plated are made the cathode and suspended in a solution that contains dissolved salts of the metal to be deposited. The anode is a suspended slab of the metal to be deposited. Other materials may be added to the electrolyte to increase its conductivity. When a dc voltage is applied, metallic ions migrate to the cathode and, upon losing their charges, are deposited as metal upon it.

Successful plating depends greatly on (1) the preparation of the surface, (2) the ability of the bath to produce coatings in recessed areas, and (3) the crystalline character of the deposited metal. The actual deposition of the plating metal is governed by the bath composition and concentration, the bath temperature, and the current density. These are interdependent and must be carefully controlled to obtain satisfactory and consistent results.

Surfaces that are to be electroplated must be prepared properly to obtain satisfactory results. All defects, pin holes, and scratches must be removed if a smooth, lustrous finish is desired. The surface must be chemically clean. Proper combinations of degreasing, cleaning, and pickling are used to assure a clean surface to which the plating material will adhere.

Plating solutions are chosen on the basis of their *throwing power,* referring to their ability to deposit sound metal in recesses. Cyanide solutions have better throwing power than acid solutions and, therefore, commonly are used, although they are more dangerous to handle.

The plating metal tends to be attracted to, and build up on, corners and protrusions (see Figure 39-7). This makes it difficult to obtain uniform plating thickness on parts of irregular shape containing recesses and interior corners. Improved results can often be obtained by using several properly spaced anodes, or anodes having a shape similar to the workpiece. Obviously, there are limitations in the use of such procedures.

Nickel plating provides good corrosion resistance but does not retain its luster and is expensive. Consequently, it has largely been replaced as an outer coating

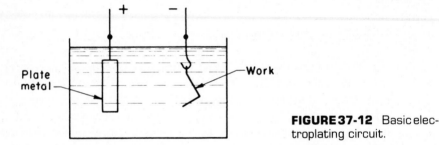

FIGURE 37-12 Basic electroplating circuit.

by chromium where appearance is important, and by cadmium for many applications where appearance is not of much importance and only moderate corrosion resistance is required.

As mentioned previously, chromium seldom is used alone. In modern practice, the first layer usually is bright acid copper, which produces a leveling effect and makes it possible to reduce the thickness of the nickel layer that follows. The nickel layer need not exceed 0.008 to 0.015 mm (0.0003 to 0.0006 inch). Chromium then is plated as the final layer to provide both protection and appearance.

A new procedure is to apply a very thin layer of stressed nickel, about 0.0005 mm (0.00002 inch), on top of the ordinary nickel layer, followed by chrome plating. The combined stresses produce fine microcracking in the chromium layer and result in improved corrosion resistance.

In mass production most electroplating is done as a continuous process, using the type of equipment shown in Figure 37-13. The parts to be plated are suspended on a conveyor and are lowered into successive plating, washing, and fixing tanks wherein the various operations are performed. Such methods make it possible to obtain economical plating where the volume of work is high. Ordinarily, only one type of workpiece can be plated at a time because the solutions, timing, and conditions of current density must be changed when different sizes and shapes are to be processed.

Hard chromium plating is used to build up worn parts to larger dimensions, to coat the face of cutting tools to reduce friction and wear, and to resist wear and corrosion. Hard chromium coatings always are applied directly to the base material and usually are much thicker than ordinary chrome plating, commonly from 0.07 to 0.25 mm (0.003 to 0.010 inch) thick. However, greater thick-

FIGURE 37-13 Modern, automatic, continuous-plating installation. Slabs of plating metal can be seen on both sides of the tank in the background. Racks of parts are lowered into the tank at the far left. (*Courtesy Udylite Corporation.*)

nesses—up to 0.76 mm (0.030 inch)—are used for such items as diesel cylinder liners. The hardness of hard chrome plating typically is from 66 to $70R_C$. Hard chrome plating does not have a leveling effect, and thus defects or roughness in the base surface will be amplified and made more apparent. Very smooth surfaces can be obtained by suitable grinding and polishing techniques.

Cadmium-titanium plating is used successfully to provide corrosion resistance for high-strength steels that are highly subject to hydrogen embrittlement when they are plated with either zinc or cadmium alone.

As will be discussed in Chapter 39, the ease and cost of obtaining satisfactory electroplating on parts can be greatly affected through design.

Electroforming. *Electroforming* is an important modification of electroplating that is used to produce metal parts by electroplating metal onto an accurately made mandrel that has the inverse contour, dimensions, and surface finish desired on the finished product. When the desired thickness of deposited metal has been obtained, the workpiece is separated from the mandrel; the method of separation depends on the mandrel material and its shape.

The mandrels are made from a variety of materials, including plastics, glass, Pyrex, and various metals (most commonly aluminum or stainless steel). The mandrel must be made electrically conductive, by a coating, if it is not made from metal.

Electroformed parts most commonly are made of nickel, iron, copper, or silver, and thicknesses up to 16 mm ($\frac{5}{8}$ inch) have been deposited successfully. Metals deposited by electroforming have their own distinct properties. Dimensional tolerances are very good, often up to 0.0025 mm (0.0001 inch), and surface finishes of 0.05 μm (2 microinches) can be obtained quite readily if the mandrel is adequately smooth.

A wide variety of parts and shapes can be made by electroforming, the principal limitation being that it must be possible to remove the part from the mandrel. In some applications a relatively thin electroformed shell is backed up with other materials, often cast into the shell, to provide the required strength. However, in most cases the wall thickness is made sufficient to provide the necessary strength.

Electroforming is particularly useful for high-cost metals and, because of the low tooling cost, for low production quantities. Some care must be taken to minimize internal stresses, which commonly may be from 13.8 to 34.5 MPa (2000 to 5000 psi), and which can be much higher if plating temperatures and current densities are not carefully controlled.

Anodizing. *Anodizing* is a process that is widely used to provide corrosion-resistant and decorative finishes to aluminum. It is somewhat the reverse of electroplating in that (1) the work is made the anode in an electrolytic circuit, and (2) instead of a layer of material being added to the surface, the reaction progresses inward, increasing the thickness of the highly protective, but thin, aluminum oxide layer that normally exists on aluminum.

One of the common forms of anodizing uses a 3% solution of chromic acid as the electrolyte at a temperature of about 38°C (100°F). The voltage is raised from about 0 to 40 at the rate of about 8 volts per minute, and then is maintained at full voltage for about 30 to 60 minutes with a current density of about 11 to 32 amperes per square meter (1 to 3 amp/ft^2). This treatment produces a converted layer that is about 0.0013 to 0.0025 mm (0.00005 to 0.0001 inch) thick. It is used primarily on aircraft materials. Because the coating is an integral part, the metal can be subjected to quite severe forming and drawing operations without destroying the coating or reducing its protective qualities. Parts that are anodized in this manner usually have a grayish-green color, resulting from the presence of reduced chromium in the coating. Other colors can be obtained by the use of suitable dye materials. The anodized surface also provides a good paint base.

More complex anodizing treatments are often used, some of which are *Alumilite* finishes. The most common of these uses a solution containing 15 to 25% sulfuric acid. This produces a transparent coating on pure aluminum and an opaque coating on alloys. These coatings are submicroscopically porous. A wide variety of colors can be obtained by the use of suitable dyes that penetrate these pores. Some of the colors will not resist sunlight, but it is possible to use a special type of dye and a sealing process that will produce good sun resistance.

Inasmuch as anodizing does not add to the dimensions, there is no necessity for providing any dimensional allowance, as must be done when electroplating is used.

ELECTROLESS PLATING

Electroless nickel plating.   Because of the almost impossibility of obtaining uniform plating thickness on even moderately complex shapes with electroplating, and because of the large amounts of energy consumed, extensive work has been done in developing *electroless plating,* with major success in plating with nickel. Basically, electroless plating is accomplished by autocatalytic reduction of the metallic ion of the plating metal in an aqueous solution in which the workpiece acts as the catalyst. In the case of electroless nickel plating, sodium hypophosphite acts as the reducing agent, reducing nickel salts to nickel metal and, incidentally, supplying a small amount of phosphorus so that the resulting plated material is a solid solution of phosphorus (about 8 per cent) nickel.

In addition to having as good corrosion resistance as electroplated nickel, electroless nickel has an as-deposited hardness of about 49 to 55R_C, which can be increased to as high as 80R_C by suitable heat treatment. Also, because it is purely a chemical process, the coatings obtained are uniform in thickness, not being affected by part complexity.

Nickel-carbide plating.   A very useful electroless plating process has been developed wherein minute particles of silicon carbide are codeposited in a nickel-alloy matrix. As shown in Figure 37-14, the particles of silicon carbide are 1 to 3 μm (0.000,04 to 0.0001 inch) in size and constitute about 25% of the volume of the deposit. Not only is the coating as corrosion resistant as nickel, because of the very high hardness of the carbide particles (about 4500 on the Vickers scale, where tungsten carbide is 1300 and hardened steel about 900, or

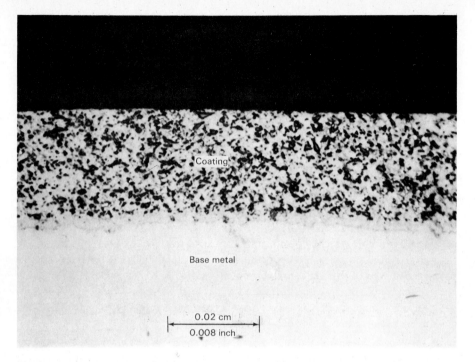

Coating

Base metal

0.02 cm
0.008 inch

FIGURE 37-14 Photomicrograph of nickel-carbide coating obtained by electroless plating. (*Courtesy Electro-Coatings, Inc.*)

$62R_C$), the wear and abrasion resistance of such coatings is outstanding. As with electroless nickel plating, the thickness of the coating is not affected by part shape. Figure 37-15 shows an example.

Coating thicknesses typically are up to 0.20 mm (0.008 inch). The process has a wide range of applications, being of outstanding value for coating molding dies for plastics that contain substantial amounts of abrasive filler materials, such as glass fibers.

A modification of this process utilizes minute artificial polycrystalline diamond particles in place of silicon carbide. Such coatings usually do not exceed 0.025 mm (0.001 inch) in thickness, and they have outstanding wear resistance.

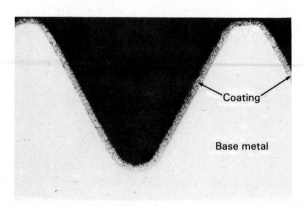

Coating

Base metal

FIGURE 37-15 Photomicrograph showing uniform deposit obtained on irregularly shaped part by electroless plating. (*Courtesy Electro-Coatings, Inc.*)

Impact plating, obtained by tumbling the parts in a tumbling barrel that contains a water slurry of very fine powder particles of the plating metal, glass spheres, and a "promoter" chemical, can be used to obtain a thin coating that is satisfactory for some purposes. The small glass balls peen the fine powder particles onto the workpieces, producing some cold welding. The deposited coatings are lamellar in structure and quite uniform in thickness. Any metal that can be obtained in a very fine powder form can be used as a plating material. One advantage of the process is that there is no danger of hydrogen embrittlement; therefore, it can be used on hardened steel.

VAPORIZED METAL COATINGS

Vacuum coating.   *Vacuum coating* is widely used to deposit thin films of metal and metal compounds on various substrate materials. The process involves the evaporation of the metal or the compound in a high vacuum and the subsequent condensation of the vapor on the cool workpiece. A pressure (vacuum) of from 0.013 to 1.33 Pa usually is required. Such coatings are used as electrical conductors and resistors in the electronics industry, as decorative coatings, and, because of their outstanding properties, as reflective surfaces. Virtually any metal can be deposited by the vacuum-coating process—aluminum, chromium, gold, nickel, silver, germanium, and platinum being very common. The coatings are usually less than 0.5 μm (0.00002 inch) in thickness, and often are as little as 0.025 μm (1 microinch).

Vaporization is a surface phenomenon and does not constitute boiling. As the metal leaves the heated surface in atomic form, it travels in line-of-sight direction to the surface of the substrate. Thus, if an entire surface of a shape is to be coated, it must be rotated to expose all the surfaces. Fortunately, in most cases only a single surface must be coated.

Because of the very thin deposits required, the process is very economical of metal, and expensive materials, such as gold or silver, often can be used economically over inexpensive part materials, such as plastics or steel.

Sputtering.   For certain applications, *radio-frequency sputtering* is used as a substitute for electroplating and vapor-deposited coatings. Its most extensive use is for depositing thin films of metals in making solid-state devices and circuits. The basic process is indicated in Figure 37-16. The substrate, upon which metal is to be deposited, and the source metal are arranged as shown in a gas-filled chamber (often argon) that is evacuated to about 10 to 50 μm pressure. The substrate is made positive, relative to the source material, by a radio-frequency power source. When the applied potential reaches the ionization energy of the gas, electrons, generated at the cathode, collide with the gas atoms, ionizing them and creating a plasma. These positively charged ions, having high kinetic energy, are accelerated toward the cathode target, overcome the binding energy of the target material, and dislodge atoms that then travel across the electrode gap and are deposited on the substrate. Because of the energy of these atoms, usually between 15 and 50 electron volts, their adherence to the substrate is

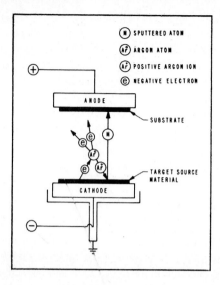

FIGURE 37-16 Schematic diagram of the radio-frequency sputtering process.

considerably better than if they were deposited by ordinary vacuum evaporation. This technique has been used to coat the edges of razor blades with chromium to prevent corrosion on the blade edges.

REVIEW QUESTIONS

1. Name three manufacturing processes that inherently result in the necessity for surface-cleaning and smoothing operations.
2. Why should a product designer be concerned about the finishing operations that may be required on the product?
3. What are three basic cleaning methods?
4. What materials commonly are used for abrasive cleaning?
5. What are the primary limiting factors in the use of abrasive cleaning?
6. Why should the barrel not be filled too full when tumbling is used for cleaning?
7. How may delicate parts be cleaned by tumbling?
8. What is the effect of wire brushing on a surface cleaned by that process?
9. On what type of surfaces is belt sanding effective?
10. What is the requisite condition for barrel finishing to be effective?
11. What is the basic difference in the action obtained in rolling and vibratory finishing barrels?
12. Basically, what type of process is buffing?
13. How is the stiffness of buffing wheels controlled?
14. Why is part design so important if buffing or belt sanding must be used to finish the part?
15. What type of surface is produced by barrel burnishing?
16. Explain how electropolishing produces a smooth surface.
17. What are three commonly used chemical cleaning methods?
18. Why is more than one of the three basic chemical cleaning processes often used on a given product?
19. Why may vapor degreasing not work well on parts made of thin aluminum?
20. What is the absolute condition that must be met for vapor degreasing to clean satisfactorily?
21. What is the purpose of pickling?
22. Under what conditions is ultrasonic cleaning usually employed?

23. Why have synthetic resin paints largely replaced oil-base paints?
24. Why are prime coats used in painting?
25. What four methods commonly are used for applying paints in manufacturing?
26. What is the reason for using the electrostatic principle in spray painting?
27. Why is rotating-disk atomization replacing air atomization in spray painting?
28. Why is dip painting limited in its use?
29. What are the principal advantages of electrocoating?
30. Why can infrared drying seldom be used for drying paints on wood products?
31. Why are paints that are excellent for finish coats usually not satisfactory for primers?
32. Why should the thickness of galvanizing be carefully controlled?
33. On what type of steel is galvanizing not satisfactory?
34. What are the differences in the purposes of Parkerizing and Bonderizing?
35. Why is it difficult to simultaneously put parts of widely differing shape through an automatic electroplating system?
36. What are two reasons why electroless plating is preferred over electroplating?
37. Why is part design especially important for parts that must be electroplated?
38. Why is vacuum coating usually preferred over electroplating when costly metals are the plating material?
39. What is the major advantage of sputtering over vapor vacuum evaporative coating?

CASE STUDY 37. The Broken Aluminum Wheel

Figure CS-37 shows a broken aluminum automobile wheel and a close-up view of the pieces that remained attached to the brake drum. It will be noted that one lug nut is missing; one relatively large piece was missing from the edge of the rim; and there is a dark abrasion on the rim almost directly opposite from the missing piece. The owner of the vehicle claimed that the wheel was defective, whereas the manufacturer claimed it had been broken as the result of abuse.

Chemical analysis revealed the metal had the following composition: silicon 7.55%; magnesium 0.32%; copper 0.31%; iron 0.53%; manganese 0.07%; titanium 0.08%; zinc 0.28%. The hardness was about 102 Brinell. Investigation found that a lug nut was missing from another wheel on the vehicle, and the tire mounted on the wheel had virtually no tread.

What is your opinion regarding the two claims, and why?

FIGURE CS-37 Photographs of broken aluminum automobile wheel with close-up view of brake drum region.

CHAPTER 38

Manufacturing Systems and Automation

In recent years, machine tools have become more powerful and versatile. Cutting tools have become more effective, capable of cutting at faster speeds with reduced cutting forces. Presses have become larger and tooling more sophisticated. However, the ability of a company to reduce costs by taking time out of the machining cycle (reducing CT or tool change time) has not done much to reduce overall manufacturing costs or make marked improvements in productivity. On an automatic screw machine producing a million-piece run, saving a couple of seconds per piece can be the difference between a healthy profit and a heavy loss. However, on a 25-piece lot, even a 10% reduction in machining time will have very little impact on total cost for the job when it takes 2 months to get the part through the shop or if the setup time for the job is 2 or 3 hours. Efficient utilization of the workers and the machines is critical to a successful organization. A good mental picture of the general utilization of machine tools is provided in Figure 38-1. Note that large blocks of time are unavailable for productive use because our plants are closed much of the time. Ultimately the productive fraction is about 6%, 525 hours per year or about $2\frac{1}{4}$ hours per day. Table 38-1 compares time utilization in numerically controlled machining centers versus machining on conventional machine tools and machining in transfer machines. Thus, we observe that for machine tools used in lot production, only about one-fourth of the time at best is spent adding value (making chips). NC

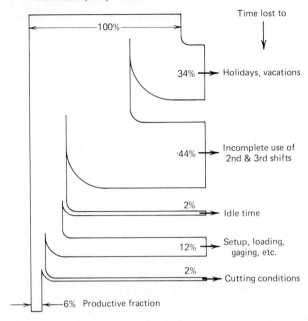

Theoretical capacity 100%

100%

Time lost to

34% → Holidays, vacations

44% → Incomplete use of 2nd & 3rd shifts

2% → Idle time

12% → Setup, loading, gaging, etc.

2% → Cutting conditions

6% Productive fraction

FIGURE 38-1 Distribution of total factory capacity of machine tools. Typical values are shown. (*From C. F. Carter, "Toward Flexible Automation,"* Manufacturing Engineering, *Aug. 1982, p. 75.*)

does reduce setup time, but considerable improvement can be made in the noncutting areas.

One of the reasons for this situation is the complexity of the manufacturing system, which requires elaborate schedules and long lead times in the planning. Superimposed on the inherent complexity are delays and disturbances caused

TABLE 38-1   Time Utilization of Conventional Equipment versus NC Machining Centers and Transfer Machines As Percentage of Overall Machining Cycle

Activity	Machining Center (Small-Medium Lots)	Conventional Machine (Small Lots)	Transfer Line (Large Lots)
Metal cutting	23% of time	20% of time	50% of time
Positioning or tool changing or transfer	27% of time	10% of time	12% of time
Gaging	8% of time	15% of time	1% of time
Loading/unloading	10% of time	15% of time	—
Setup[a]	5% of time	20% of time	—
Waiting or idle[b]	14% of time	13% of time	22% of time
Repair and technical problems	13% of time	7% of time	15% of time

[a]Actual setup of a transfer line may take 6 to 24 months.
[b]Due to stockouts, personnel allowance, slow machine cycles, etc.
Source: After C. F. Carter, "Toward Flexible Automation," *Manufacturing Engineering,* Aug. 1982, p. 75.

by engineering design changes, vendor failures, material shortages, emergency orders, machine failures, and so on. One way to cope with this has been to create a buffer at each machine to ensure that each machine has work available, but this in turn creates a large in-process inventory. Another strategy has been to use daily lists, colored tags, and people to expedite orders through the shop to conform to changes in the needs. Needless to say, this solution itself acts as a disturbance to the regularly scheduled work in progress. What is needed is effective shop floor management and organization to make sure that machines and people are utilized to the greatest possible extent. That is, more is to be gained through better overall management and planning of the manufacturing system with machines running at moderate speeds and feeds than operating at extremely fast but sporadic machining rates.

These final two chapters will try to outline how the job shop can be reorganized to take advantage of these excellent opportunities to improve productivity. We will begin with a brief discussion of the classic manufacturing systems, show how they are tied to the entire production system, and then concentrate on the most commonly used system, the job shop, indicating how it can be reorganized so that it can be more productive.

Manufacturing systems.   In manufacturing systems there have classically been three kinds of systems defined. The oldest of these is the *job shop,* a transformation process in which units for different orders follow different paths (individual routes or sequences) through processes or machines (see Figure 38-2a).

The characteristics of this kind of production are its flexibility, wide variety of product design or customer service, and the need for many highly skilled people, much indirect labor, and a great deal of manual material handling (loading, unloading, setting up, and adjusting). General-purpose machines are grouped by function and adapted to the special requirements of different orders. The price for this flexibility is long in-process times, as shown in Figure 38-3, large in-process inventories of components and materials, and the formidable task of scheduling different orders through the functional centers in which machines and people must be shared. As this kind of system grows to any size, the levels of cost and chaos increase, delivery times become long, orders get lost, and quality becomes highly variable.

For products or services where the basic design is somewhat more stable over a longer period of time resulting in larger built quantities, the manufacturing system is organized along the lines of a *flow shop* (see Figure 38-2b). The flow shop is a transformation process in which successive units of output undergo the same sequence of operations with more specialized, dedicated equipment, usually involving a production line of some sort. This is generally viewed as mass production.

The flow shop can be viewed as being either "continuous" or "intermittent." In a *continuous-flow shop,* the process goes on producing the exact same output or the same type of output in great volume. The extreme of this system is a truly continuous process, like an oil refinery. The *intermittent-flow shop* is where the process is interrupted to modify the setups to handle different specifications of the same basic design or group of designs. Typically, all the units follow the

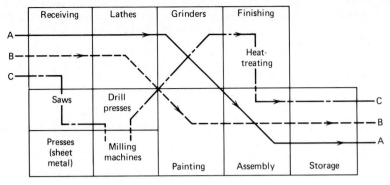

(a) Job shop — functional or process layout

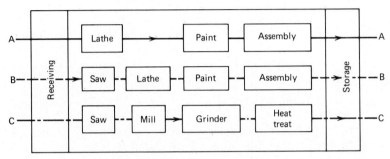

(b) Flow shop — line or product layout

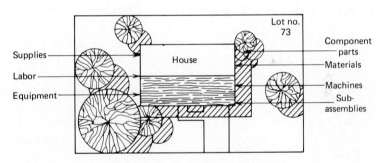

(c) Project shop — fixed position layout

FIGURE 38-2 Schematics of layouts of three classical manufacturing systems. (a) Job shop; (b) flow shop; (c) project shop. Layout for continuous processes not shown.

same sequence of operations and pass through all machines. The most automated examples of this system for machining are called *transfer lines*.

The third kind of system involves the *project shop* (see Figure 38-2c). This system is directed toward creating a product or service which is either very large (immobile) or one-of-a-kind, with a set of well-defined tasks, which typically must be accomplished in some specified sequence. The men, materials, and machines all come to the project site for assembly and processing. The project

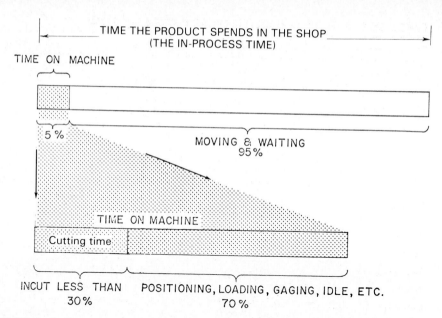

FIGURE 38-3 In the job shop, the parts spend a great deal of the in-process time moving or waiting and only about 5% of the time are they on machines. (*From C. F. Carter, "Toward Flexible Automation,"* Manufacturing Engineering, *Aug. 1982, p. 75. Used by permission of the author and the publisher.*)

shop is generally backed up by a job shop–flow shop operation to supply component parts.

Table 38-2 summarizes this introductory discussion and gives some examples of each of these classifications of production processes. Note that examples of all these types of production systems exist in both the manufacturing form (consumer or producer goods are made) and the service form, where some service is rendered to a customer.

Associated with each of the classical manufacturing systems, characterized in Figure 38-2, are systems to handle information, material movement and storage (inventory), purchasing, planning, product movement, and so forth (see Figure 38-4). Collectively, these activities represent the *production system* and the level of the activity will vary for each of the basic manufacturing systems.

The job shop represents the most common type of system, however, with estimates ranging from 30 to 50% of the systems being of this form. A typical manufacturing shop may produce 2000 to 10,000 different components per year on 100 to 500 machine tools in lot sizes ranging from 10 to 100 units, so the cost of changeovers to meet new market needs increases as the batch or lot size increases, while the unit manufacturing cost goes down as more automation and mechanization can be applied. In most job shops, there will be some items made in sufficient quantity to be made via flow-line methods, so common practice in the United States has been a system which is mixed between job and flow shops (see Figure 38-7).

In recent years two new types of manufacturing systems have emerged, one called *cellular* and the other, *flexible manufacturing systems* (FMS). These are

TABLE 38-2 **Characteristics of Basic Manufacturing Systems**

Characteristics	Job Shop	Flow Shop	Project Shop
Types of machines	Flexible, general purpose	Special purpose; single function	General purpose; mobile
Design of processes	Functional or process	Product flow layout	Project or fixed-position layout
Setup time	Long, variable	Long	Variable
Workers	Single functioned; highly skilled; one man–one machine	One function; lower skilled; one man–one machine	Single function; skilled; one man–one machine
Inventories	Large inventory to provide for large variety	Large to provide buffer storage	Variable
Lot sizes	Small to medium	Large lot	Small lot
Production time per unit	Long, variable	Short; constant	Long; variable
Examples in goods industry	Machine shop; Tool and die shop	TV set factory; automobile assembly line	Ship building; house construction
Examples in service industry	Hospital; restaurant	College registration; cafeteria	Movie; TV show; play; buffet

the sort of systems which can evolve when a functional system is analyzed and reorganized using a *group technology* (GT) approach.

Group technology is a *systems* approach to the reorganization problem, resulting in a cellular or flexible arrangement of machines which manufacture groups of similar component parts (families of parts). The machines in these systems are tooled so that one can rapidly change over from one lot of components to another. Thus, setup times are reduced to a matter of minutes or even eliminated.

Work cells can exist in either *manned* or *unmanned* forms. Figure 38-5 shows a manned version with four machine tools and four automatic inspection stations. The worker(s), who can operate all the machines, takes parts from the machines and places them in the automatic inspection devices. He takes good parts from the automatic inspection devices and places them in the next machine. Because the worker can operate all of the machines, he is called *multifunctional*. The automatic inspection system is an example of *autonomation*, automatic inspection.

Figure 38-6a shows an unmanned version of a cellular manufacturing system wherein a robot is being used to load and unload the machines, which are numerical-control-type machines. That is, all the machines in this cell are programmable. In this cell, three different parts are made and each part has six different sizes. Thus, the part family has 18 different parts.

In Figure 38-6b, a layout for a flexible manufacturing system is shown. Here again the machines are programmable, but the material handling is accomplished

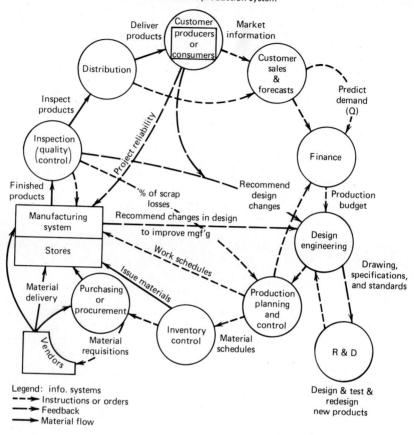

Functions within the production system

FIGURE 38-4 Typical functions within a production system which includes the manufacturing system. (*After James L. Riggs, Production Systems: Planning, Analysis and Control, 3rd ed., John Wiley & Sons, Inc., New York.*)

by a large conveyor system. The parts are usually mounted on pallets so that they can be shuttled to any of the machines, as needed. Thus, a flexible path flow of parts is developed. All the machines and the conveyor system are under the direct computer control of one computer. These systems usually have somewhere between 5 and 12 machines and are usually designed to handle larger families of parts than the cellular system. These systems are also called DNC (direct numerical control) systems.

Generally speaking, the least productive system is the job shop, resulting in products or services which are costly and in which the costs tend to keep rising with inflation. In addition, there are some social and technological trends which suggest that the number of, and need for, small-lot production systems will increase in the future. These trends are:

1. Proliferation of numbers and varieties of products and services. This results in smaller production lot sizes (as variety is increased) and decreased product

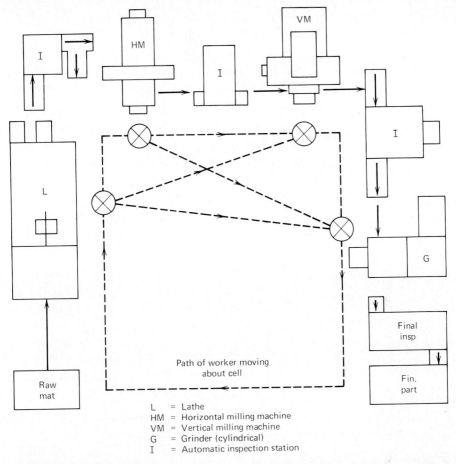

L = Lathe
HM = Horizontal milling machine
VM = Vertical milling machine
G = Grinder (cylindrical)
I = Automatic inspection station

FIGURE 38-5 Schematic of a manned work cell using conventional machine tools with autonomation.

life cycles. This means that setup costs as a percentage of total costs are often higher and the problems of managing inventories and materials become vastly greater.

2. Requirement for closer tolerances and greater precision.
3. Increased cost of product and service *liability* as consumers demand accountability on behalf of the manufacturers of products or services, putting greater emphasis on reliability and quality.
4. Increased variety in materials with greater diverse properties, requiring great flexibility and diversity in the processes used to machine or form these new materials. This goes hand-in-hand with the increased variety in the processes themselves. The 1980s will be the "age of composite materials," where even the properties of the materials become a design parameter.
5. The cost of energy needed to transform materials, the cost of capital, and the cost of materials themselves keep increasing.

On the other side of the coin, pulling against these trends are:

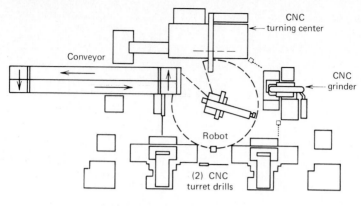

a) Cellular (unmanned)

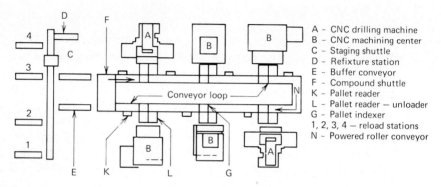

A – CNC drilling machine
B – CNC machining center
C – Staging shuttle
D – Refixture station
E – Buffer conveyor
F – Compound shuttle
K – Pallet reader
L – Pallet reader – unloader
G – Pallet indexer
1, 2, 3, 4 – reload stations
N – Powered roller conveyor

b) Flexible manufacturing system — FMS

FIGURE 38-6 Layouts for (a) cellular (unmanned) manufacturing systems (*after ASEA Inc.*); and (b) flexible manufacturing system (*after White Consolidated Industries Manufacturing Systems Division*).

1. The need to improve productivity markedly and reduce the costs of goods and services, to halt inflation and meet the prices of international competition.
2. The need to improve quality and reliability.
3. The move away from a labor-intensive manufacturing environment toward a service environment. Today, fewer people want to work in the manufacturing sector, reversing the trend that began a hundred years ago when people left the farms to go to work in the factories.
4. The need for shorter service times or lead times in production to reduce inventories.
5. Worker demands for improved quality of working life.

Overlaying these trends is the continued rapid growth of process technology, led by computer technology. No one can forecast the magnitude of this technological development, but it is clear that no segment of the production system will escape its impact. Computer-aided design, manufacturing, planning, testing, inspection, and information systems are being integrated into the factory of the future.

The United States has been slow to adopt these advanced manufacturing technologies and systems (cellular and flexible manufacturing systems). One of the best ways to reorganize a system is to embark on a GT program.

Group technology offers a systems solution to the organizational and economic problems of the functional system, allowing a significant portion of it to be reorganized into either cellular or flexible manufacturing systems. These conversions represent systems level changes, which means there will be the potential for tremendous savings but, because of the magnitude of the changes, careful planning and full cooperation from everyone involved are absolutely required.

Therefore, we conclude that because such conversions are at a systems level of *reorganization,* their greatest strength will also be their greatest difficulty. Understanding the problems and limitations of a systems conversion by all parties concerned is absolutely necessary.

Group technology. There are those who will say that group technology is a technique for coding and classification of component parts, for the purpose of determining the similarities in the parts in terms of their design and/or their manufacturing processes and sequences. But it is much, much more than that. GT provides a way to reorganize the job shop, whether it is a service or a manufacturing shop. *It is a systems approach to the redesign and reorganization of the functional shop* and it will have an impact on every segment of the job shop. *It is a manufacturing management philosophy which identifies and exploits the sameness of items and processes used in manufacturing or service industries.*

The application of this concept to a manufacturing facility results in the grouping of units or components into families wherein the components have similar design or manufacturing sequences. Machines are then collected into groups or cells (machine cells) to process the family. Portions of all of the manufacturing systems are then restructured to make families of components. The idea is to group similar components into families of parts which can be manufactured by a group or set of processes. Thus, portions of the functional system are converted, in steps, to new forms of manufacturing systems, the cellular system or the flexible manufacturing system.

This implies a great number of things. It means redesigning the entire *production system* and all functions related to it. In shifting from one type of system to another, the change will affect product design, tool design and engineering, production scheduling and control, inventories and their control, purchasing, quality control and inspection, and of course the production worker, the foreman, the supervisors, the middle managers, and so on, right up to top management. Such a conversion cannot take place overnight. These conversions must be viewed as a *long-term transformation* from one type of *production system* to another.

Note that it is unlikely that the entire shop will be able to convert into families, even in the long run (see Figure 38-7). Thus the total collection of manufacturing systems will be a mix, evolving toward a more continuous flow system over the years. This will create scheduling problems, as in-process times for components will be vastly different for products made by cellular or flexible systems versus those made under traditional job shop conditions. However, as the volume in the functional area is decreased, the total production system will become more efficient.

Finding families of parts is one of the first steps in converting the functional system to a cellular/flexible system. There are basically three ways to do this:

1. Tacit judgment or eyeballing.
2. Analysis of the production flow.
3. Coding and classification.

The eyeball method is, of course, the easiest and least expensive, but also the least comprehensive. This technique clearly works for restaurants, but not in large job shops where the number of components may approach 10,000 and the number of machines 300 to 500.

However, consider the problem of a manufacturing engineer trying to justify a NC machining center. What does he do? He "eyeballs" the shop and selects, using his experience and judgment, as many high-cost, complex parts as he can which can be machined on the NC center. He next performs an economic analysis to cost-justify the new machine. In fact, he has found a family of component parts (or at least a partial family) and the NC machining center clearly represents the simplest of machine cells—that is, one machine. However, this approach leaves much to be desired in terms of systems conversion.

The second method, *product flow analysis* (PFA), uses the information available on route cards.[1] This is explained by Figure 38-8. The idea is to sort through all the components and group them by a matrix analysis, using product routing information. This method is more analytical than tacit judgment, but not as comprehensive as coding/classification. PFA can get rather cumbersome when the numbers (of components and machines) become large. Sampling can be used (20 to 30%) to alleviate the size problem, but then one is never sure one got all the potential members into the family, or how big the family really is. However, PFA is a valuable tool in the systems reorganization problem. For example, it can be used as an "up-front" analysis. That is, the decision as to whether or not to go into group technology may require some sort of "before-the-fact" analysis which will yield some cost/benefit information for the decision maker, which has to be the chief executive officer of the company. Clearly, such information is not routinely available, but PFA can serve as such an analysis tool. The decision maker would have some information on what the company could expect in terms of the percentage of their product which could be made by cellular or flexible methods, what the families might look like, what coding/classification system would work best for them, how much money they might have to invest in new equipment, and so forth.

In short, it would greatly reduce the uncertainty in making the decision on reorganization. As part of this technique, an analysis of the flow of materials in the entire factory is performed, laying the groundwork for the new layout of the plant.

[1]See Chapter 39 for an example of a route card.

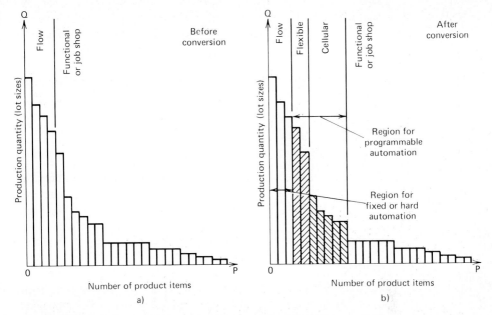

FIGURE 38-7 P-Q charts. The products are arranged in descending order of production quality (Q). P is the number of different components while Q is the *build* quantity. In (a) there are two regions: flow and functional. In (b) there are four regions: flow, flexible manufacturing, cellular, and functional.

By far the greatest number of companies converting to a cellular system have used a coding/classification method. There are design codes, manufacturing codes, and codes that cover both design and manufacture.

Classification sorts items into classes or families based on their similarities. It uses a code to accomplish this goal. Coding is the assignment of symbols (letters or numbers or both) to specific component elements based on differences in shape, function, material, size, processes, and so on.

No attempt to review coding/classification (C/C) methods will be made here. C/C systems exist in bountiful numbers in published literature and from consulting firms. Most systems are computer-compatible, so the computer sorting of the codes generates the classes or families.

Whatever C/C system is selected, it should be tailored to the production of the particular company and should be as simple as possible, so that everyone understands it. It is not necessary that old part numbers be discarded, but every component will have to be coded prior to the next step in the program, finding families of parts. This coding procedure will be costly and time consuming, but most companies opting for this conversion understand the necessity of performing this analysis.

The reader should understand that the families will not be all the same with regard to their material flow and therefore will require different designs (layouts) to process the families. In some families, every part will go to every machine in exactly the same sequence and no machine will be skipped and no back flow will be allowed. This is, of course, the purest form of a cellular system, except

Left matrix — "as found in the typical job shop"

Job Number	A	B	C	D	E	F	G	H	I	J
1								X		
2		X	X							
3				X						
4							X	X		
5	X	X	X							
6									X	X
7	X		X							
8						X		X		
9									X	X
10				X	X					
11	X	X	X				X			
12						X	X			
13									X	
14				X	X					
15									X	X
16		X				X	X	X		
17										X
18	X	X								
19						X	X	X		
20				X						

A matrix of jobs (by number) and machine tools (by code letter) as found in the typical job shop.

Right matrix — rearranged to yield families of parts

Job Number	A	B	C	D	E	F	G	H	I	J
7	X		X							
11	X	X	X				X			
2		X	X							
5	X	X	X							
18	X	X								
14				X	X					
3				X						
10				X	X					
20				X						
12						X	X			
4							X	X		
19						X	X	X		
16						X	X	X		
8	X					X		X		
1								X		
9									X	X
13									X	
6									X	X
15									X	X
17										X

Annotations on the right matrix:
- The X at G for job 11 is marked "X (exception)".
- "Cell will have 3 machines F, G, H, for manufacture of 6 jobs."
- The X at A for job 8 is marked "X (exception)".

A matrix rearranged to yield families of parts and associated groups of machines that can form a cell.

FIGURE 38-8 Schematic to explain production flow analysis, a technique to find families of parts based on the production routing.

perhaps for the single machining center, in which all parts are done on one machine.

Other families may require that not all components go to all machines, that the forward sequence of order through the machines can be modified, or that back flow in the group can exist. FMSs are designed to accommodate this situation, as are many cellular systems. This in no way alters the basic concept. It does add to the complexity of scheduling parts through the machines.

Formation of work cells and flexible systems. The formation of families of parts leads to design of FMS and work cells, but this step is by no means automatic. It is the critical step in the reorganization and must be carefully planned. Remember, the objective here is to convert the functional system with its functional layout into a flexible, group layout. The following constraints to the formation of machine groups should be noted.

1. There should be sufficient volume of work in the family to justify establishment of a FMS or work cell. Some families may be too small to form a group of machines but are good candidates for processing on a single CNC or machine center, which is the smallest form of a cell.
2. Composition of component families should permit a satisfactory utilization situation. In manned cells, this means that every machine will not be utilized totally or even that the machine utilization rate will be greater than what it was in the functional system. *The objective in manned cellular manufacturing is to improve utilization of the people.* In fact, one of the inherent results of the conversion to a cellular system is that the worker becomes multifunctional. That is, he learns to operate many machines or perform many duties or tasks. In unmanned cells and FMS, the high utilization of the equipment is more important.
3. The processes in these systems should be technologically compatible.

The conversion of the functional shop should occur in steps or stages. One system at a time should be designed and staffed. Initial cells are often of the manned variety, composed of conventional machines currently in use in the shop.

There exists little information in manufacturing texts at this time regarding the design and implementation of such arrangements of machines. Clearly, these manufacturing systems are different from flow or job shop and will require extensive, intensive study and analysis (research) on the part of industrial and manufacturing engineering scholars to discover the fundamental truths of such systems.

The selection of the first family to go flexible or cellular will always be critical. It must be carefully planned and staffed. The needed capacity of the system must be determined from quantities of parts needed and the production schedule, which determines when they are needed. Problems of balance of labor and machine utilization will be encountered along with the scheduling problems, which will be of two types:

1. Scheduling of jobs within the family through the new systems for most efficient operation of the systems themselves.
2. Scheduling of the entire manufacturing system, which will be in a state of change for some months, even years, while the transformation takes place.

Scheduling of a mixed shop will have difficulties with respect to timing of the components to arrive at assembly points on time. However, scheduling overall will be easier as the control has moved down to where the work is being done.

Some people argue that manned cellular processing can be accomplished by simply routing families through the machines without forming them into cells. This defeats many of the benefits of the cellular system and is obviously never

TABLE 38-3 Characteristics of Cellular and Flexible Manufacturing Systems

Cellular Manufacturing Systems	Flexible Manufacturing Systems
General	
Small to medium-sized lots of families of parts (1 to 200 parts)	Medium-size lots of families of parts (200 to 10,000 parts)
1 to 15 machines (A(3) and A(4) machines)	6 to 12 CNC machines (A(4) machines)
Rapid changeover—"single setup"	Rapid changeover—"single setup"
Significant reductions in inventory	Significant reductions in inventory
Greatly improved quality control through autonomation (100% inspection)	Greatly improved quality control through autonomation
Unmanned	
Flexible/programmable CNC machines	Flexible/programmable machines
Robotic integration for parts handling	Integrated conveyor system for parts and tooling
Networked computer control	Networked computer control
Manned	
General-purpose machines (A(2) machines)	
Multifunctional worker	
Natural job enlargement and job enrichment	

going to lead to conversion of the functional system to a cellular or flexible system. The primary characteristics of the cellular and flexible systems are summarized in Table 38-3.

The manned cellular system provides the worker with a natural environment for job enlargement. Greater job involvement enhances job enrichment possibilities and clearly provides an ideal arrangement for improving quality.

The similarity in shape and processes needed in the family of parts allows setup time to be reduced or even eliminated. Being able to do the setup in less than 10 minutes or in single-digit (0 to 9) minutes is called *single setup*.

In the manned cellular system, the worker is decoupled from the machine, so that the utility of the worker is no longer tied to the utility of the machine. (*Note:* This means there may be fewer workers in the cell than machines.) The objective is to improve the utilization of the people, which is accomplished by making the worker multifunctional, capable of performing all the processes in the cell.

In advanced forms of unmanned cellular layouts, the microcomputers of the CNC machine tools are networked together with a robot for material handling within a cell. It is difficult if not impossible to conceive of this kind of arrangement without resorting to some method which collects the work into compatible families. All the machines in the cell are programmable, and therefore, this kind of automation is very flexible as well as more economical than traditional job shop processing.

Benefits of conversion. The benefits of a conversion to programmable manufacturing are noted in Figure 38-9, which indicates areas for significant cost savings over a 2- to 3-year period. Specifically, manufacturing companies report significant reductions in raw materials and in-process inventories, setup and throughput times, direct labor, indirect labor, overdue orders, tooling costs, and costs to bring new designs on line.

However, this reorganization has a greater, immeasurable benefit. It prepares the "soil" so that the "seeds" of computer-aided and automated manufacturing of the future will fall on fertile ground. The progression from the functional shop to the shop with machine cells to a cluster of CNC machines to direct numerical control for the entire system must be accomplished in logical, economically justifed steps, each building from the previous stage.

Since its initiation in the USSR in the late 1950s, the group technology concept has been carried throughout the industrialized world and is well rooted

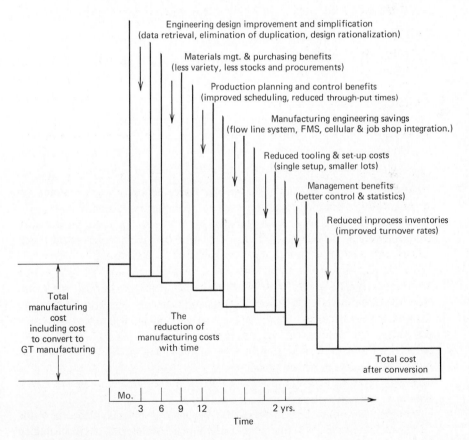

FIGURE 38-9 Reduction of manufacturing systems costs through implementation of group technology and cellular manufacturing.

in Germany, the USSR, and Great Britain, and especially Japan, where it is said to be a "way of life" in many of the manufacturing facilities and where many examples of the cellular and flexible manufacturing systems exist. One might ask why this concept has not taken stronger hold in the United States.

Clearly, it requires a major effort on the part of a business to undertake such a conversion. The constraints to implementation of a cellular conversion program into a company are as follows:

1. Systems changes are inherently difficult to implement. Changing the *entire* production system is a huge job.
2. Companies spend freely for product innovation but not for process innovation. Few machine tool companies sell integrated and coordinated systems with their machining centers.
3. Fear of the unknown. Decision making is choosing among the alternatives in the face of uncertainty. The greater the uncertainty, the more likely that the "do-nothing" alternative will be selected.
4. Faulty criteria for decision making. Decisions should be based on the ability of the company to compete (quality, reliability, delivery time, flexibility for product change or volume change) rather than output or cost alone.
5. High initial cost, plus long-term payback equals a high-risk situation in the minds of the decision makers, but is there really an alternative?
6. Short-term life of (financially oriented) middle managers versus long-term nature of program.
7. Union resistance based on fear of loss of jobs.
8. Union resistance to the multifunctional worker.
9. Lack of vertical communication in the company.
10. Lack of blue-collar involvement in the decision-making process in the company.
11. Quantification of the entire conversion in terms of dollars for management decision makers. How will we look 3 to 5 years from now?

Clearly, additional education is needed to overcome these constraints. Changing to a cellular/flexible system requires an evolutionary, dynamic philosophy but such conversions offer great potential for markedly improving the quality and productivity of our production systems while laying the organizational groundwork for CAD/CAM or CATI innovations of the future.

Automation. The term "automation" has many definitions. Apparently it was used first in the early 1950s to mean automatic materials handling, particularly equipment used to unload and load stamping equipment. It has now become a general term referring to services performed, products manufactured and inspected, information handling, materials handling, and assembly, all done automatically as an automatic operation.

In Chapter 1, Amber & Amber's Yardstick for Automation was presented. This classification is based on the concept that all work requires energy and information, and these must be provided by man or substituted for man by a machine. Whenever a machine assumes a human attribute, it is considered to have taken an "order" of automaticity. In today's factory, levels A(2), A(3), A(4), and A(5) are found, with occasional A(6) levels. This portion of the yardstick is given in greater detail in Table 38-4. Mechanization refers to the first and second orders of automaticity, which includes semiautomatic machines. Virtually all of the machines described in the previous chapters are A(2) machines. Automation as we know it today begins with the A(3) level. In recent years, this level has taken on two forms: hard or fixed-position automation, and soft or flexible or programmable automation. Instructions to the machine, telling it what to do, how to do it and when to do it, are called the program. In hard automation, as in transfer lines, the programming consists of cams, stops, slides, and hard-wired electronic circuits using relay logic. An example of this level of

TABLE 38-4   Yardstick for Automation: Levels A(3) to A(6)

Order of Automaticity	Human Attribute Mechanized	Discussion	Examples
A(3): automatic repeat cycle or open loop control	*Diligence:* carries out routine instructions without aid of man; open end or non-feedback	All automatic machines; loads, processes, unloads, repeats; system assumed to be doing okay, probability of malfunctions negligible; obeys fixed internal commands or external program; see Figure 38-24	Record player with changer; automatic screw machine, bottling machines, clock works, donut maker, spot welder, engine production lines, casting lines, newspaper printing machines, transfer machines
A(4): self-measuring and adjusting feedback or closed loop systems	*Judgment:* measures and compares result (output) to desired size or position (input) and adjusts to minimize any error	Self-adjusting devices; feedback from product position, size, velocity, etc.; multiple loops are possible	Product control, can filling, NC machine tool with position control, self adjusting grinders, windmills, thermostats, waterclock, fly ball governor on steam engine
A(5): adaptive or computer control; automatic cognition	*Evaluation:* evaluates multiple factors on process performance, evaluates and reconciles them; uses mathematical algorithms	Process performance must be expressed as equation	NC machine with AC capability, maintaining pH level, turbine fuel control
A(6): artificial intelligence or limited self-programming	*Learning by experience*	Subroutines are a form of limited self-programming; trial-and-error sequencing; develops history of usage	Phone circuits, elevator dispatching

Source: Amber & Amber, *Anatomy of Automation,* Prentice-Hall, Inc., Englewood Cliffs, N.J., 1962. Used by permission of Amber & Amber.

automation is the automatic screw machine. If the machine is programmed with a tape, a programmable controller (PC), a hand-held control box, or a microprocessor, the control means is easily changed, thus making the system or device much more flexible.

Self-adjusting and measuring machines are the A(4) level replacing human judgment and allowing these machines to be self-correcting. This is commonly known as *feedback control* or *closed loop,* meaning that information about the performance is fed back (or looped back) into the process. Figure 38-10 shows a block diagram of a machine with feedback control. A simple A(4) level is one in which the position of the workpiece or the output of the process is measured using some sort of detection device (sensor), and this information is fed back to a comparator, which makes comparisons with the desired level of operation. If the output and the input are not equal, an error signal is created and the process adjusts to reduce the error. An industrial example of an A(4) machine would be a numerical control machine or an automatic grinder which checks the size of a part and automatically repositions the grinding wheel to compensate for wheel wear and attrition.

The A(5) level, typified by *adaptive control* (AC), replaces human evaluation and is shown in the lower block diagram of Figure 38-10. Basically, A(5) machines are capable of adapting the process itself so as to optimize it in some

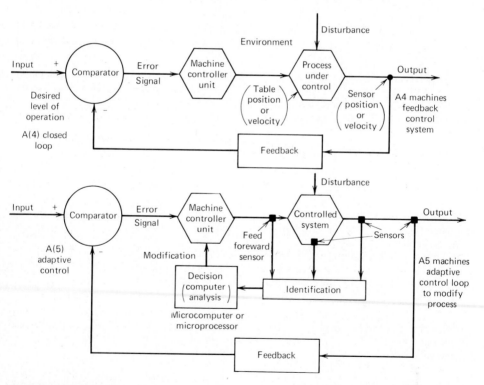

FIGURE 38-10 Block diagrams for A(4) and A(5) levels of automation, showing differences in the feedback loops.

way. This level of automation requires that the system have a computer. Programmed into the computer are models (mathematical equations) which describe how this process or system behaves, how this behavior is bounded and what aspect of the process or system is to be optimized. *This modeling obviously requires that the process be sufficiently well understood theoretically so that equations can be written which describe how the real process works.* This level of automation has been readily achieved in the realm of continuous processes (oil refineries, for example) where the theory (of heat transfer and fluid dynamics) of the process is well understood and parameters are easier to measure. Unfortunately, the theory of metal forming and metal removal is less well understood and parameters are usually difficult to measure. Consequently, these processes have resisted adequate theoretical modeling, so there are very few A(5) machines on the shop floor. The basic elements of the adaptive control loop, as shown in Figure 38-10, are:

1. *Identification:* measurements from the process itself or its output:
2. *Decision Analysis:* optimization of the process in the computer.
3. *Modification:* signal to the controller to alter the inputs.

A simple example of an A(5) machine might be a cylindrical center-type grinder on which grinding forces (as well as part size) are measured. In this process, the cutting forces tend to deflect the part more as the grinding wheel gets further from the centers. The AC program would have equations which relate deflection to grinding forces and would alter the infeed to reduce the force and minimize deflection. Notice that the overall system would still have to compensate for grit dulling and grit attrition, which will also alter the grinding forces. Thus we see that even a simple AC system will be quite complicated.

The A(6) level of automation was originally termed by Amber & Amber as "learning by experience" or limited self-programming. Today, this activity falls under the heading of artificial intelligence (AI), wherein the ability to learn (by experience) is built into the program or software in the control computer or some higher-level computer that is networked to the computer at the machine.

Programmable controllers (PCs). Modern PCs have the functional sophistication to perform virtually any control task. These devices are rugged, reliable, easy to program, and economically competitive with any alternate control device and have replaced conventional hard-wired relay panels in many applications. Relay panels are hard to reprogram, whereas PCs are very flexible. Relays have the advantage of being well understood by maintenance people and are invulnerable to electronic noise, but construction time is long and tedious. PCs allow for mathematical algorithms to be included in the closed-loop control system and are being widely used for single-axis, point-to-point control as typically required in straight-line machining, robot-handling, and robot-assembly applications. While they do not at this time challenge computer numerical controls (CNC) that are used in the control of complex multiaxis contouring machines, applications in monitoring temperature, pressure, and voltage on such machines with PCs is a growing trend. Recent applications have included on-line tool wear compensation and automatic tool setting. PCs are rapidly becoming the control device being used on transfer lines to handle complex material movement problems, gaging, and automatic inspection, giving these systems flexibility that they never had before.

Mass production, flow lines, hard automation. Ever since the birth of mass production, various approaches and techniques have been used to develop machine tools that would be highly effective in large-scale manufacturing. Their effectiveness was closely related to the degree to which the design of the products was standardized and the time over which no changes in the design were permitted. If a part or product is highly standardized and will be manufactured in large quantities, it is quite easy to develop a machine that will produce the parts with a minimum of skilled labor. A completely tooled automatic screw machine is a good example. However, such specialized machines are expensive to design and build and may not be capable of making any other products. Consequently, to be economical, such machines must be operated for considerable periods of time, making only the one product for which they are designed. These machines, although highly efficient, can be utilized only to make products in very large volume, and desired changes of design in the products must be avoided or delayed because it would be too costly to scrap the machines.

However, as we have already noted, products manufactured to meet the demands of the free-economy, mass-consumption markets need to have changes in design for improved product performance as well as style changes. Therefore, hard automation systems need to be as flexible as possible while retaining the specification needed to mass produce. This had led to the following developments.

First, the machines are constructed from basic building blocks or units that accomplish a function rather than produce a specific part. The production machine tool units are combined and interconnected to produce the desired system for making the product. These units are called self-contained power-head production machines. Then such machines are connected by automatic transfer devices to handle the material (i.e., the thing being produced), moving it automatically from one process to the next. Recently, the incorporation of computers, PCs, and other feedback control devices have made these machines more flexible.

Powerhead production units. Many machining processes involve a rotating tool, in most cases fed either longitudinally or transversely with respect to the workpiece. Powerhead production units, such as shown in Figure 38-11, provide these basic, required motions. As shown, such a unit consists essentially of (1) a powered spindle, mounted in suitable bearings, that can be used to support and drive a variety of rotating cutting tools at selected speeds through a gear box, (2) a means for power feed, and (3) a frame or bed on which the other two components are mounted. Thus it is similar to the components in the upper column and spindle assembly of a heavy-duty, upright drilling machine, and it can easily be adapted to do drilling, milling, boring, reaming, tapping, grinding, and honing with high effectiveness. Some units, in modified form, are equipped with chucks so that simple turning operations also can be done.

Powerhead production units come in a substantial range of standardized sizes

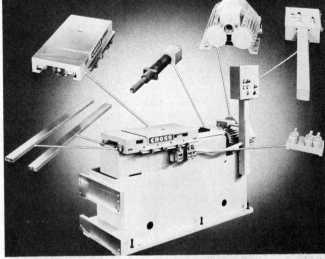

FIGURE 38-11 Powerhead production unit. (*Top*) Basic unit containing powered spindle. (*Center*) Exploded view, showing various components to provide base, spindle rotation, and feed. (*Bottom*) Complete unit. (*Courtesy Cross Company.*)

which, along with base and column units, serve as *building-block* components from which a wide variety of special production machines can be assembled. As an example, the eight standard building-block components shown in Figure 38-12 can be combined to form the different machining centers shown in Figure 38-13.

The application of this basic principle may be seen in Figures 38-14 and 38-15. In each case, although very different in size and arrangement, and for totally different products, most of each machine consists of standard components. When a design change occurs in a product, or the product is no longer to be made, the standard production units can be adapted to the new requirements. They can be regrouped, with new jigs, fixtures, and bases, if required. A unit previously used for drilling may now be used for tapping or milling on the same or entirely different product. Thus flexibility and adaptability are obtained along with high productivity. Also, an entirely new machine can be obtained in much less time because of the availability of standard components.

Transfer mechanisms. Means must be provided for moving workpieces from station to station on production machines, and for moving them from one machine to another. Machines that are provided with such automatic mechanisms are called *transfer machines*. Transferring usually is accomplished by one or more of four methods. Frequently, the work is pulled along supporting rails by means of an endless chain that moves intermittently as required. This method can be seen in Figure 38-14. In another method the work is pushed along continuous rails by air or hydraulic pistons. A third method, restricted to lighter workpieces, is to move them by an overhead chain conveyor, which may lift and deposit the work at the machining stations.

A fourth method often is employed when a relatively small number of operations—usually three to ten—are to be performed. The machining heads are arranged radially around a rotary indexing table, which contains fixtures in which the workpieces are mounted (see Figure 38-13). The table movement may be

FIGURE 38-12 Models of eight standard building-block units from which a wide variety of production machine tools can be built. (*Courtesy Heald Machine Company.*)

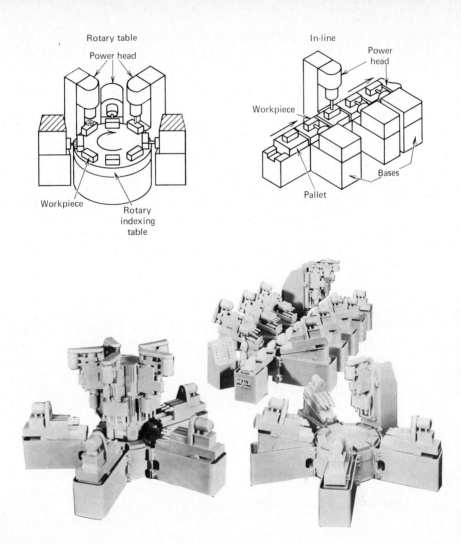

FIGURE 38-13 Examples of rotary and inline transfer units which can be constructed from components. (*Courtesy Heald Machine Company.*)

continuous or intermittent. Face milling operations sometimes are performed by moving the workpieces past one or more vertical-axis milling heads. Such circular configurations have the advantages of being compact and of permitting the workpieces to be loaded and unloaded at a single station without having to interrupt the machining.

Means must be provided for positioning the workpieces correctly as they are transferred to the various stations. One method is to attach the work to carrier pallets or fixtures that contain locating holes or points that mate with retracting pins or fingers at each work station. Excellent station-to-station precision is obtained as the fixtures thus are located and then clamped in the proper positions. This method is used in the machine shown in Figure 38-14. Obviously, carrier fixtures are costly. When possible they are eliminated and the workpiece trans-

FIGURE 38-14 (*Top*) Transfer-type production machine for machining typewriter frame (*bottom left*). (*Bottom right*) Pallet-type fixtures containing typewriter frames. (*Courtesy Cross Company*)

ported between the machines on rails, which locate the parts by self-contained holes or surfaces. This procedure, used in the machine shown in Figure 38-15, also eliminates the labor required for fastening the workpieces to the carrier pallets, as well as the pallets.

Four problems are of considerable importance when large transfer machines are used. One is the matter of the geometric arrangement of the various production units. Whether or not transfer fixtures or pallets must be used is an important factor. As was shown in Figure 38-14, these fixtures and pallets usually are quite heavy. Consequently, when they are used, a closed, rectangular arrangement, such as that shown in Figure 38-14, is often employed so that the fixtures are automatically returned to the loading point. If no fixtures or pallets are required, U- or straight-line configurations can be employed, as shown in Figure 38-15. Whether pallets or fixtures must be used is dependent primarily on precision needed as well as the size, rigidity, and the design of the workpieces. If no transfer pallet or fixture is to be used, locating bosses or points should be designed or machined into the workpiece.

The matter of tool wear and replacement is of great importance when a large number of operations are incorporated in a single production unit. In such costly machines, having high production capabilities, it is essential that they be kept

FIGURE 38-15 Complex transfer-type manufacturing system arranged in a U-pattern. This unit machines V-8 cylinder blocks at the rate of 100 per hour. There are five independent sections that perform 265 drilling, 6 milling, 21 boring, 56 reaming, 101 counterboring, 106 tapping, and 133 inspection operations. These are performed at 104 stations, including 1 loading, 53 machining, 36 visual inspection, and 1 unloading. Provision is made for banking parts between each section. (*Courtesy Cross Company.*)

operating as much of the time as possible. At the same time, tools must be replaced before they become worn and produce defective parts. Transfer machines often have more than 100 cutting tools. If the entire, complex machine had to be shut down each time a single tool became dull and had to be replaced, the resulting productivity would be very low. This is avoided by designing the tooling so that certain groups have similar lives and then utilizing control panels that record tool wear in each group and shut down the machine before the tooling has deteriorated. All the tools in the affected group are then changed so that repeated shutdowns are not necessary.

Several methods have been developed for accurately presetting tools and for changing them rapidly. Figure 38-16 shows two versions of a quick-change tool holder. These are available for a wide range of tools, and they permit a large number of tools to be changed in a few minutes, thereby reducing machine downtime. Increasingly, tools are preset in standard, quick-change holders with excellent accuracy, often to within 0.005 mm (0.0002 inch).

A third problem encountered in the use of multiple-station machines is avoiding shutting down the entire machine in case only one or two stations should become inoperative. This is usually avoided by arranging the individual units in groups, or sections with 10 to 12 stations per section, and providing for a small amount of buffer storage (*banking*) of workpieces between the sections. This permits production to continue on all remaining sections for a short time while one is shut down for tool changing or repair.

The fourth problem deals with designing the line itself so that it operates efficiently as a whole. Processes are grouped together at stations. Note in Figure 38-15 that 688 machining processes are grouped into 50 machining stations. Transfer machines, or for that matter any system wherein a sequence of processes are connected sequentially, will require that the line be *balanced*. Line balancing

FIGURE 38-16 Quick-change tool holders used in automatic machining units. (*Courtesy Bendix Corporation, Industrial Tool Division.*)

means that the process time at each station must be the same, with the total nonproductive time for all other stations minimized. Theoretically, there will be one station which will have the longest time and this station will control the cycle time for all the stations. When the line is the size of the one shown in Figure 38-15, this becomes a rather complex problem. In recent years, computer algorithms have been developed to deal with line balancing.

As we have seen, transfer machines are either A(3)- or A(4)-level machines, depending upon whether they have the built-in capacity for sensing when corrective action is required and for making such corrections. For example, an automatic broaching machine can carry out operations at a high rate without a human operator. However, if a tool wears or breaks, the machine will not make corrective adjustments or shut down to prevent defective products from being produced. To make the machine an A(4), it would have to measure the part after each operation was performed, or as it was being performed, and make any adjustments required to assure that only parts meeting the design specification were produced. Thus sensing and feedback control systems are essential requirements for the fourth level and all higher levels of automation.

Many highly mechanized mechanisms have *feed-forward* devices built into them. This means that the system takes information from the input side of the process rather than the output side of the process and uses that information to alter the process. For example, the temperature of the billet as it enters a hot rolling or hot extrusion process can be sensed and used as feed-forward information to alter the process parameters. This would be an A(5) or adaptive control example. Note that in Figure 38-10 the sensors are located in three positions— ahead of the process, in the process, and on the output side of the process. The feed-forward concept can also be applied at the A(4) level. Suppose that you have a transfer line that processes two types of flywheel housings that are similar but require different machining operations. When the housings are fed into the machine, in mixed order, a sensing device contacts a distinguishing boss that is on one type but absent from the other. The sensing and feed-forward system then sets the proper tooling for that housing and omits operations that are not required for the other type.

It is common practice to equip transfer-type machines with automated gaging or probing heads which, after each operation, determine whether the operation was performed correctly and to detect whether any tool breakage has occurred that might cause damage in subsequent operations—checking a hole after drilling to make certain it is clear prior to tapping, for example.

Automation and transfer principles also are used very successfully for assembly operations. In addition to saving labor, automatic testing and inspection can be incorporated into such machines at as many points as desired. Not only does this assure better quality, but defective, partially completed assemblies are discovered and removed for rework or scrapping without additional parts being added or additional assembly steps being completed on them.

The range of products now being completely or partially assembled automatically is very great. Figure 38-17 shows an automated assembly machine for assembling and testing electronic components. These machines are further examples of how standardized heads can be combined and converted for assembling different combinations of components.

In many cases, some manual operations are combined with some automatic operations. For example, one transfer machine for assembling steering knuckle, front wheel hub, and disk-brake assemblies has 16 automatic and 5 manual work stations. As with machining operations, automatic assembly often can be greatly facilitated by proper consideration on the part of designers.

Flexible/programmable or soft automation. The most significant development in manufacturing during the past 30 years has been the advent and wide-scale adoption of numerically, tape-, and computer-controlled machine tools. These machines bridge the gap between highly flexible, general-purpose ma-

FIGURE 38-17 Automatic assembly machine for assembling the electronic unit and components shown in the inset. (*Courtesy Bendix Corporation, Automation and Measurement Division.*)

chine tools and highly specialized, but inflexible, mass-production machines. Numerical control (NC) of machine tools not only bridges this gap, but also creates entirely new concepts in manufacturing, making routine certain operations that previously were very difficult to accomplish. And whereas in the earlier years highly trained programmers were required to do the programming required for their use, the development of low-cost, solid-state microprocessing chips has resulted in machines that can be programmed in a very short time, by personnel having only a few hours of training, using only simple machine shop language. As a consequence, there are few manufacturing facilities today, from the largest down to small job shops that do not have, and routinely use, one or more numerically controlled machine tools.

As with so many inventions, NC came into being to fill a need. The U.S. Air Force and the airframe industry were seeking a means to manufacture complex contoured aircraft components to close tolerances on a highly repeatable basis. John Parsons of the Parsons Corporation of Traverse, Michigan, had been working on a project for developing equipment that would machine templets to be used for inspecting helicopter blades. He conceived of a machine controlled by numerical data to make these templates and took his proposal to the air force, which funded developmental work for Parsons and later MIT. The first NC machine was demonstrated in 1952 at MIT in a lab located close to another lab where one of the first digital computers was being developed, the Whirlewind. It was this computer that was used to generate the digital numerical data that the first machine needed for its three-axis control system.

By 1962, NC machines accounted for about 10% of total dollar shipments in machine tools and today over 75% of the money spent for drill presses, milling machines, lathes, and machining centers goes for NC equipment.

The early machines were continuous-path or contouring machines wherein the entire path of the tool is controlled with close accuracy with regard to position and velocity. Milling machines, machining centers, and lathes are the most popular applications of continuous-path control requiring feedback control. Later, point-to-point machines were produced wherein the path taken between operations is relatively unimportant and therefore not continuously monitored. Point-to-point machines may be either open loop, A(3), or closed loop, A(4), and are chiefly used for drilling, milling straight cuts, cutoff, and punching. Automatic tool changers, which require that the tools be precisely set to a given length prior to installation in the machines, soon followed as evidenced by the first machining center, marketed by Kearney and Trecker in 1968.

During the early days of NC, programming was done manually or with a computer assist for complex workpiece geometries. The chief problems in NC programming were tool radius compensation and tool path interpolation, as shown in Figure 38-18. Tool radius offset accounts for the fact that the center of tool rotation (as held in the spindle) must be offset from the workpiece surface one wants to generate. As can be seen from the figure, the path of the cutter center line will have different dimensions than that of the surface. Interpolation refers to the fact that curved surfaces as generated by machine tools must be approximated by a series of very short, straightline, (x,y) table movements. Thus, software capability to perform linear, circular, parabolic, and so on, interpolations had to be developed.

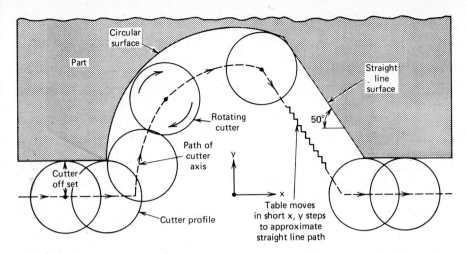

FIGURE 38-18 Two classic problems in NC programming are the determination of cutter offset and interpolation of cutter paths.

Soon it was envisioned that a large computer could be used to control directly a number of machines and so a limited number of DNC (direct numerical control) systems were developed, with the idea being that the programs were to be sent directly to the machines (eliminating paper tape handling) and the large main computer would be shared on a real-time basis by many machine tools. The machine operator would have access to the main computer via a remote terminal at his machine while management would have up-to-the-minute data on production status and machine utilization. However, this version of DNC had very few takers. Instead, NC machines soon became CNC (computer numerical control) machines through the development of small inexpensive computers, microprocessors with large memories, and programmable controllers. Now functions like program storage, tool offset and tool compensation, program edit capability, various degrees of computation, and the ability to send and receive data from a remote source were all routinely available right at the machine. Immediately it was found that the machine tool operator could readily learn how to program these machines (manually) for many component parts, often eliminating the need for a part programmer.

Now the DNC concept was revived with the small computers at the machines being networked to a larger computer to provide enhanced memory and computer computational capability. Minicomputers, supervising and controlling a flexible manufacturing system or manufacturing cell, are networked to a large central computer.

The next level of automation the A(5) level, requires that the control system perform an evaluation function of the process. CNC machines, with their on-board computers, are potentially capable of this level of automation, which requires, as shown in Figure 38-10, a feedback loop (or a feed-forward loop) from the process itself or its immediate output which can be used to modify the input process parameters. In the standard versions of NC and CNC machines in use today, speed and feed are fixed in the program unless the operator overrides them at the machine. If either speed or feed is too high, the result can be rapid tool failure, poor surface quality, or damaged parts. If the speed and feed are too low, production time is greater than desired for best productivity. An adaptive control (AC) system that can sense force, heat, torque, and the like, will use these measurements to make decisions about how the input parameters might be altered to *optimize* the process. This means that the computer must have in its software, mathematical models which describe how this process behaves and mathematical functions which state what parameter is to be optimized (i.e., cost per piece, surface quality, MRR, power consumed, etc.). The models require theory and herein lies the problem. The theory of shape-generating processes (metal forming and metal cutting) is segregated and incomplete. With incomplete theory, the models become suspect and the systems unreliable. In addition, current versions of AC systems have been very costly to develop, suffer from inadequate data banks, and produce variable cycle times. However, AC systems have the potential of improving productivity, quality, and machine utilization. Reaching this level of automation on a routine shop floor basis will require great expenditures of time and money.

Flexible manufacturing system (FMS).   The use of computers in batch manufacturing has led to the development of flexible manufacturing systems (FMS) for medium-sized batches (hundreds to thousands of units per year). These systems integrate NC machines with an automated material handling system, often incorporating computer control over all the machines and the materials handling system. Human labor is usually incorporated to load workpieces, unload finished parts, change worn tools, and perform equipment maintenance and repair. Both CNC and DNC functions can be incorporated into a single FMS. The system can usually monitor piece part counts, tool changes, and machine utilization, with the computer also providing supervisory control of the production.

Group technology is usually used to identify those families of fairly complex component parts which need to be produced in midrange volumes and which require manufacturing flexibility. The workpieces can be launched randomly into the system, which identifies each part in the family and routes it to the proper machines. The systems generally display reduced manufacturing lead time, low in-process inventory, and high machine tool utilization, with reduced indirect and direct labor. The materials handling system must be able to route any part to any machine in any order, and provide each machine with a small queue of "banked parts" waiting to be processed so as to maximize machine utilization. Convenient access for loading and unloading parts from either side, compatibility with the control system, and accessibility to the machine tools are other necessary design features for the materials handling system. Typically, an FMS contains 5 to 12 machine tools, with each machine capable of changing cutting tools to perform different tasks, such as drilling, boring, or milling. A typical computer-controlled FMS system is shown in Figure 38-19. It has three levels of computer control. The master control monitors the entire system for tool failures or machine breakdowns, schedules the work, and routes the parts to the appropriate machine. The DNC computer distributes programs to the CNC machines and supervises their operations, selecting the required programs and

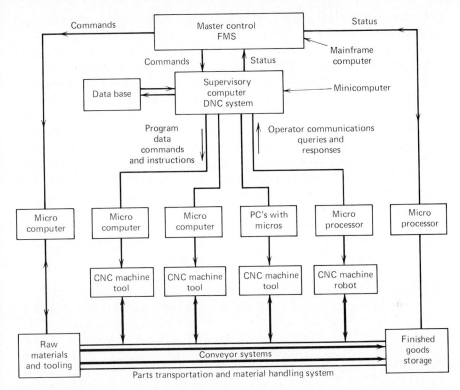

FIGURE 38-19 FMS system schematic with three levels of computer control.

transmitting them at the appropriate time. It also keeps track of the completion of the cutting programs and sends this information on to the master computer. The bottom level of computer control is at the machines themselves. Today, FMS systems employ NC or programmable machines for grinding, punching, bending, laser cutting, welding, and heat treating, as well as machining. In fact, if we refer to these machines as programmable, robots should be included.

Through the use of either tape or computer control, there can be good assurance that consecutive parts are duplicated and that a part made at some later date will be the same as one made today. Thus, repeatability and quality are improved. Workholding devices can be made more universal, setup time reduced, along with tool change time, thus making programmable machines economical for producing small lots or even a single piece. When combined with the managerial and organizational strategies of group technology (GT), programmable machines lead to tremendous improvements in productivity. GT basically leads to the creation of families of parts made in machining cells containing flexible/programmable machines. The compatibility of the components (similarity in processes and sequences of processes) greatly enhances the productivity (utility) of the programmable equipment.

It might even be said that the use of numerically controlled machines has enhanced our realization of the importance and use of preproduction planning. Such planning is an absolute requirement in the use of numerically controlled machines, and a realization of its advantages has resulted in much better planning being done even when general-purpose tools are to be used.

Another side result has been the decrease in the non-chip-producing time of machine tools, caused by the necessity for the operator to set speeds and feeds and locate the tool relative to the work. Quite simple forms of NC and digital readout equipment have provided both greater productivity and increased accuracy.

A further, but not entirely unmixed, advantage of NC has been the ability to routinely obtain greater accuracy than previously was obtainable. Most of the early NC machine tools were developed for special types of work where accuracies of as much as 0.0013 mm (0.00005 inch) might be required, and most NC machines were built to provide accuracies of at least 0.00254 mm (0.0001 inch), whether it was needed or not. While most NC machine tools today will provide greater accuracy than is required for most jobs and the tendency to specify greater accuracy on parts than is required has diminished, the matter should be monitored.

Functions involved in machining. In machining a single part on a conventional A(2) machine tool, the operator usually will perform the following functions:

1. Plan the sequence of operations.
2. Select the cutting tools and workholding devices.
3. Set and change the tools.
4. Select speeds, feeds, and depths of cut.
5. Set the speeds (rpm), feeds, and depths of cut in the machine.
6. Load and unload the workpieces.
7. Position the work relative to the tools, or vice versa.
8. Monitor or control cutter path during cutting.
9. Reposition work (or tools) after each operation for the next operation, which includes starting and stopping the machine.
10. Inspect the parts.

If the machine operator performs all these functions effectively, he or she must have a high degree of skill, yet during a substantial portion of the operation cycle she or he is idle, watching chips being made. At the same time, during a considerable portion of the total cycle, the machine is not removing chips—its functional purpose.

Items 1, 2, and 4 of the preceding list can be done before the operation begins, by someone other than the machine operator. This would lessen the skill and judgment requirements of the operator. If the number of identical parts to be produced is sufficient, and the motions required are relatively simple, items 3 and 5 through 10 can be done automatically or semiautomatically. Such procedures are not usually economically feasible for a single or a few parts, unless they are parts from a compatible family of components.

Item 7, positioning the work relative to the tool, can be greatly facilitated through the use of jigs, but this procedure is costly and cannot be justified when only one or a few parts are to be made. With a family of parts, work holders *can be made universal,* able to handle all the parts in the family with minor adjustments.

Item 8, control of the relative motion between the work and the tool, cannot be done by an operator if the motion is at all complex. For example, turning a uniform taper on a lathe by simultaneously controlling the longitudinal and cross feeds is virtually impossible, and the accurate turning of a more complex surface is completely impossible. Although irregular surfaces can be produced by such machines as copying lathes and duplicating milling machines, these require rather expensive templates or models that make the cost very high if only one or a few parts are to be made. Furthermore, such machines are quite specialized and costly, and highly skilled labor usually is required to operate them.

Historically functions 3 and 5 through 9 of the list above cannot be done economically on the basic, general-purpose machine tools when single or a few workpieces are involved. However, programmable computer control makes it possible to do all these functions automatically and economically.

Basic principles of numerical control. As the name implies, *numerical control* is a method of controlling the motion of machine components by means of numbers. It can be illustrated in simple form by its application to item 7 of the previous list—positioning. Assume that three 25.4-mm (1-inch) holes in the part shown in Figure 38-20 are to be drilled and bored on the jigmil shown in Figure 38-21. The centers of these holes must be located relative to each other and with respect to the left-hand edge of the workpiece (X direction) and the bottom

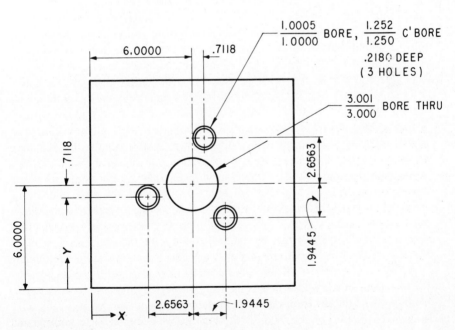

FIGURE 38-20 Part to be machined on a numerically controlled point-to-point machine tool.

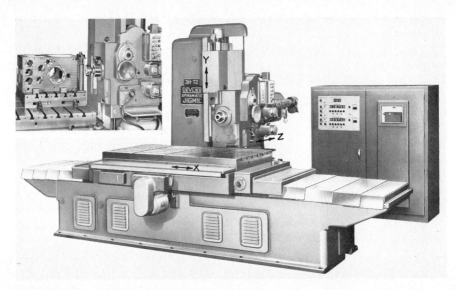

FIGURE 38-21 Tape-controlled jigmil of the type that could be used for machining the holes in the part shown in Figure 38-20. Inset shows the setup of the work for a boring operation (*Courtesy DeVlieg Machine Co.*)

edge (*Y* direction). After the workpiece is set up on the table of the machine, the work and the tool must be brought into proper relationship for each of these holes. If this is to be done automatically, means must be available for precisely specifying, measuring, and controlling the relative motions of the machine table and the spindle carrier so that the location dimensions specified on the drawing will be reproduced on the workpiece. This requires signals that will command the driving motor for the machine component, such as a table, to move the component to the desired location, where an action takes place (drill a hole). This is *open-loop* control, which is satisfactory for many point-to-point applications.

Closed-loop control, shown schematically in Figure 38-10, requires that some sort of transducer or sensing device detect machine table position (or velocity) and transmit that information back to the control unit and compare the current status with the desired state. If they are different, the control unit produces a signal to the drive motors to move the table, reducing the error signal and ultimately moving the table to the desired position at the desired velocity. Most NC controls use the closed-loop system, with the feedback signals being supplied by transducers actuated either by the feed screw or by the actual movement of the component. The transducers may provide either *digital* or *analog* information (signals).

Digital information usually is in the form of electric pulses. Two basic types of digital transducers are used. One supplies *incremental* information and tells how much motion of the input shaft or table has occurred. The information supplied is similar to telling a newsboy that he is to deliver papers to the first, fourth, and eighth houses from a given corner on one side of a block. To follow the instructions, the newsboy would have to have a means of counting the houses (pulses) as he passes them and deliver papers when he has counted 1, 4, and 8. The second type of digital information is *absolute* in character, with each pulse corresponding to a specific location of the machine component. Using the newsboy analogy, this would correspond to telling him to deliver papers to the houses having house number 2400, 2406, and 2414. In this case it would only be necessary for the newsboy (machine component) to be able to read the house numbers (addresses) and stop and deliver a paper when he has arrived at a proper address. This "address" system is a common one in numerical-control systems, because it provides absolute location information relative to a zero point.

When analog information is used, the signal is usually in the form of an electric voltage that varies as the input shaft is rotated or the machine component is moved, the variable output being a function of movement. The movement is evaluated by measuring, or matching, the voltage, or by measuring the ratios between the applied and feedback voltages; this eliminates the effect of supply-voltage variations. Again using the newsboy analogy, this method is like telling the boy to deliver papers to those houses on one side of the block that are 7.6 meters (25 feet), 68.6 meters (225 feet), and 129.5 meters (425 feet) from a given corner. For the boy to respond properly, he would need a measuring device—a tape measure—with which to measure his movement from the corner. He then would deliver papers when his traveled distance was 7.6, 68.6, and 129.5 meters. Several types of NC systems use analog information.

Figure 38-22 depicts the elements involved in the basic open-loop and closed-loop NC systems for controlling one axis of motion of a machine tool. The desired location for the workpiece for a given operation is read from a punched tape. This command signal is converted into pulses by the machine control unit (MCU), which in turn drives the servomotor. When the command counter reaches

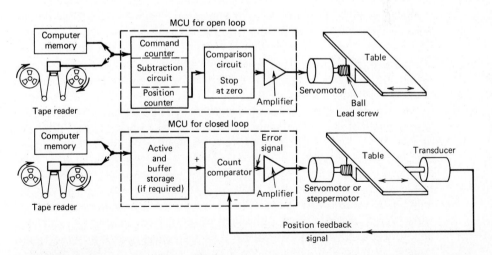

FIGURE 38-22 The elements of open loop, A(3), control for table position versus closed loop, A(4), control for position as used in NC machines. (*From Roger S. Pressman and John E. Williams,* Numerical Control and Computer-Aided Manufacturing, *John Wiley & Sons, Inc., New York.*)

zero, the correct number of pulses has been sent to move the table to the desired position. In the closed-loop system, a comparator is used to compare feedback pulses with the original value, generating an error signal. Thus when the machine control unit receives a signal to execute this command, the table is moved to the specified location, with the actual position being monitored by the feedback transducer. Table motion ceases when the error signal has been reduced to zero and the function (drill a hole) takes place. Closed-loop systems tend to have greater accuracy and respond faster to input signals but may exhibit stability problems (oscillate about a desired value instead of achieving it) not found in open-loop machines.

If the system is point-to-point (or positioning), the control system disregards the path between points. Some positioning systems provide for control of straight cuts along the machine axes and produce diagonal paths at 45° to the axes by maintaining one-to-one relationships between the motions of perpendicular axes. Contouring systems generate paths between points by interpolating intermediate coordinate positions. As many of these systems as desired can be combined to provide control in several axes—two- and three-axis controls are most common, but some machines have as many as seven. In many, conversion to either English or metric measurement is available by merely throwing a switch.

The components required for such a numerical-control system now are well-standardized items of hardware. In most cases the driving motor is electric, but hydraulic systems also are used. They usually are capable of driving the machine elements, such as tables, at high rates of speed—up to 5080 mm (200 inches) per minute being common. Thus exact positioning can be achieved much more rapidly than by manual means. Several types of transducers are used, most of them being completely electrical. In many systems the input of the transducer is connected directly to the lead screw, with special precautions being taken, such as the use of extra-large screws and ball nuts, to avoid backlash and to assure accuracy. Other systems drive the transducer from an accurate rack that is attached to the machine table. When pulse systems are used, the pulses are counted by common off–on electronic counters. Various degrees of accuracy are obtainable. Guaranteed positioning accuracies of 0.025 mm or 0.0025 mm (0.001 inch or 0.0001 inch) are most common, but greater accuracies can be obtained at higher cost. Most NC systems are built into the machines, but they can be retrofitted to some machine tools.

Initially, NC machines performed function 7 from the previous list—positioning the work relative to the tool, with the remaining functions controlled by the operator. Gradually, functions 3, 5, 8, 9, and 10 were incorporated into the control system, so the machines could change the tools automatically, change the speeds and feeds as needed for different operations, position the work relative to the tools, control the cutter path and velocity, reposition the tool rapidly between operations, start and stop the sequence as needed, and even inspect the parts. Notice that to accomplish some of these functions in programmable machines, it is necessary to have feed-forward or preset loops. Thus the machine must know in advance the rough dimensions of a casting so that it can determine how many roughing cuts are needed prior to the finishing cut. Gradually, NC machines were manufactured that had greater accuracy and repeatability and more rapid table movements than conventional machines. In addition, functions

1, 2, and 4 can, by preproduction planning, be programmed into the machine. Thus, all of the required functions can be computer controlled, permitting almost complete automation of a general-purpose tool, and for any production quantity.

Punched tape is not a recent idea—the old player-piano roll was a form of tape control, and punched cards had been used for many years for controlling complicated weaving and business machines. Thus tape control of machine tools is an extension of an existing basic concept in which holes, representing information that has been punched into the tape, are "read" by sensing devices and used to actuate relays or other devices that control various electrical or mechanical mechanisms.

Tape-controlled machine tools use a 1-inch-wide paper or Mylar tape containing eight information channels. Figure 38-23 shows an example of a "block" of such tape, punched with the information required for one operation on a turret-type drilling machine. This tape is an example of the EIA244A code, one of two symbolic codes used in NC work. The other code is called ASCII, which is used in computer and telecommunications work as well as in NC. The codes are not interchangeable even though both are based on binary numbering. The base 2 system is used because electronic circuitry responds to either of two conditions—on or off or zero or one. Thus, all the numbers, symbols, and letters that are needed to control the machine are communicated to the machine by the presence or absence of holes in the eight tracks of the tape.

Four basic types of tape format are used for NC input to communicate dimensional and nondimensional information. These are fixed-sequential format, block-address format, tab-sequential format, and word-address format. By convention, the data are usually arranged in blocks in the following sequential order, regardless of which format is being used.

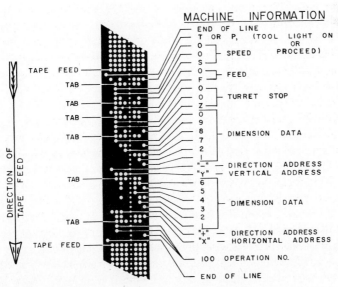

FIGURE 38-23 Block of control tape for controlling one operation on a turret-type drilling machine, using tab sequential format. (*Courtesy DeVlieg Machine Co.*)

n	sequence number: identifies the block of information.
g	preparatory function: requests different control functions.
x, y, z, a, b	dimensional data: linear and angular motion commands for the axis of the machine.
f	feed function: set feed for this operation.
s	speed function: set speed for this operation.
t	tool function: tells the machine the location of the tool in the tool holder.
m	miscellaneous function: turn coolent on or off, open spindle, reverse spindle, tool change, etc.
EOB	end of block: indicates to the MCU that a full block of information has been transmitted and the block can be executed.

When the machine has executed a block, it then reads another into the MCU.

The operator may override the tape when necessary but cannot reprogram the machine unless a new tape is prepared. If the machine is a computer NC machine, it may have the capability of reading a tape into its computer memory and the program can be modified at the machine like any other computer program.

Although the majority of NC and tape-controlled machine tools do not provide for machining contoured surfaces, many do. Most computer-assisted and direct-computer-controlled machines provide this feature. The required curves and contours are generated approximately by a series of very short, straight lines or segments of some type of regular curves, such as hyperbolas. Consequently, the program fed to the machine is arranged to approximate the required curve, within the desired accuracy. Figure 38-18 illustrated how a desired shape can be approximated by means of short, straight lines within some permissable deviation. It is apparent that the length of the straight-line or standard-curve segment must be varied in accordance with the deviation permitted. Most machine tools with contouring capability will produce a surface that is within 0.025 mm (0.001 inch) of the one desired, and many will provide considerably better performance. Most contouring machines have either two- or three-axis capability, but a good many have up to five-axis capability.

Obviously, contour machining requires that complex information be punched into the control tape by ordinary procedures, the number of straight-line or curved segments may be quite large, and manual programming of the tape can be quite laborious. This can be eliminated by the use of a computer that will translate simple commands into the complex information required by the machine.

Manual programming. Obviously, the preparation of the tape for use in NC is very important. In most cases the preparation is not difficult, since quite simple standard languages and programs have been developed. Manually prepared tapes usually are punched on a typewriter-like machine, or on devices designed specifically for punching tapes.

The basic steps can be illustrated by reference to the part shown in Figure

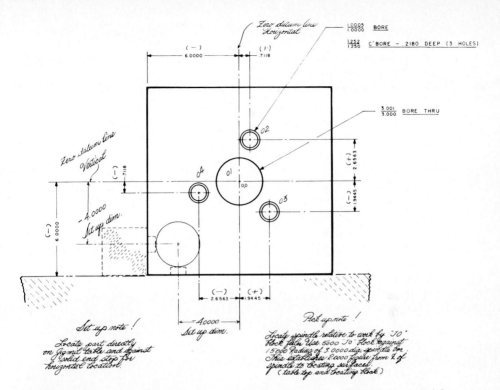

FIGURE 38-24 Drawing of a part, modified to provide information needed for preparing a tape program sheet and for setting up work on machine. (*Courtesy DeVlieg Machine Co.*)

38-20.[2] The first step is to modify the drawing to establish the zero reference axes, the *X* and *Y* directions, The dimensioning with respect with the reference axes, and the setup instructions to establish the workpiece properly on the machine table with respect to the tool. These modifications are shown in Figure 38-24. Obviously, this step can be avoided if the original drawing is made in the desired form.

The second step is to make a *program sheet,* such as shown in Figure 38-25. This sheet gives the coordinate dimensions for each operation, specifies the spindle traverse that determines the depth of the cut, the spindle speed and feed, and whether the same tool can continue the next operation or whether a tool change is required. The last four items are specified by code symbols obtained from the chart shown in Figure 38-26.

After the program sheet has been prepared, it is used to prepare the tape. Typing the information shown in Figure 38-25 produces the print copy shown in Figure 38-27 and simultaneously produces the punched tape, a section of which is shown in Figure 38-28.

Before a tape is used, it usually is checked. This can be done by running it through the tape reader which is either connected to the tape writer so as to type

---

[2]The steps required by some machines often are considerably less complicated. The method described here is used to illustrate the basic principles.

PROGRAM SHEET FOR DEVLIEG *SPIRAMATIC* JIGMIL WITH Tapac III

Part No. ABC-1001A   Part Name ___Plate___   Sheet 1 of 1

Oper. No.	X Dimension	Y Dimension	Z Turret	Spindle Feed No.	Spindle Speed No.	Tool Change Auto Step	End of Line / Carriage Ret.
000	X±000000	Y±000000	Z00	F0	S00	T / P	
000	x-040000	y-040000				t	

For set up purposes,first piece only,use .5000 Jo block against spindle bar for horizontal(end stop) and vertical (table top). Zero horizontal and vertical slides to this position and measurement.

| 001 | x+000000 | y+000000 | z01 | f6 | s05 | t | |

Core drill hole 2-7/8 dia. Hole #01 . 70 RPM .012 Feed

| 002 | | | z02 | f3 | s13 | t | |

Rough bore with tool #217 ,Hole #01;570RPM; .004 Feed

| 003 | | | z02 | f2 | s13 | t | |

Finish bore with tool # 217F; Hole #01; 570RPM; .003 FEED

| 004 | x+007118 | y+026563 | z03 | f6 | s09 | t | |

Drill 15/16 dia. thru 200 RPM .002Feed Hole #02

| 005 | x+019445 | y-019445 | | | | p | |

Same as operation #004; Hole #03

| 006 | x-026563 | y-007118 | | | | p | |

Same as operation #004; Hole #04

| 007 | x+007118 | y+026563 | z04 | f2 | s15 | t | |

Rough and finish bore and c'bore using tool #ct 10025. Hole #02 940 RPM; .003 Feed

| 008 | x+019445 | y-019445 | | | | p | |

Same as operation #007; Hole #03

| 009 | x-026563 | y-007118 | | | | p | |

Same as operation #007; Hole #04

NOTE: ....FOR MACHINE COMMAND, USE "PUNCH ON" CODE AT MARGIN
....FOR MANUSCRIPT MATTER, USE "PUNCH OFF" CODE AT MARGIN

FIGURE 38-25 Program sheet used in preparing the control tape for the part shown in Figure 38-24. (*Courtesy DeVlieg Machine Co.*)

a duplicate sheet showing the machine-information data (Figure 38-27) or to a special NC plotting machine that will trace out all the tool–work paths as they would occur on the machine tool.

On CNC, NC machines with built-in computers, the machine tool operator may perform all of these steps right at the console of the machine, programming the processing steps for the part directly into the computer memory. The program can be saved by having the machine print out a copy of the program, which can be later used for reorders of the same part. Features such as program edit, program storage, diagnostics, constant surface speed, and tape punch are common on today's CNC machines.

As more and more design work is done on the computer (CAD) using data bases and software that are compatible to the machine tools, there will be less dependency on tape for program storage and more utilization of floppy or hard disk and other typical computer storage means. For example, a machining cell designed for a family of 10 component parts may be able to do the 10 different parts without needing retooling or refixturing, but it will still have 10 different programs for these parts for each machine. If the programs are computer stored, they can be readily accessed but if stored on tape, delays will occur while dumping the different programs in and out of the control computer.

TAPAC III
SPEED & FEED CONVERSION CHART

— BAR FEED —
PROGRAM NUMBERS AND ACTUAL FEED (INCH PER REV.)

PROGRAM NO.		F 1	F 2	F 3	F 4	F 5	F 6	F 7	F 8
ACTUAL FEED	3H & 4H	.002	.003	.004	.006	.008	.012	.017	.024
	5 H	.0034	.0042	.0059	.0085	.0132	.0183	.0256	.0365

— SPINDLE SPEED —
PROGRAM NUMBERS AND ACTUAL RPM

		PROGRAM NO.	S01	S02	S03	S04	S05	S06	S07	S08	S09	S10	S11	S12	S13	S14	S15	S16				
S P E E D S	3 H	ST'D.	25	32	41	52	70	90	115	150	200	255	325	420	570	730	940	1200				
		HI-SP'D.	33	43	55	70	90	120	155	200	260	340	435	560	750	970	1250	1600				
	4 H	ST'D.	21	27	35	45	58	76	98	127	165	215	275	355	470	605	775	1000				
		HI-SP'D	26	34	44	56	70	95	120	150	210	275	355	440	575	750	950	1220				
	5 H	ST'D.	10	13	17	21	27	34	44	58	75	97	124	166	213	276	355	468	601	778	1000	N *
		PROGRAM NO.	S01	S02	S03	S04	S05	S06	S07	S08	S09	S10	S11	S12	S13	S14	S15	S16	S17	S18	S19	S20

* N=NEUTRAL

FIGURE 38-26 Speed and feed conversion code chart used in preparing the control tape for part shown in Figure 38-24. (*Courtesy DeVlieg Machine Co.*)

Manual programming of programmable machines (whether machine tools or robots) can be time consuming when the parts require complex operations, have contoured surfaces, or when multiple machine (integrated) systems are utilized. In such situations, off-line programming is used. Special programming software has been developed specifically for stand-alone machine tools and robots and more recently for integrated systems of machine tools and robots.

000	x—040000	y—040000				t
001	x+000000	y+000000	z01	f6	s05	t
002			z02	f3	s13	t
003			z02	f2	s13	t
004	x+007118	y+026563	z03	f6	s09	t
005	x+019445	y—019445				p
006	x—026563	y—007118				p
007	x+007118	y+026563	z04	f2	s15	t
008	x+019445	y—019445				p
009	x—026563	y—007118				p

FIGURE 38-27 Machine information data typed by tape-preparation unit in preparing the control tape for the part shown in Figure 38-24. (*Courtesy DeVlieg Machine Co.*)

Operation (001)

End of command

X- axis command (X + 000000)

End of command

Y-axis command (Y + 000000)

End of command

Z-axis command (Z 02)

End of command

Feed command code (f 6)

End of command

Speed command code (s 05)

End of command

Tool change code (t)

End of block

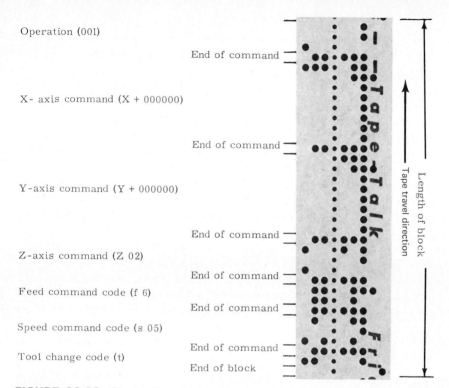

FIGURE 38-28 Block of control tape, corresponding to the machine information required for the second operation shown in Figure 38-27. The entire tape for machining the part shown in Figure 38-24 measured 28 inches. The block shown in Figure 38-28 measured $3\frac{1}{2}$ inches. (*Courtesy DeVlieg Machine Co.*)

Computer languages for NC control. The control mechanisms on NC machines do not understand ordinary shop language that people use to describe what machining, or machine–work movements, must take place. In addition, the effects of various sizes and types of cutting tools, requiring different offsets as illustrated in Figure 38-18 and the capabilities of the machine tool regarding available power, speeds, feeds, table travel, and so on, must be taken into account. This barrier has been diminished by the use of computers and special, simplified programming languages which the computer can understand and convert into the commands required by NC machine controls. One such language is APT II (Automatically Programmed Tools), a revised version of a language developed by MIT. APT is the most widely used language in the United States and is used for both positioning and continuous-path programming. APT is designed to run on large expensive computers (256K memories). ADAPT, a language developed by IBM under an Air Force contract, has many of the features of APT but is designed for smaller computers. Another APT version for smaller computers, like the PDP 11/70, is UNIAPT which delivers an EIA tape with virtually no sacrifice of features and capabilities compared to APT. AUTOSPOT was developed by IBM for point-to-point positioning but today's version can be used for contouring. Other languages (like SPLIT and COM-PACT II) have been developed on a proprietary basis for specific machine tool systems and are available on a lease basis. The basic sequence of operations for computer assisted part programming is as follows.

The part programmer prepares a manuscript which specifies the part geometry, the tool path, and the operations needed for their sequence, using English like statements of the APT language. The program can be written on punched cards or directly on a computer terminal. The computer performs an input translation, followed by the required arithmatic calculations, including things like cutter offset computations, in order to obtain the coordinate points the cutter must follow. The computer program must take into account the individualities of the machine tool, relative to the available speeds, feeds, accelerations and so forth. The machine peculiar elements are provided by a post-processor program, the output from which is the needed punched tape.

The workpiece, no matter how complex, can be conceived of as a compilation of points, straight lines, planes, circles, cylinders and other mathematically defined geometries. Historically, the part programmer's job has been to translate the component parts into basic geometric elements, defining each geometric element in terms of workpiece dimensions. In recent years, much of this work has become routine in CAD systems. However, the APT programming language is not currently compatible with many CAD programming languages, which has hindered the CAD to CAM step.

An example of how the APT language would be used on a part is shown in Figure 38-29. Essentially it is necessary to tell the computer the location of the center of the bolt hole circle or one of the holes in the circle with respect to a zero point, the number of holes in the circle, and the radius of the circle. The computer then finds the x, y coordinate values for the centers of the eight holes. The diameter of the drill defines the hole size but note that the programmer also decides on the feed rate (as well as the cutting speed). So, in addition to knowing how to program a computer using APT-type languages, the programmer must have sufficient manufacturing knowledge to make such judgments wisely. Thus, it is imperative that machine tool programmers be knowledgeable in machining fundamentals and operation sequencing. Otherwise, it is like asking someone who cannot read or write but knows how to punch the keys on a typewriter to type a letter. What comes out may not make good manufacturing sense.

Numerical control machines. Computer and numerical control are used on a wide variety of machines. These range from single-spindle drilling machines which often have only two-axis control, such as shown in Figure 38-30, and can be obtained for about $10,000, to machining centers, such as shown in Figure 38-31. The latter can do drilling, boring, milling, tapping, and so forth, with four-axis control. It can automatically select and change 32 preset tools. Such a machine can cost over $200,000. Between these extremes are numerous machine tools that do less varied work than the highly sophisticated machining centers but which combine high output, minimum setup time in changing from one job to another, and remarkable flexibility because of the number of tool motions that are provided.

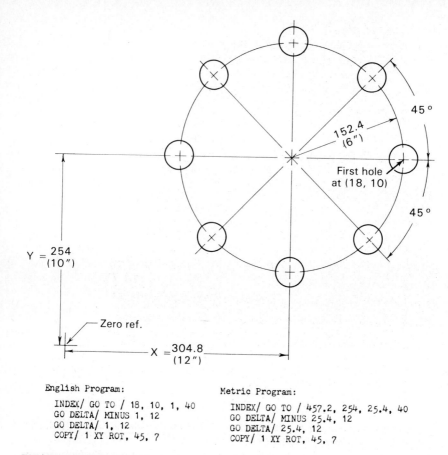

English Program:

```
INDEX/ GO TO / 18, 10, 1, 40
GO DELTA/ MINUS 1, 12
GO DELTA/ 1, 12
COPY/ 1 XY ROT, 45, 7
```

Metric Program:

```
INDEX/ GO TO / 457.2, 254, 25.4, 40
GO DELTA/ MINUS 25.4, 12
GO DELTA/ 25.4, 12
COPY/ 1 XY ROT, 45, 7
```

FIGURE 38-29 Circle of eight holes and APT program for machining these holes. Numerals 457.2, 254, and 25.4 (18, 10, and 1) are *X, Y,* and *Z* coordinates for table and tool. Numeral 40 is table movement rate. Numeral 12 is feed rate for drill. Numeral 45 is 45° of rotation. Numeral 7 is the instruction for seven duplicate holes to be drilled.

The machine shown in Figure 38-32 is an example of this trend of providing greater versatility along with high productivity. The versatility is being further increased by combining both rotary-work and rotary-tool operations—turning and milling—in a single machine. There are also numerous tape-controlled machines that provide four- and five-axis contouring capability. Table (or tool) movements commonly occur at speeds up to 6350 mm (250 inches) per minute, and tools are changed in 6 seconds or less. It is also common to provide two worktables, permitting work to be set up on one while machining is done on work mounted on the other (Figure 38-31), with the tables being interchanged automatically. Consequently, the productivity of such machines can be very high, the chip-producing time often approaching 50% of the total.

Numerical control has been applied to a wide variety of other production processes. A NC turret punch is shown in Figure 38-33 which has $X - Y$ control

FIGURE 38-30 Tape-controlled, single-spindle drilling machine. (*Courtesy Colt Industries, Pratt & Whitney Machine Tool Division.*)

on the table. NC systems are used on wire EDM machines, laser welders, flame cutters, and many other machines.

Economic considerations in tape and numerical control. NC machines are costly, but their use usually can be justified economically in from 1 to 3 years if their use factor is reasonably high, primarily through substantial savings in setup and machining time, particularly when the parts being processed are a GT family of parts. In such instances, the NC machine may be a cell unto itself with parts being loaded and unloaded with a robot. Three examples of relative costs and times are shown in Table 38-5.

Advantages and disadvantages. The advantages of programmable machine tools can be summarized as follows:

1. *Accuracy and precision.* Greater accuracy and precision is built into the machines, resulting in better quality and a high order of repeatability.
2. *High production rates.* Optimum feeds and speeds for each operation.
3. *Lower tooling costs.* Expensive jigs and templates are not needed.
4. *Less lead time.* Programs can be prepared in less time than conventional jigs and templates, and less setup time is required.
5. *Fewer setups per workpiece.* More operations can be done at each setup of the workpiece.

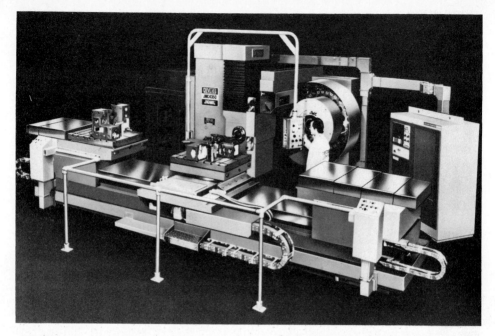

FIGURE 38-31 Horizontal spindle, four-axis CNC machining center. The table can move left/right or in/out and spindle can move up/down or in/out, with positioning accuracy in the range of 0.0003 inch in 40 inches of travel. The machine has automatic tool change and automatic work transfer so that workpieces can be loaded/unloaded while machining is in progress. (*Courtesy DeVlieg Machine Co.*)

6. *Better machine utilization.* There is less machine idle time, owing to more efficient table or tool movement between successive operations and fewer setups. Cycle time is reduced.
7. *Reduced inventory.* Less inventory needs to be carried because parts can be run economically in smaller quantities.
8. *Reduction in space required.* Greater productivity and reduced tooling lessen floor and storage space, and smaller economic lot sizes reduce storage space required for inventories.
9. *Less scrap.* Operator errors are substantially reduced.
10. *Less skill required of the operator.* Program planning in preparing tapes reduces the necessity for operator decisions.
11. *Manufacture of unique geometries.*

Through the use of computer programming of NC or CNC machines, one can generate surfaces and geometrical configurations that are not possible to make by any other method, at least not economically. Such a part is seen in Figure 38-34. This is the copper base of a cooling device. The spiral groove has constant width but constantly varying depth from start to finish. The APT programming language was used and the part machined on a three-axis CNC continuous-path vertical milling machine.

The major disadvantage of NC machine tools is their initial cost. This means

FIGURE 38-32 Tape-controlled turning machine having two indexing turrets, each capable of contouring. (*Courtesy Cincinnati Milacron, Inc.*)

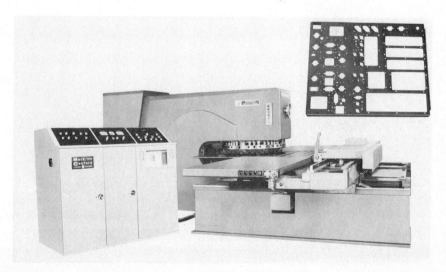

FIGURE 38-33 Tape-controlled turret-type punch press. Inset shows a typical sample of work done. (*Courtesy Warner & Swasey Company, Wiedemann Division.*)

Part	Setup Time (hr)		Cycle Time (hr)		Tooling Costs	
	Conventional	Tape-Controlled	Conventional	Tape-Controlled	Conventional	Tape-Controlled
Tool holder	5.25	None	1.86	1.30	$3140	$2350
Motor base	4.48	None	0.65	0.40	2350	1890
Bracket	4.63	None	0.64	0.21	835	260

that they must have sufficient use to justify the investment and enable the previously indicated savings to be obtained. The control equipment now is virtually all made of solid-state modules, and the reliability is excellent. Programming has been greatly simplified. Consequently, NC machines have provided a much needed solution for small- and medium-quantity production, and it is easy to see why they have been so widely adopted.

Robots: the steel-collar worker. A key element in automating manufacturing systems is the robot. As defined by the Robot Institute of America: "A *robot* is a reprogrammable, multifunctional manipulator designed to handle material, parts, tools or specialized devices through variable programmed motions for the performance of a variety of tasks." The word "robot" was coined in 1921 by Karel Capek in his play *R. U. R (Rossum's Universal Robots)*. The word "robot" is derived from the Czech word for "worker." Another famous author, Isaac Asimov, depicted robots in many of his stories and gave three laws of robotics which hold quite well for industrial robotic applications. Asimov's *Three Laws of Robotics* were:

FIGURE 38-34 Spiral groove with variable depth machined in copper plate. (*Courtesy Robert Tidmore, NASA Marshall Space Flight Center Test Laboratory, Fabrication Division.*)

1. A robot may not injure a human being, or through inaction, allow a human being to be harmed. (Safety first.)
2. A robot must obey orders given by human beings except when that conflicts with the First Law. (A robot must be programmable.)
3. A robot must protect its own existence unless that conflicts with the First or Second Law. (Reliability.)

For our purposes, if a machine is programmable, capable of automatic repeat cycles, and can perform manipulations in an industrial environment, it is an industrial robot.

All robots have the following basic components:

1. *Manipulators:* the mechanical unit, often called the "arm," that does the actual work of the robot. It is composed of mechanical linkages and joints with actuators to drive the mechanism directly or indirectly through gears, chains, or ball screws.
2. *Feedback devices:* transducers that sense the positions of various linkages and joints and transmit this information to the controllers in either digital or analog form [A(4)-level robots].
3. *Controller:* the brains of the system that direct the movements of the manipulator. In higher-level robots, computers are used for controllers. The functions of the controller are to initiate and terminate motion, store data for position and motion sequence, and interface with the "outside world," meaning other machines and human beings.
4. *Power supply:* electric, pneumatic, and hydraulic power supplies used to provide and regulate the energy needed for the manipulator's actuators.

Industrial robots manufacturers have proliferated in recent years[3] but most commercially available robots have one of four mechanical configurations, three of which are shown in Figure 38-35. Cylindrical coordinate robots have a work envelope (shaded region) that is a portion of a cylinder. Spherical coordinate robots have a work envelope that is a portion of a sphere. Jointed-spherical coordinate robots have a jointed arm and a work envelope that approximates a portion of a sphere. Rectangular coordinate robots with a rectangular work envelope have been developed for precision assembly applications. Figure 38-36 shows a jointed-arm robot with six axes of motion. That is, the arm can move up or down, in or out, and rotate about the base, as can the cylindrical or spherical coordinate robots. In addition, these robots may have two or three additional minor axes of motion at the end of the arm (commonly called the wrist). These three movements are "pitch" (vertical movement), "yaw" (horizontal motion), and "roll" (wrist rotation).

Industrial robots used in industry today fill two main functions: material handling or materials processing. They have, for the most part, very primitive motor and intelligence capabilities, with most robots being A(3)-level machines. The sensory-interactive control, decision-making, strength-to-size, dexterity, and artificial intelligence capabilities of robots are far inferior to those of human beings

[3]In 1982 there were over 50 manufacturers of robots, whereas 10 years ago there were less than 5.

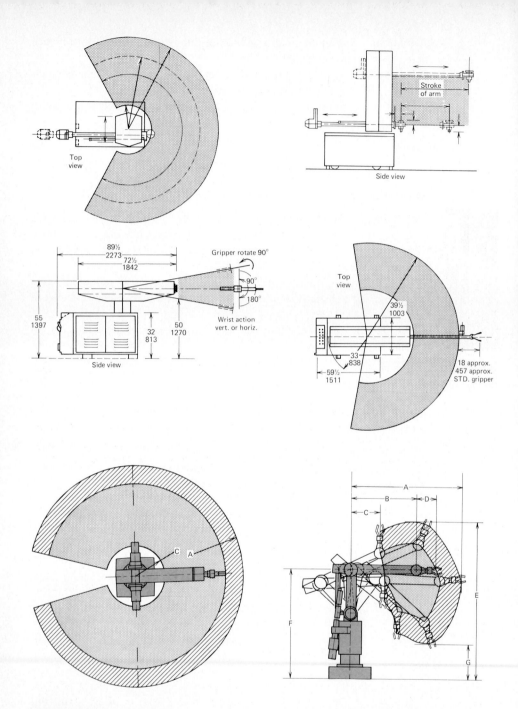

FIGURE 38-35 Cylindrical coordinate (*top*), spherical coordinate (*middle*), and jointed arm (*bottom*) work envelopes (shaded regions) for typical industrial robots.

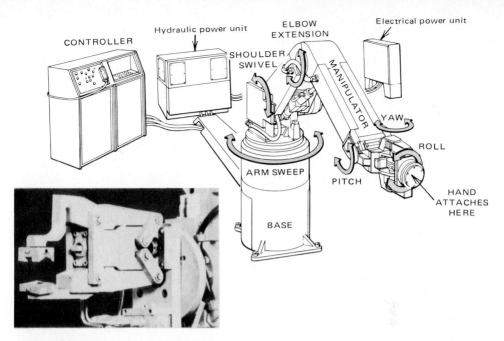

FIGURE 38-36 Typical jointed arm manipulator or robot. Inset in lower left shows hand. (*Courtesy Cincinnati Milacron, Inc.*)

at this time. Nevertheless, robots are having a strong impact in the industrial environment, often doing those jobs which are hazardous, extremely tedious, or unpleasant. Robots perform well doing paint spraying, loading and unloading small forgings or die casting machines, spot welding, and so forth.

The A(3)-level robot is usually called a "pick-and-place" machine and is capable of performing only the simplest repeat-cycle movements, on a point-to-point basis, being controlled by an electronic or pneumatic control with manipulatory movement controlled by end stops.

A(3) robots are usually small robots with relatively high speed movements, good repeatability (0.25 mm or 0.010 inch), and low cost, are simple to program, operate, and maintain, but have limited flexibility in terms of program capacity and positioning capability.

In order to raise the robot to an A(4) machine, sensory devices must be installed in the joints of the arm(s) to provide positional feedback and error signals to the servomechanisms, just as was the case for the NC machines. The addition of an electronic memory and digital control circuitry allows this level of robot to be programmed by a worker who guides the robot through the desired operations and movements using a "hand controller" (see Figure 38-37). The hand-held control box has rate control buttons for each axis of motion of the robot arm. When the arm is in the desired position, the "record" or "program" button is pushed to enter that position or operation into the memory. This is similar to point-to-point NC machines as the path of the robot arm movement is defined by selected end points when the program is played back. The elec-

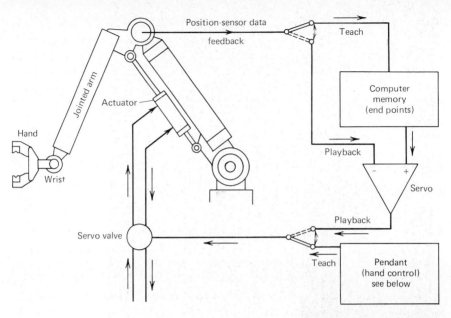

Position-sensor data
feedback

Teach

Jointed arm

Actuator

Computer
memory
(end points)

Hand

Wrist

Playback

–   +

Servo

Playback

Servo valve

Teach

Pendant
(hand control)
see below

FIGURE 38-37 Manual programming of an industrial robot is accomplished by using a hand-held "teach pendant." The robot is guided through a sequence of operations. The successive positions of all the robotic joints are stored in an electronic memory. This establishes the end points of the motions of the arm. By switching from "teach" to "playback," the stored end points are replayed. (*Courtesy Cincinnati Milacron, Inc.*)

tronic memory can usually store multiple programs and randomly access the required one, depending on the job to be done. This allows for a product mix to be handled without stopping to reprogram the machine. The addition of a computer, usually a minicomputer, makes it possible to program the robot to move its "hand" or "gripper" in straight lines or other geometric paths between given points, but the robot is still essentially a point-to-point machine.

Point-to-point servo-controlled robots have the following common characteristics: high load capacity, large working range, and relatively easy programming, but the path followed by the manipulator during operation may not be the path followed during teaching.

In order to make the A(4) robot continuous-path, position and velocity data must be sampled on a time base rather than as discretely determined points in space. Due to the high rate of sampling, many spacial positions must be stored in the computer memory, thus requiring a mass storage system. Continuous-path machines tend to be smaller than point-to-point machines, with lighter load capacity, greater precision (± 0.020 to $\pm .002$ inch), and somewhat higher end of arm speed.

Many of the industrial robots in use at this time are A(4) point-to-point machines. Most robots operate in systems wherein the items to be handled or processed are placed in precise locations with respect to the robot. Even robots with computer control which can follow a moving auto conveyor line while performing spot-welding operations have point-to-point feedback information, but this is satisfactory for most industrial applications.

To expand the capability of robots, sensors are used for obtaining information regarding position and component status. Tactile sensors provide information about force distributions in the joints and in the hand of the robot during manipulations. This information is then used to control movement rates. Visual sensors collect data on spatial dimensions by means of image recording and analysis. Visual sensors are used to identify workpieces; determine their position and orientation; check position, orientation, geometry, or speed of parts for correctness; determine the correct welding path or point; and so forth. A summary of robotic sensors is given in Table 38-6.

In order to provide a robot with tactile or visual capability, powerful computers and sophisticated software are required, but this appears to be the most logical manner to raise the robot to the A(5) level of automation, wherein it can adapt to variations in its environment. This would mean that the vision system would be used to locate parts moving past a robot on a conveyor, identify those parts which should be removed from the conveyor, and communicate this information to the robot. The robot's arm would then have to be able to track the part moving on the conveyor and orient its gripper to be able to pick up the part and move it to the desired work station.

TABLE 38-6   Sensors Used on Robots with Some Typical Applications

	Design	Application
Visual	Video pickup tubes (TV camera) Semiconductor sensors	Position detection Parts detection Consistency testing (e.g., in manipulation, welding, and assembly)
Tactile	Feelers Pin matrix Load cells (piezo, capacitance) Wire strain gauge Pressure-sensitive plastics	Position detection Tool monitoring, (e.g., in casting cleaning, grinding, manipulation, and assembly)
Electrical (inductive capacitive)	Shunt (current determination) Capacitor Coil	Position detection Status determination, (e.g., in manipulation and welding)

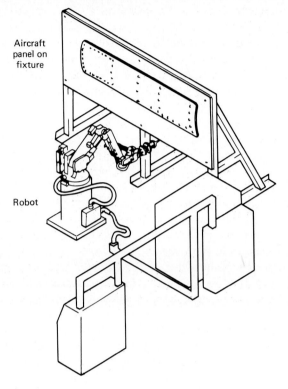

Drilling aircraft panel assemblies

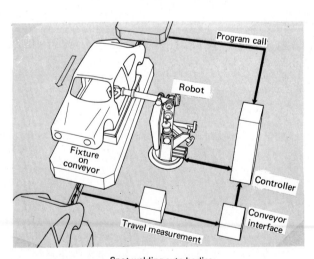

Spot welding auto bodies

FIGURE 38-38 Applications for robots in material processing. (*Courtesy KUKA Welding Systems + Robot Corp. Mi.*)

The current generation of robots is finding applications in the following areas (also see Figures 38-38 and 38-39):

Die casting. In single- or multishift operations, custom or captive shops, robots unload machines, quench parts, operate trim presses, load inserts, ladle metal, and perform die lubrication. Die life is increased because die-casting machines can be operated without breaks or shutdowns and die temperature is controlled by uniform cycle times.

Press transfer. Robots in press transfer lines guarantee consistent throughput shift after shift. Large and unwieldy parts can be handled at piece rates as high as 400 per hour with no change in cycle time due to fatigue. Robots are adaptable for long- or short-run operations. Programming for new part sizes can be accomplished in minutes.

Materials handling. Strength, dexterity, and a versatile memory allow robots to pack goods in complex palletized arrays or to transfer workpieces to (and from) moving or indexing conveyors from machines. Savings are dramatic in these labor-intensive operations.

Forging. Operating costs are reduced when robots feed forge presses and upsetters. They work continuously without fatigue or the need for relief in hot, hostile environments commonly found in forging. Robots can easily manipulate the parts in the presses.

Investment casting. Scrap rates as high as 85% have been reduced to less than 5% when molds are produced by robots. The smooth, controlled motions of the robot provide consistent mold quality impossible to achieve manually.

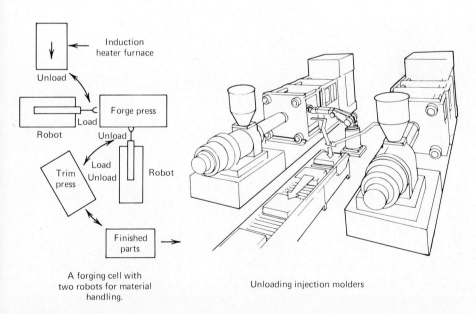

A forging cell with two robots for material handling.

Unloading injection molders

FIGURE 38-39 Application of robots for materials handling, loading, and unloading hot parts from forge and unloading two plastic injection molding machines. (*Right, from Mert Corwin, "A Computer Controlled Robot for Automotive Manufacturing," Wolfsburg, West Germany, Sept. 12–15, 1977; courtesy Cincinnati Milacron, Inc.*)

Material processing. Product quality is improved and sustained with point-to-point, continuous-path robots in jobs such as routing, flame cutting, mold drying, drilling, polishing, and grinding. Once programmed, the robot will process each part with the same high quality.

Welding. Robots spot-weld cars and trucks for almost every major manufacturer in the world, with uniformity of spot location and weld integrity. In arc welding, robots increase arc time, remove operators from hazardous environments, reduce the cost of worker protection, and improve consistency of weld quality.

Machine tool loading. The most efficient use of robots may well be achieved by integrating them into cellular arrangements of machine tools being used to process families of component parts. The robot can provide part loading and unloading of two to five CNC machines grouped properly in a machining cell. These unmanned cells facilitate maximum automation and productivity while maintaining programmable flexibility in producing small to medium-size productions lots of parts from compatible parts families. The robot can also change tools in the machines and, in the future, even the work-holding devices, thereby adding more flexibility to the cell. These cellular arrangements (or layouts) help to achieve maximum machine tool utilization by greatly increasing the percentage of time the machines are cutting which in turn increases the output for the same investment. This is the name of the game in productivity. Examples of this concept are shown in Figures 38-40 and 38-6a.

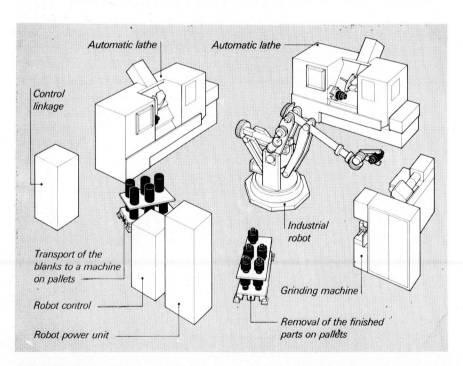

FIGURE 38-40 Schematic of unmanned machining cell with three machine tools and robot handling the parts loading and unloading of the parts from pallets and the machines. (*Courtesy KUKA Welding Systems + Robot Corp. Mi.*)

PROCESSES AND TECHNIQUES
RELATED TO MANUFACTURING

Programmable (robotic) automation economics. Debugging hard automation systems can be costly and time-consuming, taking an average of 12 months to bring a hard automation system on-stream. If products change regularly, it is quite possible the system will be obsolete before it becomes operational. Programmable automation generally takes less debugging and is less subject to obsolescence. As the robots become smarter, they will replace more manual activities. As they become cheaper, it becomes more economical for them to replace hard automation types of functions as well as human labor.

Figure 38-41 shows a comparison of the cost of the steel-collar worker versus the blue-collar worker in the U.S. automotive industry, beginning in 1961, when the first Unimate robot was used in an automotive plant, up to 1979. In 1961, labor cost about $3.80 per hour, including all fringe benefits. By 1979, this figure rose to about $14 per hour, including fringes. Through the same period, the cost of robot labor runs about $4.80 per hour. This figure includes capitalized cost for an 8-year life, maintenance, repair, installation, cost of power, and so forth. This figure is conservative, since it is based on 32,000 hours of operating life, and many robots have already exceeded 80,000 hours of in-plant service with up-times exceeding 95%.

The robot is a key element of advanced manufacturing technologies available for the 1980s. All the tools are there: computer-aided design, computer-aided manufacturing (NC, CNC, AC, DNC), computer-aided testing and inspection (CATI), automatic assembly and warehousing, robots, and much more. However, the Japanese have demonstrated that to achieve high productivity, visionary management coupled with high utilization of people as well as machines, is needed. No better example of this can be found than the Toyota Production System, developed and promoted by the Toyota Motor Company. Their production system is as unique and revolutionary as was the Taylor system of scientific management or the Ford system for mass assembly. It is significant that virtually every manufacturing system or technology cited in this chapter is practiced at Toyota and that this company has emerged as a world leader in automotive manufacturing. It will be briefly discussed in Chapter 39, but stu-

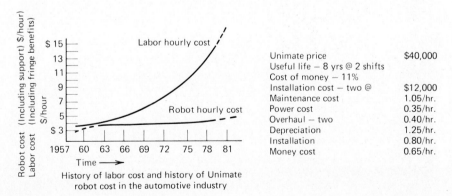

FIGURE 38-41 Cost per hour for a robot versus human labor in the automotive industry. (*Data and plot from J. F. Engelberger, The Industrial Robot, Sept. 1979, p. 115.*)

dents of manufacturing engineering are well advised to be knowledgeable of this unique system.

Instructional robots. Generally speaking, the industrial robots described in the preceding section are expensive and may be considered dangerous for student usage. Thus, a class of robots has been developed which can be called instructional robots, which make it possible for educational systems and small businesses to gain hands-on experience with this technology. These robots cost but a fraction of the industrial versions but use essentially the same electronic controls with stepping motors or low-pressure hydraulics. The microprocessors in these machines will typically have RS232 interface for connection to a small personal computer in which the program software is stored. The trade-off here is one of scaling, as these robots will have much lower speed of response, lower weight-carrying capacity, and poorer positional accuracy than their industrial counterparts but may cost only one-tenth of the industrial machine. Figure 38-42 shows three of these small instructional robots. The machine on the left has positional and velocity feedback, so it is an A(4) machine; the one on the right is open loop or A(3). The small robot in the middle simulates a "pick-and-place" device. While it has no positional feedback, it has measurement capability and a microcomputer interface, which allows it to perform as a computer-aided inspection and/or automatic sorting machine.

Many of these mini-robots are being incorporated into miniature manufacturing cells which use the small robots and scaled-down versions of machine tools. The machines in these micro-CAM setups have essentially the same minicom-

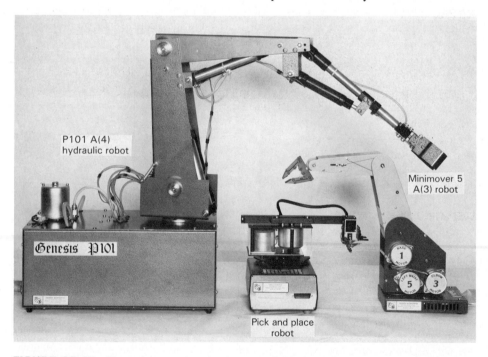

FIGURE 38-42 Three instructional robots available for laboratory work. (*Courtesy Feedback, Inc.*)

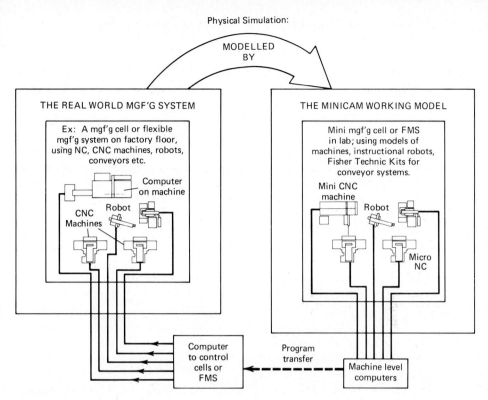

Physical Simulation:

MODELLED BY

THE REAL WORLD MGF'G SYSTEM

Ex: A mgf'g cell or flexible mgf'g system on factory floor, using NC, CNC machines, robots, conveyors etc.

Computer on machine

Robot

CNC Machines

THE MINICAM WORKING MODEL

Mini mgf'g cell or FMS in lab; using models of machines, instructional robots, Fisher Technic Kits for conveyor systems.

Mini CNC machine

Robot

Micro NC

Computer to control cells or FMS

Program transfer

Machine level computers

FIGURE 38-43 Schematic defining physical simulation, a new technique to study manufacturing systems, using scaled-down but fully functional models.

puters and software that the full-scale systems employ. This is called physical simulation (see Figure 38-43). In this way, the development of the software needed to integrate the machines and design of the cell can be done prior to the installation of the full-scale system on the shop floor. These systems are ideal for providing hands-on instruction for students in advanced manufacturing systems as both unmanned cellular and FMS can be simulated in the laboratory at quite reasonable cost.

REVIEW QUESTIONS

1. Why does the elimination or reduction of setup time greatly improve the productivity of short-run or small-lot operations?
2. If theoretical capacity of 100% represents 365 days, how many days does the typical production facility run during the year?
3. What are four classical forms for manufacturing systems?
4. Can you give a service example of each of the four classical forms for manufacturing systems?
5. What are the two new forms of manufacturing systems that have emerged in the last decade and how do these new forms differ from the classical forms?
6. What do we mean by the statement that "the manufacturing system is usually a mixed system"?

7. What are the trends that are driving companies toward smaller lot production?
8. What is meant by the statement "processes proliferate"?
9. What is group technology and how does it convert the functional job shop into a cellular/flexible shop? Do you think that 100% of the job shop processes will be converted?
10. What is a P-Q chart and how is it related to group technology?
11. What have been the benefits generally experienced by companies that have undergone conversions of their systems through GT?
12. How does the "do-nothing" alternative act as a constraint to the implementation of a GT program?
13. Can you think of examples of A(3) and A(4) systems that exist in your home? What about an A(5) or A(6) system?
14. How does an adaptive control system differ from a numerical control system?
15. In the decision analysis portion of the AC system, optimization of the process is required. What does this imply about the process?
16. How does feed forward differ from feedback in these systems?
17. Unmanned cellular and flexible manufacturing systems use NC machines with either robots or conveyors. What differentiates these systems from transfer lines?
18. Why do you think that transfer lines are moving toward the use of programmable machines?
19. Explain the problems of cutter offset and interpolation in NC programming.
20. Some of the functions performed by the operator in piece-part manufacturing (10 were listed) are going to be very difficult to completely automate. Which ones and why?
21. The first NC machines were closed loop control, but later machines were open-loop. What did this require on the part of the machine tool builders?
22. Can a continuous-path NC machine be open-loop?
23. Why don't we have NC shapers or broaches?
24. Why don't we use manual programming for continuous-path NC?
25. What are the two main areas in which robots are used?
26. What are the basic components of all robots?
27. What are the four common work envelopes for industrial robots?
28. How is positional feedback obtained in robots?
29. What are the primary differences between an instructional robot and an industrial robot?
30. What is physical simulation and how is it used?
31. Compare a rotary transfer machine with an unmanned robotic cell.
32. In Table 38-2, examples of different types of restaurants were given. That is, a sit-down restaurant with menus is a service example of a job shop. What kinds of restaurants are examples of cellular manufacturing?

CASE STUDY 38.   The Component With the Triangular Hole

The Johnstone Company estimates that its annual requirements for the socket component shown in Figure CS-38 will be at least 50,000 units, and that this volume will

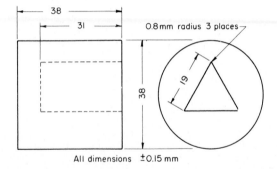

FIGURE CS-38 Component containing dead-end triangular hole.

continue for at least 5 years. Consequently, it wants to consider all practicable methods for making the component and has assigned you the task of determining these methods and of recommending which should be explored in detail to determine the most effective and economical process. The specifications call for a lightweight metal, such as an aluminum alloy, and that it must have a tensile strength of at least 138 Mpa and an elongation of at least 3%.

Determine at least five practicable methods for producing the part. Suggest which two appear most likely to be most economical and thus should be investigated fully. (Give the reasons for your selections.)

CHAPTER 39

Production Systems

Manufacturing is the utilization and management of materials, equipment, people, and money to produce products. The bulk of this text has been devoted to discussions of materials, processes, and equipment. Chapter 38 discussed how certain of these processes can be integrated together into manufacturing systems. This chapter will discuss other elements of the production system, including some of the management aspects of both the manufacturing systems themselves and the entire production system (refer back to Figure 38-4). Included in this discussion will be the relationship between design and manufacture and the planning of the production system.

For economical manufacturing to be achieved, planning must start at the design stage and continue through the selection of the materials, the processes, the tooling, the materials handling methods, and scheduling of production. Within recent years, our ability to reduce overall manufacturing costs by process improvement has decreased. This is especially true in the job shop or lot production environment. Individual processes have been improved and automated to the place where further improvements may not result in very significant gains in productivity. What will provide marked productivity improvement is effective planning and scheduling of a production system, organized to utilize the people and the machines to their best advantage. There is much to be gained from effective management and control of processes run at modest speeds and feeds.

This means better planning and scheduling. Clearly, one of the major breakthroughs in this area has been group technology, a manufacturing management philosophy which resulted in cellular and flexible manufacturing systems.

For the GT concept to be effective through the entire production system each workpiece must be coded and classified. The same codes that are used for classifying the parts into families can be used for computer-aided process planning (CAPP) and computer-aided design (CAD).

In the design area, the following kinds of items can be identified by codes: main shape, shape elements, materials, dimensions, tolerances, and even functions. Of course, nothing is produced until it has been designed and this was strictly a manual operation until the 1970's when functional computer graphics systems became available. No longer did the designer have to labor over the board. With a CRT (cathode ray tube), a keyboard, a stylus, and a list of available computer functions (called a menu), all backed by powerful computer design software, designers' productivity increased manyfold without much loss in creativity. Elements going into the designs became more standardized, reducing the complexity on the manufacturing side and assisting computer-aided manufacturing.

On the planning side, *automated process planning* schemes began to develop, utilizing the programmable machines described in Chapter 38. This activity is often called CAPP because the computer is utilized in the process planning. CAPP involves the automatic generation of the process plan (known as the route sheet) to make the part. The process routing through the manufacturing processes and automated systems is developed by recognizing the specific attributes of component parts and relating these attributes to corresponding processes and operations. Operational details such as the tool layout, speeds and feeds, estimates of machining times, and the like can be produced in such programs. Obviously, the computer must have data regarding rough and finish sizes of the part, machinability data for the part in question, specifications on cutting tools, machine tools, and work-holding devices, surface finish and tolerances required, and much more in order to be able to replace the process planner. Proponents of planning and scheduling software systems claim that there are far more savings to be gained by good planning, scheduling, and control than by faster and better machining.

However, as the elements of CAD, CAM, and CAPP draw closer together, it is evident that a unified software system that can draw the endless hardware configurations together is missing. The software programs now on the market were developed in the spirit of free enterprise, so requests for unified, integrated software have fallen on deaf ears. Also, it should be remembered that software costs money to develop and high software costs require long runs to recover, even though such costs are often hidden in the indirect or overhead costs. Thus, unless universal software is developed and widely implemented, the realization of computer-programmed, unmanned factories will be seriously crimped. Automation and computer control in the chemical and petroleum processing industries has always exceeded that of discrete-parts manufacturing industries. Gases and liquids need no shape information and have well-developed reaction rate theory with equations of state that either exist or can be found. Solids, on the other hand, require processing to bestow shape. For every processing scheme, a distinct history of stress, strain, temperature, strain rate, composition, and so on, accompanies each of these schemes. Further, each processing scheme has a distinct technology (skills not transferable), distinct processing intrastructure, distinct group of processing skills, distinct classes of optimally suited materials, and distinct methodology. The result has been a disconnected production system wherein systemized, cost-efficient organizations have emerged only for large lot sizes. Clearly, then, the integration of the design, manufacturing, planning, and control software offers great challenge and greater rewards for those who will lead the industrial world in this development.

THE RELATIONSHIP OF DESIGN TO PRODUCTION

By recognizing the systems approach, the design stage can save many dollars and much time later in production, inspection, assembly, packaging, and even distribution and marketing. Let us examine some of the simpler aspects of designing for production. This has recently been termed ''producibility'' in design.

Traditionally, a design will often develop in three phases, as was outlined in Chapter 10. In the *conceptual* or *idea* phase, the designer conceives of an idea for a device that will accomplish some function. This stage establishes the functional requirements that must be met by the device.

In the second, *functional-design* stage, a device is designed that will achieve the functional requirements established in the conceptual stage. Often more than one functional design will be made, suggesting alternative ways in which the functions can be met. At this stage, the designer is usually more concerned with materials than processes and may ignore the fact that the designed configuration cannot be produced economically utilizing the material being considered.

The third phase of design is called *production design*. Although attention should also be given to the appearance of the product at this stage, particularly if sales appeal is important, the major emphasis is on providing a design that can be manufactured economically. The *design engineer* must, of course, know that certain manufacturing processes and operations exist that can produce the desired product. However, merely knowing that feasible processes exist is not sufficient. He also needs to know their limitations, relative costs, and process capabilities (accuracy, tolerance requirements, and so forth), in order to design for producibility. An excellent example of such a situation occurred in the designing of a new automatic transmission for an automobile. Four multiple-disk clutches were required—two large and two small. By knowing not only that a blanking process could be used for producing both the friction and metal disks, but also understanding thoroughly the blanking process and the tolerance requirements, the engineers were able to design each clutch with the same number of friction and metal disks and with the small and large clutches of such diameters that each piece blanked from the center of a large clutch element was exactly the correct diameter for a small clutch element. As a result, very substantial savings in material and blanking costs were effected. This likely would not have occurred if the designers had been concerned only with function, and if the production department had then been required to manufacture whatever the designers had called for. Thus, if maximum economy is to be achieved, the designer should be aware of the intimate relationship between design details and production operations.

It is extremely important that the relationship between production and design be given careful consideration throughout the production-design phase, and it should not be forgotten even at the functional-design stage. Changes can be made for pennies in the design room that would cost hundreds and thousands of dollars to effect later in the factory. This type of consideration should be an integral and routine part of planning for manufacturing. Having the design engineers make a working prototype of every new model before any production drawings are made is one approach. If the model performs in accordance with the conceptual requirements, a second model is made, using, insofar as possible, the same manufacturing methods that will be used later for production. Any changes that will permit easier and more economical production are incorporated into this second model. If the second model meets the functional requirements of the engineering design group, it is then sent to the drafting room, and production drawings are made from it. This practice virtually eliminates costly-to-produce details from products and the need for a lot of design changes after the part has gone into production.

The designer plays a key role in determining what processes and equipment must be used to manufacture the product, although often indirectly. Clearly, one of the ways he can indirectly determine the process is through the selection of material. For example, the chassis for record players (stereos) are usually made from zinc and are die cast. Suppose, however, that the designer specifies a composite chassis, to be made from fiber-reinforced plastic; then a foam-molding process is needed instead of a die-casting machine. He specifies a particular joining process when he calls for a welded joint. These kinds of direct relationships are pretty obvious. However, other equipment and processes may be specified just as certainly in not-so-obvious ways.

One of the most common ways in which equipment and processes may be specified indirectly is through dimensional tolerances placed on a drawing. If a tolerance of ± 0.005 mm (0.0002 inch) is shown, a grinding operation may be specified just as definitely as if the word *grind* were placed on the drawing. Designers often fail to realize this fact, and specify unnecessarily close tolerances and expensive and unnecessary operations result.

On the other hand, indefinite dimensions and tolerances can lead to important requirements being ignored by the factory. In a recent large court suit, one important drawing contained the statement "Approximately not to scale," and another had an important dimension reading "$128\frac{5}{16}$ inches approx." It was not surprising that the shop ignored certain other dimensions that caused it some inconvenience but which were critical. Consequently, designers should realize that the dimensions and tolerances they place on a drawing may have implications and results far beyond what they anticipated.

Design details are directly related to the processing that will be used, making the processing easy, difficult, or impossible, and affecting the cost and/or quality. Some of these will now be considered.

Design details related to casting. Some design details that relate to casting are shown in Figure 39-1. These emphasize the basic fact that castings shrink during solidification and cooling, and that difficulties may be experienced unless uniform sections are employed and adequate provision made to avoid excessive contraction stresses. Also, by remembering this principle, one often may reduce the weight and cost of castings, as illustrated at Figure 39-1d. Figure 39-1c emphasizes that mold parting lines should be in a single plane, if practical. This also applies to forgings.

Most of the design details relating to sand castings also apply to permanent-mold castings, such as die castings. However, if permanent cores are utilized, as most often is the case, other unique problems are encountered. Figure 39-2 illustrates how relatively simple design changes can have a major effect on the cost of the die and the resulting casting.

Design details related to lettering. Lettering or decorative detail is often included on the surface of products. As shown in Figure 39-3, this should be

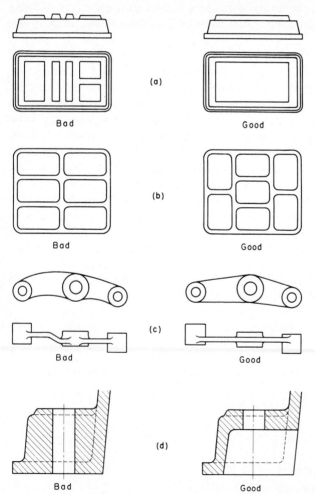

FIGURE 39-1 Good and bad details in casting design.

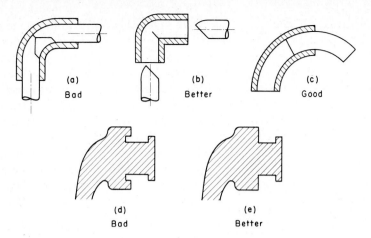

FIGURE 39-2 Good and bad design details in die castings.

handled differently for various processes. When letters or decorative devices are included on castings or forgings, the design should call for them to be above the surface of the part. On sand castings they can be produced easily in this manner by applying stock letters to the surface of the pattern, which will in turn result in their being above the surface of the casting. To have the lettering below the surface of the casting requires that the pattern be engraved. In the case of die castings and forgings, the lettering can be engraved in the dies, requiring only a minimum of machining to produce the desired, raised letters on the finished parts. On the other hand, if the design calls for depressed lettering on die castings and forgings, all the surface in the dies, except the letters, must be machined away. This is an expensive process and can be done only by use of special machines. If depressed lettering must be used on die castings and forgings, the method illustrated in Figure 39-3b sometimes can be employed, the lettering being depressed in a small, raised panel. This requires only the small amount of material in the panel, surrounding the lettering, to be removed from the dies.

When lettering must be included on a machined surface, it should be engraved into the surface. If raised lettering is essential, it should be made on a separate, small plate and attached to the machined surface. In many cases this can be done very satisfactorily by means of modern adhesives.

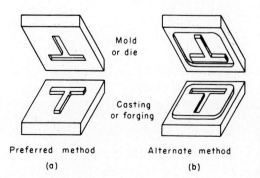

FIGURE 39-3 Preferred methods of providing lettering for different processes.

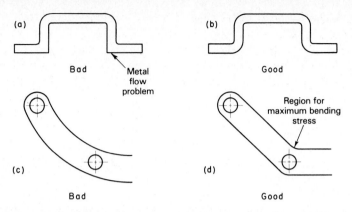

(a) Bad — Metal flow problem

(b) Good

(c) Bad

(d) Good — Region for maximum bending stress

FIGURE 39-4   Examples of good and bad design details for forgings.

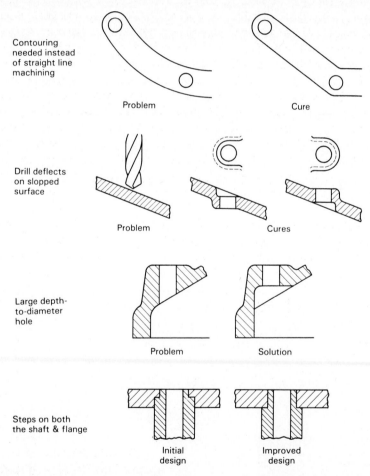

Contouring needed instead of straight line machining

Problem

Cure

Drill deflects on slopped surface

Problem

Cures

Large depth-to-diameter hole

Problem

Solution

Steps on both the shaft & flange

Initial design

Improved design

FIGURE 39-5   Examples of good and bad design details from the viewpoint of machining.

Design details related to forgings. When designing forgings, two primary factors should be kept in mind: (1) metal flow, and (2) minimizing the number of operations and dies. These two often are closely related. As illustrated in Figure 39-4a, a shape that would be very satisfactory for casting may be quite poor for forging, requiring more than one operation. This difficulty can be eliminated by the change shown in Figure 39-4b. Figure 39-4c and d illustrate how a shape that is easy to draw may be excessively costly to produce because of added difficulties in machining the forging die. If upset forging is involved, the rules illustrated in Figure 14-14 should be observed. Quite often the combining of two processes, such as forging and inertia welding, can provide an economical solution.

Design details related to machining. The ways in which design details can affect machining are almost unlimited. Figure 39-5 illustrates just a few. It is far better for the designer to visualize how the workpiece will be machined and make minor modifications that will permit easy and economical machining than to force the manufacturing department to find some way to machine a needlessly bothersome detail, usually at excessive cost. Too often the manufacturing departments will take for granted that the part has to be made as designed, rather than take the trouble to contact the designer to ascertain whether some modification can be made that will facilitate machining.

Figure 39-6 illustrates how required grinding may be made difficult, or virtually impossible, through poor design. With no provision for entry of the grinding wheel or its supporting arbor (top views), the required surfaces cannot be ground by ordinary procedures. A simple change in the design corrects the difficulty. Another common deficiency in the design of parts that require the grinding of external cylindrical surfaces is the failure to provide any means to grip or hold the workpiece during grinding. This happens most frequently when only one or a few pieces are involved.

Design details related to finishing. As was discussed in Chapter 37, the cost of required finishing operations—cleaning, smoothing, plating, and painting— is greatly affected by design. This is particularly important when large quantities

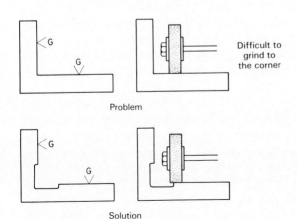

Problem

Solution

FIGURE 39-6 Method of providing proper clearance for grinding-wheel mounting and for overtravel.

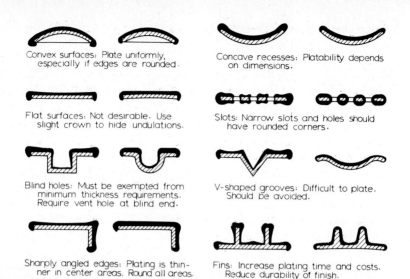

Convex surfaces: Plate uniformly, especially if edges are rounded.

Concave recesses: Platability depends on dimensions.

Flat surfaces: Not desirable. Use slight crown to hide undulations.

Slots: Narrow slots and holes should have rounded corners.

Blind holes: Must be exempted from minimum thickness requirements. Require vent hole at blind end.

V-shaped grooves: Difficult to plate. Should be avoided.

Sharply angled edges: Plating is thinner in center areas. Round all areas.

Fins: Increase plating time and costs. Reduce durability of finish.

FIGURE 39-7 Good and bad design details for electroplating.

of a product are to be manufactured. As has been pointed out in previous chapters, some manufacturing processes result in less need for finishing than others. Thus the choice of processing methods, and design to permit the use of such processes, may be very important where fine finish is required. The various finishing and decorative treatments often require special considerations and some modification of design details. Figure 39-7, for example, illustrates good and bad design details where plating is to be used.

DESIGN—PROCESSING—SERVICE RELATIONSHIPS

Satisfactory service of manufactured parts is obviously dependent on the design and the manufacturing. Most often it seems that failures of component parts in service, often resulting in product liability claims comes from a combination of factors. Some are related to poor design, some to manufacture, and some to misuse of the product. However, it is important to understand that the processes themselves impart certain properties to the materials which will influence their performance when put into service.

We have observed that the various machining processes produce widely varied surface textures (roughness, waviness, and lay) in the workpieces, but we should also be aware of the fact that different processes produce other changes in the physical or metallurgical properties in the near surfaces of the component parts. For the most part, these changes take place, in the subsurface, to a depth of 5 to 10 mils below the surface. The effect of these changes can be beneficial or detrimental, depending on the process used to create the surface, which, as we have learned, is often dictated by the material selected and the functional design.

Machining processes (both chip-forming and chipless) cause plastic deformation. Surfaces, when cut, generally are left with a tensile residual stress, microcracks, and a different hardness than that of the bulk metal. The EDM process leaves a layer of hard, recast metal on the surface which usually contains microcracks. Ground surfaces can have either residual tension or residual compression stresses, depending on the mix between chip formation and plowing and rubbing during the grinding. In some materials ground at high speed, phase transformations can occur in the subsurface.

The process known as roller burnishing, often used as a finishing process, leaves the surface with a residual compressive stress, as do many peening or tumbling processes. Welding processes leave residual stresses in the metal near the weld joint due to shrinkage of the molten weld metal as it cools and contracts. Similar shrinkage problems create residual stresses in castings. In summary, the principal causes of alterations in the subsurface state, due to material processing operations, are:

1. Plastic strain or plastic deformation.
2. Large temperature gradients or high temperatures.
3. Chemical reactions.
4. Differential shrinkage.

To understand the subtle nature of this problem, look at Figure 39-8. It shows the depth of subsurface damage in metal cutting as measured by the plastic deformation in the grain structure as a function of back rake angle of the tool, reflecting a hardness gradient in the surface. Suppose that the designer called for a design modification which, in turn, forced the manufacturing engineer to change tool materials in order to be able to increase cutting speed and thereby maintain production rates. Thus, he changed from a high-speed steel tool with a large rake (30°) to a carbide tool with a small rake (5°) but still got the same surface finish. The result was to double the depth of surface damage, and now

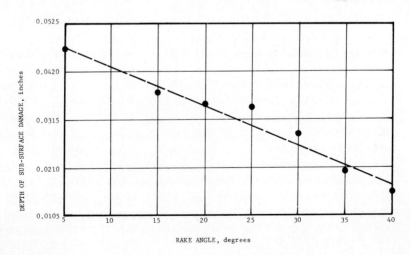

FIGURE 39-8 The depth of damage in a machined part surface increases with decreasing rake angles.

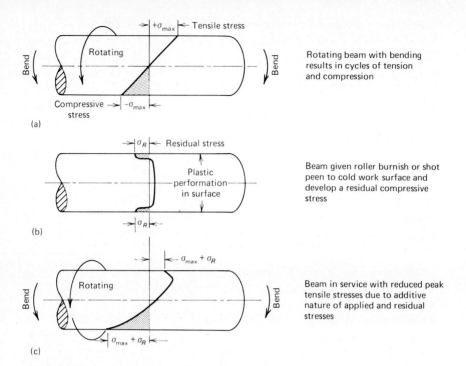

FIGURE 39-9 Lowering of peak tensile stresses in a rotating beam with bending by adding residual compressive stress during manufacture.

perhaps this part will fail in service, whereas before it performed quite admirably.

Residual stresses are often the product of nonuniform or localized plastic deformation. Since this describes most metalworking or machining processes, we can expect most components to have some residual stresses left in them after processing. This includes parts made by casting and welding, where unequal cooling and solidification rates cause residual stress and even distortion. These internal or "locked-in" stresses can be detrimental or beneficial in terms of fatigue behavior or corrosion resistance in the parts.

Suppose that we have a rotating beam which has been machined. The beam has a load on it so that it is bent while rotating. This results in a cyclic fatigue situation shown in Figure 39-9a. In order to enhance the fatigue life of the beam, it is roller-burnished or shot-peened, which cold works the surface and leaves a residual compressive stress, as shown in Figure 39-9b. The residual compressive stress has the effect of lowering the peak tensile stress in the surface, thereby enhancing the fatigue life of the part. What is important to understand here is that the final process can have a significant influence on the component's performance. For example, Table 39-1 shows some test results for fatigue strength in reverse cantilever bending as a function of eight different surface finishes. Notice that for a life of 10^7 cycles, the same part will sustain almost five times the load if the surface is prepared by ultrasonic machining rather than by EDM.

TABLE 39-1 Fatigue Strength in Reverse Cantilever Bending of Ti–5A1–2.5Sn As a Function of Surface Generation Method for a Life of 10^7 Cycles

Surface Generation Method	Strength[a] (psi)
Ultrasonic	98,000
Slab mill	86,000
Chemical mill + vacuum anneal	77,000
Shot peen	76,000
As rolled and received	61,000
Chemical mill	59,000
Ground	52,000
Electric discharge machined	21,000

[a]Average values.
Source: Data from Rooney, WADC report.

Figure 39-10 shows the results of another study where specimens were prepared by milling and turning and then either polished, shot-peened, or roller-burnished. Only the average line is shown for clarity. Suppose that the average applied stress in this situation was around 41,000 to 42,000 psi. The difference in life between a milled specimen and one which has been milled and roller-

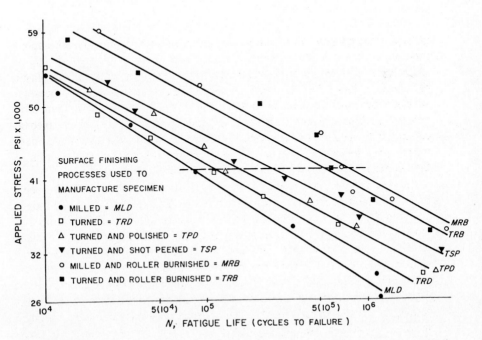

FIGURE 39-10 Variation in fatigue life of rotary bend 2024-T4 aluminum specimens as a function of surface finishing processes.

burnished is 610,000 cycles (700,000 to 90,000 cycles). To put it another way, roller burnishing demonstrated the potential of improving the life of the part seven- to eightfold! Similar results have been obtained in the area of stress corrosion resistance. Thus, the manufacturing engineer should always be aware of the functional behavior of the component part in that the processes can greatly influence the life of the part in service.

Poor design and its resulting consequence in the form of manufacturing defects are factors in product-liability claims. Unfortunately, with the experience of 20-20 hindsight, it often is easy to show how a manufacturing defect could easily have been avoided by a simple design change. Thus designers are called upon to exercise foresight in relating their designs to the required processing so that defects and difficulties will not occur. Figure 39-11 shows an example where a designer failed to take into account a very simple and elementary fact regarding casting. The company initially manufactured and marketed a relatively small woodworking tool that was supported at the end of a cast-aluminum beam having the size shown in Figure 39-11a. The product was very successful, and larger and larger models were made, each having a larger beam with an exactly proportional cross section. Finally, the beam was the size shown in Figure 39-11b, but almost the entire center was a large shrinkage cavity. One of these beams broke, resulting in a serious accident and a large damage award. The design engineer admitted that he was unaware of the problem of hot spots and shrinkage in castings!

Although, admittedly, it often is difficult for a designer to foresee all the possible uses and misuses to which a product will be subjected, there is no excuse for not properly relating the design to the processing that will be used to manufacture the product.

Although the number of examples that could be given to show the close relationship between design and production is almost limitless, those presented here should demonstrate the fact that design is an important phase of production

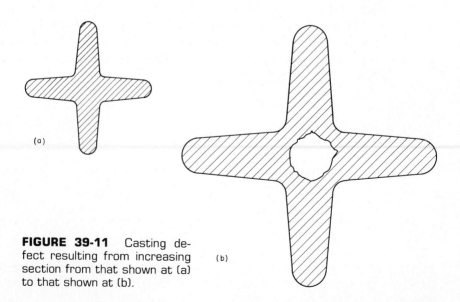

(a)

(b)

FIGURE 39-11 Casting defect resulting from increasing section from that shown at (a) to that shown at (b).

planning and that all designers should have an intimate knowledge of production processes and use this knowledge in carrying out their work.

QUANTITY–PROCESS–DESIGN RELATIONSHIPS

Most processes are not equally suitable and economical for producing a range of quantities for a given product. Consequently, the quantity to be produced should be considered, and the design should be adjusted to the process that actually is to be used before it is "finalized." As an example, consider the part shown in Figure 39-12. Assume that, *functionally,* brass, bronze, a heat-treated aluminum alloy, or ductile iron would be suitable materials. What material and process would be most economical if 1, 100, or 1000 parts were to be made?

If only one were to be made, contour sawing, followed by drilling the 19-mm ($\frac{3}{4}$-inch) hole, would be very economical. The irregular surface would be difficult to produce by other machining processes, and any casting process would require the making of a pattern, which would be about as costly to produce as to saw the desired part. It would be unlikely that a suitable piece of ductile iron or heat-treated aluminum alloy would be readily available. Because brass would be considerably cheaper than bronze, brass, contour sawing, and drilling most likely would be the best combination. For only one part, the excess cost of brass over ductile iron would not be great, and this combination would require no special consideration on the part of the designer.

For a quantity of 100 parts, the excess cost for brass would be appreciable, and machining costs would have to be minimized. Consequently, casting would probably be the most economical process, and ductile iron would be cheaper than any of the other permissible materials. Although the design requirements for casting this simple shape would be minimal, the designer would want to consider them, particularly as to whether the hole should be cored.

For 1000 parts, entirely different solutions become feasible. The use of an aluminum extrusion, with the individual 50.8-mm (2-inch) units being sawed off, might be the most economical solution.[1] All other machining would be

[1]This assumes the time required to obtain the extrusions—probably requiring the making of a special die—was not a factor.

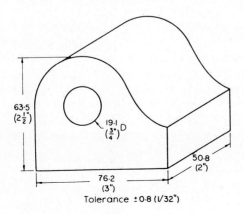

Tolerance ±0·8 (l/32")

FIGURE 39-12 Part to be analyzed for production.

eliminated. About 55 meters (180 feet) of extrusion would be required, including sawing allowance. The cost of the extrusion die would not be very high; thus the per-piece cost would not be great and likely would be more than offset by the savings in machining costs. So for this quantity, this method of production should be investigated. If it were to be used, the designer should make sure that any tolerances specified were well within commercial extrusion tolerances.

This simple example clearly illustrates how quantity can affect both material and process selection, and the selection of the process may require special considerations and design revisions on the part of the designer. Obviously, if the dimensional tolerances were changed, entirely different solutions might result. When more complex products are involved, these relationships become more complicated, but they also usually are more important and require detailed consideration by the designer.

Part analysis for basic requirements. After a satisfactory production design has been completed, the next step in planning is to determine the basic job requirements that must be satisfied. These usually are determined by analysis of the drawings and the job orders. They involve consideration and determination of the following:

1. Size and shape of the geometric components of the workpiece.
2. Tolerances.
3. Material from which the part is to be made.
4. Properties of material being machined.
5. Number of pieces to be produced.

Such an analysis for the ''threaded shaft'' shown in Figure 39-13 would be as follows:[2]

1. a. Two concentric and adjacent cylinders, having diameters of 0.877/0.873 and 0.501/0.499, respectively, and lengths of 2 inches and $1\frac{1}{2}$ inches.
   b. Three parallel, plain surfaces forming the ends of the cylinders.
   c. A $45° \times \frac{1}{8}$-inch bevel on the outer end of the $\frac{7}{8}$-inch cylinder.
   d. A $\frac{7}{8}$-inch NF-2 thread cut on the entire length of the $\frac{7}{8}$-inch cylinder.
2. The closest tolerance is 0.002 inch, and the single angular tolerance is $\pm 1°$.
3. The material is AISI 1340 cold-rolled steel, Bhn 200.
4. The job order calls for 25 parts.

A number of conclusions regarding the processing can be drawn from this analysis. First, because concentric, external, cylindrical surfaces are involved, it is apparent that turning operations are required and that the piece should be made on some type of lathe. Second, because 25 pieces are to be made, the use of a turret lathe probably would not be justified, so an NC or engine lathe will be used. Third, because the maximum, required diameter is approximately $\frac{7}{8}$-inch, 1-inch-diameter cold-rolled stock will be satisfactory; it will provide about $\frac{1}{16}$ inch for rough and finish turning of the large diameter. From this information, decisions can be made regarding the equipment and personnel that will be used and the time that will be required to accomplish the task.

[2]Because Figures 39-12 and 39-13 are dimensioned in English units, no metric equivalences are given in the discussion.

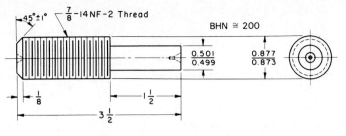

45°±1° $\frac{7}{8}$-14NF-2 Thread

BHN ≅ 200

$\frac{0.501}{0.499}$ $\frac{0.877}{0.873}$

$\frac{1}{8}$ $1\frac{1}{2}$

$3\frac{1}{2}$

Matl. AISI 1340 Medium carbon steel

Tolerance unless otherwise specified = ± $\frac{1}{64}$

FIGURE 39-13 Threaded shaft.

Routing sheets.   After the production requirements have been determined, the next step is to set up a *routing sheet.* This lists the operations that must be performed in order to produce the part, in their sequential order, and the machines or work stations and the tooling that will be required for each operation. For example, Figure 39-14 shows the drawing of a small, round "punch," and Figure 39-15 is a routing sheet for making this part. Once the routing of a part has been determined, the planning of each operation in the processing can then be done.

Operation sheets.   Although routing sheets are very useful for the general planning of manufacturing, they usually do not give sufficiently detailed information to act as instructions for the individual machine operator in carrying out an operation or for scheduling machines and personnel. Such information can be provided on *operation sheets,* such as shown in Figure 39-16, which lists, in sequence, the operations required for machining the threaded shaft shown in Figure 39-13. Commonly, a single operation sheet lists the operations that are done in sequence on a single machine. However, they may cover all the operations for a given part even though more than one machine is required.

Operation sheets vary greatly as to details. The simpler types often list only

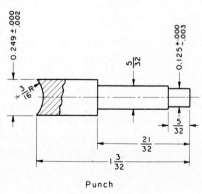

0.249 ± .002

0.125 +.000 -.003

$\frac{3}{16}$R

$\frac{5}{32}$

$\frac{5}{32}$

$\frac{21}{32}$

$1\frac{3}{32}$

Punch

Matl. - 0.250 dia. AISI 1040

H.T. to 50 R.C. on 0.249 dia.

FIGURE 39-14 Drawing of a punch.

DARVIC INDUSTRIES

ROUTING SHEET

NAME OF PART _____Punch_____   PART NO. _____2_____

QUANTITY _____1,000_____   MATERIAL _____SAE 1040_____

OPERATION NUMBER	DESCRIPTION OF OPERATION	EQUIPMENT OR MACHINES	TOOLING
1	Turn $\frac{5}{32}$, 0.125, and 0.249 diameters	J & L turret lathe	#642 box tool
2	Cut off to $1\frac{3}{32}$ length	"	#6 cutoff in cross turret
3	Mill $\frac{3}{16}$ radius	#1 Milwaukee	Special jaws in vise $\frac{3}{16}$ form cutter x 4" D
4	Heat treat. 1,700° F for 30 minutes, oil quench	Atmosphere furnace	
5	Degrease	Vapor degreaser	
6	Check hardness	Rockwell tester	

FIGURE 39-15 Routing sheet for making the punch shown in Figure 39-14.

the required operations and the machines to be used; speeds and feeds may be left to the discretion of the operator, particularly where skilled workers and small quantities are involved. However, it is common practice for complete details to be given regarding tools, speeds, and often the time allowed for completing each operation. Such data are necessary if the work is to be done on NC machines, and experience has shown that these preplanning steps are advantageous where ordinary machine tools are used.

The selection of speeds and feeds required to manufacture the part will generally require referring to handbooks in which tables of the type shown in Figure

PROCESSES AND TECHNIQUES RELATED TO MANUFACTURING

DARVIC INDUSTRIES

OPERATION SHEET

PART NAME: Threaded Shaft 1340 Cold Rolled Steel

Part No. 7358-267-10

OPER. NO.	NAME OF OPERATION	MACH. TOOL	CUTTING TOOL	CUTTING SPEED		FEED ipr	DEPTH OF CUT Inches	REMARKS
				ft/min	rpm			
10	Face end of bar	Engine Lathe		120	458	Hand		Use 3-jaw Universal chuck
20	Center Drill End	"	Combination center drill		750	Hand		
30	Cut off to $3\frac{9}{16}$ length	"	Parting tool	120	458	Hand		To prevent chattering, keep overhang of work and tool at a minimum and feed steadily. Use lubricant.
40	Face to length	"	RH facing tool (small radius point)	120	458	Hand	(R)$\frac{1}{8}$ max. .005 (F) .005	Before replacing part in 3-jaw chuck, scribe a line marking the $3\frac{1}{2}$ inch length.
50	Center Drill End	"	Combination center drill		750	Hand		
60	Place between centers, turn .501 diameter, .499 diameter, and face shoulder	"	RH turning tool (small radius point)	120 160	(R)458 (F)611	(R) .0089 (F) .0029	(R) .081(3) (F) .007	
70	Remove and replace end for end and turn .877 .873 diameter	"	RH tools (R)(small radius point) (F) Round nose tool	120 160	(R)458 (F)611	(R) .0089 (F) .0029	(R) .057 (F) .005	
80	Produce 45°-chamfer	"	RH round nose tool	120	458	Hand	(R)$\frac{1}{8}$ max. .005 (F) .005	
90	Cut $\frac{7}{8}$-14 NF-2 thread	"	Threading tool	60	208		(R) .004 (F) .001	(1) Swivel compound rest to 30 degrees. (2) Set tool with thread gage. (3) When tool touches outside diameter of work set cross slide to zero. (4) Depth of cut for roughing = .004. (5) Engage thread dial indicator on any line. (6) Depth of cut for finishing = .001 Use compound rest.
	Remove burrs and sharp edges	"	Hand file					

Date

FIGURE 39-16 Operation sheet for the threaded shaft shown in Figure 39-13.

MATERIAL	HARD-NESS Bhn	CONDITION	DEPTH OF CUT* in / mm	HIGH SPEED STEEL TOOL SPEED fpm / m/min	FEED ipr / mm/r	TOOL MATERIAL AISI / ISO	CARBIDE TOOL UNCOATED SPEED BRAZED fpm / m/min	INDEX-ABLE fpm / m/min	FEED ipr / mm/r	TOOL MATERIAL GRADE C / ISO	COATED SPEED fpm / m/min	FEED ipr / mm/r	TOOL MATERIAL GRADE C / ISO
5. ALLOY STEELS, WROUGHT (cont.)	175 to 225	Hot Rolled, Annealed or Cold Drawn	.040	135	.007	M2, M3	375	500	.007	C-7	650	.007	CC-7
			.150	105	.015	M2, M3	300	400	.020	C-6	525	.015	CC-6
Medium Carbon			.300	80	.020	M2, M3	240	315	.030	C-6	400	.020	CC-6
1340 4340 81B45			.625	65	.030	M2, M3	190	250	.040	—	—	—	—
1345 50B40 8640			1	41	.18	S4, S5	115	150	.18	P10	200	.18	CP10
4042 50B44 8642			4	32	.40	S4, S5	90	120	.50	P20	160	.40	CP20
4047 5046 8645			8	24	.50	S4, S5	73	95	.75	P30	120	.50	CP30
4140 50B46 86B45			16	20	.75	S4, S5	58	76	1.0	—	—	—	—
4142 5140 8740	225 to 275	Annealed, Normalized, Cold Drawn or Quenched and Tempered	.040	115	.007	M2, M3	350	465	.007	C-7	600	.007	CC-7
4145 5145 8742			.150	90	.015	M2, M3	280	365	.020	C-6	475	.015	CC-6
4147 5147			.300	70	.020	M2, M3	220	285	.030	C-6	375	.020	CC-6
			.625	55	.030	M2, M3	170	225	.040	—	—	—	—
			1	35	.18	S4, S5	105	140	.18	P10	185	.18	CP10
			4	27	.40	S4, S5	85	110	.50	P20	145	.40	CP20
			8	21	.50	S4, S5	67	87	.75	P30	115	.50	CP30
			16	17	.75	S4, S5	52	69	1.0	—	—	—	—
			.040	90	.007	T15, M42†	330	440	.007	C-7	575	.007	CC-7
			.150	70	.015	T15, M42†	260	340	.015	C-6	450	.015	CC-6
			.300	55	.020	T15, M42†	200	270	.020	C-6	350	.020	CC-6
			.625	—									
			27	27	.18	S9, S11†	100	135	.18	P10	175	.18	CP10
			21	21	.40	S9, S11†	79					.40	CP20
					.50								CP30

FIGURE 39-17 Portion of table from *MDC Machining Data Handbook,* 3rd ed. which gives recommended speeds and feeds for a given material, broken down by operation, material hardness, and condition according to what tool material is being used at a given depth of cut. Such data are considered to be reliable but their accuracy is not guaranteed. (*Courtesy Metcut Research Associates Inc.*)

39-17 are given. For the threaded shaft, this table will give suggested values for the turning and facing operations for either high-speed steel or carbide tools. Note that the tables are segregated by the workpiece material (medium-carbon alloy steels, wrought or cold-worked) and then by process, in this case, turning. For these materials, additional tables for drilling and threading would have to be referenced. Notice that the depth of cut dictates the speed and feed selection and that rough and finish cuts are used. That is, operation 60 requires three roughing cuts and one finishing cut. While the operation sheet does not specifically say so, high-speed steel tools are being used throughout. If the job was being done on an NC lathe, it is likely that all the cutting tools would be carbides. Notice also that the job as described for the engine lathe required two setups—

TABLE 39-2 Typical Inputs and Outputs of Computerized Machinability Data Systems in Use Today

Inputs	Outputs
Machine tool code	Recommended speed/feed
Cutting tool code	Production-time calculations
Material cost	Number and size of cuts
Operation code	Optimum speed/feed
Depth of cut	Cutting tool/grade
Material hardness	Machine tool

Source: "Machining Briefs," Metcut Research Associates Inc., Mar./Apr. 1982.

a three-jaw chuck to get the part to length and produce the centers and a between-centers setup to complete the other operations. Notice further that in each setup, it was required to stop the machine to invert the part in the setup. Do you think that this part could be manufactured more efficiently in an NC lathe?

Data of the type shown in Figure 39-17 are currently being computerized to be compatible with CAM and CAPP systems. Table 39-2 shows the most frequently reported system inputs and outputs for this kind of data.

As noted in Chapter 38, the cutting time, as computed from the machining parameters, represents only 20 to 30% of the total time needed to complete the part. Referring back to Figure 39-16, the operator will need to pick up the part after it is cut off at operation 30, stop the lathe, open up the chuck, take out the piece of metal in the chuck, scribe a line on the part marking the desired $3\frac{1}{2}$-inch length, place the part back in the chuck, change the parting tool to a facing tool, adjust the facing tool to the right height, and move the tool to the proper position for the desired facing cut in operation 40. All these operations take time and someone must estimate how much time is required for such noncutting operations if an accurate estimate of total time to make the part is to be obtained.

When an operation is machine-controlled, as in making a lathe cut of a certain length with power feed, the required time can be determined by simple mathematics. For example, if a cut is 254 mm (10 inches) in length and the turning speed is 200 rpm, with a feed of 0.127 mm (0.005 inch) per revolution, the time required will be 10 minutes. Procedures are available for determining the time required for people-controlled machining elements, such as moving the carriage of a lathe by hand back from the end of one cut to the starting point of a following cut. Such determinations of time estimates are generally considered the job of the industrial engineer and space does not permit us to cover this material in this text but the techniques are well established. Actual time studies, accumulated data from past operations, or some type of motion-time data, such as MTM,[3] can be employed for estimating such times. Each can provide accurate results that can be used for establishing standard times for use in planning. Various handbooks and books on machine-shop estimating contain tables of average times for a wide variety of elemental operations for use in estimating and setting standards. However, such data should be used with great caution, even for planning purposes, and they should never be used as a basis of wage payment. The conditions under which they were obtained may have been very different from those for which a standard is being set.

OTHER FUNCTIONAL AREAS OF THE PRODUCTION SYSTEM

As was shown in Figure 38-4 there are many functional areas within the total production system, aside from production planning, design, manufacturing, and inspection, which have not been treated in this text. Our primary purpose in discussing them is to show the kinds of interactions which can occur within the

[3]MTM stands for Methods-Time-Measurement.

production system and to show that while the processes are elemental to the system, they are not in themselves the key to economical manufacturing. The truth of the matter is that many countries have about the same level of process development when it comes to manufacturing technology and that much of the technology which exists in the world today was developed in the United States. Yet, other nations, like Japan, are making great inroads into American markets, particularly in the automotive and electronics industry. What many people have failed to recognize is that many Japanese companies have developed and promoted a totally different kind of production system that in future years will take its place with the Taylor system of scientific management and the Ford system for mass production. The Toyota production system,[4] which has been adopted by many Japanese companies, embodies many unique concepts in production organization and management which are foreign (no pun intended) to their American counterparts. At Toyota, machine layout is in accordance with the flow of processes wherein products having common or similar processes are grouped together and quick conveyance between the processes is provided, along with the means to reduce setup time. The system is simple enough that everyone understands how it works and how they can contribute to make it work better. While space in this text does not permit us to discuss this system in detail, the basic idea is to produce the kind of units needed in the quantities needed at the time needed. This is called *Just-in-Time* production (see Figure 39-18). Three key production concepts are employed.

Smoothing of production. Cornerstone of the system. Eliminates variation or fluctuation in quantities in feeder processes by eliminating fluctuation in final assembly. Small lot sizes, single unit conveyance, and very short setup times employed here.

Design of process. Manned cellular manufacturing system used with multifunctional worker who is mobile and flexible. Unmanned cellular and FMS used where economically justified.

Standardization of jobs. Every part, sequence of assembly operations, or subassembly has the same number of specified minutes. Only the number of workers needed to produce one unit of output in the cycle time are used.

The manufacturing management information system (MIS) used at Toyota is called *Kanban,* which harmoniously controls the production quantities in every process.

Cycle time is determined as follows. Suppose that the forecast is for 500 cars per day and 480 production minutes are available (60 minutes $\times$ 8 hours/day). Thus, cycle time = 0.96 minute. Every 0.96 minute a car rolls off the line. Suppose that the mix is as given in Table 39-3.

The subprocesses that feed the two-door fastback are controlled by the cycle time for this model. Every 2.4 minutes, the engine line will produce an engine for the fastback version. Every 2.4 minutes, two doors are made. Each engine needs four pistons, so every 2.4 minutes, four pistons are produced. Parts and assemblies are produced in their minimum lot sizes and delivered to the next process, under the control of Kanban. In this manner, production is smoothed.

[4]Toyota Motor Company Ltd.

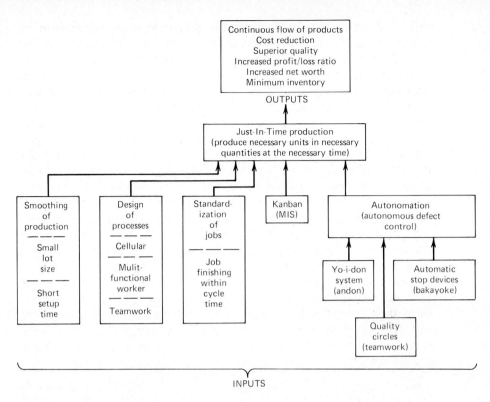

FIGURE 39-18 Elements of the Toyota Just-in-Time (JIT) production system.

Smoothing of production also requires the elimination of setup time, which in turn makes it economical to produce in small lots. This has been accomplished through the application of group technology, training workers in rapid setup, and applying Toyota's unique "single setup" concepts. Setup times of nine minutes 59 seconds or less for all operations, including the press room, have been achieved. American and European companies often spend hours making the changeovers that Toyota workers do in minutes.

Observe that one of the results of the elimination of setup time is to drive the total direct cost of the part down to where it equals the variable cost of the part.

TABLE 39-3 Example of Product Mix on Final Assembly Line Which Determines the Cycle Time by Model

Car Mix For Line		Cycle Time by Model (min)	Production Minutes by Model	Comments
Q	Model			
100	Two-door coupe	4.8	96	Every fifth car a two-door
200	Two-door fastback	2.4	192	
50	Four-door sedan	9.6	48	Every tenth car a four-door
150	Four-door wagon	3.2	144	
500			480	

480 min/500 cars = 0.96 min per car

This, in effect, means that Toyota can manufacture parts economically in very small lots, which, in turn, drastically reduces their inventories. Automobile manufacturers in the United States turn inventory over four to five times per year, whereas at Toyota, inventory is reportedly turned 70 times per year. The difference in inventory costs shows up in the cost of the final products, making them significantly cheaper.

For the Just-in-Time (JIT) system to work, 100% good units flow rhythmically to subsequent processes, without interruption. In order to accomplish this, a total quality control (TQC) program has been developed in conjunction with JIT. The responsibility for quality has been given to production and there is a company-wide attitude toward constant quality improvement. The goal is to make it right (perfect) the first time. The basic principles of this TQC concept include making quality easy to see, stopping the line when something goes wrong, and inspecting things 100%, often accomplished by autonomation.

Autonomation means to build a mechanism to prevent defective work from either being made or passed on to subsequent processes. In machines, this is accomplished with automatic checking devices. One such autonomous checking device is a mechanism called *Bakayoke*. For manual work, another system is used which is called *Andon*.

Andon is actually an electric light board which hangs high above the conveyor assembly lines so that all can see it. When a worker in the line needs help, he can turn on a yellow light. Other nearby (multifunctional) workers who have finished their jobs within the alloted cycle time move to assist workers having problems. If the problem cannot be solved within the cycle time, a red light comes on and the line automatically stops until the problem is solved. In most cases, the red lights go off within 10 seconds and the next cycle begins—a green light comes on—with all the processes beginning together. The name for this system is *Yo-i-don,* which literally means ''ready–set–go.'' Such systems are built on teamwork and a cooperative spirit among the workers, fostered by a management philosophy based on harmony and trust. In this light, it is strongly recommended that all students interested in knowing more about Japanese production systems should also explore their management philosophy.

REVIEW QUESTIONS

1. How does a production system differ from a manufacturing system?
2. In what ways are the planning of production related to the manufacturing system and the production system as a whole?
3. Why is it so important for the designer to have intimate knowledge of the available manufacturing processes at the various stages of the design activity?
4. Explain how function dictates the design with respect to the design of footwear. Use examples of different kinds of footwear (shoes, sandals, high heels, boots, etc.) to emphasize your points. For example, cowboy boots have pointed toes so that they slip into the stirrups easily and high heels to help keep the foot in the stirrup.
5. Most companies, when computing or estimating costs for a job, will add in an overhead cost, often tying that cost to some direct cost, like direct labor. What is included in this overhead cost?
6. How is this overhead cost, which is often given at 100 to 200% of direct labor cost, impacted by better planning and scheduling?

7. How does the design of the product influence the design of the manufacturing and production systems?

8. Discuss this statement: "Software can be as costly to design and develop as hardware and will require long production runs to recover, even though these costs may be hidden in the overhead costs."

9. What is meant by the statement that "all processing schemes have a distinct technology and history and call for a distinct set of optimally suited materials"?

10. If a designer called for a part to be made from cast steel, what processes has he eliminated from the possible list for casting this part?

11. Give an example of how a designer may indirectly dictate what processes or machines must be used to fabricate a part.

12. Give an example of a design (detail) that might lead to a part failure or defect.

13. In Figure 39-5, how would you have designed the shaft and flange pair if you were going to join these parts by friction welding?

14. Under what assembly conditions (again referring to Figure 39-5) might have the initial design been better than the improved redesign?

15. How might the peak tensile stresses in a part subjected to fatigue loading be enhanced rather than reduced by the fabrication of the part? Use Table 39-1 as a reference.

16. What other processes, other than shot peening and roller burnishing, could have been used to improve the fatigue life of the test specimen?

17. Outline the manufacturing processes which you think might be needed to fabricate a razor blade and estimate what you think the blade itself actually costs. Razor blades have a final edge cutting radius of 1000 angstroms or better. Did you determine the basic job requirements?

18. Figure 39-14 shows a drawing of a punch. What are the critical dimensions? Why are they critical? Why isn't overall length critical?

19. Suppose that the part in Figure 39-13 were to be made in quantities of 250 rather than 25. What processes and machine(s) would you have selected? Suppose that the quantity were 25,000; then what?

20. What is the basic philosophy of the Just-in-Time Toyota Production System?

21. What are the three key concepts?

22. What does "single setup" refer to and why is it such an important concept? Can you relate it to inventory? How?

23. To what does the word "autonomation" refer, what is the objective of autonomation, and why is this so important in this production system?

CASE STUDY 39. The Snowmobile Accident

Mary M. is suing the SnoCat Snowmobile Company and ACME Components for $250,000 over her husband's death. He was killed while racing his snowmobile through the woods in the upper peninsula of Michigan. Her lawyer claims that he was killed because a tie rod broke, causing him to lose control and crash into a tree, breaking his neck. It was impossible to determine whether the tie rod broke before the crash or as a result of the crash. The following evidence has been put forth. The tie rod was originally designed and made entirely out of low-carbon steel (heat-treated by case hardening) in three pieces, as shown in Figure CS-39. These tie rods were subcontracted by SnoCat to ACME Components. ACME Components changed the material of the tie rod bolts from steel to a heat-treated aluminum having the same UTS as the steel. They did this because aluminum rods were easier to thread-roll rather than using a thread-cutting operation. It was further found that threads on one of the tie-rod bolts were not as completely formed as they should have been. The sleeve of the tie rod in question was split open and one of the tie-rod bolts was bent. Mary's lawyer claimed that the tie rod was not assembled properly. He claimed that one rod was screwed into the sleeve too far and the other not far enough, thereby giving it insufficient thread engagement. ACME testified these tie rods are hand assembled and only checked for overall length and that such a misassembly was possible. Mary's lawyer stated that the failure was due to a combination of material change, manufacturing error, and bad assembly, all combining

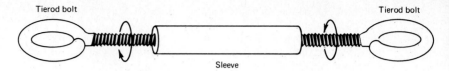

FIGURE CS-39 Assembled tie rod for snowmobile.

to result in a failure of the tie rod. A design engineer for SnoCat testified that the tie rods were "way overdesigned" and would not fail even with slightly small threads or misassembly. SnoCats' lawyer then claimed that the accident was caused by driver failure and that the tie rod broke upon impact of the snowmobile with the tree. One of the men racing with Mary's husband claimed that her husband's snowmobile veered sharply just before it crashed, but under cross-examination he admitted that they had all been drinking that night because it was so cold (he guessed $-20°F$), but since this accident took place over two years ago, he could not remember how much.

You are a member of the jury and have now been sequestered to decide if SnoCat is guilty of negligence resulting in death. The rest of the jury, knowing you are an engineer, has asked for your opinion. What do you think? Who is really to blame for this accident? What actually caused the accident?

Selected References
for Additional Study

Title	Author(s)	Publisher
1. *Anatomy of Automation*	George H. Amber & Paul S. Amber	Prentice-Hall
2. *Aluminum* (3 Vols.)		American Society for Metals
3. *Aluminum Alloys: Structure and Properties*	Mondolfo	Butterworths
4. *Aluminum Standards and Data*		The Aluminum Association
5. *ASM Metals Reference Book*		American Society for Metals
6. *Atlas of Isothermal Transformation and Cooling Transformation Diagrams*		American Society for Metals
7. *Automation in Manufacturing Systems, Processes and Computer Aids* PED Vol. 4		ASME
8. *Automation, Production Systems and Computer Aided Manufacturing*	Mikell P. Groover	Prentice Hall

Title	Author(s)	Publisher
9. *BASIC Programming Solutions for Manufacturing*	J. E. Nicks	SME
10. *CAD/CAM and the Computer Revolution*	Daniel B. Dallas	SME
11. *Case Histories in Failure Analysis*		American Society for Metals
12. *Cast Metals Handbook* (4th Ed.)		American Foundrymen's Society
13. *Computer Applications in Manufacturing Systems PED Vol. 2*		ASME
14. *Computerized Robots*		Integrated Computer Systems
15. (A) *Concise Guide to Plastics*	Simonds and Church	Van Nostrand Reinhold Company
16. *Deformation and Fracture Mechanics of Engineering Materials*	Hertzberg	John Wiley & Sons, Inc.
17. *Design of Cutting Tools*	A. Bhattacharyya and Inyong Ham	SME
18. *Dimensional Control*		Bendix Automation and Measurement Division
19. *Elements of Materials Science and Engineering* (3rd Ed.)	Van Vlack	Addison-Wesley Publishing Company, Inc.
20. *Engineering Aspects of Product Liability*	Conangelo and Thornton	American Society for Metals
21. *Engineering Cost Analysis*	Collier Ledbetter	Harper & Row
22. *Engineering Design: A Synthesis of Stress Analysis and Materials Engineering* (2nd Ed.)	Faupal and Fisher	Wiley-Interscience
23. *Engineering Materials and their Applications*	Flinn and Trojan	Houghton Mifflin
24. *Engineering Materials: Properties and Selection*	Budinski	Reston
25. *Engineering Properties of Steel*		American Society for Metals
26. *Engineering Properties of Zinc Alloys*		International Lead Zinc Research Organization
27. *Extrusion*	Laue and Stenger	American Society for Metals
28. *Forging Equipment, Materials, and Practice* (MCIC-HB-03)		Metals and Ceramics Information Center

Title	Author(s)	Publisher
29. *Forging Industry Handbook*	Jenson	Forging Industry Association
30. *Fundamentals of Metal Casting*	Flinn	Addison-Wesley Publishing Company, Inc.
31. *Fundamentals of Metal Machining and Machine Tools*	G. Boothroyd	McGraw-Hill
32. *Fundamental Principles of Polymeric Materials for Practicing Engineers*	Rosen	Barnes and Noble
33. *Fundamentals of Tribology*	Nam P. Suh and N. Saka (Eds.)	MIT Press
34. *Gear Handbook*	Dudley	McGraw-Hill
35. *Group Technology*	E. A. Arn	Springer-Verlag
36. *Group Technology*	Gallagher and Knight	London/Butterworths
37. *Handbook of Adhesives (2nd Ed.)*	Skiest	Van Nostrand Reinhold
38. *Handbook of Industrial Metrology*		Prentice-Hall
39. *Handbook of Metal Forming Processes*	Betzalel Avitzur	John Wiley-Interscience
40. *Handbook of Stainless Steels*	Peckner and Bernstein	McGraw-Hill
41. *Heat Treater's Guide, Standard Practices and Procedures for Steel*	Unterweiser	American Society for Metals
42. *Heat Treatment, Structure and Properties of Nonferrous Alloys*	Brooks	American Society for Metals
43. *High-Velocity Forming of Metals*		Society of Manufacturing Engineering
44. *Introduction to Powder Metallurgy*	Hirschhorn	American Powder Metallurgy Inst.
45. *Industrial Robots* Vols. 1 and 2		SME
46. *Instrumented Impact Testing* STP 563		ASTM
47. *(The) Introduction of Group Technology*	John L. Burbridge	Halsted Press Book John Wiley & Sons
48. *Investment Castings Handbook*		Investment Casting Institute
49. *Iron Castings Handbook*	Walton	Iron Castings Society

Title	Author(s)	Publisher
50. *ISO System of Limits and Fits, General Tolerances and Deviations*		American National Standards Institute
51. *Japanese Manufacturing Techniques*	R. J. Schonberger	Free Press Macmillan
52. *(The) Making, Shaping, and Treating of Steel* (9th Ed.)	McGannin	United States Steel Corp.
53. *Jigs & Fixtures*	Williams E. Boyes (Ed.)	SME
54. *Machining Data Handbook* (3rd Ed.)		MDC Metcut Research Inc.
55. *(The) Machine Tools That Are Building America* Vol. 218, No. 9		Iron Age
56. *Machining Fundamentals*	Daniel Follette	SME
57. *Manufacturing Engineering Processes*	Alting	Marcel Dekker
58. *Manufacturing Cost Estimating Guide* (1983 Ed.)	P. R. Ostwald	American Machinist
59. *Manufacturing Processes* (7th Ed.)	Amstead, Ostwald and Begemen	John Wiley & Sons
60. *Manufacturing Systems Engineering*	K. Hitomi	Taylor & Francis Ltd., London
61. *Material Properties and Manufacturing Processes*	Datsko	Wiley
62. *Materials* (''Scientific American'' book)		W. H. Freeman and Company
63. *Materials Handbook* (11th Ed.)	Brady and Clauser	McGraw-Hill
64. *Materials Science*	Ruoff	Prentice-Hall
65. ''Materials Selector Issue'' of *Materials Engineering* (annual)		Penton/IPC Publications
66. *Mechanical Metallurgy*	Dieter	McGraw-Hill
67. *Mechanical Processing of Metals*	Kalpakjian	Van Nostrand Reinhold
68. *Mechanics of Plastic Deformation in Metal Processing*	Thomsen, Yang, and Kobayashi	Macmillan Publishing Company
69. *Metal Cutting*	E. M. Trent	Butterworths
70. *Metal Finishing Guidebook Directory* (annual)		Metals and Plastics Publications
71. *Metallurgy and Heat Treatment of Tool Steels*	Wilson	McGraw-Hill

Title	Author(s)	Publisher
72. *Metals Handbook* (8th Ed., 11 Vols.)		American Society for Metals
73. *Metals Handbook* (9th Ed., multiple Vols.)		American Society of Metals
74. *Metals Joining Manual*	M. M. Schwartz	McGraw-Hill
75. *Metals-Principles-Treatment-Selection*	K. J. Trigger and S. Ramalingam	Campus Bookstore Champaign, IL
76. *Metalworking with Aluminum*		The Aluminum Association Inc.
77. *Mineral Facts and Problems, Bulletin 667, Bureau of Mines*		United States Department of the Interior
78. *Modern Welding Technology*	Cary	Prentice-Hall
79. *New Developments in Grinding*	M. C. Shaw (Ed.)	Carnegie Press
80. *Nondestructive Testing Handbook* (2nd Ed.)	R. C. McMaster	ASNT-ASM
81. *Numerical Control Applications*		SME
82. *Numerical Control Fundamentals*		SME
83. *Numerical Control and Computer Aided Manufacturing*	R. S. Pressman and J. E. Williams	Wiley
84. *On The Art of Cutting Metals–75 Years Later* PED Vol. 7		ASME
85. *Open Die Forging Manual* (3rd Ed.)		Forging Industry Association
86. *Powder Metallurgy, Principles and Applications*	Lenel	Metal Powder Industries Federation
87. *Powder Metallurgy Processing: New Techniques and Analysis*	Kuhn and Lawley	Academic Press
88. *Principles and Applications of Tribology*	Desmond F. Moore	Pergamon Press
89. *Principles of Heat Treatment of Steel*	Krauss	American Society for Metals
90. *Principles of Metals Casting* (2nd Ed.)	Heine, Loper and Rosenthal	McGraw-Hill
91. *Principles of Numerical Control* (3rd Ed.)	James L. Childs	Industrial Press, Inc.
92. *Proceedings of the North American Metalworking Research Conference*		Vol. II-Univ of Wisconsin Vol. III-Carnegie Press Vol. IV thru X–SME

Title	Author(s)	Publisher
93. *Product Design and Process Engineering*	B. W. Niebel and A. B. Draper	McGraw-Hill
94. *Production Systems: Planning, Analysis and Control* (3rd Ed.)	James L. Riggs	John Wiley-Interscience
95. *Products Liability and the Reasonably Safe Product*	Weinstein et al.	John Wiley-Interscience
96. *Quality Planning and Analysis* (2nd Ed.)	Juran & Gryan	McGraw-Hill
97. *(The) Role of Computers in Manufacturing Processes*	Gideon Hal	John Wiley-Interscience
98. *SAE Handbook, Part 1— Materials*		Society of Automotive Engineers
99. *(The) Science of Precision Measurement*		The DoALL Company
100. *Scientific Organization of Batch Production* AFML-TR-77-218 Vol. III	S. F. Mitrofanov	U.S. Air Force
101. *Solders and Soldering* (2nd Ed.)	Manko	McGraw-Hill
102. *Source Book on Heat Treating* (2 Vols.)		American Society for Metals
103. *Standard Handbook of Fastening and Joining*	R. O. Parmley (Ed.)	McGraw-Hill
104. *Standards Handbook: Copper, Brass, Bronze* (7 Vols.)		Copper Development Association
105. *Statistical Quality Control* (5th Ed.)	Grant and Leavenworth	McGraw-Hill
106. *Steel Castings Handbook* (5th Ed.)		Steel Founder's Society of America
107. *Steel Selection: A Guide for Improving Performance and Profits*	Kern and Suess	Wiley-Interscience
108. *Surface Preparation and Finishes for Metals*	Murphy	Society of Manufacturing Engineers
109. *Techniques of Pressworking Sheet Metal* (2nd Ed.)	Eary and Reed	Prentice-Hall
110. *(The) Testing of Engineering Materials* (4th Ed.)	Davis, Troxell and Hauck	McGraw-Hill
111. *Titanium Alloys Handbook*		Metals and Ceramics Information Center
112. *Tool Design*	Donaldson, LeGain, and Goold	McGraw-Hill

Title	Author(s)	Publisher
113. *Tool and Manufacturing Engineers Handbook* (3rd Ed.)	Dallas	McGraw-Hill
114. *Tool Steels* (4th Ed.)	Roberts and Cary	American Society for Metals
115. *Toward the Factory of the Future* PED Vol. 1		ASME
116. *Weldability of Steels*	Stout	Welding Research Council
117. *Welding Handbook* (7th Ed.)		American Welding Society
118. *Why Metals Fail*	Barer and Peters	Gordon and Breach

Chapter Number	Subject	References
1	Introduction	1, 55, 94
2	Properties of Materials	16, 19, 23, 46, 110
3	The Nature of Metals and Alloys	19, 23, 62, 66, 75
4	Production and Properties of Common Engineering Metals	23, 52, 77
5	Equilibrium Diagrams	19, 23, 72(8), 89
6	Heat Treatment	6, 41, 42, 52, 70, 71, 72(2), 73(4), 75, 89, 102, 108
7	Alloy Irons and Steels	5, 24, 25, 40, 52, 63, 64, 65, 71, 72(1), 73(1,3), 75, 98, 107, 114
8	Nonferrous Alloys	2, 3, 4, 5, 19, 24, 26, 42, 63, 72(1), 73(2), 75, 98, 104, 111
9	Nonmetallic Materials: Plastics, Elastomers, Ceramics, and Composites	15, 19, 24, 32, 62, 63, 65
10	Material Selection	11, 16, 20, 21, 22, 58, 70, 75, 80, 95, 107, 108, 118
11	Casting Processes	12, 30, 48, 49, 57, 59, 72(5), 90, 106, 113
12	Powder Metallurgy	44, 86, 87, 92, 113
13	The Fundamentals of Metal Forming	16, 39, 66, 67, 68, 92, 109, 113
14	Hot-working Processes	27, 27, 29, 30, 57, 61, 66, 67, 76, 85, 92, 113
15	Cold-working Processes	39, 43, 57, 61, 66, 67, 76, 92, 109, 113

Chapter Number	Subject	References
16	Measurement and Inspection	18, 33, 38, 50, 80, 88, 99
17	Process Capability and Quality Control	96, 105
18	Chip-type Machining Processes (Metal Cutting)	17, 31, 33, 55, 56, 57, 66, 67, 68, 69, 71, 84, 88, 92, 112
19	Shaping and Planing	54, 57, 59
20	Turning and Boring	54, 57, 59
21	Drilling and Reaming	54, 57, 59
22	Milling	54, 57, 59
23	Abrasive Machining Processes	54, 57, 79
24	Broaching	54, 57, 59
25	Sawing and Filing	54, 57, 59
26	Nontraditional Machining Processes	57
27	Thread Cutting and Forming	57
28	Gear Manufacturing	34, 57
29	Forge, Oxyfuel Gas, and Arc Welding	57, 74, 78, 113, 117
30	Resistance Welding	57, 74, 78, 113, 117
31	Miscellaneous Welding and Related Processes	57, 74, 78, 103, 113, 117
32	Torch and Arc Cutting	57, 74, 78, 113, 117
33	Brazing, Soldering, Adhesive Bonding, and Mechanical Fastening	37, 57, 72(6), 74, 78, 101, 113, 117
34	Potential Problems in Welding and Cutting	57, 74, 78, 116, 117
35	Layout	57
36	Jigs and Fixtures	17, 53, 57, 112, 113
37	Decorative and Protective Surface Treatments	57, 70, 108
38	Manufacturing Systems and Automation	1, 7, 8, 9, 10, 13, 14, 35, 36, 45, 47, 60, 81, 82, 83, 91, 92, 94, 97, 100, 115
39	Production Systems	11, 21, 51, 54, 58, 66, 92, 93, 94, 95, 97, 115

Index

Brazing, 19, 847
  definition, 847
  design of joints, 855
  flux removal, 854
  fluxes, 849
  fluxless, 855
  metals for, 848
  methods of heating, 851
  nature of joints, 847
Breaking strength, 37
Brinell hardness test, 41
Brinell hardness vs. strength in steel, 166
Brittleness, 37, 38
Broaches, 669
  classification, 670
  construction, 672
  design, 670
Broaching, 500, 529, 530, 668
  of gears, 766
  speeds, 672
Broaching machines, 673
Bronze, 180
Buffing, 909
Built-up edge, 506
Bulging, 411
Burnishing, 379
  of gears, 768
  roller, 380, 1001

C

Calender, for rubber, 224
Calipers, inside and outside, 451
  micrometer, 446
  vernier, 444
Carbon, effect in hardening steel, 117
Cast iron, 89, 118, 121
  alpha ferrite, 118
  cementite, 119
  delta ferrite, 118
  eutectoid, 118
  ferrite, 118
  gray, 122, 176
  malleable, 124
  martensite, 124
  white, 124
Casting of metals, 12, 247
  basic requirements, 250
  centrifugal, 284
  continuous, 95, 292
  defined, 12
  die, 279
  full-mold process, 270
  investment, 287
  molding machines, 262
  patterns for sand, 254

permanent-mold, 277
plaster-mold, 286
rubber-mold, 291
sand, 251
Shaw process, 288
shell, 267
Casting of plastics, 210
Castings
  design, 300, 994
  finishing, 299
  gating and risering, 252, 303
  heat treating, 300
Cellular manufacturing systems, 9, 936, 942
Cementite, 119
Center drill, combination and countersink, 593
Centers, for lathes, 562
Centrifugal casting, 284
Cermets (cast nonferrous alloys), 228, 516
Ceramics, 227
  cutting tools, 228, 517
  molecular structure, 227
  properties, 228, 517
  structure of, 227
Chaplets, 275
Charpy impact test, 49
Chemical cleaning, 912
  alkaline, 912
  emulsifiable solvent, 912
  pickling, 913
  ultrasonic, 913
  vapor, 913
Chemical machining, 696, 701, 1001
  blanking, 701
  metal removal rates for, 697
  tolerances for, 697
Chip breakers, 519, 521
Chip formation. *See* Metal cutting
Chipless machining. *See* Nontraditional machining processes
Chromium, 105, 228, 516
Chucks, drill, 596
  lathe, 565
  magnetic, 657
  vacuum, 658
Climb (or down) milling, 612
Closed loop, 945, 946, 956, 963
Coated abrasives, 660
Cobalt, 105
  alloys of, 105
Coding/Classification, 939
Coining, 378
Cold heading, 370
Cold roll forming, 385
  flanging, 386

Friction
  in metal cutting, 507
  in metal working, 326, 329
Full-mold process, 270
Furnaces, for heat treating, 152
Furnaces for melting
  air fired, 295
  arc, 295
  cupolas, 294
  induction, 296

G

Gage blocks, 432
  grades, 432
  sets, 434
Gages, 455
  air, 441, 461
  deviation-type, 459
  dial, 454, 460
  flush-pin, 458
  plug and ring, 456, 457
  small-hole, 452
  snap, 457
  step, 458
  thread plug, 457
  thread pitch, 459
Gaging, definition, 430
  see also Attributes
Galvanizing, 918
Gangue, 86
  see also Hematite
Gas metal arc welding, 788
Gas tungsten arc welding, 785
Gating, of castings, 251, 253, 304, 305
Gear(s), 743
  finishing, 767
  inspection, 768
  shaving, 767
  standard dimensions, 747
  terminology, 743
  types, illustrated, 748
Gear manufacturing, 743, 751
  bevel gear, 759
  form-cutting machines, 749
  generating machines, 755
  hobbing machines, 757
  roll-forming, 763
  template machines, 750
Goodyear, Charles, 223
Grain growth, 79
Grain structure
  development, 71
  grain boundaries, 70
  nucleation and growth, 70

Graphite
  in ductile iron, 126
  effect in cast iron, 122
  in malleable iron, 125
  properties and uses, 106
Grinding, 630
  chips, 635
  creep feed, 655
  electrochemical, 711
  gear, 768
  nature of, 635
  low stress, 638
  ratio (G-ratio), 637
  safety, 645
  speeds, 643
  thread, 655, 736
  ultrasonic, 712
  working holding devices, 656
Grinding machines, 646
Grinding (abrasive) wheels, 642
  bonding materials, 639
  dressing, 640
  grade, 636
  identification, 642
  selection, 642
  structure, 635
Group technology, 9, 932, 937, 958,
    959, 991, 1011
  benefits of conversion, 943
  coding/classification, 939
  defined, 933
  formation of parts families with, 938
  production flow analysis, 938
  relationship to CAM, 991
Guerin forming, 409

H

Hafnium, 106
Hard facing. *See* Metallizing; Surfacing
Hardenability, 143
  Jominy test for, 143
Hardening of metals, 133
  age, 133, 134
  dispersion, 33
  phase transformation, 133
  precipitation, 133, 134
  solid solution, 133
  strain, 133, 335
  surface, 149, 380
Hardness, 41, 144
  Brinell test, 41
  conversion table, 47
  Durometer test, 45
  Knoop, 45
  microhardness test, 45

Hardness *(cont.)*
  Mohs test, 46
  relationship to strength, 47, 166
  Rockwell test, 42
  Scleroscope test, 46
  Scratch test, 46
  Superficial Rockwell test, 44
  Tukon test, 45
  Vickers test, 44
Heat-affected zone (HAZ), 868
Heat treatment, 129
  annealing, 131
  definition, 21, 129
  full annealing, 131
  furnaces, 152
  normalizing, 131
  processing, 131
  roll of design in, 146
  spheroidizing, 132
  of steel, 132, 136
Height gage. *See* Gages
Hematite, 86
High-Energy-Rate forming, 412
High-speed steel, 173, 516
Hobbing
  of dies, 378
  of gears, 758
Honing, 662
Hooke's law, 35
Hot compression molding, 210, 211
Hot Isostatic Pressing (HIP), 315
Hot working, 80
  definition of, 80, 331
Hot working processes, classification, 339
  definition, 340
  drawing, 362
  extrusion, 358
  fiber structure, 333
  forging, 346
  piercing, 365
  pipe welding, 363
  rolling, 340
  spinning, 363
  swaging, 357
Hubbing, 379
Hydroform process, 410
Hypereutectoid steels, 120
Hypoeutectoid steels, 120
Hysteresis loop, 40

I

Impact tests, 49
  Charpy, 49, 866
  effect of notches, 50, 866
  Izod, 49

temperature effect on, 58
tensile, 51
Impregnation, 316
Indicator, dial, 441, 460
Induction heating, 150, 853, 857
  furnaces, 296
Inertia (friction) welding, 823
Infiltration, in powder metallurgy, 316
Ingots
  continuous casting, 91
  porosity, 96
Injection molding, of plastics, 210
Inspection, 430, 481
  factors in selecting equipment, 440
  methods of, 441
  nondestructive, 469
Interatomic distances, 67
Interchangeable manufacture, 24, 429
Intermetallic compounds, 81, 116
Internal cylindrical surfaces
  operations and machines for, 530
International meter, 430
Inventory reduction, 1011, 1012
Investment casting, 287, 983
Ions, 63
Ionitriding, 152
Iron, 85, 118, 121
  pig iron, 85
  ductile, 126
  malleable, 124
  nodular cast, 126
Iron carbide (Fe_3C), 117, 119
  see also Cementite
Iron-carbon equilibrium diagram, 117, 118, 119
Iron ore, direct reduction, 88
Ironing (see drawing), 362, 415
Isomers, 201
Isothermal forming, 337
Isothermal transformation diagrams, 136
Izod impact test, 49

J

Jet machining, 713
Jib borers, 582
Jigs and fixtures, 11, 887
  definition, 888
  economic justification, 899
  effect of reducing setup time, 894
  factors in design, 888
  types, 891, 895
  use in drilling, 608, 891
Job shop, 8, 930
Johansson, Carl E., 432
Joining processes, 19, 773
Jominy test, 143

Muller, for molding sand, 259
Multifunctional worker, 933, 941, 942

N

Nasmyth, James, 24, 25
Necking, 37, 40
Nibbling, 392
Nickel, 105, 160
 as an alloying element in steel, 160
 alloys of, 105, 195
 effect in cast iron, 122
Niobium, 106
Nitriding, 151
Nondestructive testing, 469
 table of methods, 470
Nonferrous alloys and metals, 178, 197
 aluminum metals, 183
 copper-base alloys, 179
 lead-tin alloys, 198
 magnesium-base alloys, 189
 nickel-base alloys, 195
 superalloys, 196
 zinc-base alloys, 194
Nonmetallic properties, 31
Nontraditional machining processes, 695
 electrochemical (ECM), 698, 709
 electrodischarge (EDM), 714
 mechanical, 699, 712
 thermal, 700, 714
Normalizing, 131
Notching, 392
Numerical control, 24, 25, 955, 985
 advantages and disadvantages, 959, 973
 computer languages for, 970
 continuous path (contouring), 964
 format, 966
 machines, 971
 point-to-point, 956, 961, 964, 980
 principles, 961
 programming of, 966

O

Oblique machining, 491, 509
Open-hearth process, 89
Open loop, 945, 956, 962, 963
Operation sheets, 1005
Optical contour projector, 451
Optical flat, 462
Organization (planning) for production, 27, 28
 JIT Production System, 1011
Orthogonal machining, 491, 503, 506
Oxyfuel gas cutting, 835
Oxyfuel gas welding, 777, 778

P

Painting, 20, 21, 915
 application, 915
 blackening, 919
 drying, 917
 electrocoating, 916
 electrostatic spray, 916
 hot-dip, 918
 organic finishes, 914
 phosphate coating, 919
 spray, 915
Parkerizing, 919
Parsons, John, 956
Parts family (family of parts), 933, 947
Patterns, 251
 allowances, 252
 sweeps, 258
 types, 254
 wax, 287
Pearlite, 110
Peening, 379
 shot, 1001
Percent elongation, 37
Percent reduction in area, 37
Perforating, 392
Permanent-mold casting, 277
 centrifugal, 284
 corthias, 278
 die, 279
 slush, 278
Permeability tester, 261
Phase, definition, 108
 diagrams, 109
Phase-transformation hardening, 133
Phosphate coatings, 919
 Bonderizing, 919
 Parkerizing, 919
Photochemical machining, 701
Physical properties, 32, 238
Physical simulation, 987
Pickling, 913
Piercing (hot), 365
Piercing and blanking, 391
 compound dies, 398
 design for, 398
 dies, 393
 progressive dies, 397
 steel rule dies, 395
 subpress dies, 394
Pig iron, 85
Pipe welding, 363, 781, 793, 813
 in plastic pipe, 883
Planers, 494, 540
Planing, 534, 540
Plasma arc welding, 794
Plasma torch, 842